Lecture Notes in Computer Science 13046

More information about this subseries at http://www.springer.com/series/7407

Colette Johnen · Elad Michael Schiller ·
Stefan Schmid (Eds.)

Stabilization, Safety, and Security of Distributed Systems

23rd International Symposium, SSS 2021
Virtual Event, November 17–20, 2021
Proceedings

Editors
Colette Johnen
University of Bordeaux
Bordeaux, France

Elad Michael Schiller
Chalmers University of Technology
Gothenburg, Sweden

Stefan Schmid
University of Vienna
Vienna, Austria

ISSN 0302-9743 ISSN 1611-3349 (electronic)
Lecture Notes in Computer Science
ISBN 978-3-030-91080-8 ISBN 978-3-030-91081-5 (eBook)
https://doi.org/10.1007/978-3-030-91081-5

LNCS Sublibrary: SL1 – Theoretical Computer Science and General Issues

This Springer imprint is published by the registered company Springer Nature Switzerland AG
The registered company address is: Gewerbestrasse 11, 6330 Cham, Switzerland

Preface

This volume contains the papers presented at the 23rd International Symposium on Stabilization, Safety, and Security of Distributed Systems (SSS 2021), held virtually during November 17–20, 2021.

SSS is an international forum for researchers and practitioners in the design and development of distributed systems with a focus on systems that are able to provide guarantees on their structure, performance, and/or security in the face of an adverse operational environment.

SSS started as the Workshop on Self-Stabilizing Systems (WSS), the first two of which were held in Austin, USA, in 1989 and in Las Vegas, USA, in 1995. After 1995, the workshop was held biennially until 2005 when it became an annual event. Starting from then, it broadened its scope and attracted researchers from other communities. In 2006, the name of the conference was changed to the International Symposium on Stabilization, Safety, and Security of Distributed Systems (SSS).

Due to the COVID-19 pandemic, SSS 2021 was a virtual event.

The authors were asked to align their submission with one of the four tracks:

- Track A. Self-stabilizing Systems: Theory and Practice (chaired by Lelia Blin and Fukuhito Ooshita)
- Track B. Foundations of Concurrent and Distributed Computing (chaired by Jukka Suomela and Philipp Woelfel)
- Track C. Mobile and Robot Computing (chaired by Maria Potop-Butucaru and Xavier Defago)
- Track D. Fault-tolerance, Security, and Privacy (chaired by Christian Scheideler and Moti Yung)

The conference received 48 regular submissions and 8 brief announcement submissions. From these we selected a total of 16 papers for presentation at the conference, leading to an acceptance rate of 33%. In addition, 10 submissions were accepted as brief announcements. The proceedings also include 14 invited papers.

The best paper was awarded to Louis Penet de Monterno, Bernadette Charron-Bost, and Stephan Merz for their paper "Synchronization modulo k in Dynamic Networks".

The best student paper was awarded to Archak Das, Kaustav Bose, and Buddhadeb Sau for their paper "Exploring a Dynamic Ring without Landmark".

We are also grateful to our keynote speakers who enriched the program:

- Idit Keidar, Technion - Israel Institute of Technology, Israel
- Michael Luby, University of California, Berkeley, USA
- Nancy Lynch, Massachusetts Institute of Technology, USA
- Ronitt Rubinfeld, Massachusetts Institute of Technology, USA

- Paul Spirakis, University of Liverpool, UK
- Jeffrey Ullman, Stanford University, USA

We hope that you will enjoy reading these excellent and very selective papers.

October 2021

Colette Johnen
Elad Michael Schiller
Stefan Schmid

Organization

Steering Committee

Anish Arora	Ohio State University, USA
Shlomi Dolev	Ben-Gurion University of the Negev, Israel
Sandeep Kulkarni	Michigan State University, USA
Toshimitsu Masuzawa (Chair)	Osaka University, Japan
Franck Petit	Sorbonne Université, France
Sébastien Tixeuil	Sorbonne Université, France
Elad Michael Schiller	Chalmers University of Technology, Sweden

Organization Committee

Romaric Duvignau	Chalmers University of Technology, Sweden
Carlo Brunetta	Chalmers University of Technology, Sweden

General Chair

Elad Michael Schiller	Chalmers University of Technology, Sweden

Program Co-chairs

Colette Johnen	LaBRI, Université de Bordeaux, France
Elad Michael Schiller	Chalmers University of Technology, Sweden
Stefan Schmid	University of Vienna, Austria

Publicity Co-chairs

Doina Bein	California State University, Fullerton, USA
Yuichi Sudo	Hosei University, Japan
Volker Turau	Hamburg University of Technology, Germany

Publication Co-chairs

Iosif Salem	University of Vienna, Austria
Ioannis Marcoullis	University of Cyprus, Cyprus

Program Committee

Hagit Attiya	Technion - Israel Institute of Technology, Israel
Alkida Balliu	University of Freiburg, Germany

Leonid Barenboim	Open University of Israel, Israel
Lélia Blin	Sorbonne Université, France
Borzoo Bonakdarpour	Michigan State University, USA
Silvia Bonomi	Sapienza University of Rome, Italy
Quentin Bramas	ICube, Université de Strasbourg, France
Sebastian Brandt	ETH Zurich, Switzerland
Jérémie Chalopin	LIS, CNRS, Aix-Marseille Université, and Université de Toulon, France
Gregory Chockler	University of Surrey, UK
Ashish Choudhury	International Institute of Information Technology, Bangalore, India
Gianlorenzo d'Angelo	Gran Sasso Science Institute, Italy
Xavier Defago	Tokyo Institute of Technology, Japan
Giuseppe Antonio Di Luna	Sapienza University of Rome, Italy
Shlomi Dolev	Ben-Gurion University of the Negev, Israel
Faith Ellen	University of Toronto, Canada
Laurent Feuilloley	Université de Lyon, France
Dianne Foreback	Kent State University, USA
Pierre Fraigniaud	Université de Paris and CNRS, France
Sukumar Ghosh	University of Iowa, USA
Wojciech Golab	University of Waterloo, Canada
Michiko Inoue	Nara Institute of Science and Technology, Japan
Colette Johnen	LaBRI, Université de Bordeaux, France
Sayaka Kamei	Hiroshima University, Japan
Yoshiaki Katayama	Nagoya Institute of Technology, Japan
Valerie King	University of Victoria, Canada
Fabian Kuhn	University of Freiburg, Germany
Anissa Lamani	Université de Strasbourg, France
Christoph Lenzen	Max Planck Institute for Informatics, Germany
Daniel Xiapu Luo	Hong Kong Polytechnic University, Hong Kong
Friedhelm Meyer auf der Heide	Heinz Nixdorf Institute and University of Paderborn, Germany
Antonis Michalas	Tampere University, Finland
Katerina Mitrokotsa	Chalmers University of Technology, Sweden
Krishnendu Mukhopadhyaya	Indian Statistical Institute, India
Mikhail Nesterenko	Kent State University, USA
Fukuhito Ooshita	Nara Institute of Science and Technology, Japan
Rotem Oshman	Tel Aviv University, Israel
Seth Pettie	University of Michigan, USA
Benny Pinkas	University of Haifa, Israel
Maria Potop-Butucaru	LIP6, Sorbonne Université, France
Giuseppe Prencipe	Universita' di Pisa, Italy
Stéphane Rovedakis	Laboratoire CEDRIC, France
Jared Saia	University of New Mexico, USA

Christian Scheideler	University of Paderborn, Germany
Elad Schiller	Chalmers University of Technology, Sweden
Stefan Schmid	University of Vienna, Austria
Gokarna Sharma	Kent State University, USA
Alexander Spiegelman	Technion - Israel Institute of Technology, Israel
Thorsten Strufe	Karlsruhe Institute of Technology and CeTI, TU Dresden, Germany
Yuichi Sudo	Hosei University, Japan
Jukka Suomela	Aalto University, Finland
Lewis Tseng	Boston College, USA
Xavier Urbain	Université Claude Bernard Lyon 1, France
Giovanni Viglietta	JAIST, Japan
Koichi Wada	Hosei University, Japan
Philipp Woelfel	University of Calgary, Canada
Moti Yung	Google Research, USA

Additional Reviewers

Abegg, Jean-Philippe
Aggarwal, Abhinav
Bund, Johannes
Castenow, Jannik
Chandramauli, Anirudh
Coijanovic, Christoph
D'Emidio, Mattia
Das, Shantanu
Devismes, Stéphane
Di Stefano, Gabriele
Durand, Anaïs
Faghih, Fathiyeh
Garg, Vijay
Gelashvili, Rati
Godard, Emmanuel
Gouleakis, Themistoklis
Götte, Thorsten
Habig, Jonas
Hector, Rory
Ilcinkas, David
Jayanti, Siddhartha
Kim, Yonghwan
Kinali, Attila
Kumar, Manish
Labourel, Arnaud
Liedtke, David
Manabe, Yoshifumi
Melnyk, Darya
Mostéfaoui, Achour
Nakai, Rikuo
Navarra, Alfredo
Pattanayak, Debasish
Poudel, Pavan
Rabie, Mikaël
Sangnier, Arnaud
Shahkarami, Golnoosh
Shibata, Masahiro
Srinivas, Vivek
Streit, Robert
Trahan, Jerry
van Ditmarsch, Hans
Wang, Ziyu
Wellnitz, Philip
Werthmann, Julian
Wiederhake, Ben
Xiang, Zhuolun
Young, Maxwell

Contents

Distributed Computing with the Cloud

Yehuda Afek[1], Gal Giladi[2], and Boaz Patt-Shamir[3](✉)

[1] School of CS, Tel Aviv University, 6997801 Tel Aviv, Israel
afek@tauex.tau.ac.il
[2] School of CS, Tel Aviv University, 6997801 Tel Aviv, Israel
[3] School of EE, Tel Aviv University, 6997801 Tel Aviv, Israel
boaz@tau.ac.il

Abstract. Motivated by cloud storage (à la Dropbox, Google Drive, etc.), we investigate distributed computing in message passing networks that contain a passive node that can only store and share data, and does not carry out any computations. Using basic primitives of collaborative transmission of a file from and to the cloud, we implement more complex tasks where the goal is to combine input values: e.g., each node holds a vector (or a matrix) as input and the sum (or product) of all the inputs should be stored in the cloud. We present near-optimal algorithms for these tasks. Finally we consider applications such as federated learning and file deduplication in this new model. Our results show that utilizing both node-cloud and node-node communication links can substantially speed up computation with respect to systems where processors communicate either only through the cloud or only through the network links.

1 Introduction

In 2018 Google announced that the number of users of Google Drive is surpassing one billion [25]. Earlier that year, Dropbox stated that in total, more than an exabyte (10^{18} bytes) of data has been uploaded by its users [14]. Other cloud-storage services, such as Microsoft's OneDrive, Amazon's S3, or Box, are thriving too. The driving force of this paper is our wish to let *other* distributed systems to take advantage of the enormous infrastructure that makes up the complexes called "clouds." Let us explain how.

The computational and storage capacities of servers in cloud services are relatively well advertised. A lesser known fact is that a cloud system also entails a massive component of *communication*, that makes it appear close almost everywhere on the Internet. (This feature is particularly essential for cloud-based video conferencing applications, such as Zoom, Cisco's Webex and others.) In view of the existing cloud services, our fundamental idea is to *abstract a complete cloud system as a single, passive storage node.*

To see the benefit of this approach, consider a network of the "wheel" topology: a single cloud node is connected to n processing nodes arranged in a cycle (see Fig. 1). Suppose each processing node has a wide link of bandwidth n to its

---

The original version of this chapter was revised: An error in the presentation of Gal Giladi's affiliation was corrected. The correction to this chapter is available at https://doi.org/10.1007/978-3-030-91081-5_41

C. Johnen et al. (Eds.): SSS 2021, LNCS 13046, pp. 1–20, 2021.
https://doi.org/10.1007/978-3-030-91081-5_1

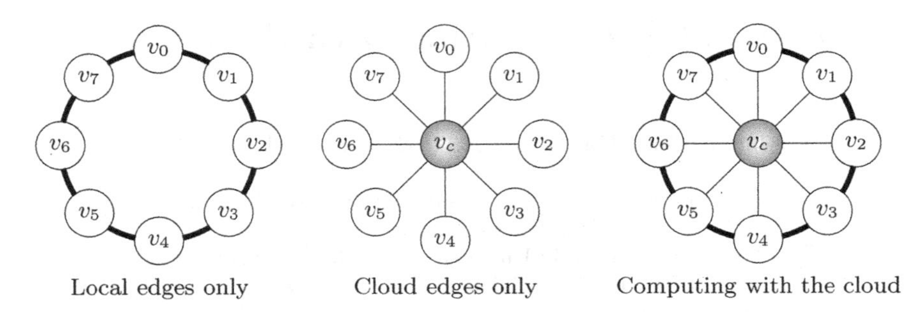

Fig. 1. Wheel topology with $n = 8$. The v_i nodes are processing nodes connected by a ring of high-bandwidth links. The cloud node v_c is connected to the processing nodes by lower-bandwidth links. All links are bidirectional and symmetric.

cycle neighbors, and a narrower link of bandwidth $\sqrt{n}$ to the cloud node. Further suppose that each processing node has an n-bit vector, and that the goal is to calculate the sum of all vectors. Without the cloud (Fig. 1, left), such a task requires at least $\Omega(n)$ rounds – to cover the distance; on the other hand, without using the cycle links (Fig. 1, middle), transmitting a single vector from any processing node (and hence computing the sum) requires $\Omega(n/\sqrt{n}) = \Omega(\sqrt{n})$ rounds – due to the limited bandwidth to the cloud. But using both cloud links and local links (Fig. 1, right), the sum can be computed in $\tilde{\Theta}(\sqrt[4]{n})$ rounds, as we show in this paper.

More generally, in this paper we initiate the study of the question of how to use an omnipresent cloud storage to speed up computations, if possible. We stress that the idea here is to develop a framework and tools that facilitate computing *with* the cloud, as opposed to computing *in* the cloud.

Specifically, in this paper we introduce the *computing with the cloud* model (CWC), and present algorithms that efficiently combine distributed inputs to compute various functions, such as vector addition and matrix multiplication. To this end, we first implement (using dynamic flow techniques) primitive operations that allow for the exchange of large messages between processing nodes and cloud nodes. Given the combining algorithms, we show how to implement some applications such as federated learning and file de-duplication (dedup).

1.1 Model Specification

The "Computing with the Cloud" (CWC) model is a synchronous network whose underlying topology is described by a weighted directed graph $G = (V, E, w)$. The node set consists of two disjoint subsets: $V = V_p \cup V_c$, where V_p is the set of *processing nodes*, and V_c is the set of *cloud nodes*. Cloud nodes are passive nodes that function as shared storage: they support read and write requests, and do not perform any other computation. We use n to denote the number of processing nodes (the number of cloud nodes is typically constant).

We denote the set of links that connect two processing nodes by E_L ("local links"), and by E_C ("cloud links") the set of links that connect processing nodes to cloud nodes. Each link $e \in E = E_L \cup E_C$ has a prescribed bandwidth $w(e)$

(there are no links between different cloud nodes). We denote by $G_p \stackrel{\text{def}}{=} (V_p, E_L)$ the graph $G - V_c$, i.e., the graph spanned by the processing nodes.

Our execution model is the standard synchronous network model, where each round consists of processing nodes receiving messages sent in the previous round, doing an arbitrary local computation, and then sending messages. The size of a message sent over a link e in a round is at most $w(e)$ bits.

Cloud nodes do not perform any computations: they can only receive requests we denote by `FR` and `FW` (file read and write, respectively), to which they respond in the following round. More precisely, each cloud node has unbounded storage; to write, a processing node v_i invokes `FW` with arguments that describe the target cloud node, a filename f, a bit string S, and the location (index) within f that S needs to be written in. It is assumed that $|S| \leq w(v_i, v_c)$ bits (longer writes can be broken to a series of `FW` operations). To read, a processing node v_i invokes `FR` with arguments that describe the cloud node, a filename f and the range of indices to fetch from f. Again, we assume that the size of the range in any single `FR` invocation by node v_i is at most $w(v_i, v_c)$.[1]

`FW` operations are exclusive, i.e., no other operation (read or write) to the same file location is allowed to take place simultaneously. Concurrent `FR` operations from the same location are allowed.

Discussion. We believe that our model is fairly widely applicable. A processing node in our model may represent anything from a computer cluster with a single gateway to the Internet, to cellphones or even smaller devices—anything with a non-shared Internet connection. The local links can range from high-speed fiber to Bluetooth or infrared links. Typically in this setting the local links have bandwidth much larger than the cloud links (and cloud downlinks in many cases have larger bandwidth than cloud uplinks). Another possible interpretation of the model is a private network (say, in a corporation), where a cloud node represents a storage or a file server. In this case the cloud link bandwidth may be as large as the local link bandwidth.

1.2 Problems Considered and Main Results

Our main results in this paper are efficient algorithms in the CWC model to combine values stored at nodes. These algorithms use building blocks that facilitate efficient transmission of large messages between processing nodes and cloud nodes. These building blocks, in turn, are implemented in a straightforward way using dynamic flow techniques. Finally, we show how to use the combining algorithms to derive new algorithms for federated learning and file de-duplication (dedup) in the CWC model.

More specifically, we provide implementations of the following tasks.

[1] For both the `FW` and `FR` operations we ignore the metadata (i.e., v_c's descriptor, the filename f and the indices) and assume that the total size of metadata in a single round is negligible and can fit within $w(v_i, v_c)$. Otherwise, processing nodes may use the metadata parameters to exchange information that exceeds the bandwidth limitations (for example, naming a file with the string representation of a message whose length is larger than the bandwidth).

Basic Cloud Operations: Let v_c denote a cloud node below.

cW$_i$ *(cloud write):* write an s-bits file f stored at node $i \in V_p$ to node v_c.
cR$_i$ *(cloud read):* fetch an s-bits file f from node v_c to node $i \in V_p$.
cAW *(cloud all write):* for each $i \in V_p$, write an s-bits file f_i stored at node i to node v_c.
cAR *(cloud all read):* for each $i \in V_p$, fetch an s-bits file f_i from node v_c to node i.

Combining and Dissemination Operations
cComb: *(cloud combine):* Each node $i \in V_p$ has an s-bits input string S_i, and there is a binary associative operator $\otimes : \{0,1\}^s \times \{0,1\}^s \to \{0,1\}^s$ (the result is as long as each operand). The requirement is to write to a cloud node v_c the s-bits string $S_1 \otimes S_2 \otimes \cdots \otimes S_n$. Borrowing from Group Theory, we call the operation $\otimes$ *multiplication*, and $S_1 \otimes S_2$ is the *product* of S_1 by S_2. In general, $\otimes$ is not necessarily commutative. We assume the existence of a unit element for $\otimes$, denoted $\tilde{\mathbf{1}}$, such that $\tilde{\mathbf{1}} \otimes S = S \otimes \tilde{\mathbf{1}} = S$ for any s-bits strings S. The unit element is represented by a string of $O(1)$ bits. Examples for commutative operators include vector (or matrix) addition over a finite field, logical bitwise operations, leader election, and the top-k problem. Examples for non-commutative operators may be matrix multiplication (over a finite field) and function composition.

cCast *(cloudcast):* All the nodes $i \in V_p$ simultaneously fetch a copy of an s-bits file f from node v_c (Similar to network broadcast).

Applications. cComb and cCast can be used directly to provide matrix multiplication, matrix addition, and vector addition. We also outline the implementation of the following.

Federated Learning (FL) [31]*:* In FL, a collection of agents collaborate in training a neural network to construct a model of some concept, but the agents want to keep their data private. Unlike [31], in our model the central server is a passive storage device that does not carry out computations. We show how elementary secure computation techniques, along with our combining algorithm, can efficiently help training an ML model in the federated scheme implemented in CWC, while maintaining privacy.

File Deduplication: Deduplication (or dedup) is a task in file stores, where redundant identical copies of data are identified (and possibly unified)—see, e.g., [32]. Using cComb and cCast, we implement file dedup in the CWC model on collections of files stored at the different processing nodes. The algorithm keeps a single copy of each file and pointers instead of the other replicas.

Special Topologies. The complexity of the general algorithms we present depends on the given network topology. We study a few cases of interest.

First, we consider *s-fat-links* network, defined to be, for a given parameter $s \in \mathbb{N}$, as the CWC model with the following additional assumptions:

- All links are symmetric, i.e., $w(u,v) = w(v,u)$ for every link $(u,v) \in E$.

- Local links have bandwidth at least s.
- There is only one cloud node v_c.

The fat links model seems suitable in many real-life cases where local links are much wider than cloud links (uplinks to the Internet), as is the intuition behind the HYBRID model [4].

Another topology we consider is the *wheel network*, depicted schematically in Fig. 1 (right). In a wheel system there are n processing nodes arranged in a ring, and a cloud node connected to all processing nodes. In the *uniform* wheel, all cloud links have the same bandwidth b_c and all local links have the same bandwidth b_l. In the uniform wheel model, we typically assume that $b_c \ll b_\ell$.

The wheel network is motivated by non-commutative combining operations, where the order of the operands induces a linear order on the processing nodes, i.e., we view the nodes as a line, where the first node holds the first input, the second node holds the second input etc. For symmetry, we connect the first and the last node, and with a cloud node connected to all—we've obtained the wheel.

Overview of Techniques. As mentioned above, the basic file operations (cW, cR, cAW and cAR) are solved optimally using dynamic flow techniques, or more specifically, *quickest flow* (Sect. 2). In the full version, we present closed-form bounds on cW and cR for the wheel topology.

We present tight bounds for cW and cR in the s-fat-links network, where s is the input size at all nodes. We then continue to consider the tasks cComb with *commutative operators* and cCast, and prove nearly-tight bounds on their time complexity in the s-fat-links network (Theorem 11, Theorem 13, Theorem 15). The idea is to first find, for every processing node i, a cluster of processing nodes that allows it to perform cW in an optimal number of rounds. We then perform cComb by combining the values within every cluster using convergecast [33], and then combining the results in a computation-tree fashion. Using sparse covers [5], we perform the described procedure in near-optimal time.

Non-commutative operators are explored in the natural wheel topology. We present algorithms for wheel networks with *arbitrary* bandwidth (both cloud and local links). We prove an upper bound for cComb (Theorem 18).

Finally, in Sect. 5, we demonstrate how the considered tasks can be applied for the purposes of Federated Learning and File Deduplication.

Paper Organization. Due to space constraints, many details and proofs are omitted from this version. They can be found in the full version [2].

1.3 Related Work

Our model is based on, and inspired by, a long history of theoretical models in distributed computing. To gain some perspective, we offer here a brief review.

Historically, distributed computing is split along the dichotomy of message passing vs shared memory [16]. While message passing is deemed the "right" model for network algorithms, the shared memory model is the abstraction of choice for programming multi-core machines.

The prominent message-passing models are LOCAL [28], and its derived CONGEST [33]. In these models, a system is represented by a connected (typically undirected) graph, in which nodes represent processors and edges represent communication links. In LOCAL, message size is unbounded, while in CONGEST, message size is restricted, typically to $O(\log n)$ bits. Thus, CONGEST accounts not only for the distance information has to traverse, but also for information volume and the bandwidth available for its transportation.

While most algorithms in the LOCAL and CONGEST models assume fault-free (and hence synchronous) executions, in the distributed shared memory model, asynchrony and faults are the primary source of difficulty. Usually, in the shared memory model one assumes that there is a collection of "registers," accessible by multiple threads of computation that run at different speeds and may suffer crash or even Byzantine faults (see, e.g., [3]). The main issues in this model are coordination and fault-tolerance. Typically, the only quantitative hint to communication cost is the number and size of the shared registers.

Quite a few papers consider the combination of message passing and shared memory, e.g., [1,12,18,19,30,35]. The uniqueness of the CWC model with respect to past work is that it combines passive storage nodes with a message passing network with restrictions on the links bandwidth.

The CONGESTED CLIQUE (CC) model [29] is a special case of CONGEST, where the underlying communication graph is assumed to be fully connected. The CC model is appropriate for computing *in* the cloud, as it has been shown that under some relatively mild conditions, algorithms designed for the CC model can be implemented in the MapReduce model, i.e., run in datacenters [20]. Another model for computing in the cloud is the MPC model [22]. Very recently, the HYBRID model [4] was proposed as a combination of CC with classical graph-based communication. More specifically, the HYBRID model assumes the existence of two communication networks: one for local communication between neighbors, where links are typically of infinite bandwidth (exactly like LOCAL); the other network is a *node-congested* clique, i.e., a node can communicate with every other node directly via "global links," but there is a small upper bound (typically $O(\log n)$) on the total number of messages a node can send or receive via these global links in a round. Even though the model was presented only recently, there is already a line of algorithmic work in it, in particular for computing shortest paths [4,10,23].

Discussion. Intuitively, our CWC model can be viewed as the classical CONGEST model over the processors, augmented by special cloud nodes (object stores) connected to some (typically, many) compute nodes. To reflect modern demands and availability of resources, we relax the very stringent bandwidth allowance of CONGEST, and usually envision networks with much larger link bandwidth (e.g., n^ϵ for some $\epsilon > 0$).

Considering previous network models, it appears that HYBRID is the closest to CWC, even though HYBRID was not expressly designed to model the cloud. In our view, CWC is indeed more appropriate for computation with the cloud. First, in most cases, global communication (modeled by clique edges in HYBRID) is limited by link bandwidth, unlike HYBRID's node capacity constraint, which seems somewhat artificial. Second, HYBRID is not readily amenable to model multiple clouds, while this is a natural property of CWC.

Regarding shared memory models, we are unaware of topology-based bandwidth restriction on shared memory access in distributed models. In some general-purpose parallel computation models (based on BSP [35]), communication capabilities are specified using a few global parameters such as latency and throughput, but these models deliberately abstract topology away. In distributed (asynchronous) shared memory, the number of bits that need to be transferred to and from the shared memory is seldom explicitly analyzed.

2 Communication Primitives in CWC

In this short section we state the complexity results for the basic operations, derived by straightforward application of *dynamic flow* techniques [34].

Intuitively, the concept of dynamic flow is a variant of maximum flow, where time is finite, links introduce delay, and the goal is to maximize the amount of flow shipped in the given time limit (the dual problem, where the amount of flow to ship is given and the goal is to minimize the time required to ship it, is called *quickest flow* [6,9,15,21]). By reduction to min-cost max-flow, strongly polynomial algorithms to these problems are known. Using these results, we can prove the following statements. Details can be found in [2].

Theorem 1. *Given any instance of the CWC model, an optimal schedule realizing cW_i or cR_i can be computed in polynomial time.*

Theorem 2. *Given any instance of the CWC model, an optimal schedule realizing cAW or cAR for one cloud node can be computed in polynomial time.*

Theorem 3. *Given any instance of the CWC model and $\epsilon > 0$, a schedule realizing cAW or cAR of length at most $(1+\epsilon)$ times the optimal can be computed in time polynomial in the instance size and ϵ^{-1}.*

3 Computing and Writing Combined Values

Flow-based techniques are not applicable in the case of writing a combined value, because the very essence of combining violates conservation constraints (i.e., the number of bits entering a node may be different than the number of bits leaving it). However, in Sect. 3.1 we explain how to implement cComb in the general case using cAW and cAR. While simple and generic, these implementations can have time complexity much larger than optimal. We offer partial remedy in Sect. 3.2, where we present our main result: an algorithm for cComb when $\otimes$ is commutative and the local network has "fat links," i.e., all local links have capacity at least s. For this important case, we show how to complete the task in time larger than the optimum by an $O(\log^2 n)$ factor.

Algorithm 1. High-level algorithm for cComb using cAW and cAR

1: $m := n$, $j := 0$
2: for all $i < n$ set $X_i^0 = S_i$, and for all $i > n$, $X_i^0 = \tilde{\mathbf{1}}$
3: **while** $m > 1$ **do**
4: run cAW with inputs $S_i = X_i^j$
5: run cAR with inputs $S_i = X_{2i}^j$
6: run cAR with inputs $S_i = X_{2i+1}^j$
7: $m := \lceil m/2 \rceil$
8: for all $i < m$ set $X_i^{j+1} = X_{2i}^j \otimes X_{2i+1}^j$, and for all $i > m$, $X_i^{j+1} = \tilde{\mathbf{1}}$
9: for all $i < m$, in parallel, node i calculates X_i^{j+1} locally
10: $j := j + 1$
11: **end while**
12: run cW from node 0 to write X_0^j to the cloud

3.1 Combining in General Graphs

We now present algorithms for cComb and for cCast on general graphs, using the primitives treated in Sect. 2. Note that with a non-commutative operator, the operands must be ordered; using renaming if necessary, we assume w.l.o.g. that in such cases the nodes are indexed by the same order of their operands.

Theorem 4. *Let T_s be the running time of cAW (and cAR) when all files have size s. Then Algorithm 1 solves cComb in $O(T_s \log n)$ rounds.*

In a way, cCast is the "reverse" problem of cComb, since it starts with s bits in the cloud and ends with s bits of output in every node. However, cCast is easier than cComb because our model allows for concurrent reads and disallows concurrent writes. We have the following result.

Theorem 5. *Let T_s be the time required to solve cAR when all files have size s. Then cCast can be solved in T_s rounds as well.*

3.2 Combining Commutative Operators in Fat Links Network

In the case of s-fat-links network (i.e., all local links are have bandwidth at least s, and all links are symmetric) we can construct a near-optimal algorithm for cComb. The idea is to use multiple cW and cR operations instead of cAW and cAR. The challenge is to minimize the number of concurrent operations per node; to this end we use sparse covers [5].

We note that if the network is s-fat-links but the operand size is $s' > s$, the algorithms still apply, with an additional factor of $\lceil s'/s \rceil$ to the running time. The lower bounds in this section, however, may change by more than that factor.

We start with a tight analysis of cW and cR in this setting and then generalize to cComb and cCast.

Implementation of cW and cR. Consider cW_i, where i wishes to write s bits to a given cloud node. The basic tension in finding an optimal schedule for cW_i

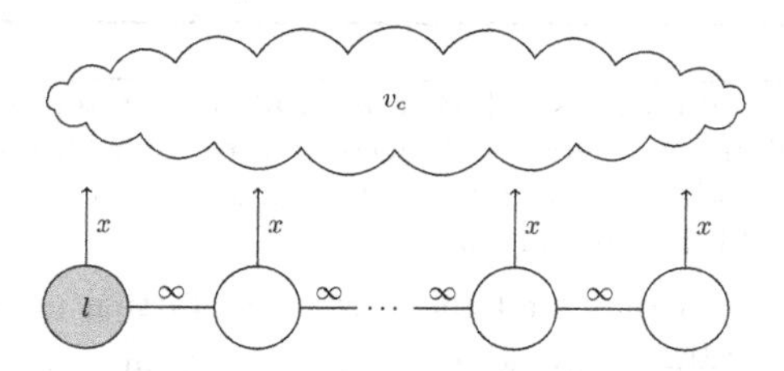

Fig. 2. A simple path example. The optimal distance to travel in order to write an s-bits file to the cloud would be $\sqrt{s/x}$.

is that in order to use more cloud bandwidth, more nodes need to be enlisted. But while more bandwidth reduces the transmission time, reaching remote nodes (that provide the extra bandwidth) increases the traversal time. Our algorithm looks for the sweet spot where the conflicting effects are more-or-less balanced.

For example, consider a simple path of n nodes with infinite local bandwidth, where each node is connected to the cloud with bandwidth x (Fig. 2). Suppose that the leftmost node l needs to write a message of s bits to the cloud. By itself, writing requires s/x rounds. Using all n nodes, uploading would take $O(s/nx)$ rounds, but $n-1$ rounds are needed to ship the messages to the fellow-nodes. The optimal solution in this case is to use only $\sqrt{s/x}$ nodes: the time to ship the file to all these nodes is $\sqrt{s/x}$, and the upload time is $\frac{s/\sqrt{s/x}}{x} = \sqrt{s/x}$, because each node needs to upload only $s/\sqrt{s/x}$ bits.

In general, we define "cloud clusters" to be node sets that optimize the ratio between their diameter and their total bandwidth to the cloud. Our algorithms for cW and cR use nodes of cloud clusters. We prove that the running-time of our implementation is asymptotically optimal. Formally, we have the following.

Definition 1. *Let $G = (V, E, w)$ be a CWC system with processor nodes V_p and cloud nodes V_c. The* ***cloud bandwidth*** *of a processing node $i \in V_p$ w.r.t. a given cloud node $v_c \in V_c$ is $b_c(i) \stackrel{\text{def}}{=} w(i, v_c)$. A* ***cluster*** *$B \subseteq V_p$ in G is a connected set of processing nodes. The* ***cloud (up or down) bandwidth*** *of cluster B w.r.t a given cloud node, denoted $b_c(B)$, is the sum of the cloud bandwidth to v_c over all nodes in B: $b_c(B) \stackrel{\text{def}}{=} \sum_{i \in B} b_c(i)$. The (strong)* ***diameter*** *of cluster B, denoted $\mathrm{diam}(B)$, is the maximum distance between any two nodes of B in the induced graph $G[B]$: $\mathrm{diam}(B) = \max_{u,v \in B} dist_{G[B]}(u, v)$.*

We use the following definition for the network without the cloud.

Definition 2. *Let $G = (V, E, w)$ be a CWC system with processing nodes V_p and cloud nodes V_c. The* ***ball of radius*** *r* ***around node*** *$i \in V_p$, denoted $B_r(i)$ is the set of nodes at most r hops away from i in G_p.*

(Note that the metric here is hop-based—w indicates link bandwidths.) Finally, we define the concept of cloud cluster of a node.

Algorithm 2. cW_i

1: Construct a BFS spanning tree of B_i rooted at node i and assign for each index $1 \le x \le |B_i|$ a unique node $v(x) \in B_i$ according to their BFS order ($v(1) = i$)
2: Broadcast S from node i to all nodes in B_i using the tree
3: **for** all $x := 1$ **to** $|B_i|$, in parallel **do**
4: Node $v(x)$ writes to the cloud the part of S starting at $s \cdot \frac{\sum_{y=1}^{x-1} b_c(v(y))}{b_c(B_i)}$ and extending for $s \cdot \frac{b_c(v(x))}{b_c(B_i)}$ bits, writing $b_c(v(x))$ bits in every round.
5: Node $v(x) \neq i$ sends an acknowledgment to i when done, and halts
6: **end for**
7: Node i halts when all acknowledgments are received. *// for* cR *reversal*

Definition 3. *Let* $G = (V, E, w)$ *be a CWC system with processing nodes* V_p *and cloud node* v_c*, and let* $i \in V_p$*. Given* $s \in \mathbb{N}$*, the* ***s-cloud radius*** *of node* i*, denoted* $k_s(i)$*, is defined to be* $k_s(i) \stackrel{\text{def}}{=} \min(\{\text{diam}(G_p)\} \cup \{k \mid (k+1) \cdot b_c(B_k(i)) \ge s\})$ *. The ball* $B_i \stackrel{\text{def}}{=} B_{k_s(i)}(i)$ *is the* ***s-cloud cluster*** *of node* i*. The* ***timespan*** *of the s-cloud cluster of* i *is denoted* $Z_i \stackrel{\text{def}}{=} k_s(i) + \frac{s}{b_c(B_i)}$*. We sometimes omit the* s *qualifier when it is clear from the context.*

In words, B_i is a cluster of radius $k(i)$ around node i, where $k(i)$ is the smallest radius that allows writing s bits to v_c by using all cloud bandwidth emanating from B_i for $k(i)+1$ rounds. Z_i is the time required (1) to send s bits from node i to all nodes in B_i, and (2) to upload s bits to v_c collectively by all nodes of B_i. Note that B_i is easy to compute. We can now state our upper bound.

Theorem 6. *Given a fat-links CWC system, Algorithm 2 solves the s-bits* cW_i *problem in* $O(Z_i)$ *rounds on* B_i*.*

Next, we show that our solution for cW_i is optimal, up to a constant factor. We consider the case of an incompressible input string: such a string exists for any size $s \in \mathbb{N}$ (see, e.g., [27]). As a consequence, in any execution of a correct algorithm, s bits must cross any cut that separates i from the cloud node, giving rise to the following lower bound.

Theorem 7. cW_i *in a fat-links CWC requires* $\Omega(Z_i)$ *rounds.*

By reversing time (and hence information flow) in a schedule of cW, one gets a schedule for cR. Hence we have the following immediate corollaries.

Theorem 8. cR_i *can be executed in* $O(Z_i)$ *rounds in a fat-links CWC.*

Theorem 9. cR_i *in a fat-links CWC requires* $\Omega(Z_i)$ *rounds.*

► *Remark:* The lower bound (Theorem 7) and the definition of cloud clusters (Definition 3) show an interplay between the message size s, cloud bandwidth, and the network diameter; For large enough s, the cloud cluster of a node includes all processing nodes (because the time spent crossing the local network is negligible relative to the upload time), and for small enough s, the cloud cluster includes only the invoking node, rendering the local network redundant.

Implementation of cComb. Below, we first show how to implement cComb using any given cover. In fact, we shall use *sparse covers* [5], which allow us to get near-optimal performance.

Intuitively, every node i has a cloud cluster B_i which allows it to perform cW_i, and calculating the combined value within every cloud cluster B_i is straightforward (cf. Algorithm 4 and Lemma 1). Therefore, given a partition of the graph that consists of pairwise disjoint cloud clusters, cComb can be solved by combining the inputs in every cloud cluster, followed by combining the partial results in a computation-tree fashion using cW and cR. However, such a partition may not always exist, and we resort to a *cover* of the nodes. Given a cover $\mathcal{C}$ in which every node is a member of at most $\text{load}(\mathcal{C})$ clusters, we can use the same technique, while increasing the running-time by a factor of $\text{load}(\mathcal{C})$ by time multiplexing. Using Awerbuch and Peleg's sparse covers (see Theorem 12), we can use an initial cover $\mathcal{C}$ that consists of all cloud clusters in the graph to construct another cover, $\mathcal{C}'$, in which $\text{load}(\mathcal{C}')$ is $O(\log n)$, paying an $O(\log n)$ factor in cluster diameters, and use $\mathcal{C}'$ to get near-optimal results.

Definition 4. *Let G be a CWC system, and let B be a cluster in G (see Definition 1). The* ***timespan of node*** i ***in*** B*, denoted $Z_B(i)$, is the minimum number of rounds required to perform cW_i (or cR_i), using only nodes in B. The* ***timespan of cluster*** B*, denoted $Z(B)$, is given by $Z(B) = \min_{i \in B} Z_B(i)$. The* ***leader*** *of cluster B, denoted $r(B)$, is a node with minimal timespan in B, i.e., $r(B) = \text{argmax}_{i \in B} Z_B(i)$.*

In words, the timespan of cluster B is the minimum time required for any node in B to write an s-bit string to the cloud using only nodes of B.

Definition 5. *Let G be a CWC system with processing node set V_p. A* ***cover*** *of G is a set of clusters $\mathcal{C} = \{B_1, \ldots, B_m\}$ such that $\cup_{B \in \mathcal{C}} B = V_p$. The* ***load of node*** i *in a cover $\mathcal{C}$ is the number of clusters in $\mathcal{C}$ that contain i, i.e., $\text{load}_{\mathcal{C}}(i) = |\{B \in \mathcal{C} \mid i \in B\}|$. The* ***load of cover*** $\mathcal{C}$ *is the maximum load of any node in the cover, i.e.. $\text{load}(\mathcal{C}) = \max_{i \in V_p} \text{load}_{\mathcal{C}}(i)$. The* ***timespan*** *of cover $\mathcal{C}$, denoted $Z(\mathcal{C})$, is the maximum timespan of any cluster in $\mathcal{C}$, $Z(\mathcal{C}) = \max_{B \in \mathcal{C}} Z(B)$. The* ***diameter*** *of cover $\mathcal{C}$, denoted $\text{diam}_{\max}(\mathcal{C})$, is the maximum diameter of any cluster in $\mathcal{C}$, $\text{diam}_{\max}(\mathcal{C}) = \max_{B \in \mathcal{C}} \text{diam}(B)$.*

We now give an upper bound in terms of any given cover.

Theorem 10. *Given a cover $\mathcal{C}$, Algorithm 3 solves cComb in a fat-links CWC in $O\left(\text{diam}_{\max}(\mathcal{C}) \cdot \text{load}(\mathcal{C}) + Z(\mathcal{C}) \cdot \text{load}(\mathcal{C}) \cdot \log |\mathcal{C}|\right)$ rounds.*

The basic strategy is to first compute the combined value in each cluster using only the local links, and then combine the cluster values using a computation tree. However, unlike Algorithm 1, we use cW and cR instead of cAW and cAR.

A high-level description is given in Algorithm 3. The algorithm consists of a preprocessing part (lines 1–2), and the execution part, which consists of the "low-level" computation using only local links (lines 3–5), and the "high-level" computation among clusters (line 6). We elaborate on each below.

Algorithm 3. High-level algorithm for cComb given a cover $\mathcal{C}$

1: For a node $i \in V_p$, let $C_i[1], C_i[2], \ldots$ be the clusters containing i.
2: For the rest of the algorithm, multiplex each round as load($\mathcal{C}$) rounds, such that each node i operates in the context of cluster $C_i[j]$ in the j-th round .
3: **for all** $B \in \mathcal{C}$, in parallel **do**
4: Compute $P_B = \bigotimes_{j \in B} S_j$ using Alg. 4 // *convergecast using local links only*
5: **end for**
6: Apply Alg. 5 // *the result is stored in the cloud*

▶ *Preprocessing.* A major component of the preprocessing stage is computing the cover $\mathcal{C}$, which we specify later (see Theorem 11). In Algorithm 3 we describe the algorithm as if it operates in each cluster independently of other clusters, but clusters may overlap. To facilitate this mode of operation, we use time multiplexing: nodes execute work on behalf of the clusters they are member of in a round-robin fashion, as specified in lines 1–2 of Algorithm 3. This allows us to invoke operations limited to clusters in all clusters "simultaneously" by increasing the time complexity by a load($\mathcal{C}$) factor.

▶ *Low levels: Combining within a single cluster.* To implement line 4 of Algorithm 3, we build, in each cluster $B \in \mathcal{C}$, a spanning tree rooted at $r(B)$, and apply Convergecast [33] using $\otimes$. Ignoring the multiplexing of Algorithm 3, we have:

Lemma 1. *Algorithm 4 computes* $P_B = \bigotimes_{i \in B} S_i$ *at node* $r(B)$ *in* $O(\mathrm{diam}(B))$ *rounds.*

To get the right overall result, each input S_i is associated with a single cluster in $\mathcal{C}$. To this end, we require each node to select a single cluster in which it is a member as its *home cluster.* When applying Algorithm 4, we use the rule that the input of node i in a cluster $B \ni i$ is S_i if B is i's home cluster, and $\tilde{\mathbf{1}}$ otherwise.

Considering the scheduling obtained by Step 2, we get the following lemma.

Lemma 2. *Steps 3–5 of Algorithm 3 terminate in* $O(\mathrm{diam}_{\max}(\mathcal{C}) \cdot \mathrm{load}(\mathcal{C}))$ *rounds, with* P_B *stored at the leader node of* B *for each cluster* $B \in C$.

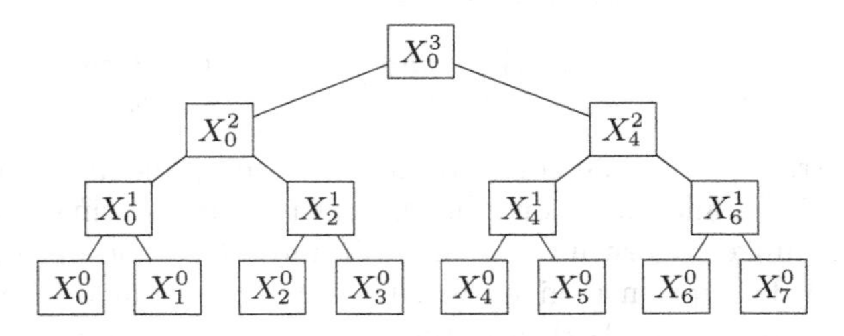

Fig. 3. Computation tree example. X_i^j denotes the result stored in i after iteration j.

Algorithm 4. Computing the combined result of cluster B at leader $r(B)$

```
1: Construct a BFS tree of B rooted at node r(B). Let h be the height of the tree.
2: for d := h to 2 do
3:    for all i ∈ B at layer d of the tree, in parallel do
4:       if i is not a leaf then
5:          i computes S'_i := S_i ⊗ ⊗_{j∈child(i)} S'_j
6:       else
7:          S'_i := S_i
8:       end if
9:       i sends S'_i to its parent node in the tree
10:   end for
11: end for
12: Node r(B) computes P_B := S_{r(B)} ⊗ ⊗_{j∈child(r(B))} S'_j
```

Algorithm 5. Computing the high level tree-nodes values

```
1: for l := ⌈log |C_1|⌉ to 1 do
2:    for all tree-nodes y in layer l of the computation tree, in parallel do
3:       Let B := cl(y)
4:       if y is not a leaf then
5:          Let y_ℓ and y_r be the left and the right children of y, respectively.
6:          r(B) invokes cR for vl(y_ℓ)
7:          r(B) invokes cR for vl(y_r)
8:          r(B) computes vl(y) := vl(y_ℓ) ⊗ vl(y_r)
9:       else
10:         vl(y) := P_B          // if y is a leaf its value is already stored at r(B)
11:      end if
12:      r(B) invokes cW for vl(y)
13:   end for
14: end for
```

► *High levels: Combining using the cloud.* When Algorithm 3 reaches Step 6, the combined result of every cluster is stored in the leader of the cluster. The idea is now to fill in a computation tree whose leaves are these values (see Fig. 3).

We combine the partial results by filling in the values of a computation tree defined over the clusters. The leaves of the tree are the combined values of the clusters of $\mathcal{C}$, as computed by Algorithm 4. To fill in the values of other nodes in the computation tree, we use the clusters of $\mathcal{C}$: Each node in the tree is assigned a cluster which computes its value using the cR and cW primitives.

Specifically, in Algorithm 5 we consider a binary tree with $|\mathcal{C}|$ leaves, where each non-leaf node has exactly two children. The tree is constructed from a complete binary tree with $2^{\lceil \log |\mathcal{C}| \rceil}$ leaves, after deleting the rightmost $2^{\lceil \log |\mathcal{C}| \rceil} - |\mathcal{C}|$ leaves. (If by the end the rightmost leaf is the only child of its parent, we delete the rightmost leaf repeatedly until this is not the case).

We associate each node y in the computation tree with a cluster $\text{cl}(y) \in \mathcal{C}$ and a value $\text{vl}(y)$, computed by the processors in $\text{cl}(y)$ are responsible to compute $\text{vl}(y)$. Clusters are assigned to leaves by index: The i-th leaf from the left is associated with the i-th cluster of $\mathcal{C}$. For internal nodes, we assign the clusters arbitrarily except that we ensure that no cluster is assigned to more than one internal node. (This is possible because in a tree where every node has two or no children, the number of internal nodes is smaller than the number of leaves).

The clusters assigned to tree nodes compute the values as follows (see Algorithm 5). The value associated with a leaf y_B corresponding to cluster B is $\mathrm{vl}(y_B) = P_B$. This way, every leaf x has $\mathrm{vl}(x)$, stored in the leader of $\mathrm{cl}(x)$, which can write it to the cloud using cW. For an internal node y with children y_l and y_r, the leader of $\mathrm{cl}(y)$ obtains $\mathrm{vl}(y_l)$ and $\mathrm{vl}(y_r)$ using cR, computes their product $\mathrm{vl}(y) = \mathrm{vl}(y_l) \otimes \mathrm{vl}(y_r)$ and invokes cW to write it to the cloud. The executions of cW and cR in a cluster B are done by the processing nodes of B.

Computation tree values are filled layer by layer, bottom up.

▶ *Remark.* We note that in Algorithm 5, Lines 6, 7 and 12 essentially compute cAR and cAW in which only the relevant cluster leaders have inputs. Therefore, these calls can be replaced with a collective call for appropriate cAR and cAW, making the multiplexing of Line 2 of Algorithm 3 unnecessary (similarly to Algorithm 1). By using optimal schedules for cAW and cAR, the running-time can only improve beyond the upper bound of Theorem 10.

Sparse Covers. We now arrive at our main result, derived from Theorem 10 using a particular flavor of covers. The result is stated in terms of the maximal timespan of a graph, according to the following definition.

Definition 6. *Let $G = (V, E, w)$ be a CWC system with fat links. $Z_{\max} \stackrel{\mathrm{def}}{=} \max_{i \in V_p} Z_i$ is the* ***maximal timespan*** *in G.*

In words, $Z_{\max}$ is the maximal amount of rounds that is required for any node in G to write an s-bit message to the cloud, up to a constant factor (cf. Theorem 7).

Theorem 11. *Let $G = (V, E, w)$ be a CWC system with fat links. Then cComb with a commutative combining operator can be solved in $O(Z_{\max} \log^2 n)$ rounds.*

To prove Theorem 11 we use sparse covers. We state the result from [5].

Theorem 12 ([5]). *Given any cover $\mathcal{C}$ and an integer $\kappa \geq 1$, a cover $\mathcal{C}'$ that satisfies the following properties can be constructed in polynomial time.*

(i) *For every cluster $B \in \mathcal{C}$ there exists a cluster $B' \in \mathcal{C}'$ such that $B \subseteq B'$.*
(ii) $\max_{B' \in \mathcal{C}'} \mathrm{diam}(B') \leq 4\kappa \max_{B \in \mathcal{C}} \mathrm{diam}(B)$
(iii) $\mathrm{load}(\mathcal{C}') \leq 2\kappa |\mathcal{C}|^{1/\kappa}$.

Proof of Theorem 11**:** Let $\mathcal{C}$ be the cover defined as the set of all cloud clusters in the system. By applying Theorem 12 to $\mathcal{C}$ with $\kappa = \lceil \log n \rceil$, we obtain a cover $\mathcal{C}'$ with $\mathrm{load}(\mathcal{C}') \leq 4 \lceil \log n \rceil$ because $|\mathcal{C}| \leq n$. By ii, $\mathrm{diam}_{\max}(\mathcal{C}') \leq 4 \lceil \log n \rceil \cdot \mathrm{diam}_{\max}(\mathcal{C})$. Now, let $B' \in \mathcal{C}'$. We can assume w.l.o.g. that there is a cluster $B \in \mathcal{C}$ such that $B \subseteq B'$ (otherwise B' can be removed from $\mathcal{C}'$). B is a cloud cluster of some node $i \in B'$, and therefore by Theorem 6 and by Definition 4, we get that $Z(B') \le Z(B) = O(Z_i) = O(Z_{\max})$. Since this bound holds for all clusters of $\mathcal{C}'$, $Z(\mathcal{C}') = O(Z_{\max})$.

An $O\left(\mathrm{diam}_{\max}(\mathcal{C}) \cdot \log^2 n + Z_{\max} \cdot \log^2 n\right)$ time bound for cComb is derived by applying Theorem 10 to cover $\mathcal{C}'$. Finally, let $B_j \in \mathcal{C}$ be a cloud cluster of diameter $\mathrm{diam}_{\max}(\mathcal{C})$. Recall that by Definition 3, $\mathrm{diam}(B_j) \le 2k(j) \le 2Z_j \le 2Z_{\max}$. We therefore obtain an upper bound of $O(Z_{\max} \log^2 n)$ rounds. ∎

We close with a lower bound.

Theorem 13. *Let $G = (V, E, w)$ be a CWC system with fat links. Then* cComb *requires $\Omega(Z_{\max})$ rounds.*

cCast. To implement cCast, one can reverse the schedule of cComb. However, a slightly better implementation is possible, because there is no need to ever write to the cloud node. More specifically, let $\mathcal{C}$ be a cover of V_p. In the algorithm for cCast, each cluster leader invokes cR, and then the leader disseminates the result to all cluster members. The time complexity for a single cluster B is $O(Z(B))$ for the cR operation, and $O(\mathrm{diam}(B))$ rounds for the dissemination of S throughout B (similarly to Lemma 1). Using the multiplexing to $\mathrm{load}(\mathcal{C})$ as in Step 2 of Algorithm 3, we obtain the following result.

Theorem 14. *Let $G = (V, E, w)$ be a CWC system with fat links. Then* cCast *can be performed in $O(Z_{\max} \cdot \log^2 n)$ rounds.*

Finally, we note that since any algorithm for cCast also solves cR_i problem for every node i, we get from Theorem 9 the following result.

Theorem 15. *Let $G = (V, E, w)$ be a CWC system with fat links. Any algorithm solving* cCast *requires $\Omega(Z_{\max})$ rounds.*

4 Non-commutative Operators and the Wheel Settings

In this section we consider cComb for non-commutative operators in the wheel topology (Fig. 1). Our description here omits many details that can be found in the full version [2].

Consider an instance with a non-commutative operator. Trivially, Algorithm 3 can be used (and Theorem 10 can be applied) if the ordering of the inputs happens to match ordering induced by the algorithm. While such a coincidence is unlikely in general, it seems reasonable to assume that processing nodes are physically connected according to their combining order. Neglecting other possible connections, assuming that the last node is also connected to the first node for symmetry, and connecting a cloud node to all processors, we arrive at the *wheel* topology, which we study in this section. We assume that all links are bidirectional and bandwidth-symmetric.

We start with the concept of intervals that refines the concept of clusters (Definition 1) to the case of the wheel topology.

Definition 7. *The* **cloud bandwidth** *of a processing node $i \in V_p$ in a given wheel graph is $b_c(i) \stackrel{\text{def}}{=} w(i, v_c)$. An* **interval** $[i, i+k] \stackrel{\text{def}}{=} \{i, i+1, \ldots, i+k\} \subseteq V$ *is a path of processing nodes in the ring. Given an interval $I = [i, i+k]$, $|I| = k+1$ is its* **size**, *and k is its* **length**. *The* **cloud bandwidth** *of I, denoted $b_c(I)$, is the sum of the cloud bandwidth of all nodes in I: $b_c(I) = \sum_{i \in I} b_c(i)$. The* **bottleneck bandwidth** *of I, denoted $\phi(I)$, is the smallest bandwidth of a link in the interval: $\phi(I) = \min\{w(i, i+1) \mid i, i+1 \in I\}$. If $|I| = 1$, define $\phi(I) = \infty$.*

(Note that bottleneck bandwidth was not defined for general clusters).

Given these interval-related definitions, we adapt the notion of "cloud cluster" (Definition 3), for problems with inputs s, this time also accounting for the bottleneck of the interval. We define I_i to be the **cloud interval** of node i, and $Z_i = |I_i| + \frac{s}{\phi(I_i)} + \frac{s}{b_c(I_i)}$ to be the **timespan** of I_i.

Similarly to fat-links, we obtain the following results for cW_i and cR_i.

Theorem 16. *In the wheel settings, cW_i can be solved in $O(Z_i)$ rounds for every node i.*

Theorem 17. *In the wheel settings, Any algorithm for cW_i requires at least $\Omega(Z_i)$ rounds for every node i.*

Our main result in this section is an algorithm for cComb for the wheel topology with arbitrary bandwidths, that works in time bounded by $O(\log n)$ times the optimal. We note that by using standard methods [24], the presented algorithm can be extended to compute, with the same asymptotic time complexity, all prefix sums, i.e., compute $\bigotimes_{i=0}^{j} S_i$ for each $0 \le j < n$.

Extending the notion of $Z_{\max}$ to the wheel case, and adapting Algorithm 3, we obtain the following theorem.

Theorem 18. *In the wheel settings, cComb can be solved in $O(Z_{\max} \log n)$ rounds by Algorithm 3.*

This is a log factor improvement over the fat-links case. The main ideas are as follows:

- In the wheel case, for any minimal cover C' of the graph, $\text{load}_{C'}(i) \le 2$ for every node i. Furthermore, a minimal cover is easy to find without resorting to Theorem 12 (see, e.g., [26]).
- Due to the limited local bandwidth, Algorithm 4 can't be used with the same time analysis as in the fat-links case in Steps 3–5 of Algorithm 3. Instead, we use pipelining to compute the inner product of every interval in the cover.

Pipelining. We distinguish between *holistic* and *modular* combining operators, defined as follows (see [2] for details). In modular combining, one can apply the combining operator to aligned, equal-length parts of operands to get the output corresponding to that part. For example, this is the case with vector (or matrix) addition: to compute any entry in the sum, all that is needed is the corresponding entries in the summands. If the operand is not modular, it is called holistic (e.g., matrix multiplication). We call the aligned parts of the operands *grains*, and their maximal size g is the *grain size*. We show that in the modular case, using pipelining, a logarithmic factor can be shaved off the running time (more precisely, converted into an additive term), as can be seen in the following theorem:

Theorem 19. *Suppose $\otimes$ is modular with grain size g, and that $w(e) \geq g$ for every link $e \in E$. Then* cComb *can be solved in $O(Z_{\max} + \log n)$ rounds, where $Z_{\max} = \max\{Z_i \mid i \in V_p\}$.*

5 CWC Applications

In this section we briefly explore some of the possible applications of the results shown in this paper to two slightly more involved applications, namely Federated Learning (Sect. 5.1) and File Deduplication (Sect. 5.2).

5.1 Federated Learning in CWC

Federated Learning (FL) [11,31] is a distributed Machine Learning training algorithm, by which an ML model for some concept is acquired. The idea is to train over a huge data set that is distributed across many devices such as mobile phones and user PCs, without requiring the edge devices to explicitly exchange their data. Thus it gives the end devices some sense of privacy and data protection. Examples of such data is personal pictures, medical data, hand-writing or speech recognition, etc.

In [8], a cryptographic protocol for FL is presented, under the assumption that any two users can communicate directly. The protocol of [8] is engineered to be robust against malicious users, and uses cryptographic machinery such as Diffie-Hellman key agreement and threshold secret sharing. We propose a way to do FL using only cloud storage, without requiring an active trusted central server. Here, we describe a simple scheme that is tailored to the fat-links scenario, assuming that users are "honest but curious."

The idea is as follows. Each of the users has a vector of m weights. Weights are represented by non-negative integers from $\{0, 1, \ldots, M-1\}$, so that user input is simply a vector in $(\mathbb{Z}_M)^m$. Let $\mathbf{x}_i$ be the vector of user i. The goal of the computation is to compute $\sum_{i=0}^{n-1} \mathbf{x}_i$ (using addition over $\mathbb{Z}_M$) and store the result in the cloud. We assume that M is large enough so that no coordinate in the vector-sum exceeds M, i.e., that $\sum_{i=0}^{n-1} \mathbf{x}_i = \left(\sum_{i=0}^{n-1} \mathbf{x}_i \bmod M\right)$.

To compute this sum securely, we use basic multi-party computation in the CWC model. Specifically, each user i chooses a private random vector $\mathbf{z}_{i,j} \in (\mathbb{Z}_M)^m$ uniformly, for each of her neighbors j, and sends $\mathbf{z}_{i,j}$ to user j. Then each user i computes $\mathbf{y}_i = \mathbf{x}_i - \sum_{(i,j)\in E} \mathbf{z}_{i,j} + \sum_{(j,i)\in E} \mathbf{z}_{j,i}$, where addition is modulo M. Clearly, $\mathbf{y}_i$ is uniformly distributed even if $\mathbf{x}_i$ is known. Also note that $\sum_i \mathbf{y}_i = \sum_i \mathbf{x}_i$. Therefore all that remains to do is to compute $\sum_i \mathbf{y}_i$, which can be done by invoking cComb, where the combining operator is vector addition over $(\mathbb{Z}_M)^m$. We obtain the following theorem from Theorem 11.

Theorem 20. *In a fat-links network, an FL iteration with vectors in* $(\mathbb{Z}_M)^m$ *can be computed in* $O(Z_{\max} \log^2 n)$ *rounds.*

Since the grain size of this operation is $O(\log M)$ bits, we can apply the pipelined version of cComb in case that the underlying topology is a cycle, to obtain the following.

Theorem 21. *In the uniform n-node wheel, an FL iteration with vectors in* $(\mathbb{Z}_M)^m$ *can be computed in* $O(\sqrt{(m \log M)/b_c} + \log n)$ *rounds, assuming that* $b_c m \log M \leq b_\ell^2$ *and* $b_c \geq \log M$.

5.2 File Deduplication with the Cloud

Deduplication, or Single-Instance-Storage (SIS), is a central problem for storage systems (see, e.g., [7,17,32]). Grossly simplifying, the motivation is the following: Many of the files (or file parts) in a storage system may be unknowingly replicated. The general goal of deduplication (usually dubbed dedup) is to identify such replications and possibly discard redundant copies. Many cloud storage systems use a dedup mechanism internally to save space. Here we show how the processing nodes can cooperate to carry out dedup without active help from the cloud, when the files are stored locally at the nodes (cf. serverless SIS [13]). We ignore privacy and security concerns here.

We consider the following setting. Each node i has a set of local files F_i with their hash values, and the goal is to identify, for each unique file $f \in \bigcup_i F_i$, a single owner user $u(f)$. (Once the operation is done, users may delete any file they do not own).

This is easily done with the help of cComb as follows. Let h be a hash function. For file f and processing node i, call the pair $(h(f), i)$ a *tagged hash*. The set $S_i = \{(h(f), i) \mid f \in F_i\}$ of tagged hashes of F_i is the input of node i. Define the operator $\tilde{\cup}$ that takes two sets S_i and S_j of tagged hashes, and returns a set of tagged hashes without duplicate hash values, i.e., if (x, i) and (x, j) are both in the union $S_i \cup S_j$, then only $(x, \min(i, j))$ will be in $S_i \tilde{\cup} S_j$. Clearly $\tilde{\cup}$ is associative and commutative, has a unit element ($\emptyset$), and therefore can be used in the cComb algorithm. Note that if the total number of unique files in the system is m, then $s = m \cdot (H + \log n)$. Applying cComb with operation $\tilde{\cup}$ to inputs S_i, we obtain a set of tagged hashes S for all files in the system, where $(h(f), i) \in S$ means that user i is the owner of file f. Then we invoke cCast to disseminate the ownership information to all nodes. Thus dedup can be done in CWC in $O(Z_{\max} \log^2 n)$ rounds.

6 Conclusion and Open Problems

In this paper we have introduced a new model that incorporates cloud storage with a bandwidth-constrained communication network. We have developed a few building blocks in this model, and used these primitives to obtain effective solutions to some real-life distributed applications. There are many possible directions for future work; below, we mention a few.

One interesting direction is to validate the model with *simulations and/or implementations* of the algorithms, e.g., implementing the federated learning algorithm suggested here.

A few algorithmic question are left open by this paper. For example, can we get good approximation ratio for the problem of combining in a general (directed, capacitated) network? Our results apply to fat links and the wheel topologies.

Another interesting issue is the case of *multiple cloud nodes*: How can nodes use them effectively, e.g., in combining? Possibly in this case one should also be concerned with privacy considerations.

Finally, *fault tolerance*: Practically, clouds are considered highly reliable. How should we exploit this fact to build more robust systems? and on the other hand, how can we build systems that can cope with varying cloud latency?

References

1. Adler, M., Gibbons, P., Matias, Y., Ramachandran, V.: Modeling parallel bandwidth: local versus global restrictions. Algorithmica **24**, 381–404 (1999)
2. Afek, Y., Giladi, G., Patt-Shamir, B.: Distributed computing with the cloud. arXiv e-prints, arXiv:2109.12930, September 2021
3. Attiya, H., Welch, J.: Distributed Algorithms. McGraw-Hill, New York (1998)
4. Augustine, J., Hinnenthal, K., Kuhn, F., Scheideler, C., Schneider, P.: Shortest paths in a hybrid network model. In: Proceedings of the SODA 2020, pp. 1280–1299 (2020)
5. Awerbuch, B., Peleg, D.: Sparse partitions (extended abstract). In: 31st FOCS, pp. 503–513. IEEE Computer Society (1990)
6. Baumann, N., Skutella, M.: Solving evacuation problems efficiently-earliest arrival flows with multiple sources. In: Proceedings of the 47th FOCS, pp. 399–410 (2006)
7. Bolosky, B., Corbin, S., Goebel, D., Douceur, J.: Single instance storage in windows 2000. In: Proceedings of the 4th USENIX Windows Systems Symposium USENIX (2000)
8. Bonawitz, K., et al.: Practical secure aggregation for privacy-preserving machine learning. In: Proceedings of the CCS 2017, pp. 1175–1191. ACM (2017)
9. Burkard, R.E., Dlaska, K., Klinz, B.: The quickest flow problem. ZOR - Methods Models Oper. Res. **37**, 31–58 (1993)
10. Censor-Hillel, K., Leitersdorf, D., Polosukhin, V.: Distance computations in the hybrid network model via oracle simulations. In: Proceedings of the 38th STACS (2021)
11. Cheng, Y., Liu, Y., Chen, T., Yang, Q.: Federated learning for privacy-preserving AI. Commun. ACM **63**(12), 33–36 (2020)
12. Culler, D., et al.: LogP: towards a realistic model of parallel computation. SIGPLAN Not. **28**(7), 1–12 (1993)
13. Douceur, J.R., Adya, A., Bolosky, W.J., Simon, P., Theimer, M.: Reclaiming space from duplicate files in a serverless distributed file system. In: Proceedings of the 22nd International Conference on Distributed Computing Systems, pp. 617–624 (2002)

14. Dropbox. Prospectus. Filing to US Securities and Exchange Commission (2018). https://www.sec.gov/Archives/edgar/data/1467623/000119312518055809/d451946ds1.htm
15. Fleischer, L., Skutella, M.: Quickest flows over time. SIAM J. Comput. **36**(6), 1600–1630 (2007)
16. Fraigniaud, P.: Distributed computational complexities: are you Volvo-addicted or NASCAR-obsessed? In: Proceedings of the 30th PODC, pp. 171–172. ACM (2010)
17. Freeman, L.: Looking beyond the hype: Evaluating data deduplication solutions. Network Appliance Inc., September 2007. http://www-download.netapp.com/edm/TT/docs/Looking_beyond_hype_Dedupe.pdf
18. Friedman, R., Kliot, G., Kogan, A.: Hybrid distributed consensus. In: Baldoni, R., Nisse, N., van Steen, M. (eds.) OPODIS 2013. LNCS, vol. 8304, pp. 145–159. Springer, Cham (2013). https://doi.org/10.1007/978-3-319-03850-6_11
19. Gibbons, P., Matias, Y., Ramachandran, V.: Can shared-memory model serve as a bridging model for parallel computation? In: Proceedings of the 9th SPAA, pp. 72–83. ACM (1997)
20. Hegeman, J.W., Pemmaraju, S.V.: Lessons from the congested clique applied to MapReduce. Theor. Comput. Sci. **608**(P3), 268–281 (2015)
21. Hoppe, B., Tardos, É.: Polynomial time algorithms for some evacuation problems. In: SODA 1994 (1994)
22. Karloff, H., Suri, S., Vassilvitskii, S.: A model of computation for MapReduce. In: Proceedings of the 21st SODA. Society for Industrial and Applied Mathematics (2010)
23. Kuhn, F., Schneider, P.: Computing shortest paths and diameter in the hybrid network model. In: Proceedings of the 39th PODC, pp. 109–118. ACM (2020)
24. Ladner, R.E., Fischer, M.J.: Parallel prefix computation. J. ACM **27**(4), 831–838 (1980)
25. Lardinois, F.: Google Drive Will Hit a Billion Users This Week. TechCrunch, July 2018
26. Lee, C.C., Lee, D.T.: On a circle-cover minimization problem. Inf. Process. Lett. **18**(2), 109–115 (1984)
27. Li, M., Vitányi, P.: An Introduction to Kolmogorov Complexity and its Applications, 4th edn. Springer, Heidelberg (2019)
28. Linial, N.: Locality in distributed graph algorithms. SICOMP **21**, 193–201 (1992)
29. Lotker, Z., Patt-Shamir, B., Pavlov, E., Peleg, D.: Minimum-weight spanning tree construction in $O(\log\log(n))$ communication rounds. SICOMP **35**(1), 120–131 (2005)
30. Mansour, Y., Nisan, N., Vishkin, U.: Trade-offs between communication throughput and parallel time. J. Complex. **15**(1), 148–166 (1999)
31. McMahan, B., Moore, E., Ramage, D., Hampson, S., Agüera y Arcas, B.: Communication-efficient learning of deep networks from decentralized data. In: PMLR, vol. 54, pp. 1273–1282 (2017)
32. Meyer, D.T., Bolosky, W.J.: A study of practical deduplication. ACM Trans. Storage **7**(4) (2012)
33. Peleg, D.: Distributed Computing: A Locality-Sensitive Approach. Society for Industrial and Applied Mathematics, Philadelphia (2000)
34. Skutella, M.: An introduction to network flows over time. In: Cook, W.J., Lovász, L., Vygen, J. (eds.) Research Trends in Combinatorial Optimization, pp. 451–482. Springer, Heidelberg (2009). https://doi.org/10.1007/978-3-540-76796-1_21
35. Valiant, L.G.: A bridging model for parallel computation. Commun. ACM **33**(8), 103–111 (1990)

Byzantine-Tolerant Reliable Broadcast in the Presence of Silent Churn

Timothé Albouy, Davide Frey, Michel Raynal(✉), and François Taïani(✉)

Univ Rennes, IRISA, CNRS, Inria, 35042 Rennes, France
raynal@irisa.fr

Abstract. This paper introduces a new reliable broadcast communication abstraction suited to n-process asynchronous message-passing systems in which up to t processes may behave arbitrarily (Byzantine processes) and where (due to transient disconnections or message losses) up to d correct processes may not receive a message broadcast by a correct (i.e., not Byzantine) process. Then the paper presents and proves correct an algorithm implementing such a communication abstraction where the system parameters n, t, and d are such that $n > 3t + 2d$.

Keywords: Asynchronous system · Byzantine processes · Churn · Message adversary · Message losses · Message-passing · Reliable broadcast · Transient disconnection

1 Introduction

Reliable Broadcast. Introduced in the mid of eighties, Reliable Broadcast is a fundamental communication abstraction that lies at the center of fault-tolerant asynchronous distributed systems. Formally defined in [2,3], it allows each process to broadcast messages in the presence of process failures, with well-defined delivery[1] properties, which allow the design of provably correct software for an upper layer based on such a broadcast abstraction.

More precisely, reliable broadcast guarantees that the non-faulty processes deliver the same set of messages, which includes at least all the messages they broadcast. This set may also contain messages broadcast by faulty processes. The fundamental property of reliable broadcast lies in the fact that no two non-faulty processes deliver different sets of messages [4,14].

In the context where some processes can commit Byzantine failures [10], the design of a reliable broadcast communication abstraction is far from being trivial. Such an algorithm is called Byzantine-tolerant reliable broadcast (BRB) and we say that a process brb-broadcasts and brb-delivers messages. The most famous BRB algorithm is due to Bracha [2] (1987). For an application message, this algorithm gives rise to 3 sequential communication steps and $(n-1)(2n+1)$

[1] The term *delivery* refers here to the application layer where a process receives and can access the content of an application message. See Sect. 3.

C. Johnen et al. (Eds.): SSS 2021, LNCS 13046, pp. 21–33, 2021.
https://doi.org/10.1007/978-3-030-91081-5_2

implementation messages. This algorithm requires $n > 3t$, which is optimal from the fault-tolerance point of view.

Recent Works Related to Reliable Broadcast. It is natural that, as it is a fundamental communication abstraction, BRB has been addressed by many authors. Here are a few recent results. Like Bracha's algorithm, all these algorithms assume an underlying fully connected reliable network.

- The versatility dimension of Bracha's algorithm has been analyzed in [8,15].
- Addressing efficiency issues, the BRB algorithm presented in [9], implements the reliable broadcast of an application message with only two communication steps and $n^2 - 1$ protocol messages. The price to pay for this gain in efficiency is a weaker t-resilience, namely $t < n/5$. Hence, this algorithm and Bracha's algorithm differ in the trade-off t-resilience versus message/time efficiency.
- Scalable BRB is addressed in [7]. The issue is here not to pay the $O(n^2)$ message complexity price. To this end, the authors use a non-trivial message-gossiping approach which allows them to design a sophisticated BRB algorithm satisfying fixed probability-dependent properties.
- BRB in *dynamic* systems is addressed in [6]. Dynamic means that a process can enter and leave the system at any time. In their article the authors present an efficient BRB algorithm for such a context. This algorithm assumes that, at any time, the number of Byzantine processes present in the system is less than one third of the total number of processes present in the system.
- An efficient algorithm for BRB with long inputs of ℓ bits using lower costs than ℓ single-bit instances is presented in [11]. This algorithm, which assumes $t < n/3$, achieves the best possible communication complexity of $\Theta(n\ell)$ input sizes. This article also presents an authenticated extension of the previous algorithm.

This article is an exploratory work on asynchronous systems in which some processes are Byzantine and, at the network level, an adversary suppresses messages (which creates a form of churn which we call *Silent Churn*). Section 2 defines the underlying computing model and the message adversary. Section 3 presents the new BRB broadcast abstraction (denoted SCB-broadcast) suited to the model. Section 4 presents an algorithm implementing SCB-broadcast. Section 5 proves it is correct. Section 6 evaluates its cost. Finally, Sect. 7 concludes the article.

Motivation. This article originated in our research on the reconciliation of process local states in distributed Byzantine money transfer systems, in which processes can disconnect for long periods of time.

2 Computing Model

Process Model. The system is composed of n asynchronous sequential processes denoted p_1, ..., p_n. Each process p_i has an identity, and all the identities are different and known by all processes. To simplify, we assume that i is the identity of p_i.

On the failure side, up to t processes can be Byzantine, where a Byzantine process is a process whose behavior does not follow the code specified by its algorithm [10,12]. Let us notice that Byzantine processes can collude to fool the non-Byzantine processes (also called correct processes). Let us also notice that, in this model, the premature stop (crash) of a process is a Byzantine failure.

Communication Model. The processes communicate through a fully connected asynchronous point-to-point communication network. Although it is assumed to be reliable in the sense it neither corrupts, duplicates, nor creates messages, as far as messages losses are concerned, the network is under the control of an adversary (defined below) that can suppress messages.

Let MSG be a message type and v the associated value. A process can invoke the unreliable operation broadcast MSG(v), which is a shorthand for "**for all** $i \in \{1, \cdots, n\}$ **do** send MSG(v) **to** p_j **end for**".

It is assumed that all the correct processes invoke broadcast() to send messages. As we can see, the operation broadcast MSG(v) is not reliable. As an example, if the invoking process crashes during its invocation, an arbitrary subset of processes receive the message implementation message MSG(v). Moreover, due to its very nature, a Byzantine process can send messages without using the macro-operation broadcast().

Terminology. From a terminology point of view, at the system/network level, we say that messages are *broadcast* and *received.*

Moreover, a message generated by the algorithm is said to be a *base* or *implementation* message, while a message generated by the application layer is said to be an *application* message.

Message Adversary. The notion of a message adversary has been implicitly introduced in [17], and then used (sometimes implicitly) in [1,5,16,18]. A short tutorial is presented in [13].

Let d be an integer constant such that $0 \leq d < n$. The communication network is under the control of a *message adversary* which eliminates messages sent by the processes, so these messages are lost. More precisely, when a correct process invokes broadcast MSG(v), the message adversary is allowed to arbitrarily suppress up to d copies of the message MSG(v). This means that, despite the fact the sender is correct, up to d correct processes can miss the message MSG(v).

As an example, let us consider a set D of correct processes, where $1 \leq |D| \leq d$, such that during some period of time, the adversary suppresses all the messages sent to them. It follows that, during this period of time, this set of processes appears as a set of correct processes input-disconnected from the other correct processes. According to the message adversary, the set D can vary with time. Let us notice that $d = 0$ corresponds to the weakest possible message adversary: it corresponds to a classical static system where some processes are Byzantine but no message is lost (the network is reliable).

Let us remark that this type of message adversary is stronger, and therefore covers, the more specific case of *silent churn*, in which processes (nodes) may decide to disconnect from the network. While disconnected, such a process silently pauses its algorithm (a legal behavior in our asynchronous model), and is implicitly moved (by the adversary) to the D adversary-defined set. Upon coming back, the node resumes its execution, and is removed by the adversary from D.

Informally, in a silent churn environment, a correct process may miss messages sent by other processes during the time it is disconnected from the network. The adjective "silent" in *silent churn* expresses the fact that no notification is sent on the network by processes whenever they leave or join the system: there is no explicit "attendance list" of connected processes, and processes are given no information on the status (connected/disconnected) of their peers. In this regard, the silent churn model strays away from the classical approach to design dynamic distributed systems, where processes send messages on the network notifying their connection or disconnection [6]. The silent churn model is a good representation of real-life large-scale peer-to-peer systems, where peers leaving the network typically do so in a completely silent manner (i.e., without warning other peers).

Let us also observe that silent churn allows us to model input-disconnections due to process mobility. When a process moves from a location to another location it is possible that the range used by the sender to broadcast messages is not large enough to ensure that the moving process remains input-connected. An even more prosaic example would be one where a user simply turns off their device, or disable its Internet connection, which entails that it would not be able to receive or send messages anymore. In this context, we consider that the message adversary removes all the incoming and outgoing messages from the corresponding process, until the device reconnects.

Computability Bound. As the algorithm presented in Sect. 4 uses signatures, it is assumed that (in that algorithm) the computability power of the adversary is bounded.

3 Silent Churn Byzantine-Tolerant Broadcast: Definition

The SCB-broadcast communication abstraction is composed of two matching operations denoted $\mathsf{scb\_broadcast}()$ and $\mathsf{scb\_deliver}()$. It considers that an identity $\langle i, sn\rangle$ (sender identity, sequence number) is associated with each application message, and assumes that any two application messages scb-broadcast by the same correct process have different sequence numbers. Sequence numbers are one of the most natural ways to design "multi-shot" reliable broadcast algorithms, that is, algorithms where the broadcast operation can be invoked multiple times.

When, at the application level, a process p_i invokes $\mathsf{scb\_broadcast}(m, i, sn)$, where m is the application message, we say it "scb-broadcasts (m, i, sn)". Similarly when the invocation of $\mathsf{scb\_deliver}$ returns (m, j, sn), we say it "scb-delivers (m, i, sn)".

We say that the application messages are *scb-broadcast* and *scb-delivered* (while, as said previously, the messages generated by an implementation algorithm are *broadcast* and *received*). The SCB-broadcast abstraction is defined by the following properties:

- Safety:
  - SCB-Validity (no spurious message). If a correct process p_i scb-delivers a message m from a correct process p_j with sequence number sn, then p_j scb-broadcast m with sequence number sn.
  - SCB-No-duplication. A correct process p_i scb-delivers at most one message m from a process p_j with sequence number sn.
  - SCB-No-duplicity. If a correct process p_i scb-delivers a message m from a process p_j with sequence number sn, then no correct process scb-delivers another message $m' \neq m$ from p_j with sequence number sn.
- Liveness:
  - SCB-Local-delivery. If a correct process p_i scb-broadcasts a message m with sequence number sn, then at least one correct process p_j eventually scb-delivers m from p_i with sequence number sn.
  - SCB-Global-delivery. If a correct process p_i scb-delivers a message m from a process p_j with sequence number sn, then at most d correct processes do not scb-deliver m from p_j with sequence number sn.[2]

If $d = 0$, the previous specification boils down to Bracha's specification [2]. Let us notice that the constraint $n > 2d$ prevents the message adversary from partitioning the system.

4 An Algorithm Implementing the SCB-Broadcast Abstraction

This section presents Algorithm 1, that implements the SCB-broadcast communication abstraction under the constraint $n > 3t + 2d > 0$.

4.1 Signatures, Local Data Structures and Message Types

The Operations sign() *and* verify() The algorithm uses an asymmetric cryptosystem to sign messages and verify their authenticity. Every node in the network has a public/private key pair. We suppose that the public keys are known by everyone, and that the private keys are kept secret by their owner. Everyone also knows the mapping between any node's index i and its public key. This signature scheme provides two operations, sign and verify:

[2] Let us observe that, as at the implementation level the message adversary can always suppress all the implementation messages send to a fixed set D of d processes, these SCB-delivery properties are the best that can be done.

- sign(MSG) creates a digest of the message MSG, signs it using the calling process' private key and returns the resulting signature. The base message MSG is made up of a triplet or a five-uplet containing an application message m, a sequence number sn, and the identity i of the sender process p_i.
- verify(MSG, sig, i) returns $\top$ (true) if the signature sig of message MSG is valid using the public key of p_i, otherwise it returns $\perp$ (false). As the system is static, the verify() operation can also check *under the hood* that the public key given in the parameter is not in the process blacklist. If it is, it also returns $\perp$.

The signatures are used to cope with the net effect of the Byzantine processes and the fact that implementation messages broadcast (sent) by correct processes can be eliminated by the message adversary. A noteworthy advantage of signatures is that, in spite of the unauthenticated nature of the point-to-point communication channels, signatures allow correct processes to verify the authenticity of messages that have not been directly received from their initial sender, but rather relayed through intermediary processes. Signatures provide us with a *network-wide* non-repudiation mechanism: if a Byzantine process issues two conflicting messages to two different subsets of correct processes, then the correct processes can detect the malicious behavior by disclosing between themselves the Byzantine signed messages.

The fact that the algorithm uses signed implementation messages does not mean that SCB-broadcast requires signatures to be implemented. The design of a signature-free SCB-broadcast algorithm (or the proof of its impossibility) is an open problem for which the techniques introduced in [19] could prove to be useful.

Message Types. The algorithm uses two message types.

- The type ECHO is associated with the base messages that allow an application message to be disseminated to all the correct processes. It implements controlled flooding.
- The type QUORUM is associated with the base messages that contain the proof testifying that a quorum of processes witnessed a given application message. This proof consists of all signatures of the quorum. The fact that this proof can be exchanged in the network as soon as one correct process witnesses a quorum incidentally guarantees the SCB-Global-delivery property.

Local Data Structures. Each (correct) process uses the following local variables.

- sn_i: integer, initialized to 0, used to generate sequence numbers.
- sig_i and sig'_i: correspond respectively to the signatures (i.e. the signed fixed-size digest of a certain data) for a triplet message and a five-uplet base message. A triplet base message contains an application message, its sequence number and its sender identity, signed and sent by the sender. The five-uplet base message is sent by a process to show that it witnessed a base triplet message.

- $echoes_i$: set, initially empty, of five-uplets representing the echoed messages that have been received by p_i.
- $quorum_i$: set containing pairs of signature/signing process identity, that constitutes proof that enough processes witnessed a given application message for it to be delivered.
- $valid_i$: set containing pairs of signature/signing process identity that are valid, i.e., signatures for which the verify() operation returned true with respect to the corresponding sender identity.
- $delivered_i$: set initially empty, that contains the identifiers (proc. id, seq. number) of the application messages scb-delivered by p_i.

4.2 Algorithm

At a high level, Algorithm 1 works by producing, forwarding, and accumulating *witnesses* of an initial broadcast operation, until a large-enough quorum is observed by at least one correct process, at which point this quorum is propagated in one final unreliable broadcast operation. Witnesses take the form of doubly signed ECHO() messages, including the process that initially invoked the scb_broadcast operation. Because the underlying broadcast is unreliable, individual ECHO() messages do not necessarily reach enough correct processes in one broadcast step to build a quorum. To overcome this weakness, correct processes greedily resend ECHO() messages they have seen for the first time.

Signatures serve to ascertain the provenance and authenticity of these propagated ECHO messages, thus providing a key ingredient to tolerate the limited reliability of the underlying network. They also authenticate the invoker of an initial scb_broadcast operation, and finally, in the last phase of our algorithm, they allow us to propagate a cryptographic proof that a quorum has been reached, ensuring that enough correct processes eventually scb_deliver the initial scb_broadcast.

In more detail, when a (correct) process p_i invokes scb_broadcast(m), it computes the next sequence number (line 1), builds the triplet $\langle m, sn_i, i\rangle$, and signs it to guarantee its non-repudiation (line 2). Then, p_i signs a second time to authenticate the wrapping base message (line 3) that will be broadcast as an ECHO() message. Next, it adds its own echo message to the $echoes_i$ set (line 4) so that it will not rebroadcast it later. Finally, p_i broadcasts the message (line 5).

When a correct process p_i receives an ECHO(m, sn, j, sig, k, sig') base message, it first checks if this message was not already received and if the inner signature is valid (line 6). If this condition is satisfied, p_i then checks if the outer signature is also valid (line 7). If it the case, then the echo is added to the $echoes_i$ set (line 8). After that, the process p_i checks if it issued an echo for the given base message (line 10). If it did not, it signs a new echo message (line 11), adds it to the $echoes_i$ set (line 12) and broadcasts it (line 13). Finally, if p_i has witnessed a quorum of strictly more than $\frac{n+t}{2}$ echoes for the same base message $\langle m, sn, j\rangle$ (line 15), it constructs a quorum message containing all the signatures of the echoes it has received so far for the base message (line 16) and broadcasts it (line 17).

```
operation scb_broadcast(m) is
(1)  sn_i ← sn_i + 1;
(2)  sig_i ← sign(⟨m, sn_i, i⟩);
(3)  sig'_i ← sign(⟨m, sn_i, i, sig_i, i⟩);
(4)  echoes_i ← echoes_i ∪ {⟨m, sn_i, i, sig_i, i, sig'_i⟩};
(5)  broadcast ECHO(m, sn_i, i, sig_i, i, sig'_i).

when ECHO(m, sn_i, j, sig, k, sig') is received do
(6)  if (⟨m, sn, j, sig, k, sig'⟩ ∉ echoes_i ∧ verify(⟨m, sn, j⟩, sig, j)) then
(7)    if (verify(⟨m, sn, j, sig, k⟩, sig', k)) then
(8)      echoes_i ← {echoes_i ∪ ⟨m, sn, j, sig, k, sig'⟩}
(9)    end if;
(10)   if (⟨m, sn, j, sig, i, −⟩ ∉ echoes_i) then
(11)     sig'_i ← sign(⟨m, sn, j, sig, i⟩);
(12)     echoes_i ← echoes_i ∪ {⟨m, sn, j, sig, i, sig'_i⟩};
(13)     broadcast ECHO(m, sn, j, sig, i, sig'_i)
(14)   end if;
(15)   if (|{⟨m, sn, j, sig, −, −⟩ ∈ echoes_i}| > (n+t)/2) then
(16)     quorum_i ← {⟨ℓ, sig''⟩ | ⟨m, sn, j, sig, ℓ, sig''⟩ ∈ echoes_i};
(17)     broadcast QUORUM(m, sn, j, sig, quorum_i)
(18)   end if
(19) end if.

when QUORUM(m, sn, j, sig, quorum) is received do
(20) valid_i ← {⟨k, sig'⟩ ∈ quorum | verify(⟨m, sn, j, sig, k⟩, sig', k)};
(21) if (|valid_i| > (n+t)/2 ∧ ⟨sn, j⟩ ∉ delivered_i) then
(22)   broadcast QUORUM(m, sn, j, sig, valid_i);
(23)   delivered_i ← delivered_i ∪ {⟨sn, j⟩};
(24)   scb_delivery of m from p_j with the sequence number sn
(25) end if.
```

Algorithm 1: Silent churn Byzantine reliable broadcast (code for p_i)

When a correct process p_i receives a QUORUM$(m, sn, j, sig, quorum)$ message, it first selects only the echoes for the base message that have a valid signature (line 20). Then, p_i checks if there are enough valid echoes to constitute a quorum and if it has not yet scb-delivered anything for the given message identifier (line 21). If this condition is satisfied, then p_i rebroadcasts the quorum message (line 22) (this is to cope with the case where the sender is Byzantine and has broadcast the message to only a subset of correct processes). Finally, p_i adds the identifier of the concerned application message m to the $delivered_i$ set (line 23) and locally scb-delivers m (line 24).

Remark. The reader can notice that the system parameters n and t appear in the algorithms, whereas the system parameter d does not. Naturally, they all explicitly appear in the proof.

5 Proof of the Algorithm

Assuming, $n > 3t + 2d > 0$, this section shows that Algorithm 1 implements SCB-broadcast. To this end it shows that it satisfies the five properties defining this communication abstraction.

Theorem 1 (SCB-Validity). *If a correct process p_i scb-delivers m from p_j, where p_j is a correct process, then p_j has previously invoked* scb_broadcast (m).

Proof. A correct process p_i scb-delivers an application m from a process p_j at line 24 when it has received a quorum of strictly more than $\frac{n+t}{2}$ echoes (due to the condition at line 21) for this message identified $\langle j, sn\rangle$. Out of these $\frac{n+t}{2}$ echoes, at most t are sent by Byzantine processes. Consequently, there remains strictly more than $\frac{n-t}{2}$ echoes from correct processes. Due to the algorithm hypothesis, we have $n > 3t + 2d \iff n - t > 2t + 2d \iff \frac{n-t}{2} > t + d \geq 0$, from which it follows that at least one echo is from a correct process p_ℓ.

We conclude that p_ℓ has checked the validity of the signature *sig* for m when it received the message ECHO$(m, sn, j, sig, -, -)$ at line 6. It follows from the sign/verify operations that the only way to create *sig* is for p_j to invoke the sign() operation at line 2, during the scb_broadcast (m) invocation. □

Theorem 2 (SCB-No-duplication). *Given a pair $\langle j, sn\rangle$, a correct process p_i scb-delivers at most one message identified $\langle j, sn\rangle$.*

Proof. Before scb-delivering m from p_j with sequence number sn at line 24, p_i adds the message identity $\langle j, sn\rangle$ to the $delivered_i$ set at line 23. However, as p_i checks if a message has not been already scb-delivered for the given message identifier at line 21, it follows that p_i can deliver a message m identified $\langle j, sn\rangle$ at most once. □

Theorem 3 (SCB-No-duplicity). *If a correct process p_i scb-delivers an application message m identified $\langle sn, k\rangle$, no correct process scb-delivers $m' \neq m$ with the identity $\langle sn, k\rangle$.*

Proof. Seeking a contradiction, let us consider two correct processes p_i and p_j which respectively scb-deliver (m, sn, k) and (m', sn, k) such that $m \neq m'$. It means that these two correct processes respectively received two quorum messages containing $\lfloor\frac{n+t}{2}\rfloor + 1$ distinct valid echoes for the (m, sn, k) and (m', sn, k) base messages (due to the condition at line 21) from two distinct sets of processes, that we respectively denote A and B. We have $|A| = |B| = \lfloor\frac{n+t}{2}\rfloor + 1$.

On another side we have $|A \cap B| = |A| + |B| - |A \cup B| \geq |A| + |B| - n \geq 2(\lfloor\frac{n+t}{2}\rfloor + 1) - n = 2\lfloor\frac{n+t}{2}\rfloor + 2 - n > 2(\frac{n+t}{2}) - n = t$. Hence, A and B have at least one correct process in common which signed and broadcast both echo messages. However, a correct process signs and broadcasts an echo only once for a given message identifier: whether it be at line 5 or line 13, a correct process always checks that it had not broadcast a conflicting echo before. For the broadcast operation at line 5, we assume that the invoking process is correct and does not invoke the scb_broadcast operation twice for the same sequence number. Before broadcasting the echo message broadcast at line 13, p_i checks at line 10 that it has not already echoed it. So, the same message with the same sequence number cannot be broadcast again at line 13. Contradiction. □

Theorem 4 (SCB-Local-delivery). *If a correct process p_i invokes* scb_broadcast(m) *with sequence number sn, at least one correct process p_j scb-delivers m from p_j with sequence number sn.*

Proof. If a correct process p_i invokes scb_broadcast(m) with sequence number sn, it broadcasts ECHO$(m, sn, i, sig_i, i, sig'_i)$ at line 5. As p_i is correct it does not invoke the scb_broadcast() operation twice with the same sequence number. It follows that every correct process that receives the ECHO$(m, sn, i, sig_i, -, -)$ message for the first time passes the conditions at lines 6 and 10, and thus broadcasts an echo message for the same base message at line 13.

Let K denote the set of correct processes p_j that broadcast the message ECHO$(m, sn, i, sig_i, j, sig_j)$, whether it is done at line 5 (if p_j is the sender p_i) or at line 13 (if p_j received a message that it did not echo yet). Let us note $k \equiv |K|$ the size of this set.

We show that we have $n - t - d \leq k \leq n - t$. To this end, let us first observe that the echo message that is broadcast by each correct process of K is eventually received by at least $(n - t - d)$ correct processes, due to the property of the message adversary. Thereby, the minimum number of echo messages that are eventually received by correct processes is $k(n-t-d)$. Let us note K' the set of correct processes that receive at least one of these $k(n - t - d)$ messages. By construction of the algorithm, $K' \subseteq K$, and therefore $|K'| \leq k$. It follows that, the (correct) processes of K' receive on average at least $k(n-t-d)/k = n-t-d$ echo messages. As the adversary does not duplicate messages, and as correct processes broadcast a given ECHO message at most once, each of the messages received by a given process of K' is further signed by a distinct correct process.

From the algorithm assumption, we have $3t + 2d < n \iff n + 3t + 2d < 2n \iff n + t < 2n - 2t - 2d \iff \frac{n+t}{2} < n - t - d$. Therefore, the average number of distinct echoes received by the processes of K' (of at least $n - t - d$ correct processes) is strictly superior to the quorum size (of $\frac{n+t}{2}$). It implies that at least one correct process passes the condition at line 15 and broadcasts a quorum message at line 17. This quorum message is then received by at least $(n-t-d)$ correct processes, which all finally scb-deliver m from p_i with sequence number sn at line 24. □

Theorem 5 (SCB-Global-delivery). *If a correct process p_i scb-delivers m from p_j with sequence number sn, then at most d correct processes do not scb-deliver it.*

Proof. If a correct process scb-delivers (m, sn, j) at line 24, then it has previously received a QUORUM message that it forwarded at line 21. From the definition of the message adversary, a quorum message can be missed by at most d correct processes. Consequently, at least $(n - t - d)$ correct processes scb-deliver the message. □

An Additional Property. The reader can check, from the previous proof, that the algorithm satisfies the following scb-delivery property. If there is a set K of k correct processes, $1 \leq k \leq d$, such that there is a finite time τ after which the message adversary never eliminates the implementation messages sent to them, then, after τ, each process of K scb-delivers all the applications messages scb-broadcast by the correct processes.

6 Cost of the Algorithm

This section assumes that the duration of local computations is negligible compared to message transfer delays. So it is assumed local computations have zero duration. Differently, for computing the time complexity of the algorithm it is assumed that the transfer of an implementation message takes one time unit.

Theorem 6 (Time-Cost). *Let $t > 0$. For the values of d such that $d < n-t-\sqrt{\frac{n^2-t^2}{2}}$, the scb-broadcast operation terminates in exactly three message rounds.*

Proof. The invocation of scb_broadcast by a correct process p_i results in the unreliable broadcast of a first ECHO message signed by p_i at line 5. This initial ECHO message is received by at least $(n-t-d-1)$ correct processes that are different from p_i, due to our assumption on the message adversary. This counts for a first message round.

In the second message round, each of these $(n-t-d-1)$ correct processes signs its own ECHO message, and broadcasts it using an unreliable broadcast (lines 11–13). At the end of the second round, in total at least $(n-t-d)$ distinct ECHO messages (counting that of p_i) have been signed and unreliably broadcast, resulting in at least $(n-t-d)^2$ receptions of said ECHO messages by correct processes. As there are at least $n-t$ correct processes, this means that on average each correct process has received $\mu = \frac{(n-t-d)^2}{n-t}$ ECHO messages by the end of the second round, and that at least one correct process, p_ℓ, receives at least this number of ECHO messages.

Using simple algebraic transformations, the assumption $d < n-t-\sqrt{\frac{n^2-t^2}{2}}$, leads to $\mu > \frac{n+t}{2}$. Since the adversary does not duplicate messages, each of these μ messages corresponds to distinct unreliable broadcast invocations (lines 5 or 13), signed by distinct correct processes (due to the test at line 10, and the use the set $echoes_i$). Consequently, at the start of the third round, the condition of line 15 has become true for p_ℓ, at which point p_ℓ has unreliably broadcast a QUORUM message, leading at least $n-t-d$ correct processes to have delivered the initial scb_broadcast message to the application by the end of the third round. The initial scb_broadcast therefore terminates in at most three rounds.

As $t \geq 1$ and $n > 3t+2d$, it follows that $n \geq 4$ and $\frac{n+t}{2} > 2$. As a result, the condition of line 15 cannot become true earlier than the end of the second communication round. With the earlier result, it follows that the scb_broadcast() broadcast operation terminates in exactly three rounds. □

Theorem 7 (Message-Cost). *The scb-broadcast of an application message by a correct process entails the sending of $O(n^2)$ implementation messages.*

Proof. The proof is easy. The broadcast of an echo message at line 5 entails its forwarding by each other correct process, and the same occurs for the quorum implementation messages. Hence the $O(n^2)$ messages complexity. □

7 Conclusion

This article has presented a new communication abstraction that extends Byzantine reliable broadcast (as defined by Bracha and Toueg [2,3]) to systems where an adversary may suppress some subset of application messages that have been broadcast. This kind of messages loss captures what we call the *silent churn* phenomenon. An algorithm implementing the corresponding Byzantine-tolerant reliable broadcast in the presence of silent churn has been presented and proven correct. This algorithm assumes $n > 3t+2d$, where n is the number of processes, t is the maximum number of Byzantine processes, and d is an upper bound on the silent churn.

Acknowledgments. This work was partially supported by the French ANR project ByBLoS (ANR-20-CE25-0002-01) devoted to the modular design of building blocks for large-scale Byzantine-tolerant multi-users applications. The authors want to thank Colette Johnen, Elad Schiller, and Stefan Schmid for their kind invitation to participate in the conference.

References

1. Afek, Y., Gafni, E.: Asynchrony from synchrony. In: Frey, D., Raynal, M., Sarkar, S., Shyamasundar, R.K., Sinha, P. (eds.) ICDCN 2013. LNCS, vol. 7730, pp. 225–239. Springer, Heidelberg (2013). https://doi.org/10.1007/978-3-642-35668-1_16
2. Bracha, G.: Asynchronous Byzantine agreement protocols. Inf. Comput. **75**(2), 130–143 (1987)
3. Bracha, G., Toueg, S.: Asynchronous consensus and broadcast protocols. J. ACM **32**(4), 824–840 (1985)
4. Cachin, Ch., Guerraoui, R., Rodrigues, L.: Reliable and Secure Distributed Programming, p. 367. Springer, Heidelberg (2011). https://doi.org/10.1007/978-3-642-15260-3. ISBN 978-3-642-15259-7
5. Charron-Bost, B., Schiper, A.: The heard-of model: computing in distributed systems with benign faults. Distrib. Comput. **22**(1), 49–71 (2009). https://doi.org/10.1007/s00446-009-0084-6
6. Guerraoui, G., Komatovic, J., Kuznetsov, P., Pignolet, P.A., Seredinschi, D.-A., Tonkikh A.: Dynamic Byzantine reliable broadcast. In: Proceedings of 24th International Conference on Principles of Distributed Systems (OPODIS'20), LIPIcs, vol. 184, Article 23, 18 p. (2020)
7. Guerraoui, G., Kuznetsov, P., Monti, M., Pavlovic, M., Seredinschi, D.-A.: Scalable Byzantine reliable broadcast. In: Proceedings of 33rd International Symposium on Distributed Computing (DISC'19), LIPIcs, vol. 146, Article 22, 16 p. (2019)

8. Hirt, M., Kastrato, A., Liu-Zhang, C.-D.: Multi-threshold asynchronous reliable broadcast and consensus. In: Proceedings of 24th International Conference on Principles of Distributed Systems (OPODIS'20), LIPICs, vol. 184, Article 6, 16 p. (2020)
9. Imbs, D., Raynal, M.: Trading t-resilience for efficiency in asynchronous Byzantine reliable broadcast. Parallel Process. Lett. **26**(4), 8 (2016)
10. Lamport, L., Shostack, R., Pease, M.: The Byzantine generals problem. ACM Trans. Program. Lang. Syst. **4**(3), 382–401 (1982)
11. Nayak, K., Ren, L., Shi, E., Vaidya, N.H., Xiang, Z.: Improved extension protocols for Byzantine broadcast and agreement. In: Proceedings of 34rd Int'l Symposium on Distributed Computing (DISC'20), LIPIcs, vol. 179, Article 28, 16 p. (2020)
12. Pease, M., Shostak, R., Lamport, L.: Reaching agreement in the presence of faults. J. ACM **27**, 228–234 (1980)
13. Raynal, M.: Message adversaries. In: Kao, M.-Y. (ed.) Encyclopedia of Algorithms, pp. 1272–1276. Springer, Heidelberg (2015). https://doi.org/10.1007/978-1-4939-2864-4
14. Raynal, M.: Fault-Tolerant Message-passing Distributed Systems: An Algorithmic Approach, p. 480. Springer, Heidelberg (2018). https://doi.org/10.1007/978-3-319-94141-7. ISBN 978-3-319-94140-0
15. Raynal, M.: On the versatility of Bracha's Byzantine reliable broadcast algorithm. Parallel Process. Lett. **31**(3), 2150006 (2021)
16. Raynal, M., Stainer, J.: Synchrony weakened by message adversaries vs asynchrony restricted by failure detectors. In: Proceedings of 32nd ACM Symposium on Principles of Distributed Computing (PODC '13), pp. 166–175. ACM Press (2013)
17. Santoro, N., Widmayer, P.: Time is not a healer. In: Monien, B., Cori, R. (eds.) STACS 1989. LNCS, vol. 349, pp. 304–313. Springer, Heidelberg (1989). https://doi.org/10.1007/BFb0028994
18. Santoro, N., Widmayer, P.: Agreement in synchronous networks with ubiquitous faults. Theoret. Comput. Sci. **384**(2–3), 232–249 (2007)
19. Srikanth, T.K., Toueg, S.: Simulating authenticated broadcasts to derive simple fault-tolerant algorithms. Distrib. Comput. **2**, 80–94 (1987). https://doi.org/10.1007/BF01667080

Building Systems of Systems with Escher

Burcu Canakci[(✉)], Lorenzo Alvisi, and Robbert van Renesse

Cornell University, Ithaca, USA
{burcu,lorenzo,rvr}@cs.cornell.edu

Abstract. This paper presents Escher, an approach to build and deploy multi-tiered cloud-based applications, and outlines the framework that supports it. Escher is designed to allow *systems of systems* to be derived methodically and to evolve over time, in a modular way. To this end, Escher includes (i) a novel authenticated message bus that hides from one another the low-level implementation details of different tiers of a distributed system; and (ii) general purpose wrappers that take an implementation of an application and deploy, for example, a sharded or replicated version of an application automatically.

Keywords: Middleware · Distributed systems · Refinement · Fault tolerance · Maintainability

1 Introduction

Large-scale distributed systems deployed today within and across datacenters are hierarchically structured. At the highest level they are decomposed into *tiers*, such as, for example, the load balancing tier, the web frontend tier, the data store tier, the caching tier, the data analysis tier, the recommendation tier, and various application logic tiers. Each tier in turn is further subdivided into smaller components, down to the executable services that run on the datacenter servers. Such a *refinement hierarchy* allows for modular development and simplifies management. Furthermore, at the lowest level applications are often containerized to simplify software distribution and deployment. Various readily available management services maintain configurations, dependencies, and deployments of such systems [7,10,22].

While this approach has produced distributed systems of unprecedented scale and sophistication, these systems of systems have introduced a new challenge: managing the interaction between tiers in this refinement hierarchy. Standards for data representation and remote invocation have made it straightforward to glue together services written in different programming languages or deployed on different operating systems [12]. However, such ease of integration stops at the implementation level in the refinement hierarchy; still lacking is a way to describe abstractly the interaction between higher-level tiers of the system.

C. Johnen et al. (Eds.): SSS 2021, LNCS 13046, pp. 34–50, 2021.
https://doi.org/10.1007/978-3-030-91081-5_3

Consider the architecture of a search engine. A client's query first goes through a query processor which is responsible for parsing and formatting the query. The request is then relayed to an online ranking tier, which computes the results. That tier interacts with an indexing tier which is responsible for crawling and indexing offline. The indexing tier itself can include different storage systems, such as an in-memory caching system for popular queries and a disk storage database system for historical indexes. Many modern applications are similarly composed of numerous interacting tiers and components [2,6,24,28]. Different components have different non-functional objectives, and so require different refinements. Additionally, these objectives may change over time, so their refinements are not static [24]. Therefore, what is needed is a clean abstraction for the interaction between *any* two refined components, one that allows them to independently evolve, or even be replaced, without affecting the rest of the system.

This paper presents *Escher*, a principled methodology to (*i*) systematically derive such systems of systems and (*ii*) continue supporting their organic evolution.

Escher specifies communication between tiers at the highest levels of their refinement hierarchy. Instead of exposing a plethora of ways for components to interact with one another, Escher offers just one: a novel message bus that is *refinement-aware*, in that it hides the low-level, refinement-specific details of how components interface with one another. For example, it transparently manages the interaction between a replicated data store and a sharded caching service, and it automatically handles how one replicated component interacts with another in a fault-tolerant fashion.

Escher's bus exposes only the *external* behavior of a system's components. Enforcing this form of isolation is critical to supporting the organic growth of the system: it allows the implementation of each component to be refined (and its correctness to be verified) modularly, and thus for it to evolve, or even be replaced, without requiring other components that are in communication with it to also be reconfigured, or recompiled with new communication libraries.

To further facilitate organic growth, Escher transparently supports modular transformations with *wrappers*, which can automatically provide components with desirable properties such as fault tolerance, load balancing, or privacy.

Escher goes well beyond conventional message streaming platforms [16,23], middleware services [14,20,22], or service-oriented-architecture solutions, such as enterprise service buses [12]. While these solutions enable interactions between objects or services implemented using different languages, data models, or communication APIs, they provide isolation only at the implementation level. Further, their ability to support a system's organic growth is limited. For example, CORBA [14] can refine a service to add fault-tolerance through replication, but does not support the same capability for clients. In Escher, all tiers can be fully and independently refined. For example, a replicated client can transparently communicate with a replicated server, regardless of the replication protocol used on either side. To our knowledge, Escher is the first system to support such two-sided transparency and generality for any refinement.

This paper covers in detail the system model underlying Escher's refinement hierarchy in Sect. 2 and presents different implementation options in Sect. 3. We discuss related work in Sect. 4.

2 Escher Design

Escher models components as *agents*; whether implemented by a single process or a collection of processes, they consist of a state machine combined with a message inbox and outbox. The Escher message bus moves messages from outboxes to inboxes. An agent can be refined by replacing it with a non-empty set of new agents, each with their own state machines, inboxes, and outboxes. A refinement mapping [1] specifies how inboxes and outboxes of the new agents map to the inbox and outbox of the original agent. Escher uses this mapping to hide refinement from other, unrelated agents.

Agents form a *refinement hierarchy*: a parent agent is refined into a set of child agents. For example, in a replicated service, the parent is a virtual, logically centralized service, while the children are the replicas that together provide that abstraction. Other refinements can be similarly modeled and applied recursively to render the refinement hierarchy.

Each agent in the refinement hierarchy has a separate management interface. In particular, each agent has its own public/private key pair. An agent that is being refined uses its private key to create public key certificates for its child agents, so they can prove that they are part of a specific refined agent. The root of the refinement hierarchy forms a *Certification Authority* (CA) for the entire system. The leaves in the hierarchy are physical agents: the Escher framework uses their private keys to sign and verify their messages.

The Escher refinement hierarchy thus forms a uniform management interface for all agents, whether virtual (refined) or physical (running code). Escher therefore solves not only the interfacing between agents, but also provides the management necessary to successfully deploy and run a system of distributed services.

Escher provides the following design properties (partially borrowed from [11]):

Transparency: The refinement of an agent is transparent to other agents. An agent only requires the high-level API (through a set of message types) and the identifier of a destination agent to communicate with it, even if the destination is refined.

Generality: Agents can be refined arbitrarily, as long as the refinement is correct.

Flexibility: Agent refinements can evolve over time.

Manageability: Agents, whether refined or not, can be deployed, executed, and monitored uniformly.

2.1 System Model: Agents and Message Bus

An Escher system is a collection of agents communicating over a message bus. Each agent a is identified by a unique identifier, denoted a.id. Each agent a includes a (possibly nondeterministic) state machine and three unordered collections of messages: a.inbox, a.outbox, and a.donebox. The inbox and done box of an agent are initially empty. The operation of a correct agent a proceeds as follows:

1. Wait until there exists a message $m \in a.\texttt{inbox} - a.\texttt{donebox}$;
2. Apply m to the state machine to produce a new state and a set of output messages;
3. Add the output messages to a.outbox;
4. Add m to a.donebox;
5. Repeat.

A message is a four-tuple: ⟨message identifier, source agent identifier, destination agent identifier, payload⟩. No two messages from the same correct source agent for and the same destination agent can have the same message identifier.

The operation of the message bus is as follows:

1. Wait until there exists a message $m = \langle \_,\ s.\texttt{id},\ d.\texttt{id},\ \_\rangle$ such that $m \in s.\texttt{outbox} - d.\texttt{inbox}$ ($\_$ denotes a wildcard);
2. Add m to d.inbox;
3. Repeat.

Logically, the contents of the three boxes of a correct agent grow monotonically. Later, we show how messages that have been handled can be garbage collected.

The system is asynchronous but fair: if a correct agent—or the message bus—can continually take a step, it eventually will. Unless refined to be fault-tolerant, agents can exhibit both crash failures and arbitrary (Byzantine) failures:

- An agent a that experiences a crash failure no longer processes messages in $a.\texttt{inbox} - a.\texttt{donebox}$;
- An agent a that experiences a Byzantine failure ignores its state machine and places any messages in a.outbox. It may also remove messages from a.outbox.

The message bus is reliable and guarantees *authenticity*: if agents s and d are correct and message $m = \langle i,\ s.\texttt{id},\ d.\texttt{id},\ p\rangle \in d.\texttt{inbox}$, then $m \in s.\texttt{outbox}$.

2.2 Refining Agents

An agent a can be refined, that is, replaced with one or more new agents $a_1, ..., a_n$ such that $a_1, ..., a_n$ collectively have the same *external behavior* as a. We call a the *parent agent* of $a_1, ..., a_n$ and $a_1, ..., a_n$ the *child agents* of agent a. We denote C_a to be the set of child agents of agent a. Child agents can be refined as well, leading to a hierarchy of agents; at the root of the refinement hierarchy is a single *system agent*, which models the entire system. Agents are then identified by *pathnames* of the form {/id}*. For example, /x/y identifies a child agent of /x. The system agent is identified by the empty path name "".

Given a set of input messages M^I for agent a, agents $a_1, ... a_n$ must produce the same set of output messages M^O that agent a might have; specifically, whether or not a correct agent a has been refined must be *invisible* to other correct agents that a communicates with. Supporting this notion of indistinguishability requires adding important features to the message bus.

First, it introduces restrictions on who can communicate with whom. A correct agent will only send messages to its siblings or siblings of its ancestors. For example, a correct agent identified by /x/y/z may send messages to agents identified by /x/y/z′, /x/y′, or /x′, but not to agents identified by /x/y/z′/w, /x/y′/z′ or /x′/y′.

Second, it requires the message bus to be refinement-aware. In particular, if an agent d is refined, the message bus attempts to deliver a message $m = \langle i,\ s.\texttt{id},\ d.\texttt{id},\ p\rangle$ from the outbox of a correct agent s to the inboxes of all child agents of agent d. The message bus guarantees that each message in the inbox of a correct agent d has a destination agent identifier that is a prefix of the identifier of d. For example, the message bus may deliver a message destined for x/y to agent x/y/z, but not a message destined for x/y/z/w nor a message destined for x/y/z'.

In a refinement of agent a, there must exist a function R_a that maps the state of the child agents to the state of the parent agent a [1]. Because the state of each agent has four components (inbox; outbox; done box; and state of its state machine), we split R_a into four corresponding components R^I_a, R^O_a, R^D_a, and R^S_a. In particular, to be able to exchange messages between refined agents, the message bus needs to understand two of these components: R^I_a and R^O_a.

R^I_d is an application-dependent function that, applied to the inboxes of d's the child agents, computes $d.\texttt{inbox}$. For example, in the case of replication, a message could be considered in $d.\texttt{inbox}$ once it is delivered to the inbox of a correct child agent of d; In the case of sharding, a message is considered delivered once it is delivered to the inbox of the child agent that can handle the request in the message.

Similarly, R^O_s is an application-dependent function that, applied to the outboxes of s's child agents, computes $s.\texttt{outbox}$. Let μ_s map a set of messages in the outboxes of child agents of s to a set of messages in the outbox of s. Formally, μ_s maps a set of messages $M_{\langle i,s,d\rangle}$ of the form $\langle i,\ c.\texttt{id},\ d.\texttt{id},\ p_c\rangle$, $c \in C_s$, to a set of messages of the form $\langle i,\ s.\texttt{id},\ d.\texttt{id},\ p_s\rangle$. Recall that we require that correct agents do not produce more than one message with the same message identifier, source agent identifier, and destination agent identifier. Therefore, the result of μ_s is either the empty set or a singleton set. Moreover, if $\mu_s(M^1_{\langle i,s,d\rangle}) = \{\langle i,\ s.\texttt{id},\ d.\texttt{id},\ p^1_s\rangle\}$ and $\mu_s(M^2_{\langle i,s,d\rangle}) = \{\langle i,\ s.\texttt{id},\ d.\texttt{id},\ p^2_s\rangle\}$, then $p^1_s = p^2_s$: messages from different subsets of child agents with the same message identifier must map to the same payload.

For example, if the refinement is "replication for crash failures", then μ_s is the identity function: a message from any of s's child agents (replicas) is a message from the parent agent s. In crash tolerant replication all replicas generate the same messages; thus, applying μ_s to non-empty sets produces singleton sets. Moreover, μ_s applied to any non-empty $M_{\langle i,s,d\rangle}$ is guaranteed to produce the *same* singleton set.

However, if the refinement is "replication for Byzantine failures" with some parameter f, then

$$\mu_s(M_{\langle i,s,d\rangle}) = \Big\{\langle i,\ s.\mathtt{id},\ d.\mathtt{id},\ p\rangle \ \Big|\ c \in C_s \ \wedge \Big|\bigcup\{\langle i,\ c.\mathtt{id},\ d.\mathtt{id},\ p\rangle \in M_{\langle i,s,d\rangle}\}\Big| > f\Big\}$$

That is, a message from s must have more than f matching messages from its child agents. Again, for Byzantine replication, the result of μ_s is either the empty set or some singleton set containing some message m. If $|M_{\langle i,s,d\rangle}| \leq f$, then the result is guaranteed to be the empty set; if $|M_{\langle i,s,d\rangle}| \geq |C_s| - f$, then the result is guaranteed to be $\{m\}$.

Let $M_{\langle i,s,d\rangle}$ be a set of messages such that $\mu_s(M_{\langle i,s,d\rangle})$ yields a singleton set containing a message from s. Then, we call $M_{\langle i,s,d\rangle}$ a *merge group* of messages. Every message in a merge group has an identical message identifier, destination agent identifier, and parent of the source agent.

Let $s.\mathtt{outbox} = R_s^O(\{c.\mathtt{outbox} \mid c \in C_s\})$ be the set of messages in the outbox of agent s according to the refinement mapping of s. We require that the μ_s function satisfies

- **Soundness**:

$$\forall M_{\langle i,s,d\rangle} \subseteq \{\langle i,\ c.\mathtt{id},\ d.\mathtt{id},\ p_c\rangle \in c.\mathtt{outbox} \mid c \in C_s\} : \quad \mu_s(M_{\langle i,s,d\rangle}) \subseteq s.\mathtt{outbox}$$

  In other words: all messages from s that can be constructed, using μ_s, from the outboxes of the child agents of s, are in fact messages produced by s.
- **Completeness**: If $m = \langle i,\ s.\mathtt{id},\ d.\mathtt{id},\ p\rangle \in s.\mathtt{outbox}$, then eventually there exists a set $M_{\langle i,s,d\rangle}$ from correct agents in C_s such that $m = \mu_s(M)$. In other words, every message in $s.\mathtt{outbox}$ is eventually retrievable from the correct child agents of s.

As mentioned above, the message bus delivers a message sent to a refined agent d to the inboxes of all its child agents. This is sufficient, but depending on the definition of R_d^I may be overkill. Escher allows agents to specify a minimum refinement "depth" to manage delivery more precisely. The minimum depth is the number of components in the destination path of the message. For example, an agent with identifier `/x/y/z` and a minimum depth of 2 would only receive messages destined to `/x/y/z` and `/x/y`, but not to `/x`. The default minimum depth is 1.

To summarize, the message bus collects messages from the child agents of s with the same message identifier and destination agent identifier, and uses

μ_s to "merge" those messages into messages from s. If the destination of such a message m is d, then the message bus delivers m to d.`inbox`. If agent d is also refined, then the message bus delivers m to the inboxes of d's child agents, constrained by the child agents' minimum depth settings.

2.3 Wrappers

In some common refinements, which include state machine replication and sharding, a child agent's state machine uses the state machine transition function of its parent as a subroutine. Escher includes a special type of agent called a *wrapper* to handle these types of refinements automatically.

For example, in the case of state machine replication, the wrappers are the replicas, and they must agree on the order in which they present incoming commands to their copies of their parent's state machine (which is required to be deterministic). Similarly, in the case of sharding, each wrapper is responsible for a range of keys and has a copy of the parent's state machine transition function. Each wrapper passes messages for keys within the shard's range to its copy of the state machine.

Escher contains a collection of general-purpose wrappers that automatically provide certain desirable *non-functional properties* to agents such as being able to survive certain classes of failures or being able to handle high load.

2.4 Deploying and Managing Agents

We separate agents into *physical agents* and *virtual agents.* Physical agents are not refined and form the leaf agents in the refinement hierarchy. Virtual agents are refined—they are the internal nodes in the refinement hierarchy. Each agent, whether physical or virtual, must be *managed.* If a is a virtual agent, its manager mgr_a is a principal that can be a person, a cluster manager, or even another agent; if a is a physical agent, mgr_a is a *runtime* that runs both the code of the leaf agent and Escher code that implements the functionality of the message bus.

Escher includes a Public Key Infrastructure (PKI) that it uses to prevent Byzantine agents from forging messages from correct agents: in particular, it ensures that if agent s is correct, then messages that claim to come from s are guaranteed to be in s.`outbox`. The PKI associates with each agent, whether physical or virtual, a private key p_a and a corresponding public key certificate P_a. For a correct agent, only mgr_a holds p_a; the private keys and public key certificates are generally transparent to the agents themselves.

The public key certificate P_a of agent a binds its identifier (a path) to its public key and is signed using the private key of the parent agent of a. Therefore, the manager of a virtual agent a is a Certification Authority (CA) to the child agents of a. The system agent is the root CA of Escher's PKI. The public key certificate of the system agent is self-signed.

Public key certificates are *chained*: they refer to the public key certificates of their parents. Thus, to refine an agent a, manager mgr_a uses its private key p_a

to generate a public key certificate for each of the child agents. Afterward, mgr_a can be offline, unless the refinement of a is updated. If the identifier of some agent x is $\langle$`path`$\rangle$`/x`, then the next certificate on the certificate chain of x is the public key certificate of the parent of x, identified by $\langle$`path`$\rangle$.

If a is a physical agent, the runtime that manages a, among other functions described in the implementation section below, signs outgoing messages using p_a and attaches to them the certificate chain (although certificates can be cached for efficiency). The runtime of destination agents can then verify the messages using the public key contained in P_a.

2.5 Compatibility with Legacy Services

To enable its own gradual deployment, Escher can incorporate legacy services.

Existing centralized microservices simply require a *shim*, which is associated with an Escher identifier and implements the inbox, outbox, and done box. The shim must define a set of message types through which other Escher agents can communicate with it. Upon receipts of such a message, the shim has to invoke the corresponding functionality in the microservice. The output of the microservice has to be similarly transformed before it can be sent on the Escher bus.

Legacy distributed services can be incorporated into Escher in a similar fashion, by defining a virtual agent that models the distributed service. The actual running components of the service can then be modeled as its child agents, each paired with its own shim.

3 Implementation

In this section, we describe several implementation options for the Escher message bus as well as the choices we favor.

3.1 Message Merging

The Escher *runtime* of a physical agent a maintains the inbox, outbox, and done box for a, as well as the state of its state machine. Each runtime also partially implements the message bus. An important functionality of the message bus is to collect messages from the child agents of a refined agent, and to merge them into the corresponding message from the parent agent before delivering the message to the destination. But which component(s) of an Escher system should be responsible for merging messages?

To frame the question, let s be a refined agent and $m_s = \langle i,\ s.\texttt{id},\ d.\texttt{id},\ p_s\rangle$ a message from s to d. Assume child agents of s are unrefined for now. To merge messages from the child agents of s into m_s, an Escher component needs to: (1) collect the necessary set $M_{\langle i,s,d\rangle}$ of messages from the child agents of s, and (2) apply a merge function μ_s to $M_{\langle i,s,d\rangle}$.

Merging at the Sender. One straightforward candidate to perform the merging would be the runtime of one of the child agents of s. This runtime would be in charge of merging outgoing messages and of sending merged messages to agents external to the refinement. After all, the runtime of a child agent of s already knows μ_s, and it can collect $M_{\langle i,s,d\rangle}$ from the runtimes of other child agents. However, this solution comes with several drawbacks: (*i*) it is not fault tolerant; (*ii*) it requires additional cryptographic support, such as multisignatures or threshold signatures, to provide message authenticity efficiently; and finally (*iii*) it requires a message from s to d to take two steps of communication: one from the runtimes of other child agents to the "merging" runtime, and one from the merging runtime to the runtime(s) of the physical descendants of d.

Merging in the Middle. A second solution would be to have a logically-centralized trusted Escher service handle all message merging in the system. This "merging" service, however, would need to be replicated for fault-tolerance, and may become a bottleneck for the entire system. Further, with this solution, as with the previous one, a message from s to d would take at least 2 steps of communication.

Merging at the Receiver. Escher opts for a third solution: it leaves the runtime of a physical destination agent d' the responsibility of merging the messages it receives. To merge messages from child agents of a refined agent s, d''s runtime needs to know μ_s. We considered endowing Escher with a trusted configuration service, which d' could query to retrieve μ_s, but we ultimately rejected this option: the service would add an extra step of communication before the runtime of d' can merge messages, would have to be replicated for fault tolerance, and may turn into a bottleneck. Further, child agents in dynamic refinements would need to synchronize their configuration with the service whenever the refinement is updated. Instead, we opted for an approach that eliminates the need for a global service and does not induce extra steps of communication: we piggyback μ_s onto all message from the child agents of s to agents that are not part of the refinement of s. When the runtime of a child agent attaches the corresponding merge function(s) to an outgoing message, we say that the runtime *tag*s the message.

This solution has minimal communication overhead, is fault tolerant, and both modular and general. Its main drawback is that running merging code in the destination agent is a potential security issue; thus, it should be done with great care, for example through the use of a safe language.

Implementing Merge Functions. Escher is intended to support any kind of refinement, and therefore specific merge functions should ideally not be built into the Escher framework itself. The most general option would be to support arbitrary programs for merge functions. However, a Byzantine agent can be refined as well, and such a Byzantine agent could issue certificates that verify a potentially malicious program μ_F. The runtime of a correct agent that received a message from a child of the Byzantine agent could be damaged if it attempted to run μ_F without any sandboxing mechanism. An obvious alternative would be for Escher to ask programmers to assemble merge functions using a safe domain-specific language (DSL), such as SQL (modelling messages as rows in a relational table and using SQL aggregation functions to merge them), or the *extended Berkeley Packet Filter* [27], which, though a verifier, ensures that an eBPF program runs only for a limited time, accesses only a restricted memory region, and calls only an allowed set of safe external functions. Message processing abstractions from specification languages for distributed protocols, such as [19], can also be used in creating a DSL for merge functions.

We leave exploring such options for future work. Escher's current merge functionality can support a variety of common quorum-based crash- and Byzantine-tolerant replication protocols, as well as sharding protocols. The functionality is implemented as a dictionary of keys to values. A key defines a specific action that a runtime should take in order to merge a message; and the value customizes this action. We currently support two such pairs. The first is ($\texttt{wait for matching}, k$), which informs the receiving runtime to wait until there are k messages with matching payloads. The second is ($\texttt{from}: a_1.\texttt{id}, ..., a_n.\texttt{id}$), which informs the runtime to only consider messages that are from a specified set of agents. For example, the merge function of a crash-tolerant replicated service with replicas $r_1, ..., r_n$ can be implemented with $\{\texttt{wait for matching}: 1, \texttt{from}: r_1.\texttt{id}, ..., r_n.\texttt{id}\}$ whereas the merge function of a Byzantine fault-tolerant replicated service can be implemented with $\{\texttt{wait for matching} : f+1, \texttt{from} : r_1.\texttt{id}, ..., r_n.\texttt{id}\}$.

3.2 Message Delivery

The second functionality of the message bus is to deliver a message, addressed to a destination agent d that is refined, to the inboxes of all interested child agents of d. This functionality could be implemented using point-to-point communication between runtimes. The runtime of the sender, however, would need to know the addresses of d's child agents, which is difficult because refinements may be dynamic and addresses may change. Another option would be to have the runtime of a specific child agent of d receive all incoming traffic, acting as an ingress proxy. The child agent could then forward each message to their appropriate destinations. Unfortunately, while simple, this is not a fault-tolerant solution.

In Escher, runtimes discover one another using a publish/subscribe messaging service. This service has the following properties:

1. Runtimes *publish* messages to *topics*. A topic is an agent identifier, either virtual or physical.
2. The runtime of an agent s can publish a message $m = \langle \_, s.\texttt{id}, d.\texttt{id}, \_ \rangle$ in $s.\texttt{outbox}$ to the topic named $d.\texttt{id}$.
3. The runtime of an agent s automatically *subscribes* to the topic named $s.\texttt{id}$; it also subscribes to prefixes of $s.\texttt{id}$ according to its minimum depth (see Sect. 2).

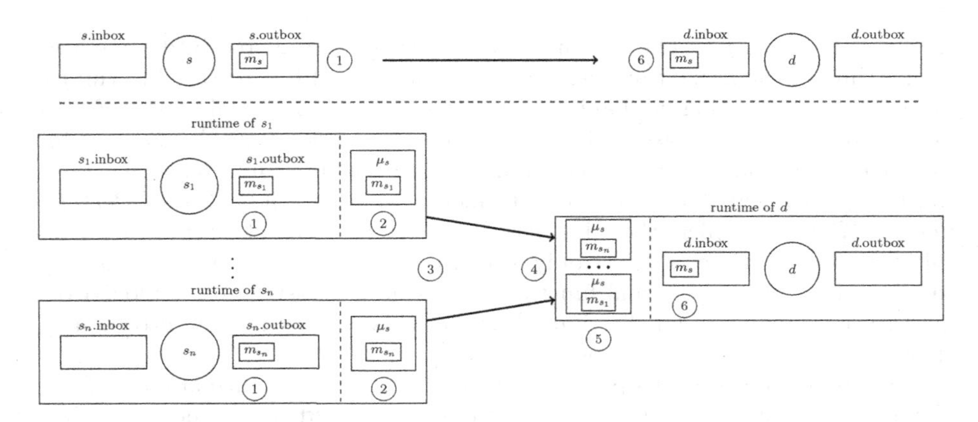

Fig. 1. Above the dashed line is a virtual representation of communication between agents s and d. (1) Virtual agent s sends a message $m_s = \langle i,\ s.\texttt{id},\ d.\texttt{id},\ p_s \rangle$ to d. This maps to a set of messages of the form $m_{s_i} = \langle i,\ s_i.\texttt{id},\ d.\texttt{id},\ p_{s_i} \rangle$ in the outboxes of the physical child agents of s. (2) The runtimes of the child agents of s tag the messages m_{s_i} with μ_s. (3) The runtimes of child agents of s send tagged messages to topic $d.\texttt{id}$. (4) The runtime of d receives tagged messages. (5) The runtime of d merges these messages into m_s. (6) Finally, the runtime of d places m_s in $d.\texttt{inbox}$.

Runtimes may opt to use the publish/subscribe messaging service for all communication. While attractive in its simplicity, this approach can negatively impact performance, as communication through pub/sub services typically results in significantly higher latencies than using point-to-point communication. Thus, the current Escher runtimes use the publish/subscribe service only to discover other agents; all ensuing communication is point-to-point. Should refinements change after this initial handshake, the runtimes need to repeat the process of discovery and connection.

3.3 Implementing Tagging and Merging

We describe now in detail (and illustrate in Fig. 1) the tagging and merging process that allows two agents s and d to communicate.

Let s be an agent refined into n child agents $C_s = \{s_1, ..., s_n\}$. Let $m_s = \langle i,\ s.\texttt{id},\ d.\texttt{id},\ p_s \rangle$ be a message in $s.\texttt{outbox}$ to be delivered to $d.\texttt{inbox}$. For example, s could be a replicated server sending a response to a client agent d. For simplicity, assume that d is physical (the merge functionality is the same for virtual destinations). Because refinement is transparent in Escher, m_s must be constructed by the message bus from messages that are physically in the outboxes of the child agents of s. The completeness property of μ_s implies that there exists a merge group $M_{\langle i,s,d \rangle}$ of messages of the form $m_{s_i} = \langle i,\ s_i.\texttt{id},\ d.\texttt{id},\ p_{s_i} \rangle$ in the outboxes of correct agents in C_s such that $\mu_s(M_{\langle i,s,d \rangle}) = m_s$. The runtimes of these child agents tag their messages with μ_s and send them to the topic $d.\texttt{id}$.

Note that messages between sibling agents (say replicas of the same server) do not need to be tagged.

The runtime of d eventually receives all messages sent by correct agents to d.id and groups together messages with the same tag. Let M be a set of messages with message identifier i, destination agent identifier d.id, and source agent identifier a child of s. The runtime of d updates M whenever a new message with matching values is received, and applies μ_s to M after each update. If this results in the empty set, the runtime waits for additional messages. If, however, the result is a non-empty set, then it is guaranteed to be a singleton set containing m_s; in this case, the runtime delivers m_s in d.inbox and discards M.

Generalizing for Multi-step Refinements. Let a be an agent that has $k+1$ ancestors with $k \geq 0$. Then, a is associated with a list of functions $\mathscr{M}_a = [\mu_k, ..., \mu_0]$, where μ_k is the merge function of a's parent and μ_0 is the merge function of the system agent. (μ_0 is not explicitly defined and never invoked.) We call $\mathscr{M}_a$ the *μ-list* of a.

If a is a correct agent with $k+1$ ancestors, its runtime may tag an outgoing message with a list of 0 to k tags, depending on the destination agent identifier. Given a message $\langle \_,$ a.id, d.id, $\_\rangle$ in a.outbox, the runtime determines the longest common prefix of a.id and d.id. Suppose the longest common prefix has length l, $0 \leq l \leq k$. The runtime tags the message with the first $k-l$ functions in the μ-list of a.

As an example, consider an agent /a/1/1. The runtime of /a/1/1 has the μ-list $[\mu_2, \mu_1, \mu_0]$. μ_2 is used to construct messages from virtual agent /a/1 and μ_1 is used to construct messages from virtual agent /a. Messages from /a/1/1 to /a/1/2 need not be tagged, since these are sibling agents. A message from /a/1/1 to /a/2 needs to be tagged only with $[\mu_2]$, since /a/2 and /a/1 are sibling agents, and the refinement of its sibling is transparent to /a/2. A message from /a/1/1 to /d is tagged with $\mathscr{M}_a = [\mu_2, \mu_1]$; it first needs to be merged using μ_2 to obtain a message from /a/1 to /d, which in turn needs to be merged using μ_1 to obtain a message from /a to /d.

Each successful application of a merge function by the runtime removes a tag from a message. Runtimes can only place in their agent's inbox messages with no tags.

Since merge functions are provided by runtimes that may be faulty, the Escher PKI must account for tag authenticity; that is, runtimes need to verify tags. Without tag authenticity, the runtime of a faulty agent could tag its outgoing messages incorrectly. For instance, Byzantine replicas of a replicated service could collude and tag their messages with a faulty μ_F such that μ_F yields a singleton set containing a bad message. To prevent this, recall that a manager of a virtual agent s issues a public key certificate to its child agents by signing their public key and identifier. We require this certificate to also includes a hash of the merge function μ_s.

3.4 Garbage Collection of Delivered Messages

An agent's inbox and outbox are specified as append-only. In practice, however, once a message is in the done box of the destination agent, it can physically be removed from the corresponding in- and outboxes. In this section, we discuss how.

We are explicitly not concerned with garbage collecting messages in an agent's done box, as the done box is outside Escher's purview. Further, while at the specification level it is another unbounded message box, an application can often implement its functionality using, say, a sequence number per client to track which messages it has handled thus far.

However, to implement garbage collection of inboxes and outboxes, a runtime must be able to find out whether a message with a given identifier is in the done box. To this end, we require each agent to expose a method $\texttt{contains}(i,\ s.\texttt{id},\ d.\texttt{id})$ that returns **true** iff there exists a message $\langle i,\ s.\texttt{id},\ d.\texttt{id},\ \_\rangle \in a.\texttt{done}$.

Garbage Collection in Outboxes. A message $m = \langle i,\ s.\texttt{id},\ d.\texttt{id},\ \_\rangle$ can be removed from $s.\texttt{outbox}$ once it is in $d.\texttt{inbox}$. To implement garbage collection of agent outboxes, Escher uses acknowledgment messages. When the runtime of an agent d delivers a message $m = \langle i,\ s.\texttt{id},\ d.\texttt{id},\ \_\rangle$ to $d.\texttt{inbox}$, it sends an ACK message $m_{\texttt{ACK}} = \langle \texttt{ACK}_i,\ d.\texttt{id},\ s.\texttt{id},\ m\rangle$ to agent s. ACK messages must be tagged the same way as regular messages. The message identifier of an ACK message is of the form $\langle ACK,\ i\rangle$, where i is the message identifier of the message being acknowledged. When the acknowledgment message is received (possibly after merging) by s or one of its descendants, the runtime removes m from its outbox. ACK messages are exchanged between runtimes only and are ephemeral: they are not placed in inboxes or outboxes and are not themselves acknowledged.

Garbage Collection in Inboxes. A message $m = \langle i,\ s.\texttt{id},\ d.\texttt{id},\ \_\rangle \in a.\texttt{inbox}$ can be removed once it is in $a.\texttt{done}$. The runtime of an agent a periodically evaluates $a.\texttt{done.contains}(i,\ s.\texttt{id},\ d.\texttt{id})$ for each such message. Also, if an unmerged *fragment* of a message arrives, the `contains` method is invoked to see if the fragment must be stored for later merging. If the message is already in the done box, an acknowledgment is sent but the fragment is otherwise dropped to avoid duplicate delivery.

4 Related Work

Many frameworks have been proposed that simplify the development of distributed systems. Here we only cover the projects most related to Escher.

Object-Oriented Middlewares. A variety of middlewares propose modeling a distributed system as a collection of objects and provide uniform interfaces to manage and access those objects, including CORBA [14], JavaBeans [26], and DCOM [20]. For example, CORBA defines an Interface Definition Language to specify object interfaces and allows applications to be written in a variety of programming languages and deployed on a variety of operating systems. An Object Request Broker allows objects to invoke interfaces of other objects transparently. To this end, both client and servers have *stubs* that automatically marshal arguments and results. Herein also lies the problem that Escher solves: these stubs do not support refinement. While CORBA does provide support for an unrefined client to securely interact with a refined server, it cannot transparently support multi-tiered distributed services in which refined clients interact with refined services. While Escher can support an object-oriented approach to application development, Escher is more general and supports other popular paradigms for distributed systems development such as streaming applications.

Reconfigurable Middlewares. Reconfigurable distributed systems support the replacement of their sub-systems. The aforementioned CORBA supports reconfiguration, see for example [8]. Ajmani et al. [3] supports multi-version systems and can gradually upgrade software of distributed systems. Escher supports agent upgrades at any level of refinement and therefore supports both upgrades of individual physical agents and upgrades of an entire distributed service.

Refinement-based Middlewares. The Ovid system [5] supports refinements (in the form of so-called *transformations*). The Escher refinement hierarchy is similar to Ovid's Logical Agent Transformation Tree, and Escher borrows the concept of using pathnames for refined agents from Ovid. However, Ovid does not support a uniform and secure way for refined agents to communicate, a key feature of Escher. The Ovid management system depends on a *controller agent*, which needs to be programmed with specific code for each type of agent and refinement and leaves open the question of who manages the control agent itself. The Escher management interface is uniform and does not require application- or refinement-specific code. Verdi [29] is another system that supports transformations. It is focused on formal verification of such transformations using a Coq-based toolchain. IronFleet [15] supports building and verifying distributed systems using the Dafny [18] language. Neither Verdi nor Ironfleet provide uniform interfaces for refined agents to communicate nor any form of application management.

The authors of [17] use stepwise refinement, to develop fault tolerant middleware for mobile agent systems. The middleware allows groups of agents to work together in isolated *scopes* and lets individual agents move around, switching scopes, while supporting interprocess communication via the Linda Tuple Space [13]. In [25], the authors describe a reflexive approach to updating communication between JavaBeans components in the face of reconfiguration events. While also based on refinement, neither approach solves communication between refined services, which is the key problem Escher addresses.

Replicated Remote Procedure Call. Perhaps the earliest work investigating how refined services can interact is *replicated procedure calls* [11,30]. The approach considers only crash-tolerant state machine replication schemes and requires that both the client and server are aware of each other's configurations. No attempt is made at making the approach transparent, and the approach does not support secure or Byzantine-tolerant replication schemes. Other proposals for specialized protocols for interacting replicated service include [2,21,28]. Aegean [4] also considers the interaction between replicated services. It considers the blocking RPC model incompatible with the state machine replication model of Paxos and proposes a solution based on speculative execution. Compared to these approaches, Escher provides a secure and general-purpose approach to interacting services that can have undergone various forms of refinement, including any of a variety of replication protocols. Moreover, Escher supports dynamic updates of a service without having to reconfigure other services that it interacts with.

Automated Synthesis. Some prior work focuses on generating low-level implementations of distributed protocols from their high-level specifications. In [9], Bonakdarpour et al. give a framework for automatedly deriving the implementation of a distributed application from its component-based "Behavior, Interaction, Priority" model. DistAlgo [19] is a high-level specification language for distributed protocols that can be compiled into Python implementations. While Escher's wrappers can be used to deploy a distributed refinement of a specific agent that has an implementation, Escher is not concerned with automatically refining a specification. Instead, Escher focuses on providing uniform communication between refined components, which is key for supporting modular and dynamic systems.

5 Conclusion

As multi-tiered cloud applications have become commonplace, the time has come to have first-class support for interacting distributed services. This paper proposes Escher, a middleware that uses service refinement to ease the development and evolution of such systems of systems. Escher manages refinement hierarchies and provides a message bus that allows services to communicate irrespective of their refinement level. Escher also provides isolation between distributed services, so that each can be developed and evolved independently of others. For certain classes of services, Escher provides built-in refinements, such as replication for fault tolerance and sharding for scalability, that can be applied without having to write or change code.

Acknowledgments. We thank the reviewers for their feedback. This work was supported in part by a Google Faculty Research Award, and by the NSF grants CSR-17620155, CNS-CORE 2008667, and CNS-CORE 2106954.

References

1. Abadi, M., Lamport, L.: The existence of refinement mappings. Theor. Comput. Sci. **82**(2), 253–284 (1991)

2. Adya, A., et al.: Farsite: federated, available, and reliable storage for an incompletely trusted environment. In: Proceedings of the 5th Symposium on Operating Systems Design and Implementation (OSDI 2002). USENIX, December 2002
3. Ajmani, S., Liskov, B., Shrira, L.: Modular software upgrades for distributed systems. In: Thomas, D. (ed.) ECOOP 2006. LNCS, vol. 4067, pp. 452–476. Springer, Heidelberg (2006). https://doi.org/10.1007/11785477_26
4. Aksoy, R.C., Kapritsos, M.: Aegean: replication beyond the client-server model. In: Proceedings of the 27th ACM Symposium on Operating Systems Principles, SOSP 2019, pp. 385–398, New York, NY, USA. Association for Computing Machinery (2019)
5. Altinbuken, D., Van Renesse, R.: Ovid: a software-defined distributed systems framework. In: 8th USENIX Workshop on Hot Topics in Cloud Computing (HotCloud 2016), June 2016
6. Ananthanarayanan, R., et al.: Photon: fault-tolerant and scalable joining of continuous data streams. In: Proceedings of the 2013 ACM SIGMOD International Conference on Management of Data, SIGMOD 2013, pp. 577–588, New York, NY, USA. Association for Computing Machinery (2013)
7. Bernstein, D.: Containers and cloud: from LXC to Docker to Kubernetes. IEEE Cloud Comput. **1**(3), 81–84 (2014)
8. Bidan, C., Issarny, V., Saridakis, T., Zarras, A.: A dynamic reconfiguration service for CORBA. In: Proceedings of the Fourth International Conference on Configurable Distributed Systems (Cat. No.98EX159) (1998)
9. Bonakdarpour, B., Bozga, M., Jaber, M., Quilbeuf, J., Sifakis, J.: A framework for automated distributed implementation of component-based models. Distrib. Comput. **25**, 10 (2012)
10. Burns, B., Grant, B., Oppenheimer, D., Brewer, E., Wilkes, J.: Borg, omega, and kubernetes. Queue **14**(1), 70–93 (2016)
11. Cooper, E.C.: Replicated procedure call. SIGOPS Oper. Syst. Rev. **20**(1), 44–56 (1986)
12. IBM Cloud Education. ESB (Enterprise Service Bus). https://www.ibm.com/cloud/learn/esb
13. Gelernter, D.: Generative communication in Linda. ACM Trans. Program. Lang. Syst. **7**(1), 80–112 (1985)
14. Object Management Group. The Common Object Request Broker: Architecture and specification, revision 2.4.1, formal/00-11-07, November 2000
15. Hawblitzel, C., et al.: IronFleet: proving practical distributed systems correct. In: Proceedings of the 25th Symposium on Operating Systems Principles, SOSP 2015, pp. 1–17, New York, NY, USA. Association for Computing Machinery (2015)
16. Apache Kafka. https://kafka.apache.org/
17. Laibinis, L., Troubitsyna, E., Iliasov, A., Romanovsky, A.B.: Fault tolerant middleware for agent systems: a refinement approach. In: 12th European Workshop on Dependable Computing (EWDC 2009), Toulouse, France, May 2009
18. Leino, K.R.M.: Dafny: an automatic program verifier for functional correctness. In: Clarke, E.M., Voronkov, A. (eds.) LPAR 2010. LNCS (LNAI), vol. 6355, pp. 348–370. Springer, Heidelberg (2010). https://doi.org/10.1007/978-3-642-17511-4_20
19. Liu, Y.A., Stoller, S.D., Lin, B.: High-level executable specifications of distributed algorithms. In: Richa, A.W., Scheideler, C. (eds.) SSS 2012. LNCS, vol. 7596, pp. 95–110. Springer, Heidelberg (2012). https://doi.org/10.1007/978-3-642-33536-5_11

20. Horstmann, M., Kirtland, M.: DCOM architecture. In: Microsoft Developer Network, July 1997
21. Netto, H.V., Lung, L.C., Correia, M., Luiz, A.F., de Souza, L.M.S.: State machine replication in containers managed by Kubernetes. J. Syst. Archit. **73** (2017)
22. Envoy Proxy. https://www.envoyproxy.io/
23. TIBCO Enterprise Message Service. https://www.tibco.com/products/tibco-enterprise-message-service
24. Shoup, R.: Service architectures at scale: Lessons from Google and eBay
25. Truyen, E., Joosen, W., Verbaeten, P., Jorgensen, B.N.: On interaction refinement in middleware. In: Workshop on Component-Oriented Programming, June 2000
26. Valesky, T.: Enterprise JavaBeans: Developing Component-Based Distributed Applications. Addison-Wesley Longman Publishing Co., Inc., USA (1999)
27. Vieira, M.A.M., et al.: Fast packet processing with EBPF and XDP: concepts, code, challenges, and applications. ACM Comput. Surv. **53**(1) (2020)
28. Wang, Y., et al.: Robustness in the Salus scalable block store. In: Proceedings of the 10th USENIX Conference on Networked Systems Design and Implementation, NSDI 2013, pp. 357–370, USA. USENIX Association (2013)
29. Wilcox, J.R., et al.: Verdi: a framework for implementing and formally verifying distributed systems. SIGPLAN Not. **50**(6), 357–368 (2015)
30. Yap, K.S., Jalote, P., Tripathi, S.: Fault tolerant remote procedure call. In: Proceedings of the 8th International Conference on Distributed Computing Systems (1988)

Deadlock and Noise in Self-Organized Aggregation Without Computation

Joshua J. Daymude[1(✉)], Noble C. Harasha[2], Andréa W. Richa[3], and Ryan Yiu[3]

[1] Biodesign Center for Biocomputing, Security and Society, Arizona State University, Tempe, AZ 85281, USA
jdaymude@asu.edu
[2] Massachusetts Institute of Technology, Cambridge, MA 02139, USA
nharasha@mit.edu
[3] School of Computing and Augmented Intelligence, Arizona State University, Tempe, AZ 85281, USA
aricha@asu.edu

Abstract. Aggregation is a fundamental behavior for swarm robotics that requires a system to gather together in a compact, connected cluster. In 2014, Gauci et al. proposed a surprising algorithm that reliably achieves swarm aggregation using only a binary line-of-sight sensor and no arithmetic computation or persistent memory. It has been rigorously proven that this algorithm will aggregate one robot to another, but it remained open whether it would always aggregate a system of $n > 2$ robots as was observed in experiments and simulations. We prove that there exist deadlocked configurations from which this algorithm cannot achieve aggregation for $n > 3$ robots when the robots' motion is uniform and deterministic. In practice, however, the physics of collisions and slipping work to the algorithm's advantage in avoiding deadlock; moreover, we show that the algorithm is robust to small amounts of noise in its sensors and in its motion. Finally, we prove that the algorithm achieves a linear runtime speedup for the $n = 2$ case when using a cone-of-sight sensor instead of a line-of-sight sensor.

Keywords: Swarm robotics · Self-organization · Aggregation

1 Introduction

The fields of swarm robotics [5,14,15,24] and programmable matter [2,18,33] seek to engineer systems of simple, easily manufactured robot modules that can cooperate to perform tasks involving collective movement and reconfiguration. Our present focus is on the *aggregation problem* (also referred to as

The authors gratefully acknowledge support from the U.S. ARO under MURI award #W911NF-19-1-0233 and from the Arizona State University Biodesign Institute.

C. Johnen et al. (Eds.): SSS 2021, LNCS 13046, pp. 51–65, 2021.
https://doi.org/10.1007/978-3-030-91081-5_4

"gathering" [8,16,19] and "rendezvous" [9,34,35]) in which a robot swarm must gather together in a compact, connected cluster [3]. Aggregation has a rich history in swarm robotics as a prerequisite for other collective behaviors requiring densely connected swarms. Inspired by self-organizing aggregation in nature [6,12,13,25,27,29], numerous approaches for swarm aggregation have been proposed, each one seeking to achieve aggregation faster, more robustly, and with less capable individuals than the last [1,11,17,26,28].

One goal from the theoretical perspective has been to identify *minimal capabilities* for an individual robot such that a collective can provably accomplish a given task. Towards this goal, Roderich Groß and others at the Natural Robotics Laboratory have developed a series of very simple algorithms for swarm behaviors like spatially sorting by size [7,23], aggregation [21], consensus [32], and coverage [31]. These algorithms use at most a few bits of sensory information and express their entire structure as a single "if-then-else" statement, avoiding any arithmetic computation or persistent memory. Although these algorithms have been shown to perform well in both robotic experiments and simulations with larger swarms, some lack general, rigorous proofs that guarantee the correctness of the swarm's behavior.

In this work, we investigate the swarm aggregation algorithm of Gauci et al. [21] (summarized in Sect. 2) whose provable convergence for systems of $n > 2$ robots remained an open question. In Sect. 3, we answer this question negatively, identifying deadlocked configurations from which aggregation is never achieved. Motivated by the need to break these deadlocks, we corroborate and extend the simulation results of [20] by showing that the algorithm is robust to two distinct forms of error (Sect. 4). Finally, we prove that the time required for a single robot to aggregate to a static robot improves by a linear factor when using a cone-of-sight sensor instead of a line-of-sight sensor; however, simulations show this comparative advantage decreases for larger swarms (Sect. 5).

2 The Gauci et al. Swarm Aggregation Algorithm

Given n robots in arbitrary initial positions on the two-dimensional plane, the goal of the *aggregation problem* is to define a controller that, when used by each robot in the swarm, eventually forms a compact, connected cluster. Gauci et al. [21] introduced an algorithm for aggregation among e-puck robots [30] that only requires binary information from a robot's (infinite range) line-of-sight sensor indicating whether it sees another robot ($I = 1$) or not ($I = 0$). The controller $x = (v_{\ell 0}, v_{r0}, v_{\ell 1}, v_{r1}) \in [-1, 1]^4$ actuates the left and right wheels according to velocities $(v_{\ell 0}, v_{r0})$ if $I = 0$ and $(v_{\ell 1}, v_{r1})$ otherwise. Using a grid search over a sufficiently fine-grained parameter space and evaluating aggregation according to a dispersion metric, they determined that the best controller was:

$$x^* = (-0.7, -1, 1, -1).$$

Thus, when no robot is seen, a robot using x^* will rotate around a point c that is 90° counter-clockwise from its line-of-sight sensor and $R = 14.45$ cm away at a speed of $\omega_0 = -0.75$ rad/s; when a robot is seen, it will rotate clockwise in place at a speed of $\omega_1 = -5.02$ rad/s. The following three theorems summarize the theoretical results for this aggregation algorithm.

Theorem 1 (Gauci et al. [21]). *If the line-of-sight sensor has finite range, then for every controller x there exists an initial configuration in which the robots form a connected visibility graph but from which aggregation will never occur.*

Theorem 2 (Gauci et al. [21]). *One robot using controller x^* will always aggregate to another static robot or static circular cluster of robots.*

Theorem 3 (Gauci et al. [21]). *Two robots both using controller x^* will always aggregate.*

Our main goal, then, is to investigate the following conjecture that is well-supported by evidence from simulations and experiments.

Conjecture 1. A system of $n > 2$ robots each using controller x^* will always aggregate.

Throughout the remaining sections, we measure the degree of aggregation in the system using the following metrics:

- *Smallest Enclosing Disc (SED) Circumference.* The smallest enclosing disc of a set of points S in the plane is the circular region of the plane containing S and having the smallest possible radius. Smaller circumferences correspond to more aggregated configurations.
- *Convex Hull Perimeter.* The convex hull of a set of points S in the plane is the smallest convex polygon enclosing S. Smaller perimeters correspond to more aggregated configurations. Due to the flexibility of convex polygons, this metric is less sensitive to outliers than the smallest enclosing disc which is forced to consider a circular region.
- *Dispersion (2nd Moment).* Adapting Gauci et al. [21] and Graham and Sloane [22], let p_i denote the (x, y)-coordinate of robot i on the continuous plane and $\overline{p} = \frac{1}{n}\sum_{i=1}^{n} p_i$ be the centroid of the system. Dispersion is defined as:

$$\sum_{i=1}^{n} ||p_i - \overline{p}||_2 = \sum_{i=1}^{n} \sqrt{(x_i - \overline{x})^2 + (y_i - \overline{y})^2}$$

  Smaller values of dispersion correspond to more aggregated configurations.
- *Cluster Fraction.* A cluster is a set of robots that is connected by means of (nearly) touching. Following Gauci et al. [21], our final metric for aggregation is the fraction of robots in the largest cluster. Unlike the previous metrics, larger cluster fractions correspond to more aggregated configurations.

We use dispersion as our primary metric of aggregation since it is the metric that is least sensitive to outliers and was used by Gauci et al. [21], enabling a clear comparison of results.

3 Impossibility of Aggregation for $n > 3$ Robots

In this section, we rigorously establish a negative result indicating that Conjecture 1 does not hold in general. This result identifies a deadlock that, in fact, occurs for a large class of controllers that x^* belongs to. We say a controller $x = (v_{\ell 0}, v_{r0}, v_{\ell 1}, v_{r1}) \in [-1, 1]^4$ is *clockwise-searching* if $v_{r0} < v_{\ell 0} < 0$. In other words, a clockwise-searching controller maps $I = 0$ (i.e., the case in which no robot is detected by the line-of-sight sensor) to a clockwise rotation about the center of rotation c that is a distance $R > 0$ away.[1]

Theorem 4. *For all $n > 3$ and all clockwise-searching controllers x, there exists an initial configuration of n robots from which the system will not aggregate when using controller x.*

Proof. At a high level, we construct a deadlocked configuration by placing the n robots in pairs such that no robot sees any other robot with its line-of-sight sensor—implying that all robots continually try to rotate about their centers of rotation—and each pair's robots mutually block each other's rotation. This suffices for the case that n is even; when n is odd, we extend the all-pairs configuration to include one mutually blocking triplet. Thus, no robots can move in this configuration since they are all mutually blocking, and since no robot sees any other they remain in this disconnected (non-aggregated) configuration indefinitely.

In detail, first suppose $n > 3$ is even. As in [21], let r denote the radius of a robot. For each $i \in \{0, 1, \dots, \frac{n}{2} - 1\}$, place robots p_{2i} and p_{2i+1} at points $(3r \cdot i, r)$ and $(3r \cdot i, -r)$, respectively. Orient all robots p_{2i} with their line-of-sight sensors in the $+y$ direction, and orient all robots p_{2i+1} in the $-y$ direction. This configuration is depicted in Fig. 1a. Due to their orientations, no robot can see any others; thus, since x is a clockwise-searching controller, all robots p_{2i} are attempting to move in the $-y$ direction while all robots p_{2i+1} are attempting to move in the $+y$ direction. Each pair of robots is mutually blocking, resulting in no motion. Moreover, since each consecutive pair of robots has a horizontal gap of distance r between them, this configuration is disconnected and thus non-aggregated.

It remains to consider when $n > 3$ is odd. Organize the first $n - 3$ robots in pairs according to the description above; since n is odd, we have that $n - 3$ must be even. Then place robot p_{n-1} at point $(3r(\frac{n}{2} - 1) + \sqrt{3}r, 0)$ with its line-of-sight sensor oriented at $0°$ (i.e., the $+x$ direction), robot p_{n-2} at point $(3r(\frac{n}{2} - 1), -r)$ with orientation $240°$, and robot p_{n-3} at point $(3r(\frac{n}{2} - 1), r)$ with orientation $120°$, as depicted in Fig. 1b. By a nearly identical argument to the one above, this configuration will also remain deadlocked and disconnected.

Therefore, in all cases there exists a configuration of $n > 3$ robots from which no clockwise-searching controller can achieve aggregation. □

[1] Note that an analogous version of Theorem 4 would hold for counter-clockwise-searching controllers if a robot's center of rotation was $90°$ clockwise rather than counter-clockwise from its line-of-sight sensor.

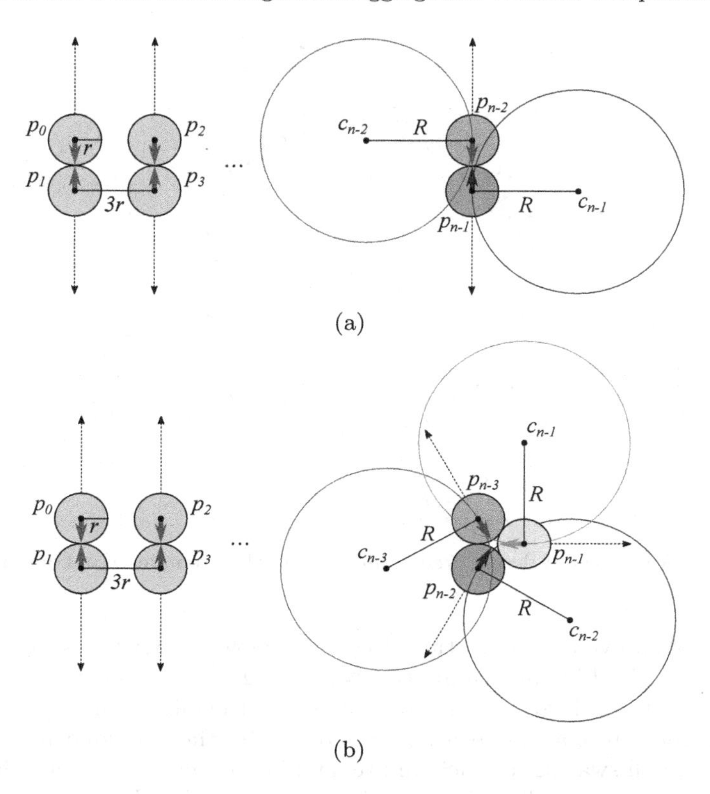

Fig. 1. The deadlocked configurations described in the proof of Theorem 4 for (a) $n > 3$ even and (b) $n > 3$ odd that remain non-aggregated indefinitely.

We have shown that no clockwise-searching controller (including x^*) can be guaranteed to aggregate a system of $n > 3$ robots starting from a deadlocked configuration, implying that Conjecture 1 does not hold in general. Moreover, not all deadlocked configurations are disconnected: Fig. 2 shows a connected configuration that will never make progress towards a more compact configuration because all robots are mutually blocked by their neighbors. Notably, these deadlocks are not observed in practice due to inherent noise in the physical e-puck robots. Real physics work to aggregation's advantage: if the robots were to ever get "stuck" in a deadlock configuration, collisions and slipping perturb the precise balancing of forces to allow the robots to push past one another. This motivates an explicit inclusion and modeling of noise in the algorithm, which we will return to in the next section.

Aaron Becker had conjectured at Dagstuhl Seminar 18331 [4] that symmetry could also lead to livelock, a second type of negative result for the Gauci et al. algorithm. In particular, Becker conjectured that robots initially organized in a cycle (e.g., Fig. 3a for $n = 3$) would traverse a "symmetric dance" in perpetuity without converging to an aggregated state when using controller x^*. However,

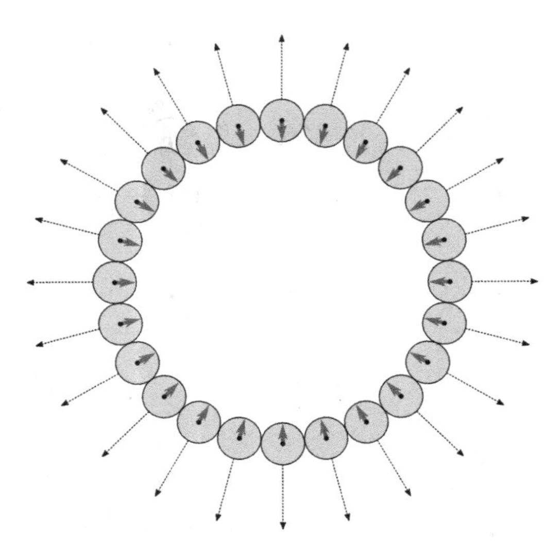

Fig. 2. A connected deadlocked configuration that remains non-compact.

simulations disprove this conjecture. Figure 3b shows that while swarms of various sizes initialized in the symmetric cycle configuration do exhibit an oscillatory behavior, they always reach and remain near the minimum dispersion value indicating near-optimal aggregation. Interestingly, these unique initial conditions cause small swarms to reach and remain in an oscillatory cycle where they touch and move apart infinitely often. Larger swarms break symmetry through collisions once the robots touch.

4 Robustness to Error and Noise

Motivated by the role of collisions and perturbations in freeing swarms from potential deadlocks, we next investigate the algorithm's *robustness* to varying magnitudes of error and noise. Our simulation platform models robots as circular rigid bodies in two dimensions, capturing all translation, rotation, and collision forces acting on the robots. Forces are combined and integrated iteratively over 5 ms time steps to obtain the translation and rotation of each robot. Figure 4 shows each of the four aggregation metrics for a baseline run on a swarm of $n = 100$ robots with no explicitly added noise. All four metrics demonstrate the system's steady but non-monotonic progress towards aggregation. Smallest enclosing disc circumference, convex hull perimeter, and dispersion show qualitatively similar progressions while the cluster fraction highlights when individual connected components join together.

We study the effects of two different forms of noise: *motion noise* and *error probability*. For motion noise, each robot at each time step experiences an applied force of a random magnitude in $[0, m^*]$ in a random direction. The parameter m^* defines the maximum noise force (in newtons) that can be applied to a robot

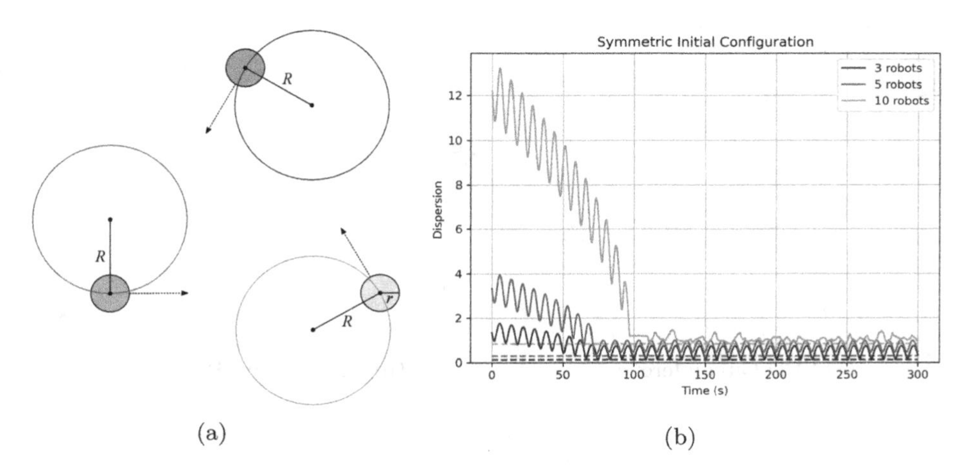

Fig. 3. (a) An example symmetric configuration of $n = 3$ robots that was conjectured to produce livelock. (b) Dispersion over time for swarms of $n = 3$ (purple), $n = 5$ (magenta), and $n = 10$ (orange) robots with symmetric initial configurations analogous to that of Fig. 3a. Dashed lines show the theoretical minimum dispersion value for the given system size. (Color figure online)

in a single time step. For error probability, each robot has the same probability $p \in [0, 1]$ of receiving the incorrect feedback from its sight sensor at each time step; more formally, a robot will receive the correct feedback I with probability $(1-p)$ and the opposite, incorrect feedback $1 - I$ with probability p.[2] The robot then proceeds with the algorithm as usual based on the feedback it receives.

In general, as the magnitude of error increases, so does the time required to achieve aggregation. The algorithm exhibits robustness to low magnitudes of motion noise with the average time to aggregation remaining relatively steady for $m^* \leq 5$ N and increasing only minimally for $5 \leq m^* \leq 20$ N (Fig. 5a). With larger magnitudes of motion noise ($m^* > 20$ N), average time to aggregation increases significantly, with many runs reaching the limit for simulation time before aggregation is reached. A similar trend is evident for error probability (Fig. 5b). The algorithm exhibits robustness for small error probabilities $p \in [0, 0.05]$ with the average time to aggregation rising steadily with increased error until nearly all runs reach the simulation time limit. Intuitively, while small amounts of noise can help the algorithm overcome deadlock without degrading performance, too much noise interferes significantly with the algorithm's ability to progress towards aggregation.

[2] Our formulation of an "error probability" p is equivalent to "sensory noise" in [21] when the false positive and false negative probabilities are both equal to p.

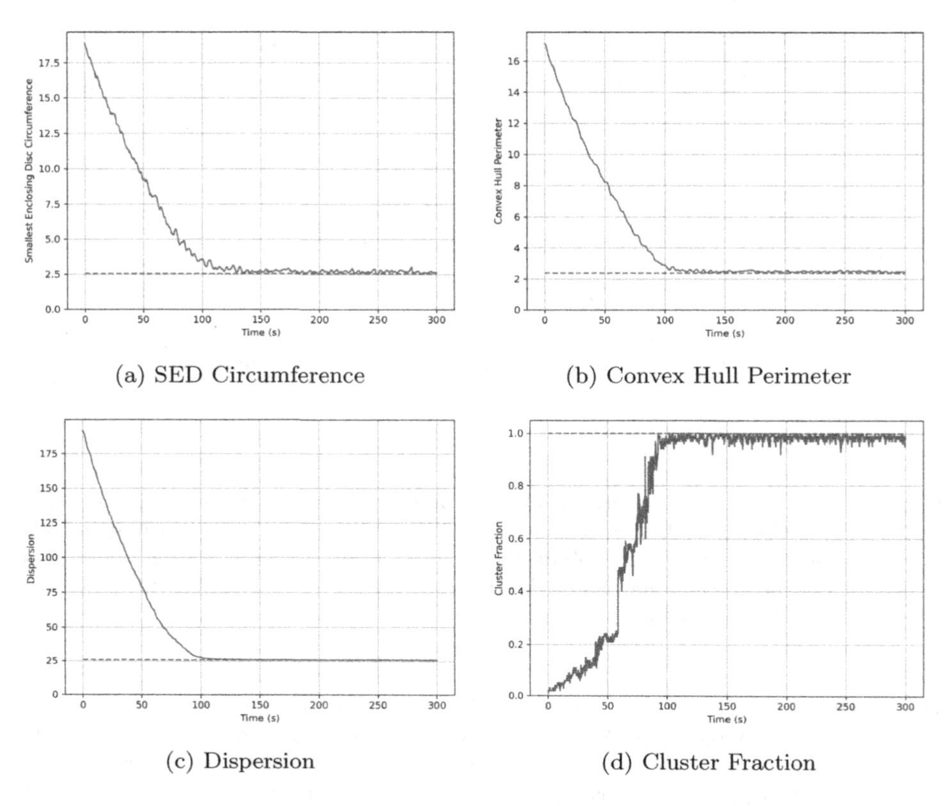

(a) SED Circumference

(b) Convex Hull Perimeter

(c) Dispersion

(d) Cluster Fraction

Fig. 4. Time evolutions of the four aggregation metrics for the same execution of x^* by a system of $n = 100$ robots for 300 s with no explicitly added noise. Dashed lines indicate the optimal value for each aggregation metric given the number of robots n.

5 Using a Cone-of-Sight Sensor

We next analyze a generalization of the algorithm where each robot has a *cone-of-sight sensor* of angle β instead of a line-of-sight sensor ($\beta = 0$). This was left as future work in [21] and was briefly considered in [20] where, for each $\beta \in \{0°, 30°, \ldots, 180°\}$, the best performing controller x_β was found via exhaustive search and compared against the others. Here we take a complementary approach, studying the performance of the original controller x^* as β varies.

We begin by proving that, in the case of one static robot and one robot executing the generalized algorithm, using a cone-of-sight sensor with size $\beta > 0$ can improve the time to aggregation by a linear factor (as a function of the initial distance between the two robots) over the original algorithm. This result follows from the fact that progress towards aggregation is achieved when the moving robot is rotating in place, moving its center of rotation closer to the static robot. With a line-of-sight sensor, the further the two robots are from each other, the smaller the moving robot's rotation in place. However, with a

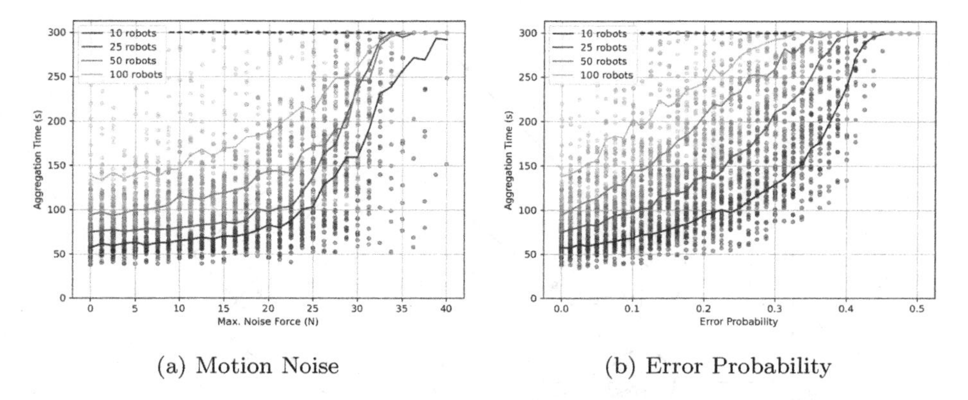

(a) Motion Noise

(b) Error Probability

Fig. 5. The time required to reach aggregation for different magnitudes of (a) motion noise and (b) error probability for systems of $n = 10$ (purple), $n = 25$ (magenta), $n = 50$ (red), and $n = 100$ (orange) robots. Each experiment for a given system size and noise strength was repeated 25 times (scatter plot); average runtimes are shown as solid lines. We consider systems that are within 15% of the minimum dispersion value as aggregated. The dashed line at 300 seconds indicates the cutoff time at which the run is determined to be non-aggregating. (Color figure online)

cone-of-sight sensor, the moving robot is guaranteed to rotate in place a fixed amount each time it sees the static robot, guaranteeing at least constant progress towards aggregation with each revolution.

Theorem 5. *One moving robot using a cone-of-sight sensor of size $\beta \in (0, \pi)$ will always aggregate with another static robot in*

$$m < \left\lceil \frac{(d_0 - R - r_i - r_j)(R + 2r_i)}{2\sqrt{3}Rr_i \sin((1 - 1/\sqrt{3}) \cdot \beta/2)} \right\rceil$$

rotations around its center of rotation, where d_0 is the initial value of $||\boldsymbol{p}_j - \boldsymbol{c}_i||$.

Proof. Consider a robot i executing the generalized algorithm at position $\boldsymbol{p}_i$ with center of rotation $\boldsymbol{c}_i$ and a static robot j at position $\boldsymbol{p}_j$. As in the proofs of Theorems 5.1 and 5.2 in [21], we first consider the scenario shown in Fig. 6 and derive an expression for $d' = ||\boldsymbol{p}_j - \boldsymbol{c}_i'||$ in terms of $d = ||\boldsymbol{p}_j - \boldsymbol{c}_i||$. W.l.o.g., let $\boldsymbol{c}_i = [0, 0]^T$ and let the axis of the cone-of-sight sensor of robot i point horizontally right at the moment it starts seeing robot j. Then the position of robot j is given by

$$\boldsymbol{p}_j = \begin{bmatrix} \frac{r_j \cos(\alpha/2+\gamma)}{\sin(\alpha/2)} \\ -\left(R + \frac{r_j \sin(\alpha/2+\gamma)}{\sin(\alpha/2)}\right) \end{bmatrix}.$$

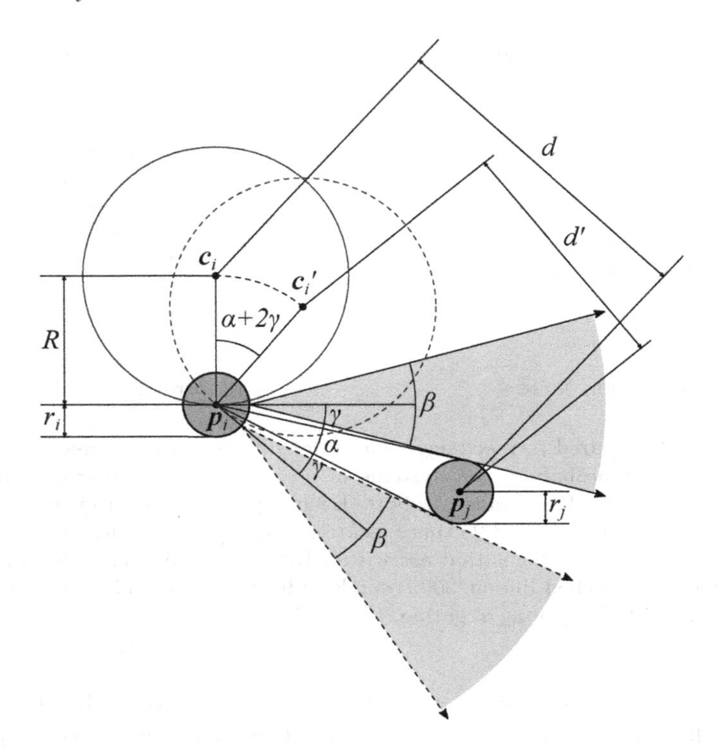

Fig. 6. The setup considered in the proof of Theorem 5. Robot i is moving and has a cone-of-sight sensor with size β while robot j is static.

Substituting this position into the distance $d = ||\boldsymbol{p}_j - \boldsymbol{c}_i||$ yields

$$d^2 = \left(\frac{r_j \cos(\alpha/2 + \gamma)}{\sin(\alpha/2)}\right)^2 + \left(R + \frac{r_j \sin(\alpha/2 + \gamma)}{\sin(\alpha/2)}\right)^2$$
$$= R^2 + \frac{2Rr_j \sin(\alpha/2 + \gamma)}{\sin(\alpha/2)} + \frac{r_j^2}{\sin^2(\alpha/2)}.$$

Using a line-of-sight sensor, robot i would only rotate α before it no longer sees robot j; however, with a cone-of-sight sensor of size β, robot i rotates $\alpha + 2\gamma$ before robot j leaves its sight, where γ is the angle from the cone-of-sight axis to the first line intersecting $\boldsymbol{p}_i$ that is tangent to robot j. With this cone-of-sight sensor, $\boldsymbol{c}_i'$ is given by

$$\boldsymbol{c}_i' = \begin{bmatrix} R\sin(\alpha + 2\gamma) \\ R(\cos(\alpha + 2\gamma) - 1) \end{bmatrix}.$$

Substituting this new center of rotation into the distance $d' = ||\boldsymbol{p}_j - \boldsymbol{c}_i'||$ yields

$$\begin{aligned}
d' &= \sqrt{\begin{aligned}&\left(\frac{r_j \cos(\alpha/2+\gamma)}{\sin(\alpha/2)} - R\sin(\alpha+2\gamma)\right)^2 \\ &+ \left(-\left(R + \frac{r_j \sin(\alpha/2+\gamma)}{\sin(\alpha/2)}\right) - R(\cos(\alpha+2\gamma)-1)\right)^2\end{aligned}} \\
&= \sqrt{\begin{aligned}&\frac{r_j^2 \cos^2(\alpha/2+\gamma)}{\sin^2(\alpha/2)} - \frac{2Rr_j \cos(\alpha/2+\gamma)\sin(\alpha+2\gamma)}{\sin(\alpha/2)} + R^2\sin^2(\alpha+2\gamma) \\ &+ \frac{r_j^2 \sin^2(\alpha/2+\gamma)}{\sin^2(\alpha/2)} + \frac{2Rr_j \sin(\alpha/2+\gamma)\cos(\alpha+2\gamma)}{\sin(\alpha/2)} + R^2\cos^2(\alpha+2\gamma)\end{aligned}} \\
&= \sqrt{R^2 + \frac{r_j^2}{\sin^2(\alpha/2)} + \frac{2Rr_j(\sin(\alpha/2+\gamma)\cos(\alpha+2\gamma) - \cos(\alpha/2+\gamma)\sin(\alpha+2\gamma))}{\sin(\alpha/2)}} \\
&= \sqrt{d^2 + \frac{2Rr_j(\sin(\alpha/2+\gamma-(\alpha+2\gamma)) - \sin(\alpha/2+\gamma))}{\sin(\alpha/2)}} \\
&= \sqrt{d^2 - \frac{4Rr_j \sin(\alpha/2+\gamma)}{\sin(\alpha/2)}}.
\end{aligned}$$

Note that this relation contains the result proven in Theorem 5.1 of [21] as a special case by setting $\beta = 0$ (and thus $\gamma = 0$), which corresponds to a line-of-sight sensor. To bound the number of $d \to d'$ updates required until $d \leq R + r_i + r_j$ (i.e., until the robots have aggregated), we write the following recurrence relation, where $\hat{d}_m = d_m^2$ and $\hat{d}_m > (R + r_i + r_j)^2$:

$$\hat{d}_{m+1} = \hat{d}_m - \frac{4Rr_j \sin(\alpha/2+\gamma)}{\sin(\alpha/2)}.$$

Observe that α is the largest when the two robots are touching, and—assuming $r_i = r_j$, i.e., the two robots are the same size—it is easy to see that $\alpha \leq \pi/3$. Also, γ is at least 0 and at most $\beta/2$; thus, by supposition, $\gamma < \pi/2$. Thus, by the angle sum identity,

$$\hat{d}_{m+1} < \hat{d}_m - \frac{4Rr_j \cos(\alpha/2)\sin(\gamma)}{\sin(\alpha/2)}.$$

Again, since $\alpha \leq \pi/3$, we have $\cos(\alpha/2) \geq \sqrt{3}/2$. By inspection, we also have $\sin(\alpha/2) < r_j/(d_m - R)$, yielding

$$\hat{d}_{m+1} < \hat{d}_m - 2\sqrt{3}R(d_m - R)\sin(\gamma).$$

Let $k_1 > 1$ be a constant such that $r_i = r_j = R/k_1$, which must exist since r_i, r_j, and R are constants and $r_i = r_j < R$. Since the robots have not yet aggregated, we have $d_m > R + r_i + r_j = (1 + 2/k_1)R$. We use this to show

$$R < \frac{d_m}{1 + 2/k_1} = k_2 d_m,$$

where $k_2 = 1/(1 + 2/k_1) < 1$ is a constant. Returning to our recurrence relation:

$$\hat{d}_{m+1} < \hat{d}_m - 2\sqrt{3}R(d_m - k_2 d_m)\sin(\gamma) = \hat{d}_m - 2\sqrt{3}R d_m(1 - k_2)\sin(\gamma)$$

Recalling that $\hat{d}_m = d_m^2$, we have

$$d_{m+1} < \sqrt{\hat{d}_m - 2\sqrt{3}R d_m(1 - k_2)\sin(\gamma)} < d_m - \sqrt{3}R(1 - k_2)\sin(\gamma),$$

where the second inequality can be verified by squaring both sides and noting that $R > 0$, $1 - k_2 > 0$, and $\gamma \in (0, \pi/2)$. As a final upper bound on d_{m+1}, we lower bound the angle γ as a function of the constant size of the cone-of-sight sensor β as $\gamma \geq (1 - 1/\sqrt{3}) \cdot \beta/2$ (see [10] for a complete derivation), yielding

$$d_{m+1} < d_m - \sqrt{3}R(1 - k_2)\sin((1 - 1/\sqrt{3}) \cdot \beta/2).$$

This yields the solution

$$d_m < d_0 - m\sqrt{3}R(1 - k_2)\sin((1 - 1/\sqrt{3}) \cdot \beta/2), \quad d_m > R + r_i + r_j$$

The number of $d \to d'$ updates required until $d \leq R + r_i + r_j$ is now given by setting $d_m = R + r_i + r_j$ in this solution and solving for m, which yields

$$m < \left\lceil \frac{d_0 - R - r_i - r_j}{\sqrt{3}R(1 - k_2)\sin((1 - 1/\sqrt{3}) \cdot \beta/2)} \right\rceil = \left\lceil \frac{(d_0 - R - r_i - r_j)(R + 2r_i)}{2\sqrt{3}R r_i \sin((1 - 1/\sqrt{3}) \cdot \beta/2)} \right\rceil,$$

concluding the proof. □

We note that this bound on the number of required updates m has a linear dependence on d_0 while the original bound proven in Theorem 5.2 of [21] for line-of-sight sensors depended on d_0^2, demonstrating a linear speedup with cone-of-sight sensors for $n = 2$ robots. However, simulation results show that as the number of robots increases, the speedup from using cone-of-sight sensors diminishes (Fig. 7). All systems benefit from small cone-of-sight sensors—i.e., $\beta \in (0, 0.5)$—reaching aggregation in significantly less time. With larger systems, however, large cone-of-sight sensors can be detrimental as robots see others more often than not, causing them to primarily rotate in place without making progress towards aggregation. This highlights a delicate balance between the algorithm's two modes (rotating around the center of rotation and rotating in place) with β indirectly affecting how much time is spent in each.

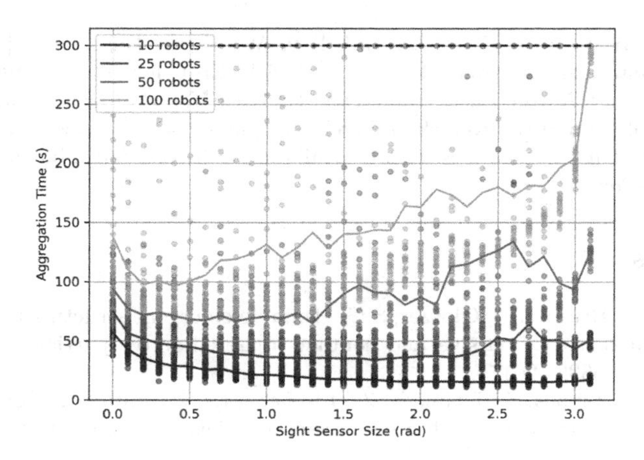

Fig. 7. The effects of cone-of-sight sensor size on the algorithm's time to aggregation for systems of $n = 10$ (purple), $n = 25$ (magenta), $n = 50$ (red), and $n = 100$ (orange) robots. Each experiment for a given system size and sensor size was repeated 25 times (scatter plot); average runtimes are shown as solid lines. We consider systems that are within 15% of the minimum dispersion value as aggregated. The dashed line at 300 seconds indicates the cutoff time at which the run is determined to be non-aggregating. (Color figure online)

6 Conclusion

In this paper, we investigated the Gauci et al. swarm aggregation algorithm [21] which provably aggregates two robots and reliably aggregates larger swarms in experiment using only a binary line-of-sight sensor and no arithmetic computation or persistent memory. We answered the open question of whether the algorithm guarantees aggregation for systems of $n > 2$ robots negatively, identifying how deadlock can halt the system's progress towards aggregation. In practice, however, the physics of collisions and slipping work to the algorithm's advantage in avoiding deadlock; moreover, we showed that the algorithm is robust to small amounts of noise in its sensors and in its motion. Finally, we considered a generalization of the algorithm using cone-of-sight sensors, proving that for the situation of one moving robot and one static robot, the time to aggregation is improved by a linear factor over the original line-of-sight sensor. Simulation results showed that small cone-of-sight sensors can also improve runtime for larger systems, though with diminishing returns.

In the full version of this work [10], we additionally introduced a noisy, discrete adaptation of the Gauci et al. algorithm in an effort to formally prove its convergence when noise is explicitly modeled as a mechanism to break deadlock. However, both the original algorithm and this discrete adaptation progress towards aggregation non-monotonically, complicating analysis techniques relying on consistent progress towards the goal state. It is possible that a proof showing convergence *in expectation* can be derived, but we leave this for future work.

Acknowledgements and Data Availability. We thank Dagstuhl [4] for hosting the seminar that inspired this research, Roderich Groß for introducing us to this open problem, and Aaron Becker and Dan Halperin for their contributions to the investigations of symmetric livelock and cone-of-sight sensors. Source code for all simulations reported in this work is openly available at https://github.com/SOPSLab/SwarmAggregation.

References

1. Agrawal, M., Bruss, I.R., Glotzer, S.C.: Tunable emergent structures and traveling waves in mixtures of passive and contact-triggered-active particles. Soft Matter **13**(37), 6332–6339 (2017)
2. Angluin, D., Aspnes, J., Diamadi, Z., Fischer, M.J., Peralta, R.: Computation in networks of passively mobile finite-state sensors. Distrib. Comput. **18**(4), 235–253 (2006)
3. Bayindir, L.: A review of swarm robotics tasks. Neurocomputing **172**, 292–321 (2016)
4. Berman, S., Fekete, S.P., Patitz, M.J., Scheideler, C.: Algorithmic foundations of programmable matter (Dagstuhl Seminar 18331). Dagstuhl Rep. **8**(8), 48–66 (2019)
5. Brambilla, M., Ferrante, E., Birattari, M., Dorigo, M.: Swarm robotics: A review from the swarm engineering perspective. Swarm Intell. **7**(1), 1–41 (2013)
6. Camazine, S., Franks, N.R., Sneyd, J., Bonabeau, E., Deneubourg, J.L., Theraula, G.: Self-Organization in Biological Systems. Princeton University Press, Princeton, NJ, USA (2001)
7. Chen, J., Gauci, M., Price, M.J., Groß, R.: Segregation in swarms of e-puck robots based on the Brazil nut effect. In: Proceedings of the 11th International Conference on Autonomous Agents and Multiagent Systems, vol. 1, pp. 163–170 (2012)
8. Cieliebak, M., Flocchini, P., Prencipe, G., Santoro, N.: Solving the robots gathering problem. In: Baeten, J.C.M., Lenstra, J.K., Parrow, J., Woeginger, G.J. (eds.) ICALP 2003. LNCS, vol. 2719, pp. 1181–1196. Springer, Heidelberg (2003). https://doi.org/10.1007/3-540-45061-0_90
9. Cortés, J., Martinez, S., Bullo, F.: Robust rendezvous for mobile autonomous agents via proximity graphs in arbitrary dimensions. IEEE Trans. Autom. Control **51**(8), 1289–1298 (2006)
10. Daymude, J.J., Harasha, N.C., Richa, A.W., Yiu, R.: Deadlock and noise in self-organizing aggregation without computation (2021). https://arxiv.org/abs/2108.09403
11. Deblais, A., et al.: Boundaries control collective dynamics of inertial self-propelled robots. Phys. Rev. Lett. **120**(18), 188002 (2018)
12. Deneubourg, J.L., Grégoire, J.C., Le Fort, E.: Kinetics of larval gregarious behavior in the bark beetle Dendroctonus micans (Coleoptera: Scolytidae). J. Insect Behav. **3**, 169–182 (1990)
13. Devreotes, P.: Dictyostelium discoideum: A model system for cell-cell interactions in development. Science **245**(4922), 1054–1058 (1989)
14. Dorigo, M., Theraulaz, G., Trianni, V.: Reflections on the future of swarm robotics. Sci. Robot. **5**(49), eabe4385 (2020)
15. Dorigo, M., Theraulaz, G., Trianni, V.: Swarm robotics: Past, present, and future. Proc. IEEE **109**(7), 1152–1165 (2021)
16. Fatès, N.: Solving the decentralised gathering problem with a reaction-diffusion-chemotaxis scheme. Swarm Intell. **4**(2), 91–115 (2010)

17. Firat, Z., Ferrante, E., Gillet, Y., Tuci, E.: On self-organised aggregation dynamics in swarms of robots with informed robots. Neural Comput. Appl. **32**(17), 13825–13841 (2020). https://doi.org/10.1007/s00521-020-04791-0
18. Flocchini, P., Prencipe, G., Santoro, N. (eds.): Distributed Computing by Mobile Entities. Springer International Publishing, Switzerland (2019)
19. Flocchini, P., Prencipe, G., Santoro, N., Widmayer, P.: Gathering of asynchronous robots with limited visibility. Theor. Comput. Sci. **337**(1–3), 147–168 (2005)
20. Gauci, M.: Swarm Robotic Systems with Minimal Information Processing. PhD Thesis, University of Sheffield, Sheffield, England (2014). https://etheses.whiterose.ac.uk/7569/
21. Gauci, M., Chen, J., Li, W., Dodd, T.J., Groß, R.: Self-organized aggregation without computation. Int. J. Robot. Res. **33**(8), 1145–1161 (2014)
22. Graham, R.L., Sloane, N.J.A.: Penny-packing and two-dimensional codes. Discret. Comput. Geom. **5**(1), 1–11 (1990). https://doi.org/10.1007/BF02187775
23. Groß, R., Magnenat, S., Mondada, F.: Segregation in swarms of mobile robots based on the Brazil nut effect. In: 2009 IEEE/RSJ International Conference on Intelligent Robots and Systems, pp. 4349–4356. IROS 2009 (2009)
24. Hamann, H.: Swarm Robotics: A Formal Approach. Springer, Heidelberg (2018)
25. Jeanson, R., et al.: Self-organized aggregation in cockroaches. Anim. Behav. **69**(1), 169–180 (2005)
26. Li, S., et al.: Programming active cohesive granular matter with mechanically induced phase changes. Sci. Adv. **7**(17), eabe8494 (2021)
27. Magurran, A.E.: The adaptive significance of schooling as an anti-predator defence in fish. Ann. Zool. Fenn. **27**(2), 51–66 (1990)
28. Misir, O., Gökrem, L.: Dynamic interactive self organizing aggregation method in swarm robots. Biosystems **207**, 104451 (2021)
29. Mlot, N.J., Tovey, C.A., Hu, D.L.: Fire ants self-assemble into waterproof rafts to survive floods. Proc. Natl. Acad. Sci. **108**(19), 7669–7673 (2011)
30. Mondada, F., et al.: The e-puck, a robot designed for education in engineering. In: Proceedings of the 9th Conference on Autonomous Robot Systems and Competitions, pp. 59–65 (2009)
31. Özdemir, A., Gauci, M., Kolling, A., Hall, M.D., Groß, R.: Spatial coverage without computation. In: International Conference on Robotics and Automation, pp. 9674–9680 (2019)
32. Özedmir, A., Gauci, M., Bonnet, S., Groß, R.: Finding consensus without computation. IEEE Robot. Autom. Lett. **3**(3), 1346–1353 (2018)
33. Woods, D., Chen, H.L., Goodfriend, S., Dabby, N., Winfree, E., Yin, P.: Active self-assembly of algorithmic shapes and patterns in polylogarithmic time. In: Proceedings of the 4th Innovations in Theoretical Computer Science Conference, pp. 353–354 (2013)
34. Yu, J., LaValle, S.M., Liberzon, D.: Rendezvous without coordinates. IEEE Trans. Autom. Control **57**(2), 421–434 (2012)
35. Zebrowski, P., Litus, Y., Vaughan, R.T.: Energy efficient robot rendezvous. In: Fourth Canadian Conference on Computer and Robot Vision, pp. 139–148 (2007)

Failure is (literally) an Option: Atomic Commitment vs Optionality in Decentralized Finance

Daniel Engel, Maurice Herlihy(✉), and Yingjie Xue

Computer Science Department, Brown University, Providence, RI 02912, USA
mph@cs.brown.edu

Abstract. Many aspects of blockchain-based decentralized finance can be understood as an extension of classical distributed computing. In this paper, we trace the evolution of two interrelated notions: failure and fault-tolerance. In classical distributed computing, a failure to complete a multi-party protocol is typically attributed to hardware malfunctions. A fault-tolerant protocol is one that responds to such failures by rolling the system back to an earlier consistent state. In the presence of Byzantine failures, a failure may be the result of an attack, and a fault-tolerant protocol is one that ensures that attackers will be punished and victims compensated. In modern decentralized finance however, failure to complete a protocol can be considered a legitimate option, not a transgression. A fault-tolerant protocol is one that ensures that the party offering the option cannot renege, and the party purchasing the option provides fair compensation (in the form of a fee) to the offering party. We sketch the evolution of such protocols, starting with two-phase commit, and finishing with timed hashlocked smart contracts.

1 Introduction

Decentralized finance (DeFi) is on the rise: between June and October 2020, the value of assets managed by DeFi protocols increased from $1 billion to $7.7 billion [26]. This paper is an informal tutorial, explaining certain basic problems in DeFi as if they were problems in fault-tolerant distributed computing. Conversely, many core problems in DeFi represent interesting and important extensions of distributed computing problems. The goal of this paper is to encourage distributed computing researchers to consider the kinds of problems and models that arise in DeFi, and conversely, to encourage DeFi researchers to benefit from the rich history of distributed computing techniques and algorithms.

The contribution of this work is simply to illustrate these claims through an extended example, presented with the hope of provoking others to take up research in this area. We explore how one core problem of distributed computing has evolved over time, gradually turning into a superficially distinct core problem

Supported by NSF grant 1917990.

C. Johnen et al. (Eds.): SSS 2021, LNCS 13046, pp. 66–77, 2021.
https://doi.org/10.1007/978-3-030-91081-5_5

of finance. We consider the problem of *atomic commitment*: how can we install updates at multiple databases or ledgers in such a way that guarantees that if all goes well, all updates are installed, but if something goes wrong, all updates are discarded. The classical challenge is, of course, tolerating failures: databases can crash or communication can be lost or delayed.

This is one of the oldest problems in distributed computing, and not surprisingly, it is central to key problems in DeFi. We explore two aspects of this problem. First, we explore the technical solutions, where DeFi has tended to borrow, whether consciously or not, from prior solutions in distributed computing. Second, we explore underlying conceptual frameworks, where DeFi extends the notion of fault-tolerance well beyond the classical models of distributed computing. We hope that our examples illustrate how each field can learn from the other.

In Sect. 3, we review the well-known *two-phase commit protocol* [2], a classical technique for making atomic updates to independently-failing databases in a distributed system. In Sect. 4, we consider the *cross-chain atomic swap* problem, where mutually-suspicious parties exchange assets atomically across distinct blockchains. The simplest atomic swap protocols are based on *hashed timelocked contracts* [11,17] (HTLCs). Technically, HTLC protocols closely resemble classical two-phase commit. The principal difference between the two protocols is in their underlying conceptual frameworks. In two-phase commit, a failure is typically an operational malfunction at a node or a network, while in atomic swap, a failure could also be a malicious action chosen by an adversarial party.

This distinction becomes more pronounced in Sect. 5. In both two-phase commit and atomic cross-chain swap protocols, fault tolerance means that if one party falls silent in the middle, the other parties are eventually made whole: database replicas are eventually restored, and escrowed assets are eventually refunded. For distributed computing's two-phase commit, the story ends there, but for DeFi's atomic swap protocol, the story has just begun. In finance, the ability to abandon or to complete an in-progress swap is called an *option*, and options themselves have value. Any party who abandons an atomic swap should compensate the other parties by paying a small fee called a *premium*. Treating failures as compensated options is alien to classical distributed computing models, where all parties implicitly are on the same team, but it opens up a range of new research challenges for distributed computing. Incorporating premiums into atomic swaps turns out to be a challenging technical problem [25], effectively requiring nesting one atomic commitment protocol within another.

Section 6 takes the notion of optionality to the next level. What if one party could sell such an option to another? Alice, who has paid for the option to complete or cancel a swap, should be able to transfer that option to Bob for a fee. Alice would relinquish her power over the swap's outcome, and Bob would assume all of Alice's power, including the power to complete or cancel the swap, and the right to be compensated if another party cancels the swap. This problem is also technically challenging, as it requires embedding yet another atomic commitment mechanism within other nested atomic commitment mechanisms.

While we advocate thinking about DeFi mechanisms as if they were distributed computing problems, we also advocate DeFi as a rich source of new problems and models for mainstream distributed computing. Originally, abandoning an atomic commitment protocol was considered a simple operational failure, and the meaning of fault-tolerance was simply to restore integrity and availability. When the parties become autonomous and potentially adversarial, however, failures can become deliberate choices, and the meaning of fault-tolerance must be extended to provide financial compensation to any victims of other parties' choices. Once failures become options (in the financial sense), then those options themselves become assets to be traded.

The questions raised here are not really about blockchains, as blockchains. Instead, they are really about the scientific and engineering problems of safely transferring value among autonomous parties. This problem will remain of enduring importance to society, independently of whether particular blockchain technologies bloom or fade, whether certain asset bubbles expand or pop, or whether regulatory agencies do or do not intervene to protect gullible investors. We believe the fault-tolerant distributed computing community has much to offer on these fundamental problems, and we encourage the community to get involved.

2 Model

A *blockchain* is a tamper-proof distributed ledger or database that tracks ownership of *assets* by *parties*. (Our discussion is mostly independent of which blockchain technology is used.) A party can be a person, an organization, or even a contract (see below). An asset can be a cryptocurrency, a token, an electronic deed to property, and so on. There are multiple blockchains managing different kinds of assets. We focus here on applications where mutually-untrusting parties trade assets among themselves, for example by swaps, loans, auctions, markets, and so on.

A *contract* is a blockchain-resident program initialized and called by the parties. A party can publish a new contract on a blockchain, or call a function exported by an existing contract. Contract code and contract state are public, so a party calling a contract knows what code will be executed. Contract code must be deterministic because contracts are typically re-executed multiple times by mutually-suspicious parties.

Multiple parties agree on a common *protocol* to execute a series of transfers, an agreement that can be monitored, but not enforced. Instead of distinguishing between faulty and non-faulty parties, as in classical distributed computing, we distinguish only between *compliant* parties who follow the agreed-upon protocol, and *deviating* parties who do not. We make no assumptions about the number of deviating parties.

We assume a *synchronous* execution model where there is a known upper bound Δ on the propagation time for one party's change to the blockchain state to be noticed by the other parties. Specifically, blockchains generate new blocks at a steady rate, and valid transactions sent to the blockchain will be included in a block and visible to participants within Δ. Our example protocols use Δ as the basis for timeouts: it is typically chosen conservatively.

We make standard cryptographic assumptions. Each party has a public key and a private key, and any party's public key is known to all. Messages are signed so they cannot be forged, and they include single-use labels ("nonces") so they cannot be replayed.

3 Classical Two-Phase Commit Protocol

Imagine we have a distributed database with a number of replicas. These replicas might be identical, or they may hold different portions of the database (so-called shards). For simplicity, assume Alice's node holds one replica, and Bob's node holds another. A node may *crash* (cease operation), and later *recover* (resume operation). Node memory is divided into *volatile* memory lost on a crash, and *stable* memory that survives crashes.

A *transaction* is a sequence of steps that modifies both replicas. As a transaction executes, Alice and Bob accumulate a list of tentative changes. If the transaction *commits*, those changes take effect, and if the transaction *aborts*, they are discarded.

The *two-phase commit protocol* [2] is a classical technique for ensuring *atomicity*: if a transaction makes tentative changes at both Alice's node and Bob's node, then the transaction either commits at both nodes or aborts at both.

Here is the simplest form of this protocol. One node, say Carol, is chosen as the *coordinator*.

1. *Prepare phase*
   - Carol, the coordinator, instructs Alice and Bob to record their tentative changes in stable storage, so they will not be lost in a crash.
   - If Alice is able to write her changes to stable storage, she sends Carol a *yes* vote. At this point, some or all of the database becomes inaccessible pending the outcome of the transaction. If for any reason, Alice cannot save her changes, she sends Carol a *no* vote. Bob does the same.
2. *Commit phase*
   - If Carol receives two *yes* votes, she instructs Alice and Bob to apply their tentative changes, committing the transaction. If Carol receives a *no* vote, or if either Alice or Bob fails to respond in time, she instructs them to discard their tentative changes, aborting the transaction. Before Carol sends her decision to Alice and Bob, she records her decision in stable memory, in case she herself crashes.
   - Alice follows Carol's instructions. If Alice crashes after preparing but before Carol decides, Alice must learn the transaction's outcome from Bob or Carol before resuming use of her database.

This description is vastly simplified, and omits many practical considerations, but it serves as a baseline for the more complex DeFi commitment protocols considered in later sections. The key pattern is that commitment requires that each party agrees to lock up a set of tentative changes, thereby freezing something of value (here, the database) until the outcome of the protocol becomes known.

4 Cross-Chain Atomicity

Alice has invested in the guilder cryptocurrency, while Bob has invested in the florin cryptocurrency. Alice and Bob would both like to diversify: Alice wants to trade some her guilders for florins, and Bob wants the opposite trade. Such an exchange would be almost trivial if both cryptocurrencies reside on the same chain, but florins reside on the Florin blockchain, and guilders on the Guilder blockchain. Naturally, Alice and Bob do not trust one another, so we are presented with a more difficult version of last section's atomic commitment problem: is there a safe way to guarantee that either both transfers happen, or neither happens, *given untrusting participants*.

The two-phase commit protocol is a good start, but it assumes that all parties are acting in good faith. Each node reports honestly whether it was able to prepare, and the coordinator does not lie about the votes it received. Nevertheless, we can build an atomic cross-chain swap by "hardening" the classical two-phase commit protocol.

We assume each blockchain supports contracts, and each party can inspect the state of each blockchain. We make use of a technical gadget called a *hashlock*. Alice creates a secret value s, called the *hashkey*. She then applies a cryptographic hash function H to s, yielding a (public) *hashlock* $h = H(s)$. It is effectively impossible to reconstruct s from h, or to find another value s' such that $h = H(s')$.

The notion of *escrow* plays the role of stable storage: an escrow contract is given custody of Alice's coins, along with a hashlock h and a timeout. If s is presented to the contract before the timeout, then Alice's coins are transferred to Bob, and if not, those coins are refunded to Alice. Bob creates a symmetric escrow contract, only with Alice's hashlock and a different timeout.

Here is the hardened two-phase commit protocol.

1. *Prepare phase*
   - Alice transfers her guilders to her escrow contract with timeout 2Δ.
   - When Bob verifies that Alice's coins have been escrowed, he transfers his florins to his escrow contract with timeout Δ.
2. *Commit phase*
   - When Alice verifies that Bob has put his florins in escrow, she sends her secret to the escrow contract on the Florin blockchain, unlocking and collecting Bob's florins. Alice has now recorded her hashkey on the Florin blockchain.
   - As soon as Alice's hashkey appears on the Florin blockchain, Bob forwards that hashkey to the escrow contract on the Guilder blockchain, unlocking and collecting Alice's guilders.

Placing coins in escrow is the analog of writing updates to stable storage and then voting to commit: each party gives up the ability to back out. For two-phase commit, it does not matter which party writes first to stable storage. For the atomic swap, however, Alice must escrow first, and Bob second, because Alice controls the hashkey, and she could steal Bob's coins if he escrowed first. The choice of timeouts is critical: if Bob's timeout were 2Δ instead of Δ, then Alice could wait until the timeout was about to expire to claim Bob's florins, leaving Bob without enough time to claim Alice's guilders. Atomic swap is less forgiving than two-phase commit: if Bob falls asleep and fails to claim Alice's guilders before 2Δ timeout, then Bob loses the coins on both chains.

A full analysis of this protocol, including failure paths, is beyond the scope of this paper. This protocol is called a *hashed timelock contract* protocol. It was invented by Nolan [17], generalized to multiple parties [11], and used on a number of blockchains [3,4,6,18,28].

5 Cross-Chain Atomicity with Optionality

In the previous section, we argued that one can solve the atomic cross-chain swap problem by "hardening" an existing solution to the atomic cross-chain commitment problem. In this section, we argue that the transition from a system where agents cooperate with one another despite failures, to a system where agents are potentially adversarial changes the conceptual framework underlying common coordination problems.

The HTLC protocol in the last section is safe in the sense that no compliant party's coins can be stolen. Each party either completes the swap, or gets its coins back. Nevertheless, the HTLC protocol introduces a new problem that could not have been formulated in the classical distributed computing model: the *sore loser* attack [25].

Suppose that after Alice escrows her coins, but before Bob escrows his, the market shifts, and Alice's florins lose value with respect to Bob's guilders. Bob now has the *option* to walk away from the deal, leaving Alice's coins locked up for a long time, while Bob is free to use his coins as he pleases. This problem did not arise in the classical two-phase commit protocol where all parties' interests were assumed to be aligned.

Premiums. The problem of optionality is well-understood in the financial world. If Bob has the option to walk away, leaving Alice temporarily unable to access her coins, called her *principal*, then Bob should compensate Alice by paying her a small fee, called a *premium.* There are well-known formulas for computing fair premiums given asset volatility and escrow duration [9]. In practice, a 2% premium is often appropriate.

The problem of adding premiums to atomic swaps is tricky, because it involves nesting one kind of atomic commitment (the premium deposit) inside another (the swap). If the premium is deposited before the principal, then the principal is protected from sore loser attacks. But the premium itself is now exposed to a reverse sore loser attack: what if Alice walks away immediately after Bob escrows his premium? The way to resolve this "chicken-and-egg" problem is to observe that the value of the premium is much lower than the value of the principal, and while Alice might not be willing to risk locking up 100 coins, Bob is probably willing to risk locking up 1 coin. For very large principals, Alice and Bob can bootstrap their premiums: Bob risks 1 coin, Alice escrows 100 coins protected by Bob's 1-coin premium, Bob escrows 1000 coins protected by Alice's 100-coin premium, and so on.

Two-Party Swap with Premiums. Here we present a simple two-party swap protocol with premiums, taken from Xue and Herlihy [25]. Let p_a be the compensation Alice should pay to Bob if Bob is a victim, and let p_b be the compensation from Bob to Alice. A contract on the guilder blockchain accepts Alice's principal and Bob's premium, and a symmetric contract on the florin blockchain accepts Bob's escrow and Alice's premium. The timeout for the first step is Δ from the start of the protocol, and subsequent timeouts increase by Δ.

A straightforward idea is to let Alice deposit premium p_a and Bob p_b. However, if Alice does not redeem Bob's principal, Bob will not be able to redeem Alice's principal, so as a result, Bob pays a premium to Alice, and Alice to Bob. Therefore, Alice should pay $p_a + p_b$ to Bob in case she does not redeem Bob's principal. Here is the protocol, where each step is labeled with its timeout. See Fig. 1.

- Δ Alice deposits her premium $p_a + p_b$ on the florin blockchain's escrow contract with timelock $t_B = 5\Delta$.
- 2Δ Bob deposits his premium on the guilder blockchain's escrow contract with timelock $t_A = 6\Delta$.
- 3Δ Alice escrows her principal on guilder blockchain's escrow contract. If she fails to do so, the premium p_b is refunded to Bob. Otherwise, the premium remains in the contract.
- 4Δ Bob escrows his principal on florin blockchain's escrow contract. If he fails to do so, the premium $p_a + p_b$ is refunded to Alice. Otherwise, the premium remains in the contract.
- 5Δ Alice sends a secret x where $H(x) = h$ to redeem Bob's principal. If she fails to do so, the premium $p_a + p_b$ in the contract is paid to Bob. If she redeems Bob's principal, the premium is refund to her.
- 6Δ Bob sends a secret s where $H(x) = h$ to redeem Alice's principal. If he fails to do so, the premium p_b in the contract is paid to Alice. If he redeems Alice's principal, the premium is refund to him.

After Alice escrows her principal, if Bob reneges, Alice can get p_b as compensation. If the swap fails after Bob escrows his principal due to Alice, Bob is compensated p_a.

The goal of this chapter is to illustrate the progression from the classical two-phase commit protocol to atomic cross-chain swap, to atomic cross-chain swap with premiums. The techniques are recognizably similar: move the item of value to a safe place, check that everything is ok, and if so, pull the trigger. The nature of the problem has shifted in interesting ways: protocol failures are no longer external events beyond the parties' control, they have become potentially rational choices requiring nested atomic commitment mechanisms for protection. In the following section, we take optionality to the next level.

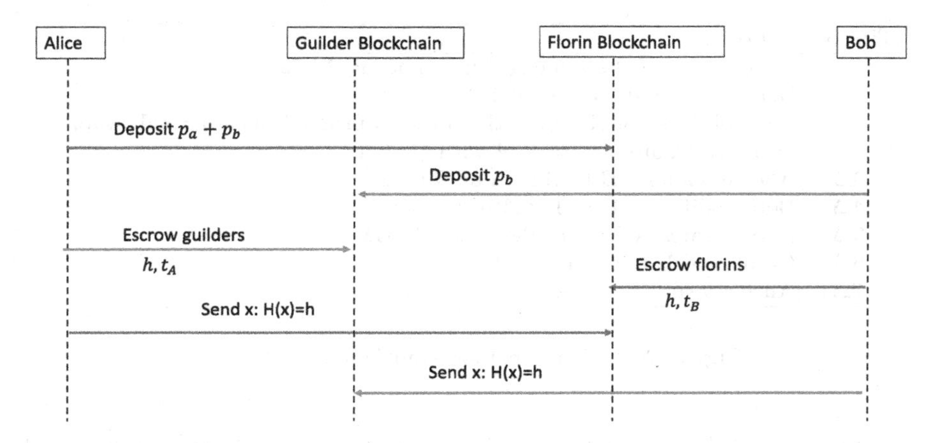

Fig. 1. Two-party Swap with Premiums

6 Cross-Chain Atomicity with Transferable Optionality

At this point, we have shifted the protocol from one where Alice and Bob agree to trade guilders for florins to one where Alice buys the *option* to make that trade. If she exercises the option, the swap happens, and if she declines to do so, she pays Bob a premium for his troubles.

While the option is capable of being exercised, it has value. It is standard in traditional finance to trade option contracts: Alice should be able to sell her option with Bob to a third party, Carol. If Carol buys the option, she acquires Alice's right to exercise the option before it expires, and Alice relinquishes all her rights. As usual, it should be possible for Alice to sell the option to Carol without placing any compliant party at risk.

Why might Alice want to transfer her option to Carol? Perhaps Alice has private information suggesting that the relative value of florins to guilders will change in the near future. If she does not plan to exercise the option, then selling it will help pay for her lost premium.

Why might Carol be willing to by an option from Alice? Perhaps Alice and Carol have asymmetric information: one thinks florins will increase in value and the other disagrees. In an illiquid options market, Carol might have trouble finding a way to buy florins, so Alice would be a natural counterparty. In a highly liquid market, Alice might be willing to offer a discount to dump her option.

Even if Alice and Carol have symmetric information, they might have different risk tolerances. Consider the price of a florin expressed in guilders at time $t = 0$. At $t = 1$, both Alice and Carol believe that with equal probability, the price will either increase by dx or decrease by dx. If Alice is risk-averse or indifferent, but Carol is risk-seeking, then Carol will want to buy that option from Alice, and Alice will want to sell.

Timeout	Action
	Alice creates AB swap edge with timeout $A : 7\Delta$.
Δ	Bob creates BA with $A : 6\Delta$.
2Δ	If Carol does not show up, the protocol proceeds as a normal swap. Otherwise Carol creates CA with $C : 9\Delta$.
3Δ	Alice modifies AB to $A : 7\Delta$ or $C : 8\Delta$.
4Δ	Bob modifies BA to $A : 6\Delta$ or $C : 7\Delta$.
5Δ	Alice creates AC swap edge with $C : 7\Delta$.
6Δ	Carol reveals C on both BA, AC.
7Δ	Alice reveals C on CA.

Fig. 2. Partial Protocol for Transferable Option

A full protocol for transferable cross-chain options is beyond the scope of this paper, and appears elsewhere [8]. Instead, we present a naive protocol that almost solves the problem, but the ways in which it falls short are instructive.

Here is a naïve Transfer Protocol. For simplicity, we address the easier problem: how to transfer a position in a 2-party swap (without premiums). The protocol is shown in Fig. 2. Initially Alice creates a swap with Bob. If Carol offers to buy the option and Alice agrees, Alice transfers her position to Carol. If Alice does not agree, Alice proceeds with the protocol as normal. Alice has a secret A and Carol has a secret C.

For brevity, we use *edge* XY as shorthand for a tentative (escrowed) transfer from party X to party Y. The notation $X : k\Delta$ means that the asset on that edge is transferred if triggered by X's secret before $k\Delta$ time after the start of the protocol. "$X : k\Delta$ or $Y : \ell\Delta$" means the asset is transferred if either X or Y triggers the transfer by revealing a secret before the respective timeouts. While this naïve protocol conveys the flavor of a full protocol, there are several reasons it is unsatisfactory.

First, there is no clear distinction between when Carol buys the swap from Alice, and when she exercises the swap. Alice just wants to sell her option and have Carol assume Alice's role immediately. Here, however, Alice she has to wait for Carol to make up her mind. Alice should be able to walk away as soon as decides she wants to buy the option.

Whether Carol does nor does not decide to participate, the ability to sell the option adds 3Δ extra rounds to the original swap protocol. An ideal protocol would behave like a typical 2-party swap if Carol never participates, taking the usual 4Δ rounds at most.

Because Alice is entangled in the protocol until Carol decides to exercise it, Alice has to escrow more than she would otherwise. That is, she has to escrow assets on AC in addition to the original assets she escrowed on AB. Alice should only have to escrow what she had in the original swap protocol.

These observations illustrate the challenges of designing transferable options for even a simple two-party swap option. In general, we would like to be able to transfer more complex, linked options. For example, in a *cross-chain deal* [13], parties can set up a complex network of swaps to be executed atomically, and a mature DeFi system would allow any party to sell their position in that network to another party. Similar challenges arise with types of cross-chain commerce such as bonds, stocks, and derivatives.

7 Related Work

The use of HTLCs for two-party cross-chain swaps is generally attributed to Nolan [17]. HTLCs have adapted to several uses [3,4,6,18]. Herlihy [11] extended HTLCs to support multi-party swaps on directed graphs.

Herlihy *et al.* [14] introduce the notion of *cross-chain deals*. They focus on how conventional notions of atomicity are inadequate for an adversarial environment, and give protocols using both HTLCs and a central coordinating blockchain. Zakhary *et al.* [27] propose a cross-chain swap protocol for proof-of-work blockchains using a *witness blockchain* as a central coordinator.

The BAR (byzantine, altruistic and rational) model [1,5] supports cooperative services spanning autonomous administrative domains that are resilient to Byzantine and rational manipulations. BAR-tolerant systems assume a bounded number of Byzantine faults, and as such do not fit our adversarial model, where any number of parties may be Byzantine, rationally or not.

In finance, *optionality* [15] is the notion that there is value in acquiring the right, but not the obligation, to invest in something later. Atomic swap based on HTLCs exposes such optionality to both parties. However, multiple researchers [9,10,16] have observed that both parties are exposed to sore loser attacks where the counterparty reneges at critical points in the protocol. Robinson [21] proposes to reduce vulnerability to sore loser attacks by splitting each swap into a sequence of very small swaps, an approach that works only for fungible, divisible tokens.

Xue and Herlihy [25] show how to incorporate premiums into multi-party swaps, auctions, and brokered sales. Prior work was focused exclusively on two-party swaps, and proposed *asymmetric* protocols, meaning that only one party pays a premium to the other, protecting only that side of the swap from a sore loser attack. These protocols include Han *et al.* [9], Eizinger *et al.* [7], Liu [16], the Komodo platform [19], Eizinger et al. [7], and the Arwen protocols [10].

Xu *et al.* [24] analyze the success rate of cross-chain swaps using HTLCs. Liu [16] proposed an atomic swap protocol that protects both parties from sore loser attacks, structured so that Alice explicitly purchases an option from Bob, and her premium is never refunded. There is no obvious way to extend this protocol to applications other than two-party swaps. Tefagh *et al.* [22] propose a similar protocol based on an options model.

8 Conclusions

We have argued elsewhere [12] that some early blockchain work recapitulated ideas and algorithms from distributed computing, sometimes falling prey to familiar pitfalls [20,23]. Here, we argue that blockchain and DeFi open up new opportunities for distributed computing research. In this paper, we outlined how atomic commitment, a classical distributed computing problem, lies at the heart of several DeFi challenges. At the same time, moving from a hardware failure model to a Byzantine failure model opens up rich new research possibilities. Dealing with optionality requires nesting one atomic commitment mechanism inside another (to support premiums), and fully embracing optionality requires nesting yet another atomic commitment mechanism (to support option sale and transfer). We hope that this paper will help draw the attention of our community to these intriguing questions.

References

1. Aiyer, A.S., Alvisi, L., Clement, A., Dahlin, M., Martin, J.P., Porth, C.: BAR fault tolerance for cooperative services. In: Proceedings of the Twentieth ACM Symposium on Operating Systems Principles. SOSP 2005, pp. 45–58. ACM, New York (2005). https://doi.org/10.1145/1095810.1095816, http://doi.acm.org/10.1145/1095810.1095816
2. Bernstein, P.A., Hadzilacos, V., Goodman, N.: Concurrency Control and Recovery in Database Systems. Addison-Wesley Longman Publishing Co., Inc., Boston (1986)
3. bitcoinwiki: Atomic cross-chain trading. https://en.bitcoin.it/wiki/Atomic_cross-chain_trading
4. Bowe, S., Hopwood, D.: Hashed time-locked contract transactions. https://github.com/bitcoin/bips/blob/master/bip-0199.mediawiki
5. Clement, A., Li, H., Napper, J., Martin, J.P.M., Alvisi, L., Dahlin, M.: BAR primer. In: Proceedings of the international conference on dependable systems and networks (DSN), DCC symposium (2008), place: Anchorage, AK
6. DeCred: Decred cross-chain atomic swapping. https://github.com/decred/atomicswap
7. Eizinger, T., Fournier, L., Hoenisch, P.: The state of atomic swaps (2018). http://diyhpl.us/wiki/transcripts/scalingbitcoin/tokyo-2018/atomic-swaps/
8. Engel, D., Herlihy, M., Xue, Y.: Transferrable cross-chain options (2021)
9. Han, R., Lin, H., Yu, J.: On the optionality and fairness of Atomic Swaps. In: Proceedings of the 1st ACM Conference on Advances in Financial Technologies. pp. 62–75. ACM, Zurich, October 2019. https://doi.org/10.1145/3318041.3355460, https://dl.acm.org/doi/10.1145/3318041.3355460
10. Heilman, E., Lipmann, S., Goldberg, S.: The Arwen trading protocols, January 2019. https://www.arwen.io/whitepaper.pdf
11. Herlihy, M.: Atomic cross-chain swaps. In: Proceedings of the 2018 ACM Symposium on Principles of Distributed Computing. PODC 2018, pp. 245–254. ACM, New York (2018). https://doi.org/10.1145/3212734.3212736, http://doi.acm.org/10.1145/3212734.3212736, number of pages: 10 Place: Egham, United Kingdom tex.acmid: 3212736
12. Herlihy, M.: Blockchains from a distributed computing perspective. Commun. ACM **62**(2), 78–85 (2019). https://doi.org/10.1145/3209623
13. Herlihy, M.: Cross-chain deals and adversarial commerce. CoRR abs/1905.09743 (2019). http://arxiv.org/abs/1905.09743

14. Herlihy, M., Liskov, B., Shrira, L.: Cross-chain deals and adversarial commerce. Proc. VLDB Endow. **13**(2), 100–113 (2019). https://doi.org/10.14778/3364324.3364326, http://arxiv.org/abs/1905.09743
15. Higham, D.J.: An Introduction to Financial Option Valuation: Mathematics, Stochastics and Computation, 4th edn. Cambridge University Press, Cambridge (2009)
16. Liu, J.A.: Atomic swaptions: cryptocurrency derivatives. arXiv:1807.08644 [cs, q-fin], March 2020
17. Nolan, T.: Atomic swaps using cut and choose, February 2016. https://bitcointalk.org/index.php?topic=1364951
18. Organization, T.K.: The BarterDEX whitepaper: a decentralized, open-source cryptocurrency exchange, powered by atomic-swap technology. https://supernet.org/en/technology/whitepapers/BarterDEX-Whitepaper-v0.4.pdf
19. Platform, K.: Advanced blockchain technology, focused on freedom, July 2019. https://docs.komodoplatform.com/basic-docs/start-here/core-technology-discussions/introduction.html#note-on-changes-since-whitepaper-creation-cr-2019
20. Popper, N.: A venture fund with plenty of virtual capital, but no capitalist. New York Times (man 2016). https://www.nytimes.com/2016/05/22/business/dealbook/crypto-ether-bitcoin-currency.html
21. Robinson, D.: Htlcs considered harmful (2019). http://diyhpl.us/wiki/transcripts/stanford-blockchain-conference/2019/htlcs-considered-harmful/
22. Tefagh, M., Bagheri, F., Khajehpour, A., Abdi, M.: Capital-free futures arbitrage, October 2020. https://doi.org/10.13140/RG.2.2.31609.90729/1, https://www.researchgate.net/profile/Mojtaba-Tefagh-2/publication/344886866_Capital-free_Futures_Arbitrage/links/5fdc88e3a6fdccdcb8d89ee1/Capital-free-Futures-Arbitrage.pdf
23. Vigna, P.: Chiefless company rakes in more than $100 million. Wall Street Journal, May 2016. https://www.wsj.com/articles/chiefless-company-rakes-in-more-than-100-million-1463399393
24. Xu, J., Ackerer, D., Dubovitskaya, A.: A game-theoretic analysis of cross-chain atomic swaps with HTLCs. arXiv:2011.11325 [cs], April 2021. http://arxiv.org/abs/2011.11325
25. Xue, Y., Herlihy, M.: Hedging against sore loser attacks in cross-chain transactions. In: ACM Symposium on Principles of Distributed Computing (2021)
26. Young, J.: Defi explosion: Uniswap surpasses coinbase pro in daily volume (2020)
27. Zakhary, V., Agrawal, D., El Abbadi, A.: Atomic commitment across blockchains. CoRR abs/1905.02847 (2019). http://arxiv.org/abs/1905.02847 tex.bibsource: dblp computer science bibliography, https://dblp.org tex.biburl: https://dblp.org/rec/bib/journals/corr/abs-1905-02847 tex.timestamp: Mon, 27 May 2019 13:15:00 +0200
28. Zyskind, G., Kisagun, C., FromKnecht, C.: Enigma Catalyst: a machine-based investing platform and infrastructure for crypto-assets. https://www.enigma.co/enigma_catalyst.pdf

Privacy-Preserving Data Sharing for Medical Research

Michael J. Fischer[1(✉)], Jonathan E. Hochman[2], and Daniel Boffa[3]

[1] Computer Science, Yale University, New Haven, USA
michael.fischer@yale.edu
[2] UNS Project and Hochman Consultants, Hartford, USA
jonathan@unsproject.com
[3] School of Medicine, Yale University, New Haven, USA
daniel.boffa@yale.edu

Abstract. Electronic patient medical records contain vast amounts of information of potential value to researchers striving to increase understanding of diseases, treatments, and outcomes. Effective use of such data is limited by privacy and technical concerns. Privacy laws require the removal of Personally Identifiable Information (PII) from the released data. Technical concerns are that the data must be abstracted for consistency across different providers. To be most useful, data from different providers for the same patient must be linked together. This paper applies cryptographic techniques to the problem of privacy-preserving linking of medical records.

Keywords: Medical research · Privacy · Cryptographic data linking

1 Introduction

The goal of this work is to design a national system for medical data sharing that meets several criteria:

1. Understandable to stakeholders
2. Supportable by stakeholders
3. Simple and explainable
4. Actionable with minimal startup investment
5. Sustainable
6. Secure

Our contribution includes a new cryptographic primitive, called a *blinding-completion pair*, which addresses the practical problem of linking anonymous medical records. Blinding-completion pairs provide a method for generating a multitude of anonymous pseudonyms for an entity, to be used by data sources, and then consolidating that multitude into a single anonymous pseudonym at the destination database. We also describe a distributed system and protocol for

C. Johnen et al. (Eds.): SSS 2021, LNCS 13046, pp. 78–89, 2021.
https://doi.org/10.1007/978-3-030-91081-5_6

sharing medical data that can preserve privacy when confronted with occasional data breaches.

This paper is organized as follows: Sect. 2 summarizes related work. Section 3 describes the stakeholders and sketches the existing flow of clinical data from patient to researcher. Section 4 focuses on the problem of privacy-preserving linking of patient data. Section 5 suggests possible modifications of the workflow to shift responsibility for maintaining privacy to specialized "security nodes". Section 6 provides a brief threat model and discloses known limitations of the system. Section 7 summarizes our contribution and directions for future research.

2 Related Work

The problem of connecting records from multiple databases is called *record linking*. What we aim to achieve is called *privacy-preserving record linking (PPRL)*, where the goal is to link patient records while protecting the PII of the patient. The interest in PPRL of health records goes back at least 16 years to the paper of Demuynck and De Decker [8], who propose a complicated multi-stakeholder protocol that uses cryptographic techniques to achieve PPRL. The idea of using pseudonym identifiers to substitute for PII linkage appears in Alhaqbani and Fidge in 2008 [1].

Many other linkage techniques have subsequently been studied. Vatsalan, Christen, and Verykios [20] give a taxonomy of PPRL techniques in a 2013 paper containing 143 references to the extensive literature on the subject! They observe that preserving the privacy of shared data such as medical records is a difficult problem and that existing approaches have a variety of drawbacks. For example, some systems have focused on joining records from just two sources, which would not satisfy our design requirements.

Recent work aims to address one or more of the shortcomings of prior work. Camenisch and Lehmann [6] add user-auditability to pseudonym systems. The PRIMAT system [9] handles multiple sources of medical data, but it assumes the existence of a "trusted linkage unit (LU) [that] performs the actual linkage of encoded records submitted".

Our approach is different because we leverage existing universal identifiers and anonymize them by generating two levels of pseudonyms. This results in a system with resilience to limited breaches that is easy to understand by stakeholders. Moreover, our system does not put sensitive medical data in the hands of any third party. Medical data flows directly from a health care provider to a research database.

Our proposed system also differs from many others in the location of the three types of data: *identifier* is a segment of coded information that is unique for each person, *identifying information* enables a specific person to be identified, and *sensitive information* is desired to be kept private and not shared publicly as being attributable to an individual.

An identifier by itself is meaningless and is just a code. For example, any random combination of nine numbers very well may be a social security number, but without identifying information, there is no relevance, utility or vulnerability. Identifying information alone is not overly relevant, because it simply notes the existence of a person, without any detail of that person. For example, names (and addresses and phone numbers), have historically been distributed in phone books. Finally, sensitive information that cannot be linked to a specific person poses no risk to privacy and is the principal that allows large databases to exist for medical research. Any medical textbook could contain what most would consider sensitive information, such as data related to the treatment of specific patients. When such information is not attributed to any person, it poses no vulnerability. An important caveat is that if the anonymized sensitive information is sufficiently detailed, it may serve as a fingerprint that can be correlated with publicly available data and used to identify the subject [15].

3 Medical Information Workflow

Data enters the health care system when a patient contacts a provider, whether a primary care physician or a hospital. At that point, the provider determines and records the patients PII and begins or updates the patient's chart.

To curate data for statistical and research purposes, trained registrars extract select data elements from the medical record according to specific data field definitions, resulting in highly structured data sets. The datasets are then stripped of PII and exported in a deidentified manner to one or more of several national databases. Importantly, not every health care entity submits to every database, and each database only requests certain fragments of the patient's medical information [5]. As a result, each patient's care is captured by the databases in a piecemeal fashion. Because the databases do not collect PII, there is no way to consistently reunite the fragments of the health care data back together to create a complete picture of a patient's journey through the diagnosis, treatment, and outcome of their medical condition.

The analysis of patient outcomes captured within the databases has led to dramatic improvements in the safety and effectiveness of care for almost every medical condition [18]. However, the ability of the database research to characterize relationships between variables and outcomes in medical care is critically dependent on the breadth of information available for analysis (e.g., to control for bias and confounding effects). Because databases are only capturing fragments of the medical journey, there are limitations to the types of improvements that currently can be made with database research [16].

For example, the database that best captures cancer stage does not capture the specific type of chemotherapy that patients received [5]. If there were a way to reunite all the fragments of data back together, medical research using existing databases would become far more powerful, and many more improvements would be possible. (See Daniel Boffa, *Comparing Comparisons*, in comments to [16]).

A simple but unacceptable "solution" to the linking problem is to give each patient a universal health identifier to be included with the patient's record in each database. This is used in some other countries (e.g., Norway [2]). However, in the United States, the topic of a national identifier has become highly polarizing, making this a less feasible option. Moreover, anyone with access to the curated databases would be able to join the several databases into one master health record for the patient with that identifier. This is often sufficient, when combined with other information readily available on the internet, to deanonymize the health record and reveal the patient's PII.

4 Privacy-Preserving Linking of Patient Data

We describe a new cryptographic primitive for generating identifiers that allows a database to link records from different data providers while preserving privacy in the face of many kinds of breaches.

4.1 Blinding-Completion Pairs

Let x be the identifier used by a data provider h to identify an entity. Let $b(x)$ be a one-way hash function. A *blinding-completion pair* for h is a pair of one-way functions $(b_h(x), c_h(y))$ such that $b(x) = c_h(b_h(x))$ for all x, and $b(x)$ is also one-way. Like a cryptosystem $(E_h(x), D_h(y))$, the composition of the second function in the pair with the first yields the same function for all keys h. In our case, the composition is the fixed blinding function $b(x)$, which defines the alias $y = b(x)$ for x. While neither x nor y can be recovered from $y_h = b_h(x)$, y *can* be recovered from *any* single value y_h if the corresponding completion function c_h is available, since $y = c_h(y_h)$.

4.2 Implementation

There are several ways to implement blinding-completion pairs. One way is to use cryptographic accumulators. Let $Q = \{b_1, \ldots, b_N\}$ be a set of *quasi-commutative cryptographic hash functions* [3]. They have the property that the N-way composition of these functions in any order yields the same function B. Hence, for any subset $S \subseteq Q$, the composition of those functions in S, call it b_S, can be used as the first element of a blinding-completion pair, and the composition of $b_{(Q-S)}$ becomes the completion function c_S. The drawback of this scheme is that N must be known in advance, and the time complexity of finding b_S and c_S grows with N.

We use a different scheme based on discrete logarithms. First we introduce some standard number theory. For positive integer n, let $\mathbf{Z}_n^*$ be the set of positive integers less than n that are relatively prime to n. The size of $\mathbf{Z}_n^*$ is given by Euler's totient function $\phi(n)$.

In the special case that n is a prime p, $\mathbf{Z}_p^* = \{1, \ldots, p-1\}$, so $\phi(p) = p - 1$. Also, p has primitive roots. We say g is a *primitive root* of p if every number $a \in \mathbf{Z}_p^*$ can be expressed as $a = g^k \bmod p$ for some $k \in \mathbf{Z}_p^*$. The number k is called the *discrete logarithm of a modulo p*. Computing the discrete logarithm is believed to be computationally difficult when p and g are chosen carefully.

For our purposes, we choose $p = 2q + 1$, where q is a *Sophie Germain prime* and p is called a *safe prime*. Such prime pairs are widely used in cryptography, so a suitable supply exists for our purposes. An estimate of the number of Sophie Germain primes less than n is $\Theta(n/(\log n)^2)$ [19, pp.123–124].

There are $\phi(\phi(p)) = \phi(p-1)$ primitive roots in $\mathbf{Z}_p^*$. This makes it possible to find a primitive root g by a guess-and-check method. Guess a number $g \in \mathbf{Z}_p^*$ and check that $g^q \equiv -1 \pmod{p}$. The expected number of guesses required to find g is $(p-1)/\phi(p-1) = \mathcal{O}(\log\log p)$ [14, p. 391]. How big is $\phi(p-1)$? Because we've chosen $p - 1 = 2q$, then $\phi(p-1) = \phi(2)\phi(q) = q - 1 = (p-3)/2$.

Let r, u be positive integers in $\mathbf{Z}_{\phi(p)}^* = \mathbf{Z}_{p-1}^*$, and let v be a positive integer less than $\phi(p)$ such that $r = (u+v) \bmod \phi(p)$. Define $b_u(x) = xg^u \bmod p$ and let $c_v(y) = yg^v \bmod p$. Then (b_u, c_v) is a blinding-completion pair for the blinding function $b(x) = xg^r \bmod p$. This follows since

$$c_v(b_u(x)) \equiv c_v(xg^u) \equiv (xg^u)g^v \equiv xg^{u+v} \equiv xg^r \pmod{p}.$$

The last identity follows from Euler's Theorem, which states that for $a \in \mathbf{Z}_p^*$, $a^{\phi(p)} \equiv 1 \pmod{p}$.

The parameters p, q matter both for convenience and security. To choose r from $\mathbf{Z}_{\phi(p)}^*$, we need to find an r that is relatively prime to $(p-1)$. But an arbitrary $p-1$ might have many small factors, e.g., $p = 71$. However, since we choose p, q so that $p - 1 = 2q$, we know the only factors of $p-1$ are 2 and q. Choosing a safe prime p makes it easy to find random numbers in $\mathbf{Z}_{\phi(p)}^*$. As for security, the discrete logarithm problem is hard in general, but a solution may be feasible via the *Pohlig–Hellman algorithm* when $p-1$ has no large prime factors [17]. A safe prime does not have this weakness.

As explained in Sect. 5 below, we will allow security nodes to independently choose random values of $r, u \in \mathbf{Z}_{\phi(p)}^*$ and calculate $v = (r-u) \bmod \phi(p)$.

Because r and u may be chosen independently by different security nodes, there is a theoretical risk of $v = 0$, which would produce the undesirable result $y_h = y$. From the point of view of a cryptographer, such a result is not a problem, but to satisfy our design requirements we want to provide an unqualified guarantee that the identifier used by a data provider h does not appear in a database that links its records.

The value r is secret and may not be shared, therefore the security node choosing u cannot "peek" at r to make sure it chooses a safe value for u. In practice, when p is very large, the probability of $r = u$ is vanishingly small. Should this ever happen, a simple remedy is to choose a new value for r via key rotation, as explained next.

4.3 Key Rotation

The values of r, u, v should be rotated periodically in case they are ever compromised. We provide a sketch of how such rotation could be implemented.

To rotate the value of r, choose a random $0 < s < \phi(p)$ and calculate $r' = (r+s) \bmod \phi(p)$. Check that $r' \in \mathbf{Z}_{\phi(p)}^*$, and if not, choose a different random

s and try again. Then recalculate $v' = (r' - u) \bmod \phi(p)$. Finally, to update the blinded value $y = xg^r \bmod p$, calculate a new blinded value $y' = yg^s \bmod p$. Note that x is not needed for this calculation. Then y' is the new blinded value for the same x, since $y' \equiv xg^r g^s \equiv xg^{r'} \pmod{p}$.

To rotate the value of u, choose a random $0 < s < \phi(p)$ and calculate $u' = (u + s) \bmod \phi(p)$. Check that $u' \in \mathbf{Z}^*_{\phi(p)}$, and if not, choose a different random s and try again. To update the blinded value $y_h = xg^u \bmod p$, calculate a new blinded value $y'_h = y_h g^s \bmod p$. Then y'_h is the new blinded value for the same x, since $y'_h \equiv xg^u g^s \equiv xg^{u'} \pmod{p}$.

5 Proposed Workflow for Enhanced Security

We propose three additions to the existing workflow to maintain security while still permitting research data sharing.

1. We envision a system of restricted *local patient identifiers* (LPIDs) that can be used to identify the medical records of a given patient within the context of a single health care provider. Local identifiers are obtained from a patient's PII via a one-way cryptographic function. This prevents the local identifier from being reverse engineered to obtain PII.
2. Using the cryptographic technique of *blinding-completion functions*, the local patient identifiers for different health care providers can be used to calculate an anonymized patient identifier (APID). The APID allows a medical database to link patient records across providers while still providing no clear path to finding the corresponding PII.
3. To further protect patient anonymity and privacy, we propose to separate the security services from the servers and databases holding the actual PII (in the case of hospitals) and medical data (in the case of curated database).

5.1 Trust

Our model of trust has two dimensions: whether the party has *good intentions* to keep sensitive information private, and whether the party is *competent* to do so. For example, while we may trust health care systems to do their best to keep sensitive patient information private, they are not always good at cybersecurity, as evidenced by the large number of cyber-attacks against health care organizations.[1] Even when a health care system has a central information technology department capable of maintaining network security, that competency may be a scarce resource.

Our model of trust is different from traditional adversarial models that consider the worst possible outcomes from an untrusted party. Our model is informed by one author's experience in analyzing dozens of actual lawsuits related to online

[1] In a recent survey of health care organizations, 70% of respondents reported that their organizations had experienced significant security incidents in the prior 12 months [12].

identity and privacy. While malice is sometimes present, incompetence is much more likely. In the modern era this idea has been called Hanlon's razor, which states, "Never attribute to malice that which is adequately explained by stupidity" [21]. Earlier such attributions go back at least to Goethe [10]: "I have realized once again that misunderstandings and lethargy can cause more going wrong in the world than cunning and wickedness do. At least, those two are certainly less common".

While health care providers are expected to be competent at medical treatment, there is no reason to expect them to be competent at cryptography (nor would we trust the average cryptographer to perform surgery). To mitigate the risk of health care providers performing cryptographic functions insecurely or leaking secret keys, we restrict access to certain functions and the secret keys that power them. To keep everyone safe, we introduce additional parties to the transaction, called "security nodes", which have demonstrated technical competence. Each health care provider will choose a security node to work with, and so will each medical database.

A security node is a network service that can be trusted to implement cryptographic functions correctly and to hold secret keys without leaking them. Security nodes could be independently operated, or they might be operated by a department within a medical organization with the required competence. Importantly, a security node isolates the secret keys used by cryptographic functions or for signing messages in a single location. This makes it easier to protect secret keys by storing them in specialized computing hardware, such as a hardware security module (HSM). Security nodes also provide authentication services to health care providers. Each security node has a public-private key pair it can use to sign and authenticate messages for other security nodes. A special "executive" security node keeps a list of all security nodes and their public keys. This list may be periodically updated and distributed, enabling security nodes to reliably authenticate each other's messages.

5.2 Parties

A transaction at minimum includes six parties:

1. A patient w who is identified with a *user identifier* (UID),
2. A health care provider h who treats patients and gathers medical data,
3. The health care provider's security node that provides LPIDs that can be attached to medical data in lieu of UIDs,
4. The database's security node that attaches an APID to medical data in lieu of LPIDs,
5. A database d that collects anonymous patient profiles, identified only by APID,
6. Researchers that receive anonymous patient profiles.

5.3 Identifiers

There are three levels of identifiers, each with distinct properties:

1. UID is an invariant identifier, such as a name-birthday pair or a Social Security number. The UID is readily available to the patient and widely used by health care providers. A patient's UID must never be shared because it constitutes PII.
2. LPID identifies patients relative to a health care provider and has no apparent connection to any PII. A patient's LPID is different at every provider and is used for sending anonymized records to a database.
3. APID identifies patients relative to a database and has no apparent connection to any PII or to any LPID. Anonymized records sent to a database by different health care providers for the same patient are associated with the same APID, which enables record linking.

The LPID and APID identifiers are rotated periodically to frustrate any attacker who manages to breach the system. Rotation can be done if a breach is detected, or on a regular schedule to limit the damage from an undetected breach, and to provide other benefits. Key rotation is a widely accepted good practice in cloud computing [11].

5.4 Initialization

Initially, one or more databases join our proposed system, which provides the values p, q, and g. Each database d chooses a security node which generates a random value r_d such that $r_d \in \mathbf{Z}^*_{\phi(p)}$. The value r_d is used to generate blinding-completion function pairs and must be kept secret.

Each health care provider h joining the system chooses a security node. The health care provider obtains a public-private key pair for signing messages (e.g., an X.509 security certificate [4]), using a digital signature algorithm such as DSA or ECDSA, and verifying the health care provider's identity to its security node. The public key is registered, or "pinned", to the security node. Upon registration, the health care provider's security node will chose a random value u_h such that $u_h \in \mathbf{Z}^*_{\phi(p)}$. The value u_h must be kept secret and is used to generate a blinding function $b_h()$.

To join a database d, a provider h causes its security node to send u_h to the security node of database d. The security node of d then calculates a value $v_d = (r_d - u_h) \bmod \phi(p)$. The value v_d must be kept secret and is used to generate a completion function $c_h()$.

These blinding-completion functions are constructed in such a way that:

1. Each blinding function for each provider h produces a different pseudorandom identifier LPID_h for each patient.
2. Each completion function for each provider h to each database d maps each LPID_h to APID_d.

If a health care provider h participates in multiple databases, it uses the same LPID_h identifiers, but each database d will generate different APID_d identifiers. Conversely, when multiple providers contribute medical data to a database, each provider h has different LPID_h identifiers and the database d has the same APID_d identifiers. To preserve privacy, no provider knows any of the APID_d values, and no database knows any of the LPID_h values. The patient identifier equivalence pairs (LPID_h, APID_d) are only known to, or computed by, security nodes.

5.5 Contribution of Patient Profiles

Medical providers may contribute patient profiles to a database. A profile contains demographic and medical information of interest to researchers. For example, a patient profile might include age, medical diagnosis codes and dates, occupation, ethnicity, treatment history, and other target characteristics. Existing standards for storing digital medical records can be used.

To contribute a profile to a database d, a health care provider h performs several steps.

1. The provider hashes the patient's UID, w, with a standard, widely available hash function such as SHA256 [7], to generate a value x that it sends to its security node. The security node then applies the blinding function for that health care provider to x, resulting in the value LPID_h. The security node returns LPID_h to the health care provider, and h adds it to the patient's medical record.
2. The provider generates a random transaction number t. The relevant profile data m is then composed into a message (h, t, m, d) and sent to the database d. The medical data will only be added to the database after it is authenticated by the database's security node.
3. The provider creates a token (such as a JSON web token [13]) containing the quadruple $(\mathrm{LPID}_h, h, t, d)$ and signs it using its secret key. The provider sends the signed token to the provider's security node, which then authenticates the signature using the health care provider's public key and appends its own signature to the token.

The provider's security node sends token $(\mathrm{LPID}_h, h, t, d)$ to the security node of database d which does the following steps:

1. Authenticates the signature of the health care provider's security node.
2. Verifies that the health care provider's name h in the token matches the name in the message.
3. Applies the appropriate completion function $c_h()$ to LPID_h to generate APID_d.
4. Creates a new token $(\mathrm{APID}_d, p, t, d)$, signs, and sends it to database d.

Database d receives the token $(\mathrm{APID}_d, h, t, d)$ and then performs these steps:

1. Authenticates the signature of its own security node.
2. Finds the message (h, t, m, d) with the same transaction number t.
3. Verifies that the health care provider's name h in the message matches the health care provider's name in the token.

4. Adds the medical data m, the provider h, and APID_d to its data store. If there is an existing record with APID_d, the new data is linked to the existing record.

5.6 Accessing Medical Data for Research

A researcher can connect to a database and search for patient profiles that match desired criteria for the study. Upon approval by an appropriate medical research ethics board, the researcher can then requisition specific medical data from the database that is relevant to the research being conducted.

Upon receiving an approved request for medical data related to a patient profile, the database then retrieves from its database the patient medical data that meets the researcher's specific criteria.

When releasing data to a researcher, the identifier for each record, APID_d, should be removed or hashed with a one way function such as SHA256. Researchers are not security experts. Therefore, they should not be trusted to keep the APID_d identifiers private.

6 Threat Analysis

The system we describe, like all such systems, does not confer perfect security. If a security node were compromised, an attacker might learn the secret values r_d, u_h, or v_h. These secret values could enable an attacker to recover some or all of the blinding-completion functions $b()$, $b_h()$, or $c_h()$ and their inverses. Having one or more of these inverse functions could give an attacker who possesses APID_d the ability to calculate LPID_h or x, the hash of the patient's UID. While it is not practical to invert the hash function used to generate x, an attacker could test whether a known UID value, when hashed, equals x.

Assume an attacker gathers medical data from researchers. This data would contain hashes of the APID_d for each record. If the attacker additionally compromises health care providers and security nodes, it is conceivable that they could eventually link a UID to anonymous research data. Given UID and r_d, an attacker can calculate APID_d, hash this value and then compare it to the data collected from researchers. The difficulty of such an attack is high because it requires compromising at least one health care provider and at least one security node of a database containing data from that health care provider within a limited time frame (the key rotation period). Moreover, if such an attack were to succeed, it would likely deanonymize only a limited number of medical records, especially if there are many independent health care providers, databases, and security nodes.

The security nodes described in this system only need to communicate with other security nodes and with the medical providers or databases they serve. Consequently, a firewall can protect each security node so that it only communicates with systems on an "allow" list. This type of protection increases the difficulty of breaching a security node because even if the system has vulnerabilities, an attacker needs to gain access to a system on a security node's "allow" list even to commence a remote attack on the security node.

We believe that the difficulty of attacking our proposed system is sufficiently high, and the profitability sufficiently low, that attackers would prefer to attack health care providers directly and aggregate data via UID. Therefore, our proposed system does not materially increase the risk of private medical data being exposed in a data breach versus the status quo.

7 Conclusion

We have presented a new cryptographic technique called blinding-completion pairs and demonstrated how they could be used to enable the sharing of private data without revealing personally identifiable information (PII).

Based upon blinding-completion pairs maintained by security nodes, we have drawn a sketch of how health care providers could supply medical data to one or more databases that would aggregate data for each patient and then make those consolidated records available as anonymous data to researchers. Our system could release data for medical research in a way that protects patient PII while still enabling qualified researchers to identify records from different health care providers that belong to the same patient.

Possible areas for future work include constructing a prototype system, developing new blinding-completion functions with improved security properties, and investigating alternative sharing protocols that may offer stronger privacy guarantees in the event of data breaches.

Acknowledgments. We are grateful to Ewa Syta of Trinity College (Connecticut) for a thorough reading of an early draft of this paper and for providing many helpful comments and pointers to the relevant literature. We thank Bonnie Kaplan of the Yale School of Medicine for sharing her vast knowledge of the world of electronic health data with us. Lastly, we are indebted to Alice Fischer from the University of New Haven, who scrutinized our nearly final draft.

References

1. Alhaqbani, B., Fidge, C.: Privacy-preserving electronic health record linkage using pseudonym identifiers. In: 10th International Conference on e-health Networking, Applications and Services, HealthCom 2008, pp. 108–117 (2008). https://doi.org/10.1109/HEALTH.2008.4600120
2. Bakken, I.J., Ariansen, A.M.S., Knudsen, G.P., Johansen, K.I., Vollset, S.E.: The Norwegian Patient Registry and the Norwegian Registry for Primary Health Care: Research potential of two nationwide health-care registries. Scand. J. Public Health **48**(1), 49–55 (2020)
3. Benaloh, J., de Mare, M.: One-way accumulators: a decentralized alternative to digital signatures. In: Helleseth, T. (ed.) EUROCRYPT 1993. LNCS, vol. 765, pp. 274–285. Springer, Heidelberg (1994). https://doi.org/10.1007/3-540-48285-7_24

4. Boeyen, S., Santesson, S., Polk, T., Housley, R., Farrell, S., Cooper, D.: Internet X.509 Public Key Infrastructure Certificate and Certificate Revocation List (CRL) Profile. RFC 5280, May 2008. https://doi.org/10.17487/RFC5280
5. Boffa, D.J., et al.: Using the national cancer database for outcomes research: a review. JAMA Oncol. **3**(12), 1722–1728 (2017)
6. Camenisch, J., Lehmann, A.: Privacy-preserving user-auditable pseudonym systems. In: 2017 IEEE European Symposium on Security and Privacy (EuroS&P), pp. 269–284 (2017). https://doi.org/10.1109/EuroSP.2017.36
7. Dang, Q.: Secure hash standard, 04 August 2015. https://doi.org/10.6028/NIST.FIPS.180-4
8. Demuynck, L., De Decker, B.: Privacy-preserving electronic health records. In: Dittmann, J., Katzenbeisser, S., Uhl, A. (eds.) CMS 2005. LNCS, vol. 3677, pp. 150–159. Springer, Heidelberg (2005). https://doi.org/10.1007/11552055_15
9. Franke, M., Sehili, Z., Rahm, E.: PRIMAT: a toolbox for fast privacy-preserving matching. Proc. VLDB Endow. **12**(12), 1826–1829 (2019). https://doi.org/10.14778/3352063.3352076
10. von Goethe, J.W.: The Sorrows of Young Werther. Oxford World's Classics, Oxford, (tr.) David Constantine, online edn., December 2020. https://doi.org/10.1093/owc/9780199583027.001.0001, Accessed 25 Sept 2021
11. Google Cloud Key Management Service: Key rotation. https://cloud.google.com/kms/docs/key-rotation. Accessed 26 Sept 2021
12. HIMSS Cybersecurity Survey (2020). https://www.himss.org/sites/hde/files/media/file/2020/11/16/2020_himss_cybersecurity_survey_final.pdf
13. Jones, M., Bradley, J., Sakimura, N.: JSON Web Token (JWT). RFC 7519, May 2015. https://doi.org/10.17487/RFC7519
14. Knuth, D.E.: The Art of Computer Programming, vol. 2: Seminumerical Algorithms, 3rd edn. Addison-Wesley Professional (1998)
15. Kolata, G.: Your Data were 'Anonymized'? These Scientists Can Still Identify You. The New York Times, 23 July 2019. https://www.nytimes.com/2019/07/23/health/data-privacy-protection.html
16. Kumar, A., et al.: Evaluation of the use of cancer registry data for comparative effectiveness research. JAMA Netw. Open **3**(7), e2011985 (2020). https://doi.org/10.1001/jamanetworkopen.2020.11985
17. Pohlig, S., Hellman, M.: An improved algorithm for computing logarithms over GF(p) and its cryptographic significance (Corresp.). IEEE Trans. Inf. Theory **24**(1), 106–110 (1978). https://doi.org/10.1109/TIT.1978.1055817
18. Salazar, M.C., et al.: Association of delayed adjuvant chemotherapy with survival after lung cancer surgery. JAMA Oncol. **3**(5), 610–619 (2017). https://doi.org/10.1001/jamaoncol.2016.5829
19. Shoup, V.: A Computational Introduction to Number Theory and Algebra, chap. 5.5.5 Sophie Germain Primes, 2nd edn., pp. 123–124. Cambridge University Press, February 2009. ISBN 9780521516440
20. Vatsalan, D., Christen, P., Verykios, V.S.: A taxonomy of privacy-preserving record linkage techniques. Inf. Syst. **38**(6), 946–969 (2013). https://doi.org/10.1016/j.is.2012.11.005
21. Wikipedia contributors: Hanlon's razor – Wikipedia, The Free Encyclopedia (2021). https://en.wikipedia.org/w/index.php?title=Hanlon's_razor&oldid=1045571584. Accessed 24 Sept 2021

How Do Mobile Agents Benefit from Randomness?

Pierre Fraigniaud(✉)

Institut de Recherche en Informatique Fondamentale,
Université de Paris and CNRS, Paris, France
pierre.fraigniaud@irif.fr

Abstract. This paper focuses on mobile agents modeling biological entities such as foraging insects. It compares the behavior of randomized mobile agents with the behavior of deterministic agents subject to probabilistic perturbations of their actions caused by the environment, e.g., a gust of wind deviating the trajectory of a flying insect. We show that, for a large class of scenarios, the two types of settings cannot be distinguished on the sole basis of external observations. On the other hand, we also identify specific scenarios for which particular observable behaviors can only result from agents benefiting from their own individual sources of randomness. That is, in these scenarios, probabilistic perturbations of the environment are not sufficient for allowing deterministic agents to behave, from an external observer perspective, the same as randomized agents.

Keywords: Mobile agents · Mobile computing · Random search · Graph exploration · ANTS' problem · Navigation problems

1 Introduction

In the context of system biology, understanding the behavior of biological entities (e.g., insects, birds, fishes, etc.) is often addressed by modeling each entity as an autonomous computing *agent*, typically a finite state machine. The action of the agent at each time t is determined by the internal state of the agent at time t, and by its perception of its local environment at that time. Probabilistic effects help significantly for solving problems inspired by tasks accomplished by biological entities (e.g., foraging for food, looking for a new nest, escaping from a predator, etc.), and some problems may even be impossible to solve without randomization. For instance, even very powerful machines such as Cook and Rackoff's JAGs [4] cannot explore all graphs[1]. Indeed, biological entities seem to take benefit from probabilistic effects [8], whether it be thanks to internal

[1] A *jumping automaton for graphs* (JAG) can be viewed as a team of finite automata that cooperate constantly, and such that any member of the team can jump instantly to any vertex occupied by another member.

Additional support from ANR projects DUCAT and QuDATA.

C. Johnen et al. (Eds.): SSS 2021, LNCS 13046, pp. 90–107, 2021.
https://doi.org/10.1007/978-3-030-91081-5_7

capabilities of these entities allowing them to generate random values, and/or thanks to random noise generated by the environment in which the entities navigate (e.g., atmospheric variations, presence of physical micro-structures on the terrain, encounters with other agents, etc.). This paper is aiming at identifying which of these two phenomenons predominates. Coming back to the computing agents modeling the biological entities, the question addressed in this paper is whether the agents must have the ability to generate random values for solving navigation problems efficiently, or whether the presence of random noises that is external to deterministic agents is actually sufficient.

Statement of the Problem. We adopt a very generic and flexible model for a mobile agent, based on three functions: the *action* function, the *perception* function, and the *transition* function. The action function returns which action (e.g., go straight, turn right, etc.) is to be performed by the agent. The perception function provides the agents with information about their current location (e.g., presence of obstacles, list of available directions, etc.). The transition function enables the agent to change state after each movement. Similarly, we adopt a simple and flexible model for the environment in which the agent evolves. It is essentially characterized by a single function: the *movement* function. Given a current position, and an action performed by the agent at that position, the movement function merely returns the new position where the agent is moved.

Specifically, we consider an agent $\mathcal{A}$ modeled by a finite state machine moving in a discrete environment $\mathcal{E}$, i.e., the set X of agent's positions is discrete. At each discrete time step, the agent currently located at a position $x \in X$, gets as input a local *perception* $\rho(x)$ of x. Based on this input, and on its current state s, the agent computes an *action* $a = \alpha(s, \rho(x))$ where the function α is specific to the agent. For instance, an action may be "rotate 90° clockwise" or "perform one step ahead". The action a results in a *movement* from position x to position $y = \mu(a, x) \in X$, where the function μ is depending on the environment in which the agent evolves—see Algorithm 1.

Algorithm 1: Search Protocol

```
1 s ← s0 ;                               // initial state
2 x ← x0 ;                               // initial position
3 repeat forever
4 |   a ← α(s, ρ(x)) ;                   // agent selects action
5 |   y ← μ(a, x) ;                      // movement is performed
6 |   s ← δ(s, ρ(y)) ;                   // agent updates its state
7 |   x ← y
8 end
```

For instance, an insect facing north at x, and executing the action "perform one step ahead" may move one millimeter north, but it may as well move half a millimeter north-west, or even remain at x, if an obstacle prevents it from moving straight, or if the atmospheric conditions cause a deviation of the agent's trajectory. Once at y, the agent updates its internal state from s to $\delta(s, \rho(y))$, where δ is a transition function specific to the agent, and $\rho(y)$ is the perception of position y by the agent. Note that the agent is aware of its actions, but not of its actual movements. For instance, in our illustrative example, the agent aiming at moving straight does not necessarily perceive whether this action was actually executed, depending on its perception $\rho(y)$ of the new position y versus its perception $\rho(x)$ of the previous position x.

We focus on *search* tasks modeled as navigation problems in an environment $\mathcal{E}$, that is, the ability to travel from a source $s \in X$ to a target $t \in X$, governed by the action function α and the transition function δ of the agent, and by the movement function μ of the environment. We study the ability to solve search tasks by randomized agents (i.e., agents using randomized action functions) moving in a deterministic environment (i.e., specified by a deterministic movement function), and we compare it with the ability to solve search tasks by deterministic agents (i.e., the action function is deterministic) moving in a probabilistic environment (i.e., in which the movement function is randomized, that is, subject to probabilistic perturbations). We consider two comparison criteria, one at a small scale, and one other weaker criterium, at a large scale.

- The small scale measure refers to the distribution $\Pi = \big(\Pr[y|x]\big)_{x,y}$ describing the probability of moving from position x to position y, for all pairs $(x, y) \in X \times X$;
- The large scale measure refers to the performances of the agents, quantified as the expected number of steps performed by the agents for solving search problems.

These two measures allow us to address questions such as the following. Given a randomized agent $\mathcal{A} = (\alpha, \rho, \delta)$ navigating in a deterministic environment $\mathcal{E} = (X, \mu)$, is there a probabilistic perturbation μ' of the movement function μ, and a deterministic action function α' for $\mathcal{A}$ such that

$$\Pi = \big(\Pr_{\alpha,\mu}[y|x]\big)_{x,y}$$

and

$$\Pi' = \big(\Pr_{\alpha',\mu'}[y|x]\big)_{x,y}$$

are identical? Note that if the answer to this question is positive, then an external observer examining the random movements of the agent in its environment cannot be certain that these movements are caused by random effects internal to the agent, and not by probabilistic effects caused by the environment.

Another question addressed in this paper is, given a randomized agent $\mathcal{A} = (\alpha, \rho, \delta)$ navigating form position s to position t in expected time T in a deterministic environment $\mathcal{E} = (X, \mu)$, is there a probabilistic perturbation μ' of the movement function μ, and a deterministic action function α' for $\mathcal{A}$ enabling the agent to navigate from s to t in expected time $T' = T$? Note that is the answer to this question if negative, and, in particular, if $T' \gg T$ for every deterministic agent and every probabilistic perturbation of the environment, then an external observer witnessing agents traveling from s to t in expected time T can unambiguously conclude that the random effects must be caused by a source of randomness internal to the agents, i.e., by random variables produced by the agent itself.

1.1 Our Results

First, we show that external randomness, i.e., random effects caused by probabilistic perturbations of the environment, can always be simulated by internal randomness, i.e., by random values produced by the agent itself. Specifically, we show that the behavior of any deterministic or randomized agent with action function α, subject to a probabilistic perturbation μ' of a deterministic movement function μ, can be simulated by an agent with randomized action function α', and moving according to the deterministic movement function μ, in the sense that the two distributions $\left(\Pr_{\mu',\alpha}[y|x]\right)_{x,y}$ and $\left(\Pr_{\mu,\alpha'}[y|x]\right)_{x,y}$ are identical. In other words, an external observer cannot unambiguously measure the impact of random noise while observing solely the movements of an agent, as this agent may well be the source of all the probabilistic effects.

Next, we address the main question motivating our work: under which circumstances the behavior of an agent must unambiguously witness the fact that the agent uses randomness produced by the agent itself? The answer to this question is contrasted, and depends on the space complexity of the agent.

We show that if the agent does not possess a sufficiently large memory, i.e., a sufficiently large set of internal states, with respect to the number of possible actions, then internal randomness can systematically be simulated by probabilistic noise. Specifically, we show that, unless the number of states of the agent is strictly larger than the number of its possible actions, then the behavior of any randomized agent with action function α, and subject to any deterministic or probabilistic movement function μ, may also be produced by a deterministic agent with action function α' subject to a probabilistic movement function μ', in the sense that the two distributions $\left(\Pr_{\alpha,\mu}[y|x]\right)_{x,y}$ and $\left(\Pr_{\alpha',\mu'}[y|x]\right)_{x,y}$ are identical. In particular, an external observer cannot be unambiguously certain that a *stateless* agent is randomized by observing solely the movements of the agent, as all probabilistic effects may actually be caused by random noise due to the environment. We show that this also holds for agents with a small set of states.

In contrast, we exhibit a setting in which the behavior of the agent is unambiguously the result of random effects produced by the agent itself, and cannot be caused by probabilistic perturbations of the environment. Specifically, we describe an environment in which a randomized agent can travel from a position s to a position t in expected time T while any deterministic agent traveling from s to t takes expected time $T' \gg T$ for any probabilistic perturbation of the movement function. An external observer witnessing agents traveling from s to

t in expected time T in our setting can thus unambiguously conclude that the agents are randomized.

To sum up, it was known that randomization helps for solving navigation problems with finite state machines. However, we show that external randomness (i.e., noise caused by the environment) is not necessarily required, as such probabilistic effects can be simulated internally by a randomized agent. On the other hand, we show that while internal randomness is not required from agents with small memory, as noise can simulate such internal randomness, there are settings involving agents with large memory for which *internal randomness helps.* In these settings, randomized agents achieve performances that cannot be obtained from deterministic agents, even if the environment helps the latter thanks to random noise.

1.2 Related Work

A recent trend in system biology aims at studying biological entities through the lens of computer science, by modeling these entities as computing machines executing an algorithm whose objective is to allow these machines to complete a specific task (see, e.g., [2,8,14]). This approach offers the advantage of decoupling the search strategy performed by the biological entity (i.e., the algorithm) from the environment in which these entities evolve (i.e., the constraints), and from the task to be achieved (i.e., the problem). This allows one to separate the different computational factors impacting the performances of the biological entities, including their computing capabilities, their ability to communicate and to cooperate with other entities, their perception of the environment, their ability to memorize events, etc.

Studying biological entities through the lens of computer science was popularized within the TCS community by Feinerman and Korman [7] in 2012, who introduced and investigated the *ants nearby treasure search* (ANTS) problem. This seminal paper inspired several subsequent work, which revisited the ANTS' problem, as well as other problems inspired by biological tasks, under different assumptions—see, e.g., [5,6,9,11–13]. All these papers consider mobile agents modeled as randomized finite state machines.

A parallel line of research is modeling the behavior of biological entities by random models, including Lévy walks [15], random walks with heterogeneous step lengths [10], power laws [3], preferential attachment [1], etc.

The objective of this paper is inspired by the general framework introduced by Feinerman and Korman [7]. The models considered for approaching the ANTS' problem assume randomized agents. In this paper, we are decoupling the random effects caused by the agents themselves from the random effects potentially caused by their environment, and we aim at studying the relative impact of these two sources of randomness. The long term objective of this study is to figure out which problems (e.g., the ANTS' problem) and which environments require randomized agents for the problems be solved in these environments, and for which problems and environments deterministic agents suffice as long as the environments are probabilistic.

2 Model

2.1 Deterministic Systems

A search *environment* is defined as a 4-tuple $\mathcal{E} = (X, x_0, A, \mu)$ where X and A are non-empty finite sets, $x_0 \in X$, and $\mu : A \times X \to X$. The elements of X are called *positions*, and the elements of A are called *actions*. The position x_0 is called *initial* position, and the map μ is called a *movement* function. Given a search environment $\mathcal{E}$, an *agent for* $\mathcal{E}$ is a 5-tuple $\mathcal{A} = (\Sigma, s_0, \rho, \alpha, \delta)$ where Σ is the non-empty finite set of *states* of the agent, $s_0 \in \Sigma$, and $\rho : X \to Y = \rho(X)$, $\alpha : \Sigma \times \rho(X) \to A$, and $\delta : \Sigma \times \rho(X) \to \Sigma$. The function α is called the *action* function of the agent, and δ is the agent's *transition* function. The state s_0 is the initial state of the agent. The map ρ is called the *perception* function of the agent $\mathcal{A}$, and the value $\rho(x)$ of a position $x \in X$ is called the perception of x by the agent. This setting is quite general, and the only requirement imposed on the agent and its environment is to be consistent in the following sense.

Definition 1 (Consistency). *A movement function $\mu : A \times X \to X$ and a perception function $\rho : X \to Y$ are* consistent *if, for every $(x, x') \in X \times X$ with $\rho(x) = \rho(x')$, and for every $(a, b) \in A \times A$:*

$$\mu(a, x) = \mu(b, x) \iff \mu(a, x') = \mu(b, x').$$

The consistency condition merely says that if the agent perceives two different positions x and x' the same, then the movements resulting from its actions must also share similarities, in the sense that if two actions a and b performed at x lead to the same position y, then the same two actions a and b performed at x' must also lead to the same position y'. For instance, in a 2-dimensional square mesh modeling a flat terrain, if the two nodes x and x' are not distinguishable by an agent provided with a compass, and if the actions "go north" and "move down" lead the agent to the same node y from x, then these two actions also lead the agent to the same node y' from x'. The rationale is that if it was not the case, then the down direction would not face north, and the agent (provided with a compass) could detect this fact, and could therefore distinguish x from x'.

Definition 2 (System). *An environment $\mathcal{E}$ and an agent $\mathcal{A}$ for $\mathcal{E}$ form a* system *$(\mathcal{A}, \mathcal{E})$ if the movement function of $\mathcal{E}$ and the perception function of $\mathcal{A}$ are consistent.*

An execution of an agent $\mathcal{A}$ in the environment $\mathcal{E}$ is displayed as Algorithm 1. Starting from position $x_0 \in X$, in the prescribed initial state s_0, the agent evolves in X. At every execution of the loop in the algorithm, the agent selects an action $a \in A$ based on its current state $s \in \Sigma$, and its perception $\rho(x)$ of the current position x. Then the agent is moved from its current position x to a new position, thanks to the movement function μ applied to the action a performed by the agent at x. Finally, the agent updates its internal state based on its current state s, and on its perception of the new position reached after its movement.

2.2 Randomized Systems

The action function α of an agent may be *randomized*, in which case $\alpha : \Sigma \times \rho(X) \times \Omega \to A$ for some discrete probability space (Ω, p) where $p : \Omega \to (0, 1]$ denotes the probability mass function. Given a current state s and a current position x of the agent, the probability that the agents performs action a is $\sum_{r \in E} p(r)$ where $E = \{r \in \Omega : \alpha(s, x, r) = a\}$. For the sake of avoiding confusion between deterministic and randomized agents, we will systematically denote by α_{D} the action function of deterministic agents, and by α_{R} the action function of randomized agents, keeping α solely for generic statements applying to both types of functions.

The environment may also be subject to probabilistic perturbations, and we thus consider randomized movement functions $\mu : A \times X \times \Omega' \to X$ for some discrete probability space (Ω', p'). We adopt the same convention as for the action functions, that is, deterministic and randomized movement functions are denoted by μ_{D} and μ_{R}, respectively. For the sake of simplifying the notations, the probability spaces (Ω, p) and (Ω', p') will be omitted when clear from the context. A randomized movement function must satisfy two basic properties for being a probabilistic perturbation of a system. These properties are called *reachability* and *statistical consistency*.

Definition 3 (Reachability). *A randomized movement function $\mu_{\text{R}} : A \times X \to X$ satisfies* reachability *with respect to a deterministic movement function $\mu_{\text{D}} : A \times X \to X$ if, for every $(x, y) \in X \times X$ and every $a \in A$:*

$$\Pr[\mu_{\text{R}}(a, x) = y] > 0 \implies \exists b \in A : y = \mu_{\text{D}}(b, x).$$

The reachability condition specifies that the randomized movement function μ_{R} cannot lead an agent from position x to some position y that is not reachable from x thanks to the deterministic movement function μ_{D}. That is, we focus on small scale effects, and ignore large scale effects like landslides, floods, or gales susceptible to move a biological entity far from its current position in an instant.

Definition 4 (Statistical consistency). *A randomized movement function $\mu_{\text{R}} : A \times X \to X$ satisfies* statistical consistency *with respect to a perception function $\rho : X \to Y$ and a deterministic movement function $\mu_{\text{D}} : A \times X \to X$ if, for every $(x, x') \in X \times X$ with $\rho(x) = \rho(x')$, and for every $(a, b) \in A \times A$:*

$$\Pr[\mu_{\text{R}}(a, x) = \mu_{\text{D}}(b, x)] = \Pr[\mu_{\text{R}}(a, x') = \mu_{\text{D}}(b, x')].$$

The statistical consistency condition specifies that, at two positions x and x' that are perceived the same by the agent, if the agent aims at performing action a, then it is moved by μ_{R} from x to a position $y = \mu_{\text{D}}(b, x)$ with the same probability as it might be moved from x' to a position $y' = \mu_{\text{D}}(b, x')$. In other words, the randomized movement function μ_{R} acts statistically the same at two positions that are perceived the same by the agent. For instance, revisiting the example of an agent in a 2-dimensional square mesh, if the two nodes x and x'

are not distinguishable by the agent, the probability that the agent is moved from x to the node y north of x' when performing the action "go down" is equal to the probability that the agent is moved from x' to the node y' north of x' when performing the action "go down". The rational behind this definition is that if these two probabilities were different, then the positions x and x' must be sufficiently different (e.g., different wind conditions, different slopes, etc.), and therefore be distinguishable by the agent.

Definition 5 (Perturbation). *Given a system $(\mathcal{A}, \mathcal{E})$, a randomized movement function μ_{R} is a* probabilistic perturbation *the system if μ_{R} satisfies reachability w.r.t. the movement function μ_{D} of $\mathcal{E}$, and statistical consistency w.r.t. the perception function ρ of $\mathcal{A}$.*

Note that, for any system $(\mathcal{A}, \mathcal{E})$, its own movement function $\mu_{\mathrm{D}} : A \times X \to X$ is a probabilistic perturbation of itself. This fact directly follows from the consistency condition imposed on systems (cf. Definitions 1 and 2).

Notation. Given a (possibly randomized) action function α, and given a (possibly randomized) movement function μ, we denote

$$\Pr_{\alpha,\mu}[y|s,x] \triangleq \Pr[\mu(\alpha(s,\rho(x)),x) = y]$$

for every $(x, y) \in X \times X$ and $s \in \Sigma$, where the probabilities are taken oven the random values drawn in the probability spaces of α and μ. Similarly, for every action $a \in A$, we denote

$$\Pr_{\alpha}[a|s,x] \triangleq \Pr[\alpha(s,x) = a], \text{ and } \Pr_{\mu}[y|a,x] \triangleq \Pr[\mu(a,x) = y].$$

To compare different action functions α and α' under different movement functions μ and μ', we study and compare the two collections of probability distributions $\big(\Pr_{\alpha,\mu}[y|s,x]\big)_{s,x,y}$ and $\big(\Pr_{\alpha',\mu'}[y|s,x]\big)_{s,x,y}$.

Remark. Given a system $(\mathcal{A}, \mathcal{E})$ with $\mathcal{E} = (X, x_0, A, \mu)$, and $\mathcal{A} = (\Sigma, s_0, \rho, \alpha, \delta)$, an external observer can observe the trajectory of $\mathcal{A}$, that is, the sequence of positions in X generated by $\mathcal{A}$ and μ. On the other hand, the sets Σ and A, as well as the functions ρ, α, and δ are a priori not observable. Nevertheless, depending on the context (e.g., the specimen of a biological entity at hand), an external observer may have a good estimate of the type of actions that may be produced by an agent (e.g., thanks to physiological characteristics of the biological entity). The same holds for the perception function and the set of states, i.e., it may be possible to characterize those that are plausible, depending on the context. As a consequence, unless specified otherwise, when simulating a system $(\mathcal{A}, \mathcal{E})$ by another system $(\mathcal{A}', \mathcal{E}')$ the two systems differ only in their movement and action functions.

3 Indistinguishability

This section focuses on settings for which an external observer cannot tell whether the behavior of the agent results from a probabilistic perturbation of the environment, or from an internal source of randomness. In particular, Theorem 1 states that any given probabilistic perturbation can be simulated by a randomized agent, and Theorem 2 states that, under some conditions, any given randomized agent can be simulated by a probabilistic perturbation. In Sect. 4, we shall show that the conditions in Theorem 2 are actually tight.

Theorem 1. *Let $(\mathcal{A}, \mathcal{E})$ be a system, with $\mathcal{E} = (X, x_0, A, \mu_{\text{D}})$, and where $\mathcal{A} = (\Sigma, s_0, \rho, \alpha_{\text{R}}, \delta)$ may be randomized. For every probabilistic perturbation μ_{R} of the movement function μ_{D}, there exists a randomized agent $\mathcal{A}' = (\Sigma, s_0, \rho, \alpha'_{\text{R}}, \delta)$ for $\mathcal{E}$ such that, for all $(x, y) \in X \times X$ and $s \in \Sigma$, $\Pr_{\alpha'_{\text{R}},\mu_{\text{D}}}[y|s,x] = \Pr_{\alpha_{\text{R}},\mu_{\text{R}}}[y|s,x]$.*

Proof. For $x \in X$ and $s \in \Sigma$, we define $\alpha'_{\text{R}}(s, \rho(x))$ as follows. For every action $a \in A$, let $\widehat{a} = \{b \in A : \mu_{\text{D}}(b, x) = \mu_{\text{D}}(a, x)\}$. We set

$$\Pr[\alpha'_{\text{R}}(s, \rho(x)) = a] = \frac{1}{|\widehat{a}|} \cdot \sum_{b \in A} \Pr[\alpha_{\text{R}}(s, \rho(x)) = b] \cdot \Pr[\mu_{\text{R}}(b, x) = \mu_{\text{D}}(a, x)].$$

By construction, we have $\sum_{a \in A} \Pr[\alpha'_{\text{R}}(s, \rho(x)) = a] = 1$. To show that α'_{R} is well defined, let $x, x' \in X$ be two positions with $\rho(x) = \rho(x')$. Obviously, for every $b \in A$ and $s \in \Sigma$, $\Pr[\alpha_{\text{R}}(s, \rho(x)) = b] = \Pr[\alpha_{\text{R}}(s, \rho(x')) = b]$. Moreover, by the statistical consistency property, we also have

$$\Pr[\mu_{\text{R}}(b, x) = \mu_{\text{D}}(a, x)] = \Pr[\mu_{\text{R}}(b, x') = \mu_{\text{D}}(a, x')].$$

It follows that, for every $a \in A$ and $s \in \Sigma$, the equality

$$\Pr[\alpha'_{\text{R}}(s, \rho(x)) = a] = \Pr[\alpha'_{\text{R}}(s, \rho(x')) = a]$$

holds, which guarantees that α'_{R} depends on the perception of the positions, and not the positions themselves.

We now show that the pairs $(\alpha'_{\text{R}}, \mu_{\text{D}})$ and $(\alpha_{\text{R}}, \mu_{\text{R}})$ are indistinguishable from an external observer perspective. On the one hand, by definition, for every $y \in X$, we have

$$\Pr_{\alpha_{\text{R}},\mu_{\text{R}}}[y|s,x] = \sum_{a \in A} \Pr_{\alpha_{\text{R}}}[a|s,x] \cdot \Pr_{\mu_{\text{R}}}[y|a,x] = \sum_{a \in A} \Pr[\alpha_{\text{R}}(s, \rho(x)) = a] \cdot \Pr[\mu_{\text{R}}(a, x) = y].$$

On the other hand, by construction, for every $y \in X$, we have

$$\Pr_{\alpha'_{\text{R}},\mu_{\text{D}}}[y|s,x] = \sum_{a \in A} \Pr_{\alpha'_{\text{R}}}[a|s,x] \cdot \mathbf{1}(\mu_{\text{D}}(a, x) = y), \tag{1}$$

where $\mathbf{1}(\text{false}) = 0$, and $\mathbf{1}(\text{true}) = 1$. Let $\mu_{\text{D}}^{-1}(x, y) = \{a \in A : \mu_{\text{D}}(a, x) = y\}$. Observe that if $\mu_{\text{D}}^{-1}(x, y) = \varnothing$, i.e., if $\mathbf{1}(\mu_{\text{D}}(a, x) = y) = 0$ for all $a \in A$, then, by the reachability property, it must be the case that $\Pr_{\mu_R}[y|a,x] = 0$ for all $a \in A$, and thus $\Pr_{\alpha_R,\mu_R}[y|s,x] = \Pr_{\alpha'_R,\mu_D}[y|s,x]$ holds (they are both equal to 0). Therefore, we assume now that $\mu_D^{-1}(x,y) \neq \varnothing$. Equation (1) can then be rewritten as

$$\begin{aligned}\Pr_{\alpha'_R,\mu_D}[y|s,x] &= \sum_{a\in\mu_D^{-1}(x,y)} \Pr_{\alpha'_R}[a|s,x]\\ &= \sum_{a\in\mu_D^{-1}(x,y)} \frac{1}{|\widehat{a}|}\cdot\sum_{b\in A}\Pr[\alpha_R(s,\rho(x))=b]\cdot\Pr[\mu_R(b,x)=\mu_D(a,x)]\\ &= \sum_{a\in\mu_D^{-1}(x,y)} \frac{1}{|\widehat{a}|}\cdot\sum_{b\in A}\Pr[\alpha_R(s,\rho(x))=b]\cdot\Pr[\mu_R(b,x)=y].\end{aligned}$$

Since, for every $a \in \mu_D^{-1}(x,y)$, the equality $\widehat{a} = \mu_D^{-1}(x,y)$ holds, we get that

$$\Pr_{\alpha'_R,\mu_D}[y|s,x] = \sum_{b\in A}\Pr[\alpha_R(s,\rho(x))=b]\cdot\Pr[\mu_R(b,x)=y] = \Pr_{\alpha_R,\mu_R}[y|s,x],$$

as desired, which completes the proof. □

We now focus on determining under which conditions the reciprocal of Theorem 1 holds. We first show that if the action set A can be altered, then the behavior of a randomized agent can systematically be simulated by a deterministic agent provided with an appropriate perturbation of its movements. Note that, as mentioned before, altering the action set is not desirable, as such modification may be detected by an external observer. Nevertheless, as we shall see later, the following lemma has interesting consequences, even to systems in which the action sets of the agents involved in the simulation are kept identical.

Lemma 1. *Let $(\mathcal{A},\mathcal{E})$ be a system, with $\mathcal{E} = (X, x_0, A, \mu_D)$, and where $\mathcal{A} = (\Sigma, s_0, \rho, \alpha_R, \delta)$ is randomized. For every perturbation μ_R of $(\mathcal{A},\mathcal{E})$, there exists an environment $\mathcal{E}' = (X, x_0, A', \mu_D)$, a deterministic agent $\mathcal{A}' = (\Sigma, s_0, \rho, \alpha_D, \delta)$ for $\mathcal{E}'$, and a perturbation μ'_R of $(\mathcal{A}',\mathcal{E}')$ such that, for every $(x,y) \in X\times X$, and for every $s \in \Sigma$, $\Pr_{\alpha_D,\mu'_R}[y|s,x] = \Pr_{\alpha_R,\mu_R}[y|s,x]$.*

Proof. Let $A' = \Sigma$. Let us define $\alpha_D : \Sigma\times\rho(X) \to A'$ as $\alpha_D(s,\rho(x)) = s$, and $\mu'_R : A'\times X \to X$ as $\mu'_R(s,x) = \mu_R(\alpha_R(s,\rho(x)),x)$. The function μ'_R satisfies reachability because, whenever applied at a position x, μ'_R outputs only positions y that can be outputted by μ_R at x. To check statistical consistency, let x and x' be two positions perceived the same by the agent, i.e., for which $\rho(x) = \rho(x')$, let t be an action in A' (i.e., a state in Σ), and let $y = \mu_D(t,x)$ and $y' = \mu_D(t,x')$. For every $s \in A' = \Sigma$, we have

$$\begin{aligned}\Pr[\mu'_R(s,x)=y] &= \Pr[\mu_R(\alpha_R(s,\rho(x)),x)=y]\\ &= \textstyle\sum_{a\in A}\Pr[\mu_R(a,x)=y]\cdot\Pr[\alpha_R(s,\rho(x))=a]\\ &= \textstyle\sum_{a\in A}\Pr[\mu_R(a,x')=y']\cdot\Pr[\alpha_R(s,\rho(x'))=a] \text{ (by stat. consistency)}\\ &= \Pr[\mu_R(\alpha_R(s,\rho(x')),x')=y']\\ &= \Pr[\mu'_R(s,x')=y'].\end{aligned}$$

It follows that μ'_{R} satisfies statistical consistency, and thus it is a perturbation of the system $(\mathcal{E}', \mathcal{A}')$, as desired. Now, given the current position $x \in X$, and the current state $s \in \Sigma$ of the agent at x, the probability for $y \in X$ to be the next position using $\mathcal{A}'$ under perturbation μ'_{R} is

$$\begin{aligned}\Pr_{\alpha_{\mathrm{D}},\mu'_{\mathrm{R}}}[y|s,x] &= \Pr[\mu'_{\mathrm{R}}(\alpha_{\mathrm{D}}(s,\rho(x)),x) = y] \\ &= \Pr[\mu'_{\mathrm{R}}(s,x) = y] \\ &= \Pr[\mu_{\mathrm{R}}(\alpha_{\mathrm{R}}(s,\rho(x)),x) = y] \\ &= \Pr_{\alpha_{\mathrm{R}},\mu_{\mathrm{R}}}[y|s,x].\end{aligned}$$

The latter equality completes the proof. □

The proof of Lemma 1 can be used for establishing simulations results involving systems with identical action sets. In particular, the next result shows that if the set of actions is large compared to the set of agent's states, then internal randomness (i.e., produced by the agent) can be simulated by external randomness (i.e., caused by probabilistic perturbations). Note that $|\Sigma|$ must not be interpreted as, say, the "size of a brain" (even if some micro-animals have no more than a couple of hundreds of neurons), but merely as the amount of memory dedicated to navigating in an environment.

Theorem 2. *Let $(\mathcal{A}, \mathcal{E})$ be a system, with $\mathcal{E} = (X, x_0, A, \mu_{\mathrm{D}})$, and where $\mathcal{A} = (\Sigma, s_0, \rho, \alpha_{\mathrm{R}}, \delta)$ is randomized. If $|\Sigma| \leq |A|$ then, for every perturbation μ_{R} of $(\mathcal{A}, \mathcal{E})$, there exists a deterministic search agent $\mathcal{A}' = (\Sigma, s_0, \rho, \alpha_{\mathrm{D}}, \delta)$ for $\mathcal{E}$ and a perturbation μ'_{R} of $(\mathcal{A}', \mathcal{E})$ such that, for every $(x, y) \in X \times X$, and for every $s \in \Sigma$, $\Pr_{\alpha_{\mathrm{D}},\mu'_{\mathrm{R}}}[y|s,x] = \Pr_{\alpha_{\mathrm{R}},\mu_{\mathrm{R}}}[y|s,x]$.*

Proof. Let $f : \Sigma \to A$ be any one-to-one mapping from the set of agent's states to the set of actions. Such a mapping exists, merely because $|A| \geq |\Sigma|$. We then set $\alpha_{\mathrm{D}} : \Sigma \times \rho(X) \to A$ as $\alpha_{\mathrm{D}}(s, \rho(x)) = f(s)$, and $\mu'_{\mathrm{R}} : A \times X \to X$ as

$$\mu'_{\mathrm{R}}(a,x) = \begin{cases} \mu_{\mathrm{R}}(a,x) & \text{if } a \notin f(\Sigma) \\ \mu_{\mathrm{R}}(\alpha_{\mathrm{R}}(s,\rho(x)),x) & \text{if } a = f(s) \text{ for some } s \in \Sigma. \end{cases}$$

We just need to show statistical consistency for actions $a \notin f(\Sigma)$. This is straightforward. Indeed, let x and x' with $\rho(x) = \rho(x')$, let $b \in A$ be an action, and let $y = \mu_{\mathrm{D}}(b, x)$ and $y' = \mu_{\mathrm{D}}(b, x')$. Let $a \in A$. If $a \notin f(\Sigma)$, then

$$\Pr[\mu'_{\mathrm{R}}(a,x) = y] = \Pr[\mu_{\mathrm{R}}(a,x) = y] = \Pr[\mu_{\mathrm{R}}(a,x') = y'] = \Pr[\mu'_{\mathrm{R}}(a,x') = y'].$$

The rest of the proof is identical to the proof of Lemma 1. □

The following is a direct consequence of Theorem 2 on memoryless agents, i.e., agents for which $|\Sigma| = 1$, and $\delta = Id$, where $Id(s, \rho(x)) = s$ for every $(s, x) \in \Sigma \times X$.

Corollary 1. *Let $(\mathcal{A}, \mathcal{E})$ be a system with movement function μ_{D}, where $\mathcal{A}$ is memoryless and randomized, with action function α_{R}. For every perturbation μ_{R}*

of $(\mathcal{A}, \mathcal{E})$, *there exists a deterministic memoryless agent* $\mathcal{A}'$ *with action function* $\alpha_{\rm D}$, *and a perturbation* $\mu'_{\rm R}$ *of* $(\mathcal{A}', \mathcal{E})$ *such that, for every* $x, y \in X$,

$$\Pr_{\alpha_{\rm D},\mu'_{\rm R}}[y|x] = \Pr_{\alpha_{\rm R},\mu_{\rm R}}[y|x].$$

4 Distinguishability

This section aims at exhibiting a system $(\mathcal{A}, \mathcal{E})$ for which the behavior of the agent can solely result from internal randomness, and not from a probabilistic perturbation of the movement function. By Theorem 2, the set Σ of agent states must be larger than the set of actions A. We show that $|\Sigma| = |A| + 1$ suffices. Specifically, we consider environments $\mathcal{E} = (X, x_0, A, \mu_{\rm D})$ for which

- there exists a unique position $y \in X$ such that $\mu_{\rm D}(a, y) = y$ for every $a \in A$;
- for every $x \in X$, there exists a sequence of positions $x_1, \ldots, x_k$, $k \geq 1$, such that $x_1 = x$, $x_k = y$ and, for every $i \in \{1, \ldots, k-1\}$, $x_{i+1} = \mu_{\rm D}(a_i, x_i)$ for some $a_i \in A$.

A system $(\mathcal{A}, \mathcal{E})$ where $\mathcal{E}$ satisfies the above is called a *rooted* system, $\mathcal{E}$ is called a rooted environment, and y is called the root of the environment. Given a (possibly randomized) agent $\mathcal{A}$ for a rooted environment $\mathcal{E}$, with action function α, and given a perturbation μ of $(\mathcal{A}, \mathcal{E})$, we denote by $T_{\alpha,\mu}$ the random variable equal to the number of positions (with multiplicity) traversed by the agent from the initial position x_0 until reaching the root y of $\mathcal{E}$. The expectation of $T_{\alpha,\mu}$ is denoted by $\mathbb{E}T_{\alpha,\mu}$. Note that $\mathbb{E}T_{\alpha,\mu} = +\infty$ may occur, typically when both α and μ are deterministic. The following result shows that there exists a rooted system $(\mathcal{A}, \mathcal{E})$ in which no deterministic agents can meet the performances of the randomized agent $\mathcal{A}$, for every probabilistic perturbation of the system. Note that, by Theorem 2, the number of agent's states $k + 1$ is minimum as a function of the size k of the action set for such a result to hold. Note also that the size of the environment is quasi-linear in k.

Theorem 3. *For every* $\epsilon > 0$, *and every integer* $k \geq 1$, *there exists a rooted system* $(\mathcal{E}, \mathcal{A})$ *where* $\mathcal{E} = (X, x_0, A, \mu_{\rm D})$ *is an environment with* $|A| = k$ *and* $|X| = O(k(\log k + \log \frac{1}{\epsilon}))$, *and* $\mathcal{A} = (\Sigma, s_0, \rho, \alpha_{\rm R}, \delta)$ *is a randomized agent with* $|\Sigma| = k+1$, *such that, for every perturbation* $\mu_{\rm R}$ *of* $\mu_{\rm D}$, *and for every deterministic search agent* $\mathcal{A}' = (\Sigma, s_0, \rho, \alpha_{\rm D}, \delta)$ *for* $\mathcal{E}$, *we have* $\mathbb{E}T_{\alpha_{\rm R},\mu_{\rm D}} < \epsilon \cdot \mathbb{E}T_{\alpha_{\rm D},\mu_{\rm R}}$.

Proof. Let $\epsilon > 0$, and let $k \geq 1$. First, we describe the system $(\mathcal{A}, \mathcal{E})$. The set of actions is let to be $A = \{1, \ldots, k\}$. The set X of positions, and the movement function $\mu_{\rm D} : A \times X \to X$ are as follows. Intuitively, X can be viewed as a directed path $x_0, \ldots, x_{n-1}$ where, at each position x_i, only one action $a_i \in A$ leads to x_{i+1} thanks to the movement function $\mu_{\rm D}$, while all other $k-1$ actions lead back to x_0. Moreover, the action a_i belongs to a pair $P_{i \bmod (k+1)}$ of actions, where $P_0, \ldots, P_k$ are $k+1$ distinct pairs of actions, which are fixed. The movement function $\mu_{\rm D}$ is such that, for every $j \in \{0, \ldots, k\}$, for half of the positions x_i with

$i \bmod (k+1) = j$, one action in P_j leads to x_{i+1}, and, for the other half, the other actions in P_j leads to x_{i+1}.

A randomized agent with $k+1$ states can keep track of $j = i \bmod (k+1)$, and then pick its action at x_i uniformly at random in P_j, i.e., each with probability $\frac{1}{2}$. Instead, a deterministic agent, even one keeping track of $j = i \bmod (k+1)$, will systematically err at half of the positions x_i, in the sense that it does not return the right action for μ_{D} at these positions. We shall show that this cannot be balanced by a perturbation μ_{R} of the movement function.

Roughly, since $|A| = k$ is smaller than the number of pairs of actions $P_0, \dots, P_k$, there are two distinct pairs P_i and P_j for which the deterministic action function returns the same action. It follows that the perturbation μ_{R} has to randomly choose which movement to perform among those corresponding to actions in $P_i \cup P_j$, which has a cardinality of at least three (instead, the randomized agent has the ability to either randomly choose in P_i, or randomly choose in P_j). As a consequence, the probabilistic perturbation cannot do better than leading the deterministic agent to err with probability roughly $\frac{1}{3}$ (actually, with probability at least $1/(2\sqrt{2})$) at numerous positions where the randomized agent err with probability at most $\frac{1}{2}$. For sufficiently many positions in which this phenomenon occurs, the ratio between the performances of the randomized agent in its deterministic environment, and the performances of the deterministic agent its probabilistic environment can be made as small as desired, and in particular smaller than ϵ.

Formally, for defining X and μ_{D}, let us fix a set

$$\{P_0, \dots, P_k\} \subseteq \binom{A}{2}$$

of $k+1$ distinct pairs of distinct actions. Let $w = 00\dots0$ and $\overline{w} = 11\dots1$ be the binary words consisting of $k+1$ consecutive 0s and 1s, respectively. Let $m \geq 1$ be an integer, and let us consider the binary sequence $\mathbf{w} = w\overline{w}w\overline{w}\dots w\overline{w}$ consisting of m occurrences of the pair $w\overline{w}$. The sequence $\mathbf{w}$ thus includes $n = 2m(k+1)$ bits, and we denote $\mathbf{w} = \mathbf{w}_0 \dots \mathbf{w}_{n-1}$, with $\mathbf{w}_i \in \{0,1\}$ for $i = 0, \dots, n-1$.

There are $n + k + 1$ positions in X (see Fig. 1). Specifically,

$$X = \{y_1, \dots, y_k, x_0, \dots, x_n\},$$

where the positions $x_0, \dots, x_{n-1}$ will be placed in direct correspondence with the n bits in the sequence $\mathbf{w}$. For every $i \in \{1, \dots, k\}$ and every $a \in A$, we set $\mu_{\text{D}}(a, y_i) = x_0$. Given an action $a \in A$ and a position $x_i \in X$, the movement function μ_{D} is set as

$$\mu_{\text{D}}(a, x_i) = \begin{cases} x_{i+1} \text{ if } i < n,\ a = \min\{b \in P_{i \bmod (k+1)}\}, \text{ and } \mathbf{w}_i = 0 \\ x_{i+1} \text{ if } i < n,\ a = \max\{b \in P_{i \bmod (k+1)}\}, \text{ and } \mathbf{w}_i = 1 \\ x_n \quad \text{if } i = n \\ y_a \quad \text{otherwise.} \end{cases}$$

We now define an agent $\mathcal{A} = (\Sigma, s_0, \rho, \alpha_{\text{R}}, \delta)$ for $\mathcal{E}$. We set $\Sigma = \{0, \dots, k\}$ and $s_0 = 0$. We assume that the agent perceives all positions x_i the same, but

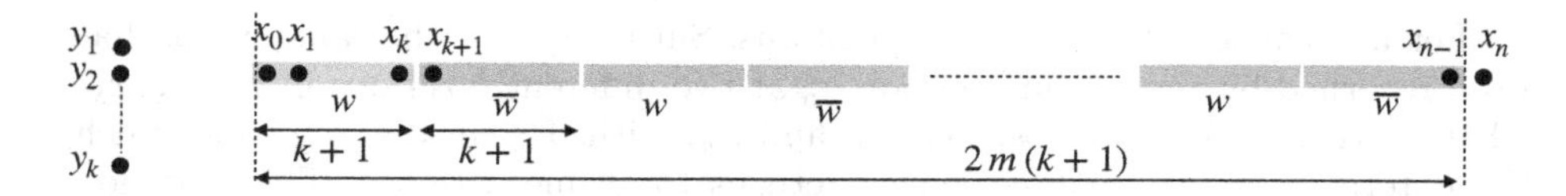

Fig. 1. Construction in the proof of Theorem 3

the final position x_n, and that the agent perceives all positions y_j as the same. That is,

$$\begin{cases} \rho(x_i) &= go \quad \text{for all } 0 \leq i \leq n-1 \\ \rho(x_n) &= end \\ \rho(y_j) &= back \text{ for all } 1 \leq j \leq k \end{cases}$$

Note that the consistency condition (cf. Definition 1) is satisfied. The transition function δ merely increases by 1 the current state, modulo $k+1$, except when the agent is back at a position y_j, $j = 1, \ldots, k$, in which case the state is reset. More specifically,

$$\delta(s, \rho(x)) = \begin{cases} k & \text{if } \rho(x) = back \\ s + 1 \bmod (k+1) & \text{otherwise.} \end{cases}$$

This ensures, in particular, that the agent is systematically in state 0 at position x_0. Finally, the randomized action function is set as follows:

$$\alpha_{\text{R}}(s, \rho(x)) = \begin{cases} \max\{b \in P_s\} \text{ with probability } 1/2 \\ \min\{b \in P_s\} \text{ with probability } 1/2 \end{cases}$$

Let us analyze the behavior of agent $\mathcal{A}$ in $\mathcal{E}$, and the time $T_{\alpha_{\text{R}},\mu_{\text{D}}}$ it takes for the agent starting at position x_0 to reach the final position $y = x_n$. By the setting of the transition function δ, for every $i \in \{0, \ldots, n-1\}$, the state s of the agent at node x_i satisfies $s = i \bmod (k+1)$. Therefore, the agent chooses uniformly at random among one of the two actions in P_s that are susceptible to bring it to x_{i+1}. This will succeed with a probability $1/2$. It follows that the agent traverses the sequence $x_0, x_1, \ldots, x_n$ with probability $1/2^n$. As a consequence,

$$\mathbb{E}T_{\alpha_{\text{R}},\mu_{\text{D}}} \leq n\, 2^n$$

since, during each failing attempt to reach x_n, the agent traverses at most n positions.

We now consider a deterministic agent $\mathcal{A}' = (\Sigma, s_0, \rho, \alpha_{\text{D}}, \delta)$. The action function $\alpha_{\text{D}} : \Sigma \times \rho(X) \to A$ boils down to a function $\alpha_{\text{D}} : \Sigma \to A$ at positions $x_0, \ldots, x_{n-1}$ since all these positions are perceived the same by the agent. Let μ_{R} be a probabilistic perturbation of μ_{D}. Recall that μ_{R} must satisfy the reachability and statistical consistency condition. In particular, the latter condition specifies that the probabilistic distribution of the movements returned by μ_{R} must be identical at all positions $x_0, \ldots, x_{n-1}$ since the perception function ρ of the agent does not distinguish these positions.

Let us focus on the first $k+1$ positions. Since $|A| = k$, the same action has been returned by the action function α_{D} at two different positions in $x_0, \ldots, x_k$. That is, there exist two positions x_u and x_v, with $0 \leq u < v \leq k$, for which the deterministic action function α_{D} returns the same action c. This common action c is then returned by α_{D} at all positions $x_{u+i(k+1)}$ and $x_{v+i(k+1)}$ for all $i \geq 0$. We call these positions *critical*. Note that a critical position may actually not be visited by the agent during all attempts to go from x_0 to x_n, but if the agent was visiting this position, then it would return action c at this position. We show that the critical positions are slowing down the traversal of $x_0, \ldots, x_n$ by the deterministic agent $\mathcal{A}'$ compared to the randomized agent $\mathcal{A}$, for any probabilistic perturbation μ_{R} of μ_{D}.

For every position x_t, $0 \leq t \leq k$, all positions $x_{t+i(k+1)}$ for $i \geq 0$ are called the *twins* of x_t. At every twin position x_i of x_t, the agent $\mathcal{A}'$ returns the same action c_t, and only one of the two actions in $P_{t \bmod (k+1)}$ enables to make progress toward x_n. For maximizing the probability that the agent does not fail at any of these positions (i.e., is not moved back to x_0 via some y_j), the best is that the probabilistic perturbation guarantees that

$$\Pr[\mu_{\text{R}}(c_t, x_t) \in \{\mu_{\text{D}}(a, x_t) : a \in P_{t \bmod (k+1)}\}] = 1.$$

Let $J_t = \{t + i(k+1) : 0 \leq i \leq 2m-1\}$. Our construction of the environment $\mathcal{E}$ satisfies the following.

Fact 1. *For every* $0 \leq t \leq k$, $|\{i \in J_t : \mathbf{w}_i = 0\}| = |\{i \in J_t : \mathbf{w}_i = 1\}|$.

Fact 1 implies that the best scenario for x_t and its twins is that, for every $a \in P_{t \bmod (k+1)}$,

$$\Pr[\mu_{\text{R}}(c_t, x_t) = \mu_{\text{D}}(a, x_t)] = \frac{1}{2}.$$

This is because the maximum of xy under the constraint of non-negative variables x, y satisfying $x+y=1$ is reached for $x = y = \frac{1}{2}$, for an optimum of $\frac{1}{4}$. We conclude that the probabilistic perturbation of the environment must guarantee that, for every non-critical position x_i, the probability of reaching x_{i+1} from x_i is equal to $\frac{1}{2}$.

However, this scenario, which enables the deterministic agent $\mathcal{A}'$ to perform as well as the randomized agent $\mathcal{A}$ at non-critical positions, cannot hold at critical positions. Indeed, the action function α_{D} of $\mathcal{A}'$ returns the same action c at critical positions x_u and x_v, where $0 \leq u < v \leq k$. For maximizing the probability of reaching x_n, the best setting is that that

$$\Pr[\mu_{\text{R}}(c, x_t) \in \{\mu_{\text{D}}(a, x_t) : a \in P_{u \bmod (k+1)} \cup P_{v \bmod (k+1)}\}] = 1.$$

If $P_{u \bmod (k+1)} \cap P_{v \bmod (k+1)} = \varnothing$, then Fact 1 implies that the best scenario for x_u, x_v, and their twins is that, for every $a \in P_{u \bmod (k+1)} \cup P_{v \bmod (k+1)}$,

$$\Pr[\mu_{\text{R}}(c, x_t) = \mu_{\text{D}}(a, x_t)] = \frac{1}{4}.$$

This is because the maximum of $xyzt$ under the constraint of non-negative variables x, y, z, t satisfying $x + y + z + t = 1$ is reached for $x = y = z = t = \frac{1}{4}$, for an optimum of $\frac{1}{256}$.

It may however be the case that the two pairs of actions intersect, in which case $P_{u \bmod (k+1)} \cup P_{v \bmod (k+1)}$ contains only three actions a_1, a_2, and a_3. By Fact 1, and by the setting of μ_{D}, the action common to the two pairs, say a_1, occurs half of the time at the twins of either x_u or x_v, while each the other two actions, a_2 and a_3, occurs only a quarter of the time. The best option for μ_{R} is then to set

$$\Pr[\mu_{\text{R}}(c, x_u) = \mu_{\text{D}}(a_1, x_u)] = \frac{1}{2},$$

and

$$\Pr[\mu_{\text{R}}(c, x_u) = \mu_{\text{D}}(a_2, x_u)] = \Pr[\mu_{\text{R}}(c, x_u) = \mu_{\text{D}}(a_2, x_u)] = \frac{1}{4}.$$

This is because the maximum of x^2yz under the constraint of non-negative variables x, y, z satisfying $x + y + z = 1$ is reached for $x = \frac{1}{2}$, and $y = z = \frac{1}{4}$, for an optimum of $\frac{1}{64} > \frac{1}{256}$.

The case of two intersecting pairs $P_{u \bmod (k+1)}$ and $P_{v \bmod (k+1)}$ is therefore more favorable, as far as the probability of making progress toward x_n is concerned. Note however that $(\frac{1}{64})^{1/4} = \frac{1}{2\sqrt{2}}$, which is to say that the amortized probability of making progress at a critical position is to the best $\frac{1}{2\sqrt{2}}$, which is smaller than $\frac{1}{2}$.

The following fact directly follows from our construction.

Fact 2. *There are at least* $4m$ *critical positions.*

By Fact 1, for one-half of these critical positions, the success probability at each position is to the best $\frac{1}{2}$, while, for the other half, the success probability at each position is to the best $\frac{1}{4}$. Overall, we get that

$$\mathbb{E}T_{\alpha_{\text{D}},\mu_{\text{R}}} \geq 2^{n-4m} \cdot 2^{2m} \cdot 4^{2m} = 2^{n+2m}.$$

It follows that

$$\mathbb{E}T_{\alpha_{\text{R}},\mu_{\text{D}}} < \epsilon \mathbb{E}T_{\alpha_{\text{D}},\mu_{\text{R}}}$$

whenever $n < \epsilon\, 2^{2m}$. This latter inequality is satisfied by picking m sufficiently large. In particular, $m = \Theta(\log k + \log \frac{1}{\epsilon})$ suffices. □

Remark. An interesting question is finding the simplest deterministic agent $\mathcal{A}'' = (\Sigma', s'_0, \rho, \alpha'_{\text{D}}, \delta')$ for which there is a perturbation μ_{R} of μ_{D} such that $\mathbb{E}T_{\alpha'_{\text{D}},\mu_{\text{R}}}$ is at least as good as $\mathbb{E}T_{\alpha_{\text{R}},\mu_{\text{D}}}$, where $\mathbb{E}T_{\alpha'_{\text{D}},\mu_{\text{R}}}$ and $\mathbb{E}T_{\alpha_{\text{R}},\mu_{\text{D}}}$ respectively denote the expected number of steps required by $\mathcal{A}''$ and the agent $\mathcal{A}$ defined in the proof of Theorem 3 to search the environment $\mathcal{E}$ also defined in that proof. A deterministic agent $\mathcal{A}''$ with $2(k+1)$ states can achieve that with $\mu_{\text{R}} = \mu_{\text{D}}$, as such an agent can maintain a pair (s_1, s_2) of counters with $s_1 \in \{0, \ldots, k\}$ and $s_2 \in \{0, 1\}$, which is sufficient for the agent to be perpetually aware of which action to perform at each step. What about deterministic agents with less than $2(k+1)$ states? And what about deterministic agents with at most $k+1$ states?

5 Conclusion

This paper has studied the relative power of (1) deterministic agents subject to probabilistic perturbations of their movements, and (2) randomized agents moving in a deterministic environment. We have established a sharp threshold governing the respective power of these two settings. Specifically, if the set of states Σ of the agent is not larger than its set of actions A, i.e., if $|\Sigma| \leq |A|$, then the power provided to the agent by giving it access to its own individual source of randomness can be simulated by probabilistic perturbations of its movements. In contrast, there exists a setting where $|\Sigma| = |A| + 1$ in which a randomized agent outperforms any deterministic agent, even if the latter is helped by probabilistic perturbations of the movements.

This work opens several directions for future research. In particular, it would be interesting to study the case of randomized transition functions, and to figure out how much such a form of randomization helps compared to agents for which solely the action function is randomized. Another possible direction is to consider a group of agents, possibly cooperating thanks to some light form of communication, and to analyze their ability to coordinate for solving specific search tasks, as a function of their sources of randomness: internal vs. external.

Acknowledgements. The author is thankful to Amos Korman for exciting and motivating early discussions about the possible causes and effects of macroscopic random phenomenons in the context of system biology.

References

1. Avin, C., Cohen, A., Fraigniaud, P., Lotker, Z., Peleg, D.: Preferential attachment as a unique equilibrium. In: 27th ACM Conference on World Wide Web (WWW), pp. 559–568 (2018)
2. Boczkowski, L., Natale, E., Feinerman, O., Korman, A.: Limits on reliable information flows through stochastic populations. PLoS Comput. Biol. **14**(6), e1006195 (2018)
3. Chaintreau, A., Fraigniaud, P., Lebhar, E.: Networks become navigable as nodes move and forget. In: Aceto, L., Damgård, I., Goldberg, L.A., Halldórsson, M.M., Ingólfsdóttir, A., Walukiewicz, I. (eds.) ICALP 2008. LNCS, vol. 5125, pp. 133–144. Springer, Heidelberg (2008). https://doi.org/10.1007/978-3-540-70575-8_12
4. Cook, S.A., Rackoff, C.: Space lower bounds for maze threadability on restricted machines. SIAM J. Comput. **9**(3), 636–652 (1980)
5. Emek, Y., Langner, T., Stolz, D., Uitto, J., Wattenhofer, R.: How many ants does it take to find the food? Theor. Comput. Sci. **608**, 255–267 (2015)
6. Emek, Y., Langner, T., Uitto, J., Wattenhofer, R.: Solving the ANTS problem with asynchronous finite state machines. In: Esparza, J., Fraigniaud, P., Husfeldt, T., Koutsoupias, E. (eds.) ICALP 2014. LNCS, vol. 8573, pp. 471–482. Springer, Heidelberg (2014). https://doi.org/10.1007/978-3-662-43951-7_40
7. Feinerman, O., Korman, A.: The ANTS problem. Distrib. Comput. **30**(3), 149–168 (2016). https://doi.org/10.1007/s00446-016-0285-8

8. Gelblum, A., Fonio, E., Rodeh, Y., Korman, A., Feinerman, O.: Ant collective cognition allows for efficient navigation through disordered environments. eLife **9**(e55195) (2020)
9. Ghaffari, M., Musco, C., Radeva, T., Lynch, N.A.: Distributed house-hunting in ant colonies. In: 34th ACM Symposium on Principles of Distributed Computing (PODC), pp. 57–66 (2015)
10. Guinard, B., Korman, A.: Tight bounds for the cover times of random walks with heterogeneous step lengths. In: 37th International Symposium on Theoretical Aspects of Computer Science (STACS), pp. 28:1–28:14 (2020)
11. Langner, T., Uitto, J., Stolz, D., Wattenhofer, R.: Fault-tolerant ANTS. In: Kuhn, F. (ed.) DISC 2014. LNCS, vol. 8784, pp. 31–45. Springer, Heidelberg (2014). https://doi.org/10.1007/978-3-662-45174-8_3
12. Lenzen, C., Lynch, N., Newport, C., Radeva, T.: Searching without communicating: tradeoffs between performance and selection complexity. Distrib. Comput. **30**(3), 169–191 (2016). https://doi.org/10.1007/s00446-016-0283-x
13. Musco, C., Su, H., Lynch, N.A.: Ant-inspired density estimation via random walks. In: 35th ACM Symposium on Principles of Distributed Computing (PODC), pp. 469–478 (2016)
14. Radeva, T., Dornhaus, A.R., Lynch, N.A., Nagpal, R., Su, H.: Costs of task allocation with local feedback: effects of colony size and extra workers in social insects and other multi-agent systems. PLoS Comput. Biol. **13**(12), e1005904 (2017)
15. Reynolds, A.M.: Current status and future directions of Lévy walk research. Biol. Open **7**(1), bio030106 (2018)

A Lattice Linear Predicate Parallel Algorithm for the Housing Market Problem

Vijay K. Garg(✉)

Department of Electrical and Computer Engineering, University of Texas at Austin, Austin, TX 78712, USA
garg@ece.utexas.edu

Abstract. It has been shown that Lattice Linear Predicate (LLP) algorithm solves many combinatorial optimization problems such as the shortest path problem, the stable marriage problem and the market clearing price problem. In this paper, we give an LLP algorithm for the Housing Market problem. The Housing Market problem is a one-sided matching problem with n agents and n houses. Each agent has an initial allocation of a house and a totally ordered preference list of houses. The goal is to find a matching between agents and houses such that no strict subset of agents can improve their outcome by exchanging houses with each other rather than going with the matching. Gale's celebrated Top Trading Cycle algorithm to find the matching requires $O(n^2)$ time. Our parallel algorithm has expected time complexity $O(n \log^2 n)$ with and expected work complexity of $O(n^2 \log n)$.

1 Introduction

The housing market problem proposed by Shapley and Scarf [1] is a matching problem with one-sided preferences. There are n agents and n houses. Each agent a_i initially owns a house h_i for $i \in \{1, n\}$ and has a completely ranked list of houses. There are variations of this problem when the agents do not own any house initially. In this paper, we focus on the version with the initial endowment of houses for the agents. The list of preferences of the agents is given by $pref[i][k]$ which specifies the kth preference of the agent i. Thus, $pref[i][1] = j$ means that a_i prefers h_j as his top choice. The goal is to come up with an optimal house allocation such that each agent has a house and no subset of agents can improve the satisfaction of agents in this subset by exchanging houses within the subset. It can be shown that there is a unique such matching called the *core* for any housing market. The standard algorithm for this problem is Gale's Top Trading Cycle Algorithm that takes $O(n^2)$ time. This algorithm is optimal in terms of the time complexity since the input size is $O(n^2)$. Our interest in this paper is to design parallel algorithms for this problem.

Supported by the NSF Grant CCR-1812351 and Cullen Trust Endowed Professorship.

C. Johnen et al. (Eds.): SSS 2021, LNCS 13046, pp. 108–122, 2021.
https://doi.org/10.1007/978-3-030-91081-5_8

The housing market problem has been studied by many researchers [1–8]. Possible applications of the housing market problem include: assigning virtual machines to servers in cloud computers, allocating graduates to trainee positions, professors to offices, and students to roommates. In this paper, we apply the Lattice Linear Predicate (LLP) method [9] to give a parallel algorithm for the housing market problem. This problem has also recently been studied by Zheng and Garg [10] where it is shown that the problem of verifying that a matching is a core is in NC, but the problem of computing the core is CC-hard[1] The paper [10] also gives a *distributed* message-passing algorithm to find the core with $O(n^2)$ messages. In this paper, we focus on computing the core and give a parallel algorithm for finding the core that is nearly linear in the number of agents. Our algorithm takes expected $O(n \log^2 n)$ time and expected $O(n^2 \log n)$ work.

Another goal of this paper is to show applications of the Lattice Linear Predicate (LLP) algorithm for the problem. It has been shown that the Lattice Linear Predicate (LLP) algorithm solves many combinatorial optimization problems such as the shortest path problem, the stable marriage problem and the market clearing price problem [9]. In [11], we show that the LLP algorithm also solves many dynamic programming problems in parallel. These problems include the longest subsequence problem, the optimal binary search tree problem, and the knapsack problem.

The lattice-linear predicate detection method to solve a combinatorial optimization problem is as follows. The first step is to define a lattice of vectors L such that each vector is *assigned* a point in the search space. For the stable matching problem, the vector corresponds to the assignment of men to women (or equivalently, the choice number for each man). For the shortest path problem, the vector assigns a cost to each node. For the housing problem studied in this paper, the vector corresponds to the assignment of agents to houses. The comparison operation ($\leq$) is defined on the set of vectors such that the least vector, if feasible, is the extremal solution of interest. For example, in the stable marriage problem if each man orders women according to his preferences and every man is assigned the first woman in the list, then this solution is the man-optimal solution whenever the assignment is a matching and has no blocking pair. Similarly, in the shortest path problem, the zero vector would be optimal if it were feasible. For the housing problem, each agent orders the list of houses in order of its preference giving us the comparison operator. For two vectors G and H in the lattice, $G \leq H$ if and only if each agent prefers the house assigned to them in G at least as much as the house assigned to them in H.

The second step in our method is to define a boolean predicate B that models the feasibility of the vector. For the stable matching problem, an assignment is feasible iff it is a matching and there is no blocking pair. For the shortest path problem, the vector G only gives the lower bound on the cost of a path and there may not be any path to vertex v_i with cost $G[i]$. To capture that an assignment

[1] The class CC (Comparator Circuits) is the complexity class containing decision problems which can be solved by comparator circuits of polynomial size.

is feasible, we define *feasibility* which requires the notion of a *parent*. We say that v_i is a parent of v_j in G iff there is a direct edge from v_i to v_j and $G[j]$ is at least $(G[i]+w[i,j])$. For the shortest path problem, an assignment is feasible iff every reachable node except the source node has a parent. For the housing problem, we say that a housing assignment is feasible if no subset of agents can improve the satisfaction of agents in this subset by exchanging houses within the subset. Figure 1 gives the feasibility predicate for each of these problems.

Problem	**Feasibility Predicate** B
Shortest Path	every reachable vertex other than the source has a parent
Stable Marriage	the assignment is a matching and there is no blocking pair
Housing market	the assignment is a matching and there is no break away coalition

Fig. 1. The feasibility predicate for various problems

The third step is to show that the feasibility predicate is a lattice-linear predicate [9,12]. Lattice-linearity property allows one to search for a feasible solution efficiently. If any point in the search space is not feasible, it allows one to make progress towards the optimal feasible solution without any need for exploring multiple paths in the lattice. Moreover, multiple processes can make progress towards a feasible solution in a *parallel* fashion. In a finite distributive lattice, it is clear that the maximum number of such advancement steps before one finds the optimal solution or reaches the top element of the lattice is equal to the height of the lattice. In this paper, we derive a parallel LLP algorithm that solves the housing market problem using this approach.

This paper is organized as follows. Section 2 gives background on Gale's Top Trading Cycle Algorithm and the LLP method. Section 3 applies LLP method to the unconstrained Housing market problem and derives a high-level parallel algorithm. Section 4 gives a parallel Las Vegas algorithm for the Housing market problem.

2 Background

In this section, we cover the background information on Gale's Top Trading Cycle Algorithm and the LLP Algorithm [9]. Consider the housing market instance shown in Fig. 2. There are four agents a_1, a_2, a_3 and a_4. Initially, the agent a_i holds the house h_i. The preferences of the agents is shown in Fig. 2.

2.1 Gale's Top Trading Cycle (TTC) Algorithm for Housing Market

The Top Trading Cycle (TTC) algorithm attributed to Gale by Shapley and Scarf [1] works in stages. At each stage, it has the following steps:

Agents' Preferences	Initial Allocation	Matching returned by the TTC algorithm
$a_1 : h_2, h_3, h_1, h_4$	$a_1 : h_1$	$a_1 : h_2$
$a_2 : h_1, h_4, h_2, h_3$	$a_2 : h_2$	$a_2 : h_1$
$a_3 : h_1, h_2, h_4, h_3$	$a_3 : h_3$	$a_3 : h_4$
$a_4 : h_2, h_1, h_3, h_4$	$a_4 : h_4$	$a_4 : h_3$

Fig. 2. Housing market and the matching returned by the top trading cycle algorithm

Step 1. We construct the *top choice* directed graph $G_t = (A, E)$ on the set of agents A as follows. We add a directed edge from agent $a_i \in A$ to agent $a_j \in A$ if a_j holds the current top house of a_i. Figure 3 shows the directed graph at the first stage.
Step 2. Since each node has exactly one outgoing edge in G_t, there is at least one cycle in the graph (possibly, a self-loop). All cycles are node disjoint. We find all the cycles in the top trading graph and implement the trade indicated by the cycles, i.e., each agent which is in any cycle gets its current top house.
Step 3. Remove all agents which get their current top houses and remove all houses which are assigned to some agent from the preference list of remaining agents.

The above steps are repeated until each agent is assigned a house. At each stage, at least one agent is assigned a final house. Thus, this algorithm takes $O(n)$ stages in the worse case and needs $O(n^2)$ computational steps.

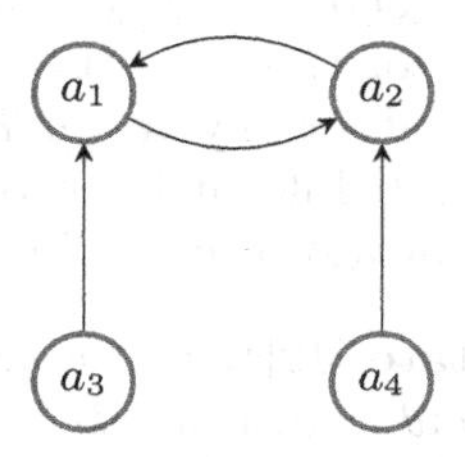

Fig. 3. The top choice graph at the first stage.

2.2 LLP Algorithm

Let L be the lattice of all n-dimensional vectors of reals greater than or equal to zero vector and less than or equal to a given vector T where the order on the vectors is defined by the component-wise natural $\leq$. The lattice is used to model the search space of the combinatorial optimization problem. The combinatorial optimization problem is modeled as finding the minimum element in L that satisfies a boolean *predicate* B, where B models *feasible* (or acceptable) solutions. We are interested in parallel algorithms to solve the combinatorial optimization problem with n processes. We will assume that the systems maintains as its state, the current candidate vector $G \in L$ in the search lattice, where $G[i]$ is maintained at process i. We call G, the global state, and $G[i]$, the state of process i.

Figure 4 shows a finite poset corresponding to n processes (n equals two in the figure), and the corresponding lattice of all eleven global states.

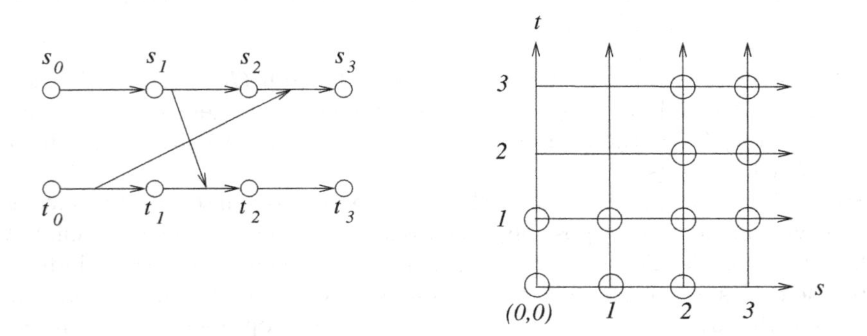

Fig. 4. A poset and its corresponding distributive lattice L

Finding an element in lattice that satisfies the given predicate B, is called the *predicate detection* problem. Finding the *minimum* element that satisfies B (whenever it exists) is the combinatorial optimization problem. A key concept in deriving an efficient predicate detection algorithm is that of a *forbidden* state. Given a predicate B, and a vector $G \in L$, a state $G[j]$ is *forbidden* (or equivalently, the index j is forbidden) if for any vector $H \in L$, where $G \leq H$, if $H[j]$ equals $G[j]$, then B is false for H. Informally, this means that any global state $H \geq G$ which satisfies B must be advanced on index j. Formally,

Definition 1 (Forbidden State [12]). *Given any distributive lattice L of n-dimensional vectors of $\mathbf{R}_{\geq 0}$, and a predicate B, we define* $\text{forbidden}(G, j, B) \equiv \forall H \in L : G \leq H : (G[j] = H[j]) \Rightarrow \neg B(H)$.

We define a predicate B to be *lattice-linear* with respect to a lattice L if for any global state G, B is false in G implies that G contains a *forbidden state*. Formally,

Definition 2 (lattice-linear Predicate [12]). *A boolean predicate B is* lattice-linear *with respect to a lattice L iff* $\forall G \in L : \neg B(G) \Rightarrow (\exists j : \text{forbidden}(G, j, B))$.

Once we determine j such that $forbidden(G, j, B)$, we also need to determine how to advance along index j. To that end, we extend the definition of forbidden as follows.

Definition 3 (α-forbidden). *Let B be any boolean predicate on the lattice L of all assignment vectors. For any G, j and positive real $\alpha > G[j]$, we define $forbidden(G, j, B, \alpha)$ iff*

$$\forall H \in L : H \geq G : (H[j] < \alpha) \Rightarrow \neg B(H).$$

Given any lattice-linear predicate B, suppose $\neg B(G)$. This means that G must be advanced on all indices j such that $forbidden(G, j, B)$. We use a function $\alpha(G, j, B)$ such that $forbidden(G, j, B, \alpha(G, j, B))$ holds whenever $forbidden(G, j, B)$ is true. With the notion of $\alpha(G, j, B)$, we have the Algorithm LLP. The algorithm LLP has two inputs—the predicate B and the top element of the lattice T. It returns the least vector G which is less than or equal to T and satisfies B (if it exists). Whenever B is not true in the current vector G, the algorithm advances on all forbidden indices j in parallel. This simple parallel algorithm can be used to solve a large variety of combinatorial optimization problems by instantiating different $forbidden(G, j, B)$ and $\alpha(G, j, B)$.

ALGORITHM LLP. To find the minimum vector at most T that satisfies B

```
1 vector function getLeastFeasible(T: vector, B: predicate)
2 var G: vector of reals initially ∀i : G[i] = 0;
3     while ∃j : forbidden(G, j, B) do
4         for all j such that forbidden(G, j, B) in parallel:
5             if (α(G, j, B) > T[j]) then return null;
6             else G[j] := α(G, j, B);
7     endwhile;
8     return G; // the optimal solution
```

The following Lemma is useful in proving lattice-linearity of predicates.

Lemma 1 [9,12]. *Let B be any boolean predicate defined on a lattice L of vectors.*
(a) Let $f : L \rightarrow \mathbf{R}_{\geq 0}$ be any monotone function defined on the lattice L of vectors of $\mathbf{R}_{\geq 0}$. Consider the predicate $B \equiv G[i] \geq f(G)$ for some fixed i. Then, B is lattice-linear.
(b) If B_1 and B_2 are lattice-linear then $B_1 \wedge B_2$ is also lattice-linear.

We now give an example of lattice-linear predicates for the scheduling of n jobs. Each job j requires time t_j for completion and has a set of prerequisite jobs, denoted by $pre(j)$, such that it can be started only after all its prerequisite jobs have been completed. Our goal is to find the minimum completion time for each job. We let our lattice L be the set of all possible completion times. A completion vector $G \in L$ is feasible iff $B_{jobs}(G)$ holds where $B_{jobs}(G) \equiv \forall j : (G[j] \geq t_j) \wedge (\forall i \in pre(j) : G[j] \geq G[i] + t_j)$. B_{jobs} is lattice-linear because if it is false, then there exists j such that either $G[j] < t_j$ or $\exists i \in pre(j) : G[j] < G[i] + t_j$. We claim that $forbidden(G, j, B_{jobs})$. Indeed, any vector $H \geq G$ cannot be feasible with $G[j]$ equal to $H[j]$. The minimum of all vectors that satisfy feasibility corresponds to the minimum completion time.

As an example of a predicate that is not lattice-linear, consider the predicate $B \equiv \sum_j G[j] \geq 1$ defined on the space of two dimensional vectors. Consider the vector G equal to $(0,0)$. The vector G does not satisfy B. For B to be lattice-linear either the first index or the second index should be forbidden. However, none of the indices are forbidden in $(0,0)$. The index 0 is not forbidden because the vector $H = (0,1)$ is greater than G, has $H[0]$ equal to $G[0]$ but it still satisfies B. The index 1 is also not forbidden because $H = (1,0)$ is greater than G, has $H[1]$ equal to $G[1]$ but it satisfies B.

2.3 Notation

We now go over the notation used in the description of our parallel algorithms. Figure 5 shows a parallel algorithm for the job-scheduling problems.

The **var** section gives the variables of the problem. We have a single variable G in the example shown in Fig. 5. G is an array of objects such that $G[j]$ is the state of thread j for a parallel program.

The **input** section gives all the inputs to the problem. These inputs are constant in the program and do not change during execution.

The **init** section is used to initialize the state of the program. All the parts of the program apply to all values of j. For example, the *init* section of the job scheduling program in Fig. 5 specifies that $G[j]$ is initially $t[j]$. Every thread j would initialize $G[j]$.

The **always** section defines additional variables which are derived from G. The actual implementation of these variables are left to the system. They can be viewed as macros. We will show its use later.

The LLP algorithm gives the desirable predicate either by using the **forbidden** predicate or **ensure** predicate. The *forbidden* predicate has an associated *advance* clause that specifies how $G[j]$ must be advanced whenever the forbidden predicate is true. For many problems, it is more convenient to use the complement of the forbidden predicate. The *ensure* section specifies the desirable predicates of the form $(G[j] \geq expr)$ or $(G[j] \leq expr)$. The statement *ensure* $G[j] \geq expr$ simply means that whenever thread j finds $G[j]$ to be less than $expr$; it can advance $G[j]$ to $expr$. Since $expr$ may refer to G, just by setting $G[j]$ equal to $expr$, there is no guarantee that $G[j]$ continues to be equal to $expr$—the value of $expr$ may change because of changes in other components. We use *ensure* statement whenever $expr$ is a monotonic function of G and therefore the predicate is lattice-linear.

3 Applying LLP Algorithm to the Housing Market Problem

We model the housing market problem as that of predicate detection in a computation. There are n agents and n houses. Each agent proposes to houses in the

```
Pj: Code for thread j
// common declaration for all the programs below
var G: array[1..n] of 0..maxint;// shared among all threads
input: t[j] : int, pre(j): list of 1..n;
init: G[j] := t[j];

job-scheduling:
    forbidden: G[j] < max{G[i] + t[j] | i ∈ pre(j)};
        advance: G[j] := max{G[i] + t[j] | i ∈ pre(j)};

job-scheduling:
    ensure: G[j] ≥ max{G[i] + t[j] | i ∈ pre(j)};
```

Fig. 5. LLP Parallel Program for (a) job scheduling problem using forbidden predicate (b) job scheduling problem using ensure clause

decreasing order of preferences. These proposals are considered as events executed by n processes representing the agents. Thus, we have n events per process. Each event is labeled as (i, h, k), which corresponds to the agent i proposing to the house h as his choice number k.

The global state corresponds to the number of proposals made by each of the agents. Let $G[i]$ be the number of proposals made by the agent i. We will assume that in the initial state every agent has made his first proposal. Thus, the initial global state $G = [1, 1, .., 1]$. We extend the notation of indexing to subsets $J \subseteq [n]$ such that $G[J]$ corresponds to the subvector given by indices in J.

We now model the possibility of reallocation of houses based on any global state. Recall that $pref[i][k]$ specifies the kth preference of the agent a_i. Let $wish(G, i)$ denote the house that is proposed by a_i in the global state G, i.e.,

$$wish(G, i) = pref[i][G[i]]$$

A global state G satisfies *matching* if every agent proposes a different house, i.e.,

$$matching(G) \equiv \forall i, j : i \neq j : wish(G, i) \neq wish(G, j).$$

We generalize *matching* to refer to a subset of agents rather than the entire set.

Definition 4 (submatching). *Let $J \subseteq [n]$. Then, $submatching(G, J)$ iff $wish(G, J)$ is a permutation of indices in J.*

Intuitively, if $submatching(G, J)$ holds, then all agents in J can exchange houses within the subset J.

For any G, it is easy to show that

Lemma 2. *For all G, there always exists a nonempty J such that submatching(G, J).*

Proof: Given any G, we can create a directed graph as follows. The set of vertices is agents and there is an edge from i to j if $wish(G, i) = j$. There is exactly one outgoing edge from any vertex in $[n]$ to $[n]$ in this graph. This implies that there is at least one cycle in this graph (possibly, a self-loop). The indices of agents in the cycle gives us such a subset J. ■

We now show that

Lemma 3. *$submatching(G, J_1)$ and $submatching(G, J_2)$ implies that $submatching(G, J_1 \cup J_2)$.*

Proof: Any index $i \in J_1 \cup J_2$ is mapped to J_1 if $i \in J_1$ and J_2, otherwise. ■

Hence, there exists the biggest submatching in G. Note that $matching(G)$ is equivalent to $submatching(G, [n])$.

Definition 5 (Feasible Global State). *A global state G is feasible for the housing market problem iff it is a matching and for all global states $F < G$, there does not exist any submatching which is better in F than in G. Note that if there exists a submatching J which is better in F than G, then the agents in J can improve their allocation by just exchanging houses within the subset J. Formally, let*

$$B_{housing}(G) \equiv matching(G) \wedge (\forall F < G : \forall J \subseteq [n] : submatching(F, J) \Rightarrow F[J] = G[J]).$$

We show that $B_{housing}(G)$ is a lattice-linear predicate. This result will let us use the lattice-linear predicate detection algorithm for the housing market problem.

Theorem 1. *The predicate $B_{housing}(G)$ is lattice-linear.*

Proof: Suppose that $\neg B_{housing}(G)$. This implies that either G is not a matching or it is a matching but there exists a smaller global state F that has a submatching better than G.

First, consider the case when G is not a matching. Let J be the largest set such that $submatching(G, J)$. Consider any index $i \notin J$ such that $wish(G, i) \in J$. We claim that $forbidden(G, i, B_{housing})$. Let H be any global state greater than G such that $G[i] = H[i]$. We consider two cases.

Case 1: $H[J] > G[J]$.
Then, from the second conjunct of $B_{housing}$, we know that $\neg B_{housing}(H)$ because $submatching(G, J)$ and $H[J] \neq G[J]$.
Case 2: $H[J] = G[J]$.
Since $wish(H, i) = wish(G, i)$, $wish(G, i) \in J$, and $G[J] = H[J]$, we get that H is not a matching because the house given by $wish(G, i)$ is also in the wish list of some agent in J.

Now consider the case when G is a matching but $\neg B_{housing}(G)$. This implies

$$\exists F < G : \exists J \subseteq [n] : submatching(F, J) \wedge F[J] < G[J]).$$

However, the same F will also result in guaranteeing $\neg B_{housing}(H)$ for any $H \geq G$. ■

It is also easy to see from the proof that if an index is part of a submatching, then it will never become forbidden.

This theorem gives us the algorithm shown in Fig. 6. Let G be the initial global state. Let $S(G)$ be the biggest submatching in G. All agents such that they are not in $S(G)$ and wish a house which are part of $S(G)$ are forbidden and can move to their next proposal. The algorithm terminates when no agent is forbidden. This algorithm is a parallel version of the top trading cycle (TTC) mechanism attributed to Gale in [1].

```
Algorithm Housing-Market:
var
    G: array[1..n] of int initially 1;// every agent starts with the top choice
    T = (n, n, ..., n); //maximum number of proposals at a_i
always
    S(G) = largest J such that submatching(G, J)
    forbidden(G, j, B) ≡ (j ∉ S(G)) ∧ (wish(G, j) ∈ S(G))

while ∃j : forbidden(G, j, B) do
    for all j such that forbidden(G, j, B) in parallel:
        if (G[j] = T[j]) then return null;
        else G[j] := G[j] + 1;
endwhile;
return G; // the optimal solution
```

Fig. 6. A high-level parallel algorithm to find the optimal house market

We now show that

Theorem 2. *There exists at least one feasible global state G such that $B_{housing}(G)$.*

Proof: Every agent has his own house in the list of preferences. If he ever makes a proposal to his own house, he forms a submatching. That particular event is never forbidden because it is a part of a submatching. Hence, lattice-linear predicate detection algorithm will never mark that event as forbidden. Since such an event exists for all processes, we are guaranteed to never go beyond this global state. ■

The above proof also shows that agents can never be worse-off by participating in the algorithm. Each agent will either get his own house back or get a house that he prefers to his own house.

4 An Efficient Parallel Algorithm for the Housing Market Problem

We now present an efficient parallel algorithm for the housing market problem. We note here that [10] gives a distributed algorithm with $O(n^2)$ messages for the housing market problem. In this paper, we focus on computing the core and give a parallel algorithm for finding the core that is nearly linear in the number of agents. Our algorithm takes expected $O(n \log^2 n)$ time and expected $O(n^2 \log n)$ work.

By renumbering houses, if necessary, we assume that initially agent a_i has the house h_i. We assume that the preference list is provided as two data structures: $prefList$ and $prefPointer$. The variable $prefList$ is an array of doubly linked list such that $prefList[i]$ points to the list of preferences of agent i. As the algorithm executes, we advance on $prefList$ and the head of the $prefList[i]$ corresponds to the variable $wish$ for agent a_i in Fig. 6.

To facilitate the quick deletion of houses from this list, we also have a data structure $prefPointer$. The variable $prefPointer$ is a two dimensional array such that $prefPointer[i][j]$ points to the node corresponding to house h_j in the doubly-linked list of agent a_i. If at any stage in the algorithm, we find out that the house h_j has been permanently allocated to some other agent than a_i, then we need to remove the house h_j from the preference list of a_i. Since $prefPointer[i][j]$ points to that node in the doubly linked list $prefList[i]$, we can delete the house in $O(1)$ time. Due to these deletions, we maintain the invariant that the head of $prefList[i]$ always corresponds to the top choice of the agent a_i. Note that if the input is given as the two dimensional array $pref$, where $pref[i][j]$ is the top j^{th} choice for the agent a_i, then it can be converted into $prefList$ and $prefPointer$ in $O(n)$ time with $O(n)$ processors.

We keep the array $fixed$ such that $fixed[i]$ indicates that the agent i has been assigned its final house. If an agent i is fixed, then it can never be forbidden in Fig. 6. Once all agents are fixed, we get that no agent is forbidden and the algorithm terminates.

At every iteration, we keep the array $inCycle[i]$ that indicates agents that are in Top Trading Cycle at that iteration. In Fig. 6, these agents correspond to $S(G)$ in the global state G. Algorithm LLP-TTC uses a *while* loop to fix some number of agents in every iteration. At least one agent is fixed in every iteration, and therefore there are at most n iterations of the while loop.

Each iteration has four steps. In the first step, we initialize *inCycle* to be false by default. In the second step (function *markRoots*) we use symmetry breaking via randomization and pointer jumping to mark one node called *root* in every cycle as belonging to a cycle. The reader is referred to [13] for symmetry breaking and pointer jumping. During the process of pointer jumping, we also construct a tree rooted at a vertex such that it consists of all the nodes in the cycle. In the third step (function *informTree*), we inform all the agents that are in some rooted tree that they are in a cycle. In the fourth step, we fix all the agents that are in cycles and remove their houses from $prefList$. This step corresponds to advancing G in Fig. 6.

ALGORITHM LLP-TTC. Parallel LLP Top Trading Cycle Algorithm

```
1  // By renumbering houses, ensure that initially agent a_i is assigned house h_i
2  var
3      prefList:array[1..n] of list initially ∀i : prefList[i] has preferences for a_i;
4      prefPointer:array[1..n, 1..n] of pointer to the node in prefList;
5      fixed: array[1..n] of booean initially ∀i : fixed[i] = false;
6      inCycle: array[1..n] of booean initially ∀i : inCycle[i] = false;
7      children: array[1..n] of set of nodes that a_i traversed initially {};
8  while (∃i : ¬fixed[i])
9      // Step 1: initialize inCycle
10     forall i : ¬fixed[i] in parallel do: inCycle[i] := false;
11     // Step 2: Mark one node in every cycle as the root
12     markRoots();
13     // Step 3: inform all the agents in any rooted tree that they are in a cycle
14     informTree();
15     // Step 4: Now delete all the agents that are in cycle
16     forall i : ¬fixed[i] ∧ inCycle[i], j : ¬[fixed[j] ∧ ¬inCycle[j] in parallel do
17             delete the node prefPointer[j][i] from the linked list prefList[j];
18     forall i : ¬fixed[i] ∧ inCycle[i] in parallel do
19         fixed[i] := true;
20 endwhile
21 return prefList; //prefList[i] points to the house assigned to the agent a_i
```

The function *markRoots* uses variable *active* to denote agents that are active. Initially, all agents are active. The variable $succ[i]$ is used to point to the next active agent. Initially, $succ[i]$ points to the agent who has the top choice house of agent i. The variable $done[i]$ indicates whether a cycle has been discovered in the subgraph that agent i is pointing to. Once, a cycle has been discovered then any active agent knows that it cannot be part of any cycle and it becomes inactive.

The function *markRoots* uses a *while* loop at line 5 to run while there is any active node. Every active agent flips a coin at line 7. If its own coin is a head and its successor gets a tail, then this agent becomes inactive at line 9. It is clear that two consecutive agents can never become inactive in the same round because we require an agent to get "head" and its successor to get "tail" to become inactive. It is also clear that the number of active agents is reduced by a constant fraction in every round of coin toss in expectation. Thus, the outer while loop at line 5 is executed expected $O(\log n)$ times.

If an agent is active, it traverses its *succ* pointer till it reaches the next *active* node. This is done using the *while* loop at line 11. This traversal has a length of one or zero because there cannot be two consecutive inactive agents due to the rule of becoming active.

If agent i reaches itself as the next active node at line 15, it marks *inCycle* to be true. It also sets $done[i]$ to be true so that any active node j that points to i knows that a cycle has been found and that the node j can stop looking for the cycle. If the successor of the node is different, then we check if the successor is done. If the successor is done, then this node is not part of the cycle and can therefore make itself inactive and also mark itself as done. Since all agents execute the statements in *forall* in parallel, we get that the function *markRoots()* has parallel expected time complexity of $O(\log n)$. Also, for every cycle in the graph, there is exactly one node that sets its *inCycle* to be true.

The function *informTree* uses variable *rootSet* to initially include all the roots found in the function *markRoots*. Once all the nodes in any rooted tree have been informed, the root is deleted from the *rootSet*. To inform agents in the tree, we follow the usual method of broadcasting a value from the root to its children. To detect that all agents in the tree have been notified, we let any subtree that has finished informing its subtree to leave the tree by deleting itself from the children set of its parent. If the agent is a root, then it deletes itself from the *rootSet*. Once all roots have deleted themselves, the function terminates. Since the height of any tree is expected to be $O(\log n)$ and the number of children of any node is also $O(\log n)$, we get that the algorithm takes $O(\log^2 n)$ time.

ALGORITHM markRoots. Function markRoots for the Parallel LLP Top Trading Cycle Algorithm

```
1  function markRoots()
2      succ: array[1..n] of 1..n initially ∀i : succ[i] = prefList[i].head(); //successor
   of a_i which is active
3      active: array[1..n] of booean initially ∀i : active[i] = true;
4      done: array[1..n] of booean initially ∀i : done[i] = false;
5      while (∃i : active[i])
6          forall i : ¬fixed[i] ∧ active[i] in parallel do
7              coin[i] := "head" or "tail" // based on the flip of a coin
8              if (coin[i] = "head") ∧ (coin[succ[i]] = "tail") then
9                  active[i] := false;
10             else // node i is active
11                 while ¬active[succ[i]] do
12                     children[i] := children[i] ∪ succ[i]
13                     succ[i] := succ[succ[i]]
14                 endwhile
15                 if (succ[i] = i) // found a cycle
16                     done[i] := true
17                     inCycle[i] := true
18                     active[i] := false
19                 else if done[succ[i]] then
20                     active[i] := false
21                     done[i] := true
22         endforall
23     endwhile
```

ALGORITHM informTree. Function informTree for the Parallel LLP Top Trading Cycle Algorithm

```
1  function informTree()
2      informed: array[1..n] of booean initially ∀i : informed[i] = false;
3      parent: array[1..n] of 1..n initially ∀i : parent[i] = i;
4      rootSet: set of 1..n initially {i | inCycle[i]}
5      while (rootSet ≠ {}) do
6          forall i : ¬fixed[i] in parallel do
7              if (inCycle[i] ∧ ¬informed[i]) then
8                  informed[i] := true
9                  for (j ∈ children[i]) do
10                     inCycle[j] := true
11                     parent[j] := i
12                 endfor
13             if (inCycle[i] ∧ informed[i] ∧ (children[i] = {})) then
14                 if (parent[i] = i) then rootSet.remove(i);
15                 else  children[parent[i]] := children[parent[i]] − {i}
16         endforall
17     endwhile
```

We first show the correctness of the parallel Algorithm LLP-TTC.

Theorem 3. *The Algorithm LLP-TTC returns the core of the housing market problem.*

Proof: It is sufficient to show that the Algorithm LLP-TTC finds all top trading cycles in each iteration. Consider any top trading cycle of size 1 at node i. The function markRoot can never mark node i in the cycle as inactive due to the requirement of the coin turning at node i as head and its successor, itself, as tail. Furthermore, since $succ[i]$ equals i, node i is marked as *inCycle*. Now, consider any top trading cycle of size $k > 1$. Since we require the successor of the node to have a different toss to turn inactive, all nodes cannot turn inactive. The active nodes keep the inactive nodes following it as its children. After every coin toss, the length of the cycle for active nodes is expected to shrink by a constant factor. Hence, in expected $O(\log n)$ coin tosses, the cycle reduces to size 1 and the former case applies.

Now consider any node i that is not in any top trading cycle. Since our graph is functional (every vertex has out-degree exactly one), node i leads to a cycle by following the *succ* edge. By previous discussion in $O(\log n)$ expected time, one of the nodes in that cycle, say j will set $inCycle[j]$ and $done[j]$ to be true. Since any path of active nodes reduces by a constant factor, in $O(\log n)$ expected time node i will point to a node that is *done* and will also mark itself as *done*.

The function *informTree* simply sets the variable *inCycle* of all nodes in the cycle to be true. Finally, step 4 removes all houses and agents that are in any cycle and thus implements the top trading cycle mechanism. ■

We now analyze the time and work complexity of LLP-TTC.

Theorem 4. *LLP-TTC takes expected* $O(n \log^2 n)$ *time and expected* $O(n^2 \log n)$ *work.*

Proof: Since every functional graph has at least one cycle, there exists at least one new node that finds itself in a cycle in every iteration of the while loop. Hence, there are at most n iterations of the while loop. In each iteration, Step 1 takes $O(1)$ time and $O(n)$ work. Step 2 takes expected $O(\log n)$ time and expected $O(n \log n)$ work. Step 3 takes $O(\log^2 n)$ time and $O(n)$ work. Let α_k be the number of agents that are fixed in the kth iteration of the while loop. Step 4 takes $O(\alpha_k)$ time and $O(n\alpha_k)$ work. Adding up over all iterations, we get the desired time and work complexity. ■

Acknowledgments. I thank Changyong Hu, Robert Streit, and Xiong Zheng for various discussions on the housing allocation problem. I also thank the anonymous reviewers for comments on the paper.

References

1. Shapley, L., Scarf, H.: On cores and indivisibility. J. Math. Econ. **1**(1), 23–37 (1974)
2. Hylland, A., Zeckhauser, R.: The efficient allocation of individuals to positions. J. Polit. Econ. **87**(2), 293–314 (1979)
3. Zhou, L.: On a conjecture by Gale about one-sided matching problems. J. Econ. Theory **52**(1), 123–135 (1990)
4. Abdulkadiroğlu, A., Sönmez, T.: Random serial dictatorship and the core from random endowments in house allocation problems. Econometrica **66**(3), 689–701 (1998)
5. Abdulkadiroğlu, A., Sönmez, T.: House allocation with existing tenants. J. Econ. Theory **88**(2), 233–260 (1999)
6. Roth, A.E., Postlewaite, A.: Weak versus strong domination in a market with indivisible goods. J. Math. Econ. **4**(2), 131–137 (1977)
7. Roth, A.E.: Incentive compatibility in a market with indivisible goods. Econ. Lett. **9**(2), 127–132 (1982)
8. David, M.: Algorithmics of Matching Under Preferences, vol. 2. World Scientific (2013)
9. Garg, V.K.: Predicate detection to solve combinatorial optimization problems. In: Scheideler, C., Spear, M. (eds.) SPAA 2020: 32nd ACM Symposium on Parallelism in Algorithms and Architectures, Virtual Event, USA, 15–17 July 2020, pp. 235–245. ACM (2020)
10. Zheng, X., Garg, V.K.: Parallel and distributed algorithms for the housing allocation problem. In: Felber, P., Friedman, R., Gilbert, S., Miller, A. (eds.) 23rd International Conference on Principles of Distributed Systems, OPODIS 2019, Neuchâtel, Switzerland, 17–19 December 2019. LIPIcs, vol. 153, pp. 23:1–23:16. Schloss Dagstuhl - Leibniz-Zentrum für Informatik (2019)
11. Garg, V.K.: A lattice linear predicate parallel algorithm for the dynamic programming problems. CoRR, abs/2103.06264 (2021)
12. Chase, C.M., Garg, V.K.: Detection of global predicates: techniques and their limitations. Distrib. Comput. **11**(4), 191–201 (1998)
13. JáJá, J.: An Introduction to Parallel Algorithms, vol. 17. Addison-Wesley Reading, Boston (1992)

Security in Asynchronous Interactive Systems

Ivan Geffner(✉) and Joseph Y. Halpern

Cornell University, Ithaca, NY 14850, USA
ieg8@cornell.edu, halpern@cs.cornell.edu

Abstract. Secure function computation has been thoroughly studied and optimized in the past decades. We extend techniques used for secure computation to simulate arbitrary protocols involving a mediator. The key feature of our notion of simulation is that it is bidirectional: not only does the simulation produce only outputs that could happen in the original protocol, but the simulation produces all such outputs. In asynchronous systems there are also new subtleties that arise because the scheduler can influence the output. Thus, these requirements cannot be achieved by the standard notion of secure computation. We provide a construction that is secure if $n > 4t$, where t is the number of malicious agents, which is provably the best possible. We also show that our construction is secure in the *universal composability* model and that it satisfies additional security properties even if $3t < n \leq 4t$.

1 Introduction

In a distributed system, agents often want to be able to carry out a computation without revealing any private information. There has been a great deal of work showing how and to what extent this can be done. We briefly review the most relevant work here.

Ben-Or, Goldwasser and Widgerson [3] (BGW from now on) and Chaum, Crépeau, and Damgard [8] showed that, if $n > 3t$, then every function f of n inputs can be t-securely computed by n agents in a synchronous system with private communication channels, where "t-securely computed" means that no coalition of at most t malicious agents can either (a) prevent the honest agents from correctly computing the output of f given their inputs (assuming some fixed inputs for malicious agents who do not provide inputs) or (b) learn anything about the inputs of the honest agents (beyond what can be concluded from the output of f). The notion of an agent "not learning anything" is formalized by comparing what happens in the actual computation to what could have happened had there been a trusted third party (which we here call a *mediator*) who will calculate $f(x_1, \ldots, x_n)$ after being given the input x_i by agent

Supported in part by NSF grants IIS-1703846 and IIS-0911036, ARO grant W911NF-17-1-0592, MURI grant W911NF-19-1-0217 from the ARO, and a grant from Open Philanthropy.

C. Johnen et al. (Eds.): SSS 2021, LNCS 13046, pp. 123–140, 2021.
https://doi.org/10.1007/978-3-030-91081-5_9

i, for $i = 1, \ldots, n$. Then, roughly speaking, the malicious agents do not learn anything if the distribution of outputs in the actual computation could have also resulted in the computation with a mediator if the malicious agents had given the appropriate input to the mediator.

Ben-Or, Canetti and Goldreich [2] (BCG from now on) proved analogous results in the asynchronous case. Asynchrony raises new subtleties. For example, agent i cannot tell if the fact that he has received no messages from another agent j (which means that i cannot use j's input in computing f) is due to the fact that j is malicious or that its messages have not yet arrived. Roughly speaking, when defining secure function computation in an asynchronous setting, BCG require that for every scheduler σ_e and set T of malicious agents, no matter what the agents in T do, the resulting distribution over outputs could have also resulted in the computation with a mediator if the malicious agents had given the appropriate input to the mediator.

BCG show that, in asynchronous systems, if $n > 4t$, the malicious agents cannot prevent the honest agents from correctly computing the output of f given their inputs, nor can the malicious agents learn anything about the inputs of the honest agents. Ben-Or, Kelmer, and Rabin [4] (BKR from now on) then showed if we are willing to tolerate a small probability $\epsilon > 0$ that the agents do not correctly compute f or that the malicious agents learn something, then we can achieve this if $n > 3t$. BCG and BKR also prove matching lower bounds for their results, showing that we really need to have $n > 4t$ (resp., $n > 3t$).

We can view secure function computation as a one-round interaction with a trusted mediator: each agent sends its input to the mediator, the mediator waits until it receives enough inputs, applies f to these inputs (again, replacing missing inputs with a default value), and sends the output back to the agents, who then output it. We generalize BCG and BKR's results for function computation to a more general setting. Specifically, we want to simulate arbitrary interactions with a mediator, not just function computation. Also, unlike previous approaches, we want the simulation to be "bidirectional": the set of possible output distributions that arise with the mediator must be the same as those that arise without the mediator, even in the presence of malicious parties. More precisely, we show that, given a protocol profile $\boldsymbol{\pi}$ for n agents[1] and a protocol π_d for a mediator, we can construct a protocol profile $\boldsymbol{\pi}'$ such that for all sets T of fewer than $n/4$ malicious agents, the following properties hold:

(a) For all protocols $\boldsymbol{\tau}'_T$ for the malicious agents and all schedulers σ'_e in the setting without the mediator, there exists a protocol $\boldsymbol{\tau}_T$ for the agents in T and a scheduler σ_e in the setting with the mediator such that, for all input profiles $\boldsymbol{x}$, the output distribution in the computation with $\boldsymbol{\pi}'$, $\boldsymbol{\tau}'$, and σ'_e with input $\boldsymbol{x}$ is the same as the output distribution with $\boldsymbol{\pi} + \pi_d$, $\boldsymbol{\tau}$, and σ_e with input $\boldsymbol{x}$.

[1] In the economics literature, the term "profile" is used to denote a tuple, so, for example, a protocol profile is a tuple of protocols, one for each agent. In this paper, we refer to protocol profiles as just "protocols", as is standard in the distributed computing literature.

(b) For all protocols $\boldsymbol{\tau}_T$ for the malicious agents and all schedulers σ_e in the setting with the mediator, there exists a protocol $\boldsymbol{\tau}'_T$ for the agents in T and a scheduler σ'_e in the setting without a mediator such that, for all input profiles $\boldsymbol{x}$, the output distribution in the computation with $\boldsymbol{\pi}'$, $\boldsymbol{\tau}'$, and σ'_e with input $\boldsymbol{x}$ is the same as the output distribution with $\boldsymbol{\pi} + \pi_d$, $\boldsymbol{\tau}$, and σ_e with input $\boldsymbol{x}$.

We use the notation $\boldsymbol{\pi} + \pi_d$ to indicate that the agents use protocol $\boldsymbol{\pi}$ and the mediator uses protocol π_d (we use the subscript d to denote the mediator); we view the mediator as just another agent here. This result implies that arbitrary distributed protocols that work in the presence of a trusted mediator can be compiled to protocols that work without a mediator, as long as there are less than $n/4$ malicious agents. And, just as BKR, if we allow a probability ϵ of error, we can get this result while tolerating up to $n/3$ malicious agents. BCG proved the analogue of (a) for secure function computation, which is enough for security purposes: if there is any bad behavior in the protocol without the mediator, this bad behavior must already exist in the protocol with the mediator. However, (b) also seems like a natural requirement; if a protocol satisfies this property, then all behaviors in the protocol with the mediator also occur in the protocol without the mediator.

Property (b) is typically not required in security papers. It plays a critical role in our work on implementing mediators [1], but we believe it of independent interest. Requiring only (a) may result in protocols where outcomes that may be likely in the mediator setting do not arise at all. This is especially relevant in asynchronous systems, since by requiring only (a) we are implicitly assuming that the adversary has total control over the scheduler. However, it may be the case that the scheduler acts randomly or that is even influenced by honest agents. For instance, suppose that a group of n agents wants to check who has the fastest internet connection. To do this, each agent pings the server and waits for the server's response. The server (who we are viewing as the mediator) waits until the first ping arrives, then sends a message to each agent saying which agent's ping was received first. In this example, the scheduler determines the lag in the system. If we wanted to simulate this interaction without the mediator but requiring only property (a), even with no malicious agents, a protocol profile in which every honest agent does nothing and outputs 1 would suffice. But, intuitively, this implementation does not capture the behavior of the server. Similar examples exist even in the case of function evaluation. Suppose a group of n congressmen vote remotely (by sending a vote to a trusted third party) to either pass or not a bill that requires support from at least 90% of them. We can view this as a multiparty computation of a function f in which each agent has input 0 (vote against) or 1 (vote for), and the output is either 0 (reject the bill) or 1 (pass the bill) depending on how many agents had input 1 (agents that do not submit input count as 0). In this case, a protocol in which every agent does nothing and outputs 0 securely computes f while tolerating up to $n/4$ malicious agents. To see this, note that regardless of the adversary, the scheduler can delay $n/4$ of the players until everyone else has finished the computation. This is indistinguishable from $n/4$ agents deviating from the protocol and submitting no input. However, again, this protocol does not capture the intended behavior of the voting process. By way of contrast, a protocol that bisimulates f would come closer to capturing the intended behavior of the voting process.

Clearly, the results of BCG and BKR are special cases of our result. However, in general, our results do not follow from those of BCG/BKR, as is shown in Sect. 3.2. Specifically, the results of BCG/BKR do not give us property (b), since the outcome can depend on the behavior of the scheduler. For example, consider protocols for two agents and a mediator m in which each agent sends its input to the mediator, the mediator m sends to each agent the first message it receives, and each agent outputs whatever they receive from the mediator. Let σ_e^i be the scheduler that delivers the message from agent i first, for $i = 1, 2$. It is easy to check that if the agents have inputs 0 and 1, respectively, and play with mediator σ_e^1, then they both output 0, while if they play with σ_e^2, then they both output 1. This means that, unlike secure function computation, even if all the agents are honest, the distribution over the agents' outputs can depend on the scheduler's protocol, not just the agents' inputs.

Even though our results do not follow from those of BCG/BKR, our proofs very much follow the lines of those of BCG/BKR. However, there are some new subtleties that arise in our setting. In particular, as the example above shows, when we try to implement the setting with the mediator, the agents must somehow keep track of the scheduler's possible behaviors. Doing this adds nontrivial complexity to our argument. We also show that our construction satisfies an analogue of (a) and (b) in the *universal composability* framework [7], which intuitively means that if a set of agents runs a distributed protocol that requires calls to a subroutine that can be implemented with a mediator, then the agents can implement that subroutine using our construction instead (with no need of a mediator), and the resulting protocol would preserve its original security properties.

Besides the main result, we also show that our protocol without the mediator has two additional security properties, which may be of independent interest. Specifically, we show that the following two properties hold for coalitions of malicious agents of size at most $t < n/3$.

(P1) The only way malicious agents can disrupt the computation is by preventing honest agents from terminating; if an honest agent terminates, then its output is correct.

(P2) If $2t + 1$ or more honest agents terminate, then all honest agents terminate. That is, either all the honest agents terminate or a nontrivial number of honest agents (more than $n - 2t$) do not terminate.

If we allow an ϵ probability of error, we get analogous results if we have $n > 2t$ rather than $n > 3t$. We remark that these two properties are in fact also satisfied by BCG's and BKR's implementations, but they do not prove this (or even state the properties explicitly).

Our interest in these properties stems in part from a game-theoretic variant of the problem that is considered by Abraham et al. [1], where agents get utility for various outcomes, and, in addition to honest and malicious agents, there are *rational agents*, who will deviate from a protocol if (and only if) it is in their interest to do so. We also assume that honest agents can leave "wills", so that if sufficiently many honest agents do not terminate, the remaining agents will be punished. Property P2 guarantees that either all the honest agents terminate, or sufficiently many of them do not terminate so as to guarantee that rational agents will not try to prevent honest agents from terminating (due to the threat of punishment). Property P1 guarantees that if all the honest agents terminate, their output will be correct. Thus, using these results allows us to obtain results stronger than those of this paper in the game-theoretic setting.

The focus of this paper is on upper bounds. Since our algorithms have the same upper bounds as those of BCG and BKR, despite the results of BCG and BKR being special cases of our results, and BCG and BKR prove lower bounds that match their upper bonds on the number of malicious agents that can be tolerated, we immediately get lower bounds that match our upper bounds from the results of BCG and BKR.

2 The Model

The model used throughout this paper is that of an asynchronous network in which every pair of agents can communicate through a private and reliable communication channel. For most of our results, we assume that all messages sent through any of these channels are eventually received, but they can be delayed arbitrarily. The order in which these messages are received is determined by the *environment* (also called the *scheduler*), which is an adversarial entity. The scheduler also chooses the order in which the agents are scheduled. For some of the results of this paper, we drop the condition that all messages must be eventually delivered. We call these more general schedulers *relaxed schedulers*.

Whenever an agent is scheduled, it reads all the messages that it has received since the last time it was scheduled, sends a (possibly empty) sequence of messages, and then performs some internal actions. We assume that the scheduler does not deliver any message or schedule other agents during an agent's turn. Thus, although agent i does not send all its messages simultaneously when it is scheduled, they are sent atomically, in the sense that no other agent is scheduled while i is scheduled, nor are any messages delivered while i is scheduled. Note that the atomicity assumption is really a constraint on the scheduler's protocol.

More precisely, consider the following types of *events*:

- $sch(i)$: Agent i gets scheduled.
- $snd(\mu, j, i)$: Agent i sends a message μ to agent j.
- $rec(\mu, j, i)$: Message μ sent by j is received by i. The message μ must be one sent at an earlier time to i that was not already received.
- $comp(v, i)$: Agent i locally computes value v.
- $out(s, i)$: Agent i outputs string s.
- $done(i)$: i is done sending messages and performing computations (for now).

For simplicity, we assume that agents can output only strings in $\{0,1\}^*$. Note that all countable sets can be encoded by such strings, and thus we can freely talk about agents being able to output any element of any countable set (for instance, elements of a finite field $\mathbb{F}_q$) by assuming that they are actually outputting an encoding of these elements. We also assume that at most one event occurs at each time step. Let $h(m)$ denote a *global view* up to time m: a sequence that starts with an input profile $\boldsymbol{x}$, followed by the ordered sequence of events that have occurred up to and including time m. We assume that the only events between events of the form $sch(i)$ and $done(i)$ are ones of the form $snd(\mu, j, i)$ and $comp(v, i)$. This captures our atomicity assumption. We do not include explicit events that correspond to reading messages. (Nothing would change if we included them; they would simply clutter the notation.) Message delivery (which is assumed to be under the control of the scheduler) occurs at times between when agents are scheduled. We can also consider the subsequence involving agent i, namely, i's initial state, followed by events of the form $sch(i)$, $snd(\cdot,\cdot,i)$, $comp(\cdot,i)$, $rec(\cdot,\cdot,i)$, and $done(i)$. This subsequence is called i's *local view* . We drop the argument m if it can be deduced from context or if it is not relevant (for instance, when we consider the local view of an agent after a particular event).

Agent i moves only after a $sch(i)$ event. What it does (in particular, the order in which i sends messages) is determined by i's protocol, which is a function of i's local view. The scheduler moves after an action of the form $done(i)$ or $rec(\cdot,\cdot,i)$. It is convenient to assume that the scheduler is also running a protocol, which is also a function of its local view. Since the scheduler does not see the contents of messages, we can take its view to be identical to $h(m)$, except that $comp$ events and the contents of the messages in snd and rec events are removed, although we do track the index of the messages delivered; that is, we replace events of the form $snd(\mu, i, j)$ and $rec(\mu, i, j)$ by $snd(i, j)$ and $rec(i, j, \ell)$, where ℓ is the index of the message sent by i to j in $h(m)$. For instance, $rec(i, j, 2)$ means that the second message sent by i to j was delivered to j. Note that the scheduler does see events of the form $done(i)$; indeed, these are signals to the scheduler that it can move, since i's turn is over. Since we view the agents (and the mediator) as sending messages atomically, in the sequel, we talk about an agent's (or the mediator's) *turn*. An agent's kth turn takes place the kth time it is scheduled. During its turn, the agent sends a block of messages and performs some local computation.

It is more standard in the literature to assume that agents perform at most one action when they are scheduled. We can view this as a constraint on agents' protocols. A *single-action* protocol for agent i is one where agent i sends at most one message before performing the $done(i)$ action. As we show in Appendix 3.6, we could have restricted to single-action protocols with no loss of generality as far as our results go; allowing agents to perform a sequence of actions atomically just makes the exposition easier.

Even though it might appear that malicious agents and the scheduler act independently, it is shown by Abraham et al. [1, Section A.1] that we can assume without loss of generality that they can coordinate their actions, even when there is no direct communication channel between them. In fact, we can assume without loss of generality that they are all under the control of a single entity that is aware of all their local views at all times. We call this entity the *adversary*.

Definition 1. *An* adversary *is a triple* $(T, \boldsymbol{\sigma}_T, \tau_e)$, *consisting of a set* T *of malicious agents, the protocol* $\boldsymbol{\tau}_T$ *used by the agents in* T*, and a protocol* σ_e *for the scheduler. An adversary where the scheduler is relaxed is a* relaxed adversary.

In this paper, we consider protocols that involve a *mediator*, typically denoted d, using a protocol denoted π_d. In protocols that involve a mediator, we assume that honest agents' protocols are always such that the honest agents communicate only with the mediator and not with each other, as opposed to malicious agents that can do both. As far as the scheduler is concerned, the mediator is like any other agent, so the scheduler (and the mediator's protocol) determine when the mediator sends and receives messages. However, the mediator is never malicious, and thus never deviates from its announced protocol.

We deal only with bounded protocols, where there is a bound N on the number of messages that an honest agent sends. Of course, there is nothing to prevent malicious agents from spamming the mediator and sending an arbitrary number of messages. We assume that the mediator reads at most N messages from each agent i, ignoring any further messages sent by i.

For our results involving termination, specifically, (P2), it is critical that agents know when the mediator stops sending messages. For these results, we restrict the honest agents and the mediator to using protocols that have the following *canonical form*: Using a canonical protocol, each honest agent tags its ℓth message with label ℓ and all honest agents are guaranteed to send at most N messages regardless of their inputs or the random bits they use. Whenever the mediator receives a message from an agent i, it checks its tag ℓ; if $\ell > N$ or if the mediator has already received a message from i with tag ℓ, it ignores the message. The mediator is guaranteed to eventually terminate. Whenever this happens, it sends a special "STOP" message to all agents and halts. Whenever an honest agent receives a "STOP" message, it terminates.

Even though canonical protocols have a bound N on the number of messages that honest agents and the mediator can send, the mediator's local view in a canonical protocol can be arbitrarily long, since it can be scheduled an arbitrary number of times. We conjecture that, in general, since the message space is finite, the expected number of messages required to simulate the mediator is unbounded. However, we can do better if the mediator's protocol satisfies two additional properties. Roughly speaking, the first property says that the mediator can send messages only either at its first turn or in response to an agent's message; the second property says that the mediator ignores *empty turns*, that is, turns where it does not receive or send messages. More precisely, the first property says that whenever the mediator π_d is scheduled with view h_d, then if $h_d \neq ()$ (i.e., if h_d is not the initial view) or if the mediator has not received any messages in h_d since the last time it was scheduled, then $\pi_d(h_d) = done(d)$. The second property says that $\pi_d(h_d) = \pi_d(h'_d)$, where h'_d is the result of removing consecutive $(done(d), sch(d))$ pairs in h_d (e.g., if $h_d = (sch(d), snd(\mu, j, d), done(d), sch(d), done(d), rec(\mu', i, d), sch(d), done(d), sch(d))$, then $h'_d = (sch(d), snd(\mu, j, d), done(d), rec(\mu', i, d), sch(d)))$. A protocol for the mediator that satisfies these two properties is called *responsive*. In the full paper [9, Section 4.4], we show that if the mediator uses a responsive protocol π_d that can be represented using a circuit with c gates, then we can simulate all protocol profiles $\boldsymbol{\pi} + \pi_d$ in such a way that the expected number of messages sent by honest agents during the simulation is polynomial in n and N and linear in c.

3 Secure Computation in Interactive Settings

In this section, we present the main results of this paper and show how they extend and generalize other well-known results.

3.1 The BGW/BCG Notion of Secure Computation

Secure computation is concerned with jointly computing a function f on n variables, where the ith input is known only to agent i. For instance, if we want to compute the average salary of the people from the state of New York, then n would be New York's population, the input x_i is i's salary, and $f(x_1, \dots, x_n) = \frac{\sum_{i=1}^{n} x_i}{\sum_{x_i \neq 0} 1}$. (For the denominator we count only people who are actually working.) Ideally, a secure computation protocol that computes f would be a protocol in which each agent i outputs $f(x_1, \dots, x_n)$ and gains no information about the inputs x_j for $j \neq i$. In our example, this amounts to not learning other people's salaries.

Typically, we are interested in performing secure computation in a setting where some of the agents might be malicious and not follow the protocol. In particular, they might not give any information about their input or might just pretend that they have a different input (for instance, they can lie about their salary). What output do we want the secure computation of f to produce in this case? To make precise what we want, we use notation introduced by BGW and BCG.

Let $\boldsymbol{x}$ be a vector of n components; let C be a subset of $[n]$ (where we use the notation $[n]$ to denote the set $\{1, \dots, n\}$, as is standard); let $\boldsymbol{x}_C$ denote the vector obtained by projecting $\boldsymbol{x}$ onto the indices of C; and if $\boldsymbol{z}$ is a vector of length $|C|$, let $\boldsymbol{x}/_{(C,\boldsymbol{z})}$ denote the vector obtained by replacing the entries of x indexed by C with $\boldsymbol{z}$. Given a set C of indices, a default value, which we take here to be 0, and a function f, we take f_C to be the function results from applying f, but taking the inputs of the agents not in C to be 0; that is, $f_C(\boldsymbol{x}) = f(\boldsymbol{x}/_{(\overline{C},\boldsymbol{0})})$. Roughly speaking, if only the agents in C provide inputs, we want the output of the secure computation to be $f_c(\boldsymbol{x})$.

What about agents who lie about their inputs? A malicious agent i who lies about his input x_i and pretends to have some other input y_i is indistinguishable from an honest agent who has y_i as his actual input. We can capture this lie using a function $L : D^{|T|} \to D^{|T|}$, where D is the domain of the inputs and T is the set of malicious agents. The function L encodes the inputs malicious agents pretend to have given their actual inputs. BCG require that all the honest agent output the same value and that the output has the form $(C, f_C(\boldsymbol{y}))$, where $\boldsymbol{y} = \boldsymbol{x}/_{(T,L(\boldsymbol{x}_T))}$. They allow C to depend on $\boldsymbol{x}_T$, since malicious agents can influence the choice of C. They also allow the choice of C and the function L to be randomized. Since the choice of L and C can be correlated, L and C are assumed to take as input a common random value $r \in \mathcal{R}$, where $\mathcal{R}$ denotes the domain of random inputs. That is, $C = c(\boldsymbol{x}_T, r)$ for some function c, and the malicious agents with actual input $\boldsymbol{x}_T$ pretend that their input is $L(\boldsymbol{x}_T, r)$.

BCG place no requirements on the output of malicious agents, but they do want the inputs of honest agents to remain as secret as possible. Hence, in an ideal scenario, the outputs of malicious agents can depend only on $\boldsymbol{x}_T$, $f_C(\boldsymbol{y})$, and possibly some randomization. Taking O_i to denote the output function of a malicious agent i, we can now give BCG's definitions.

Definition 2. *An* ideal t-adversary *A is a tuple $(T, c, L, \boldsymbol{O})$ consisting of a set $T \subseteq [n]$ of malicious agents with $|T| \leq t$ and three randomized functions $c : D^{|T|} \times \mathcal{R} \to \mathcal{P}([n])$ with $|c(\boldsymbol{z}, r)| \geq n - t$ for all input profiles $\boldsymbol{z}$ and r, $L : D^{|T|} \times \mathcal{R} \to D^{|T|}$ and $\boldsymbol{O} : D^{|T|} \times D \times \mathcal{R} \to (\{0,1\}^*)^{|T|}$. The* ideal *output $\boldsymbol{\rho}$ of A given function f, input profile $\boldsymbol{x}$, and a value $r \in \mathcal{R}$ is*

$$\rho_i(\boldsymbol{x}, A, r; f) = \begin{cases} (c(\boldsymbol{x}_T, r), f_{c(\boldsymbol{x}_T,r)}(\boldsymbol{x}/_{(T,L(\boldsymbol{x}_T,r))})) & \text{if } i \notin T \\ O_i(\boldsymbol{x}_T, f_{c(\boldsymbol{x}_T,r)}(\boldsymbol{x}/_{(T,L(\boldsymbol{x}_T,r))}), r) & \text{if } i \in T. \end{cases}$$

Note that an ideal t-adversary is somewhat different from the adversary as defined in Definition 1, although they are related, as we show in Sect. 3.2. We use variants of A to denote both types of adversary.

Let $\boldsymbol{\rho}(\boldsymbol{x}, A; f)$ denote the distribution induced over outputs by the protocol profile $\boldsymbol{\rho}$ on input x given the ideal t-adversary A. We can now give the BCG definition of secure computation. Let $\boldsymbol{\pi}(\boldsymbol{x}, A)$ be the distribution of outputs when running protocol $\boldsymbol{\pi}$ on input $\boldsymbol{x}$ with adversary $A = (T, \boldsymbol{\tau}_T, \sigma_e)$.

Definition 3 (Secure computation). *Let $f : \mathcal{D}^n \to \mathcal{D}$ be a function on n variables and $\boldsymbol{\pi}$ a protocol for n agents. Protocol $\boldsymbol{\pi}$ t-securely computes f if, for every adversary $A = (T, \boldsymbol{\tau}_T, \sigma_e)$, the following properties hold:*

SC1. For all input profiles $\boldsymbol{x}$, all honest agents terminate with probability 1.
SC2. There exists an ideal t-adversary $A' = (T, c, L, \boldsymbol{O})$ such that, for all input profiles $\boldsymbol{x}$, $\boldsymbol{\rho}(\boldsymbol{x}, A'; f)$ and $\boldsymbol{\pi}(\boldsymbol{x}, A)$ are identically distributed.

Note that BCG just require that *some* ideal t-adversary A gives the same distribution over the outputs of $\boldsymbol{\pi}$. This captures the idea that all ways that malicious agents can deviate are modeled by adversaries. Also note that SC1 follows from SC2 if we view non-termination as a special kind of output.

BCG prove the following result:

Theorem 1 (BCG). *Given n and t such that $n > 4t$ and a function $f : D^n \to D$, there exists a protocol $\boldsymbol{\pi}^f$ that t-securely computes f.*

The construction of $\boldsymbol{\pi}^f$ is sketched in [2] and [9, Section 3.2.7]; most of the primitives used in this construction are also used in ours.

3.2 Secure Computation and Mediators

Even though it is not explicitly proven by BCG, their construction of $\boldsymbol{\pi}^f$ satisfies an additional property that we call SC3, which is essentially a converse of SC2.

SC3. For all ideal t-adversaries $A = (T, c, L, \boldsymbol{O})$, there exists an adversary $A' = (T, \boldsymbol{\tau}_T, \sigma_e)$ such that, for all input profiles $\boldsymbol{x}$, $\boldsymbol{\rho}(\boldsymbol{x}, A; f)$ and $\boldsymbol{\pi}(\boldsymbol{x}, A')$ are identically distributed.

Lemma 1. *Given a function $f : D^n \to D$, protocol $\boldsymbol{\pi}^f$ satisfies SC3.*

Proof (Proof (sketch)). Given a trusted-party adversary $A = \{T, c, L, \boldsymbol{O}\}$ and an input profile $\boldsymbol{x}_T$, the adversary A' runs $\boldsymbol{\pi}_T^f$ with input $L(\boldsymbol{x}_T)$, except that if a malicious agent i would output a tuple of the form (S, z) (note that all outputs of honest players have this form), it outputs $O_i(\boldsymbol{x}_T, z)$ instead. Meanwhile, the scheduler delays all messages from agents not in $c(\boldsymbol{x}_T)$ until all honest players finish their part of the computation. We can show that the outputs of A and A' are identically distributed. Since the full proof requires the actual implementation of $\boldsymbol{\pi}^f$, the details are given in [9, Section 3.3]

We next show how secure computation relates to simulating a mediator. Consider the following protocol $\boldsymbol{\tau}^f + \tau_d^f$ for n agents and a mediator: Agents send their inputs to the mediator the first time that they are scheduled. The mediator waits until it has received a valid input from all agents in a subset $C \subseteq [n]$ with $|C| \geq n - t$. The mediator then computes $y = f_C(\boldsymbol{x})$ and sends each agent the pair (C, y). When the agents receive a message from the mediator, they output that message and terminate.

Clearly $\boldsymbol{\tau}^f + \tau_d^f$ satisfies SC1. It is easy to see that it also satisfies SC2: Given a set T of malicious agents, a deterministic protocol profile $\boldsymbol{\tau}_T$ for the malicious agents, and a deterministic scheduler σ_e, define $L(\boldsymbol{x}, r)$ to be whatever the malicious agents send to the mediator with input $\boldsymbol{x}$, let $c(\boldsymbol{x})$ be the set of agents from whom the mediator has received a message the first time it is scheduled after having received a message from a least $n - t$ agents (given σ_e, $\boldsymbol{\tau}_T$, and input $\boldsymbol{x}$), and let $O(\boldsymbol{x})$ be the output function that malicious agents use in $\boldsymbol{\tau}^f + \tau_d^f$ (note that they receive a single message with the output of the computation, so their output depends only on $\boldsymbol{x}$, $\boldsymbol{\tau}_T$, and σ_e). Clearly SC2 holds with this choice of t-ideal adversary. Randomized functions $\boldsymbol{\tau}_T$ and σ_e can be viewed as resulting from sampling random bits r according to some distribution and then running deterministically; the protocols c, h, and O can sample r from the same distribution and then proceed as above with respect to the deterministic $\boldsymbol{\tau}_T(r)$ and $\sigma_e(r)$.

The protocol $\boldsymbol{\tau}^f + \tau_d^f$ satisfies SC3 as well. Given $A = (T, c, L, O)$, the definition of $\boldsymbol{\tau}_T$ and σ_e is straightforward: the agents in T choose a random input $r \in \mathcal{R}$ and then each agent $i \in T$ sends $L(x_i, r)$ to the mediator. The scheduler

σ_e delivers all messages from the agents in $c(\boldsymbol{x}_T, r)$ first, and then schedules the mediator. It then delivers all the other messages.

Since both $\boldsymbol{\tau}^f + \tau_d^f$ and $\boldsymbol{\pi}^f$ satisfy SC2 and SC3, for all adversaries A, there exists an adversary A' (resp., for all adversaries A' there exists an adversary A) such that $(\boldsymbol{\tau}^f + \tau_d^f)(\boldsymbol{x}, A)$ and $\boldsymbol{\pi}^f(\boldsymbol{x}, A')$ are identically distributed.

Unfortunately, given a protocol $\boldsymbol{\pi}_d$ for the mediator, there might not exist a function f such that SC2 and SC3 hold, as the example given in the introduction shows (where the mediator sends to the agents the first message it receives). Note that, in this example, the output of the agents is not a function of their input profile; thus, there is no function f for which SC2 and SC3 hold. Nevertheless, we are still interested in securely computing the output of the protocol with the mediator. That is, we are interested in getting analogues to SC2 and SC3 for arbitrary interactive protocols. This is captured by the following definition:

Definition 4. *Protocol $\boldsymbol{\pi}'$ t-*bisimulates *$\boldsymbol{\pi}$ if the following two properties hold:*

(a) For all adversaries $A = (T, \boldsymbol{\tau}_T, \sigma_e)$ with $|T| \leq t$, there exists an adversary $A' = (T, \boldsymbol{\tau}'_T, \sigma'_e)$ such that for all input profiles $\boldsymbol{x}$, $\boldsymbol{\pi}(\boldsymbol{x}, A)$ and $\boldsymbol{\pi}'(\boldsymbol{x}, A')$ are identically distributed.

(b) For all adversaries $A' = (T, \boldsymbol{\tau}'_T, \sigma'_e)$ with $|T| \leq t$, there exists an adversary $A = (T, \boldsymbol{\tau}_T, \sigma_e)$ such that all input profiles $\boldsymbol{x}$, $\boldsymbol{\pi}(\boldsymbol{x}, A)$ and $\boldsymbol{\pi}'(\boldsymbol{x}, A')$ are identically distributed.

Note that the first clause is analogous to SC2, while the second clause is analogous to SC3. There is no clause analogous to SC1 since we allow agents not to terminate. In any case, since we can view non-termination as a special type of output (i.e., we can view an agent that does not terminate as outputting $\bot$), so SC2 already guarantees that non-termination happens with the same probability in $\boldsymbol{\pi}'$ and $\boldsymbol{\pi}$ (In the setting of BGW, since all functions terminate, with this viewpoint, SC2 implies SC1, a point already made by Canetti [6].)

The following proposition follows from Theorem 1 and Lemma 1.

Proposition 1. *$\boldsymbol{\pi}^f$ t-bisimulates $\boldsymbol{\tau}^f + \tau_d^f$ if $n > 4t$.*

3.3 Beyond Secure Computation

Although BCG make claims for their protocol only if $n > 4t$, variants of some of the properties that they are interested in continue to hold even if $n < 4t$. The first of these properties is that if $n > 3t$, then the only way that the adversary can affect $\boldsymbol{\pi}^f$ is by preventing some honest agents from terminating. We can capture this notion as follows.

Definition 5. *A scheduler is* relaxed *if it can decide not to deliver some of the messages. A protocol $\boldsymbol{\pi}'$ (t, t')-*bisimulates *$\boldsymbol{\pi}$ if it t-bisimulates $\boldsymbol{\pi}$ but the schedulers σ'_e and σ_e of the first and second clause of Definition 4 respectively may be relaxed for $t \geq |T| > t'$.*

Proposition 2. *$\boldsymbol{\pi}^f$ (t, t')-bisimulates $\boldsymbol{\tau}^f + \tau_d$ and $t \geq t'$.*

This means that if $3t + t' < n$, then adversaries of size between t' and t have the same power to affect the outcome with $\boldsymbol{\pi}^f$ as with $\boldsymbol{\tau}^f + \tau_d$ as long as schedulers are allowed to discard messages, so that they never reach their recipient. In particular, this means that the adversary cannot influence the outcome in any other way than by preventing some honest players from terminating. However, we can show that the BCG protocol has the property that if at least $2t+1$ honest agents terminate, then all the remaining honest agents terminate. This observation motivates the following definition:

Definition 6. *A protocol* $\boldsymbol{\pi}$ (t,k)-coterminates *if, all adversaries* $A = (T, \boldsymbol{\tau}_T, \sigma_e)$ *with* $|T| \leq t$ *and all input profiles* $\boldsymbol{x}$*, in all executions of* $\boldsymbol{\pi}$ *with adversary* A *and input* $\boldsymbol{x}$*, either all the agents not in* T *terminate or strictly fewer than* k *agents not in* T *do.*

Proposition 3. $\boldsymbol{\pi}^f$ *$(t, 2t+1)$-coterminates.*

We do not prove Proposition 2 or 3 here, since we prove a generalization of them below (see Theorem 2).

3.4 Simulating Arbitrary Protocols

The goal of this paper is to show that we can securely implement any interaction with a mediator, and do so in a way that ensure the two properties discussed in Sect. 3.3. This is summarized in the following theorem:

Theorem 2. *For every protocol* $\boldsymbol{\pi} + \pi_d$ *for* n *agents and a mediator, there exists a protocol* $\boldsymbol{\pi}'$ *for* n *agents such that* $\boldsymbol{\pi}'$

(a) (t,t')*-bisimulates* $\boldsymbol{\pi}$ *if* $n > 3t + t'$ *and* $t \geq t'$*, and*
(b) $(t, 2t+1)$*-coterminates if* $n > 3t$ *and* $\boldsymbol{\pi} + \pi_d$ *is in canonical form.*

Moreover, if π_d *is responsive, the expected number of messages sent in an execution of* $\boldsymbol{\pi}'$ *is polynomial in* n *and* N*, and linear in* c*, where* N *is the expected number of messages sent when running* $\boldsymbol{\pi} + \pi_d$ *and* c *is the number of gates in an arithmetic circuit that implements the mediator's protocol.*

The construction of $\boldsymbol{\pi}'$ is sketched in Sect. 4 and given in full detail in given in the full paper [9, Section 4.2] and, not surprisingly, uses many of the techniques used by BCG. And, like BKR, if we allow an ϵ probability of error we get stronger results. We define ϵ-t-bisimulation just like t-bisimulation (Definition 4), except that, in both clauses, the distance between $(\boldsymbol{\pi} + \pi_d)(\boldsymbol{x}, A)$ and $\boldsymbol{\pi}'(\boldsymbol{x}, A')$ is less than ϵ, where the distance d between probability measures ν and ν' on some finite space S is defined as $d(\nu, \nu') = \sum_{s \in S} |\nu(s) - \nu'(s)|$. The definition of ϵ-t-bisimulation and ϵ-(t,t')-bisimulation are analogous. A protocol ϵ-(t,k)-*coterminates* if it (t,k)-coterminates with probability $1 - \epsilon$.

Theorem 3. *For every protocol* $\boldsymbol{\pi} + \pi_d$ *for* n *agents and a mediator and all* $\epsilon > 0$*, there exists a protocol* $\boldsymbol{\pi}'$ *for* n *agents such that* $\boldsymbol{\pi}'$

(a) ϵ*-*(t, t')*-bisimulates* $\boldsymbol{\pi} + \pi_d$ *if* $n > 2t + t'$ *and* $t \geq t'$*, and*
(b) ϵ*-*$(t, t+1)$*-coterminates if* $n > 2t$ *and* $\boldsymbol{\pi} + \pi_d$ *is in canonical form.*

Moreover, if π_d *is responsive,* $\boldsymbol{\pi}'$ *can be implemented in such a way that the expected number of messages when running* $\boldsymbol{\pi}' + \pi_d$ *is polynomial in* n *and* N*, and linear in* c*, where* N *is the expected number of messages sent when running* $\boldsymbol{\pi} + \pi_d$*.*

3.5 Universally Composable Security

Both the definition of secure computation (Definition 3) and of bisimulation (Definition 4) capture only the intended security properties in the *stand-alone* model, where only a single execution of a given protocol is run. However, in many cases, it is important that these properties are satisfied even when a protocol is run several times in succession, or even when these executions are performed concurrently. For this purpose, the standard approach is to prove that the given protocol is secure in the *universal composability model* [7]. Kushilevitz, Lindell and Rabin [10] showed that every protocol that is perfectly secure in the stand-alone model and has a *straight-line black-box simulator* is also secure in the UC model. Having a black-box straight-line simulator means that the adversary is able to simulate what its view would be when running the protocol without the mediator (resp., with the mediator), given its view in the protocol with the mediator (resp., without the mediator), and that it is able to do so without having to *rewind*, which means to go back to a previous state and interact with the other agents in a different way. This is exactly the approach we take when showing that the protocol presented in [9, Section 4.2] satisfies the properties of Theorem 2 (see [9, Section 4.3] for details). Therefore, Theorem 2 holds with perfect universally composable security as well; that is, if a protocol uses the mediator as a subroutine, and we replace the subroutine with our implementation, then all the desired properties would still hold, even if the protocol is ran concurrently with another protocol.

3.6 Variant Models

In this section, we show that the choices made in our formal model are essentially being made without loss of generality. We start by considering our assumption that agents perform a sequence of actions atomically when they are scheduled. We next show that we would get theorems equivalent to the ones that we are claiming if we had instead assumed that agents perform just a single action when they are scheduled. To prove this, we first need the following notion:

Definition 7. *A protocol* $\boldsymbol{\pi}$ *is* N-message bounded *if, for all inputs, no agent ever sends more than* N *messages in a single turn. A protocol is* message bounded *if it is* N*-message bounded for some* N*.*

Proposition 4. *There exist a function* H *from message-bounded protocols to single-action protocols such that for all protocols* $\boldsymbol{\pi}$*, the following holds:*

(a) For all schedulers (resp., relaxed schedulers) σ_e there exists a scheduler (resp., relaxed scheduler) σ_e' such that, for all input profiles $\boldsymbol{x}$, $\boldsymbol{\pi}(\boldsymbol{x}, \sigma_e)$ and $H(\boldsymbol{\pi})(\boldsymbol{x}, \sigma_e')$ are identically distributed, where $H(\boldsymbol{\pi}) := (H(\pi_1), \ldots, H(\pi_n))$ and we view σ_e and σ_e', respectively, as the adversaries (i.e., we take $T = \emptyset$).

(b) For all schedulers (resp., relaxed schedulers) σ_e' there exists a scheduler (resp., relaxed scheduler) σ_e such that, for all input profiles $\boldsymbol{x}$, $\boldsymbol{\pi}(\boldsymbol{x}, \sigma_e)$ and $H(\boldsymbol{\pi})(\boldsymbol{x}, \sigma_e')$ are identically distributed.

The converse of Proposition 4 is trivial, since single-action protocols are protocols. It follows from Proposition 4 that Theorem 2 holds even if we restrict agents to using single-action protocols (note that canonical protocols are message bounded).

Proof. Intuitively, $H(\pi_i)$ is identical to π_i, except that rather than sending a sequence of messages when it is scheduled, i sends the messages one at a time. The scheduler σ_e' is then chosen to ensure that i is scheduled so that it sends all of its messages as if they were sent atomically. In addition to keeping track of the messages it has sent and received, i uses the variable U_i whose value is a sequence of messages (intuitively, the ones that i would have sent at this point in the simulation of π_i that it has not yet sent), initially set to the empty sequence, and a binary variable *next*, originally set to 1. When i is scheduled by σ_e', $H(\pi_i)$ proceeds as follows: If $next = 1$, then i sets U_i to the sequence of messages that it would send with π_i given its current view. (If π_i randomizes, then $H(\pi_i)$ does the same randomization. If U_i is the empty sequence (so π_i would not send any messages at that point), i performs the action $done(i)$, and outputs whatever it does with π; otherwise, i sets *next* to 0, sends the first message in U_i to its intended recipient, and removes this message from U_i. If $next = 0$, then if U_i is empty, i sets *next* to 1, sends $done(i)$, and outputs whatever it does with π; otherwise, i sends the first message in U_i to its intended recipient and removes it from U_i.

Since $\boldsymbol{\pi}$ is message bounded, there exists an N such that $\boldsymbol{\pi}$ is N-message bounded. For part (a), given σ_e, we construct σ_e' so that it simulates σ_e, except that if σ_e schedules i, σ_e' schedules i repeatedly until either it observes $done(i)$ or until i sends messages in $N+1$ consecutive turns. Since $\boldsymbol{\pi}$ is N-message bounded, it is clear that $\boldsymbol{\pi}(\boldsymbol{x}, \sigma_e)$ and $H(\boldsymbol{\pi})(\boldsymbol{x}, \sigma_e')$ are identically distributed. Note that it is necessary for $\boldsymbol{\pi}$ to be N-message bounded, since if the scheduler schedules each agent i repeatedly until it stops sending a message during its turn, an agent that keeps sending messages would be scheduled indefinitely, and so would prevent other agents from being scheduled.

For part (b), given σ_e', we construct σ_e so that it simulates σ_e. There is one issue that we have to deal with. Whereas with σ_e, an agent i can send k messages each time it is scheduled, with σ_e', it can send only one message when it is scheduled. The scheduler σ_e' constructed from σ_e in part (a) scheduled i repeatedly until it sent all the messages it did with σ_e. But we cannot assume that the scheduler σ_e' that we are given for part (b) does this. Thus, σ_e must keep track of how many of the messages that each agent i was supposed to send the last time it was scheduled by σ_e have been sent so far. To do this, σ_e uses variables mes_i, one for each agent i, initially set to 0, such that mes_i keeps track of how many of the messages that agent i sent with σ_e still need to be sent by σ_e'. As we observed above, given a local view h of the scheduler where the agents use $\boldsymbol{\pi}$ and the scheduler uses σ_e, there is a corresponding local view h' of the scheduler where the agents use $\boldsymbol{\pi}'$ and the scheduler uses σ_e'. If, given h', σ_e' schedules agent i with probability α_i, then with the same probability α_i, σ_e proceeds as follows: if $mes_i = 0$ (which means that all the messages that i sent the last time it was scheduled have been delivered in h'), then σ_e schedules i, sees how many messages i delivers according π_i, and sets mes_i to this number; if $mes_i \neq 0$, then mes_i is decremented by 1 but no agent is scheduled. Again, it is clear that that $\boldsymbol{\pi}(\boldsymbol{x}, \sigma_e)$ and $H(\boldsymbol{\pi})(\boldsymbol{x}, \sigma_e')$ are identically distributed.

BCG put further constraints on the scheduler. Specifically, they assume that, except possibly for the first time that agent i is scheduled, i is scheduled immediately after receiving a message and only then. That is, in our terminology, BCG assume that a $rec(\cdot, \cdot, i)$ event must be followed by a $sch(i)$ event, and all $sch(i)$ events except possibly the first one occur after a $rec(\cdot, \cdot, i)$ event. We call the schedulers that satisfy this constraint *BCG schedulers.*

We now prove a result analogous to Proposition 4, from which it follows that we could have obtained our results using a BCG scheduler.

Proposition 5. *There exist a function H from protocols to protocols such that for all protocols $\boldsymbol{\pi}$ the following holds:*

(a) For all schedulers (resp., relaxed schedulers) σ_e there exists a BCG scheduler (resp., relaxed BCG scheduler) σ_e' such that, for all input profiles $\boldsymbol{x}$, $\boldsymbol{\pi}(\boldsymbol{x}, \sigma_e)$ and $H(\boldsymbol{\pi})(\boldsymbol{x}, \sigma_e')$ are identically distributed.
(b) For all BCG schedulers (resp., relaxed schedulers) σ_e' there exists a scheduler (resp., relaxed scheduler) σ_e such that, for all input profiles $\boldsymbol{x}$, $\boldsymbol{\pi}(\boldsymbol{x}, \sigma_e)$ and $H(\boldsymbol{\pi})(\boldsymbol{x}, \sigma_e')$ are identically distributed.

Proof. As in Proposition 4, the idea is that σ_e' simulates σ_e, but since σ_e can schedule an agent only when it delivers a message, we have each agent i send itself special messages, denoted $proceed_i$, to ensure that there are always enough messages in the system. In more detail, $H(\pi_i)$ works as follows. When it is first scheduled, agent i sends itself a $proceed_i$ message. Since we are considering BCG schedulers, agent i is scheduled subsequently only when it receives a message. If it receives a message other than $proceed_i$, it does nothing (although the message is added to its view). If it receives a $proceed_i$ message, then it does whatever it would do with π_i given its current view with the $proceed_i$ messages and the $sch(i)$ events not preceded by a $proceed_i$ message removed, and sends itself another $proceed_i$ message.

For part (a), given σ_e, σ_e' first schedules each agent once (in some arbitrary order), to ensure that that each of them has sent a $proceed_i$ message that is available to be delivered. Given a view h', σ_e' considers what σ_e would do in the view h that results from h' by removing the initial $sch(i)$ event for each agent i, the last message that each agent i sends when it is scheduled if it sends a message at all, and the receipt of these messages. If h' is a view that results where the agents are running $H(\boldsymbol{\pi})$, then the send and receive events removed are precisely those that involve $proceed_i$. If σ_e delivers a message with some probability, then σ_e' delivers the corresponding message with the same probability; if σ_e schedules an agent i with some probability, σ_e' delivers the last $proceed_i$ that i sent and schedules agent i with the same probability. If there is no $proceed_i$ message to deliver, then σ_e' does nothing, but our construction of $H(\pi_i)$ guarantees that if h' is a view that results from running $H(\boldsymbol{\pi})$, then there will be such a message that can be delivered. Again, it is clear that $\boldsymbol{\pi}(\boldsymbol{x}, \sigma_e)$ and $H(\boldsymbol{\pi})(\boldsymbol{x}, \sigma_e')$ are identically distributed.

For part (b), given σ_e', the construction of σ_e is similar to that of Proposition 4. Again, given a local view h of σ_e where the agents use $\boldsymbol{\pi}$, there is a corresponding view h' of σ_e' where the agents use $H(\boldsymbol{\pi})$. If, given input h', σ_e' delivers a message with some probability p and the messages is not a $proceed_i$ message, then σ_e delivers the corresponding message with probability p. If the message is a $proceed_i$ message, then σ_e also schedules agent i. If σ_e' schedules an agent i with probability p, and in h' this is the first time that i is scheduled, then σ_e schedules i with probability p and otherwise does nothing with probability p. Yet again, it is straightforward to show that $\boldsymbol{\pi}(\boldsymbol{x}, \sigma_e)$ and $H(\boldsymbol{\pi})(\boldsymbol{x}, \sigma_e')$ are identically distributed.

4 Proof of Theorem 2

In this section we sketch the construction of a protocol $\boldsymbol{\pi}'$ that satisfies Theorem 2. We provide the full construction and the proof of correctness in the full paper [9, Sections 4.3 and 4.5].

The construction uses a number of primitives, some of which were defined by BCG and some of which go back much further. Among other, it uses *secret-sharing*, which goes back to Shamir [11], a broadcast protocol due to Bracha [5], and a *circuit computation* protocol. Recall that with secret sharing, a sender can distribute a secret s (which is just an element of the field $\mathbb{F}_p$) among n agents in such a way that no subset of t agents can guess the value of s with better probability than $1/p$ (the cardinality of $\mathbb{F}_p$), but such that any subset of $t+1$ agents can compute s with no probability of error. This is done by having the sender choose a polynomial $p_s \in \mathbb{F}_p[X]$ of degree t uniformly at random such that $p_s(0) = s$ and sending each agent i i's *share* of s, which is $s_i := p_s(i)$. Bracha's broadcast protocol allows a sender to broadcast a message to a group in an asynchronous system so that, if $n > 3t$, all honest players will eventually get it.

The circuit computation protocol [2] allows agents to compute their shares of the output of a circuit f from their shares of the inputs of f without learning anything about the shares of other players.

The construction proceeds as follows. Intuitively, agents simulate $\boldsymbol{\pi} + \pi_d$ by jointly computing the mediator's state, which messages it receives, and which messages it would send to each agent. To keep the mediator's computation secret, instead of computing the mediator's state directly, agents just compute their share of it. Each agent i sends j its share of each of the messages that i sends in the protocol with the mediator; each agent j then uses these shares to update the state of the mediator. Agents use the verifiable secret sharing and circuit computation protocols provided by BCG in order to tolerate malicious behavior when distributing the shares of their messages and when computing the shares of the mediator's state every time it is updated. They also use Bracha's consensus protocol to agree on what messages the mediator receives every time it is scheduled and in which order it does so. The properties of all of these primitives are given in [5], [2], and [9, Section 3.2]

While this is the outline of the protocol, there are still a number of subtleties that have to be dealt with. For example, we must decide when each agent and the mediator is scheduled in the simulation of $\vec{\pi} + \pi_d$, or how agents update the mediator's state. All of these details are provided in the full version [9, Section 4.2].

5 Conclusion

We have shown how to simulate arbitrary protocols securely in an asynchronous setting in a "bidirectional" way (as formalized by our notion of bisimulation). This bidirectionality plays a key role in the application of these results in [1]; we believe that it might turn out to be useful in other settings as well. While this property holds for BCG's secure computation implementation, proving that we can simulate arbitrary protocols so that it holds seems to be nontrivial.

Our construction may not be message-efficient in the general case. However, for responsive mediators, a small modification (see [9, Section 4.4]) allows us to bound the expected number of messages by a function that is polynomial in the number of agents n and the maximum number of messages N sent in the setting with the mediator, and linear in c, the number of gates in a circuit that implements the mediator's protocol. It is still an open problem whether all protocols $\vec{\pi} + \pi_d$ can be implemented in a way that the expected number of messages sent by honest agents is bounded by some function of n, N, and c.

References

1. Abraham, I., Dolev, D., Geffner, I., Halpern, J.Y.: Implementing mediators with asynchronous cheap talk. In: Proceedings 38th ACM Symposium on Principles of Distributed Computing, pp. 501–510 (2019)
2. Ben-Or, M., Canetti, R., Goldreich, O.: Asynchronous secure computation. In: STOC 1993: Proceedings of the 25 Annual ACM Symposium on Theory of Computing, pp. 52–61. ACM Press, New York (1993). http://doi.acm.org/10.1145/167088.167109
3. Ben-Or, M., Goldwasser, S., Wigderson, A.: Completeness theorems for non-cryptographic fault-tolerant distributed computation. In: Proceedings of the 20th ACM Symposium on Theory of Computing, pp. 1–10 (1988)

4. Ben-Or, M., Kelmer, B., Rabin, T.: Asynchronous secure computations with optimal resilience (extended abstract). In: Proceedings of the 13th ACM Symposium Principles of Distributed Computing, pp. 183–192. ACM Press, New York (1994). http://doi.acm.org/10.1145/197917.198088
5. Bracha, G.: An asynchronous $[(n-1)/3]$-resilient consensus protocol. In: Proceedings of the 3rd ACM Symposium on Principles of Distributed Computing, pp. 154–162 (1984)
6. Canetti, R.: Studies in secure multiparty computation and applications. Ph.D. thesis, Technion (1996). http://citeseer.nj.nec.com/canetti95studies.html
7. Canetti, R.: Universally composable security: a new paradigm for cryptographic protocols. In: Proceedings of the 42nd IEEE Symposium Foundations of Computer Science, p. 136 (2001)
8. Chaum, D., Crépeau, C., Damgård, I.: Multi-party unconditionally secure protocols. In: Proceedings of the 20th ACM Symposium on Theory of Computing, pp. 11–19 (1988)
9. Geffner, I., Halpern, J.Y.: Security in asynchronous interactive systems. https://arxiv.org/abs/1906.02069
10. Kushilevitz, E., Lindell, Y., Rabin, T.: Information-theoretically secure protocols and security under composition. SIAM J. Comput. **39**(5), 2090–2112 (2010)
11. Shamir, A.: How to share a secret. Commun. ACM **22**, 612–613 (1979)

A New Problem Setting for Mobile Robots Based on Backscatter-Based Communication and Sensing

Teruo Higashino[1,2(✉)], Akira Uchiyama[2], Hirozumi Yamaguchi[2], Shunsuke Saruwatari[2], Takashi Watanabe[2], and Toshimitsu Masuzawa[2]

[1] Kyoto Tachibana University, Kyoto 607-8175, Japan
higashino-t@tachibana-u.ac.jp
[2] Osaka University, Osaka 565-0871, Japan
{higashino,uchiyama,h-yamagu,saru,watanabe,masuzawa}@ist.osaka-u.ac.jp

Abstract. In this paper, we introduce a new problem setting for mobile robots based on backscatter-based communication and sensing. *Ambient backscatter communication* is a technology that transmits/receives data only by switching the impedance of the antenna at high speed without creating a carrier wave on the transmitting side (target backscatter tag). It modulates and transmits data by turning on/off radio waves and reflecting/absorbing radio waves such as Wi-Fi existing in the environment. Data transmission and backscatter tags' sensing can be done with several tens of μW of power consumption while general Wi-Fi communication requires several tens of mW of power consumption. We have developed a software defined radio (SDR) system for backscatter-based communication and advanced sensing of humans and objects. By equipping each SDR system with multiple antennas and implementing a mechanism to estimate the direction of backscatter communication with high accuracy, and by using multiple SDR systems, our SDR systems can not only transmit/receive data of ambient backscatter communication but also analyze signals obtained from backscatter tags and estimate their positions concurrently with an error of a few centimeters. Computation by a swarm of autonomous *mobile robots* is one of the most active fields in the distributed computing community. By arranging the multiple SDR systems in a target area, it may be possible to give new environmental conditions to the time-series position estimation of mobile robots. Such environmental conditions can be used for finding a new problem setting for the mobile robots and/or context recognition of humans and objects using mobile robots and backscatter-based communication.

Keywords: Mobile robots · Ambient backscatter · Software defined radio

C. Johnen et al. (Eds.): SSS 2021, LNCS 13046, pp. 141–153, 2021.
https://doi.org/10.1007/978-3-030-91081-5_10

1 Introduction

Computation by a swarm of autonomous *mobile robots* [1,2] is one of the most active fields in the distributed computing community. There also exist commercial mobile robots such as Kilobot [3]. In order to reduce the energy consumption of mobile robots, a model without memory nor communication function is set as a basic model of the mobile robots. There are several variations for the visibility and compasses of the mobile robots. Here, we focus on ultra-low power wireless communication technology called *ambient backscatter communication* and introduce a new problem setting for mobile robots based on backscatter-based communication and sensing.

Generally, IoT devices consume power for three processes: sensing, process, and communication, but the power required for communication is extremely high, and the key technology for connecting to the Internet of IoT devices is the spread of ultra-low power consumption communication methods. In recent years, a wireless communication technology called *ambient backscatter communication*, which can reduce power consumption to about 1/10,000 (about 10 μW) of the conventional one by using radio waves existing in the environment such as Wi-Fi, is being developed [4,5]. Ambient backscatter communication does not generate any radio waves on the transmitting side (target backscatter tag). It reflects or absorbs radio waves such as Wi-Fi existing in the environment by switching the impedance of the antenna and increases or decreases their signal strengths. It sends 0/1 bits by utilizing the fact that the signal strengths can be increased or decreased. In general, many existing sensing devices are powered by button batteries and so on, but due to the time and efforts of battery replacement and the complexity of voltage control circuits, many IoT devices cannot be connected to the Internet. If we can develop a mechanism to sense the strength (fluctuation) of wireless radio waves by ambient backscatter communication, it is one of the promising ways to popularize battery-less and maintenance-free small IoT devices for sensing the movement and stop of humans and objects, seating and standing, temperature change, shape change, and so on.

Wi-Fi based backscatter communication technology capable of transmitting and receiving data at the distance of several tens of meters at several Mbps and RFID communication technology capable of transmitting and receiving data from the distance of several meters are being developed [6]. In addition, sensing elements that use only the electric power obtained from energy harvesting and sensing technologies for grasping human behavior have been devised. However, most of the existing sensing technologies based on ambient backscatter communication are limited to the development of relatively simple situational awareness technologies for humans and objects, such as the existence of a person and the presence or absence of movement at the target location [3]. It is less powerful than the situational awareness technology assuming general Wi-Fi and enough power supply. At present, we still need intellectual situation recognition technology for humans and objects using battery-less and maintenance-free small IoT devices that can be applied to practical purpose.

For solving this problem, we have been developing a software defined radio (SDR) system that can send and receive information obtained by ambient backscatter communication as an existing IEEE 802.11 compatible frame (a wireless access point that mediates between regular Wi-Fi and backscatter). In the developed SDR system, by applying reflection, absorption and modulation to target backscatter devices at different frequencies Δf, the IDs of those backscatter devices can be known even in an environment where multiple backscatter devices coexist. By equipping each SDR system with multiple antennas and implementing a mechanism to estimate the direction of ambient backscatter communication with high accuracy, our SDR systems can estimate their positions concurrently with an error of a few centimeters. Using these SDR systems, we can recognize the positions of multiple persons in a target area and estimate the trajectories of multiple backscatter tags concurrently.

In the developed SDR systems, all the mobile robots in their wireless ranges can know their positions concurrently. By giving ultra-low power communication functions to the mobile robots, a model that knows only its own position can be considered, or a model that can know the positions of a part of nearby mobile robots can be considered. Different assumptions for ID recognition of nearby mobile robots can be also considered. Since ambient backscatter communication can be used for the ultra-low power context recognition of humans and objects, several types of problems for context recognition of humans and objects using mobile robots can be also considered. We hope the developed SDR systems can be used for finding a new problem setting for the mobile robots and/or creating ultra-low power methods for context recognition of humans and objects using the mobile robots.

In this paper, Sect. 2 introduces the recent research trends for ambient backscatter communication and the context recognition of humans and objects using ambient backscatter communication. In Sect. 3, we introduce an overview of the developed software defined radio (SDR) systems. In Sect. 4, we introduce our recent research work for the context recognition of humans and objects using the SDR systems. In Sect. 5, we present the basic model of the mobile robots and discuss how our SDR systems can be used for finding a new problem setting for the mobile robots and/or context recognition of humans and objects. Section 6 concludes the paper.

2 Ambient Backscatter Communication and Sensing

2.1 Ambient Backscatter Communication

Backscatter communication is a technology that can reduce the power consumption of wireless transmission to 1/1000–1/10000. Generally, wireless communication consumes significant power. For example, the Silex SX-SDCAG 802.11a/b/g SDIO card module for Wi-Fi consumes 525 mW for the transmission and 195 mW for the reception, and the Texas Instruments CC2530 for IEEE 802.15.4 (ZigBee) consumes 58 mW for the transmission and 48 mW for the reception. Fig. 1(a) shows how an IEEE 802.15.4 transmitter works: the baseband signal, generated as a 2 MHz signal, is multiplied by a 2.4 GHz carrier signal generated by a local oscillator (LO) using a mixer and converted to a 2.402 GHz modulated signal. The modulated signal is then amplified by a power amplifier (PA) and emitted as radio waves into space. The transmitter consumes a large amount of power in the LO that generates the 2.4 GHz carrier signal and the PA that amplifies the transmitted signal.

Backscatter communication drastically reduces the transmission power by transmitting data without using the LO and PA in Fig. 1(a). The basic idea of backscatter communication is similar to that of the heliograph. Fig. 1(b) shows an example of a heliograph. When sending information from the ground to an airplane, information such as Morse code is sent by reflecting the sunlight with a mirror. At this time, the only energy consumed by the transmitter is the energy to move the mirror. Backscatter communication is the realization of heliograph by radio waves. Fig. 1(c) shows the circuit diagram of backscatter communication. In backscatter communication, data is transmitted by superimposing a signal on the flying radio wave by switching between reflecting and absorbing the radio wave by changing the impedance at high speed. This backscatter mechanism eliminates the need for a LO and PA in a typical radio transmitter and thus significantly reduces power consumption.

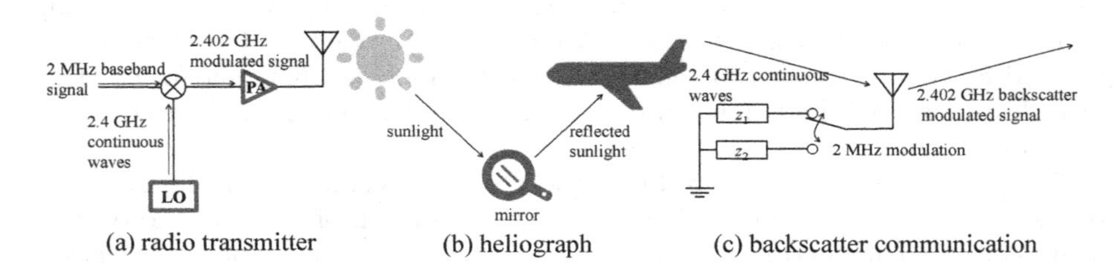

Fig. 1. The basis of backscatter communication.

As shown in Fig. 2, there are three types of communication models for backscatter communication [5]. The first is monostatic backscatter, which has a long history and corresponds to RFID communication [7]. The second is bistatic backscatter [8, 9, 10]. Bistatic backscatter is a model that uses carrier signals emitted by a carrier emitter such as a wireless LAN AP. Bistatic backscatter assumes a clean carrier signal such as continuous waves. The use of the clean carrier signal enables its flexibility in modulation, such as creating IEEE 802.11 frames [8] and IEEE 802.15.4 frames [9], and its more extended transmission range than monostatic backscatter and ambient backscatter. By combining one carrier emitter and multiple backscatter receivers, multiple backscatter communications can be realized simultaneously in single continuous waves [10]. The third model is ambient backscatter [4,11]. While monostatic backscatter and bistatic backscatter use continuous waves for backscatter communication, the ambient backscatter shares the same signals used in other communications such as TV [4] and wireless LAN [11]. Although the communication distance is lower than that of monostatic backscatter and bistatic backscatter, it has the advantage that backscatter communication can be performed without a carrier emitter.

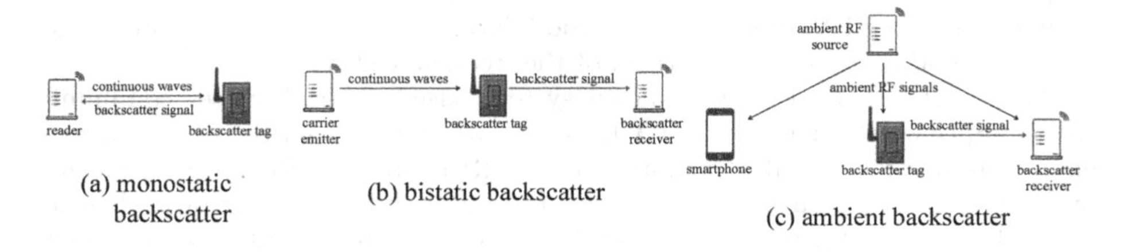

Fig. 2. Three types of backscatter communication.

2.2 Backscatter Sensing of Humans and Objects

As we reviewed in Sect. 2.1, backscatter is a key enabler for ultra-low-power wireless communication. Nevertheless, it still requires digital modulation for communication. This means it needs processing by computation modules such as microcontroller units. This leads to additional costs for additional components as well as additional energy consumption, which can limit the application design space. However, interestingly, backscatter is also useful for sensing by directly observing the change of the signal backscattered from tags. For making signal changes, backscatter sensing leverages analog sensors or physical movements owing to contexts such as wind, water flow, acoustic vibration, and human interaction (e.g., touch).

As for battery-less approaches, Printed Wi-Fi [12] is an interesting concept which converts physical movement owing to various contexts into the fluctuation of Wi-Fi signal without any IC chip and battery. The authors presented prototypes of a wind speed meter, liquid flow meter, slider bar, and so on. Similarly, LiveTag [13] is a printed tag with antennas and resonators which backscatters Wi-Fi signal with absorption of specific frequency. Touching the printed resonator cancels the corresponding frequency absorption, which is used for battery-less touchpads.

On the other hand, researchers have also investigated another research direction using ultra-low-power backscatter tags. Because of its ultra-low-power nature, we can deploy a large number of backscatter tags. This enables us to observe fluctuation of signal backscattered from the tags due to human motion in the proximity of the tags. BARNET [14] proposed Backscatter Channel State Information (BCSI) between backscatter tags to obtain activity-related signal change information, increasing the number of wireless links for sensing. Furthermore, we can directly attach the tags to the targets (i.e., humans and objects) for sensing of their states such as movement. RF Bandaid [15] takes such an approach for ultra-low-power sensing of various phenomena such as temperature, force, respiration, and heart rate. RF Bandaid employs a micropower precision programmable oscillator from Linear Technology LTC6906 which is a key component for converting the resistance or capacitance of analog sensors into a frequency shift in the backscattered signal. UbiquiTouch [16] designs a touchpad by backscattering FM radio signal for modulating a touch point on a surface to its corresponding time-series pattern of the frequency shift.

Meanwhile, RFID is a well-known system based on backscatter communication. Specifically, passive RFID tags operate without battery by charging energy from radio signal transmitted by an RFID reader. Receiving Continuous Wave (CW) signal from the reader, the tags send back their own identification by backscatter communication. RFID readers provide information about replied tags' identification with Received Signal Strength (RSS) and phase of the backscattered signal. Especially the phase information is useful for measurement of distance change between a tag and a reader. Also, RFID readers and tags are available on the market. Therefore, many researchers have proposed a wide range of applications such as respiration monitoring [17], heart rate variability sensing [18], pose estimation [19], and so on. In [20], we have also proposed some backscatter communication-based sensing methods.

Different from RFID-based sensing, backscatter sensing further enhances the capability of wireless sensing by directly converting context of humans and objects into ambient RF signal change such as Wi-Fi and BLE. This is achieved by designing backscatter tags for changing antenna impedance according to target context.

3 Developed Software Defined Radio (SDR) System

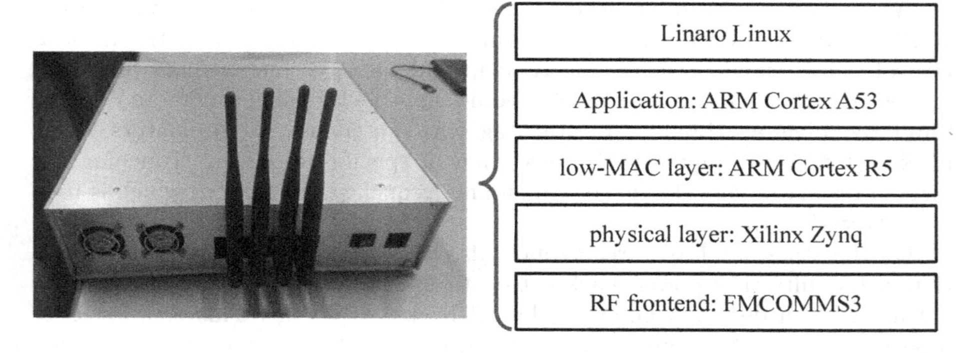

Fig. 3. Developed software defined radio (SDR) system called SD-WiFi.

Figure 3 shows the overview of the developed software defined radio (SDR) system called "SD-WiFi". In SD-WiFi, we have developed a protocol stack that allows existing communication protocols and backscatter communication to coexist in the MAC layer so that the obtained information in a backscatter tag can be communicated in backscatter as existing IEEE 802.11 compatible frames or IEEE 802.15.4 frames. We implemented the frontend using FMCOMMS3, the physical layer using FPGA, the MAC layer using ARM Coretex R5, a real-time CPU, and the data link layer to the application layer implemented using Coretex A53. The MAC layer can be programmed exclusively by the Realtime CPU, which simplifies the implementation of the MAC protocol. The Coretex A53 runs Debian-based Linaro Linux, which makes it easy to install various software and services using apt commands.

Furthermore, receiving the same signal with two antennas makes it possible to obtain the phase difference. From both antennas' obtained radio wave strength, it becomes possible to know in real-time from which direction and how strong the radio wave is being transmitted. By installing several of the developed software defined radio (SDR) systems in the target area and estimating the direction of radio wave transmission from the radio wave strength information of each antenna, the location of the backscatter device can be determined with high accuracy.

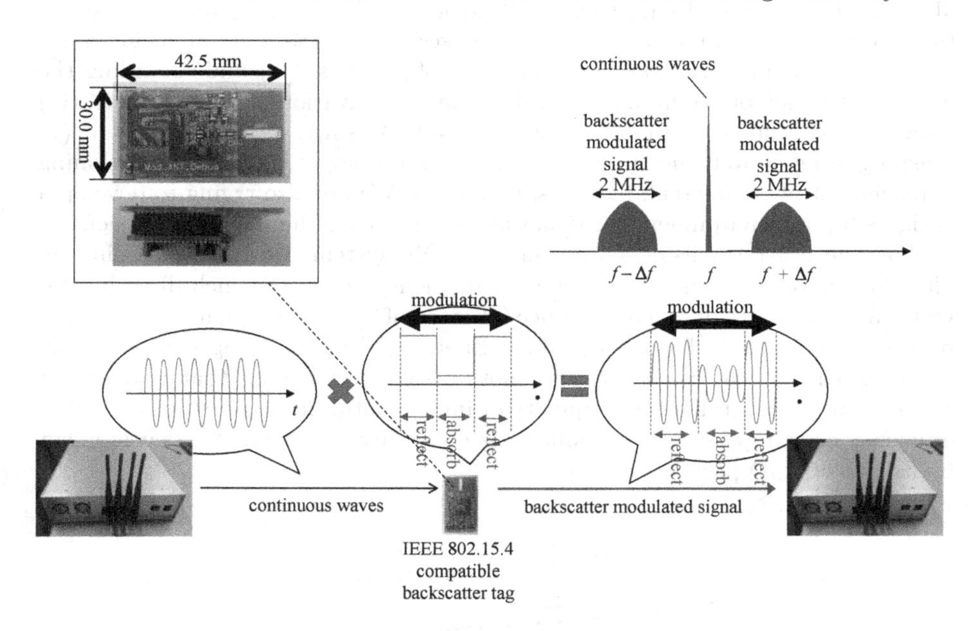

Fig. 4. SD-WiFi and IEEE 802.15.4 compatible backscatter tag.

Figure 4 shows an example of the operation of SD-WiFi in combination with an IEEE 802.15.4 compatible backscatter tag. The IEEE 802.15.4 compatible backscatter tag consists of two parts: the MCU and the RF switch. We used STMicroelectronics STM32F446RE as an MCU and single-pole double-throw (SPDT) switch as an RF switch. Reference [9] shows the communication range was 3 m, but later improvements were made to enable communication at 15 m. It has also been confirmed that the IEEE 802.15.4 chip, Texas Instruments CC1352, can receive the backscatter modulated signal. All these circuit data are available to GitHub [21,22], so you can run it by yourself. When continuous waves of frequency f are switched at a rate of Δf at an IEEE 802.15.4 compatible backscatter tag, the backscattered carrier signals appear at $f + \Delta f$ and $f - \Delta f$. In Fig. 4, two frames with $f + \Delta f$ and $f - \Delta f$ as center frequencies are generated by backscatter modulating the 2 MHz baseband signal with IEEE 802.15.4 modulation in the backscatter tag.

4 Context Recognition of Humans and Objects Using SDR System

We have been designing and developing battery-less and maintenance-free IoT sensing devices (human sensor, accelerometer, camera, temperature) that can be applied to the situation recognition of humans and objects by using backscatter communication technology and electronic circuit design technology using 3D printers together. Using those IoT devices, we are trying to create situation recognition technologies that can be used for (i) watching at facilities for the elderly, (ii) understanding the activities of athletes, (iii) understanding the movement trajectory of humans and detecting the invasion of wild animals, (iv) building sociograms for understanding the relationships between children, (v) grasping wind power and ground fluctuations on slopes, (vi) air conditioning management of commercial facilities, and so on. We are also trying to develop a design support environment for situation recognition of humans and objects.

For backscatter sensing based on the SDR system, we are developing an ultra-low-power backscatter tag called a frequency shift tag which directly converts target context into the existence/absence of specific frequency shift in the backscattered signal of Bluetooth Low Energy (BLE). The tag consists of an antenna, RF switch, oscillator, and motion switch. We note that it does not require a microcontroller for simplicity of implementation and further reduction of energy consumption. The simplicity is especially important for development by users using 3D printers, and so on.

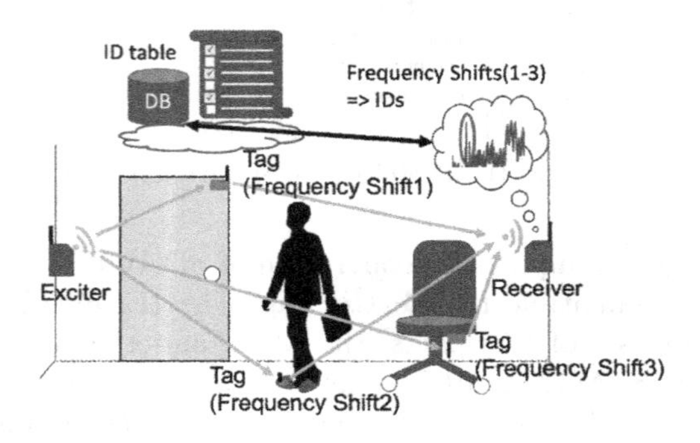

Fig. 5. Backscatter sensing by frequency shift tags.

Figure 5 illustrates the overview of backscatter sensing by the frequency shift tags. An exciter emits a carrier wave signal for the tags attached to humans and objects such as shoes, doors, and chairs. A receiver then observes a frequency spectrum to detect the existence of the tags and the state of the motion switches coupled with corresponding contexts. We use our SDR system as exciters and receivers.

If the backscatter tag exists in the target environment with its motion switch turned on, the frequency shift corresponding to the frequency of the carrier signal f_c appears in the backscattered signal. We assume a pair of the oscillation frequency f_i of tag$_i$ and the corresponding context c_i of the tag is registered in a database. The receiver detects peaks in the frequency spectrum and determines whether there is a frequency shift f_i or not. If the peak at f_i is detected, we recognize the occurrence of c_i.

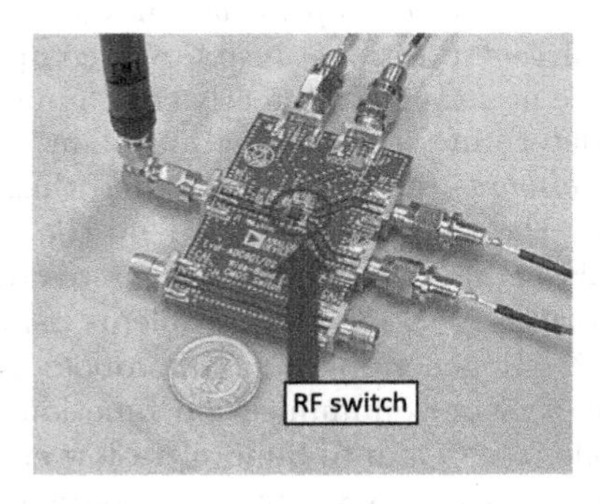

Fig. 6. Frequency shift tag prototype.

We have implemented a prototype of the frequency shift tag as shown in Fig. 6. Currently, we have confirmed that the backscatter sensing is achievable within 3 m from an exciter and a receiver by using pseudo continuous waves created by BLE Interscatter [23] as the backscatter signal source. In our preliminary experiment, we used BLE Interscatter for the experiment in the real environment (i.e., outside an RF shielded tent). We note that we are working on obtaining Technical Standards Compliance Certification to use our SDR system in the real environment. Also, the energy consumption of the frequency shift tag is 19.8 μW at 1.8 V of the operating voltage. This means the lifetime of the tag is more than 2 years with a CR2032 coin cell without any sleep mode. Therefore, we can usually expect a longer lifetime since the energy is intermittently supplied according to the state of the motion switch. By using the prototype, we have implemented a seating sensor, pedometer, and door sensor as concept applications. To further enhance the sensing range, we are planning to leverage computation power and energy source on the infrastructure side (i.e., SDR AP).

5 Backscatter-Based Communication and a New Problem Setting for Mobile Robots

Computation by a swarm of autonomous *mobile robots* is one of the most active fields in the distributed computing community. Following the seminal work by Suzuki and Yamashita [1], computability and complexity for several problems have been extensively executed. Many fundamental results are surveyed in [2].

Here, we introduce commonly used models of mobile robots in a continuous two-dimensional space. The robots are *anonymous*; they execute the same algorithm, and they cannot be distinguished by their appearances. Each robot has its own x-y coordinate system, called a *compass*, with the origin that is the current position of the robot. The compasses of different robots may be different in direction of the x-axis, unit length, and chirality. Each robot operates in *Look-Compute-Move (LCM)* cycles, each consisting of a Look, a Compute, and a Move operation in this order. A *Look operation* allows the robot to take a snapshot consisting of the current positions, in its own compass, of all the robots. Note that the robots have no means for directly communicating with each other. A *Compute operation* determines the destination point of the robot in its own compass. The robot is *oblivious* (or *memoryless*) and thus the destination point is determined only from the snapshot obtained by the Look operation in the same LCM cycle. In a *Move operation*, the robot moves to the destination.

Several types of distributed cooperated problems using mobile robots have been discussed so far [1,2]. In general, the mobile robots assume that each robot does not have any memory nor communication function. This means that the researchers for the mobile robots aim to build ultra-low energy distributed cooperated systems. As we discussed above, wireless communication requires relatively a large amount of energy. On the other hand, ambient backscatter communication requires relatively a very small amount of energy. The developed SDR system can also provide a rather accurate location for each mobile robot within the communication range of each SDR system.

We can develop mobile robots with a backscatter communication function where each mobile robot can sense a specific frequency shift for the backscattered signal. If we assume that each mobile robot has its own specific frequency for the frequency shift, each mobile robot can only know its position with high accuracy at regular intervals while it cannot know the positions of its neighboring mobile robots. On the other hand, if multiple mobile robots have a common specific frequency for the frequency shift and if those mobile robots ask their own locations to a nearby SDR system asynchronously, those mobile robots can know their positions in parallel although packet collisions might occur. In this case, those mobile robots can know the locations and IDs of all the neighboring mobile robots which use the same frequency for the frequency shift. Since each SDR system has its own communication range and a large number of mobile robots cannot have the common frequency for the frequency shift in real situations, each mobile robot can only know the locations and IDs of a small part of its neighboring mobile robots at different times. Also, the set of such neighboring mobile robots is not fixed. Since the range of the radio is not so great, we need multiple SDR systems for covering a target area in general. It is not always possible for each mobile robot to communicate with at least one of those SDR systems. In such communication environments, we might be able to give different environmental conditions for the mobile robots and find different distributed cooperative problems.

In the backscatter, by detecting the directions of the radio waves from two base stations (SDR systems), the exact position of a target backscatter tag is known at the intersection of the two directions of the radio waves under the assumption that we know the exact position of two base stations (SDR systems). For two base stations, it is relatively easy to measure the distance between them while it requires some effort to know their absolute positions (e.g., usage of accurate GNSS systems, anchors, and accurate compasses, and so on). The problem of finding the exact positions of a group of a large number of base stations when arranging them is not so easy, but it is possible to find their rough positions. There exist some iterative closest point (ICP) algorithms that calculate adequate positions for multiple nodes (e.g., [24]). Actually, the following assumptions are adequate: for each pair of two base stations, the distance between them is known, their positions and directions are not exactly known, and it is possible to use the premise that each mobile robot operates in an environment where there are multiple sets of such base stations.

In such environments, we can assume that robots that can use the same base stations have almost (but not exactly) common x-y coordinate systems. Moreover, the robots can get the positions of each other when they use the common frequency. These assumptions enhance the model of mobile robots. It is interesting to investigate the impact of the assumptions on the complexity or solvability of problems. In case that multiple frequencies are available, each robot can change the robots that it can recognize by changing the frequency it uses. In such an environment, there might be a problem in which a kind of scheduling of how to dynamically change the frequency is the key. Also, if this scheduling is made probabilistic, groups can be constructed with appropriate sampling, and it may be possible to come up with an efficient solution for some problems.

As we explained in Sect. 4, we can create several types of techniques for context recognition of humans and objects based on ambient backscatter communication. Thus, there might exist problems for finding or surrounding backscatter tags with specific features using autonomous mobile robots. Also, there might exist problems for collecting the locations and/or IDs of backscatter tags with specific features using autonomous mobile robots.

6 Conclusion

In this paper, we introduced a new problem setting of the mobile robots based on backscatter-based communication and sensing. By equipping each SDR system with multiple antennas and implementing a mechanism to estimate the direction of backscatter communication with high accuracy, and by using multiple

SDR systems, our SDR systems can not only transmit/receive data of ambient backscatter communication but also analyze signals obtained from backscatter tags and estimate their positions concurrently with an error of a few centimeters. For a swarm of autonomous mobile robots, it may be possible to give new environmental conditions to the time-series position estimation of mobile robots as we discussed in Sect. 5. Since the mobile robots assume ultra-low-power consumption for moving and sensing, ambient backscatter communication-based environments might be able to provide new types of problem settings for a swarm of autonomous mobile robots. We hope our research could give some hints to the distributed computing community.

Acknowledgements. The research in this paper is partly supported by JSPS Grants-in-Aid for Scientific Research (Grant Numbers: 19H05665, 20K20398, 19K11941, 18H03231, and 19H01101). The research is also partly supported by JST PRESTO, Japan (Grant Numbers: JPMJPR1932 and JPMJPR2032).

References

1. Suzuki, I., Yamashita, M.: Distributed anonymous mobile robots. In: 3rd International Colloquium on Structural Information and Communication Complexity (SIROCCO 1996), pp. 313–330 (1996)
2. Flocchini, P., Prencipe, G., Santoro, N. (eds.): Distributed Computing by Mobile Entities, Current Research in Moving and Computing. Theoretical Computer Science and General Issues, vol. 11340. Springer, Heidelberg (2019). https://doi.org/10.1007/978-3-030-11072-7
3. Kilobot: https://www.k-team.com/mobile-robotics-products/kilobot. Accessed 08 Sept 2021
4. Liu, V., Parks, A., Talla, V., Gollakota, S., Wetherall, D., Smith, J.R.: Ambient backscatter: wireless communication out of thin air. SIGCOMM Comput. Commun. Rev. **43**(4), 39–50 (2013)
5. Huynh, N.V., Hoang, D.T., Lu, X., Niyato, D., Wang, P., Kim, D.I.: Ambient backscatter communications: a contemporary survey. IEEE Commun. Surv. Tutor. **20**(4), 2889–2921 (2018)
6. Uchiyama, A., Saruwatari, S., Maekawa, T., Ohara, K., Higashino, T.: Context recognition by wireless sensing: a comprehensive survey. J. Inf. Process. (JIP) **29**, 46–57 (2021)
7. Want, R.: An introduction to RFID technology. IEEE Pervasive Comput. **5**(1), 25–33 (2006)
8. Kellogg, B., Talla, V., Gollakota, S., Smith, J. R.: Passive Wi-Fi: bringing low power to Wi-Fi transmissions. In: 13th USENIX Symposium on Networked Systems Design and Implementation (USENIX NSDI 2016), pp. 151–164 (2016)
9. Konishi, Y., Ueda, T., Kizaki, K., Fujihashi, T., Saruwatari, S., Watanabe, T.: Experimental evaluation on IEEE 802.15.4 compatible backscatter. In: 2020 IEEE Global Communications Conference (IEEE GLOBECOM 2020), pp. 1–6 (2020)
10. Zeba, O., Saruwatari, S., Watanabe, T.: QuadScatter for simultaneous transmissions in a large-scale backscatter network. In: 2020 IEEE International Conference on Communications (IEEE ICC 2020), pp. 1–6 (2020)

11. Kellogg, B., Parks, A., Gollakota, S., Smith, J.R., Wetherall, D.: Wi-Fi backscatter: internet connectivity for RF-powered devices. In: 2014 ACM Conference on SIGCOMM (ACM SIGCOMM 2014), pp. 607–618 (2014)
12. Iyer, V., Chan, J., Gollakota, S.: 3D printing wireless connected objects. ACM Trans. Graph. **36**(6), pp. 1–13 (2017). Article no. 242
13. Gao, C., Li, Y., Zhang, X.: LiveTag: sensing human-object interaction through passive chipless WiFi tags. In: 15th USENIX Symposium on Networked Systems Design and Implementation (USENIX NSDI 2018), pp. 533–546 (2018)
14. Ryoo, J., Karimi, Y., Athalye, A., Stanaćević, M., Das, S.R., Djurić, P.: BARNET: towards activity recognition using passive backscattering tag-to-tag network. In: 16th ACM International Conference on Mobile Systems, Applications, and Services (MobiSys 2018), pp. 414–427 (2018)
15. Ranganathan, V., Gupta, S., Lester, J., Smith, J.R., Tan, D.: RF Bandaid: a fully-analog and passive wireless interface for wearable sensors. Proc. ACM Interact. Mob. Wearable Ubiquitous Technol. **2**(2), 1–21 (2018). Article no. 79
16. Waghmare, A., et al.: UbiquiTouch: self sustaining ubiquitous touch interfaces. Proc. ACM Interact. Mob. Wearable Ubiquitous Technol. **4**(1), 1–22 (2020). Article no. 27
17. Yang, Y., Cao, J.: Robust RFID-based respiration monitoring in dynamic environments. In: 17th Annual IEEE International Conference on Sensing, Communication and Networking (SECON 2020), pp. 1–9 (2020)
18. Wang, C., Xie, L., Wang, W., Chen, Y., Bu, Y., Lu, S.: RF-ECG: heart rate variability assessment based on COTS RFID tag array. Proc. ACM Interact. Mob. Wearable Ubiquitous Technol. **2**(2), 1–26 (2018). Article no. 85
19. Jin, H., Yang, Z., Kumar, S., Hong, J.I.: Towards wearable everyday body-frame tracking using passive RFIDs. Proc. ACM Interact. Mob. Wearable Ubiquitous Technol. **1**(4), 1–23 (2018). Article no. 145
20. Higashino, T., Uchiyama, A., Saruwatari, S., Yamaguchi, H., Watanabe, T.: Context recognition of humans and objects by distributed zero-energy IoT devices. In: 2019 IEEE 39th International Conference on Distributed Computing Systems (ICDCS 2019), pp. 1787–1796 (2019)
21. IEEE 802.15.4 Compatible Backscatter Tag. https://github.com/watalabo/backscatter-15.4. Accessed 08 Sept 2021
22. 4 GHz Backscatter Communication Tools. https://github.com/watalabo/backscatter-tools. Accessed 08 Sept 2021
23. Iyer, V., Talla, V., Kellogg, B., Gollakota, S., Smith, J.: Inter-technology backscatter: towards internet connectivity for implanted devices. In: 2016 ACM SIGCOMM Conference (ACM SIGCOMM 2016), pp. 356–369 (2016)
24. Yoshisada, H., Yamada, Y., Hiromori, A., Yamaguchi, H., Higashino, T.: Indoor map generation from multiple LIDAR point clouds. In: 2018 IEEE International Conference on Smart Computing (SMARTCOMP 2018), pp. 73–80 (2018)

Round-Oblivious Stabilizing Consensus in Dynamic Networks

Manfred Schwarz[1] and Ulrich Schmid[2(✉)]

[1] Schwarz Energietechnik and Geoinformatik, Mils, Austria
[2] TU Wien, Vienna, Austria
s@ecs.tuwien.ac.at

Abstract. In this paper, we study deterministic consensus in practical directed dynamic networks. We show that it is straightforward to remove the dependence on round numbers even for quite complex algorithms like the optimal terminating consensus algorithm for the short-lived vertex-stable root components message adversary (Winkler et al., Distributed Computing, 2019). The resulting algorithm is inherently resilient against synchronization-related errors, and can be proved correct by a simple simulation equivalence. Moreover, our approach naturally leads to a novel optimal solution for stabilizing consensus for this message adversary. Finally, we negatively answer the question of whether such algorithms could benefit from a stronger communication model, where senders get immediate notification of successful transmission. The typical strive for implementing low-level bidirectional connectivity, or implicit acknowledgments as in the JAG protocol (Boano et al., RTSS'12), in practical implementations hence appears to be an overkill for solving consensus.

1 Introduction

We consider variants of deterministic consensus in dynamic networks, where a potentially unknown number $n \leqslant N$ of fault-free[1] processes with unique ids that may possibly join the system at different times communicate over unreliable point-to-point links. Consensus, a pivotal service in truly distributed applications, is the problem of irrevocably computing a common output value based on local input values of all the processes. Stabilizing consensus is the weaker problem of computing an eventually stabilizing common output value, without the need to irrevocably decide on a value once and forever, however.

An execution of a distributed algorithm in our system proceeds in a sequence of lock-step synchronous rounds, where message loss is controlled by an omniscient *message adversary* [1] that determines the directed *communication graph*

[1] Nevertheless, a crash of a process p in a round could easily be modelled by p sending no messages in a later round.

M. Schwarz—Supported by the Austrian Science Fund (FWF) under project ADynNet (P28182).

C. Johnen et al. (Eds.): SSS 2021, LNCS 13046, pp. 154–172, 2021.
https://doi.org/10.1007/978-3-030-91081-5_11

$\mathcal{G}^r$ for each round r. A directed edge $(p \to q)$ present in $\mathcal{G}^r$ means that the message sent by p in round r is successfully received by q in the same round.

In most work in this area, e.g. [1,11,20,21], the message adversary is considered *oblivious*, i.e., may choose each $\mathcal{G}^r$ from a fixed *set* of admissible graphs arbitrarily in each round. For instance, the classic result from Santoro and Widmayer [20] states that consensus is impossible if the set of admissible graphs comprises all graphs that suppress no more than $n-1$ messages in every round.

By contrast, we explore consensus solvability under message adversaries that support *eventual stabilization* [3–5,22,24]: Here, the set of admissible choices for $\mathcal{G}^r$ may change with evolving round numbers r. Rather than constraining the set of admissible graphs, we hence constrain admissible *graph sequences* here. For example, the eventual vertex-stable root message adversary from [24] generates graph sequences consisting of graphs that are single-rooted, i.e., have exactly one root component (a strongly connected component without incoming edges from outside of the component), and contain a subsequence of $x = D + 1$ consecutive graphs whose root component is formed by the same set of nodes ("vertex-stable root component"). Herein, the dynamic diameter $D \leqslant n - 1$ is the number of rounds needed by the members of the root component to reach all processes in the system. In [24], Winkler et al. proved that it is impossible to solve consensus for $x = D$, and provided an optimal algorithm for $x = D + 1$. Note that it is impossible to solve consensus under an oblivious message adversary that may choose any of the graphs generated by the eventual vertex-stable root message adversary [11,20].

Since the algorithm of [24], like the alternative algorithms [3–5,22] we are aware of, send messages containing round numbers and trigger actions in specific rounds, they are obviously vulnerable w.r.t. synchronization errors: Messages containing a corrupted round number, not matching the receiver's current round counter, might be delivered in the wrong round. Unfortunately, unlike the simultaneous round switching itself, which can be implemented robustly, e.g., based on low-level pulse synchronization mechanisms [23] and even self-stabilizing protocols [12,14], transiently erroneous round counters and, hence, round numbers cannot easily be avoided in clock synchronization protocols like the popular IEEE 1588 or FTSP [19]. Hence, the question arises whether and how consensus algorithms for dynamic networks can be made *round-oblivious*, i.e., resilient against such faults.

In addition, most of the existing work on consensus in directed dynamic networks under message adversaries assumes that all processes start their execution simultaneously in round $r = 1$. This is inherently not required by stabilizing consensus algorithms, like the strikingly simple *MinMax* algorithms proposed in [9], however, which does not even need unique ids. Whereas the MinMax algorithm has been shown to work correctly under message adversaries that guarantee rootedness with bounded delay[2] in every round, it does not always work in graph sequences where multiple root components can occur.

Another question, which has also been inspired by round obliviousness, is whether algorithms for directed dynamic networks would benefit from a somewhat stronger communication model: There are network protocols that provide the sender with immediate feedback, i.e., in the same round, on whether a sent message has been successfully delivered or not. This is obviously the case in systems that enforce bidirectional low-level communication between processes. A more lightweight alternative is the popular CSMA/CA protocol as e.g. used in the CANbus [15], where one can reasonably infer successful reception from the absence of a collision in the packet header. Another interesting example is the JAG protocol for wireless sensor networks [7], which has been explicitly designed for guaranteeing sender-receiver agreement on messages by using jamming in some clever way.

Obviously, if the sender of a message knows whether it has been received or not, the lossy-link impossibility of consensus among 2 processes [20] disappears immediately. The question remains, however, whether consensus among more than 2 processes would also benefit from this feature in general.

Contributions: In this paper, we show how to convert the optimal consensus algorithm for the vertex-stable root message adversary $\lozenge\mathsf{STABLE}_{\leqslant N,D}(D+1)$ from [24] into a round-oblivious algorithm, and provide a simple simulation-based correctness proof. The resulting algorithm, unlike the original one, hence tolerates an arbitrary number of processes with erroneous round clocks, as long as the lock-step synchrony of the round switching itself remains intact. Furthermore, we prove that immediate acknowledgments do not allow reducing the required stability interval of $D + 1$ rounds if at least 3 processes are present in the system. Consensus does hence not benefit from bidirectional low-level communication in our setting.

Moreover, for a considerably stronger message adversary, which is also too strong for applying the MinMax algorithm [9], we develop an optimal stabilizing consensus algorithm and prove its optimality using a matching impossibility result. Note that our algorithm and MinMax differ fundamentally, in the sense that the former (resp. latter) work under message adversaries that relax a safety (resp. a liveness) property in $\lozenge\mathsf{STABLE}_{\leqslant N,D}(D+1)$.

Paper Organization: After providing some additional related work in Sect. 2, we present our system model in Sect. 3 and a set of basic message adversaries in Sect. 4. Section 5 provides a round-oblivious variant of the optimal consensus algorithm from [24], along with the centerpiece of a simulation-based correctness proof. In Sect. 6, we add immediate acknowledgments to our basic communication model and show that this does not provide an advantage in the worst case. In Sect. 7, we provide the message adversary $\lozenge\mathsf{WEAKSTAB}_{\leqslant N}(D+1, D)$ and a

[2] Essentially, rootedness with bounded delay means that at least one process must have a path to all other processes in the product graph $\mathcal{G}^t \circ \cdots \circ G^{t+T-1}$, for every $t \geqslant 1$ and some bounded T.

suitable stabilizing consensus algorithm, and prove it to be correct and optimal. Some conclusions in Sect. 8 complete our paper.

2 Additional Related Work

Research on consensus in synchronous message passing systems subject to link failures dates back at least to the seminal paper [20] by Santoro and Widmayer; generalizations have been provided in [6,8,10,11,21]. In all these papers, consensus, resp. variants thereof, are solved in systems where, in each round, a digraph is picked from a set of possible communication graphs. The term message adversary was coined by Afek and Gafni in [1].

The first instance of a stabilizing message adversary that allows solving consensus has been provided in [3], namely, the $4D + 1$-vertex-stable root message adversary. It has been improved in [22], where a consensus algorithm for stability periods of at least $2D + 1$ rounds has been provided. An interesting property of this algorithm is that the communication graphs outside the stable period need not be single-rooted, in which case it is impossible to solve consensus for a stability period of at most $2D$. If all graphs are single-rooted, a $D + 1$-vertex stable root component has been proved necessary and sufficient for solving consensus in [24].

Considerably less is known about stabilizing consensus, which seems to have been first introduced in [2]. The authors studied the problem in an asynchronous system of fully connected processes with uids, which are subject to crashes and even byzantine faults. A randomized solution has been proposed in [13]. In [9], Charron-Bost and Moran considered deterministic stabilizing consensus in directed dynamic networks controlled by message adversaries. In sharp contrast to [2] (and our work), however, they consider anonymous processes. The authors showed that the problem cannot be solved if the message adversary can generate graph sequences where the set of processes that can eventually reach all others infinitely often, i.e., can successfully broadcast infinitely often, is empty. For message adversaries that guarantee bounded rootedness in all rounds, they provided a very elegant and non-full-history algorithm called MinMax and proved it correct. This algorithm may fail, however, if the message adversary can generate graphs with multiple root components.

Regarding stronger communication models, we are not aware of any related work, except for the existing work on non-directed dynamic networks (where all links are bidirectional). A prominent example of the latter is the work on T-interval connectivity [17], which guarantees a common subgraph in the communication graphs of every T consecutive rounds. Kuhn, Oshman and Moses [18] study agreement problems in this setting, which are easy to solve here: 1-interval-connectivity, the weakest form of T-interval connectivity, corresponds to all nodes constituting a perpetually constant set of source nodes.

3 Basic System Model

We use the same system model and notations as [24], which we briefly summarize below. Our system consists of a set of fault-free processes Π with unique ids, taken from $\{1, \dots, n\}$ with $n = |\Pi| \leqslant N$, which communicate via message passing over unreliable directed point-to-point links. The processes execute a deterministic algorithm and execute in lock-step synchronous rounds $r = 1, 2, 3, \dots$, each consisting of a message exchange phase and a simultaneous computing step. We conveniently assume that all operations of round r take place strictly within time $r-1$ and time r, which results in a well-defined *configuration* C^r, consisting of the vector of local states of all processes at time r, i.e., at the end of round r, with C^0 denoting the initial configuration.

An *admissible execution* $\langle C^0, \sigma \rangle$ (also called a *run*) is uniquely determined by C^0 and a *graph sequence* σ, which is an infinite sequence $\sigma = \mathcal{G}^1, \mathcal{G}^2, \dots$ of directed round-r *communication graphs* $\mathcal{G}^r = \langle \Pi, E^r \rangle$: The round-$r$ message of p sent to q is received by q in the same round if and only if (p, q) in $\mathcal{G}^r$. Since every process p always successfully receives from itself, $(p, p) \in \mathcal{G}^r$ for every $r \geqslant 1$. The *in-neighborhood* of p in $\mathcal{G}^r$, $In_p(\mathcal{G}^r) = \{q \mid (q, p) \in \mathcal{G}^r\}$ is the set of processes from which p received a message in round r.

A contiguous subsequence σ' of σ, abbreviated $\sigma' \subseteq \sigma$, ranging from round a to round b, is denoted as $\sigma' = (\mathcal{G}^r)_{r=a}^{b}$, with $|\sigma'| = b - a + 1$. Given two graphs $\mathcal{G} = \langle V, E \rangle$, $\mathcal{G}' = \langle V, E' \rangle$ with the same vertex set V, let the *compound graph* $\mathcal{G} \circ \mathcal{G}' := \langle V, E'' \rangle$, where $(p, q) \in E''$ if and only if there exists a $p' \in V$ such that $(p, p') \in E$ and $(p', q) \in E'$. Information propagation in an execution can be crisply described by the well-known concept of the causal past:

Definition 1 Causal past [24, **Def. 3**]. *Given a sequence σ of communication graphs that contains rounds a and b, the causal past of process p from time b down to time a is $CP_p^b(a) = \emptyset$ if $a \geqslant b$ and $CP_p^b(a) = In_p(\mathcal{G}^{a+1} \circ \dots \circ \mathcal{G}^b)$ if $a < b$.*

A *message adversary* MA [1] is just a set of infinite graph sequences, which are called *admissible*. Consequently, we can compare different message adversaries via set inclusion: A message adversary A is stronger than B iff $A \supseteq B$. Conceptually, we assume that processes know *a priori* the specification[3] of the message adversary. Sometimes, it is convenient to make the system size, and hence the size of the vertex set of the graph sequences generated by a message adversary, explicit: MA_n states that every graph sequence of MA_n has a vertex set of size exactly n, while $\mathsf{MA}_{\leqslant N}$ denotes that this size is at most N. Note that an algorithm designed for $\mathsf{MA}_{\leqslant N}$ must be able to cope with any system size $n \leqslant N$. For a given $\sigma \in \mathsf{MA}_{\leqslant N}$, we will sometimes use Π_σ to denote the set of vertices of the graphs in σ.

In the *consensus problem*, each process p starts with some input value $x_p \in \mathbb{N}$, contained in the initial configuration C^0, and has a dedicated write-once output variable y_p, where $y_p = \perp$ initially; eventually, every process needs to irrevocably decide, i.e., assign a value to y_p (*termination*) that is the same for every process

[3] Whereas this does not mean that a solution algorithm is tied to one specific messages adversary B, one cannot simply assume that the algorithm will also work under any $A \subseteq B$. Typically, however, the processes require only knowledge of some key parameters of B, see Sect. 5 for an example, which may also be respected by A.

(*agreement*) and was the input of a process (*validity*). A given algorithm $\mathcal{A}$ solves consensus under message adversary MA if, for every $\sigma \in \mathsf{MA}$ and every input assignment C^0, validity, agreement, and termination are all satisfied in the execution $\langle C^0, \sigma \rangle$ of $\mathcal{A}$.

As usual, we write $\varepsilon \sim_p \varepsilon'$ if the finite or infinite executions ε and ε' are *indistinguishable* to p (i.e., the state of p at time r is the same in both executions) until p decides. When establishing our lower bounds, we will often exploit that, as outlined above, the configuration at time r is uniquely determined by the initial configuration C^0 and the sequence of communication graphs until round r. As one of our impossibility proofs relies on a bivalence argument, we briefly rephrase the terminology from [16]: Consider an algorithm $\mathcal{A}$ that solves the binary consensus problem, where, for every process p, the initial value $x_p \in \{0, 1\}$. Given a message adversary MA, we call a configuration $C = \langle C^0, \sigma \rangle$ of $\mathcal{A}$ *univalent* or, more specifically, *v-valent*, if all processes decide v in $\langle C, \sigma' \rangle$ for all σ' satisfying that the concatenated sequence $\sigma \circ \sigma' \in \mathsf{MA}$. We call C *bivalent*, if it is not univalent.

4 Basic Message Adversaries

We will now provide the definitions of some basic message adversaries, which have been introduced in [24] and will be used in this paper as well. They rely on the pivotal notion of a *root component* R, often called root for brevity, which denotes the vertex set of a strongly connected component of a graph where there is no edge from a process outside of R to a process in R.

Definition 2 Root Component [24, **Def. 1**]. *$R \neq \emptyset$ is a root (component) of graph $\mathcal{G}$, if it is the set of vertices of a strongly connected component $\mathcal{R}$ of $\mathcal{G}$ and $\forall p \in \mathcal{G}, q \in R : (p \to q) \in \mathcal{G} \Rightarrow p \in R$. A graph $\mathcal{G}$ that has a* single *root component is called* rooted*; its root component is denoted by* $\mathrm{Root}(\mathcal{G})$.

Note that a rooted graph $\mathcal{G}$ is weakly connected and contains a directed path from every node of $\mathrm{Root}(\mathcal{G})$ to every other node of $\mathcal{G}$.

If a (sub)sequence of rooted communication graphs has the same root component R, possibly with varying interconnect topology, it is called the *stable root component* of the sequence.

Definition 3 Stable Root Component [24, **Def. 2**]. *We say that a non-empty sequence $(\mathcal{G}^r)_{r \in I}$ of graphs has a stable root component R, if and only if each $\mathcal{G}^r$ of the sequence is rooted and $\forall i, j \in I : \mathrm{Root}(\mathcal{G}^{\mathrm{i}}) = \mathrm{Root}(\mathcal{G}^{\mathrm{j}}) = \mathrm{R}$. We call such a sequence an R-rooted sequence.*

Note that while "rooted" is a graph property, "R-rooted" is a property of a sequence of graphs.

We are now ready to introduce the message adversary that adheres to *dynamic network depth D*, which gives a bound on the duration of the information propagation from a stable root component to the entire network.

Definition 4 From [24, Def. 4]. $\mathsf{DEPTH}_n(D)$ *is the set of all infinite communication graph sequences* σ *s.t.* $|\Pi_\sigma| = n$ *and, for all finite rounds* r_1*, for all subsequences* $\sigma' = (\mathcal{G}^{r_1}, \dots, \mathcal{G}^{r_1+D-1})$ *of* σ*, if* σ' *is R-rooted, then* $R \subseteq CP_p^{r_1+D-1}(r_1 - 1)$ *for all* $p \in \Pi_\sigma$.

Note that D has been shown in [5, Cor. 1] to satisfy $D \leqslant n-1$, so $\mathsf{DEPTH}_n(D)$ for $D \geqslant n-1$ is void. However, there are also other relevant settings: For example, $D = O(\log n)$ if all $\mathcal{G}^r$ are expander graphs.

The following liveness property, *eventual stability*, ensures that eventually every graph sequence σ has an R-rooted subsequence $\sigma' \subseteq \sigma$ of length x.

Definition 5 From [24, Def. 5]. $\Diamond\mathsf{GOOD}_n(x)$ *is the set of all infinite communication graph sequences* σ *such that* $|\Pi_\sigma| = n$ *and there exists a set* $R \subseteq \Pi_\sigma$ *and an R-rooted* $\sigma' \subseteq \sigma$ *with* $|\sigma'| \geqslant x$.

For finite x, $\Diamond\mathsf{GOOD}_n(x)$ alone is insufficient for solving consensus: Arbitrarily long sequences of graphs that are not rooted before the stability phase occurs can fool any consensus algorithm to make wrong decisions. For this reason, we introduce a safety property in the form of the message adversary that generates only rooted graphs.

Definition 6 From [24, Def. 6]. ROOTED_n *is the set of all infinite sequences* σ *of* rooted *communication graphs such that* $|\Pi_\sigma| = n$.

5 Round-Oblivious Consensus

In this section, we will provide a round-oblivious version of the consensus algorithm from [24], which also works under the short-lived eventually stabilizing message adversary $\Diamond\mathsf{STABLE}_{\leqslant N,D}(D+1)$ defined below. Theorem 1 will prove that our algorithm actually simulates the original algorithm in a one-to-one fashion. Consequently, the correctness proof given in [24] carries over immediately.

5.1 The Message Adversary $\Diamond\mathsf{STABLE}_{\leqslant N,D}(D+1)$

The short-lived eventually stabilizing message adversary $\Diamond\mathsf{STABLE}_{n,D}(D+1)$ is defined as follows:

Definition 7 From [24, Def. 7]. *We call* $\Diamond\mathsf{STABLE}_{n,D}(x) = \mathsf{ROOTED}_n \cap \Diamond\mathsf{GOOD}_n(x) \cap \mathsf{DEPTH}_n(D)$ *the eventually stabilizing message adversary with stability period* x*. For a fixed* D*, we consider the following generalizations:*

- $\Diamond\mathsf{STABLE}_{<\infty,D}(x) = \bigcup_{n \in \mathbb{N}\setminus\{0,1\}} \Diamond\mathsf{STABLE}_{n,D}(x)$
- $\Diamond\mathsf{STABLE}_{\leqslant N,D}(x) = \bigcup_{n=2}^{N} \Diamond\mathsf{STABLE}_{n,D}(x)$

Note that it has been proved in [24] that it is impossible to solve consensus if $x \leqslant D$.

5.2 A Round-Oblivious Consensus Algorithm for $\Diamond\mathsf{STABLE}_{\leqslant N,D}(D+1)$

The round-oblivious consensus algorithm given in Algorithm 2, which uses the helper functions Algorithm 1, is a very simple transformation of [24, Alg. 2].

Algorithm 1: Helper functions for process p

```
1  Function update(q, P_q, S_q, A_q):
2  |   P ← P ∪ {q} ∪ P_q
3  |   S ← S ∪ S_q
4  |   A ← A ∪ A_q
5  |   A ← A ∪ {(1, q, p)}
6  Function shiftLayerNumbers:
7  |   add −1 to s in every tuple (s, i, j) ∈ A as well as to s and l ≠ ⊥ in every
   |     (p, s, x, l) ∈ S
8  Function searchRoot(s):
9  |   V ← {v ∈ P | ∃(s, ∗, v) ∈ A or ∃(s, v, ∗) ∈ A}
10 |   E ← {(u, v) ∈ P² | ∃(s, u, v) ∈ A}
11 |   Let SCC(V, E) denote the set of vertex sets of the strongly connected
   |     components (SCCs) of ⟨V, E⟩. A single node q may constitute a SCC only
   |     if (q, q) ∈ E.
12 |   foreach C ∈ SCC(V, E) do
13 |   |   if ∄v ∈ V \ C: (v, u) ∈ E for some u ∈ C then
14 |   |   |   return C
15 |   return ∅
16 Function L(q, s):
17 |   if ∃(q, s, ∗, ℓ) ∈ S then return ℓ
18 |   else return −∞
19 Function X(q, s):
20 |   if ∃(q, s, x, ∗) ∈ S then return x
21 |   else return −1
22 Function latestRefutation(a, b, x):
23 |   T ← {i ∈ [a, b] | ∃q ∈ P: L(q, i) = ⊥ or X(q, i) ∉ {−1, x}}
24 |   if T ≠ ∅ then return max(T)
25 |   else return −∞
26 Function uniqueCandidate(a, b):
27 |   if ∃k ∈ ℕ: ∀u ∈ P, ∀i ∈ [a, b]: L(u, i) ∉ {−∞, ⊥} ⇒ X(u, i) = k and
   |     ∃q ∈ P, ∃j ∈ [a, b]: L(q, j) ∉ {−∞, ⊥} then
28 |   |   return k
29 |   else
30 |   |   return −1
31 Function allGood(a, b, x):
32 |   return (∀q ∈ P, ∀i ∈ [a, b]: L(q, i) ≠ ⊥ and X(q, i) ∈ {−1, x})
```

Algorithm 2: Consensus algorithm, code for process p. Uses function definitions from Algorithm 1.

Initialization:

1 $\mathsf{x} \leftarrow x_p$, $\ell \leftarrow \bot$, $\mathsf{A} \leftarrow \emptyset$, $\mathsf{P} \leftarrow \emptyset$
2 $\mathsf{S} \leftarrow \{(p, 0, \mathsf{x}, \ell)\}$,

Round r communication, $r \geqslant 1$:

3 Attempt to send $(\mathsf{P}, \mathsf{S}, \mathsf{A})$ to all
4 Receive m_q from all q with $(q, p) \in \mathcal{G}^r$

Round r computation, $r \geqslant 1$:

5 **foreach** m_q *s.t. p received* $m_q = (\mathsf{P}_q, \mathsf{S}_q, \mathsf{A}_q)$ *in round* r **do**
6 | `update`$(q, \mathsf{P}_q, \mathsf{S}_q, \mathsf{A}_q)$
7 `shiftLayerNumbers`()
8 $\mathsf{R} \leftarrow$ `searchRoot`$(-D)$
9 **if** $\mathsf{R} \neq \emptyset$ *and* ($\ell = \bot$ *or* $\mathsf{R} \neq$ `searchRoot`$(-D-1)$) **then**
10 | $\mathsf{x} \leftarrow \max\{\mathtt{X}(q, -D) \mid q \in \mathsf{R}\}$
11 | $\ell \leftarrow 0$
12 **else if** $r > N$ **then**
13 | **if** `latestRefutation`$(-N, -1, \mathsf{x}) \geqslant \ell$ **then**
14 | | $\ell \leftarrow \bot$
15 | **if** `uniqueCandidate`$(-N, -1) \neq -1$ **then**
16 | | $\mathsf{x} \leftarrow$ `uniqueCandidate`$(-N, -1)$
17 **if** $r > N(D + 2N)$, $y_p = \bot$, $\ell \neq \bot$, *and* `allGood`$(-N(D+2N), -1, \mathsf{x}) = \text{TRUE}$ **then**
18 | $y_p \leftarrow \mathsf{x}$
19 $\mathsf{S} \leftarrow \mathsf{S} \cup (p, 0, \mathsf{x}, \ell)$
20 $\ell \leftarrow \ell - 1$

Following the high-level description of the algorithm from [24], the operation of Algorithm 2 can be summarized as follows:

Every process maintains a local estimate of the current graph sequence and the history of the states of the other processes, in particular, their current decision value estimate (proposal value) x, according to its local view. If a process p detects a root component R that might have been stable for $D+1$ rounds in its graph approximation (the eventual occurrence of which is guaranteed by $\lozenge\mathsf{STABLE}_{\leqslant N,D}(D+1)$), it locks on to the maximum of the proposal values of the members of R. Subsequently, p waits for contradictory evidence, i.e., information from some other process that disproves p's assumed stability. If p gets such information within a suitable chosen (quadratic in N) number of rounds, it clears its locked-on state, otherwise, it decides on its current proposal value. To ensure a safe decision, i.e., one that does not violate an earlier decision of some other process q, process p always adopts the proposal value of every process q that might be convinced to hold the correct proposal value, without locking on to it, however.

The original [24, Alg. 2] organizes all the recorded history, in particular, the graph approximation A, indexed by absolute round numbers. As all processes start simultaneously at round $r = 0$ and round numbers are assumed to be correct, this is perfectly feasible. In order to make the algorithm round-oblivious, however, all that needs to be done is to index the recorded history *relative to the current round.* We call these indices *layers*, with layer 0 referring to the current round r, -1 to the round before the current round $r - 1$, etc. Processes merge knowledge from other processes with their own layer-by-layer, prepend the local history by a temporary layer 1, and finally shift all layer-related information (that is, the round number s and the lock round ℓ) by -1. Since round-oblivious algorithms can rely upon the fact that only the round numbers may be erroneous, but not the synchrony of the current round at all processes, layer -1 at different processes still refers to the same round ($r - 1$ in the previous example).

All that needs to be done in the consensus algorithm [24, Alg. 2] to arrive at Algorithm 2 is to replace every helper function call parameter $r - s$ associated with some round number (where r represents the current round) by $r-s-r = -s$. For example, `searchRoot(`r`-`D`)` translates to `searchRoot(-`D`)`. Note that the smallest layer number accessed in Algorithm 2 is $-N(D + 2N)$ (in Line 17), which implies that the maintained history can be pruned appropriately.

In addition, as [24, Alg. 2] uses the values 0 and -1, which cannot occur as round numbers but are legitimate layer numbers, as special values for the lock round ℓ and the return value of the functions `latestRefutation(`a, b`)` and `L(`q, s`)`, we just replace those by $\perp$ and $-\infty$, respectively.

The following Theorem 1 establishes a one-to-one correspondence of layers and round numbers, which in conjunction with the correctness of [24, Alg. 2] implies the correctness of Algorithm 2. It shows by induction that, for every process p in the r-th round, the tuples $(\mathsf{P}, \mathsf{S}, \mathsf{A})$ in the history of the new algorithm indexed by layer $-\lambda$, $0 \leqslant \lambda \leqslant r$, are the same as the tuples indexed in the history of the original one with round number $r - \lambda$.

Theorem 1 Layer/round number correspondence. *Let $r \geqslant 0$ be the r-th round in any run of Algorithm 2, and $0 \leqslant \lambda \leqslant r$. Then, at every process p in the current round, both*

(i) every tuple $(*, -\lambda, *, -l) \in \mathsf{S}$ of Algorithm 2 if and only if $(*, r - \lambda, *, l') \in \mathsf{S}$ of [24, Alg. 2], where $l' = 0$ iff $l = \perp$ and $l' = r - l$ otherwise.
(ii) every tuple $(-\lambda, *, *) \in \mathsf{A}$ of Algorithm 2 if and only if $(r - \lambda, *, *) \in \mathsf{A}$ of [24, Alg. 2].

Proof. We will show our theorem by induction on $r \geqslant 0$. For the induction base $r = 0$, both algorithms have initialized their history identically in the initial configuration, except for the different initial values $\ell = \perp$ resp. $\ell' = 0$ that we will consider identically in our proof. For the induction step $r \to r + 1$, we assume identical histories of all processes according to (i) and (ii) during rounds $r' \in \{0, \dots, r\}$, for $\lambda \leqslant r'$. In the $r + 1$-st round of Algorithm 2 at process p, every tuple $m_q = (\mathsf{P}_q, \mathsf{S}_q, \mathsf{A}_q)$ received from process q, which originates from the previous round r, is first integrated as-is into p's history $(\mathsf{P}, \mathsf{S}, \mathsf{A})$ by calling `update`$(q, \mathsf{P}_q, \mathsf{S}_q, \mathsf{A}_q)$ in Line 6. In addition, this function adds a tuple $\{(1, q, p)\}$ at layer 1 to A (see Line 5). Finally, `shiftLayerNumbers()` is called in Line 7, which decreases the layer numbers (i.e., s and $l \neq \perp$) in all tuples in $(\mathsf{P}, \mathsf{S}, \mathsf{A})$ of p by one. Since each of the tuples $m_q = (\mathsf{P}_q, \mathsf{S}_q, \mathsf{A}_q)$ originated from round r thus ends up being integrated at layer -1, applying the induction hypothesis reveals that the correspondences (i) and (ii) hold for $r + 1$ for all $1 \leqslant \lambda \leqslant r + 1$; for $\lambda = 0$, they hold since Algorithm 2 and [24, Alg. 2] update $(\mathsf{P}, \mathsf{S}, \mathsf{A})$ identically in the current round $r + 1$. □

Whereas there are only a few simple changes required to make the original algorithm of [24] round-oblivious, they make the algorithm significantly more attractive in practice: Given the quite significant probability of observing erroneous round counters e.g. in cheap wireless sensor network nodes, this additional robustness (that comes free of charge) is an important asset. Moreover, our findings paved the way for the development of the novel stabilizing consensus algorithm presented in Sect. 7.

6 Impossibility of Consensus with Immediate Acknowledgments

In this section, we will address the question of whether consensus algorithms like Algorithm 2 would benefit from a stronger communication model, where the sender of a round-r message gets immediate feedback, i.e., in the same round, of whether the transmission was successful or not. As argued in Sect. 1, there are wireless network protocols like JAG [7] that provide this feature.

In essence, immediate feedback implies that sender p and receiver q automatically agree at the end of round r of whether $(p, q) \in \mathcal{G}^r$ or not. Obviously, this immediately prevents the lossy-link consensus impossibility for $n = 2$ processes [20]. Surprisingly, for larger n, it turns out that immediate feedback does not change the solvability/impossibility border for consensus under stabilizing message adversaries like $\lozenge\mathsf{STABLE}_{n,D}(x)$ (Definition 7): We will prove that $x = D+1$ is also necessary and sufficient here.

Indeed, the statement of Theorem 3 below is almost the same as the one of [24, Thm. 2], except for the increase of the minimum system size from $D+1$ to $D+2$. Its proof, however, albeit it also uses a bivalence argument, is complicated by the fact that changing a single edge in some $\mathcal{G}^r$ (which is traditionally used to prove "valence connectivity" of the successor configurations of a bivalent configuration, see [21]) changes the state of *two* processes, i.e., not just the receiver's but also the sender's state.

For proving our main theorem, we need a technical lemma from [24]. It shows, for $n > 2$, that by adding/removing a single edge at a time, we can arrive at a desired rooted communication graph when starting from any other rooted communication graph. Furthermore, during this construction, we can avoid graphs that contain a certain "undesirable" root component R''.

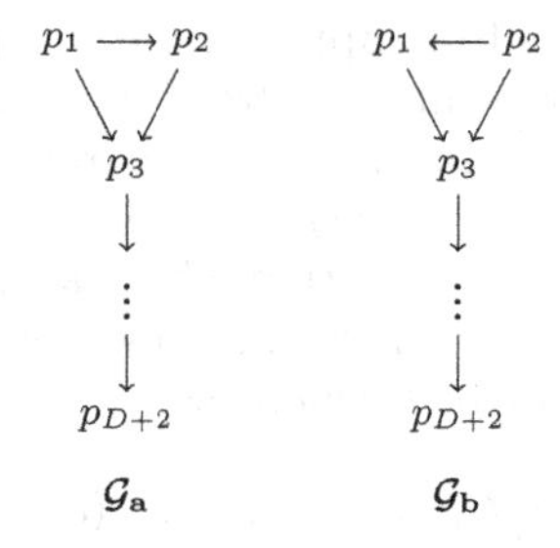

Fig. 1. Communication graphs for Theorem 3. We assume there is an edge from every process depicted in the graph to every process not depicted in the graph.

Lemma 2 From [24, **Lem. 1**]. *Let $n > 2$, let $\mathcal{G}$ be a rooted communication graph with* $\mathrm{Root}(\mathcal{G}) = \mathrm{R}$, *let $\mathcal{G}'$ be a rooted communication graph with* $\mathrm{Root}(\mathcal{G}') = \mathrm{R}'$, *and let R'' be a root component with $R'' \neq R$ and $R'' \neq R'$. Then, there is a sequence of communication graphs, $\mathcal{G} = \mathcal{G}_1, \ldots, \mathcal{G}_k = \mathcal{G}'$ s.t. each $\mathcal{G}_i$ in the sequence is rooted,* $\mathrm{Root}(\mathcal{G}_\mathrm{i}) \neq \mathrm{R}''$, *and, for $1 \leqslant i < k$, $\mathcal{G}_i$ and $\mathcal{G}_{i+1}$ differ only in a single edge.*

We can now prove the main result of this section:

Theorem 3. *There is no consensus algorithm for $\Diamond\mathsf{STABLE}_{n,D}(x)$ with $1 \leqslant x \leqslant D$ and $n \geqslant 3$, even if the adversary guarantees that the first x rounds are R-rooted.*

Proof. Since $\Diamond\mathsf{STABLE}_{n,D}(x) \supset \Diamond\mathsf{STABLE}_{n,D}(D)$ for $x \leqslant D$, it suffices to show the impossibility of consensus under $\Diamond\mathsf{STABLE}_{n,D}(D)$: If the execution ε where consensus cannot be solved is admissible under $\Diamond\mathsf{STABLE}_{n,D}(D)$, it is also admissible under $\Diamond\mathsf{STABLE}_{n,D}(x)$.

Our bivalence proof is similar to the one of [24, Thm. 2], but uses the graph sequences shown in Fig. 1 and a different argument for ensuring valence connectivity in the induction step: We first prove that, for any consensus algorithm, there is a bivalent configuration at the end of round $r = D$, and proceed by induction to show that every bivalent configuration has a bivalent successor configuration. Hence, any consensus algorithm permits a perpetually bivalent execution under $\Diamond\mathsf{STABLE}_{n,D}(D)$, where consensus cannot be solved.

We actually show that such a bivalent execution is even contained in the slightly weaker message adversary $\Diamond\mathsf{STABLE}'_{n,D}(D) \subseteq \Diamond\mathsf{STABLE}_{n,D}(D)$, which consists of those graph sequences of $\Diamond\mathsf{STABLE}_{n,D}(D)$ where already the first D rounds are R-rooted.

For the induction base $r = D$, we show that not all configurations of $\mathcal{A}$ at the end of round D can be univalent: Assume that an algorithm $\mathcal{A}$ solves consensus under $\Diamond\mathsf{STABLE}'_{n,D}(D)$ and suppose that all configurations of $\mathcal{A}$ at time D were univalent.

Let C^0 be an initial configuration of $\mathcal{A}$ with $x_{p_1} = 0$ and $x_{p_2} = 1$ and consider the graphs $\mathcal{G}_a$ and $\mathcal{G}_b$ from Fig. 1. For $i \in \{a, b\}$ let $C_i^D = \langle C^0, (\mathcal{G}_i)_{r=1}^D \rangle$ denote the configuration which results from applying $\mathcal{G}_i$ D times to C^0. Let $\mathcal{S}(p)$ denote the star-like graph where there is an edge from the center vertex p to every other vertex and from every vertex to itself but there are no other edges in the graph. Clearly, C_a^D is 0-valent since $\langle C_a^D, (\mathcal{S}(p_1))_{D+1}^{\infty}\rangle \in \Diamond\mathsf{STABLE}'_{n,D}(D)$ and for p_1 this is indistinguishable from the situation where all processes p have $x_p = 0$. A similar argument shows that C_b^D is 1-valent. However, this provides the required contradiction, since C_a^D cannot be 0-valent as

$$\langle C_a^D, (\mathcal{S}(p_{D+2}))_{D+1}^{\infty}\rangle \sim_{p_{D+2}} \langle C_b^D, (\mathcal{S}(p_{D+2}))_{D+1}^{\infty}\rangle. \tag{1}$$

Hence, not all configurations at the end of round D can be univalent.

For the induction step, let us assume that there exists a bivalent configuration C^r for a time $r \geqslant D$. For a contradiction, assume that all configurations at time $(r+1)$ reachable from C^r are univalent. Thus, there exists a 0-valent configuration $C_0^{r+1} = \langle C^r, \mathcal{G}_0\rangle$ that results from applying a communication graph $\mathcal{G}_0$ to C^r. Moreover, there is a 1-valent configuration $C_1^{r+1} = \langle C^r, \mathcal{G}_1\rangle$ that results from applying a communication graph $\mathcal{G}_1$ to C^r.

First, let us show that for $\mathcal{G} \in \{\mathcal{G}_0, \mathcal{G}_1\}$, it holds that, if $\text{Root}(\mathcal{G}) = \text{Root}(\mathcal{G}^r)$, there is an applicable graph $\mathcal{G}'$ s.t. $\langle C^r, \mathcal{G}'\rangle$ has the same valency as $\langle C^r, \mathcal{G}\rangle$ and $\text{Root}(\mathcal{G}) \neq \text{Root}(\mathcal{G}')$. The reason for this is that we can construct $\mathcal{G}'$ from $\mathcal{G}$ by simply adding an edge $(p \to q)$ for a $q \neq p$, $p \notin \text{Root}(\mathcal{G})$, $q \in \text{Root}(\mathcal{G})$ if $|\text{Root}(\mathcal{G})| = 1$, respectively, by removing $(p \to q)$ for a $p \in \text{Root}(\mathcal{G})$ and all $p \neq q \in \text{Root}(\mathcal{G})$ if $|\text{Root}(\mathcal{G})| > 1$. This yields a graph $\mathcal{G}'$ with the desired property, as $\langle C^r, \mathcal{G}, (\mathcal{S}(p))_{r+1}^{\infty}\rangle \sim_p \langle C^r, \mathcal{G}', (\mathcal{S}(p))_{r+1}^{\infty}\rangle$. The applicability of $\mathcal{G}'$ follows because $\mathcal{G}'$ is rooted and $\text{Root}(\mathcal{G}') \neq \text{Root}(\mathcal{G}^r)$ ensures that the resulting subsequence is a prefix of a sequence of $\mathsf{DEPTH}_n(D)$ for all $D > 1$, because, for these choices of D, a changing root component trivially satisfies Definition 4.

Hence, there are graphs $\mathcal{G}'_0, \mathcal{G}'_1$ such that $\text{Root}(\mathcal{G}'_0) \neq \text{Root}(\mathcal{G}^r)$, $\text{Root}(\mathcal{G}'_1) \neq \text{Root}(\mathcal{G}^r)$, and $\langle C^r, \mathcal{G}'_0\rangle$ is 0-valent while $\langle C^r, \mathcal{G}'_1\rangle$ is 1-valent. Because of $n > 2$, we can hence apply Lemma 2 to go from $\mathcal{G}'_0$ to $\mathcal{G}'_1$ by adding/removing a single edge at a time, without ever arriving at a graph that has more than one root component or has the same root component as $\mathcal{G}^r$. Somewhere during adding/removing a single edge, we transition from a graph $\mathcal{G}_i$ to a graph $\mathcal{G}_{i+1}$, by modifying an edge $(p \to q)$, where the valency of $C = \langle C^r, \mathcal{G}_i\rangle$ differs from the valency of $C' = \langle C^r, \mathcal{G}_{i+1}\rangle$. Nevertheless, $\mathcal{G}_i$ and $\mathcal{G}_{i+1}$ can be applied to C^r, because they are rooted and have a different root component as $\mathcal{G}^r$, hence guarantee the membership of the sequence in $\mathsf{DEPTH}_n(D)$ for all $D > 1$. However, C and C' cannot have a different valency, since $n > 2$ implies that $\exists u \neq p, q$ such that $\langle C, (\mathcal{S}(u))_{r+1}^{\infty}\rangle \sim_u \langle C', (\mathcal{S}(u))_{r+1}^{\infty}\rangle$. This is a contradiction, hence not all configurations at time $(r+1)$ can be univalent. □

7 Stabilizing Consensus

In this section, we will turn our attention to the weaker stabilizing consensus problem, which is essentially consensus without irrevocable decisions.

7.1 Extensions of the Basic Model

Among the advantages of stabilizing consensus is the fact that participants need not start their execution simultaneously. In order to accommodate this in our system model, we allow processes to be *active* or *passive* and allow them to start *passive*. The environment is in control of this state and can switch the state of a process at most once, from passive to active.

Only active processes participate in the execution of a distributed algorithm, and the message adversary guarantees that there is no edge $(p, q) \in \mathcal{G}^r$ going out from a non-active process p, not even the self-loop (p, p), in any round r before p's *starting round* $a_p \geqslant 1$, at the beginning of which p has been activated for the first time. For a given run of our system, let $\Pi_a = \{p \in \Pi : a_p < \infty\}$ denote the set of active processes. Note that we assume that the message adversary does not generate any incoming edge $(q, p) \in \mathcal{G}^r$ for $r < a_p$ either. This is feasible, since a passive process q would ignore any message sent to it in such a round anyway. Thanks to these assumptions, the causal past (see Definition 1) of an active process p only dates back until the beginning of round a_p.

In the *stabilizing consensus* problem, every active process $p \in \Pi_a$ has an input value x_p and output value x_p, initially $\mathsf{x}_p = x_p$. Each process must assign a value to x_p, possibly multiple times, such that there exists some round s where $\forall r \geqslant s$, $\forall p \in \Pi : \mathsf{x}_p^r = v$ (*agreement*), and x_p is the input x_q of some active process q (*validity*). The input value x_p for each active process p is specified in the configuration $C^{a_p - 1}$ at the beginning of p's starting round a_p; recall that we assume that every process initially sets $\mathsf{x}_p^{a_p - 1} = x_p$ as well.

7.2 The Message Adversary $\Diamond\mathsf{WEAKSTAB}_{\leqslant N}(D + 1, D)$

We start with the definition of the message adversary, under which we will solve stabilizing consensus. It is essentially the MA from Definition 7, albeit without incorporating ROOTED_n, but with the additional constraint that the stable root guaranteed by $\Diamond\mathsf{GOOD}_n$ occurs after the last participating process became active.

Definition 8 Weakly stabilizing MA. *Let $\Diamond\mathsf{WEAKSTAB}_n$ be the message adversary defined by $\Diamond\mathsf{WEAKSTAB}_n(x, D) = \Diamond\mathsf{GOOD}_n(x) \cap \mathsf{DEPTH}_n(D)$, restricted by the property that the R-rooted subsequence σ' guaranteed by $\Diamond\mathsf{GOOD}_n(x)$ starts at or after round $r_0 = \max_{p \in \Pi_a}\{a_p\}$. For a fixed upper bound N on the number of processes, we define $\Diamond\mathsf{WEAKSTAB}_{\leqslant N}(x, D) = \bigcup_{n=2}^{N} \Diamond\mathsf{WEAKSTAB}_n(x, D)$.*

7.3 A Stabilizing Consensus Algorithm for $\Diamond\mathsf{WEAKSTAB}_{\leqslant N}(D + 1, D)$

We will now show that Algorithm 4, which has been derived from the round-oblivious Algorithm 2 by dropping the termination-related part, solves stabilizing consensus under $\Diamond\mathsf{WEAKSTAB}_{\leqslant N}(D+1, D)$. Note that this also allowed us to get rid of the lock round variable ℓ altogether, even in the state S of the processes.

Algorithm 3: Helper functions for process p for stabilizing consensus

```
1  Function update(q, P_q, S_q, A_q):
2  |  P ← P ∪ {q} ∪ P_q
3  |  S ← S ∪ S_q
4  |  A ← A ∪ A_q
5  |  A ← A ∪ {(1, q, p)}
6  Function shiftLayerNumbers:
7  |  add −1 to s in every tuple (s, i, j) ∈ A and every (p, s, x) ∈ S
8  Function searchRoot(s):
9  |  V ← {v ∈ P | ∃(s, ∗, v) ∈ A or ∃(s, v, ∗) ∈ A}
10 |  E ← {(u, v) ∈ P² | ∃(s, u, v) ∈ A}
11 |  Let SCC(V, E) denote the set of vertex sets of the strongly connected
   |    components (SCCs) of ⟨V, E⟩. A single node q may constitute a SCC only
   |    if (q, q) ∈ E.
12 |  foreach C ∈ SCC(V, E) do
13 |  |  if ∄v ∈ V \ C: (v, u) ∈ E for some u ∈ C then
14 |  |  |  return C
15 |  return ∅
16 Function X(q, s):
17 |  if ∃(q, s, x) ∈ S then return x
18 |  else return −1
```

Algorithm 4: Stabilizing consensus algorithm, code for process p. Uses function definitions from Algorithm 3

```
   Initialization:
1  x ← x_p, A ← ∅, P ← ∅
2  S ← {(p, 0, x)}
   Round r communication, r ⩾ 1:
3  Attempt to send (P, S, A) to all
4  Receive m_q from all q with (q, p) ∈ G^r
   Round r computation, r ⩾ 1:
5  foreach m_q s.t. p received m_q = (P_q, S_q, A_q) in round r do
6  |  update(q, P_q, S_q, A_q)
7  shiftLayerNumbers()
8  R ← searchRoot(−D)
9  if R ≠ ∅ and R ≠ searchRoot(−D − 1) then
10 |  x ← max {X(q, −D) | q ∈ R}
11 S ← S ∪ (p, 0, x)
```

For the proof of our main Theorem 6, we need (part of) two technical lemmas from [24], which will be re-stated below. Albeit they originally refer to $\Diamond\mathsf{STABLE}_{\leqslant N,D}(D+1)$ rather than $\Diamond\mathsf{WEAKSTAB}_{\leqslant N}(D+1,D)$ (and are not formulated in a round-oblivious fashion), they are applicable here since their proofs (that is, of the actually required parts) do not depend on the missing ROOTED_n. Applying the substitution $r-s \rightarrow -s$ and $s \rightarrow r-s$ for translating absolute round numbers into the corresponding layers, as introduced in Sect. 5.2, and restricting their applicability to processes in Π_a and to rounds $\geqslant r_0$, these lemmas read as follows:

Lemma 4 Adapted from [24, **Lem. 5**]. *Pick* $\sigma \in \Diamond\mathsf{WEAKSTAB}_{\leqslant N}(D+1, D)$, *fix a round* $s \geqslant r_0$ *and let the current round be* $r > s$. *If* $(\mathcal{G}^i)_{i=s}^{s+D}$ *is* R*-rooted and* $r \geqslant s + D$, *then* $\mathtt{searchRoot}_p^0(s - r) = R$ *for every process* $p \in \Pi_a$.

Lemma 5 Adapted from [24, **Lem. 11**]. *If* $p \in \Pi_a$ *enters Line 10 in the current round* r *with* $r - D \geqslant r_0$, *then* $\mathsf{x}_p^0 = \max\{\mathsf{x}_q^{-D} \mid q \in \mathrm{Root}(\mathcal{G}^{\mathrm{r-D}})\}$.

The following Theorem 6 proves that Algorithm 4 indeed solves stabilizing consensus under $\Diamond\mathsf{WEAKSTAB}_{\leqslant N}(D + 1, D)$, with bounded stabilization time (counted from the beginning of the stable root guaranteed by $\Diamond\mathsf{GOOD}_n$):

Theorem 6. *Algorithm 4 solves stabilizing agreement under the message adversary* $\Diamond\mathsf{WEAKSTAB}_{\leqslant N}(D + 1, D)$, *with stabilization time* $D + 1$.

Proof. Since $\Diamond\mathsf{WEAKSTAB}_{\leqslant N}(D+1, D)$ still contains $\Diamond\mathsf{GOOD}_n(D+1)$, we can be sure that there exists a R-rooted sequence σ of length $\geqslant D+1$; let b be the round where σ ends. Moreover, $\Diamond\mathsf{WEAKSTAB}_{\leqslant N}(D + 1, D)$ guarantees that σ starts in round $a \geqslant r_0$. By Lemma 4, it holds that at R is consistently detected by every process in Π_a in round $a + D$, and by Lemma 5, that *every* process $p \in \Pi_a$ generates the same $\mathsf{x}_p^0 = v$ based on R. Therefore, from round $a + D + 1$ on, x_p^0 cannot be changed at any process $p \in \Pi_a$ in the system, which guarantees the agreement property of stabilizing consensus: Line 9 guarantees that Line 10 is only executed after R changes. The earliest round after b at which any process can detect such a change is $r = b+D+1$, at which, however, $\forall p, q \in \Pi_a : \mathsf{x}_p^0 = \mathsf{x}_q^0 = v$ already holds.

Since validity is trivially fulfilled, as only input values $x_p^{a_p}$ are ever used to generate proposal values by our algorithm, Theorem 6 follows. □

We conclude this subsection by briefly comparing our algorithm to the MinMax algorithm from [9]. To be able to do so, we first note that we can easily get rid of the assumption that the stability interval must occur after r_0 in Definition 8, by just assuming that the stability interval recurs infinitely often. After all, according to the proof of Theorem 6, stability intervals that occur before round r_0 are irrelevant in the sense that they either "froze" the proposal value (i.e., caused all processes in the system to adopt a unique value already), or are overridden by the freeze done in the stability interval that occurs after r_0. Further stability intervals after r_0 do not have any effect at all, as the proposal value is already frozen.

The MinMax algorithm works in a system model with anonymous processes, but with a property ("rootedness with bounded delay", [9, Lem. 4]) that essentially guarantees the bounded information propagation $\mathsf{DEPTH}_n(D)$ to hold not just for the (beginning(s) of the) stability interval(s), as in our case, but rather for *every* round. Obviously, this property does not hold for our message adversary $\Diamond\mathsf{WEAKSTAB}_n$, which may generate arbitrarily long subsequences where there are multiple root components, isolated processes etc. that violate rootedness with bounded delay. Consequently, MinMax cannot be used in conjunction with $\Diamond\mathsf{WEAKSTAB}_n$.

Conversely, even if one would add process ids to the system model of [9] and strengthen the assumption of rootedness with bounded delay to rootedness with some known delay, our Algorithm 4 could not be used either: The resulting system would not necessarily guarantee the stability property $\Diamond\mathsf{GOOD}_n(D+1)$, which is mandatory for the correctness of our algorithm. We note, however, that a suitably parametrized version of Algorithm 4 would work in the counterexample setting of [9, Thm. 13], which cannot be handled by any safe MinMax algorithm.

7.4 Impossibility of Stabilizing Consensus with Insufficient Stability

A natural question to ask is whether the stability interval of $D+1$ used in $\Diamond\mathsf{WEAKSTAB}_n(D+1, D)$ is necessary. We can answer this question in the affirmative, by re-using part of the proof of Theorem 3 to show the following impossibility result. Note carefully that it covers both our standard communication model and immediate acknowledgments.

Theorem 7. *There is no stabilizing consensus algorithm for* $\Diamond\mathsf{WEAKSTAB}_n(x, D)$ *with* $1 \leqslant x \leqslant D$ *and* $n \geqslant 3$*, even if immediate acknowledgments are available,* $r_0 = 1$ *and the adversary guarantees that the first* x *rounds are R-rooted.*

Proof. Our claim follows from the bivalence of some configuration at the end of round D, as established in the induction base in the proof of Theorem 3: Recall that p_{D+2} could not distinguish C_a^D and C_b^D by Eq. 1, so must have the same state in C_a^D and C_b^D. Since $\Diamond\mathsf{WEAKSTAB}_n(x, D)$ allows us to continue after round D with an infinite suffix of empty graphs $\mathcal{E}$ (that only contain self-loops) from both configurations, we arrive at $\langle C_a^D, (\mathcal{E})_{D+1}^{\infty}\rangle \sim_{p_{D+2}} \langle C_b^D, (\mathcal{E})_{D+1}^{\infty}\rangle$. Consequently, p_{D+2} must eventually set its output to some value $\mathsf{x}_{p_{D+2}} = v$, for the last time, in both executions. We distinguish two cases: If $v = 0$, since process p_2 started with $x_{p_2} = 1$ and $\langle C_b^D, (\mathcal{E})_{D+1}^{\infty}\rangle$, it must eventually set its output $\mathsf{x}_{p_2} = 1$ by validity, as it only knows its own input value. This contradicts the agreement property of stabilizing consensus, however. Analogously, if $v = 1$, we observe that p_1 started with $x_{p_1} = 0$ and never gets any message in $\langle C_a^D, (\mathcal{E})_{D+1}^{\infty}\rangle$, so must output $\mathsf{x}_{p_1} = 0$ by validity, which again provides the required contradiction. □

8 Conclusions

We showed that existing algorithms for consensus in directed dynamic networks controlled by message adversaries could easily be made round-oblivious. The resulting algorithms are hence inherently resilient to an arbitrary number of processes suffering from erroneous round numbers in messages and erroneous round counters of the processes, which is important in practice. On the other hand, it turned out that, in general, consensus algorithms do not benefit from the stronger communication model of immediate acknowledgments for a successful transmission. The round-oblivious version of our consensus algorithm also allowed us to derive a novel stabilizing consensus algorithm, which does not require an irrevocable decision and allows the processes to start their execution arbitrarily. We accomplished this by dropping the rootedness assumption ROOTED_n, but retaining the stability requirement $\Diamond\mathsf{GOOD}_n(D+1)$ of the message adversary. By contrast, the anonymous MinMax stabilizing consensus algorithm by Charron-Bost and Moran works for a message adversary that essentially guarantees ROOTED_n but not $\Diamond\mathsf{GOOD}_n(D+1)$. Hence, consensus solvability can be obtained on top of stabilizing consensus solvability by fundamentally different means.

Acknowledgments. We are grateful to Christoph Echtinger-Sieghart for his feedback on an earlier version of our paper.

References

1. Afek, Y., Gafni, E.: Asynchrony from synchrony. In: Frey, D., Raynal, M., Sarkar, S., Shyamasundar, R.K., Sinha, P. (eds.) ICDCN 2013. LNCS, vol. 7730, pp. 225–239. Springer, Heidelberg (2013). https://doi.org/10.1007/978-3-642-35668-1_16
2. Angluin, D., Fischer, M.J., Jiang, H.: Stabilizing consensus in mobile networks. In: Gibbons, P.B., Abdelzaher, T., Aspnes, J., Rao, R. (eds.) DCOSS 2006. LNCS, vol. 4026, pp. 37–50. Springer, Heidelberg (2006). https://doi.org/10.1007/11776178_3
3. Biely, M., Robinson, P., Schmid, U.: Agreement in directed dynamic networks. In: Even, G., Halldórsson, M.M. (eds.) SIROCCO 2012. LNCS, vol. 7355, pp. 73–84. Springer, Heidelberg (2012). https://doi.org/10.1007/978-3-642-31104-8_7
4. Biely, M., Robinson, P., Schmid, U., Schwarz, M., Winkler, K.: Gracefully degrading consensus and k-set agreement in directed dynamic networks. In: Bouajjani, A., Fauconnier, H. (eds.) NETYS 2015. LNCS, vol. 9466, pp. 109–124. Springer, Cham (2015). https://doi.org/10.1007/978-3-319-26850-7_8
5. Biely, M., Robinson, P., Schmid, U., Schwarz, M., Winkler, K.: Gracefully degrading consensus and k-set agreement in directed dynamic networks. Theor. Comput. Sci. **726**, 41–77 (2018)
6. Biely, M., Schmid, U., Weiss, B.: Synchronous consensus under hybrid process and link failures. Theor. Comput. Sci. **412**(40), 5602–5630 (2011)
7. Boano, C.A., Zamalloa, M.A.Z., Römer, K., Voigt, T.: JAG: reliable and predictable wireless agreement under external radio interference. In: Proceedings of the 33rd IEEE Real-Time Systems Symposium, RTSS 2012, San Juan, PR, USA, 4–7 December 2012, pp. 315–326 (2012). https://doi.org/10.1109/RTSS.2012.82
8. Charron-Bost, B., Függer, M., Nowak, T.: Approximate consensus in highly dynamic networks: the role of averaging algorithms. In: Halldórsson, M.M., Iwama, K., Kobayashi, N., Speckmann, B. (eds.) ICALP 2015. LNCS, vol. 9135, pp. 528–539. Springer, Heidelberg (2015). https://doi.org/10.1007/978-3-662-47666-6_42
9. Charron-Bost, B., Moran, S.: Minmax algorithms for stabilizing consensus. Distrib. Comput. **34**(3), 195–206 (2021)
10. Charron-Bost, B., Schiper, A.: The heard-of model: computing in distributed systems with benign faults. Distrib. Comput. **22**(1), 49–71 (2009). https://doi.org/10.1007/s00446-009-0084-6

11. Coulouma, É., Godard, E., Peters, J.G.: A characterization of oblivious message adversaries for which consensus is solvable. Theor. Comput. Sci. **584**, 80–90 (2015). https://doi.org/10.1016/j.tcs.2015.01.024
12. Daliot, A., Dolev, D., Parnas, H.: Self-stabilizing pulse synchronization inspired by biological pacemaker networks. In: Huang, S.-T., Herman, T. (eds.) SSS 2003. LNCS, vol. 2704, pp. 32–48. Springer, Heidelberg (2003). https://doi.org/10.1007/3-540-45032-7_3
13. Doerr, B., Goldberg, L.A., Minder, L., Sauerwald, T., Scheideler, C.: Stabilizing consensus with the power of two choices. In: Rajaraman, R., auf der Heide, F.M. (eds.) SPAA 2011: Proceedings of the 23rd Annual ACM Symposium on Parallelism in Algorithms and Architectures, San Jose, CA, USA, 4–6 June 2011 (Co-located with FCRC 2011), pp. 149–158. ACM (2011). https://doi.org/10.1145/1989493.1989516
14. Dolev, D., Hoch, E.N.: Byzantine self-stabilizing pulse in a bounded-delay model. In: Masuzawa, T., Tixeuil, S. (eds.) SSS 2007. LNCS, vol. 4838, pp. 234–252. Springer, Heidelberg (2007). https://doi.org/10.1007/978-3-540-76627-8_19
15. Farsi, M., Ratcliff, K., Barbosa, M.: An overview of controller area network. Comput. Control Eng. J. **10**(3), 113–120 (1999). https://doi.org/10.1049/cce:19990304
16. Fischer, M.J., Lynch, N.A., Paterson, M.S.: Impossibility of distributed consensus with one faulty process. J. ACM **32**(2), 374–382 (1985)
17. Kuhn, F., Lynch, N.A., Oshman, R.: Distributed computation in dynamic networks. In: Proceedings of the Forty-Second ACM Symposium on Theory of Computing (STOC), pp. 513–522 (2010)
18. Kuhn, F., Oshman, R., Moses, Y.: Coordinated consensus in dynamic networks. In: Proceedings of the 30th Annual ACM SIGACT-SIGOPS Symposium on Principles of Distributed Computing (PODC). pp. 1–10. ACM (2011)
19. Maróti, M., Kusy, B., Simon, G., Lédeczi, A.: The flooding time synchronization protocol. In: 2nd International Conference on Embedded Networked Sensor Systems (SenSys), pp. 39–49. ACM, New York (2004). https://doi.org/10.1145/1031495.1031501
20. Santoro, N., Widmayer, P.: Time is not a healer. In: Monien, B., Cori, R. (eds.) STACS 1989. LNCS, vol. 349, pp. 304–313. Springer, Heidelberg (1989). https://doi.org/10.1007/BFb0028994
21. Schmid, U., Weiss, B., Keidar, I.: Impossibility results and lower bounds for consensus under link failures. SIAM J. Comput. **38**(5), 1912–1951 (2009). http://www.vmars.tuwien.ac.at/documents/extern/2554/paper.pdf
22. Schwarz, M., Winkler, K., Schmid, U.: Fast consensus under eventually stabilizing message adversaries. In: Proceedings of the 17th International Conference on Distributed Computing and Networking (ICDCN), pp. 7:1–7:10. ACM (2016). https://doi.org/10.1145/2833312.2833323
23. Widder, J., Schmid, U.: The theta-model: achieving synchrony without clocks. Distrib. Comput. **22**(1), 29–47 (2009)
24. Winkler, K., Schwarz, M., Schmid, U.: Consensus in rooted dynamic networks with short-lived stability. Distrib. Comput. **32**(5), 443–458 (2019)

Towards a Robust Distributed Framework for Election-Day Voter Check-In

Alexander A. Schwarzmann(✉)

Augusta University, Georgia, USA
aschwarzmann@augusta.edu

Abstract. Electronic poll books are computerized distributed systems that replace paper-based voter lists used to enable eligible voters to cast their ballots on the Election Day. These systems have the potential for speeding up voter check-in at the polling place, and making voter records more accurate by reducing human errors in dealing with printed voter lists and post-election transcription. At the same time, electronic poll books are non-trivial distributed computing systems, and ensuring correctness, security, integrity, fault-tolerance, and performance of such systems is a challenging problem. In fact we are not aware of a single commercially available system that does not contain major deficiencies and risk factors. This paper focuses on the distributed system aspects of electronic poll book solutions and identifies the obstacles that are inherent in any distributed system that must deal with failure and asynchrony while providing a consistent and dependable service. We review several requirements that need to be satisfied by electronic poll book systems, we discuss selected important results from distributed computing research that the commercial developers of electronic poll book systems appear to not be aware of. We then present a wider landscape, including social and political science aspects, we survey broader research issues, and discuss system implementation considerations. This paper brings for the first time to the attention of the research community an in-depth presentation of an important new problem of immediate relevance. Moreover, the electronic poll book technology is an attractive application domain for the research results in dependable and secure distributed computing.

1 Introduction

Electronic voting systems are an integral component of the modern electoral procedures and an essential part of any democratic society. Such systems are composed of several entities working in concert: e.g., "Voter Registration Systems", "Election Management Systems", "Voting Terminals", and "Electronic

Supported in part by NSF Award 2131538, with the senior personnel A. Aleroud, C. Busch, R. DeFrancisco, D. Kowalski, G. Murray, N. Panwar, R. Rahaeimehr, A. Schwarzmann, E. Tremel, J. Heslen, C. Albert (Augusta Univ.), and Sh. Dolev (Ben Gurion Univ.). This paper also cites results obtained in collaboration with L. Michel, A. Russell, and M. Desmarais (Univ. of Connecticut).

C. Johnen et al. (Eds.): SSS 2021, LNCS 13046, pp. 173–193, 2021.
https://doi.org/10.1007/978-3-030-91081-5_12

Poll Books". The first of these has been computerized for some time using well-understood database technologies. In the United States, the HAVA act of 2002 [39] mandated a nationwide modernization of the voting infrastructure and led to the broad adoption of computerized solutions for the next two components. Consequently, all 50 States initiated efforts to select digital voting solutions. The private sector rushed to market with hastily created and often inadequately engineered solutions. The result was an adoption of products that suffered from severe flaws: poor engineering, lack of resilience against the most elementary tampering or simple mis-configuration, the illusion of safety from misuse of cryptography, and, in general, an under-appreciation for the complexity of the electronic election systems that need to be built as sophisticated and secure distributed systems. The third component, Electronic Poll Books, provides the key function that ensures that all eligible voters can cast their vote, and cast it at most once, thus the imperative of "one voter, one vote." These systems are the focus of this presentation. They are relatively new, and they are being widely adopted to replace the old manual processes. The broad concerns surrounding the integrity and security of electronic election systems used in the 2016 and 2020 Presidential Elections underscore the need for a rigorous scientific approach to designing and implementing all such systems.

Contributions. This paper brings for the first time to the attention of the research community an in-depth view of an important new multidimensional problem of immediate relevance. The electronic poll book technology is an attractive and worthy application domain for synergistic research encompassing distributed computing, fault tolerance, cryptography and authentication, networking and network security, consisted distributed storage, and human factors. The research directions suggested here have the promise of high impact in an important area of political elections that are so essential for any democratic society. Unlike some research advances whose applicability to solving real life problems may need to wait for many years before having impact, in the space of electronic check-in systems we have a rewarding situation where research advances can directly benefit a crucial aspect of our increasingly digital civilization, viz. the integrity and security of the electronic electoral process.

On the Broad State of Technology. Dislocation of theory and practice in computing can lead to significant (and even sensational) problems with basic computing infrastructure, and indeed, a number of striking examples arises in the context of voting technology. Numerous technical reports have documented serious issues with electronic election systems, Electronic election systems [7,40,43,44] are perceived to be inherently complex from a security and verification standpoint. Our own work had exposed a number of vulnerabilities in the design and implementation of electronic voting systems, in their use of cryptography, their logging capabilities and communication software that an attacker could exploit to alter the result of an election, e.g., [13,14,27,29,30,37,45]. Technical issues have been reported by essentially all researchers and evaluators who have conducted independent assessment of security and integrity of electronic election systems, ranging from the well-known problems with the use of paperless DRE systems (direct recording, electronic) and the security vulnerabilities in all examined stand-alone systems to severe problems in Internet voting systems [48]. In addition to raising concerns about the dependability of electronic voting systems and suitability of adopted solutions, the ensuing reexamination led a number of states to abandon their chosen solutions and switch technology altogether at a remarkable cost to the U.S. taxpayer [42], e.g., in Maryland [18], where the decision was made to migrate to Optical Scan technology with voter-verified paper ballots, which emerged as the safest option due to its reliance on voter generated paper audit trail.

Whereas researchers exposed a number of vulnerabilities in the design, implementation, use of cryptography, logging capabilities, and communication software of voting systems that an attacker could exploit to alter the result of an election, it is not surprising that the bulk of these problems can be traced back to the divergence from, and the ignorance of, established results in algorithmics, verification, validation, cryptography, and sound engineering practices.

In this work we focus on the problem of voter check-in in elections, the area that received surprisingly little attention from the community of researchers working on electronic election systems. While substantial research has been dedicated to electronic tabulation of votes, ballot processing, vote aggregation, and audits, the very first step that enables voters to cast their votes on the election day and that ensures the "one voter, one vote" imperative has not been the target of sufficient research to our knowledge.

Document Structure. In Sect. 2 we describe the e-pollbook landscape in the US. In Sect. 3 we present the high-level requirements that address the distributed nature of e-pollbooks as replacement for the manual process, and provide specific examples of jurisdictional requirements. In Sect. 4 we presented selected distributed systems facts that make it challenging to implement e-pollbooks. In Sect. 5 we present a wide landscape that encompasses social and political science dimension, broader research needs, and implementation considerations. We conclude in Sect. 6.

2 Electronic Poll Books

It is interesting that while the premature deployment of immature technology resulted in numerous documented cases of serious problems with voting terminals and election management systems, another component of the electoral process—the use of registered voter lists to allow voters to cast votes—still broadly relies on perilously inadequate manual solutions using paper pollbooks, i.e., printed voter lists. The use of a questionable manual process can cause poll opening delays, long lines, and errors, possibly disenfranchising numerous voters [5]. According to the National Conference of State Legislatures, the majority of jurisdictions nationwide are using the manual process, however the usage of electronic solutions is growing rapidly. In the 2020 Presidential Elections, already millions of voters were checked-in electronically. The emerging computerized solutions, "electronic poll books", are a key component of an electronic election system whose basic purpose is to ensure "One voter, one vote." Electronic pollbooks, or "e-pollbooks", are available from a number of vendors, and while the integrity and security of these systems is paramount to the entire electoral process, vendors are again rushing to produce e-pollbooks based on naive premises and software fraught with deficiencies (cf. patent [26]). Naturally, the integrity and security of this component is paramount to the entire electoral process. Without the scientific approach, the desire to modernize and ease the administrative process can again lead to the premature adoption of severely deficient solutions. Indeed, there were a number of reports in the press on the failures of e-pollbook systems during the 2020 US Presidential Election.

It is noteworthy that while the electronic tabulators used on the election day are stand-alone systems, an e-pollbook system is conceptually more complex because it is inherently a dynamic distributed system where multiple check-in devices must operate in concert in providing "one voter, one vote" guarantee, with security and integrity, and despite possible failures and the resulting need to dynamically reconfigure this distributed system on the fly.

We have worked for over a decade with several jurisdictions in New England and New York on security and integrity of electronic election systems. In particular, we provided substantial contributions to the definition of e-pollbook requirements in the States of Connecticut [36] and New Hampshire [20]. We have also participated in evaluating several e-pollbook systems, and while non-disclosure and legal considerations prevent is from identifying specific systems, we must report that the state of the technology in e-pollbook implementations leaves much to be desired. To start with, some systems fail to synchronize properly, leaving a non-trivial window of vulnerability that allows a speedy and mischievous vote to cast more than one ballot. Also, despite the claims of reconfigurability, it is generally impossible to replace crashed devices, and it is not easy to introduce new devices to deal with long lines. Typically, reconfiguration is only possible offline, and even then the systems do not implemented controls to identify and authenticate configured devices adequately. Other systems do not secure wireless communications adequately, allowing nefarious actors to eavesdrop and even launch an impersonation attacks. In one system the network traffic is authenticated and encrypted, yet the keys are self-signed by a third party using a personal email address (@gmail.com). (Here we also note that for practical reasons wired connection are almost never used, thus keeping the digital doors open for attempted attacks, including denial of service, which can be trivially accomplished by inexpensive wireless jamming devices.) Some system implementations also insist on relying on remote servers, where communication over the Internet on the election days creates the potential for obvious problems. Existing implementations do not perform sufficient input checks, e.g., when scanning a barcode on an identity card, and some systems can be crashed with the help of a maliciously designed barcode, or even allow injection attacks that can disenfranchise voters.

In 2017, to address the challenges associated with the use of technology in the elections, the National Academies of Sciences, Engineering, and Medicine

(NASEM) appointed an ad-hoc committee, the Committee on the Future of Voting: Accessible, Reliable, Verifiable Technology. The committee held several meetings, where we have also reported on the topic of security challenges, concerns, and increasing vulnerabilities associated with e-pollbooks [2]. NASEM published an extensive committee report "Securing the Vote: Protecting American Democracy" in 2018 [2]. The report summarizes its findings regarding e-pollbooks as follows.

> "Eligible voters may be denied the opportunity to vote a regular ballot if pollbooks are inaccurate. Internet access to e-pollbooks increases the risks associated with the use of e-pollbooks to manage elections. Cyberattacks can alter the voter registration databases used to generate and update pollbooks. If pollbooks are altered by external actors, eligible citizens might, on election days, be denied the right to vote or ineligible individuals might be permitted to vote. Cyberattacks could also compromise the record of who actually voted on Election Day-or disrupt an election in numerous other ways. If an e-pollbook is connected to a remote voter registration database and there is no offline backup, a denial-of-service cyberattack could force voting to be halted."

Among its recommendations, the report states that "Congress should authorize and fund" the development of "security standards and verification and validation protocols for electronic pollbooks." Our own work in this area is funded by a NSF award within the Secure and Trustworthy Cyberspace (SaTC) program.

The distributed computing community is well positioned to address the challenges inherent in e-pollbook solutions. Indeed, there are a number of approaches to the underlying problem that revolves around a collection of computing devices reliably and securely maintaining distributed and replicated databases (i.e., "voter lists") in the face of equipment failures. Equally importantly, the research in distributed computing identified a number of problems that are notoriously difficult to solve or that even cannot be solved in the most general setting. The lack of adoption of existing techniques and certain obliviousness of the non-trivial challenges in providing solutions could be attributed to a lack of awareness from practitioners or, equally likely, to the difficulty of specializing and implementing the known approaches in this application domain.

The main part of this paper focuses specifically on the distributed system aspects of e-pollbook solutions. We also include a broader discussion of an approach to researching and developing such systems that in our opinion needs to include a sound computer science foundation, including the cybersecurity dimension, advanced system development, and also a social and political dimension that reflects the societal and legal aspects.

3 Electronic Pollbooks as a Distributed System

In this section we describe the basic setting for poll books, the technological challenges and specific questions regarding e-pollbook implementations.

3.1 The Manual Process

Consider the objectives of officials running an election at the scale of a precinct. Prior to the election, officials are accumulating in a database the collection of registered voters throughout the precinct. On Election Day, a paper listing is printed for each polling station indicating which voters are expected. If electoral procedures permit voters to register "on site", the official must also record individuals who desire to vote but are not on the voter list. As ballots are issued to eligible voters, their names get crossed off. Additionally, in some jurisdictions, absentee voters who were previously issued a ballot must be accommodated if they decide to vote in person. The process is meant to help achieve "one voter, one vote". Naturally, a rogue voter can go from polling station to polling station and attempt to vote several times in this way. In such a case, since the authorities collect voter credentials for on-site voting, they would have a legal recourse. To deal with high voter turn-out, the voter lists are partitioned either by street addresses or names, thus allowing for multiple check-in lines and increased "throughput" at a polling place. Multiple voting is impossible because a voter's name appears in only one partition of the voter list. Clearly absent are failures due to technology. From the procedural and safety standpoints, no further significant weaknesses exist with a paper solution. Yet, a paper process is slow, prone to human error, and work intensive before, during and after the election as the voter information must be collated and re-encoded in the voter database. High voter turn-out still results in long lines because the partition of the voter lists cannot be balanced dynamically. An electronic solution is appealing to streamline the process, but it also creates many difficulties.

3.2 Distribution and Consistency: Immediate Challenge

Shared storage services are at the core of most information-age systems and e-pollbooks are not an exception. Imagine an implementation that is based on a central server storage system. The server accepts requests from check-in devices to perform operations on its data objects (e.g., voter records) and returns responses. While this is conceptually simple, this approach already presents two major problems: (1) the central server is a performance bottleneck, and (2) the server is a single point of failure. The quality of service in such an implementation may degrade as the number of users grows, and the service becomes unavailable if the server crashes. Thus the system must, first of all, be available. This means that it must provide its services despite failures within the scope of its specification. In particular, the system must be able to mask device and communication failures up to a limit. The system must also support multiple concurrent accesses without imposing unreasonable degradation in performance. The only way to guarantee availability is through redundancy, that is, to use multiple check-in devices and servers and to replicate the data among these devices.

It is also critically important to ensure data longevity. A storage system may be able to tolerate failures of some servers, but over a long period it is conceivable that all servers may need to be replaced, because no servers are infallible. Additionally, it may be necessary to provide migration of data from one collection of servers to another as the needs dictate. The storage system must provide seamless runtime migration of data: one cannot stop the world and reconfigure the system in response to failures and changing environment. A major problem that comes with replication is consistency. How does the system find the latest voter record if the data is replicated? What the users (people and high level system components) should expect to see is the illusion of a single-copy object that serializes all accesses so that each operation that reads the object returns the value of the preceding write or update operation, and that this value is at least as recent as that returned by any preceding operation [23,38]. This notion of consistency is formalized as *atomicity* [32] or, equivalently, as *linearizability* [25], and we aim to provide this, most desirable, notion of consistency.

3.3 Specific Technical Questions

Here we present some immediate questions that any e-pollbook solution must address as a distributed system. These questions pose non-trivial challenges in implementing functional and dependable solutions. We primarily focus on the dynamic distributed system view of the required solutions, and in Sect. 5 we deal with a broader landscape.

1. Multiple front-end devices are needed to allow concurrent check-in. How does one ensure that device failures do not impact system functionality? How are the misbehaving devices removed from the system? How are new devices introduced to replace faulty devices and to deal with high voter turnout?
2. If electronic devices can fail, this raises questions about the status of voter information accumulated in each device. How is the information recovered? Passed on to other devices? Replicated in real-time? How does one transfer on the fly the relevant state to a "spare device"?
3. Where is the information (voter lists) held? On each device? On a server nearby on a LAN? In the (local) cloud for more reliability? Depending on the answer, one must question what to do in case of network failure. What to do if the network connectivity is lost? Experiences delays? How does one deal with transient (or prolonged or even permanent) network partitions?
4. Any dynamic replication involves some protocol that may itself be subjected to perturbations, e.g., denial of service attacks. How does one guarantee the legitimacy of messages exchanged among the participating devices?
5. When multiple devices add voters or check-off voters upon handing out ballots, how does one maintain a coherent view of the world guaranteeing consistency and enforcing the rule of voting at most once?
6. Last, but not least is the question of functionality of the system with respect to the voters and the election officials that the system is interacting with. Does the system comply with the law? Does it favor one type of voter and might it disenfranchise another? Is it is easy to use? Does it allow for all acceptable forms of identification to be used to check in a voter? Does it allow a manual override by a qualified election official in case of a problem?

Given the generally non-complimentary state of the extant electronic election systems, there is the real concern that existing and emerging commercial offerings for electronic poll-books might side-step the majority of these questions leaving jurisdictions with brittle systems that suffer from major shortcomings and that perform adequately only in benign and friendly environments. In the next section we present some existing requirements for e-pollbooks systems. We note that documenting such requirements is quite challenging. On one hand they must be readable by people without a technical background, e.g., citizens and election officials, on the other hand they must specific requirements that are most useful for technologists who will perform the needed research and development. Thus it is likely that the requirements officially adopted by jurisdictions are viewed as incomplete by the technologists.

3.4 Requirements for Electronic Poll Books as Distributed Systems

Some of the more comprehensive requirements for e-pollbook systems have been published by the States of Connecticut [36] and New Hampshire [20]. We extract and present several requirements from [36] that specifically address the necessarily distributed nature of e-pollbook solutions. We begin by stating several definitions for the terms used in the requirements.

Electronic poll book system (EPBS) – A collection of hardware and software including at least one *electronic poll book* and aiming to implement e-pollbook functionality that satisfies the requirements (in [36]).

Electronic poll book (EPB) – A component of the *electronic poll book system* that includes a user interface device and that is to be used by a poll worker to view and update voter registration records.

EPB system configuration (EPBSC) – A physical instance of an EPBS with all its components configured for use. An EPBS *configuration* consists of peripherals (e.g., printers, scanners, etc.) and a set of configured, networked EPBs. An EPBSC may contain auxiliary servers.

Voter record – The *voter registration record* and *voter activity record* of a voter.

Local voter database - A collection of all *voter records* specific to a jurisdiction. The initial state of the *local voter database* is compiled and certified by the relevant authority. Poll workers make updates to the *local voter database* throughout the election by using the EPBS to reflect ongoing voter activity.

Voter list – A printable, exportable, and human-readable representation of the *local voter database*.

Completed update – An update to a *voter registration record* is *completed* if a query for said *voter registration record* on any active EPB within the EPBS returns the same data.

Quiescent – The EPBS is *quiescent* if all user-initiated updates have completed at all *electronic poll books*.

Reconfigure EPBSC – Configuring, adding, or removing any of the *electronic poll book configuration's* peripherals, *electronic poll books*, or auxiliary servers.

Requirements Relevant to the Distributed Nature of the System. We now present the most relevant requirements [36] that specifically deal with the distributed nature of any comprehensive e-pollbook solution. Broadly speaking, the requirements are formulated to ensure the following.

- Fault-tolerance: The system must not contain a single point of failure, and failures (up to a design limit) must not prevent the system from operating. Main types of failures are the failures of the physical system components or the software in these components, and communication failures.
- Service availability: The system must be able to provide the required service in the face of adversity and perturbations (again, up to its design limits).
- Data consistency: The data contained within the system (e.g., voter records) must be viewed consistently following any changes to the data.
- Data survivability: No data may be lost if certain components of the system fail (up to its design limit).
- System reconfigurability: The system must enable faulty components to be removed/replaced without requiring halting or restarting the overall system.

We now state the most relevant requirements [36] in an abridged form.

AR-2: No single point of failure: The EPBS must be designed to tolerate any single point of failure scenarios.

AR-1: At least three EPBs in an EPBS: An EPBS must support at least three (3) electronic poll books in a single polling location. Each of the electronic poll books must be usable concurrently. Should one of the electronic poll books become inoperable, the operation of the remaining electronic poll books must not be affected.

FR-1: Adding a new EPB to the EPBS: The EPBS must provide means for the integration of an additional EPB into its configuration at any point throughout the election without requiring a shutdown or a restart.

FR-2: Removing an EPB from the EPBS: The EPBS must provide means for the exclusion of an EPB from its configuration at any point throughout the election without requiring a shutdown, or restart of the EPBS. This action does not require physical access to the EPB that is to be excluded.

FR-21: One voter/one vote within EPBS: The EPBS must guarantee that within an EPBS configuration a voter can be checked in at most once during normal connectivity.

RR-1.1: Voter check-in during interruption of connectivity: In the event of a temporary interruption of connectivity within an EPBS, the EPBS must permit a voter to check-in.

RR-1.2: Upon restoration of connectivity: In the event of a temporary interruption of connectivity within an EPBS, the EPBS must automatically restore voter list consistency across the EPBs after connectivity is restored.

RR-1.3: Identify double voting: In the event of a temporary interruption of connectivity within an EPBS, the system must identify voters that have been checked in more than once during the interruption of connectivity.

RR-5: Local voter database replicas: Within the EPBS there must exist at least two replicas (logical or physical) of the local voter database. These replicas must be stored in distinct physical storage components. (Note: together with AR-1 and RR-1.1 the number of replicas may need to be higher.)

RR-6: Local voter database replica consistency: If the EPBS is in a quiescent state all replicas of the local voter database must be logically consistent.

RR-7: Operational consistency: Any update to a voter record or to any other data pertaining to the election completed on one EPB must be seen as complete on all other EPBs.

These requirements are quite intuitive, and we consider them necessary for any implementation of an e-pollbook system. Next we identify several results that make it challenging to satisfy these requirements.

4 E-Pollbooks and the Distributed Systems Theory

We cite results from the distributed systems theory that stress the need for careful design in developing e-pollbook solutions. *A well-informed reader will be familiar with some if not most of these results. Some of these results are quite venerable, and they have been known for some time. This is why it is particularly surprising and troubling that all existing commercial e-pollbook systems appear to be oblivious of these results, and some claims regarding the capabilities of these systems are in conflict with the known facts.*

At first glance, the facts we cite here cast a pessimistic view on the ability to build dependable systems that coordinate their activities in non-trivial ways or that maintain replicated shared data with guaranteed consistency (e.g., if a data object is changed, then the following read of the object value must reflect the change). However, this does not mean that one cannot build reliable and usable e-pollbook systems. In order to succeed, one needs to understand the theoretical limitations and to make sensible assumptions about the nature of failures, communication, and asynchrony. The main point here is that any claims about a system that provides a solution that is able to deal with the requisite adversity and perturbations in the computation and networking medium, but that are not aware of these known results, is to be suspect.

In what follows we do not cross-reference the requirements from Sect. 3, but, as indicated earlier, the selected set of requirements deals collectively with the issues of fault-tolerance and availability (AR-1, AR-2), communication (RR-1.1, RR-1.2), agreement and consistency (FR-21, RR-1.3, RR-6, RR-7), and survivability (FR-1, FR-2, RR-5) of the shared data. (See [36] for details.)

Here we focus on the negative (impossibility) research results. Although specialized practical solutions exist for certain modified versions of the problems given here, we do not present them: not only the solution space is very large, but more importantly, it is the duty of responsible system designers to investigate relevant solutions when they will have started gaining the necessary insight.

4.1 Consistent Data Store with Device Crashes

Any e-pollbook system must be able to tolerate benign failures of individual devices, specifically, a failure where the faulty device stops at an arbitrary instant of time and does not perform any further actions. Such benign failures are known as crashes. E.g., a polling place may have several devices used to check in voters. A crash of such a device must not prevent other working devices from functioning, and the crash must not destroy the consistency of the shared data maintained by the system (of course the data must be replicated for survivability). E.g., if a voter was successfully checked in, then all operating devices must agree that this is the case, regardless of a crash. If this cannot be guaranteed, then an ill-motivated voter may attempt to vote more than once. Suppose the type of data we are interested in is consistent read/write data.[1] This is a basic data type, much simpler than data types that support more complicated read-modify-write operations. It turns out that any system of devices implementing such objects can tolerate the crashes of only a minority of the devices, e.g., [6,11]. *A system of N processors cannot implement a consistent read/write object where all object access operations terminate (complete) in the presence of F crashes if $N \leq 2F$.*

This means that to tolerate a single crash, three replicas are needed. To tolerate two crashes requires replication at five devices, etc. Any poll book solution that replicates its data in two locations and that claims to tolerate a single runtime crash cannot possibly be correct.

4.2 Coordinated Action with Link Failures

Given that multiple devices are necessary in any e-pollbook system, they must must provide their service consistently. Suppose several devices (say, individual poll book devices) need to agree on a common course of action, e.g., by deciding on a value that indicates what action to take. This is known as the agreement problem, and the correctness conditions for a solution are as follows: *(a) Agreement*: no two devices agree on different values, *(b) Validity*: if all devices propose the same value, then this is the only possible agreement value, *(c) Termination*: all non-faulty devices eventually decide. Now suppose that there are just two devices that never fail, but communication can be unreliable, e.g., messages can be lost because of failures or interference. If this is the case, one of the oldest results in distributed computing tells us that there is no protocol that always solves this agreement problem [24]. *There is no algorithm that solves the coordinated action problem for two processors that communicate using unreliable messaging.*

Needless to say, if the problem cannot be solved for two devices, it cannot be solved for any larger number of devices. Of course, this problem still needs to be solved in real systems. This is normally done by strengthening the assumptions

[1] Recall that here we are interested in an implementation of a data object that is consistent, i.e., atomic or linearizable, if the users that access the object are presented with an illusion that there is a single copy of the object that is accessed sequentially regardless of how the object is implemented in the underlying distributed system.

about the model of computation or by relaxing the problem requirements [34]. For example, this can be done by limiting the types of failure that the system tolerates and by stating guarantees probabilistically, thus allowing a system to be incorrect with very small probability. (Similar approaches can be applied in solving other problems we describe in the sequel.) Incidentally, this problem is known as the "Two Generals Problem" in the literature. Here, two generals must launch a coordinated attack, lest they be defeated one at a time by the opposing force. The generals communicate by messengers that can be intercepted or destroyed. The commercially available e-pollbook solutions are routinely silent about the system behavior when communication may be unreliable or lossy.

4.3 Availability, Consistency, and Network Partitions

All devices in the e-pollbook system must present a consistent view of the underlying data shared by the system. Because there must not be any single points of failure, the system must be distributed. If the system implementation is distributed, it must rely on some network for communication among its components. The implementation cannot assume that communication is always reliable; in particular, network failures may isolate some of the devices in the system. Clearly it is desirable for the service to be available and consistent, however, the well-known "Brewer's conjecture" posits that it is not possible to simultaneously guarantee consistency, availability, and partition-tolerance [12]. *It is impossible for any distributed service implementation to provide the guarantees of (i) consistency, (ii) availability, and (iii) partition-tolerance.*

The above statement can be made more specialized for read/write objects as follows [22]. *It is impossible for any distributed service implementation of shared read/write data objects to guarantee (i) consistency, and (ii) availability, if the underlying asynchronous messaging system allows for message loss.*

This means, in particular, that if messages can be lost (e.g., due to jamming or denial-of-service attack), then either the data (e.g., voter records) may appear inconsistent or the service may be unavailable. The above result holds even if the system becomes synchronous, with known delays on the messages.

4.4 Reaching Agreement in the Presence of Crashes and Asynchrony

A polling place with several devices used to check in voters must be able to tolerate crashes. A crash must not prevent the overall system from taking coordinated actions. As before, the e-pollbook devices may not be in perfect synchrony with each other, e.g., processing delays and arbitrary timing of actions by the poll officials is likely to introduce some measure of asynchrony. Unfortunately, reaching agreement in the presence of even a single crash may be impossible in all cases for an asynchronous system, even if no message is ever lost. A seminal and venerable result from the distributed computing theory states the following [19].

For an asynchronous system of processors that communicate using reliable channels there is no algorithm that solves the agreement problem and that guarantees termination in the presence of a single crash.

This generally means that if a system relies on solving the agreement problem as part of its implementation, there may be some operations that never complete (or that are very slow). Thus, any system that claims that all of its devices are always in some type of agreement on certain values and that has good performance for all operations must make several non-trivial assumptions about the nature of failures and the constraints on asynchrony. If these assumptions are not explicitly stated, then the claims are to be taken with a grain of salt.

4.5 Agreement in the Presence of Malicious Failures

Given the plethora of malware and viruses that may affect a computer system, an e-pollbook system may also need to tolerate malicious failures of individual devices. Such malicious failures are called *byzantine* failures [33,41]. Here if a device fails, it does not stop as in a crash, but instead starts behaving arbitrarily, and in particular, it may perform malicious actions. This will be the case if a device is maliciously tampered with, or if it is infected with malware. For this setting, another seminal result states that a system of three devices (processors) cannot tolerate even a single byzantine fault [41]. *A system of three processors cannot solve the agreement problem in the presence of a single byzantine failure.*

Note that this result holds even if the processors are in a complete synchrony with each other and if there are no other perturbations, such as message delay or loss. For e-pollbook systems this means that a system with less than four devices cannot tolerate even a single malicious failure. The more general result dictates that any system of processors cannot tolerate malicious failures of even a third of the processors [41]. *A system of N processors cannot solve the agreement problem in the presence of F byzantine failures if $N \leq 3F$.*

Thus, in any system where the devices must reach agreement, the correct devices must outnumber the faulty devices by more than a factor of three-to-one. If this is not logistically feasible for a real installation, then the system cannot possibly claim to tolerate tampering with (or theft of) even one device.

4.6 The Problem of Reconfiguration in Dynamic Systems

Thus far we looked at static systems, i.e., a system where the universe of devices is fixed in the initial state. Providing e-pollbook solutions only for static systems is inadequate. Consider a polling place with three initial check-in devices. On the election day everything proceeds smoothly for a while. However the voter turnout is much higher than expected and the lines are getting long. Now suppose one of the three devices crashes. Even if the remaining two devices are operational and are able to provide the needed services, the lines of voters are getting really long now. A well-designed system must always allow for additional check-in devices to be introduced in order to cope with the faulty devices and the higher-than-expected voter turnout. Needless to say, this must be accomplished without halting the check-in process and without restarting any devices in the system.

The general problem of removing devices from a system and introducing new devices is known as the *reconfiguration* problem. In our context, we are not concerned with simply adding and removing devices: we need to also make sure that no data is lost and that the new devices are brought up to date with respect to the state of the data. Here all devices must have a consistent view of the state of the system. For a distributed system that is charged with maintaining consistent shared data the reconfiguration operation is described as follows [21].

Reconfiguration is the process of replacing one set of devices in a distributed system with an updated set of devices. In this process, the data is propagated from the old devices to the new set, and allowing devices that are not in the new configuration to safely leave the system. This changeover has no effect on data-access operations, which may continue to store and retrieve the shared data.

Development of algorithms implementing reconfiguration while providing uninterrupted and consistent access to data is an active area of research. A discussion of approaches to reconfiguration, including the use of Consensus and Group Communication Systems [9] can be found in [23,38]. The solutions to the reconfiguration problem are going to be difficult and fraught with impossibility results akin to those we discussed earlier. Regardless of how the reconfiguration of the set of devices is done, any practical implementation of e-pollbooks must address the challenge of deciding when to reconfigure. One approach is to leave this decision to the environment, e.g., the users of the system. Access to the reconfiguration service should be available to system administrators to enable reconfiguration based on policies, such as introducing new device in case of high voter turnout or removing misbehaving devices. For larger installations, reconfiguration could be enacted automatically when a failure of a certain number of devices is detected. This is a more complicated solution, but it has the potential of providing superior quality of service. Given that solutions to the reconfiguration problem are not routine, one must exercise caution. Any e-pollbook system that claims to provide reconfiguration features without supporting documentation and without rigorous arguments about the system's correctness must be carefully examined before any use on the election day.

5 A Broader Look at E-Pollbook Landscape

While this presentation focuses on the distributed systems aspects of e-pollbook systems, a comprehensive solution needs to consider a much broader landscape. There are additional dimensions of this problem that include not only the technological issues, but also societal and legal issues that impose separate requirements and that need to be incorporated. Furthermore, a substantial effort is needed to research, design, and implement a robust framework with rigorous semantics and security guarantees that provides a sound foundation for implementing e-pollbook solutions meeting the relevant requirements. To be successful, an approach to a solution needs to weave seamlessly key insights and contributions in distributed computing and cryptography into a general and implementable framework that is flexible, reusable and promotes the adoption of proven techniques with strong theoretical underpinnings.

Thus, to ensure that the technological approach encompassing theory and systems addresses meaningfully the societal, political, and legal needs it has envelop three interwoven areas: (1) Social and Political Science Dimension, (2) Computer Science as a Foundation for Software, and (3) Systems Implementation and Evaluation. This is illustrated in Fig. 1.

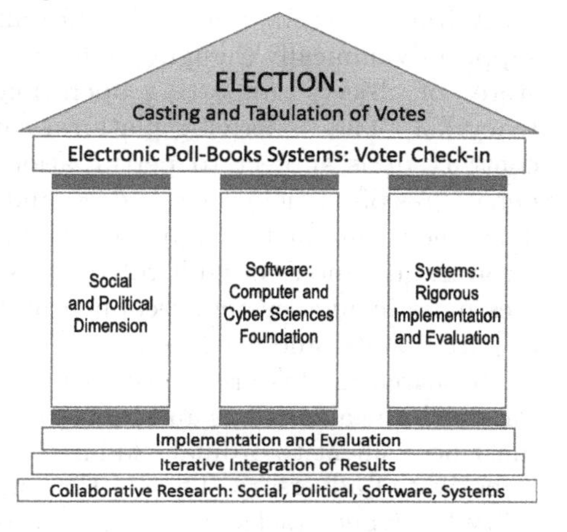

Fig. 1. Voters participate in the election proper, i.e., they cast their votes, only after they are admitted by means of the electronic poll-book system. The development of the systems rests on the three integrated pillars as shown.

5.1 Social and Political Science

A 2020 poll conducted by well-regarded polling firm Ipsos found among US voters that 17% had waited in line more than one hour to vote, 4% were told their name was not on the registered voter list, 4% could not physically access the polling location, and 3% were told they did not have the correct identification [46]. Voter disenfranchisement can lead to a loss of system and government legitimacy [1]. Democracies are sustained and driven by the general principle of one person one vote. When that principle is challenged, whether due to social or administrative barriers, such as with a flawed voter check-in system, negative systemic and governing outcomes can result. Pew survey following the 2020 US Election found that more than 40% of respondents believed the election was not run well, and 15% were not confident that their vote was accurately counted.

Any e-pollbook system must offer features that enhance system legitimacy and efficiency. To accomplish this, it is important to conduct survey research, including elections administrators and citizens, that assesses the features of check-in systems. The results need to be incorporated into any e-pollbook system, and then evaluated, while keeping in mind the efficiency, usability, and availability of the implementation. Software artifacts need to be open-source and publicly available to assuage voter concerns of security and propriety.

5.2 Computing Theory Foundation

Any usable e-pollbook system is an inherently distributed system, consisting of multiple check-in devices and any supporting servers at the polling place (we take the position that within the current state-of-the-art it is unwise to rely on the Internet, or extend the system beyond a single polling place). In this section we survey some of the research topics that can be considered "theory." The problems that are more engineering in nature are deferred to the following section.

A framework for building secure and dependable e-pollbook systems must support dynamically changing collections of physical devices and provide resilient stores of objects supporting operations with provable semantics and consistency [23]. The underlying platform can not be assumed to be "nice." On the contrary, it is subject to perturbations, such as failures, delays, arrivals and departures of participating devices, and–ultimately–even malicious participants. The algorithms included and implemented in the framework must come with rigorous guarantees of fault-tolerance, security, and performance. These guarantees must include system reconfigurability to deal with the changing collection of participants, and consistency of the provided object storage.

Research in this area must include a definition of threat and failure model for which the algorithms and protocols are to be designed. An e-pollbook system must be resilient to ordinary failures of devices and networks, and to attacks by a motivated adversary. The election officials and voters are aware of the possibility for cyberattacks on election infrastructure and the States have mandated that polling places must not allow election-related devices to connect to the Internet [20,36], and the NASEM Committee on the Future of Voting recommends avoiding Internet-dependent election systems [2]. An e-pollbook system must operate in isolation within a polling place. Yet the system must tolerate attacks that can occur in the proximity to the polling place. One needs to assume that an adversary can monitor all network traffic at the polling place, and inject arbitrary packets into this network. The adversary can also cause traffic in the network to be dropped or delayed for an arbitrary, but finite, amount of time. Moreover, the adversary can compromise some of the devices, turning them into malicious participants that can also leak information to the adversary. All of these attacks can be launched by an attacker in close geographic vicinity of the public section of a polling place, especially if the devices comprising the system communicate with the over a wireless network (which is usually the case).

While the distributed computing research has a wealth of relevant results, much work remains in several ares. The data storage must be resilient and provide consistency guarantees supporting the at-most-once update semantics [28] (at most one vote). The imperfect biological units (i.e., voters and officials) must be able to interact with the system without destroying its semantics, here we note that election officials must be legally authorized to override that automated operation of the e-pollbook system. It is also fruitful to explore the provision of lightweight blockchain-style services [17], e.g., to implement audit logs and the monotone collection of data (yes, the reader expected us to say "blockchain" at least once). It is also relevant to extend the realm of byzantine fault tolerance to the current context (cf. [8,16]). While the byzantine model is heavy-handed, it does incorporate all malicious behaviors. Auditability is a must, and so the provision of immutable audit logs and efficient tools for examining such logs [4,37] is required. It is also necessary to adapt cryptographic techniques to deal efficiently with the security requirements, although much of this can be addressed through sound software engineering principles with respect to security, including authentication and encryption.

Catastrophic failures present another challenge. What should happen if the system completely fails? In the traditional distributed systems work one specifies an upper bound on the number of failures, then all bets are off when the limit is exceeded. This is unacceptable in the elections context, and solutions must deal with such scenarios. Here an approach based on self-stabilization needs to be explored (cf. [10,15]). But even this may not be enough, and for this reason there may be no way to avoid having an up to date printed record as the election progresses. In fact several states adopt the policy of a parallel use of the manual systems alongside the e-pollbook system to avoid disasters.

5.3 Systems: Development, Implementation, and Evaluation

In tandem with the theoretical efforts in the previous section, it is also necessary to develop and implement a middleware substrate, providing consistency guarantees, and to showcase this to solve a pressing implementation challenge in e-pollbook systems. The solutions to the research problems must also be shown be practical. The middleware must support dynamic storage [23] with at-most-once updates [28], providing a tamper-proof append-only audit log [4,37], and limiting the effect of compromised devices without imposing undue cryptographic overhead [35,47]. The middleware needs to also manage a reconfigurable collection of devices while tolerating benign and malicious failures. It must maintain a consistent view of the membership of devices, ensuring that only the authorized devices can participate in the e-pollbook operation and submit check-in requests. The reconfiguration of the system must be in response to failures and on demand (by election officials) without disrupting the check-in process. Note that a theft or an unauthorized access must cause or enable detection so that an immediate action can be taken. The system must have acceptable performance and must reasonably scale (in some jurisdictions there will be several tens of devices). This performance must not unduly degrade or degrade gracefully in the face of failures of devices or the network. This includes providing a reasonably short response time for check-in requests regardless of how many client devices are participating.

These goals must be achieved while also remaining resilient to adversarial behavior. The system must minimize the effect of failed or compromised devices, ensuring that malicious records they may introduce are not accepted by the system components. If a system deploys local servers in addition to the check-in device (usually tablets) the system must also tolerate malfunctions among the servers, including byzantine failures, and a few compromised servers should not spoil the integrity of the election. Finally, the system must implement the generation of trustworthy, auditable logs of all actions taken by client and server devices, and these logs must be recoverable despite any number of device failures.

The middleware must faithfully implement the guarantees of the underlying algorithms, including the guaranteed performance that depends on the perturbations in the underlying platform—the system must provide safety in all executions and conditional performance guarantees. The focus here is not on high volume data movers, but more on the integrity of the system state. Indeed, in the case of e-pollbooks, the information being exchanged by the devices boils down to the occasional on-site voter registration record creation and the "crossing-off" of identified voters whose arrival rate at the polling station is slow relative to the processing speed. However, resilience of the system is paramount as no information can be lost, corrupted, unavailable or inconsistent because of the failures (or disappearance, e.g., theft) of individual devices. It is equally important to retain the ability to inject new devices into the active pool to offset the loss of existing devices and without incurring any down time for *any* devices.

Because it is unavoidable that e-pollbook devices communicate wirelessly, it is important to incorporate traffic analysis to detect possible network attacks. Here a machine learning-based intrusion detection approach can be explored [3,31]. An e-pollbook system must also have means to communicate with the state-wide voter registration systems prior to the start of an election. This communication can be electronic and/or by means of removable media, in all cases the security, authenticity, and integrity of the voter data must be guaranteed.

The system must implement a rigorous framework for authenticating devices, software components, and communication, and for encrypting the relevant information (but note that audit logs must be authenticated but not be encrypted because the public must be able to inspect logs without relying on any system for decryption). Much of this can be accomplished by a careful application of known techniques, and intelligent the use of cryptographic tools (lest we create only a false sense of security [14]).

Lastly, research on and development of e-pollbook systems must be evaluated. To evaluate an implementation framework and the underlying algorithms, the following criteria must be kept in mind: (1) Simplicity: The system should be clear, with succinct, mathematically clear properties, and easy to understand and to use. (2) Applicability: It should be directly useful in developing a real e-pollbook application, such that a jurisdiction might be tempted to use. (3) Feasibility: It should be implementable with acceptably good performance. The underlying algorithms and protocols must be evaluated according to their correctness in with respect to their specifications, their simplicity, and their degree of security, performance and fault-tolerance.

6 Discussion

We presented a view of e-pollbook solutions as distributed systems. We cited requirements that need to be satisfied by any robust e-pollbook system in order to guarantee fault-tolerance, availability, and consistency. We also present a broader view that societal issues, additional research directions, and the need for reference implementations. While it is also necessary to address certain security issues, including the security of underlying physical platforms, and catastrophic failure issues, these are outside of the intended scope. We provide key results from the distributed system research showing that it is challenging to build adequate poll book solutions, and that without being grounded in relevant research any solution is likely to be incorrect and provide only an illusion of fault-tolerance, consistency, and safety. Indeed, we are not aware of a single commercially available implementation that satisfies the overall requirements and that are consistent with the relevant research. An important conclusion is that e-pollbook development is an attractive application domain for the research in dependable distributed computing.

References

1. Deepening democracy: a strategy for improving the integrity of elections worldwide. In: Global Commission on Elections, Democracy and Security. Kofi Annan Foundation (2012). https://aceproject.org/ace-en/topics/ei/onePage
2. Securing the Vote: Protecting American Democracy. National Academies of Sciences, Engineering, and Medicine September 2018
3. AlEroud, A., Karabatis, G.: Bypassing detection of url-based phishing attacks using generative adversarial deep neural networks. In: Proceedings of the Sixth International Workshop on Security and Privacy Analytics, pp. 53–60 (2020)
4. Antonyan, T., et al.: Automating voting terminal event log analysis. In: EVT/WOTE (2009)
5. Aponte, A., Cruz, J., Jennings, C., MacDonald, D., Wooden, S.: Committee of Inquiry Report of Factual Findings. Tech. Rep, City of Hartford Court of Common Council (2015)
6. Attiya, H., Bar-Noy, A., Dolev, D., Peleg, D., Reischuk, R.: Renaming in an asynchronous environment. J. ACM **37**(3), 524–548 (1990)
7. Bernhard, M., et al.: Public evidence from secret ballots. In: Electronic Voting, pp. 84–109 (2017)
8. Binun, A., et al.: Self-stabilizing byzantine-tolerant distributed replicated state machine. In: Bonakdarpour, B., Petit, F. (eds.) Stabilization, Safety, and Security of Distributed Systems, SSS 2016. LNCS, vol. 10083, pp. 36–53. Springer, Cham (2016). https://doi.org/10.1007/978-3-319-49259-9_4
9. Birman, Ken: A history of the virtual synchrony replication model. In: Charron-Bost, B., Pedone, F., Schiper, A. (eds.) Replication. LNCS, vol. 5959, pp. 91–120. Springer, Heidelberg (2010). https://doi.org/10.1007/978-3-642-11294-2_6
10. Blanchard, P., Dolev, S., Beauquier, J., Delaët, S.: Practically self-stabilizing Paxos replicated state-machine. In: Noubir, G., Raynal, M. (eds.) Networked Systems, NETYS 2014. LNCS, vol. 8593, pp. 99–121. Springer, Cham (2014). https://doi.org/10.1007/978-3-319-09581-3_8
11. Bracha, G., Toueg, S.: Asynchronous consensus and broadcast protocols. J. ACM **32**(4), 824–840 (1985)
12. Brewer, E.A.: Towards robust distributed systems (abstract). 19th ACM Symposium on Principles District Computer, 7, ACM, 16-19 July 2000
13. Davtyan, S., et al.: Taking total control of voting systems: firmware manipulations on an optical scan voting terminal. In: 2009 ACM Symposium on Applied Computing, SAC 2009, pp. 2049–2053. ACM (2009)

14. Davtyan, S., Kiayias, A., Michel, L., Russell, A., Shvartsman, A.A.: Integrity of electronic voting systems: fallacious use of cryptography. In: Proceedings of 27th Annual ACM Symposium on Applied Computing, SAC 2012, pp. 1486–1493. ACM, March 2012
15. Dolev, S., Georgiou, C., Marcoullis, I., Schiller, E.M.: Self-stabilizing byzantine tolerant replicated state machine based on failure detectors. In: Dinur, I., Dolev, S., Lodha, S. (eds.) Cyber Security Cryptography and Machine Learning, CSCML 2018. LNCS, vol. 10879, pp. 84–100. Springer, Cham (2018). https://doi.org/10.1007/978-3-319-94147-9_7
16. Dolev, S., Liber, M.: Toward Self-stabilizing blockchain, reconstructing totally erased blockchain (preliminary version). In: Dolev, S., Kolesnikov, V., Lodha, S., Weiss, G. (eds.) Cyber Security Cryptography and Machine Learning, CSCML 2020. LNCS, vol. 12161, pp. 175–192. Springer, Cham (2020). https://doi.org/10.1007/978-3-030-49785-9_12
17. Dolev, S., Wang, Z.: Sodsbc: Stream of distributed secrets for quantum-safe blockchain. In: IEEE International Conference on Blockchain, pp. 247–256 (2020)
18. Cox, E.: New voting machines finally on horizon: seven years later, Maryland finally buying voting machines with a paper trail. Baltimore Sun (2014)
19. Fischer, M.J., Lynch, N.A., Paterson, M.S.: Impossibility of distributed consensus with one faulty process. J. ACM **32**(2), 374–382 (1985)
20. Gardner, W.M.: New Hampshire Electronic Poll Book System Request for Information V0.1. New Hampshire Department of State (2017)
21. Gilbert, S., Lynch, N., Shvartsman, A.: RAMBO: a robust, reconfigurable atomic memory service for dynamic networks. Distrib. Comput. **23**(4), 225–272 (2010)
22. Gilbert, S., Lynch, N.A.: Brewer's conjecture and the feasibility of consistent, available, partition-tolerant web services. SIGACT News **33**(2), 51–59 (2002)
23. Gramoli, V., Nicolaou, N., Schwarzmann, A.A.: Consistent Distributed Storage. Morgan & Claypool Publishers (2021)
24. Gray, J.N.: Notes on data base operating systems. In: Bayer, R., Graham, R.M., Seegmüller, G. (eds.) Operating Systems. LNCS, vol. 60, pp. 393–481. Springer, Heidelberg (1978). https://doi.org/10.1007/3-540-08755-9_9
25. Herlihy, M.P., Wing, J.M.: Linearizability: a correctness condition for concurrent objects. ACM Trans. Prog. Lang. Syst. **12**(3), 463–492 (1990)
26. Iredale, T., Clark, K.: System and method for synchronizing electronic poll book voter databases. U.s. patent us8812594 b2, 19 August 2014
27. Jancewicz, R., Kiayias, L., Michel, Russell, A., Shvartsman, A.: Malicious takeover of voting systems: arbitrary code execution on optical scan voting terminals. In: 28th Annual ACM Symposium on Applied Computer, pp. 1816–1823 (2013)
28. Kentros, S., Kiayias, A., Nicolaou, N., Shvartsman, A.A.: At-most-once semantics in asynchronous shared memory. In: Keidar, I. (ed.) Distributed Computing, DISC 2009. LNCS, vol. 5805, pp. 258–273. Springer, Heidelberg (2009). https://doi.org/10.1007/978-3-642-04355-0_27
29. Kiayias, A., Michel, L., Russell, A., Sashidar, N., See, A., Shvartsman, A.:. An authentication and ballot layout attack against an optical scan voting terminal. In: Electronic Voting Technology Workshop, EVT 2007. USENIX Association (2007)
30. Kiayias, A., et al.: Tampering with special purpose trusted computing devices: a case study in optical scan e-voting. In: Computer Security App-s Conference, ACSAC 2007. 23rd Annual, pp. 30–39 (2007)
31. Kolias, C., Kambourakis, G., Stavrou, A., Gritzalis, S.: Intrusion detection in 802.11 networks: empirical evaluation of threats and a public dataset. IEEE Commun. Surv. Tutor. **18**(1), 184–208 (2015)

32. Lamport, L.: On interprocess communication. part i: basic formalism. Distrib. Comput. **2**(1), 77–85 (1986). https://doi.org/10.1007/BF01786228
33. Lamport, L., Shostak, R., Pease, M.: The byzantine generals problem. ACM Trans. Program. Lang. Syst. **4**(3), 382–401 (1982)
34. Lynch, N.A.: Distributed Algorithms. Morgan Kaufmann Publishers Inc. (1996)
35. Maleki, H., Rahaeimehr, R., Jin, C., van Dijk, M.: New clone-detection approach for RFID-based supply chains. In: 2017 IEEE International Symposium on Hardware Oriented Security and Trust (HOST), pp. 122–127 (2017)
36. Merrill, D.W.: Connecticut Electronic Poll Book System: Requirement Specification V1.0. Office of the Connecticut Secretary of the State (2015)
37. Michel, L.D., Shvartsman, A.A., Volgushev, N.: A systematic approach to analyzing voting terminal event logs. USENIX J. Election Technol. Syst. (JETS) **2**(2), 34–53 (2014)
38. Musial, P.M., Nicolaou, N.C., Shvartsman, A.A.: Implementing distributed shared memory for dynamic networks. Commun. ACM **57**(6), 88–98 (2014)
39. One Hundred Seventh Congress of the United States of America. Help America Vote Act of 2002 (2002)
40. Pawlak, M., Guziur, J., Poniszewska-Marańda, A.: Towards the blockchain technology for system voting process. In: Cyberspace Safety and Security, pp. 209–223 (2018)
41. Pease, M.C., Shostak, R.E., Lamport, L.: Reaching agreement in the presence of faults. J. ACM **27**(2), 228–234 (1980)
42. Peisch, P.: Procurement and the polls: how sharing responsibility for acquiring voting machines can improve and restore confidence in American voting systems. Georgetown Law J. **97**(877), 877–915 (2009)
43. Rivest, R.: Clipaudit: a simple risk-limiting post-election audit. ArXiv, abs/1701.08312 (2017)
44. Rivest, R.L., Stark, P.B.: When is an election verifiable? IEEE Secur. Priv. **15**(3), 48–50 (2017)
45. Shvartsman, A.A., Kiayias, A., Michel, L., Russell, A.: On the security and integrity issues of optical scan voting systems. In: County of Nassau Board of Elections against State of New York, New York State Board of Elections, pp. 1–23. Supreme Court of the State of New York, County of Nassau, 19 March 2010
46. Thomson-DeVeaux, A., Mithani, J., Bronner, L.: Why many Americans don't vote. In: FiveThirtyEight, 26 October 2020
47. van Dijk, M., Jin, C., Maleki, H., Ha Nguyen, P., Rahaeimehr, R.: Weak-unforgeable tags for secure supply chain management. In: Meiklejohn, S., Sako, K. (eds.) Financial Cryptography and Data Security, FC 2018. LNCS, vol. 10957, pp. 80–98. Springer, Heidelberg (2018). https://doi.org/10.1007/978-3-662-58387-6_5
48. Wolchok, S., Wustrow, E., Isabel, D., Halderman, J.A.: Attacking the Washington, D.C. internet voting system. In: Keromytis, A.D. (ed.) Financial Cryptography and Data Security, FC 2012. LNCS, vol. 7397, pp. 114–128. Springer, Heidelberg (2012). https://doi.org/10.1007/978-3-642-32946-3_10

Asynchronous Proof-of-Stake

Jakub Sliwinski(✉) and Roger Wattenhofer

ETH Zurich, Zürich, Switzerland
{jsliwinski,wattenhofer}@ethz.ch

Abstract. We introduce a new permissionless blockchain architecture called Cascade (Consensusless, Asynchronous, Scalable, Deterministic and Efficient). The protocol is completely asynchronous, and does rely on neither randomness nor proof-of-work. Transactions exhibit finality within one round trip of communication.

Cascade is consensusless and only satisfies a relaxed form of consensus by introducing a weaker termination property. Without full consensus, the protocol does not support certain applications, such as general smart contracts. However, many important applications do not require general smart contracts, and Cascade is an advantageous solution for these applications. In particular, the architecture can implement the functionality of a cryptocurrency such as Bitcoin, replacing Bitcoin's energy-hungry proof-of-work with a proof-of-stake validation.

1 Introduction

Nakamoto's Bitcoin protocol [12] has taught the world how to achieve trust without a designated trusted party. The Bitcoin architecture provides an interesting deviation from classic distributed systems approaches, for instance by using proof-of-work to allow anonymous participants to join and leave the system at any point, without permission.

However, Bitcoin's proof-of-work solution comes at serious costs and compromises. The security of the system is directly related to the amount of investments in designated proof-of-work hardware, and to spending energy to run that hardware. Since the system's participants that provide the distributed infrastructure (often called miners) bear significant costs (hardware, energy), the protocol compensates them with Bitcoins. However, adversaries might disrupt this scheme by bribing the miners to behave untruthfully or disrupt the reward payments.

Irrespectively of how costly Bitcoin's proof-of-work gets, this solution can only process a fixed amount of transactions in a given time period, hampering adoption and often making it infeasible to use Bitcoin at all.

To make matters worse, proof-of-work protocols assume critical requirements related to the communication between the participants regarding message loss and timing guarantees. In other words, such protocols are vulnerable to attacks on the underlying network.

C. Johnen et al. (Eds.): SSS 2021, LNCS 13046, pp. 194–208, 2021.
https://doi.org/10.1007/978-3-030-91081-5_13

In the decade since the original Bitcoin publication, researchers have tried to address the wastefulness and ineffectiveness of proof-of-work. One of the most prominent research directions is replacing Bitcoin's proof-of-work with a proof-of-stake approach. In proof-of-stake designs, miners are replaced with participants who contribute to running the system according to the amounts of cryptocurrency they hold. Alas, proof-of-stake protocols require similar communication guarantees as proof-of-work, and thus can also be attacked by disrupting the network. Moreover, proof-of-stake introduces some of its own problems. Prominently, existing proof-of-stake designs critically rely on randomness. To achieve consensus, the participants of such systems repeatedly choose a leader among themselves. Despite being random, this choice needs to be taken collectively and in a verifiable way, which complicates the problem.

Due to the way blockchains typically process transactions, participants have to wait a significant amount of time before they can be confident that their transactions are accepted by the system. For example, it usually takes around an hour for merchants to accept Bitcoin transactions as confirmed, which is unacceptable for time-sensitive applications.

In his seminal paper, Nakamoto made the crucial assumption that his system has to be able to totally order the transactions submitted to the system in order to reject the fraudulent ones. However, meeting this requirement is equivalent to solving the problem known as consensus. Nakamoto's assumption has shaped the design of blockchain systems to this day. Thus, many blockchain systems achieve consensus while not taking advantage of this powerful property, but suffering the associated costs.

Our Contribution. We relax the usual notion of consensus to extract the requirements necessary for an efficient cryptocurrency. Thus we introduce a blockchain design that is Consensusless, Asynchronous, Scalable, Deterministic, and Efficient (Cascade). We claim the protocol to offer a host of exciting properties:

Permissionless: Most importantly, Cascade offers its advantages without relying on permissioned participation. The protocol is permissionless in the same way as other proof-of-stake systems, where participants of the system freely exchange cryptocurrency tokens. Token holders run the system by validating new transactions. Additionally, any token holder can delegate the validation role to other participants, but preserving his ownership of the associated tokens.

Parallelizable: In Cascade, validators running the system can parallelize the processing of transactions. There is no limit to the number of transactions a validator can process by parallelization.

Asynchronous: Cascade does not require the messages to be delivered within any known period of time. Thus the protocol is fully resilient to all network-related threats, such as delaying messages, denial-of-service, or network eclipse attacks. An adversary having complete control of the network always

Table 1. Comparison of cascade to selected BFT/blockchain protocols. Permissioned protocols are on the left, permissionless protocols on the right. We mark all protocols providing full consensus as supporting general smart contracts, even though particular implementations might not feature smart contracts.

	PBFT [3]	HoneyBadger BFT [11]	Broadcast-based [7]	Bitcoin and Ethereum [16]	Ouroboros [8]	Algorand [4]	Cascade
Permissionless				✓	✓	✓	✓
Proof-of-work free	✓	✓	✓		✓	✓	✓
Finality	✓	✓	✓			✓	✓
Asynchronous		✓	✓				✓
Deterministic	✓		✓				✓
Parallelizable			✓				✓
General smart contracts	✓	✓		✓	✓	✓	

can delay the progress of the system (by simply disabling communication), but otherwise cannot interfere with the protocol or trick the participants in any way. Previously approved transactions cannot be invalidated and impermissible transactions cannot be approved.

Final: Under normally functioning network communication, transactions in Cascade are instantly confirmed. Confirmation is final and impossible to revert. This is in stark contrast to systems such as Bitcoin, where the confidence in a transaction being confirmed only probabilistically increases over time.

Deterministic: We assume the functionality provided by asymmetric encryption and hashing. Apart from these cryptographic necessities, Cascade is completely deterministic and surprisingly simple.

Efficient: Unlike proof-of-work, the security of the system does not depend on the amount of devoted resources such as energy, computational power, memory, etc. Instead, similarly to proof-of-stake protocols, Cascade requires that more than two-thirds of the system's cryptocurrency is held by honest participants.

Cascade does not support consensus. This prevents the protocol from supporting applications that involve smart contracts open for interaction with anybody. For example, the smart contract functionality of Ethereum cannot be directly implemented with Cascade. Many important applications (e.g., cryptocurrencies or IoT systems), do not require consensus, and Cascade offers an advantageous solution for these applications.

Table 1 compares the properties of Cascade with some of the most relevant existing BFT/blockchain paradigms. Many more protocols exist that improve some aspects, for example, many protocols improve upon PBFT. While many of these protocols are more performant and efficient than the original PBFT, they share the fundamental disadvantages of PBFT: They are not permissionless, they are not parallelizable, and in order to make progress ("liveness"), they need synchronous communication.

Table 1 shows the close relation of Cascade with broadcast-based protocols. One may argue that Cascade brings the simplicity, robustness, and efficiency of broadcast-based protocols to the permissionless domain.

In this paper we focus on the basic correctness properties of the protocol and leave in depth discussion of the scalability aspect to future work.

1.1 Relaxing Consensus

In the context of a cryptocurrency, consensus is used to solve the problem of double-spending. Suppose Alice holds one cryptocurrency coin. Now Alice sets up a transaction that transfers her coin to Bob (in exchange for a good or service). However, Alice wants to cheat, trying to simultaneously spend the same coin in another transaction to Carol. Upon receiving one (or both) of Alice's transactions, honest agents need to agree on what happens to Alice's coin, preventing Alice from doubling her money. In this context, achieving consensus consists of the following requirements:

Definition (Consensus).
Each honest agent observes some transaction from a pairwise conflicting set of transactions $\{t_0, t_1, \dots\}$.

Agreement: *If some honest agent accepts a transaction* t_i*, every honest agent will accept* t_i*. No conflicting transaction can be accepted.*
Validity: *If all honest agents observe only one transaction* t_i*, only* t_i *can be accepted by honest agents.*
Termination: *One of the transactions* t_i *will be accepted by honest agents.*

The insight leading to the relaxation, is that malicious agents do not need to enjoy any guarantees. Alice tried to cheat by issuing two conflicting transactions. Cascade does not guarantee that any of Alice's conflicting transactions will be accepted.

On the other hand, an honest agent will only create one transaction spending her coin. Thus, every honest agent will see the same candidate transaction. Hence we relax consensus to guarantee termination only for honest agents:

Definition (Cascade Consensus).

Agreement: *As above.*
Validity: *As above.*
Honest-Termination: *If all honest agents observe only one transaction* t_i*,* t_i *will be accepted by honest agents.*

Under this relaxed notion of consensus, if Alice tries to cheat, it is possible that neither Bob nor Carol will accept Alice's transaction. Some honest agents might see one of the transactions first, while others might see the other first. Then the requirement of Honest-Termination does not apply, and the transactions might stay without a resolution forever. This turn of events can be seen as Alice losing her coin due to misbehaviour.

Otherwise, Consensus and Cascade Consensus do not differ. Agreed upon results are final, conflicting results are precluded and honest transactions are accepted. Despite the difference being insignificant with respect to the functioning of a cryptocurrency, this relaxation allows Cascade to combine a large set of advantages.

1.2 Intuition

For simplicity of presentation, we describe Cascade in the terminology of a cryptocurrency and refer to the cryptocurrency managed by the protocol as the money. A more formal description follows in Sect. 3.

Transactions. A transaction transfers money from one or more inputs to one or more outputs. Inputs and outputs are money amounts paired with keys required to spend them.

Validators. In proof-of-stake systems, the agents that own some of the money in the system also run the system. These agents (validators) stay online and participate in validating transactions. In Cascade, we do not require agents to stay online and participate, but allow agents to delegate this responsibility to other agents. Every agent can be a validator. Validators sign correct transactions. The system works correctly as long as agents holding more than two-thirds of the system's money delegate to honest validators.

Confirmations. A transaction t is confirmed by the system if enough validators ack (acknowledge by signing) t. If a transaction receives enough acks, no other transaction conflicting with t can become confirmed. If a cheating Alice attempts to issue two conflicting transactions t and t' at roughly the same time, it is possible that (a) either t or t' gets confirmed (but not both), or (b) neither t nor t' are ever confirmed. Case (b) happens if some validators see and sign t, while others see and sign t'. The system might stay in this state forever with the validators' approval split between t and t'. The result is equivalent to Alice losing the money she attempted to double-spend, and does not constitute any threat to the system.

It is intuitive to verify that such a system does work correctly, if the validating power amounts are statically assigned to the validators, and a set of validators controlling more than two-thirds of the cryptocurrency obeys the protocol. Our system still works correctly when the agents can freely exchange the cryptocurrency and change the appointed validators, even in the harsh conditions of an asynchronous network. Thus, we establish a system with the participation model similar to proof-of-stake protocols, but much simpler than known proof-of-stake protocols.

2 Model

Agents and Adversary. Our blockchain is used and maintained by its participants called agents. Agents who follow the protocol are called honest. The set of agents who do not follow the protocol is controlled by the adversary. The adversary behaves in an arbitrary (adversarial) way.

We make a standard assumption pertaining to proof-of-stake systems that the adversary always controls less than one-third of the cryptocurrency in the system. The assumption is the equivalent of assuming that the adversary controls less than one-third of the permissions in a BFT protocol, or half of the hashing power in a proof-of-work system such as Bitcoin. The idea behind the assumption is that an agent owning a large stake in a system is heavily invested in the system. While sufficiently deep pockets make it possible to disrupt any system, the proof-of-stake assumption ensures that an attack is costly and self-destructive. We introduce more concepts to state this requirement precisely in Sect. 3.4.

Asynchronous Communication. All agents are connected by a virtual network supporting a message diffusion mechanism (such as Bitcoin's network), where agents can broadcast their messages to all other agents. Like in Bitcoin, new agents can join this network to receive new and prior messages.

The network is asynchronous: The adversary controls the network, dictating when messages are delivered and in what order. Messages are required to arrive *eventually*, without any bound on the time it might take. Under such weak requirements, an adversary delaying the delivery of messages can delay the progress of an agent, but otherwise will not be able to interfere with the protocol.

Cryptographic Primitives. We assume the functionality of asymmetric encryption where a public key allows every agent to verify a signature of the associated secret key. Agents can freely generate public/secret key pairs.

We also assume cryptographic hashing, where for every message a succinct, unique hash can be computed. Whenever we say that a transaction t_2 refers to a transaction t_1, we mean that t_2 includes a hash of t_1, and as such uniquely identifies t_1.

Apart from these two cryptographic primitives, the Cascade protocol is completely deterministic.

3 Protocol

3.1 Transactions

Outputs. Outputs are the basic unit of information. Outputs are included in transactions and identify who owns how much money after the transaction was confirmed by the system.

Definition 1 (Output). *An output contains:*

- *Value: A number representing the amount of money.*
- *Owner key: A public key. The agent holding the associated secret key is the owner of the money.*

Agents can reuse their keys for multiple outputs, but for simplicity of presentation we assume that the owner key uniquely identifies a single output.

Transactions. A transaction is a request issued by an agent (or a set of agents) to transfer money to other agent(s). Outputs of a transaction identify recipients of the transaction. The transaction also indicates a validator – some agent devoted to maintaining the system.

Outputs can be associated with some identifying number, but for simplicity of presentation we assume that outputs uniquely identify the originating transaction.

Definition 2 (Transaction). *A transaction t contains:*

- *A set of inputs, where each input is an output of some previous transaction. Transaction t is said to spend these inputs.*
- *A set of outputs. The sum of values of the outputs equals the sum of values of the inputs. This sum is called the value of the transaction.*
- *Validator key: A public key. The value of the transaction is delegated to the agent holding the associated secret key (validator).*

The transaction is signed by all secret keys associated with the inputs.

The validator cannot spend the transaction outputs. After t is confirmed, the validator's signing stake increases until t's outputs are spent.

Genesis. The *genesis* is a special transaction without inputs. The genesis is hard-coded in the protocol and known upfront to every agent. The genesis describes the initial distribution of money among the original agents and the initial validators (which could or could not be the same as the original agents).

The values of all genesis outputs sum up to M, so M is the total money in the system. In this paper, we assume that M never changes.

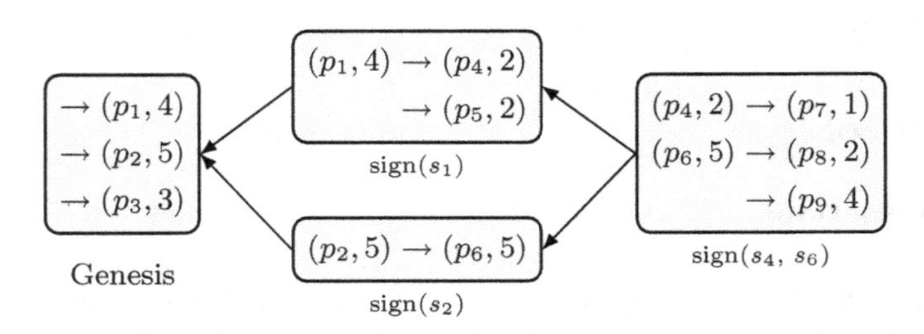

Fig. 1. Example DAG of transactions, validator keys are omitted. The p_i's are owner keys, and s_i's are the corresponding secret keys.

3.2 Validators

Validators are agents processing transactions in the system. Validators listen for transactions being broadcast, and sign them if they have not observed a conflict. An honest validator signs all non-conflicting transactions.

After a transaction t with a value of m is confirmed by the system (explained below), the "signing power" of the validator v indicated in transaction t increases by m (at the cost of the validators indicated in transactions that have output the inputs of t). To spend an output of t, the owner of an output must later broadcast a new transaction, as v cannot spend the outputs of t. An owner of an output of t can change the appointed validator v to any other validator by spending t's output (for instance by self-sending the money), when including a different validator key. Any agent can also indicate themselves as the validator.

The validator v signs transactions in the system to contribute to their confirmation, and the contribution is proportional to the amount delegated to v.

Number of Validators. Similarly to Bitcoin mining pools, the number of validators in Cascade might naturally be relatively small, such that a small number of validator's signatures is needed to confirm a transaction. The protocol can also enforce or encourage the number of validators to form groups, for example by an appropriate fee structure. In contrast to Bitcoin mining pools, the validators forming a group can maintain trustlessness with respect to each other by using an aggregatable signature scheme such as BLS [1]. In this way, a few validator pools would preserve the agency of individual validators. We believe that this is more decentralized than for example Bitcoin. Due to the page limit, we leave these aspects of the protocol to future work.

3.3 Confirmations

A validator broadcasts an *ack* message to communicate a new set of transactions the validator signed.

Definition 3 (ack). *An ack contains:*

- *A reference to the previous ack issued by the same validator.*
- *A set of references to transactions the validator signs.*

The ack is signed by the validator's secret key.

All messages can only reference previously created messages with hashes. Cyclic hash references are impossible and hence all messages form a directed acyclic graph (DAG), with the genesis being the only root. Messages are processed in any order respecting references. Agents do not process a transaction t until they have fully received *past*(t).

Definition 4 (past). *The set of messages reachable by following references from t is called past(t). For a set of messages T, past(T) $= \bigcup_{t \in T}$ past(t).*

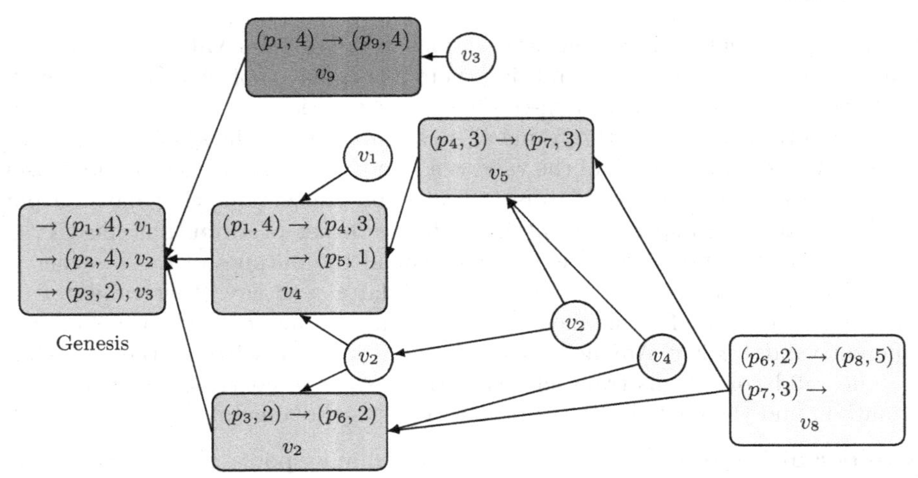

(a) Example transaction DAG, p_i's represent the owners and v_i's the validators. Circle nodes are acks labelled by the issuing validators. Acks point to the transactions being signed and (if available) the previous acks of the same validator. Light blue transactions are confirmed based on the acks. When issuing an ack, validators have to point to the previously issued ack, as exhibited by v_2. The dark grey transaction is an attempt at double-spending; it conflicts with a confirmed transaction and will never be confirmed. The white transaction is not yet confirmed.

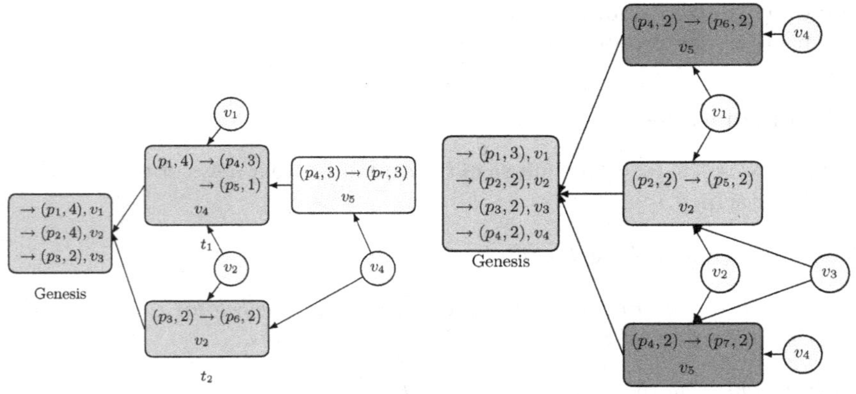

(b) A subview of the transaction DAG from Figure 2a. The set A_{t_1} consisting of the acks of validators v_1 and v_2 is proof that t_1 is confirmed. The set A_{t_2} consisting of the acks of validators v_1, v_2 and v_4 is proof that t_2 is confirmed.

(c) Example attempt at double-spending. The validator v_4 is adversarial, does not reference previous acks in new acks and attempts to confirm conflicting transactions. Honest validators are split between conflicting transactions such that neither will ever be confirmed.

Fig. 2. Example DAGs.

Transactions can be confirmed by the system, and confirmation is permanent. A transaction t becomes confirmed when enough validators broadcast an ack signing it. After a transaction is confirmed, the *stake delegated* to the validator indicated in t increases by the value of t (and appropriately decreases for the validators to whom the inputs were delegated). Thus we define transaction confirmation and the stake delegated to a validator inductively (from genesis) with respect to each other. Genesis is confirmed from the start.

Definition 5 (delegated stake). *Given a set of acks A, let T_A be the set of transactions confirmed in past(A) that indicate v as the validator. The stake delegated to v in past(A) is equal to the sum of values of outputs of transactions in T_A that are delegated to v and that are unspent in past(A).*

Definition 6 (confirmed). *A transaction t is confirmed if the transactions that output the inputs of t are confirmed, and there exists a set of acks A_t such that:*

- *some validators $v_1, \ldots, v_k$ with respective delegated stake $m_1, \ldots, m_k$ in past(A_t) sign t, and $\sum_{i=1}^{k} m_i > \frac{2}{3}M$;*
- *no transaction $t' \in$ past(A_t) shares any input with t.*

Honest agents do not spend their outputs more than once, i.e. every output becomes an input at most once. Assume that t is a transaction by an honest agent. Then we will never see a transaction t' which tries to spend the same outputs as inputs of t. In this case, it is straightforward to collect validator acks for t, and eventually t will have enough acks to be confirmed.

On the other hand, if some t' is sharing inputs with t is also present in the transaction DAG, it is unclear if there can be a set A_t such that t is confirmed. It is only the misbehaving agent's concern to find an appropriate A_t and prove to the recipient of t that t is confirmed.

3.4 Adversary

The adversary behaves in an arbitrary way, and thus might create conflicting transactions, transmit acks that do not reference previously issued acks, send different messages to different recipients, etc.

Any message sent by an honest agent is immediately seen by the adversary. The delivery of each message from an honest agent to an honest agent can be delayed by the adversary for an arbitrary amount of time.

Stake. As explained in Sect. 2, we assume that the value of genesis outputs delegated to the adversary sums up to less than $M/3$. In every transaction, a new validator is indicated. Hence the stake delegated to the adversary shifts over time.

Definition 7 (adversary stake). *Let m_t^h and m_t^a be the sums of values of inputs of t that are outputs of transactions delegated to honest agents and the adversary respectively.*

When transaction t delegated to an honest agent is **confirmed** *(i.e. any A_t exists), then we subtract m_t^a from the amount we count as delegated to the adversary. When transaction t delegated to the adversary is* **issued**, *then we add m_t^h to the amount we count as delegated to the adversary.*

4 Correctness

In this section we outline the proof that the Cascade protocol upholds Cascade Consensus as defined in Sect. 1.1. The proof is available in the online version of the paper [14].

The difficulty lies in the complete asynchrony of the system. In an orthodox blockchain, all confirmed transactions are totally ordered. Such a total order does not exist in Cascade. Moreover, the stake distribution among validators is constantly shifting. The protocol prevents problems by requiring honest validators to reference previous acks. Moreover, when some transaction t shifts the stake from a validator v_1 to a validator v_2, the stake is retracted from v_1 as soon as t is observed, but only credited to v_2 when t is referenced by many other validators and confirmed.

Theorem 8 is the main result we want to prove.

Theorem 8. *The Cascade protocol satisfies Cascade Consensus.*

Under our assumption from Sect. 3.4, more than two-thirds of the money is always delegated to honest validators. Hence, if there is no double-spend alternative to a transaction t, honest validators will sign t and t will be confirmed by the system. Thus Validity and Honest-Termination of Definition 1.1 hold. Whenever any agent observes a transaction t as confirmed, the acks A_t serve as the proof that t is confirmed to any other agent. Therefore, to show that Agreement holds, it suffices to show that no pair of conflicting transactions is ever confirmed. Then the Cascade protocol satisfies Cascade consensus.

For contradiction, assume that some transaction DAG can be produced by the protocol where two conflicting transactions t_x and t_y are confirmed. Consider the instance of such a DAG G that is minimal in terms of the number of transactions.

Consider some transaction t_0 confirmed in G during the protocol's execution based solely on the stake distribution specified in genesis. We show that for any other confirmed transaction t, either $t_0 \in \mathit{past}(A_t)$ or $t \in \mathit{past}(A_{t_0})$ holds in DAG G. We conclude that t_0 cannot conflict with any transaction. Then t_0 does not serve a purpose for the construction of DAG G, as t_0's inputs could be replaced in the genesis with t_0's outputs for a smaller DAG. This contradicts with the choice of G, and Theorem 8 summarizes that under our assumptions, conflicting transactions cannot be confirmed in a single DAG.

As we mention in Sect. 5, in practice agents running Cascade would not need to precisely compute $\mathit{past}(A_t)$ for normal workloads. Every agent would confirm almost all transactions based on a lower-bound of the stake delegated to other agents computed from the observed confirmed and yet-to-be-confirmed transactions. Precise $\mathit{past}(A_t)$ might need to be computed only for some contentious transactions when there is a conflict.

5 Future Work

Due to space constraints we focussed on the basic properties of Cascade in this paper. In this section we briefly outline the aspects of Cascade we plan to discuss and expand on in the future to exhibit the advantages of the protocol.

Parallelization. Provided the topology of the workload is not inherently impossible to parallelize (such as all transactions passing the same token in a chain of transactions), validators can parallelize the signing and processing of transactions. Thus, if we increase the number of machines (with constant bandwidth each) at the validator's disposal, the throughput of Cascade increases without limit. To exclude the corner cases inherently resistant to parallelization, we state Assumption 1.

Assumption 1. *If xM is the value of honest transactions not determined to be confirmed by some honest validators yet, honest validators control more than $(\frac{2}{3}+x)M$ of the stake.*

For example, if some 5% of the system's money is being moved and unconfirmed at some instant, about 71.7% of validators need to be active to process transactions in parallel efficiently.

Signing in Parallel. Each validator v can split the space of keys between multiple servers, for example based on the first few characters of the key. The servers can independently store the spent inputs corresponding to the assigned key space.

To issue an ack signing a lot of transactions in parallel, the implementation might support splitting an ack into multiple parallel messages marked with a message count number and the same ack sequence number.

Determining Confirmation in Parallel. To determine transaction confirmation in parallel, the key space is similarly split between machines that listen to messages being broadcast in the network.

Since bandwidth is limited for individual machines, the network might simply be split into a number of subnetworks corresponding to the key space splitting.

The validators maintain the sets of confirmed and yet-to-be-confirmed transactions in their view. The set of outputs of confirmed transactions that are unspent in the set of confirmed and unprocessed transactions gives a lower bound of the stake delegated to each validator in the view. By Assumption 1, these lower bounds are enough to determine transactions as confirmed without identifying the exact sets of confirming acks A_t or the exact corresponding stake amounts.

Smart Contracts. To support smart contracts callable by arbitrary parties Cascade needs to be augmented with a consensus mechanism ordering inputs. However, such consensus mechanism would be invoked only for the inputs requiring it, where traditional BFT/blockchain protocols totally order *all* transactions, and hence introduce an inherent bottleneck in the design of a system.

The consensus overhead is only necessary for some smart contracts, only when conflicting inputs are issued at the same time, and only with respect to such relevant inputs. Thus, a system processing mostly parallelizable content could enjoy the properties of Cascade for the most part, while resorting to consensus for the contents that require it.

Pruning the DAG. In contrast to standard blockchain systems, Cascade naturally supports checkpoints and pruning old, redundant data from the blockchain, which we discuss in the full version of the paper [14].

6 Related Work

Permissioned Systems. Traditionally, distributed ledgers [3,9] operate with a carefully selected committee of trusted machines. Such systems are called permissioned. The committee repeatedly decides which transactions to accept, using some form of consensus: The committee agrees on a transaction, votes on and commits that transaction, and only then moves forward to agree on the next transaction.

Gupta [7] proposes a permissioned transaction system that does not rely on consensus. In this design, a static set of validators is designated to confirm transactions. Our concepts (such as the use of parallelization) do work in the permissioned setting as well, and could be applied to this work.

The authors of [6] show that the consensus number of a Bitcoin-like cryptocurrency is 1, or in other words, that consensus is not needed. The paper provides an analysis and discussion of which applications rely on consensus and to what extent, all of which is directly relevant to Cascade. The authors draw parallels between permissioned consensusless transaction systems and Byzantine consistent broadcast [2,10].

HoneyBadger BFT [11] provides an asynchronous permissioned system by relying on advanced cryptographic techniques with full consensus. Again, the main differences from Cascade are that the system is permissioned, much more involved, and reliant on randomization.

The authors of [5] introduce a protocol based on reliable broadcast that allows participants to join and leave the system. However, the adversary is required to control a limited number of participants (as opposed to hashing power or stake), so the protocol cannot be applied in permissionless contexts where unknown participants can join freely. The protocol consists of a few rounds of communication to agree on nodes joining or leaving the system.

Permissionless Systems. Bitcoin [12] radically departed from the established model and became the first permissionless blockchain. In the Bitcoin system, there is no fixed committee; instead, everybody can participate. Bitcoin achieves this by using proof-of-work. Proof-of-work is a randomized process tying computational power and spent energy to the system's security, while also requiring synchronous communication. However, Bitcoin's form of consensus hardly satisfies the traditional consensus definition. Instead of terminating at any point, the extent to which the consensus is ensured raises over time, approaching but never reaching certainty. More precisely, in Bitcoin transactions are never finalized, and can be reverted with ever decreasing probability.

Similar to Bitcoin, Cascade allows permissionless participation. In contrast to Bitcoin, Cascade does not rely on proof-of-work or randomization, features parallelizability and finality, and works under full asynchrony.

To address the problems associated with proof-of-work, proof-of-stake has been suggested, first in a discussion on an online forum [13]. Proof-of-stake blockchains are managed by participants holding a divisible and transferable digital resource, as opposed to holding hardware and spending energy. Academic works proposing proof-of-stake systems include designs such as Ouroboros [8] or Algorand [4]. Proof-of-stake blockchains solve consensus and thus do not parallelize without compromises. The reliance on synchronous communication and randomization in proof-of-stake are potential security risks. Despite avoiding these pitfalls, Cascade is also simpler.

DAG Blockchains. To increase the relatively modest throughput of Bitcoin, some proof-of-work protocols employ directed acyclic graphs in the place of Bitcoin's single chain. SPECTRE [15] is likely the closest relative of Cascade among such protocols, as it relaxes consensus similarly to Cascade. However, the similarities are largely superficial, as SPECTRE remains a proof-of-work protocol, employs different techniques, and does not share the other of Cascade's advantages. SPECTRE improves many aspects of Bitcoin, but with respect to the harsh criteria of Table 1, SPECTRE can only earn a tick at permissionless.

ABC. We have been working on the idea of building a consensusfree permissionless DAG blockchain for a few years already. A predecessor of this work [14] discusses related topics not developed here due to space constraints, such as pruning the transaction DAG, fees and money creation.

7 Conclusions

In this paper we presented Cascade, a permissionless and parallelizable blockchain protocol. Cascade provides the functionality of a cryptocurrency without consensus, without proof-of-work, without requiring synchronous communication, without relying on randomness. The protocol is scalable and exhibits finality. The design of Cascade is arguably the simplest possible design for a variety of blockchain applications.

Cascade provides an advantageous solution for applications like cryptocurrencies, where honest participants do not generate conflicting status updates. Supporting general smart contracts would require performing consensus some of the time. Adding this functionality would check the last box in Table 1.

References

1. Boneh, D., Lynn, B., Shacham, H.: Short signatures from the Weil pairing. In: Boyd, C. (ed.) Advances in Cryptology—ASIACRYPT 2001, ASIACRYPT 2001.

LNCS, vol. 2248, pp. 514–532. Springer, Heidelberg (2001). https://doi.org/10.1007/3-540-45682-1_30
2. Cachin, C., Guerraoui, R., Rodrigues, L.: Introduction to Reliable and Secure Distributed Programming. Springer Science & Business Media (2011)
3. Castro, M., Liskov, B., et al.: Practical byzantine fault tolerance. OSDI **99**, 173–186 (1999)
4. Gilad, Y., Hemo, R., Micali, S., Vlachos, G., Zeldovich, N.: Algorand: Scaling byzantine agreements for cryptocurrencies. In: Proceedings of the 26th Symposium on Operating Systems Principles, pp. 51–68. ACM (2017)
5. Guerraoui, R., Komatovic, J., Seredinschi, D.A.: Dynamic byzantine reliable broadcast [technical report]. arXiv preprint arXiv:2001.06271 (2020)
6. Guerraoui, R., Kuznetsov, P., Monti, M., Pavlovič, M., Seredinschi, D.A.: The consensus number of a cryptocurrency. In: Proceedings of the 2019 ACM Symposium on Principles of Distributed Computing, pp. 307–316. ACM (2019)
7. Gupta, S.: A Non-Consensus Based Decentralized Financial Transaction Processing Model with Support for Efficient Auditing. Master's Thesis (2016)
8. Kiayias, A., Russell, A., David, B., Oliynykov, R.: Ouroboros: a provably secure proof-of-stake blockchain protocol. In: Katz, J., Shacham, H. (eds.) CRYPTO 2017. LNCS, vol. 10401, pp. 357–388. Springer, Cham (2017). https://doi.org/10.1007/978-3-319-63688-7_12
9. Lamport, L.: The part-time parliament. ACM Trans. Comput. Syst. (TOCS) **16**(2), 133–169 (1998)
10. Malkhi, D., Merritt, M., Rodeh, O.: Secure reliable multicast protocols in a wan. In: Proceedings of 17th International Conference on Distributed Computing Systems, pp. 87–94. IEEE (1997)
11. Miller, A., Xia, Y., Croman, K., Shi, E., Song, D.: The honey badger of BFT protocols. In: Proceedings of the 2016 ACM SIGSAC Conference on Computer and Communications Security, pp. 31–42. ACM (2016)
12. Nakamoto, S.: Bitcoin: a peer-to-peer electronic cash system (2008)
13. QuantumMechanic (2011). https://bitcointalk.org/index.php?topic=27787.0
14. Sliwinski, J., Wattenhofer, R.: ABC: proof-of-stake without consensus (2019). http://arxiv.org/abs/1909.10926
15. Sompolinsky, Y., Lewenberg, Y., Zohar, A.: Spectre: a fast and scalable cryptocurrency protocol. IACR Cryptology ePrint Arch. **2016**, 1159 (2016)
16. Wood, G., et al.: Ethereum: a secure decentralised generalised transaction ledger. Ethereum Proj. Yellow Pap. **151**(2014), 1–32 (2014)

Lack of Quorum Sensing Leads to Failure of Consensus in *Temnothorax* Ant Emigration

Jiajia Zhao[1(✉)], Lili Su[2], and Nancy Lynch[1]

[1] Massachusetts Institute of Technology, Cambridge, MA 02139, USA
{jiajiaz,lynch}@csail.mit.edu
[2] Northeastern University, Boston, MA 02115, USA
l.su@northeastern.edu

Abstract. We investigate the importance of quorum sensing in the success of house-hunting of emigrating *Temnothorax* ant colonies. Specifically, we show that the absence of the quorum sensing mechanism leads to failure of consensus during emigrations. We tackle this problem through the lens of distributed computing by viewing it as a natural distributed consensus algorithm. We develop an agent-based model of the house-hunting process, and use mathematical tools such as conditional probability, concentration bounds and Markov mixing time to rigorously prove the negative impact of not employing the quorum sensing mechanism on emigration outcomes. Our main result is a high probability bound for failure of consensus without quorum sensing in a two-new-nest environment, which we further extend to the general multiple-new-nest environments. We also show preliminary evidence that appropriate quorum sizes indeed help with consensus during emigrations. Our work provides theoretical foundations to analyze why *Temnothorax* ants evolved to utilize the quorum rule in their house-hunting process.

Keywords: Bio-inspired algorithms · Distributed consensus · Stochastic dynamical systems

1 Introduction

Social insect colonies are motivated to move the locations of their nesting site as a functional response to various selected forces, such as colony growth, competition, foraging efficiency, microclimate, nest deterioration, nest quality, parasitism, predation, and seasonality [18]. Through constant adaptation to a changing environment, many social insect species such as ants, termites, and bees have evolved robust algorithms to accomplish the task of collective nest relocation [32]. In this paper, we study one such algorithm observed in colonies of *Temnothorax* ants.

Temnothorax ant colonies have many biological constraints: individuals with limited memory and computational power, limited communication, and no central control. Despite that, colonies as a whole can reach various global goals

C. Johnen et al. (Eds.): SSS 2021, LNCS 13046, pp. 209–228, 2021.
https://doi.org/10.1007/978-3-030-91081-5_14

such as nest-site selections and foraging [10]. Their remarkable collective intelligence is not only an interesting problem for biologists, but also inspiring for the computer science community. In particular, from the distributed computing perspective, the collective house-hunting behavior is closely related to the fundamental problem of consensus. Building a theoretical understanding of the key mechanisms in the house-hunting process can thus shed light on the designs of novel distributed consensus algorithms.

Colonies consist of active ants who move the remaining passive workers, the queen, and brood items (immature ants) [4,25]. All workers are female ants. At the beginning of an emigration event, individual active ants independently search for new nest sites. If an ant finds one, she evaluates the site's quality according to various metrics [7,12]. Quality evaluation is relative to the old home nest [3]. If she is not satisfied with the site, she keeps searching. Otherwise if she is satisfied with the site, she returns to the home nest after some time interval that is inversely related to the new nest site quality; during this interval she might continue searching for other new potential nest sites [15,22]. If she returns to the old nest, she recruits another active ant to the site by leading a slow *tandem run* from the old nest to the new site [19,26]. This is done by the leader ant directing the follower ant along a pheromone trail (Fig. 1(a)). Upon arriving at the nest, the follower ant also evaluates the nest's quality independently of the leader ant. Both ants then continue monitoring the quality of the nest and repeat the process of quality estimation, wait interval/continued search, and further recruitment [29].

Fig. 1. From [14] (Fig. 2). (a) Recruitment via tandem running in the ant of genus *Temnothorax*. The worker at the front is leading a tandem run, and the follower behind is about to signal its presence by tapping with its antennae on the gaster of the leader. (b) Recruitment by transport in *Temnothorax* ants. One worker is simply carrying another quickly to the new nestsite. (Both photographs by S. C. Pratt.)

An ant continues leading tandem runs until she perceives that the new nest's population has exceeded a threshold, or quorum [23]. At this point, she ceases tandem runs and instead starts transporting other ants by picking one up and carrying her from the old home nest to the new nest (Fig. 1(b)). These transports are much faster than tandem runs, and they are largely directed at the passive workers and brood items, hence they serve to quickly move the entire colony to the new nest [22,25]. The transporter rarely drops out of transporting other ants, and hence is considered fully committed to the new nest as the colony's home [29].

Both tandem runs and transports are forms of recruitment to accelerate the emigration process, but the marginal benefits of transports in ensuring consensus remain relatively poorly understood. Previous studies have regarded the quorum sensing mechanism as a way to tune the speed-accuracy trade-off [9,16,17,20,24,31], where a smaller quorum prompts ants to commit sooner (higher speed) to a nest that has accumulated enough population, although that nest could be inferior to another nest that is discovered later in the process (lower accuracy). However, these studies generally equate accuracy with consensus or cohesion [5,6], when all or most ants commit to the same nest. The difference between accuracy and consensus is that the former evaluates the individuals' ability to choose the best option in the environment, but the latter is concerned only with their ability to agree with each other. The ability to stay in a single group is not only an interesting algorithmic question, but also highly beneficial for the survival of these ant colonies [6,9,14,30]. However, consensus during emigrations has been comparatively understudied. Such studies require examinations of both consensus cases and split cases, and the latter is difficult to induce experimentally. Therefore, in this paper, we conduct one of the first theoretical studies of the role of quorum sensing in emigration consensus.

At the outset, quorum sensing significantly benefits consensus because once enough ants make their choice, that choice is "locked in" and has a higher chance of becoming the final choice. This helps to ensure consensus when there are many choices and the search effort is dispersed. However, a closer look reveals that the quorum size must be carefully chosen. If the quorum size is too large, it would be very unlikely to be reached by any nest; if it is too small, multiple nests will likely reach quorum (a split), incurring significant additional costs in time and risk of exposure of the emigration [1,2,23]. These trade-offs pose the question of whether quorums help with consensus at all. In this paper, we aim to answer this question partially by investigating the probability of emigration consensus *without* the quorum sensing mechanism.

We start by modeling individual active ants as coupled random processes without considering the quorum sensing mechanism. Unlike in most classical distributed algorithms, the ants in our model do not receive initial input preferences, but must determine these preferences through exploration. Another difference is that our consensus requirement exempts a small portion of ants from committing to the same nest. Intuitively, we expect that the distribution of ant states converges to a limiting distribution in the long run. However, due to the probabilistic modeling, there is a non-zero probability that an emigration deviates greatly from this expectation, and this probability depends also on how many ants can be exempted by the requirement. Therefore, detailed calculations are needed to quantify the probability of deviations that satisfy the consensus requirement. Using probability tools such as conditional probability, concentration bounds and Markov mixing time, we then show that without quorum sensing, the probability of consensus is small and decays to zero exponentially fast as the colony size grows. In addition, we show preliminary evidence that appropriate quorum sizes indeed help with consensus during emigrations.

The rest of the paper is organized as follows. In Sect. 2, we present our model of individual ants, of the entire colony, and of an execution, for two-nest environments. In Sect. 3, we formally state the definition of consensus, and the metrics to measure a model's performance in terms of consensus. In Sect. 4, we show that with a high probability, emigrations cannot eventually reach consensus without quorum sensing. In Sect. 5, we extend our results to general $k-$nest environments where $k > 2$. Then, in Sect. 6, we consider the addition of the quorum sensing mechanism to the emigration process in two-nest environments, and show simulation results on the quorum sizes that are sufficient for consensus.

2 Model

2.1 Timing Model and the Environment

We divide time into discrete rounds. Individual active ants are modeled as identical probabilistic finite state machines and their dynamics are coupled through recruitment actions, as described later in Sect. 2.2. Let N denote the total number of active ants in the colony. Note that passive ants, the queen, and brood items can only be transported and have no states. For ease of exposition, in the sequel, by an "ant" we mean an "active ant". Each ant starts a round with its own state. During each round, ants can perform various state transitions and have new states, before all entering the next round at the same time. Throughout the paper, the state of an ant at round t refers to her state at the end of round t.

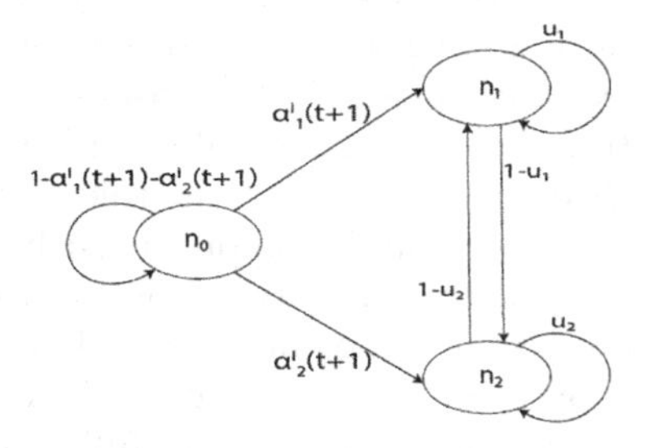

Fig. 2. State transition diagram for ant a_i during round $t + 1$ before/without quorum attainment. $\alpha_1^i(t+1)$ and $\alpha_2^i(t+1)$ are composite functions each including the probabilities of an ant taking different paths (independent discovery or tandem running) to transition out of n_0 into n_1 and n_2, respectively.

The environment contains the original home nest n_0 and two new nests n_1 and n_2. The new nests n_1 and n_2 have qualities q_1 and q_2 respectively, *relative* to the home nest quality. For the convenience of our analysis, we let $0 < q_2 < q_1 \leq 1$, where a higher value corresponds to a better nest. Each nest is also associated with a *population* that changes from round to round. We use $x_0(t)N$, $x_1(t)N$ and $x_2(t)N$, where $x_0(t) + x_1(t) + x_2(t) = 1$, to denote (active) ant populations in nest n_0, n_1 and n_2 respectively at the end of round t. Initially, individual ants have no information on q_1 and q_2.

2.2 Model of Individual Ants Without Quorums

In this subsection, we describe the dynamics of an ant without quorums (a.k.a. without performing state transitions based on seeing a quorum), compactly illustrated in Fig. 2 and Eq. (1)–(6). Though these dynamics are not Markovian as the state transition of an ant is influenced by other ants during recruitments (tandem runs), we prove (in Sect. 4) that after a finite time, the state transitions of an ant become independent of the others' states.

Individual State. The set of possible states of an ant is denoted as $\mathcal{S} \triangleq \{n_0, n_1, n_2\}$. Each state n_i refers to the ant being at nest n_i, and thus in the sequel we use "in state n_i" and "in nest n_i" interchangeably. Denote the state of ant a_i at the end of round t as $\boldsymbol{s}_i(t)$ with $\boldsymbol{s}_i(0) = n_0$ for all a_i, i.e., initially all ants locate at the home nest n_0.

Transitions out of the Home Nest. In a round, an ant a_i in n_0 can be recruited by following a tandem run to either n_1 or n_2. If a_i is not recruited, she discovers nest n_1 or n_2 for the first time through independent discovery with probability $\alpha \in (0, 1/2]$ for either nest and a total discovery probability of 2α. Note that the biological meaning of the parameter α is that it encodes the home nest quality - the higher the home nest quality, the less likely a_i is to search for a new nest during any round t and the smaller α is. Recruitment takes priority over her performing a probabilistic state transition to either n_1 or n_2 through independent discovery.

Formally, at the end of round t, if ant a_i is in n_0, let $TR_1^i(t+1), TR_2^i(t+1)$ be the event that ant a_i is recruited to n_1 and n_2 respectively during round $t+1$. Let $\tau_1^i(t+1), \tau_2^i(t+1)$ represent their respective conditional probabilities during round $t+1$, i.e.,

$$\mathbb{P}\left\{TR_m^i(t+1) \mid \boldsymbol{s}_i(t) = n_0\right\} = \tau_m^i(t+1), \text{ for } m \in \{1, 2\}.$$

Note that for any ant a_i, the two events are mutually exclusive, and $\tau_1^i(t+1) + \tau_2^i(t+1) \leq 1$. The exact expressions for $\tau_1^i(t+1)$ and $\tau_2^i(t+1)$ are very complex and affect the time that ant a_i transitions out of n_0, which is an important milestone time for the proofs in this paper. Fortunately, we manage to circumvent calculating the exact expressions of $\tau_1^i(t+1)$ and $\tau_2^i(t+1)$ by deriving a bound on this time using a coupling argument (Proposition 2). We found that this bound was sufficient for proving our main theorem.

With this notation, conditioning on an ant a_i being at state n_0 at time t, the probability of her transitioning to n_1 in the nest round, denoted by $\alpha_1^i(t)$ can be expressed as

$$\begin{aligned}
&\mathbb{P}\{\boldsymbol{s}_i(t+1) = n_1 \mid \boldsymbol{s}_i(t) = n_0\} \triangleq \alpha_1^i(t+1) \\
&= \mathbb{P}\left\{TR_1^i(t+1) \mid \boldsymbol{s}_i(t) = n_0\right\} \\
&\quad + \mathbb{P}\left\{\boldsymbol{s}_i(t+1) = n_1 \mid (\boldsymbol{s}_i(t) = n_0 \wedge \neg(TR_1^i(t+1) \vee TR_2^i(t+1)))\right\} \\
&\quad\quad \cdot \mathbb{P}\left\{\neg(TR_1^i(t+1) \vee TR_2^i(t+1)) \mid \boldsymbol{s}_i(t) = n_0\right\} \\
&= \tau_1^i(t+1) + \alpha(1 - \tau_1^i(t+1) - \tau_2^i(t+1)),
\end{aligned} \tag{1}$$

where α can be formally expressed as

$$\alpha = \mathbb{P}\left\{\boldsymbol{s}_i(t+1) = n_1 \mid (\boldsymbol{s}_i(t) = n_0 \wedge \neg(TR_1^i(t+1) \vee TR_2^i(t+1)))\right\}$$

. It is easy to see that $\alpha_1^i(t+1)$ sums up the probability of her getting recruited to n_1 and the probability of independent discovery of n_1 in the case that she does not get recruited to either n_1 or n_2. Similarly, we define $\alpha_2^i(t+1)$ as the probability of her transitioning to n_2 during round $t+1$, i.e.,

$$\begin{aligned} & \mathbb{P}\{\boldsymbol{s}_i(t+1) = n_2 \mid \boldsymbol{s}_i(t) = n_0\} \\ \triangleq\ & \alpha_2^i(t+1) = \tau_2^i(t+1) + \alpha(1 - \tau_1^i(t+1) - \tau_2^i(t+1)). \end{aligned} \quad (2)$$

Correspondingly,

$$\mathbb{P}\{\boldsymbol{s}_i(t+1) = n_0 \mid \boldsymbol{s}_i(t) = n_0\} = 1 - \alpha_1^i(t+1) - \alpha_2^i(t+1). \quad (3)$$

Transitions Between New Nests

When $\boldsymbol{s}_i(t) = n_m$ for $m \in \{1,2\}$, at the beginning of round $t+1$, with probability $(1 - u_m)$, ant a_i chooses to search her environment and discover the new nest she is not currently at, i.e.,

$$\mathbb{P}\{\boldsymbol{s}_i(t+1) = n_{3-m} \mid \boldsymbol{s}_i(t) = n_m\} = 1 - u_m, \quad \forall\ m \in \{1,2\}; \quad (4)$$

with probability u_m, ant a_i tries to recruit another ant from state n_0 through a tandem run and comes back to n_m, i.e.,

$$\mathbb{P}\{\boldsymbol{s}_i(t+1) = n_m \mid \boldsymbol{s}_i(t) = n_m\} = u_m, \quad \forall\ m \in \{1,2\}. \quad (5)$$

If there is no more ant left in n_0 to recruit, the leader ant a_i simply returns to nest n_m without recruiting another ant. The recruiting probability u_m is determined by the quality of new nest n_m as

$$u_m \triangleq \frac{1}{1 + \exp(-\lambda q_m)}, \quad \forall\ m \in \{1,2\}, \quad (6)$$

where the parameter $\lambda > 0$ represents the noise level of individual decision making to evaluate the quality of a nest n_m for $m \in \{1,2\}$. A larger λ means a less noisy decision rule, and thus a higher probability of recruitment to the superior site n_m. Also note that $u_1, u_2 \in [0.5, 1]$ and $u_1 > u_2$.

Our choice of the sigmoid function is rooted in empirical evidence. The decision making mechanism for individual ant recruitment has been shown by a number of experimental and modeling studies to be both quality-dependent [15,21,24,25] and threshold-based (individuals compare the perceived nest quality to a fixed threshold) [27,28]. The sigmoid function we chose here is thus a common choice that incorporates both dependencies into the modeling of noisy individual decision making. Intuitively, when n_m has a quality higher than that of n_0's, n_m is the better choice and it is beneficial for ants to recruit to it. When a nest n_m is strongly superior to n_0, i.e., the quality difference surpasses a threshold, the probability of an individual ant recruiting to n_m should thus be very high (close to 1 in our model). The sigmoid function is a "smooth" representation of this threshold-based rule. On the other hand, when the quality difference is small, the probability of recruitment has stronger dependencies on the quality difference. This case is modeled by a near-linear segment in the sigmoid function.

Remark 1 (Non-markovian dynamics of an individual ant). The state $\boldsymbol{s}_i(t)$ of any individual ant a_i during round t has dependencies on 1) her own state in the previous round $\boldsymbol{s}_i(t-1)$, and 2) the recruitment actions of other ants.

2.3 Dynamics of the Entire Colony

We now describe what happens in an arbitrary execution, or emigration. Throughout the paper, we use "an execution" and "an emigration" interchangeably, referring to an emigration event.

Let $\boldsymbol{s}(t) = \{\boldsymbol{s}_1(t), \cdots, \boldsymbol{s}_N(t)\}$ for $t = 0, 1, \cdots$ denote the random process of the entire colony state, represented by a vector of dimension N that stacks the states of individual ants in the colony. Although $\boldsymbol{s}_i(t)$ for any i is not Markovian, it is easy to see that $\boldsymbol{s}(t)$ is a Markov chain, since for any tandem leader in round t, the choice of a follower only depends on $\boldsymbol{s}(t-1)$ and not on any history prior to round $t-1$. An emigration starts from round 1, with $\boldsymbol{s}_i = n_0$ for all $i = 1, \cdots, N$. During each round, each ant not in n_0 performs one state transition in random order, followed by each ant in n_0 performing one state transition in random order. At the beginning of a round t, each ant has her own state $\boldsymbol{s}_i(t-1)$ and the colony has state $\boldsymbol{s}(t-1)$. If at the beginning of round t she is in nest n_1 or n_2, respectively, the population at that nest at the beginning of round t is also available to a_i. During a round t, each individual ant performs one state transition according to the individual models in Sect. 2.2, which results in a transition of the colony state as well during this round. At the end of round t, each ant has a new state $\boldsymbol{s}_i(t)$ and the colony has state $\boldsymbol{s}(t)$. All ants then enter the next round $t+1$ with their new states.

3 The Consensus Problem

Here we define what it means for an emigration to reach consensus. We say that an emigration has reached Δ-consensus (where $\Delta \in [0, \frac{1}{2}]$) if there exists $\widetilde{t}$ such that for all $t \geq \widetilde{t}$ and a nest $m \in \{1, 2\}$, the proportion of the population at nest n_m at time t is greater than or equal to $1 - \Delta$, i.e., $x_m(t) \geq (1 - \Delta)$.

The metric to evaluate a model's performance is the *consensus probability* C, which is the probability that an emigration reaches consensus as defined above.

Remark 2. Note that Δ represents the proportion of ants that can be *exempted* from the consensus requirement. We can see that the smaller Δ is (lowest value is 0), the larger $(1 - \Delta)N$ is, and hence the more ants are required for an emigration to reach consensus. In other words, the smaller Δ is, the more "strict" the consensus metric is and the more challenging it is for an emigration to reach consensus.

4 Failure of Consensus in Two-Nest Environments

In this section, we explore colony emigration behavior *only* with individual transition rules and tandem runs defined above (i.e., without quorum sensing). Equivalently, we consider the case where the quorum size is N, so that the quorum sensing mechanism never has any effect. We show an upper bound on the consensus probability C for a given Δ and colony size N. This upper bound decreases to 0 exponentially fast as $N \to \infty$.

Next we introduce two quantities, denoted by H and π^*, that will be used in the statement of our main result. It is easy to see from Eq. (4) and (5) that if an ant a_i jumps out of the home nest n_0 at some time, then from that time onward, the state transition of a_i becomes Markovian and is governed by the following transition matrix

$$H = \begin{bmatrix} u_1 & 1-u_1 \\ 1-u_2 & u_2 \end{bmatrix}. \tag{7}$$

The transition in H is also illustrated in Fig. 4. It can also be seen (which we will formally show later) that the state of each ant has an identical limiting distribution, denoted by $\pi^* \triangleq \frac{1}{2-u_1-u_2}[1-u_2, 1-u_1] \in \mathbb{R}^2$, with support on $\{n_1, n_2\}$ only.

Theorem 1. *For any* $\Delta \in [0, 1-\pi^*(n_1)]$, *let* $\epsilon_0 = \frac{1-\pi^*(n_1)-\Delta}{2} > 0$. *Then it holds that*

$$\mathbb{P}\left\{\sum_{i=1}^{N} \mathbb{1}\{s_i(t) = n_1\} \geq (\pi^*(n_1) + 2\epsilon_0)\, N = (1-\Delta)N\right\} \leq 2\exp\left(-\frac{\epsilon_0^2 N}{2}\right),$$

for any $t > \left(\frac{1}{\ln(1-2\alpha)} + \frac{1}{\ln(1-R(H))}\right)\ln\frac{\epsilon_0}{2}$, *where* $R(H) = 2 - u_1 - u_2$ *is Dobrushin's coefficient of ergodicity ([11, Chapter 6.2]) of* H.

Remark 3. Theorem 1 is stated for n_1. A similar result holds for n_2. Theorem 1 says that for any t greater than $\left(\frac{1}{\ln\beta} + \frac{1}{\ln(1-R(H))}\right)\ln\frac{\epsilon_0}{2}$, the probability of $x_1(t)$ reaching $(1-\Delta)$ is upper bounded by $2\exp\left(-\frac{\epsilon_0^2 N}{2}\right)$. Thus, the total consensus probability C for the given Δ is upper bounded by $4\exp\left(-\frac{\epsilon_0^2 N}{2}\right)$, which decreases to 0 exponentially fast as N increases. It is worth noting that real ant colonies often need Δ to be very small or even zero for survival. From the theorem expression, we can see that the smaller Δ is, i.e., the more stringent the consensus, the lower is the upper bound of the consensus probability. Therefore, Theorem 1 implies that extra mechanisms, such as the quorum rule are necessary to help the emigration reach consensus.

Later in Sect. 5, we also show that the proofs in this section and related results can easily extend to environments with multiple nests.

4.1 Analysis of Main Result

Despite the fact that the dynamics of the entire ant colony is a Markov chain, analyzing this Markov chain is highly non-trivial because the state is quite involved and the state space is huge – it contains all the possible partitions of ants into three groups, with each group representing one nest as the state of an individual ant. In this section we analytically show that despite the fact that the emigration behaviors of individual ants are *interactive*, the dynamics of any individual ant are independent of other ants shortly after she leaves the original home nests either through discovery or through recruitment. Moreover, we show that this independence manifests itself in a non-trivial way after a few rounds – suggesting that a large portion of ants quickly rely only on individual intelligence. Then we show that this independence is harmful to realizing social cohesion.

Several intermediate results are derived in proving Theorem 1. The connections of the supporting lemmas and corollaries with respect to Theorem 1 are shown in Fig. 3. Please note that due to space constraints, we show the proof details of only Theorem 1 in this paper. Those of all other intermediate results can be found in [33].

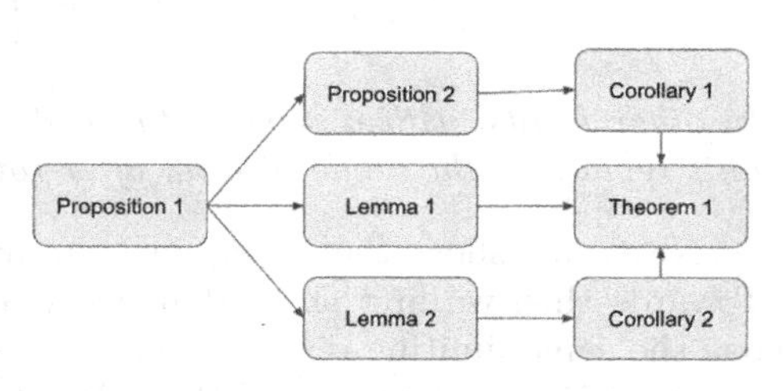

Fig. 3. Flowchart of the proofs.

Definition 1. *For each $i \in [N]$, define random variable $T_i^1 \triangleq \inf\{t : \boldsymbol{s}_i(t) \neq n_0\}$ as the first round at the beginning of which ant a_i has transitioned out of the n_0 state in any arbitrary execution of the emigration.*

Remark 4. It can be shown that T_i^1 is finite with probability 1 [33]. It follows immediately from Definition 1 that $\mathbb{P}\left\{\boldsymbol{s}_i(t) = n_0 \mid t \geq T_i^1\right\} = 0$ for any ant a_i.

It turns out that ant a_i's state transitions become independent of other ants after T_i^1, the time that a_i leaves n_0, formally stated in the following proposition.

Proposition 1. *For every $i, j \in [N], i \neq j$ and every $t > T_i^1$, the state transitions of ant a_i are independent from a_j, i.e.,*

$$\begin{aligned} &\mathbb{P}\left\{\boldsymbol{s}_i(t+1) = s_1' \mid (\boldsymbol{s}_i(t) = s_1) \wedge (\boldsymbol{s}_j(t) = s_2) \wedge (t > T_i^1)\right\} \\ &= \mathbb{P}\left\{\boldsymbol{s}_i(t+1) = s_1' \mid (\boldsymbol{s}_i(t) = s_1) \wedge (t > T_i^1)\right\}, \end{aligned}$$

where $s_1, s_2, s_1' \in \mathcal{S}$ and $s_1' \neq n_0$.

The next proposition is devoted to showing that after a few rounds, many ants have left the home nest n_0. Consider N random indicator variables $\mathbb{1}\{T_i^1 > t\}$ for any t, each variable taking values in the $\{0, 1\}$. Using stochastic dominance and Hoeffding's inequality [13], we show a high probability upper bound on the number of ants still in n_0 at round t. Here stochastic dominance is used to tackle the challenges caused by the dependency among the N indicator random variables.

Proposition 2. *Let $\beta \triangleq 1 - 2\alpha$. For $t \geq 1$ and any number $d \in [0, 1]$, it holds that*

$$\mathbb{P}\left\{\sum_{i=1}^{N} \mathbb{1}\{T_i^1 > t\} < N\left(\beta^t + d\right)\right\} > 1 - \exp\left(-2Nd^2\right),$$

i.e., with a probability of at least $(1 - \exp(-2Nd^2))$*, the number of ants staying at home nest beyond time* t *is at most* $N(\beta^t + d)$*.*

Corollary 1. *For any given* $\epsilon \in (0, 1)$*, for any* $t \geq \log_\beta(\frac{\epsilon}{2})$*, it holds that*

$$\mathbb{P}\left\{\sum_{i=1}^{N} \mathbb{1}\{T_i^1 > t\} < \epsilon N\right\} > 1 - \exp\left(-N\epsilon^2/2\right).$$

In other words, with a probability of at least $(1 - \exp(-N\epsilon^2/2))$*, at most* ϵN *ants remain in the home nest* n_0 *after round* $\log_\beta(\frac{\epsilon}{2})$*.*

Next, we show that every ant a_i has an identical limiting distribution. Towards this, we first show that every ant a_i that has transitioned out of n_0 has the same limiting distribution. Furthermore, we show that all ants eventually transition out of n_0 and thus all ants share the same limiting distribution. The proof of Lemma 1 uses the quantity $Q(t)$, defined as

$$Q(t) \triangleq \{a_i : \boldsymbol{s}_i(t) \neq n_0\} \tag{8}$$

which is a random variable representing the set of ants that have transitioned out of n_0 by the end of round t, in an arbitrary emigration. $Q(t)$ is thus a function of an execution. It is easy to see that w.r.t. this emigration, $Q(t-1) \subseteq Q(t)$ for any $t \geq 1$.

Lemma 1. *For each* a_i*, its limiting distribution, denoted by* π_i*, is well-defined, and can be expressed as*

$$\pi_i \triangleq \frac{1}{2 - u_1 - u_2}\left[1 - u_2, 1 - u_1\right]. \tag{9}$$

For ease of exposition, we define $\pi^* = \pi_i$. From Lemma 1 it can be seen that the probability ratio $\frac{\pi^*(n_1)}{\pi^*(n_2)} = \exp(\lambda(q_1 - q_2))$ is very sensitive to the nest quality gap $(q_1 - q_2)$ and λ.

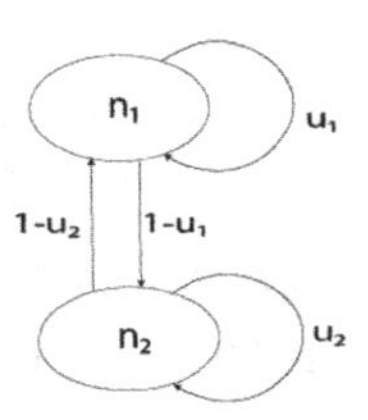

Fig. 4. State transition diagram for individual ants after they leave n_0, before/without quorum attainment.

It turns out that for t large enough, any ant that has transitioned out of n_0 has state distributions "close" to the stationary distribution π^*, formally stated next.

Lemma 2. *For any ant* a_i*, let* $\pi_{i,t}$ *denote the probability distribution of her state over the possible states depicted in Fig. 4 at time* $t \geq T_i^1$*. Then for any number of rounds* $\ell > 0$*, it holds that*

$$\|\pi_{i,T_i^1+\ell} - \pi^*\|_1 \leq 2\left(1 - R(H)\right)^{\ell}.$$

Using Lemma 2, the following corollary immediately follows:

Corollary 2. *Fix any* $\delta \in (0,1)$. *For any ant* a_i *and* $t > T_i^1 + \ell$, *where* $\ell \triangleq \log_{(1-R(H))} \frac{\delta}{2}$, *it holds that*

$$\|\pi_{i,t} - \pi^*\|_1 \leq \delta.$$

Combined with Corollary 1, we are now ready to prove Theorem 1.

Proof of Theorem 1
We first give the intuition and a proof sketch to show an upper bound on the probability of the population at n_1 being higher than a certain number C_0, for t large enough.

We break down the problem into two cases. In the first case, by a certain milestone-round k_1, the number of ants that have transitioned out of n_0 is low. In the second case, that number is high. Now, by applying concentration bounds, we show that the first case has a low probability. We thus subsequently focus on analyzing the second case which has a high probability. From Corollary 2 we know that after a certain number k_2 of rounds, most of the ants that have left n_0 will have distributions that are very close to the limiting distribution π^*. Thus, at any round $t \geq k_1 + k_2$, with high probability, the proportion of ants in n_1 is also close to $\pi^*(n_1)$ among ants that have left n_0 (at most N ants). In other words, after $k_1 + k_2$ rounds the probability of n_1's population being much higher than $\pi^*(n_1)N$ should be quite low. Summing up the bounds for the first and second cases gives us an overall upper bound on this probability, proving the theorem.

For ease of exposition, let $B_i(t) = \mathbb{1}\{\boldsymbol{s}_i(t) = n_1\}$ for each $i \in [N]$ and $t \geq 0$. Let C_0 be an arbitrary positive number, $C_0 \in [0, N]$. Let $C_1 = (1 - \epsilon_0)N$. Recall that $\beta = 1 - 2\alpha$.

$$\begin{aligned}
&\mathbb{P}\left\{\sum_{i=1}^{N} \mathbb{1}\{\boldsymbol{s}_i(t) = n_1\} \geq C_0\right\} = \mathbb{P}\left\{\sum_{i=1}^{N} B_i(t) \geq C_0\right\} \\
&= \mathbb{P}\left\{\sum_{i=1}^{N} B_i(t) \geq C_0 \mid \left|Q(\log_\beta \frac{\epsilon_0}{2})\right| < C_1\right\} \mathbb{P}\left\{\left|Q(\log_\beta \frac{\epsilon_0}{2})\right| < C_1\right\} \\
&\quad + \mathbb{P}\left\{\sum_{i=1}^{N} B_i(t) \geq C_0 \mid \left|Q(\log_\beta \frac{\epsilon_0}{2})\right| \geq C_1\right\} \mathbb{P}\left\{\left|Q(\log_\beta \frac{\epsilon_0}{2})\right| \geq C_1\right\} \\
&\leq \mathbb{P}\left\{\left|Q(\log_\beta \frac{\epsilon_0}{2})\right| < C_1\right\} + \mathbb{P}\left\{\sum_{i=1}^{N} B_i(t) \geq C_0 \mid \left|Q(\log_\beta \frac{\epsilon_0}{2})\right| \geq C_1\right\}. \qquad (10)
\end{aligned}$$

We bound the two terms in the RHS of Eq. (10) separately.

Bounding the 1st term: For any $t \geq \log_\beta \frac{\epsilon_0}{2}$, we have

$$\begin{aligned}
\mathbb{P}\{|Q(t)| < C_1\} &= \mathbb{P}\{|Q(t)| < (1-\epsilon_0)N\} \\
&= \mathbb{P}\left\{\sum_{i=1}^{N} \mathbb{1}\{\boldsymbol{s}_i(t) \neq n_0\} < (1-\epsilon_0)N\right\} \\
&= \mathbb{P}\left\{\sum_{i=1}^{N} \mathbb{1}\{T_i^1 \leq t\} < (1-\epsilon_0)N\right\} \\
&= \mathbb{P}\left\{\sum_{i=1}^{N} \mathbb{1}\{T_i^1 > t\} > \epsilon_0 N\right\} \\
&\leq \exp\left(-\frac{\epsilon_0^2 N}{2}\right),
\end{aligned}$$

where the last inequality follows from Corollary 1.

Bounding the 2nd Term: Note that

$$\sum_{i=1}^{N} B_i(t) = \sum_{a_i \in Q(t)} B_i(t) + \sum_{a_i \notin Q(t)} B_i(t).$$

It is easy to see that

$$\sum_{a_i \notin Q(t)} B_i(t) = 0. \tag{11}$$

In addition, we have

$$\begin{aligned}
&\mathbb{P}\left\{\sum_{a_i \in Q(t)} B_i(t) - \sum_{a_i \in Q(t)} \mathbb{E}\left[B_i(t)\right] \geq \epsilon_0 \left|Q(t)\right| \quad \middle| \quad \left|Q(t)\right| \geq (1-\epsilon_0)N\right\} \\
&= \mathbb{P}\left\{\sum_{a_i \in Q(t)} B_i(t) - \sum_{a_i \in Q(t)} \pi_{i,t}(n_1) \geq \epsilon_0 \left|Q(t)\right| \quad \middle| \quad \left|Q(t)\right| \geq (1-\epsilon_0)N\right\} \\
&\leq \exp\left(-2\left|Q(t)\right|\epsilon_0^2\right) \\
&\leq \exp\left(-2(1-\epsilon_0)\epsilon_0^2 N\right).
\end{aligned}$$

Conditioning on $\left|Q(\log_\beta \frac{\epsilon_0}{2}))\right| \geq (1-\epsilon_0)N$, from Corollary 2, we know that for each $a_i \in Q(\log_\beta \frac{\epsilon_0}{2})$, for any $t > \log_\beta \frac{\epsilon_0}{2} + \ell$, where $\ell = \log_{(1-R(H))} \frac{\epsilon_0}{2}$, it holds that $\pi_{i,t}(n_1) \leq \pi^*(n_1) + \epsilon_0$. Hence we get

$$\begin{aligned}
\sum_{a_i \in Q(t)} \pi_{i,t}(n_1) + \epsilon_0 \left|Q(t)\right| &\leq \left(\pi^*(n_1) + \epsilon_0\right)\left|Q(t)\right| + \epsilon_0 \left|Q(t)\right| \\
&\leq \left(\pi^*(n_1) + 2\epsilon_0\right) N.
\end{aligned}$$

Thus,

$$\begin{aligned}&\mathbb{P}\left\{\sum_{a_i\in Q(t)} B_i(t) \geq (\pi^*(n_1)+2\epsilon_0)\,N\right\}\\&\leq \mathbb{P}\left\{\sum_{a_i\in Q(t)} B_i(t) - \sum_{a_i\in Q(t)} \mathbb{E}\left[B_i(t)\right] \geq \epsilon_0\,|Q(t)| \quad \middle| \quad |Q(t)| \geq (1-\epsilon_0)N\right\}\\&\leq \exp\left(-2(1-\epsilon_0)\epsilon_0^2\,N\right). \qquad (12)\end{aligned}$$

Combining Eq. (11) and (12), we conclude that

$$\mathbb{P}\left\{\sum_{i=1}^{N} B_i(t) \geq (\pi^*(n_1)+2\epsilon_0)\,N \mid |Q(t)| \geq (1-\epsilon_0)N\right\} \leq \exp\left(-2(1-\epsilon_0)\epsilon_0^2\,N\right).$$

Combining the probability bounds on the first and second terms of Thoerem 1, we have

$$\begin{aligned}&\mathbb{P}\left\{\sum_{i=1}^{N} \mathbb{1}\{s_i(t)=n_1\} \geq (\pi^*(n_1)+2\epsilon_0)\,N\right\}\\&\leq \exp\left(-2(1-\epsilon_0)\epsilon_0^2\,N\right) + \exp\left(-\frac{\epsilon_0^2 N}{2}\right)\\&\leq 2\exp\left(-\frac{\epsilon_0^2 N}{2}\right) \quad \text{as } \epsilon_0 \in (0,1/2),\end{aligned}$$

proving Theorem 1.

5 Extension: Failure of Consensus in More-Nest Environments

Both the results on asymptotic independence and its negative impact presented in Sect. 4 can be extended to the general k-new-nest environments where $k > 2$. On a high level, the necessary additions to the individual transition model (Fig. 2) are: 1) a new state for each new nest, each similar to the n_1 and n_2, 2) all new nests can exchange ants with each other, and 3) all new nests can receive ants from n_0 through recruitment or discovery. The model for timing, environment, and execution of the whole colony remain the same as the two-nest case, where n_1 has the highest quality. After adjusting quantities H and π^*, one can derive results similar to Theorem 1: without quorum sensing, the probability of consensus can be arbitrarily low. We detail these changes below in this section.

Figure 5 shows the transition diagram for an individual ant before/without her seeing a quorum at any nest, and Eq. (13)–(16) define transition probabilities among the four states. Similar to the two-nest case, $\mathbb{P}\left\{TR_l^i(t+1)\right\} = \tau_l^i(t+1)$ for each $l \in [1,k]$ is defined as the probability of the event that ant a_i transitions

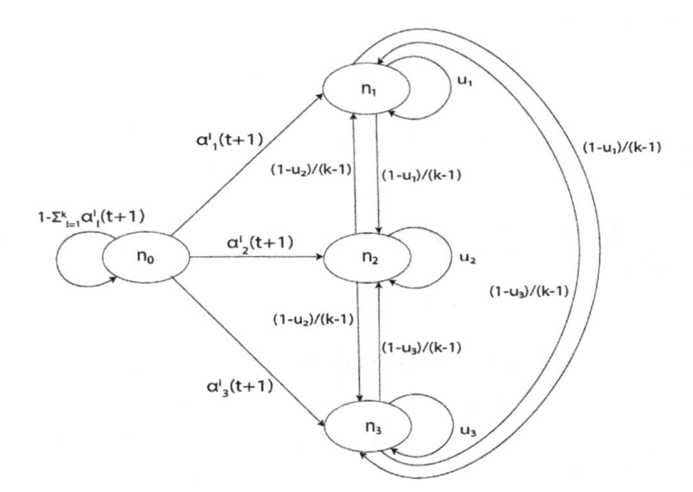

Fig. 5. State transition diagram for ant a_i during round $t+1$ in a k-nest environment before/without quorum attainment. Probabilities $\alpha_l^i(t+1), l \in [1,k]$ are composite functions each including the probabilities of an ant taking different paths (independent discovery or tandem running) to transition out of n_0 into n_l. Compared to Fig. 2, this figure shows the addition of one more new nest n_3; any more new nests can be added in the same way.

from n_0 to n_l during round $t+1$ by following a tandem run. Figure 5 displays only the addition of a third new nest, n_3, and any more new nest can be added in the same way. The addition of n_3 requires that during round t, an ant at n_0 transitions to n_3 with probability $\alpha_3(t)$; an ant at n_3 stays in n_3 with probability u_3; and an ant at a new nest l transitions to any other new nest $m \neq l$ with transition probability $\frac{1-u_l}{k-1}$.

$$\mathbb{P}\{\boldsymbol{s}_i(t+1) = n_l \mid \boldsymbol{s}_i(t) = n_0\} = \alpha_l^i(t) \text{ for } l \in [1,k] \tag{13}$$

$$\mathbb{P}\{\boldsymbol{s}_i(t+1) = n_0 \mid \boldsymbol{s}_i(t) = n_0\} = 1 - \sum_{l=1}^{k} \alpha_l^i(t) \tag{14}$$

$$\mathbb{P}\{\boldsymbol{s}_i(t+1) = n_l \mid \boldsymbol{s}_i(t) = n_l\} = u_l \text{ for } l \in [1,k] \tag{15}$$

$$\mathbb{P}\{\boldsymbol{s}_i(t+1) = n_m \mid \boldsymbol{s}_i(t) = n_l\} = \frac{1-u_l}{k-1} \text{ for } l,m \in [1,k] \text{ and } m \neq l \tag{16}$$

where

$$\begin{aligned}\alpha_l^i(t) &\triangleq \mathbb{P}\left\{TR_l^i(t)\right\}\\ &\quad+\mathbb{P}\left\{\boldsymbol{s}_i(t+1)=n_l \mid (\boldsymbol{s}_i(t)=n_0\wedge\neg(TR_1^i(t)\vee TR_2^i(t)\vee\ldots\vee TR_k^i(t)))\right\}\\ &\quad\cdot\mathbb{P}\left\{\neg(TR_1^i(t)\vee TR_2^i(t)\vee\ldots\vee TR_k^i(t))\mid \boldsymbol{s}_i(t)=n_0\right\}\\ &=\tau_l^i(t)+\alpha(1-\sum_{l=1}^{k}\tau_l^i(t)),\ \text{for } l\in[1,k],\\ u_l &\triangleq \frac{1}{1+\exp(-\lambda q_l)}\quad \text{for } l\in[1,k].\end{aligned}$$

The two quantities used in the main theorem for a k-nest environment, H and π^*, are also different, as shown below.

- H, a $k\times k$ transition matrix of an arbitrary ant a_i's state $\boldsymbol{s}_i$ after she transitions out of n_0, as specified in Eq. (17).
- $\pi^*\in\mathbb{R}^k$, a vector representing the limiting distribution of an arbitrary ant a_i (Eq. (18)). The l-th element is the limiting distribution of state n_l, for $l\in[1,k]$.

$$H=\begin{bmatrix} u_1 & \frac{1-u_1}{k-1} & \frac{1-u_1}{k-1} & \frac{1-u_1}{k-1} & \cdots & \frac{1-u_1}{k-1}\\ \frac{1-u_2}{k-1} & u_2 & \frac{1-u_2}{k-1} & \frac{1-u_2}{k-1} & \cdots & \frac{1-u_2}{k-1}\\ \frac{1-u_3}{k-1} & \frac{1-u_3}{k-1} & u_3 & \frac{1-u_3}{k-1} & \cdots & \frac{1-u_3}{k-1}\\ \cdots & \cdots & \cdots & \cdots & \cdots & \cdots\\ \frac{1-u_k}{k-1} & \frac{1-u_k}{k-1} & \frac{1-u_k}{k-1} & \frac{1-u_k}{k-1} & \frac{1-u_k}{k-1} & u_k \end{bmatrix}. \tag{17}$$

Solving the equation system $\pi_i=\pi_i H$, we also obtain that

$$\pi^*(l)=\frac{\prod_{\forall m\in[1,k],m\neq l}^{k}(1-u_m)}{\sum_{w=1}^{k}\left(\prod_{\forall m\in[1,k],m\neq w}^{k}(1-u_m)\right)},\ \text{for } l\in[1,k]. \tag{18}$$

Main Theorem for k-Nests

Theorem 2. *For any* $\Delta\in[0,1-\pi^*(n_1)]$, *let* $\epsilon_0=\frac{1-\pi^*(n_1)-\Delta}{2}>0$. *Then it holds that*

$$\mathbb{P}\left\{\sum_{i=1}^{N}\mathbb{1}\{\boldsymbol{s}_i(t)=n_1\}\geq(\pi^*(n_1)+2\epsilon_0)\,N=(1-\Delta)N\right\}\leq 2\exp\left(-\frac{\epsilon_0^2 N}{2}\right),$$

for any $t>\left(\frac{1}{\ln(1-k\alpha)}+\frac{1}{\ln(1-R(H))}\right)\ln\frac{\epsilon_0}{2}$, *where* $R(H)$ *is Dobrushin's coefficient of ergodicity ([11, Chapter 6.2]) of* H.

Remark 5. Theorem 2 is stated for n_1. A similar result holds for n_l for $l>1$. Like in the two-nest case, Theorem 2 again implies that extra mechanisms, such as the quorum rule are necessary to help the emigration reach consensus.

6 Consensus with Quorum Sensing in Two-Nest Environments

An important work in progress is analyzing the probability of consensus when the quorum rule is in effect. In this section, we show as a work in progress the addition of the quorum sensing mechanism to our model, and our current results on the quorum sizes that are sufficient for consensus of average emigrations in two-nest environments.

Fig. 6. State transition diagram for individual ants committed to n_1 and n_2, on the left and right, respectively.

Note that the dynamics shown in Fig. 2 are also accurate here before an ant a_i sees a quorum for the first time at either nest. Thus, she starts her transitions according to Fig. 2 before seeing any quorum. The evaluation of whether n_m has reached quorum happens whenever a_i is in n_m at the beginning of a round t. Before she performs any transitions during round t, she compares the nest population to a quorum size, if a_i has not yet seen a quorum at n_m (or at any other nest). Once she detects that the population is at least as high as the quorum size, she becomes "committed" to n_m. After that, she no longer monitors the nest's population. We model an ant's commitment by disallowing her to transition out of n_m. This means she has to perform a transport action and stay in the n_m state at any round after n_m's quorum attainment. As a result, once a nest reaches the quorum, it never drops out of the quorum and every ant that transitions to that nest gets "stuck" in that nest. We thus model a "committed" ant with a separate Markov chain that essentially only has one possible state, as shown in Fig. 6 and Eq. (19)–(22). For a committed ant a_i, let

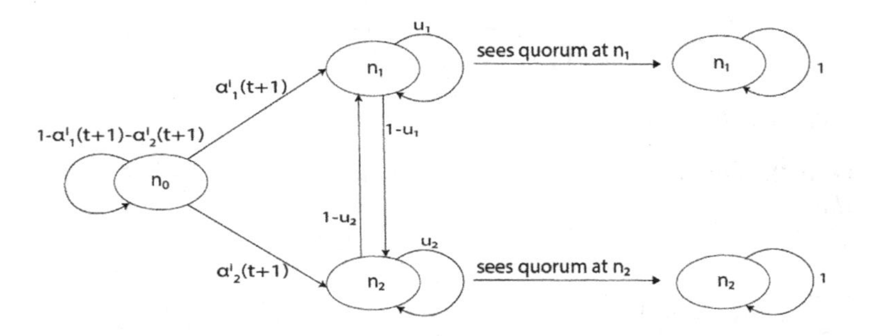

Fig. 7. State transition diagram for ant a_i during round $t+1$ with the quorum sensing mechanism. She first starts transitioning according to the left part of the figure, identical to Fig. 2. Then once she sees a quorum at either n_1 and n_2 (but not both), she commits to that nest and can only stay in that nest, as shown on the right part of the figure, identical to Fig. 6.

n_m be the nest that she is committed to where $m \in \{1, 2\}$. Then the other new nest she is not committed to is n_{3-m}.

$$\mathbb{P}\{\boldsymbol{s}_i(t+1) = n_m \mid \boldsymbol{s}_i(t) = n_m\} = 1 \tag{19}$$
$$\mathbb{P}\{\boldsymbol{s}_i(t+1) = n_{3-m} \mid \boldsymbol{s}_i(t) = n_m\} = 0 \tag{20}$$
$$\mathbb{P}\{\boldsymbol{s}_i(t+1) = n_m \mid \boldsymbol{s}_i(t) = n_{3-m}\} = 0 \tag{21}$$
$$\mathbb{P}\{\boldsymbol{s}_i(t+1) = n_{3-m} \mid \boldsymbol{s}_i(t) = n_{3-m}\} = 0. \tag{22}$$

Individual Model With Quorums. We show the full model in Fig. 7. The addition of transporting as a possible recruitment method thus has two impacts in the full model:

- An ant a_i in n_0 can get recruited by being transported to either n_1 or n_2, in addition to following a tandem run.
- an ant a_i in either state n_1 or n_2 choosing to stay in the same state tries to recruit another ant from state n_0 through a tandem run if the quorum is not reached (Fig. 2, Eq. (5)), or through a transport otherwise (Fig. 6, Eq. (19)). Whether the recruitment is successful or not still has no effect on a_i's own state transitions during this round. Otherwise, if she does not recruit, she searches her environment and discovers the new nest she is not currently at (Eq. (4)).

It still holds that during any given round t, if an ant a_i at n_0 does not get recruited, her transitions are Markovian and independent (Fig. 2, Fig. 6). The whole colony dynamics are the same as shown in Sect. 2.3 and the whole colony state retains its Markovian properties.

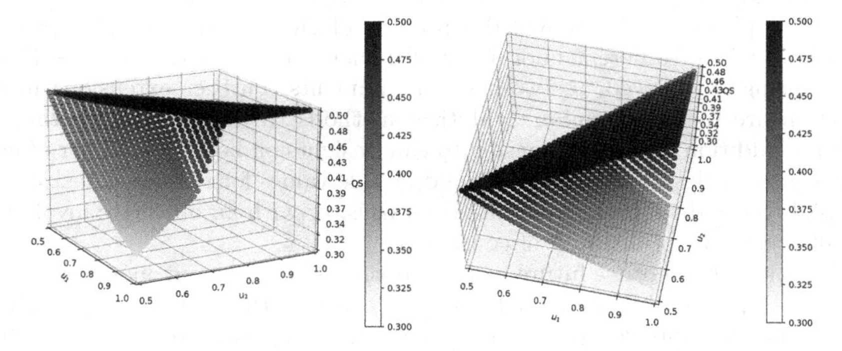

Fig. 8. 3D plots demonstrating quorum sizes that are sufficient for consensus, when $\alpha \leq \frac{1}{3}$. Views from two angles.

Current Work: In our work in progress, through theoretical analysis and simulation work, we are striving to derive quorum sizes that are sufficient for consensus. Our preliminary results in Fig. 8 show such quorum sizes for two-nest environments. In these emigrations, we enforce that $\Delta = 0$ to model the most challenging requirement of consensus. In Fig. 8, for the full ranges of u_1 and u_2 in the frequent case that $\alpha < \frac{1}{3}$, we visualize the quorum sizes (QS) in the range $[0.25, 0.5]$ that are expected to lead to consensus. The desirable values of the quorum size show general consistency with experimental findings of the observed quorum size employed by *Temnothorax* ant colonies [8,22]. However, we are still working on deriving the mathematical expressions for quorum sizes that are sufficient for consensus, as well as on extending these results to k-nest environments ($k > 2$). We plan to show all related details of this effort in a follow-up manuscript.

7 Discussion and Future Work

In this paper, we used analytical tools to show that without quorum sensing, the collective nest site selection process by *Temnothorax* ants has a limited probability to reach consensus. And this probability can be arbitrarily low for a colony size arbitrarily large. Conversely, we obtain a high probability bound for failure of consensus. Without quorum sensing, the only form of recruitment, tandem runs, does speed up the emigration process, but our results show that emigrations would still have a high probability of splitting among multiple new sites, imposing significant risks to the colony's survival. We first analyze a model of a two-new-nest environment, and then extend our results to environments with more nests. Our results provide insights into the importance of extra mechanisms, such as the quorum sensing mechanism, for emigrations to reach consensus in an unpredictable environment with multiple nests.

In this paper we also provided a preview of an important work in progress investigating how different quorum sizes influence emigration outcomes if quorum sensing is involved, in two-nest environments. Further extensions in this direction are to apply similar analytical methods to the general environment with the addition of quorum sensing to gain insights on how the number of nests and their qualities might influence the desirable values for the quorum size, with the goal to avoid splits, or to ensure consensus, or with an objective involving a specific degree or probability of consensus.

Additionally, another future work direction is to make our model more bio-plausible. Specifically, our model does not consider the very small probability that committed ants "drop out" of the nest they are committed to, and go back to searching. Adding this into the model could make it biologically more realistic.

Finally, one more way to strengthen our theoretical results is by adding a time bound metric to our consensus problem. Our current consensus metric, the consensus probability C, only requires that at least $(1 - \Delta)N$ ants keep staying at either n_1 or n_2 after a finite number of rounds. By adding a time bound metric as well, we would be able to better characterize the consensus probability (even if lower than a given C) of an emigration by a certain time t.

Acknowledgements. J. Zhao and N. Lynch are supported by NSF Awards CCF-2003830, CCF-1461559 and CCF-0939370.

References

1. Doering, G.N., Pratt, S.C.: Queen location and nest site preference influence colony reunification by the ant temnothorax rugatulus. Insectes Soc. **63**(4), 585–591 (2016). https://doi.org/10.1007/s00040-016-0503-1
2. Doering, G.N., Pratt, S.C.: Symmetry breaking and pivotal individuals during the reunification of ant colonies. J. Exp. Biol. **222**(5), jeb194019 (2019)
3. Doran, C., Newham, Z.F., Phillips, B.B., Franks, N.R.: Commitment time depends on both current and target nest value in temnothorax albipennis ant colonies. Behav. Ecol. Sociobiol. **69**(7), 1183–1190 (2015). https://doi.org/10.1007/s00265-015-1932-y
4. Dornhaus, A., Holley, J.A., Pook, V.G., Worswick, G., Franks, N.R.: Why do not all workers work? Colony size and workload during emigrations in the ant temnothorax albipennis. Behav. Ecol. Sociobiol. **63**(1), 43–51 (2008)
5. Franks, N.R., Dechaume-Moncharmont, F.X., Hanmore, E., Reynolds, J.K.: Speed versus accuracy in decision-making ants: expediting politics and policy implementation. Philos. Trans. R. Soc. Lond. Ser. B Biol. Sci. **364**(1518), 845–852 (2009)
6. Franks, N.R., Dornhaus, A., Fitzsimmons, J.P., Stevens, M.: Speed versus accuracy in collective decision making. Proc. R. Soc. Lond. Ser. B: Biol. Sci. **270**(1532), 2457–2463 (2003)
7. Franks, N.R., Mallon, E.B., Bray, H.E., Hamilton, M.J., Mischler, T.C.: Strategies for choosing between alternatives with different attributes: exemplified by house-hunting ants. Anim. Behav. **65**(1), 215–223 (2003)
8. Franks, N.R., Stuttard, J.P., Doran, C., Esposito, J.C., Master, M.C., Sendova-Franks, A.B., Masuda, N., Britton, N.F.: How ants use quorum sensing to estimate the average quality of a fluctuating resource. Sci. Rep. **5**(1), 11890 (2015)
9. Franks, N., et al.: Speed-cohesion trade-offs in collective decision making in ants and the concept of precision in animal behaviour. Anim. Behav. **85**(6), 1233–1244 (2013)
10. Gordon, D.M.: The ecology of collective behavior in ants. Ann. Rev. Entomol. **64**(1), 35–50 (2019). pMID: 30256667
11. Hajek, B.: Random Processes for Engineers. Cambridge University Press, Cambridge (2015)
12. Healey, C.I.M., Pratt, S.C.: The effect of prior experience on nest site evaluation by the ant temnothorax curvispinosus. Anim. Behav. **76**, 893–899 (2008)
13. Hoeffding, W.: Probability inequalities for sums of bounded random variables. J. Am. Stat. Assoc. **58**(301), 13–30 (1963)
14. Johnstone, R.A., Dall, S.R.X., Franks, N.R., Pratt, S.C., Mallon, E.B., Britton, N.F., Sumpter, D.J.T.: Information flow, opinion polling and collective intelligence in house-hunting social insects. Philos. Trans. R. Soc. Lond. Ser. B: Biol. Sci. **357**(1427), 1567–1583 (2002)
15. Mallon, E., Pratt, S., Franks, N.: Individual and collective decision-making during nest site selection by the ant leptothorax albipennis. Behav. Ecol. Sociobiol. **50**(4), 352–359 (2001). https://doi.org/10.1007/s002650100377
16. Marshall, J.A.R., Bogacz, R., Dornhaus, A., Planqué, R., Kovacs, T., Franks, N.R.: On optimal decision-making in brains and social insect colonies. J. R. Soc. Interface **6**(40), 1065–1074 (2009)

17. Marshall, J.A., Dornhaus, A., Franks, N.R., Kovacs, T.: Noise, cost and speed-accuracy trade-offs: decision-making in a decentralized system. J. R. Soc. Interface **3**(7), 243–254 (2006)
18. McGlynn, T.P.: The ecology of nest movement in social insects. Ann. Rev. Entomol. **57**(1), 291–308 (2012). pMID: 21910641
19. Moglich, M.H.J.: Social organization of nest emigration in leptothorax (hym., form.) (1978)
20. Planqué, R., Dornhaus, A., Franks, N.R., Kovacs, T., Marshall, J.A.R.: Weighting waiting in collective decision-making. Behav. Ecol. Sociobiol. **61**(3), 347–356 (2007). https://doi.org/10.1007/s00265-006-0263-4
21. Pratt, S.C.: Behavioral mechanisms of collective nest-site choice by the ant temnothorax curvispinosus. Insectes Soc. **52**(4), 383–392 (2005). https://doi.org/10.1007/s00040-005-0823-z
22. Pratt, S., Mallon, E., Sumpter, D., et al.: Quorum sensing, recruitment, and collective decision-making during colony emigration by the ant leptothorax albipennis. Behav. Ecol. Sociobiol. **52**(2), 117–127 (2002). https://doi.org/10.1007/s00265-002-0487-x
23. Pratt, S.C.: Quorum sensing by encounter rates in the ant Temnothorax albipennis. Behav. Ecol. **16**(2), 488–496 (2005)
24. Pratt, S.C., Sumpter, D.J.T.: A tunable algorithm for collective decision-making. Proc. Natl. Acad. Sci. **103**(43), 15906–15910 (2006)
25. Pratt, S.C., Sumpter, D.J., Mallon, E.B., Franks, N.R.: An agent-based model of collective nest choice by the ant temnothorax albipennis. Anim. Behav. **70**(5), 1023–1036 (2005)
26. Richardson, T.O., Sleeman, P.A., Mcnamara, J.M., Houston, A.I., Franks, N.R.: Teaching with evaluation in ants. Curr. Biol. **17**(17), 1520–1526 (2007)
27. Robinson, E.J.H., Franks, N.R., Ellis, S., Okuda, S., Marshall, J.A.R.: A simple threshold rule is sufficient to explain sophisticated collective decision-making. PLOS ONE **6**(5), 1–11 (2011)
28. Robinson, E.J., Smith, F.D., Sullivan, K.M., Franks, N.R.: Do ants make direct comparisons? Proc. R. Soc. B: Biol. Sci. **276**(1667), 2635–2641 (2009)
29. Sasaki, T., Colling, B., Sonnenschein, A., Boggess, M.M., Pratt, S.C.: Flexibility of collective decision making during house hunting in temnothorax ants. Behav. Ecol. Sociobiol. **69**(5), 707–714 (2015). https://doi.org/10.1007/s00265-015-1882-4
30. Stroeymeyt, N., Giurfa, M., Franks, N.R.: Improving decision speed, accuracy and group cohesion through early information gathering in house-hunting ants. PLOS ONE **5**(9), 1–10 (2010)
31. Sumpter, D.J., Pratt, S.C.: Quorum responses and consensus decision making. Philos. Trans. R. Soc. B: Biol. Sci. **364**(1518), 743–753 (2009)
32. Visscher, P.K.: Group decision making in nest-site selection among social insects. Ann. Rev. Entomol. **52**(1), 255–275 (2007). pMID: 16968203
33. Zhao, J., Su, L., Lynch, N.: (2021). https://github.com/snowbabyjia/QuorumSensingConsensus

Location Functions for Self-stabilizing Byzantine Tolerant Swarms

Yotam Ashkenazi[1(✉)], Shlomi Dolev[1], Sayaka Kamei[2], Yoshiaki Katayama[3], Fukuhito Ooshita[4], and Koichi Wada[5]

[1] Department of Computer Science, Ben-Gurion University of the Negev, Beersheba, Israel
{yotamash,dolev}@post.bgu.ac.il

[2] Graduate School of Advanced Science and Engineering, Hiroshima University, Higashihiroshima, Japan
s10kamei@hiroshima-u.ac.jp

[3] Graduate School of Engineering, Nagoya Institute of Technology, Nagoya, Japan
katayama@nitech.ac.jp

[4] Graduate School of Science and Technology, Nara Institute of Science and Technology, Ikoma, Japan
f-oosita@is.naist.jp

[5] Department of Applied Informatics, Faculty of Science and Engineering, Hosei University, Koganei, Japan
wada@hosei.ac.jp

Abstract. This paper proposes a novel framework to realize self-stabilizing Byzantine tolerant swarms. In this framework, non-Byzantine robots execute tasks while satisfying location functions, that is, the robots use a policy for their location choice, which restricts their location to satisfy the functions. We give a general Byzantine-resilient self-stabilizing algorithm based on the location function, and then provide an efficient implementation of the self-stabilizing algorithms for special classes of tasks, called polynomial-based tasks and shape-based tasks. We also demonstrate the usefulness of the proposed framework by implementing typical tasks of robots.

Keywords: Mobile robots · Byzantine faults · Self-stabilization

1 Introduction

Often, a swarm of robots must perform a task with a specific structure defined in a two- or three-dimensional space. For example, a line of marching robots may

This work was supported in part by JSPS KAKENHI No. 19K11828, 20H04140, 20K11685, and 21K11748 the Ministry of Science and Technology, Israel & JST SICORP (Grant#JPMJSC1806), Lynne and William Frankel Center for Computer Science, the Rita Altura Trust Chair in Computer Science and the German Research Funding (DFG, Grant#8767581199). It was also supported in part by the Helmsley Charitable Trust through the Agricultural, Biological and Cognitive Robotics Initiative, the Marcus Endowment Fund both at Ben-Gurion University of the Negev.

C. Johnen et al. (Eds.): SSS 2021, LNCS 13046, pp. 229–242, 2021.
https://doi.org/10.1007/978-3-030-91081-5_15

clean a field, and drones may try to reduce air resistance by creating a three-dimensional aerodynamic shape, possibly mimicking a swarm of fish moving within a higher-level dimensional polynomial structure.

When dealing with a swarm of robots in practice, we have to consider the presence of Byzantine, faulty, or malicious robots. These robots may not follow a given algorithm, possibly, due to experiencing a fault or an intentional malicious takeover that may intentionally interrupt or disturb other robots. For example, the behavior is caused by a bug in the software of honest robots or malicious malware injection. These possibly Byzantine robots may be temporarily or constantly controlled by an adversary. Traditionally, a swarm is designed to cope with a certain upper bound on the number of Byzantine robots [2–4, 10]. However, the number of such Byzantine robots may exceed (at least temporally) any given threshold up to all robots being faulty. Still, we would like the robots to execute their task once enough of them are correct(ed). In addition, there are cases where a global fault, such as momentary lightning, effects all members in the swarm. In these situations, automatic recovery in the form of self-stabilization [6, 7] in the presence of a minority of Byzantine robots is very important. Nevertheless only a few works consider self-stabilization and Byzantine-fault tolerance at the same time for a swarm of robots, see [1, 5] and the references therein.

This paper proposes a novel framework to realize self-stabilizing Byzantine tolerant swarms. In this framework, non-Byzantine robots execute tasks while satisfying *location functions*, that is, the robots use a policy for their location choice, which restricts their location to satisfy the functions. For example, let us consider a marching task where robots move while forming a line. In this task, all non-Byzantine robots move so that their locations satisfy some line function. For function $f : \mathbb{R} \times \mathbb{R} \rightarrow \{true, false\}$, we say non-Byzantine robots form f if location (x, y) of every non-Byzantine robot satisfies f (i.e., $f(x, y) = true$). In the marching task, non-Byzantine robots should form f such that, for some a, b and c ($ab \neq 0$), $f(x, y) = true$ if and only if $ax + by + c = 0$. To define more generally, let us consider a two-dimensional environment where n synchronous robots exist and t of them are Byzantine. Let $\mathcal{T}$ be a task executed by robots. Function $f : \mathbb{R} \times \mathbb{R} \rightarrow \{true, false\}$ is a *location function* of task $\mathcal{T}$ if non-Byzantine robots form f at some legitimate configuration of task $\mathcal{T}$. We characterize task $\mathcal{T}$ by a set of location functions that cover all legitimate configurations of task $\mathcal{T}$. We refer to such a set of location functions as a *location function set* $\mathcal{F}_\mathcal{T}$. For example, in the case of the above marching task, the location function set consists of functions corresponding to lines formed by robots. The location function set $\mathcal{F}_\mathcal{T}$ is useful for Byzantine identification if $n - t$ non-Byzantine locations identify a unique function in $\mathcal{F}_\mathcal{T}$ regardless of t Byzantine locations.

The location functions enable robots to execute tasks in a self-stabilizing Byzantine-tolerant fashion. If robots identify a unique function that current $n-t$ locations satisfy, they are in correct configurations and move to the next locations based on the task definition (closure). Otherwise, robots can detect the incorrect configurations and move to the default or corrected locations depending on the current configuration (convergence).

Although the framework gives a self-stabilizing Byzantine-tolerant algorithm, finding a unique location function from the current locations is not easy. In a trivial way, for every combination of $n - t$ locations, robots must compute whether the combination defines a unique location function or not. This requires $\Omega(n^t)$ local computation time and it is not practical. For this reason, this paper proposes efficient methods for cases where the location functions are given as polynomials or shape structures.

First, we focus on the case where location functions in $\mathcal{F}_T$ are given as polynomials, that is, any location function $f_i \in \mathcal{F}_T$ is defined as $f_i(x, y) = true \Leftrightarrow y = P_i(x)$ for some polynomial $P_i(x)$. That is, non-Byzantine robots move so that they form polynomial $y = P_i(x)$ for some i. In this case, the proposed framework exploits Berlekamp-Welch algorithm [12] (referred in the sequel as the BW algorithm). The BW algorithm ensures that, when a polynomial P is of degree at most $n-2t-1$ and at least $n-t$ non-Byzantine robots stand on points of P, then the non-Byzantine robots find the polynomial P with an efficient calculation. Hence, if every location function in $\mathcal{F}_T$ is a polynomial of degree at most $n - 2t - 1$, robots can efficiently identify a unique location function by the BW algorithm.

Next, we focus on the case where location functions in $\mathcal{F}_T$ are given as shape structures, that is, any location function $f_i \in \mathcal{F}_T$ is defined by a shape such as a circle, an ellipse, and so on. For example, in case of a circle, location function $f_i \in \mathcal{F}_T$ is defined such that $f_i(x, y) = true \Leftrightarrow (x - a_i)^2 + (y - b_i)^2 = r_i^2$ for some a_i, b_i and r_i. Such shapes can be defined by a relatively small set of points, and hence the proposed framework uses the majority of such (sub)sets of the robot locations to decide on Byzantine robots that do not obey the majority agreement on the shape.

In the last part, this paper demonstrates that the proposed framework is useful to implement typical tasks of robots. In particular, we demonstrate self-stabilizing Byzantine-tolerant implementations of convergence, marching and exploration.

2 Preliminaries

2.1 A Robot Model

The system consists of a set of n robots. We define R as a set of all robots. Each robot is modeled as a point on a two-dimensional Euclidean space. Robots agree on a global Cartesian coordinate system. We represent a position of a robot by coordinate (x, y). We assume that, if multiple robots share the same coordinate, they can recognize its local order among the robots, that is, they can change their behaviors based on the local order. Robots are anonymous, that is, they have no identifiers and execute the same algorithm. Robots are oblivious, that is, they have no memories to record past information. Robots cannot communicate with other robots explicitly, however they can communicate implicitly by observing coordinates of all other robots.

We assume the fully-synchronous (FSYNC) model [8,9,11]. That is, every robot repeats a cycle composed of Look, Compute and Move phases synchronously. In the Look phase, the robot obtains a set of coordinates of all robots and the number of robots in each of the coordinates. In the Compute phase, depending on the observation in the previous Look phase, the robot decides on its target coordinate. If the robot decides to move, it moves to the target coordinate in the Move phase. We assume a rigid movement, that is, if a robot moves, it arrives at the target coordinate during the Move phase. We say a robot executes a *step* if it executes one cycle composed of Look, Compute and Move phases.

In this paper, some robots may be *Byzantine*. Byzantine robots can make arbitrary movements that do not obey the algorithm choice. We define $R_B \subset R$ as a set of all Byzantine robots. We also define $R_N = R \setminus R_B$ as a set of all non-Byzantine robots. Throughout the paper, we assume that the number of Byzantine robots is at most t.

A *configuration* c is defined as a combination of coordinates of all robots. A sequence of configurations $E = c_1, c_2, \ldots$ is an *execution* starting from initial configuration c_1 if, for every $i \geq 1$, c_{i+1} is reached from c_i by a step of all robots. For a configuration c and a subset of robots R', we define partial configuration $c|_{R'}$ as a combination of coordinates of robots in R' in configuration c. Similarly, for an execution $E = c_1, c_2, \ldots$ and a subset of robots R', we define partial execution $E|_{R'}$ as a sequence of partial configurations $c_1|_{R'}, c_2|_{R'}, \ldots$. From the definition, $E|_{R_N}$ is an execution in case that we focus on only non-Byzantine robots in E. A *task* $\mathcal{T}$ is defined as a predicate $\mathcal{L}_\mathcal{T}$ on executions without Byzantine robots. We say an execution E achieves task $\mathcal{T}$ if and only if $\mathcal{L}_\mathcal{T}(E|_{R_N}) = \mathit{true}$ holds. A configuration c is legitimate if c appears in some execution E that achieves task $\mathcal{T}$. Note that in a legitimate configuration Byzantine robots can stay at arbitrary points.

Definition 1. (t-Byzantine-resilient self-stabilization)
An algorithm $\mathcal{A}$ is a t-Byzantine-resilient self-stabilizing *algorithm for $\mathcal{T}$ if and only if, when $|R_B| \leq t$ holds, there exists a set of* safe configurations C_{safe} *such that both of the following properties hold.*

- Convergence*: Starting in any arbitrary configuration the system eventually reaches a configuration in C_{safe}.*
- Closure*: For any execution E that starts from a configuration in C_{safe}, $\mathcal{L}_\mathcal{T}(E|_{R_N}) = \mathit{true}$ holds.*

2.2 Location Functions

In this subsection, we introduce *location functions*. The location function specifies possible coordinates of non-Byzantine robots in legitimate configurations for a task. For a set of coordinates S and a function $f : \mathbb{R} \times \mathbb{R} \rightarrow \{\mathit{true}, \mathit{false}\}$, we say S satisfies function f if $f(x, y) = \mathit{true}$ holds for any $(x, y) \in S$. For a set of robots R', we define $S_{R'}(c)$ as a set of coordinates of all robots in R' in configuration c.

Definition 2. (Location function)
Function $f : \mathbb{R} \times \mathbb{R} \to \{true, false\}$ *is a* location function *of task* $\mathcal{T}$ *if and only if there exists a legitimate configuration* c *of task* $\mathcal{T}$ *such that* $S_{R_N}(c)$ *satisfies* f.

Definition 3. (Location function set)
A location function set *of task* $\mathcal{T}$, *denoted by* $\mathcal{F}_\mathcal{T}$, *is a set of location functions of* $\mathcal{T}$ *such that, for any legitimate configuration* c *of* $\mathcal{T}$, $S_{R_N}(c)$ *satisfies some location function* $f \in \mathcal{F}_\mathcal{T}$.

Consider the case that a task $\mathcal{T}$ is given with its location function set $\mathcal{F}_\mathcal{T}$. Fix a legitimate configuration c of task $\mathcal{T}$. Since $|R_B| \leq t$ holds, at least $n-t$ coordinates (i.e., coordinates of non-Byzantine robots) in c satisfy some location function $f \in \mathcal{F}_\mathcal{T}$. Here, if no set of $n-t$ coordinates in c satisfies $f' \in \mathcal{F}_\mathcal{T} \setminus \{f\}$, robots can recognize the location function f corresponding to the current configuration c and hence identify at most t Byzantine robots. If such a property holds for every legitimate configuration of task $\mathcal{T}$, robots can always use the location function to identify Byzantine robots. Formally we define such a task as a *t-Byzantine-identifiable task* in Definition 4.

Definition 4. (t-Byzantine-identifiable task)
A task $\mathcal{T}$ *with location function set* $\mathcal{F}_\mathcal{T}$ *is t-Byzantine-identifiable if and only if the following property holds: For any legitimate configuration* c *of task* $\mathcal{T}$, *if* $S_{R_N}(c)$ *satisfies location function* $f \in \mathcal{F}_\mathcal{T}$, *no subset* $S \subset S_R(c)$ *with* $|S| = |R| - t$ *satisfies any location function* $f' \in \mathcal{F}_\mathcal{T} \setminus \{f\}$.

Consider a t-Byzantine-identifiable task $\mathcal{T}$ with $\mathcal{F}_\mathcal{T}$. Definition 4 implies that, if c is a legitimate configuration of $\mathcal{T}$, there exists exactly one location function $f \in \mathcal{F}_\mathcal{T}$ such that $S \subset S_R(c)$ with $|S| = n - t$ satisfies f. If the number of such location functions is more than one or zero in configuration c, c is not legitimate.

2.3 Function-Based Tasks

We can define some tasks by specifying location functions that robots form during an execution. For example, a marching task with a line can be regarded as a task such that robots form some linear functions successively. In general, we define *function-based tasks* in Definition 5. Note that, to make robots form some function with multiple coordinates, we force non-Byzantine robots not to share the same coordinate.

Definition 5. (Function-based task)
Consider task $\mathcal{T}$, *its location function set* $\mathcal{F}_\mathcal{T}$, *and a function* $\psi_\mathcal{T} : \mathcal{F}_\mathcal{T} \to \mathcal{F}_\mathcal{T}$. *Task* $\mathcal{T}$ *is a function-based task with* $(\mathcal{F}_\mathcal{T}, \psi_\mathcal{T})$ *if and only if the following condition holds: For an execution* $E = c_0, c_1, \ldots$, $\mathcal{L}_\mathcal{T}(E) = true$ *holds if and only if there exists a sequence of functions* $f_0, f_1, \ldots$ *starting from* $f_0 \in \mathcal{F}_\mathcal{T}$ *such that* $|S_{R_N}(c_i)| = |R_N|$ *holds,* $S_{R_N}(c_i)$ *satisfies* f_i, *and* $f_{i+1} = \psi_\mathcal{T}(f_i)$ *holds for any* $i \geq 0$.

As an example, let us consider a marching task $\mathcal{T}$ such that robots move to direction (δ_x, δ_y) while keeping a circle. In this case, its location function set $\mathcal{F}_\mathcal{T}$ contains any circle function, that is, $\mathcal{F}_\mathcal{T} = \{f_{a,b,c} \mid a, b, c \in \mathbb{R}, c > 0\}$, where $f_{a,b,c}$ is a function such that $f_{a,b,c}(x, y) = true$ holds if and only if $(x - a)^2 + (y - b)^2 = c^2$ holds. Then, transition function $\psi_\mathcal{T}$ is defined as $\psi_\mathcal{T}(f_{a,b,c}) = f_{a+\delta_x, b+\delta_y, c}$. Hence, we can define such marching task $\mathcal{T}$ as a function-based task with $(\mathcal{F}_\mathcal{T}, \psi_\mathcal{T})$.

3 t-Byzantine-resilient Self-stabilizing Algorithms Based on Location Functions

In this section, we describe how location functions can be used to realize t-Byzantine-resilient self-stabilizing algorithms for t-Byzantine-identifiable function-based tasks. In Sect. 3.1, we give a general t-Byzantine-resilient self-stabilizing algorithm. However, a trivial implementation of the algorithm is not efficient because it requires a vast amount of local computation. Hence, in Sects. 3.2 and 3.3, we focus on polynomial-based and shape-based tasks, respectively, and give more efficient implementation of the algorithm.

3.1 A General Algorithm

If $\mathcal{T}$ is a t-Byzantine-identifiable function-based task with $(\mathcal{F}_\mathcal{T}, \psi_\mathcal{T})$, we have a simple t-Byzantine-resilient self-stabilizing algorithm for $\mathcal{T}$ based on location functions.

Algorithm 1 gives the pseudocode of the algorithm. In the Look phase, each robot r obtains the set of coordinates of all robots S. In the Compute phase, r computes the coordinate that r moves to in the Move phase. To compute it, r tries to find $f \in \mathcal{F}_\mathcal{T}$ such that some set $S' \subset S$ with $|S'| = n - t$ satisfies f. If r finds exactly one such location function f, it can identify the next location function f_{next} by transition function $\psi_\mathcal{T}$. Note that, since $\mathcal{T}$ is t-Byzantine-identifiable, in any legitimate configuration r can identify such f and consequently the next location function f_{next}. Then, r computes its next coordinate (x, y) that satisfies $f_{next}(x, y) = true$. Here, to satisfy the condition of a function-based task, r should choose (x, y) such that other non-Byzantine robots do not choose the same coordinate (x, y). This is possible because robots can observe all coordinates and break a tie by a local order even if multiple robots stay at the same coordinate. If r does not find exactly one location function f, it understands that the current configuration is not legitimate. Hence, r uses a known default location function f_{def}, and moves to the next coordinate (x, y) that satisfies $f_{def}(x, y) = true$. Here r also should choose (x, y) such that other non-Byzantine robots do not choose the same coordinate (x, y).

Theorem 1. *If $\mathcal{T}$ is a t-Byzantine-identifiable function-based task, Algorithm 1 is a t-Byzantine-resilient self-stabilizing algorithm for $\mathcal{T}$.*

Algorithm 1. A t-Byzantine-resilient self-stabilizing algorithm for t-Byzantine-identifiable function-based task $\mathcal{T}$ with $(\mathcal{F}_\mathcal{T}, \psi_\mathcal{T})$. Function $f_{def} \in \mathcal{F}_\mathcal{T}$ is a default location function known to all robots.

A cycle of robot r
Look phase:
1: $S \leftarrow$ the set of coordinates of all robots
Compute phase:
2: **if** there exists exactly one $f \in \mathcal{F}_\mathcal{T}$ such that some set $S' \subset S$ with $|S'| = n - t$ satisfies f **then**
3: $f_{next} \leftarrow \psi_\mathcal{T}(f)$
4: $(x_{next}, y_{next}) \leftarrow (x, y)$ s.t. $f_{next}(x, y) = true$ and (x, y) is unique among robots
5: **else**
6: $(x_{next}, y_{next}) \leftarrow (x, y)$ s.t. $f_{def}(x, y) = true$ and (x, y) is unique among robots
7: **end if**
Move phase:
8: Move to coordinate (x_{next}, y_{next})

Proof. Assume that $\mathcal{T}$ is a t-Byzantine-identifiable function-based task and $|R_B| \leq t$ holds. Let C be a set of configurations such that $c \in C$ holds if and only if $|S_{R_N}(c)| = n - t$ holds and $S_{R_N}(c)$ satisfies f for some $f \in \mathcal{F}_\mathcal{T}$. We prove that C is a set of safe configurations in Definition 1.

First we consider the convergence property. Consider an arbitrary initial configuration c not in C. That is, there exists no location function $f \in \mathcal{F}_\mathcal{T}$ such that $S_{R_N}(c)$ satisfies f. By considering the following two cases, we prove that the next configuration is in C.

- Case that there exists exactly one function $f \in \mathcal{F}_\mathcal{T}$ and a set of robots $R' \subset R$ such that $|S_{R'}(c)| = n - t$ holds and $S_{R'}(c)$ satisfies f. Note that this case happens when $R' \neq R_N$ holds. In this case, all non-Byzantine robots identify the same f in line 2. Consequently they compute the same f_{next} in line 3, and obtain (x_{next}, y_{next}) that satisfies $f_{next}(x_{next}, y_{next}) = true$ in line 4. Hence all non-Byzantine robots move to different coordinates that satisfy the same location function f_{next} in line 8, and thus the next configuration is in C.
- Other case. In this case, all non-Byzantine robots move to different coordinates that satisfy the same default location function f_{def}. Hence, the next configuration is in C.

Next we consider the closure property. Consider an execution $E = c_0, c_1, \ldots$ starting from $c_0 \in C$. From $c_0 \in C$, $|S_{R_N}(c_0)| = n - t$ holds and there exists function $f_0 \in \mathcal{F}_\mathcal{T}$ such that $S_{R_N}(c_0)$ satisfies f_0. Let $f_0, f_1, \ldots$ be a sequence of functions that satisfies $f_{i+1} = \psi_\mathcal{T}(f_i)$ for any i $(i \geq 0)$. We prove that $|S_{R_N}(c_i)| = n - t$ holds and $S_{R_N}(c_i)$ satisfies f_i for any $i \geq 0$ by induction. The base case is clear. Assume that $|S_{R_N}(c_i)| = n - t$ holds and $S_{R_N}(c_i)$ satisfies f_i for some $i \geq 0$. Since $\mathcal{T}$ is t-Byzantine-identifiable, all non-Byzantine robots identify f_i in line 2. Hence all non-Byzantine robots move to different coordinates that satisfy $f_{i+1} = \psi_\mathcal{T}(f_i)$, and thus $|S_{R_N}(c_{i+1})| = n - t$ holds and $S_{R_N}(c_{i+1})$ satisfies f_{i+1}.

Since Algorithm 1 satisfies the convergence and closure properties, it is a t-Byzantine-resilient self-stabilizing algorithm for $\mathcal{T}$. □

Note that, although robots must choose, in line 4 (resp., line 6), different coordinates that satisfy f_{next} (resp., f_{def}), Algorithm 1 does not describe which coordinates robots actually choose. When implementing the algorithm, we can specify the coordinates by considering other factors. For example, robots can choose coordinates sufficiently close to each other, or choose coordinates that are sufficiently distributed, as long as location functions allow such choices.

Although Algorithm 1 is general, it requires a vast number of local computation. This is because, in line 2 of Algorithm 1, robots must search for f such that some subset $S' \subset S$ satisfies f. From $|S'| = n - t$ and $|S| = n$, the number of possible sets of S' is $\binom{n}{t} = \Omega(n^t)$. This implies that each robot requires an $\Omega(n^t)$ amount of local computation. For this reason, in the following subsections, we focus on polynomial-based and shape-based tasks, and give more efficient implementation of the algorithm.

3.2 Implementation for Polynomial-Based Tasks

In this subsection, we focus on a special case of function-based tasks, called a *polynomial-based task*. A polynomial-based task is a function-based task such that every location function is represented by a polynomial. That is, robots move so that they form polynomials specified by the task. We give the formal definition in Definition 6.

Definition 6. (Polynomial-based task)
A function-based task $\mathcal{T}$ with $(\mathcal{F}_\mathcal{T}, \psi_\mathcal{T})$ is called a polynomial-based task with $(\mathcal{F}_\mathcal{T}, \psi_\mathcal{T})$ if and only if any location function $f \in \mathcal{F}_\mathcal{T}$ is defined as $f(x,y) = true \Leftrightarrow y = P(x)$ for some polynomial $P(x)$. By abuse of terminologies, we use polynomial P to indicate location function f.

In line 2 of Algorithm 1, robots search for a polynomial that $n - t$ coordinates satisfy. Different from general function-based tasks, if the degree of the polynomial is at most $n - 2t - 1$, robots can efficiently execute this by using the Berlekamp-Welch (BW) algorithm.

Theorem 2. *[12]* **(The BW algorithm).** *Let n, t, d be positive integers such that $d \leq n-2t-1$ holds. Let $S = \{(x_1, y_1), \ldots, (x_n, y_n)\}$ be a set of n coordinates such that $x_i \neq x_j$ for $i \neq j$. The BW algorithm takes S as an input, and computes a single polynomial P with degree at most d such that at least $n - t$ coordinates in S are on $y = P(x)$, if such polynomial P exists. If such polynomial P does not exist, the BW algorithm fails to output. The BW algorithm requires an $O(n^3)$ amount of computation.*

Note that, when robots obtain a set of coordinates $S = S_R(c)$ in configuration c, S can contain less than n coordinates if multiple robots share the same coordinate. In addition, S can contain multiple coordinates whose x-coordinates are identical. However, as we prove in the following lemma, we can still use the BW algorithm because at least $n - t$ non-Byzantine robots stay at different coordinates.

Lemma 1. *Consider a t-Byzantine-identifiable polynomial-based task $\mathcal{T}$ with $(\mathcal{F}_\mathcal{T}, \psi_\mathcal{T})$ such that $\mathcal{F}_\mathcal{T}$ consists of polynomials with degree at most $d = n-2t-1$. Consider a legitimate configuration c of $\mathcal{T}$ and polynomial $P \in \mathcal{F}_\mathcal{T}$ such that $S_R(c)$ satisfies P. In configuration c, robots can compute polynomial P by the BW algorithm in an $O(n^3)$ amount of computation.*

Proof. Let $S = S_R(c)$. Since c is legitimate, there exists polynomial $P \in \mathcal{F}_\mathcal{T}$ with degree at most d such that at least $n - t$ coordinates in S are on $y = P(x)$. Since all non-Byzantine robots stay at different coordinates, $n - |S|$ Byzantine robots share coordinates with some other robots if $n > |S|$ holds. In addition, since all non-Byzantine robots are on $y = P(x)$, at most $t-(n-|S|)$ coordinates in S are not on $y = P(x)$.

Let $x_1, \ldots, x_k$ be x-coordinates that duplicate in S, and let α_i $(1 \le i \le k)$ be the number of coordinates in S whose coordinates are x_i. We have $\alpha_i \ge 2$ $(1 \le i \le k)$ from the definition. We construct set S' by removing from S all coordinates that have duplicated x-coordinates, that is, $S' = \{(x, y) \in S \mid x \notin \{x_1, \ldots, x_k\}\}$. Let $n' = |S'| = |S| - \sum_{i=1}^{k} \alpha_i$. Let t' be the number of coordinates in S' that are not on $y = P(x)$. Now S' does not contain duplicated x-coordinates and $n' - t'$ coordinates in S' are on polynomial $y = P(x)$ with degree at most d. For each i $(1 \le i \le k)$, since at most one coordinate with x-coordinate x_i in S is on $y = P(x)$, at least $\alpha_i - 1$ coordinates with x-coordinate x_i in S are not on $y = P(x)$. Hence, $t' \le t - (n - |S|) - \sum_{i=1}^{k}(\alpha_i - 1)$ holds. This implies that $n' - 2t' - 1 \ge n - 2t - 1 + (n - |S|) + (\sum_{i=1}^{k} \alpha_i - 2k) \ge n - 2t - 1 \ge d$. From Theorem 2, we can apply the BW algorithm to S' and obtain polynomial P. This requires an $O(n'^3) = O(n^3)$ amount of computation. □

From Lemma 1, we can apply the BW algorithm to search for a location function in line 2 of Algorithm 1. Clearly, we have the following theorem.

Theorem 3. *Consider a t-Byzantine-identifiable polynomial-based task $\mathcal{T}$ with $(\mathcal{F}_\mathcal{T}, \psi_\mathcal{T})$ such that $\mathcal{F}_\mathcal{T}$ consists of polynomials with degree at most $n-2t-1$. In this case, robots can execute Algorithm 1 for $\mathcal{T}$ with an $O(n^3)$ amount of local computation.*

3.3 Implementation for Shape-Based Tasks

In this subsection, we focus on a special case of function-based tasks, called a *shape-based task*. A shape-based task is a function-based task such that every location function is represented by a shape. That is, robots move so that they form some shapes specified by the task. As examples of shapes, we can consider a circle, an ellipse, a square, a hexagonal shape, and so on. Formally shape H is defined as a set of (possibly an infinite number of) coordinates. We give the formal definition in Definition 7.

Definition 7. (Shape-based task)
A function-based task $\mathcal{T}$ *with* $(\mathcal{F}_\mathcal{T}, \psi_\mathcal{T})$ *is called a shape-based task with* $(\mathcal{F}_\mathcal{T}, \psi_\mathcal{T})$ *if and only if any location function* $f \in \mathcal{F}_\mathcal{T}$ *is defined as* $f(x, y) = true \Leftrightarrow ((x, y)$ *is a coordinate on some shape* $H)$. *By abuse of terminologies, we use shape* H *to indicate location function* f.

In line 2 of Algorithm 1, robots search for a shape such that $n-t$ coordinates are on the shape. Let x be the minimum number of coordinates that can identify a single shape H in $\mathcal{F}_\mathcal{T}$. Here x is smaller than $n-t$ for many shapes. For example, if $\mathcal{F}_\mathcal{T}$ contains only circles, $x = 3$ holds. Hence, for any configuration, we have only $\binom{n}{x}$ candidate shapes, and so, by checking all candidates, robots can find a shape such that at least $n - t$ coordinates are on the shape. Since robots check, for $\binom{n}{x}$ candidate shapes, whether at least $n - t$ coordinates are on the shape, this requires an $O(n^{x+1})$ amount of local computation.

Here we consider the condition that robots can identify a single shape H. In any legitimate configuration $n - t$ non-Byzantine robots form a single shape. On the other hand, t Byzantine robots can form another shape by using $x - 1$ coordinates from non-Byzantine robots. Hence, if $n - t > t + x - 1$ holds, robots can identify a correct shape.

From the above discussion, clearly we have the following theorem.

Theorem 4. *Consider a t-Byzantine-identifiable shape-based task* $\mathcal{T}$ *with* $(\mathcal{F}_\mathcal{T}, \psi_\mathcal{T})$ *such that* x *coordinates can identify a shape in* $\mathcal{F}_\mathcal{T}$ *and* $x \leq n-2t$ *holds. In this case, robots can execute Algorithm 1 for* $\mathcal{T}$ *with an* $O(n^{x+1})$ *amount of local computation.*

4 Task Scheme

In this section we demonstrate that location functions (polynomials and shapes) can support executing basic missions in self-stabilizing fashion, in the presence of Byzantine robots. We consider basic tasks in [9].

4.1 Convergence

Under the global view and general movement assumptions convergence can be done very easy. The convergence task $\mathcal{T}$ can be defined by using the shape-based task with $(\mathcal{F}_\mathcal{T}, \psi_\mathcal{T})$ such that robots move to very small circle with some center (δ_x, δ_y). In this case, its location function set $\mathcal{F}_\mathcal{T}$ contains any function that represents a circle, that is, $\mathcal{F}_\mathcal{T} = \{f_{a,b,\epsilon} \mid a, b, \epsilon \in \mathbb{R}\}$ where $f_{a,b,\epsilon}$ is a function such that $f_{a,b,\epsilon}(x, y) = true$ holds if and only if $(x - \delta_x)^2 + (y - \delta_y)^2 = \epsilon^2$ holds. The transition function $\psi_\mathcal{T}$ is defined as $\psi_\mathcal{T}(f_{a,b,\epsilon}) = f_{a,b,\epsilon/2}$.

Instead of choosing a specific location (δ_x, δ_y), we can choose the location to be a function of the current location, say, the average location of all (non-Byzantine, if identifiable) robots. At this point all robots move to the same coordinate and the shape is reduced to be a small circle. In Fig. 1, all non-Byzantine robots converge to a specific small circle.

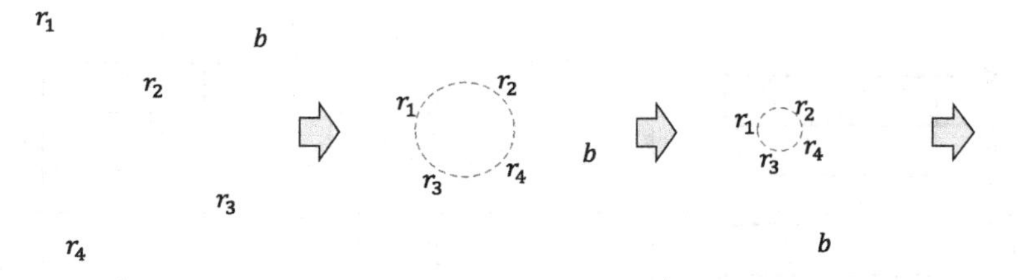

Fig. 1. Non-Byzantine robots (denoted by r_1 to r_4) converge to a specific small circle. A Byzantine robot (denoted by b) moves to another location and is identified as Byzantine.

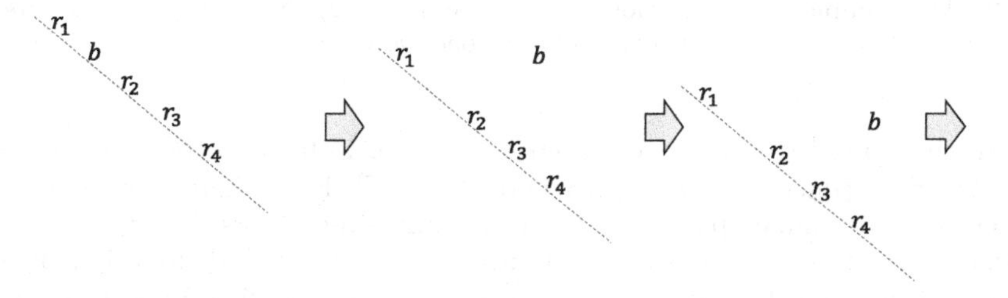

Fig. 2. Non-Byzantine robots (denoted by r_1 to r_4) march down ignoring a Byzantine robot (denoted by b).

4.2 Marching

In Sect. 2.3 an example is presented to the marching task using a circle function. The marching task can be done based on polynomial or shape structure.

Now, let us consider a marching task $\mathcal{T}$ such that robots move to direction (δ_x, δ_y) while keeping their formation parallel to a polynomial $y = P(x)$. The marching task is defined as a polynomial-based task, and its location function set $\mathcal{F}_\mathcal{T}$ contains function $P_{a,b} = P(x-a) + b$ for any a and b. The marching can be defined by increasing the x-value by δ_x and the y-value by δ_y, that is, its transition function $\psi_\mathcal{T}$ is defined as $\psi_\mathcal{T}(P_{a,b}) = P_{a+\delta_x, b+\delta_y}$. In Fig. 2, all non-Byzantine robots march together keeping the polynomial with degree 1.

4.3 Exploration

To consider the exploration task, we regard a bounded two-dimensional Euclidean environment as a bounded board. For integers x and y, we associate coordinate (x, y) in the two-dimensional Euclidean environment to a tile (x, y) in the board. The goal of the exploration task is to make robots visit every tile in the board. We assume $n \geq 2t + 1$.

Based on location functions, we can develop a self-stabilizing t-Byzantine-resilient self-stabilizing algorithm for the exploration task. Let H be a set of all

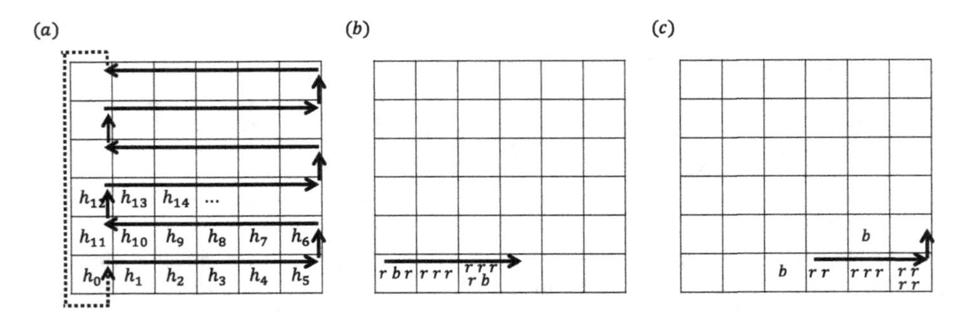

Fig. 3. An example of exploration ($n = 11$ and $t = 2$). Non-Byzantine robots and Byzantine robots are denoted by r and b, respectively.

tiles in the bounded board. We assign sequential indices to tiles and represent them by $H = \{h_0, h_1, \ldots, h_{m-1}\}$, where $m = |H|$. For example, we can assign indices using a "snake" path: We regard a right direction as a forward direction in every even row, and regard a left direction as a forward direction in every odd row (Fig. 3(a)). For simplicity, we consider mathematical operations on tile indices as operations modulo m. Let $l = \lfloor (n - 2t - 1)/(t + 1) \rfloor$.

We formulate the exploration task in the following manner. In legitimate configurations, we put robots on $l + 1$ successive tiles such that the head tile contains at least $t + 1$ robots and each of other l tiles (called a tail) contains at least one robot. To do this, non-Byzantine robots try to make a configuration such that the head tile contains at least $2t + 1$ robots and each of l tail tiles contains at least $t + 1$ robots (Fig. 3(b)). That is, for some i, non-Byzantine robots move so that, if Byzantine robots also move correctly, at least $2t+1$ robots stay on h_i and at least $t + 1$ robots stay on each of tiles $h_{i-l}, h_{i-l+1}, \ldots, h_{i-1}$. Of course, t Byzantine robots can move to arbitrary tiles. However, the head tile h_i contains at least $t+1$ non-Byzantine robots and Byzantine robots cannot counterfeit such a head. In addition, each of l tail tiles contains at least one non-Byzantine robot. Hence, non-Byzantine robots can easily identify the current legitimate configuration (for example see Fig. 3(c)).

Robots move their formation forward by $l + 1$ tiles per cycle and explore the board along the tile indices. That is, if robots identify the current head h_i with tail $h_{i-l}, h_{i-l+1}, \ldots, h_{i-1}$, they next construct the head h_{i+l+1} with tail $h_{i+1}, h_{i+2}, \ldots, h_{i+l}$. For example, robots move from configuration in Fig. 3(b) to Fig. 3(c) in one cycle. More concretely, robots move as follows. First robots assign sequential numbers 0 to $n - 1$ to all robots along the tile indices; that is, robots on tiles with small indices obtain small sequential numbers. Since we assume that robots can recognize their local orders when they share the same coordinate, they can assign different numbers even if they share the same tile. Then, for every x ($0 \le x \le l - 1$), robots with numbers $x(t + 1)$ to $x(t + 1) + t$ move to tile h_{i+x+1}, and robots with numbers $(l + 1)(t + 1)$ to $n - 1$ move to h_{i+l+1}. Since every tile of the head and the tail contains at least one non-

Byzantine robot, robots can achieve exploration by repeating this behavior in $O(m/l) = O(mt/n)$ cycles.

To describe the behavior, we can define location function $f_i \in \mathcal{F}_T$ ($0 \leq i \leq m-1$) as follows: $f_i(x,y) = \mathit{true}$ if and only if coordinate (x,y) corresponds to a tile in $\{h_{i-l}, h_{i-l+1}, \ldots, h_i\}$. Similarly to Algorithm 1, robots can achieve exploration in the self-stabilizing manner. If robots identify a unique location function f_i corresponding to the current configuration, they can move to the next configuration corresponding to f_{i+l+1} as described above. Otherwise, they can move to the configuration that satisfies the default function (for example f_l in Fig. 3(b)).

5 Conclusions

We demonstrated the usage of error correcting of functions and in particular polynomials in coping with Byzantine robots in swarms. The implementations of these fundamental tasks are self-stabilizing in spite of (a bounded number of) Byzantine robots. Moreover, one can design a sequence of swarm tasks, moving from one task to the next. Typically Byzantine tolerant algorithms cope with a threshold on the number of Byzantine robots, as consistency of the swarm cannot be preserved when too many (e.g., all) robots are Byzantine. Attackers that can overtake several robots will surely try to exceed the declared threshold on the number of compromised robots. Self-stabilization in spite of Byzantine robots tolerates periods in which even all participants are Byzantine. Once enough robots recover the self-stabilization property ensures that the entire swarm starts convergence to act as desired again. We believe that we have established a rich and useful framework for the designing and practical implementation of a swarm of robots.

References

1. Ashkenazi, Y., Dolev, S., Kamei, S., Ooshita, F., Wada, K.: Forgive and forget: self-stabilizing swarms in spite of byzantine robots. Concurr. Comput. Pract. Exp., 188–194 (2020). CANDAR Workshops 2019
2. Attiya, H., Welch, J.L.: Distributed Computing - Fundamentals, Simulations, and Advanced Topics. Wiley Series on Parallel and Distributed Computing, 2nd edn. Wiley, Hoboken (2004)
3. Bouzid, Z., Gradinariu Potop-Butucaru, M., Tixeuil, S.: Byzantine convergence in robot networks: the price of asynchrony. In: Abdelzaher, T., Raynal, M., Santoro, N. (eds.) OPODIS 2009. LNCS, vol. 5923, pp. 54–70. Springer, Heidelberg (2009). https://doi.org/10.1007/978-3-642-10877-8_7
4. Bouzid, Z., Potop-Butucaru, M.G., Tixeuil, S.: Optimal byzantine-resilient convergence in uni-dimensional robot networks. Theoret. Comput. Sci. **411**(34–36), 3154–3168 (2010)
5. Défago, X., Potop-Butucaru, M., Raipin-Parvédy, P.: Self-stabilizing gathering of mobile robots under crash or Byzantine faults. Distrib. Comput. **33**(5), 393–421 (2019). https://doi.org/10.1007/s00446-019-00359-x

6. Dijkstra, E.W.: Self-stabilizing systems in spite of distributed control. Commun. ACM **17**(11), 643–644 (1974)
7. Dolev, S.: Self-Stabilization. MIT Press, Cambridge (2000)
8. Flocchini, P., Prencipe, G., Santoro, N.: Distributed Computing by Oblivious Mobile Robots. Synthesis Lectures on Distributed Computing Theory. Morgan & Claypool Publishers (2012)
9. Flocchini, P., Prencipe, G., Santoro, N.: Distributed Computing by Mobile Entities: Current Research in Moving and Computing. Springer, Heidelberg (2019). https://doi.org/10.1007/978-3-030-11072-7
10. Souissi, S., Izumi, T., Wada, K.: Byzantine-tolerant circle formation by oblivious mobile robots. In: Proceedings of the International Conference on Communications, Computing and Control Applications, pp. 1–6 (2011)
11. Suzuki, I., Yamashita, M.: Distributed anonymous mobile robots: formation of geometric patterns. SIAM J. Comput. **28**(4), 1347–1363 (1999)
12. Welch, L., Berleklamp, E.: Error correction for algebraic block codes. US Patent 4 633 470 (1986)

Applications and Implications of a General Framework for Self-Stabilizing Overlay Networks

Andrew Berns(✉)

Department of Computer Science, University of Northern Iowa, Cedar Falls, IA, USA
andrew.berns@uni.edu

Abstract. From data centers to IoT devices to Internet-based applications, overlay networks have become an important part of modern computing. Many of these overlay networks operate in fragile environments where processes are susceptible to faults which may perturb a node's state and the network topology. Self-stabilizing overlay networks have been proposed as one way to manage these faults, promising to build or restore a particular topology from any initial configuration or after the occurrence of any transient fault. To date there have been several self-stabilizing protocols designed for overlay networks. These protocols, however, are either focused on a single specific topology, or provide very inefficient solutions for a general set of overlay networks.

In this paper, we analyze an existing algorithm and show it can be used as a general framework for building many other self-stabilizing overlay networks. Our analysis for time and space complexity depends upon several properties of the target topology itself, providing insight into how topology selection impacts the complexity of convergence. We then demonstrate the application of this framework by analyzing the complexity for several existing topologies. Next, using insights gained from our analysis, we present a new topology designed to provide efficient performance during convergence with the general framework. Our process demonstrates how the implications of our analysis help isolate the factors of interest to allow a network designer to select an appropriate topology for the problem requirements.

Keywords: Topological self-stabilization · Overlay networks · Fault-tolerant distributed systems

1 Introduction

Distributed systems have become an ubiquitous part of modern computing, with systems continuing to grow in size and scope. As these systems grow larger, the need for topologies that allow for efficient operations like search and routing increases. To this end, many systems use *overlay networks* to control the network topology. In overlay networks, connections are made using logical links, each of

C. Johnen et al. (Eds.): SSS 2021, LNCS 13046, pp. 243–257, 2021.
https://doi.org/10.1007/978-3-030-91081-5_16

which consists of zero or more physical links. This use of logical links means program actions can add and delete edges in the network, allowing the system to maintain an arbitrary logical topology even when the physical topology may be fixed.

Many of these large systems operate in environments where faults are commonplace. Servers may crash, communication links may be damaged, and processes may join or leave the system frequently. This reality has increased the demand for fault tolerant overlay networks. One particularly strong type of fault tolerance is *self-stabilization*, where a legal configuration is guaranteed to be reached after any transient fault. For overlay networks, this means a correct topology can always be built when starting from any configuration provided the network is not disconnected.

1.1 Problem Overview

Our current work focuses on self-stabilizing overlay networks. A self-stabilizing overlay network guarantees that program actions will build a legal topology even when the system starts in *any* weakly-connected topology. Our interest, then, is in the design and analysis of algorithms that, when executed on an arbitrary initial weakly-connected topology, add and delete edges with program actions until a legal *target topology* is reached.

To date, most work has focused on algorithms for a single topology, or has been inefficient in terms of time or space complexity. Work focused on a specific topology is hard to generalize and derive insights from for expanding to other overlay network applications, while general frameworks with high time and space complexity may be too inefficient to be useful in practice. Our interest is in general frameworks for overlay network creation that allow efficient algorithms to be built while still being general enough to provide insights into the stabilization of arbitrary topologies.

1.2 Main Results and Significance

In this paper, we build upon the work of Berns [2] to present a general algorithm for creating self-stabilizing overlay network protocols for a variety of target topologies, and provide several examples of the application of our general algorithm, including with a new overlay network topology. More specifically, our contributions are as follows:

- We update the analysis of Berns [2] to show how their algorithm can be extended into an algorithm for *any* target topology. Our updated analysis is the first to show a general framework for self-stabilizing overlay network creation that allows for efficient stabilization in both time and space.
- As part of this updated analysis, we introduce several measures of complexity that are properties of the target topology itself. These measures are useful for two reasons. First, they allow us to analyze the general algorithm easily for a variety of topologies. Perhaps just as important, however, is that they

provide valuable insight into how the selection of the target topology affects convergence in terms of both time and space.
- We analyze several existing overlay network topologies, providing the necessary metrics for analysis in our general framework. This analysis provides an example of how our framework can be applied, and also helps provide concrete insight into factors affecting convergence and demonstrate how diameter, degree, and robustness are balanced when designing self-stabilizing overlay networks.
- Using the insight gained from our earlier contribution, we define a new network topology which stabilizes with sublinear time and space complexity in our framework. This demonstrates how the framework can provide network designers with guidance to help them build new topologies that can stabilize efficiently with our approach. The design and analysis of new topologies targeted for efficient stabilization in this framework could be an interesting area of future study.

The key idea of our work is the extension of the algorithm of Berns [2] to work with other topologies by defining the target topology and analyzing several relevant measures regarding this topology. This definition allows the creation and analysis of many self-stabilizing overlay networks without having to design the algorithm from scratch. Furthermore, our framework highlights the factors of the network that affect stabilization, allowing a designer to tune the topology to meet their needs.

1.3 Related Work and Comparison

The past few decades have seen a large body of work develop on overlay networks. Early work focused on defining *structured* overlay networks, where a single correct configuration existed for a particular set of nodes. One such example of a structured network is CHORD [12]. Early work on structured networks often did not consider fault tolerance, or considered a weak model with limited possible failures.

As work expanded in overlay networks, so did work in various types of fault tolerance. One category of work considered *self-healing* networks, where a particular network property could be maintained even during limited node deletions [7]. Several examples of this work used virtual nodes [13], although they were not used to create a specific embedding as done in this work. In DCONSTRUCTOR [5], the authors present a framework for building overlay networks. DCONSTRUCTOR works by forming clusters and merging these clusters together. However, DCONSTRUCTOR is not self-stabilizing as it assumes all nodes begin in a single node cluster. Said in another way, DCONSTRUCTOR assumes an arbitrary initial *topology*, but not an arbitrary initial state. This is also the same assumption in the work of Götte et al. [6], who presented an algorithm for transforming a constant-degree network into a tree in $\mathcal{O}(\log n)$ rounds.

Our work considers *self-stabilizing overlay networks*, where the correct configuration is reached after an arbitrary number of transient faults that do not disconnect the network. There are several examples of these as well. The SKIP+ graph [9] presents a self-stabilizing variant of the SKIP graph [1] with polylogarithmic convergence time, although the space requirements are linear for some configurations. Another example of a self-stabilizing overlay network is RE-CHORD [10], a CHORD variant with virtual nodes and a convergence time of $\mathcal{O}(n \log n)$ rounds.

To date, most work has been focused on the convergence to a particular topology. One exception to this is the *Transitive Closure Framework* [4], which presents a general algorithm for creating any locally-checkable overlay network. They identify a general measure of interest for stabilization time which they call the *detector diameter*. Unfortunately the space requirement of their algorithm is $\Theta(n)$ for any topology, limiting the applicability.

In AVATAR [2], Berns presented both a locally-checkable definition of a network embedding for arbitrary topologies, and a self-stabilizing algorithm for building an embedded binary search tree with polylogarithmic time and space requirements. The work only considered a single topology, however, and did not offer insight into measures for arbitrary topologies. Our goal with this work is to build upon AVATAR to address these issues.

2 Preliminaries

2.1 Model of Computation

We model our distributed system as an undirected graph $G = (V, E)$, with n processes in V communicating over the edges E. Each node $u \in V$ has a unique identifier $u.id \in \mathbb{N}$, which is stored as immutable data in u. Where clear from the context, we will use u to represent the identifier of u.

Each node $u \in V$ has a *local state* consisting of a set of variables and their values, along with its immutable identifier $u.id$. A node executes a *program* whose actions modify the values of the variables in its local state. All nodes execute the same program. Nodes can also communicate with their neighbors. We use the *synchronous message passing* model of computation [8], where computation proceeds in synchronous rounds. During each round, a node receives messages sent to it in the previous round from any node in its neighborhood $N(u) = \{v \in V : (u, v) \in E\}$, executes program actions to update its local state, and sends messages to any of its neighbors. We assume reliable communication channels with bounded delay, meaning a message is received by node u in some round i if and only if it was sent to u in round $i - 1$.

In the overlay network model, nodes communicate over logical links that are part of a node's state, meaning a node may execute actions to create or delete edges in G. In any round, a node may delete any edge incident upon it, as well as create any edge to a node v which has been "introduced" to it from some neighbor w, such that (u, w) and (w, v) are both in E. Said in another way, in a particular round a node may connect its neighbors to one another by direct logical links.

The goal for our computation is for nodes to execute actions to update their state (including modifying the topology by adding and deleting edges to other nodes) until a legal configuration is reached. A *legal configuration* can be represented as a predicate over the state of the nodes in the system. In the overlay network model, links are part of a node's state, meaning a legal configuration is defined at least in part by the overlay network topology. The *self-stabilizing overlay network problem* is to design an algorithm $\mathcal{A}$ such that when executing $\mathcal{A}$ on each node in a connected network with nodes in an arbitrary state, and allowing $\mathcal{A}$ to add and delete edges, eventually a legal configuration, including a predicate defined at least in part by the network's logical topology, is reached. This means that a self-stabilizing overlay network will always automatically restore a legal configuration (including reconfiguring the network topology) after *any* transient failure so long as the network remains connected.

2.2 Complexity Measures

When designing self-stabilizing overlay network protocols, there are two measures of interest: the time required to build a correct configuration, and the space required to do so (in terms of a node's degree). In our model, we are concerned with the number of synchronous rounds that are required to reach a legal state. In particular, the maximum number of synchronous rounds required to build a legal topology when starting from any arbitrary configuration is called the *convergence time*.

When measuring the space requirements, we use the *degree expansion* measure from the original Avatar work [2], which is defined as the ratio of the maximum node degree of any node during convergence over the maximum node degree from the initial or final configuration. This measure is based upon the idea that if a node begins with a large degree in the initial configuration, or ends with a large degree in the final configuration, the overall algorithm cannot be expected to have a low degree during convergence. Instead, we are interested in the "extra" degree growth caused by the algorithm during convergence.

3 Generalizing Avatar

The original Avatar work [2] provided two things: a definition of a locally-checkable embedding from any set of real nodes to a particular target topology, and a self-stabilizing algorithm for creating a specific binary tree topology. Below, we review these contributions and expand the analysis of the algorithm to show it can work for arbitrary topologies.

3.1 Avatar Definition

The Avatar network definition is simply a dilation-1 embedding between a *guest network* and a *host network*. More specifically, let $\mathcal{F}$ be a family of graphs such that, for each $N \in \mathbb{N}$, there is exactly one graph $F_N \in \mathcal{F}$ with node set

$\{0, 1, \ldots, N-1\}$. We call $\mathcal{F}$ a *full graph family*, capturing the notion that the family contains exactly one topology for each "full" set of nodes $\{0, 1, \ldots, N-1\}$ (relative to the identifiers). For any $N \in \mathbb{N}$ and $V \subseteq \{0, 1, \ldots, N-1\}$, $\text{AVATAR}_{\mathcal{F}}(N, V)$ is a network with node set V that realizes a dilation-1 embedding of $F_N \in \mathcal{F}$. The specific embedding is given below. We also show that, when given N, AVATAR is locally checkable (N can be viewed as an upper bound on the number of nodes in the system). It is on this full graph family *guest network* that our algorithms shall execute, as we will show later.

Definition 1. *Let $V \subseteq [N]$ be a node set $\{u_0, u_1, \ldots, u_{n-1}\}$, where $u_i < u_{i+1}$ for $0 \leq i < n-1$. Let the* range *of a node u_i be $range(u_i) = [u_i, u_{i+1})$ for $0 < i < n-1$. Let $range(u_0) = [0, u_1)$ and $range(u_{n-1}) = [u_{n-1}, N)$. $\text{AVATAR}_{\mathcal{F}}(N, V)$ is a graph with node set V and edge set consisting of two edge types:*

Type 1: $\{(u_i, u_{i+1}) | i = 0, \ldots, n-2\}$
Type 2: $\{(u_i, u_j) | u_i \neq u_j \wedge \exists (a, b) \in E(F_N), a \in range(u_i) \wedge b \in range(u_j)\}$

When referring to a general AVATAR network for any set of nodes, we will omit the V and simply refer to $\text{AVATAR}_{\mathcal{F}}(N)$.

With AVATAR, we consider two "networks": a *host network* consisting of the *real* nodes in V, and a *guest network* consisting of the N *virtual* nodes from the target topology. Each real node in V is the host of one or more virtual nodes in N. This embedding provides several advantages. First, it allows us to make many networks locally-checkable provided all nodes know N in advance. Second, it provides a simple mechanism for which to reason about network behavior in the guest network. As the target N-node topology is fixed regardless of the actual set of real nodes V, the design and analysis of our algorithms is simplified by executing them on the guest network. Since we are using a dilation-1 embedding, most metrics for performance regarding the guest network (e.g. diameter) still apply to the host network. We shall assume that the virtual nodes also use the same synchronous message passing model described for the real nodes.

Note the work of AVATAR does require all nodes know N, an upper bound on the number of nodes in the system. From a practical standpoint, even in cases where N and n are significantly different, a polylogarithmic convergence time in N may still be small enough (e.g. for IPv6, $\mathcal{O}(\log N)$ is only 128).

3.2 The Avatar Algorithm

The original AVATAR work was focused on the creation of a specific topology (a binary search tree) as the *target topology*. Their algorithm followed a divide-and-conquer approach, separating nodes into clusters and then merging them together. One can think of their self-stabilizing algorithm as involving three different components:

1. *Clustering*: The first step in the algorithm is for nodes to form clusters. These clusters begin as a single host node hosting a full N node guest network of the

target topology (CBT in the original work). In the initial configuration, nodes may not be a part of a cluster, but since AVATAR is locally checkable, all faulty configurations contain at least one node which detects the faulty configuration and will begin forming the single-node clusters. This fault detection and cluster creation will propagate through the network until eventually all nodes are members of N node clusters of the target topology.

2. *Matching*: The second step of the algorithm is to match together clusters so that they may merge together. To do this, the root node of a spanning tree defined on the cluster repeatedly polls the nodes of its cluster, asking them to either find neighboring clusters that are looking for merge partners (called the leader role), or to look at neighboring clusters that can assign them a merge partner (called the follower role). The role of leader or follower is randomly selected. Leader clusters will match together all of their followers for merging by adding edges between the roots of each cluster, creating a matching between clusters that may not be direct neighbors. This ability to create edges to match non-neighboring clusters allows more matches to occur, and thus more merges, and thus a faster convergence time.
3. *Merging*: The algorithm then deals with the merging of matched clusters. To prevent degrees from growing too large, a cluster is only allowed to merge with at most one other cluster at a time. Once two clusters have matched from the previous step, the roots of the clusters connect as "partners" and update their successor pointers based upon the identifier of the host of the root of the other cluster. One node will have its responsible range become smaller, and this node will send all virtual nodes that were in its old responsible range to its partner in the other cluster. The children of the root nodes are connected, and then they repeat the process of updating successor pointers and passing along virtual nodes outside their new responsible range. Eventually this process reaches the leaves, at which point all nodes in both clusters have updated their responsible ranges and now form a new legal cluster of the target topology.

As it turns out, the algorithm components from the original AVATAR work do not depend upon the specific topology that is being built (the *target topology*). While the analysis of complexity assumes a complete binary search tree, the algorithm components themselves simply rely upon an arbitrary target topology and a spanning tree defined upon that topology on which to execute PIF waves. We can therefore extend this algorithm to other topologies if we update the analysis and include several additional metrics. We define these metrics next after discussing the algorithm's intuition.

3.3 Relevant Metrics

To update the analysis of the original AVATAR work for any target topology requires two measures of the target topology. The first of these is diameter of the target topology, which will be a factor in determining the convergence time. The second of these is a measure of a real node's degree inside the embedded target topology, which will be a factor in determining the degree expansion.

Spanning Tree Diameter. In the original AVATAR algorithm, a spanning tree embedded onto the target topology was used to communicate and coordinate between nodes in a particular cluster. For our work, we will simply use a spanning tree with a root of (virtual) node 0 and consisting of the shortest path from node 0 to all other nodes. Obviously the diameter of this spanning tree is at most the diameter of the target topology, and we therefore shall use the diameter of the target topology as our first metric of interest. We denote the diameter of a particular target topology T with N nodes as $D(T_N)$.

As we shall see, this diameter measure will be key in determining the stabilization time. Intuitively, a low-diameter spanning tree results in faster communication within clusters, and therefore faster convergence than a higher diameter spanning tree.

Maximum Degree of Embedding. The other measure of interest has to do with the degree of the *real* nodes when embedding the target topology. More formally, let the *maximum degree of embedding* T_N *in* AVATAR be defined as the maximum degree of any node in $\text{AVATAR}_T(N, V)$ for any node set $V \subseteq N$. Where clear from context, we will refer to this simply as the *maximum degree of embedding* and denote it as $\Delta_{\text{A}}(T_N)$. Note the maximum degree of embedding is almost entirely determined by the target topology T, as there are only 2 edges in $\text{AVATAR}_T(N)$ per node that are not present to realize a dilation-1 embedding of T.

The maximum degree of embedding is a critical measure for degree expansion as it determines how many additional edges a node may receive during the various stages of the algorithm. The clusters in the AVATAR algorithm are N node instances of the target topology T, and using metrics defined on T_N is acceptable for running time. However, each time an edge is added to a virtual node within a cluster, we must consider the effects on the degree in the host network, not just the guest network.

Note that the definition of maximum degree of embedding considers *any* possible subset of real nodes V. This differs from typical (non-stabilizing) overlay network results, where it is common to assume that identifiers are uniformly distributed, meaning that the ranges of each real node are of similar size. However, since we are building a self-stabilizing protocol, and each cluster is by itself an N node instance of the target topology, the ranges of hosts inside clusters during convergence may be quite skewed, even when the final distribution of node identifiers is not.

3.4 Overall Complexity

If we are given the diameter and the maximum degree of embedding of an arbitrary target topology, we can then determine the convergence time and degree expansion of Berns' algorithm for the arbitrary target topology, as shown below. The full analysis, modified from the original by Berns, can be found in the full version of the paper [3]. Our contribution is the observation that the algorithm works for any topology, the updated metrics for the analysis, and the examples and discussion that follow.

Theorem 1. *The algorithm of Berns [2] defines a self-stabilizing overlay network for* AVATAR_T*, for some full graph family target topology* T*, with convergence time of* $\mathcal{O}(D(T_N) \cdot \log N)$ *in expectation, where* $D(T_N)$ *is the diameter of the* N *node topology* T_N*.*

Proof Sketch. A sketch of the steps for proving this theorem are as follows:

- In at most $\mathcal{O}(D(T_N))$ rounds, every node is a member of a cluster.
- For any cluster, in an expected $\mathcal{O}(D(T_N))$ rounds, the cluster has completed a merge with another cluster, meaning the number of clusters has decreased by a constant fraction in $\mathcal{O}(D(T_N))$ rounds.
- Reducing the number of clusters by a constant fraction needs to be done $\mathcal{O}(\log N)$ times before a single cluster remains.

Theorem 2. *The algorithm of Berns [2] defines a self-stabilizing overlay network for* AVATAR_T*, for some full graph family target topology* T*, with degree expansion of* $\mathcal{O}(\Delta_{\text{A}}(T_N) \cdot \log N)$ *in expectation, where* $\Delta_{\text{A}}(T_N)$ *is the maximum degree of embedding of the* N *node target topology* T_N*.*

Proof Sketch. To prove this, we consider the actions that might increase a node's degree.

- Regardless of the number of merges a node participates in, the node's degree will grow to at most $\mathcal{O}(\Delta_{\text{A}}(T_N))$ as the result of merge actions. By definition, a node's degree within its cluster after a merge cannot exceed $\Delta_{\text{A}}(T_N)$.
- During the process of matching clusters together, a node's degree may grow by one for every child it has in the spanning tree. Since the node has at most $\Delta_{\text{A}}(T_N)$ children from other nodes, each time the node participates in the matching its degree grows by $\mathcal{O}(\Delta_{\text{A}}(T_N))$. In expectation, this matching happens $\mathcal{O}(\log N)$ times (see proof of Theorem 1).

The implications of Theorems 1 and 2 are that we can simply define a target topology and analyze its diameter and maximum degree of embedding to have a self-stabilizing protocol for our target topology.

4 Examples

In this section, we demonstrate how the selection of the target topology affects the complexity of our algorithm by considering several different topologies: the LINEAR network, a complete binary search tree (CBT, taken from [2]), and CHORD [12].

4.1 Linear

As the name suggests, the LINEAR network consists of a line of nodes sorted by identifier. The formal desired end topology for the LINEAR network is given next.

Definition 2. *The* LINEAR(N) *network, for* $N \in \mathbb{N}$, *consists of nodes* $V = \{0, 1, \ldots, N-1\}$ *and edges* $E = \{(i, i+1), i \in [0, N-2]\}$.

Lemma 1. *The diameter of an N node* LINEAR *network is* $\mathcal{O}(N)$.

Each virtual node has a degree of at most 2, and it is easy to see that each real node also has a degree of at most 2. Therefore, the maximum degree of embedding for LINEAR is $\mathcal{O}(1)$.

Lemma 2. *The maximum degree of embedding of the* LINEAR *topology is* $\mathcal{O}(1)$.

Proof. Note that for any particular $range(u) = [x, y]$, there are at most two external edges: $(x-1, x)$ and $(y, y+1)$. All other edges are between virtual nodes inside the range.

The above lemmas combined with Theorems 1 and 2 give us the following corollary.

Corollary 1. *The self-stabilizing* AVATAR *algorithm builds* $\text{AVATAR}_{\text{LINEAR}}(N)$ *in an expected* $\mathcal{O}(N \cdot \log N)$ *rounds with an expected degree expansion of* $\mathcal{O}(\log N)$.

Note the convergence time of this algorithm is a logarithmic factor slower than previous results [11]. This logarithmic factor comes from the "cost of coordination", as edges are only added when clusters have matched.

4.2 Complete Binary Search Tree

In the first AVATAR paper, the author defined and analyzed an algorithm for one specific topology, the complete binary search tree (called CBT). We formally define the desired end topology for CBT, list the relevant measures for this topology below and omit the proofs, as those are contained in the work of Berns [2].

Definition 3. *For* $a \leq b$, *let* CBT$[a, b]$ *be a binary tree rooted at* $r = \lfloor (b+a)/2 \rfloor$. *Node r's left cluster is* CBT$[a, r-1]$, *and r's right cluster is* CBT$[r+1, b]$. *If* $a > b$, *then* CBT$[a, b] = \perp$. *We define* CBT$(N) =$ CBT$[0, N-1]$.

Lemma 3. *The diameter of an N node* CBT *network is* $\mathcal{O}(\log N)$.

Lemma 4. *The maximum degree of embedding of an N node* CBT *network is* $\mathcal{O}(\log N)$.

The above lemmas combined with Theorems 1 and 2 give us the following corollary.

Corollary 2. *The self-stabilizing* AVATAR *algorithm builds a target topology of* $\text{AVATAR}_{\text{CBT}}(N)$ *in expected* $\mathcal{O}(\log^2 N)$ *rounds with* $\mathcal{O}(\log^2 N)$ *expected degree expansion.*

Note the above corollary matches with the detailed proofs given in the original AVATAR work. Unlike the original, however, we reached our conclusions based simply upon the metrics we defined earlier. This corollary, then, serves as a nice "sanity check" on the accuracy of our results.

4.3 Chord

Both LINEAR and CBT are tree topologies, meaning they are fragile: a single node or link failure may partition the network. In this section, we consider the more robust CHORD network [12], defined as follows.

Definition 4. *For any* $N \in \mathbb{N}$*, let* $\text{CHORD}(N)$ *be a graph with nodes* $[N]$ *and edge set defined as follows. For every node* i*,* $0 \leq i < N$*, add to the edge set* (i, j)*, where* $j = (i+2^k) \mod N$*,* $0 \leq k < \log N - 1$*. When* $j = (i+2^k) \mod N$*, we say that* j *is the* k*-th finger of* i*.*

The original CHORD paper proves the following lemma regarding the network's diameter.

Lemma 5. *The diameter of an* N *node* CHORD *network is* $\mathcal{O}(\log N)$*.*

While the logarithmic diameter means we can efficiently build CHORD in terms of time complexity, the results are not so hopeful in terms of degree complexity. In an N-node CHORD network, every node has $\mathcal{O}(\log N)$ neighbors, some of which have identifiers up to $N/2$ away. The result of this is that if a real node has a range of size $N/2$, each virtual node in the range may potentially have a connection to a virtual node on a different host, leading to the following result.

Lemma 6. *The maximum degree of embedding of the* CHORD *topology is* $\mathcal{O}(N)$*.*

Proof. To see this, consider a specific n node embedding of CHORD where $range(u) = [0, N/2)$. Note each node in u's range is incident on at least one edge whose other endpoint is outside $range(u)$ – specifically, the $\log N - 1$ CHORD finger. As there are $N/2$ nodes in u's range, we have at least $N/2$ edges with exactly one endpoint outside $range(u)$, and our lemma holds.

Note this result is not a concern in the *final* configuration if we assume identifiers are uniformly distributed (a common assumption with overlay networks). However, we are working in a self-stabilizing setting where *any* initial configuration is possible, and therefore one could imagine a scenario where the system reaches a configuration where a single cluster consisting of node 0 is matched with an $N/2$ node cluster consisting of all real nodes from the range $[N/2, N)$.

Combining the above two lemmas with Theorem 1 and Theorem 2 gives us the following corollary.

Corollary 3. *The self-stabilizing* AVATAR *algorithm builds* $\text{AVATAR}_{\text{CHORD}}(N)$ *in expected* $\mathcal{O}(\log^2 N)$ *rounds and with expected degree expansion of* $\mathcal{O}(N \cdot \log N)$.

Note we can actually improve the bound of the degree expansion, as it is at most $\mathcal{O}(N)$. As our theorems provide an upper bound, however, we leave them stated as is for simplicity.

5 SkipChord

We showed above the role the diameter and maximum degree of embedding play in determining performance of the AVATAR algorithm. One of the benefits of the general analysis of the AVATAR algorithm is that it highlights the factors of the *topology* that will affect convergence, allowing a network designer to select an existing topology, or design new topologies, based upon the problem requirements while weighing the impact of the topology on convergence time and degree expansion.

In this section, we present a new network topology designed specifically for embedding in the AVATAR framework which we call SKIPCHORD. This ring-based network is more robust than LINEAR and CBT while avoiding the high degree requirements of the CHORD network.

5.1 Definition

As trees, the LINEAR and CBT networks make poor choices for many fault-prone applications as they are easily disconnected by node or link failure. CHORD represents a more robust choice, but suffers from a high maximum degree of embedding due to each of the N nodes having a long link (to a neighbor with identifier $\mathcal{O}(N)$ away from itself). Our SKIPCHORD network tries to balance this by limiting the number of fingers while still maintaining a topology more robust than a simple tree. We give the formal definition of SKIPCHORD below.

Definition 5. *For any* $N \in \mathbb{N}$, *let* SKIPCHORD(N, s) *be a graph with nodes* $[N]$ *and* skip factor s. *The edge set for* SKIPCHORD(N, s) *is defined as follows:*

- Ring edges*: For every node* i, $0 \leq i < N$, *add edges* $(i-1 \mod N, i)$ *and* $(i, i+1 \mod N)$
- Finger edges*: For every node* j, *where* $j = sk$, *for* $k = 0, 1, \ldots, (N-s)/s$, *add edge* $(j, j + 2^{k \mod \log N} \mod N)$. *We say the* size *of this finger edge is* $2^{k \mod \log N}$.

This construction basically takes the $\log N$ fingers from each node in the original CHORD and distributes them out over a range of nodes as determined by the *skip factor* s. As we shall show, by "skipping" the fingers, fewer virtual nodes in a real node's range have long (to a neighbor $\mathcal{O}(N)$ away) outgoing links, and therefore the number of edges to other real nodes is limited while not compromising efficient routing.

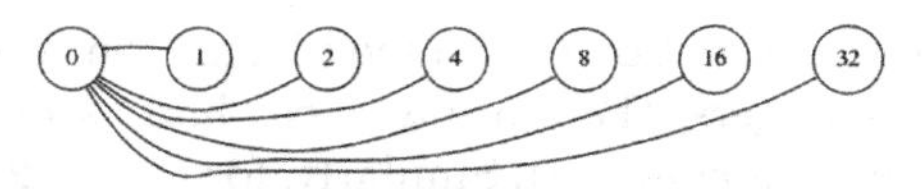

(a) Six Fingers of Node 0 in CHORD(64)

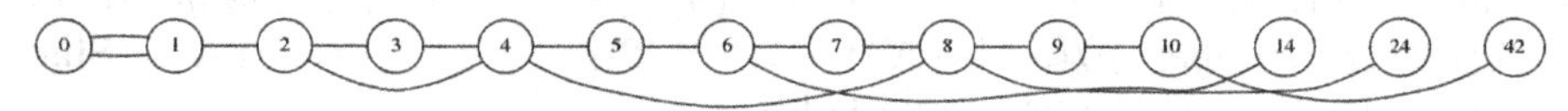

(b) First Six "Skipped" Fingers for SKIPCHORD(64, 2)

Fig. 1. (a) The neighborhood of node 0 in CHORD, and (b) the corresponding six fingers in SKIPCHORD. Note how the fingers in SKIPCHORD are no longer all incident on node 0 but instead have "skipped" ahead.

To better understand SKIPCHORD, consider Fig. 1. The network on the top shows node 0's neighborhood for CHORD(64), while the network on the bottom shows the corresponding fingers "skipped" with a skip factor of 2 (i.e. a subset of the edges for SKIPCHORD(64, 2)). The first six fingers in CHORD are all incident upon node 0, while the first six fingers in SKIPCHORD are distributed amongst nodes 0, 2, 4, 6, 8, and 10. As a result of this, each node in SKIPCHORD has a much smaller degree than in CHORD.

5.2 Metrics

We begin with a proof of the diameter of SKIPCHORD. The intuition behind our result is simple: any node is at most $s \cdot \log N$ away from an edge that at least halves the distance from itself to any other node.

Lemma 7. *The diameter of an N-node* SKIPCHORD *network with skip factor s (*SKIPCHORD(N, s)*) is $\mathcal{O}(s \cdot \log^2 N)$.*

Proof. Consider the hops required to halve the distance between two arbitrary nodes u and v. In at most $\mathcal{O}(s \cdot \log N)$ hops from u using ring edges, at least one finger of every size is reachable. One of these fingers will at least halve the distance to the other node v. As the distance to v can be halved in $\mathcal{O}(s \cdot \log N)$ hops, and this halving will occur $\log N$ times before reaching v, our lemma holds.

Next we see how spreading the fingers out over a set of nodes results in a lower maximum degree of embedding.

Lemma 8. *The maximum degree of embedding of an N-node* SKIPCHORD *with a skip factor of s (*SKIPCHORD(N, s)*) is $\mathcal{O}(\frac{N}{s \cdot \log N})$.*

Proof. Let an edge be called an *external edge* for node u if and only if it has exactly one endpoint in $range(u)$. Note that there are at most 2 external ring edges for any possible range for node u. We consider then the external finger edges in $range(u)$. Let k be the largest finger that has exactly one endpoint in $range(u)$. There are at most $|range(u)|/s \cdot \log N$ such fingers, where $|range(u)|$

denotes the size of the range (for our embeddings, the number of nodes u is hosting). For the $k-1$ fingers, there are at most half as many as the k fingers with exactly one endpoint in $range(u)$. Similarly, for the $k-2$ fingers, there are at most 1/4 as many as the k fingers with exactly one endpoint in $range(u)$, and so on. Summing these together, we get $(|range(u)|/(s \cdot \log N)) + 1/2(|range(u)|/(s \cdot \log N)) + 1/4(|range(u)|/(s \cdot \log N)) + \ldots = 2(|range(u)|/(s \cdot \log N))$. Since $|range(u)|$ is at most N, our lemma holds.

The above lemmas, combined with Theorem 1 and Theorem 2 give us the following corollary.

Corollary 4. *The* AVATAR *algorithm builds the* SKIPCHORD(N, s) *target topology in an expected* $\mathcal{O}(s \cdot \log^3 N)$ *rounds with an expected degree expansion of* $\mathcal{O}(N/s)$.

Note that we can select a skip factor in such a way as to have efficient time and space complexity. For instance, if we select a skip factor of $\log N$, we have polylogarithmic convergence time, sublinear degree expansion, and a target topology that is more robust than the tree topologies of LINEAR and CBT.

Given the fact that SKIPCHORD can be built efficiently whereas CHORD cannot, one can imagine several applications for its use. Like CHORD, SKIPCHORD can be used as a distributed hash table for storing and retrieving files, particularly in settings where transient failures may cause node and link failures, such as with unreliable Internet connections, or in cases where the nodes join and leave in large numbers frequently.

6 Discussion and Future Work

Besides providing a simple way to build and analyze self-stabilizing overlay networks, our analysis provides a set of parameters that a designer can tune to achieve a target level of efficiency. One application of our work, then, is in guiding the creation of new topologies that use the AVATAR embedding and strive for low diameter *and* low maximum degree of embedding while still maintaining other desirable properties like robustness to node or link failure. We have demonstrated this process with the creation of the SKIPCHORD topology. It would be interesting to see how other topologies perform in this framework.

Our framework can also be used to better understand the upper and lower bounds for the work or degree expansion of self-stabilizing overlay network protocols. While our results deal entirely with network embeddings, it would be interesting to see if the provided insights help find general bounds for any network.

References

1. Aspnes, J., Shah, G.: Skip graphs. In: SODA '03: Proceedings of the Fourteenth Annual ACM-SIAM Symposium on Discrete Algorithms, pp. 384–393. Society for Industrial and Applied Mathematics, Philadelphia (2003)

2. Berns, A.: Avatar: a time- and space-efficient self-stabilizing overlay network. In: Pelc, A., Schwarzmann, A.A. (eds.) SSS 2015. LNCS, vol. 9212, pp. 233–247. Springer, Cham (2015). https://doi.org/10.1007/978-3-319-21741-3_16
3. Berns, A.: Applications and implications of a general framework for self-stabilizing overlay networks (2021). https://arxiv.org/abs/2109.14125
4. Berns, A., Ghosh, S., Pemmaraju, S.V.: Building self-stabilizing overlay networks with the transitive closure framework. In: Défago, X., Petit, F., Villain, V. (eds.) SSS 2011. LNCS, vol. 6976, pp. 62–76. Springer, Heidelberg (2011). https://doi.org/10.1007/978-3-642-24550-3_7
5. Gilbert, S., Pandurangan, G., Robinson, P., Trehan, A.: Dconstructor: efficient and robust network construction with polylogarithmic overhead. In: Proceedings of the 39th Symposium on Principles of Distributed Computing, PODC 2020, pp. 438–447. Association for Computing Machinery, New York (2020). https://doi.org/10.1145/3382734.3405716
6. Götte, T., Hinnenthal, K., Scheideler, C., Werthmann, J.: Time-optimal construction of overlay networks. In: Proceedings of the 2021 ACM Symposium on Principles of Distributed Computing, PODC'21, pp. 457–468. Association for Computing Machinery, New York (2021). https://doi.org/10.1145/3465084.3467932
7. Hayes, T.P., Saia, J., Trehan, A.: The forgiving graph: a distributed data structure for low stretch under adversarial attack. In: Proceedings of the 28th ACM Symposium on Principles of Distributed Computing, PODC 2009, pp. 121–130. Association for Computing Machinery, New York (2009). https://doi.org/10.1145/1582716.1582740
8. Herlihy, M., Rajsbaum, S., Tuttle, M.R.: Unifying synchronous and asynchronous message-passing models. In: Proceedings of the Seventeenth Annual ACM Symposium on Principles of Distributed Computing, PODC 1998, pp. 133–142. Association for Computing Machinery, New York (1998). https://doi.org/10.1145/277697.277722
9. Jacob, R., Richa, A., Scheideler, C., Schmid, S., Täubig, H.: A distributed polylogarithmic time algorithm for self-stabilizing skip graphs. In: PODC '09: Proceedings of the 28th ACM Symposium on Principles of Distributed Computing, pp. 131–140. ACM, New York (2009). https://doi.org/10.1145/1582716.1582741
10. Kniesburges, S., Koutsopoulos, A., Scheideler, C.: Re-chord: a self-stabilizing chord overlay network. In: Proceedings of the 23rd ACM symposium on Parallelism in algorithms and architectures, SPAA 2011, pp. 235–244. ACM, New York (2011). https://doi.org/10.1145/1989493.1989527. http://doi.acm.org/10.1145/1989493.1989527
11. Onus, M., Richa, A.W., Scheideler, C.: Linearization: locally self-stabilizing sorting in graphs. In: ALENEX. SIAM (2007)
12. Stoica, I., Morris, R., Karger, D., Kaashoek, M.F., Balakrishnan, H.: Chord: a scalable peer-to-peer lookup service for internet applications. SIGCOMM Comput. Commun. Rev. **31**(4), 149–160 (2001). http://doi.acm.org/10.1145/964723.383071
13. Trehan, A.: Self-healing using virtual structures. CoRR abs/1202.2466 (2012)

Network Scaffolding for Efficient Stabilization of the CHORD Overlay Network

Andrew Berns(✉)

Department of Computer Science, University of Northern Iowa, Cedar Falls, IA, USA
andrew.berns@uni.edu

Abstract. Overlay networks, where nodes communicate with neighbors over logical links consisting of zero or more physical links, have become an important part of modern networking. From data centers to IoT devices, overlay networks are used to organize a diverse set of processes for efficient operations like searching and routing. Many of these overlay networks operate in fragile environments where faults that perturb the logical network topology are commonplace. Self-stabilizing overlay networks offer one approach for managing these faults, promising to build or restore a particular topology from any weakly-connected initial configuration.

Designing efficient self-stabilizing algorithms for many topologies, however, is not an easy task. For non-trivial topologies that have desirable properties like low diameter and robust routing in the face of node or link failures, self-stabilizing algorithms to date have had at least linear running time or space requirements. In this work, we address this issue by presenting an algorithm for building a CHORD network that has polylogarithmic time and space complexity. Furthermore, we discuss how the technique we use for building this CHORD network can be generalized into a "design pattern" for other desirable overlay network topologies.

Keywords: Topological self-stabilization · Overlay networks · Fault-tolerant distributed systems

1 Introduction

As computers and network connectivity have become an ubiquitous part of society, the size and scope of distributed systems has grown. It is now commonplace for these systems to contain hundreds or even thousands of computers spread across the globe connected through the Internet. To better facilitate common operations for applications, like routing and searching, many distributed systems are built using *overlay networks*, where connections occur over logical links that consist of zero or more physical links. Overlay networks allow nodes to embed

An early version of this work appeared as a Brief Announcement in SPAA 2021.

C. Johnen et al. (Eds.): SSS 2021, LNCS 13046, pp. 258–272, 2021.
https://doi.org/10.1007/978-3-030-91081-5_17

a predictable topology onto their (usually fixed) physical topology, selecting the best network for the application's particular needs.

Complicating the use of these overlay networks, however, is the reality that systems composed of such a wide variety and distribution of devices are more prone to failures caused from problems with the devices or physical links. For instance, fiber optic cables can be severed, power outages can cause machines to disconnect without warning, and even intentional user actions like joining or leaving the system on a predictable schedule can result in an incorrectly-configured overlay network causing the client application to fail.

One approach for managing these faults and preventing failures is to design protocols which are resilient to a targeted set of specific system faults, such as nodes joining or leaving the system. However, the unpredictable nature of these distributed systems makes it difficult to identify and control for every possible fault. It is for this reason that researchers have turned to *self-stabilizing overlay networks*. A self-stabilizing overlay network guarantees that after *any* transient fault, a correct topology will eventually be restored. This type of network can ensure autonomous operation of distributed systems even in the face of a variety of unforeseen transient faults.

The Problem. Our focus is on building robust self-stabilizing overlay networks efficiently. More specifically, we are interested in creating *efficient* algorithms that add and delete logical edges in the network to transform an arbitrary weakly-connected initial topology into a correct *robust* topology. By *efficient*, we mean these algorithms have a time *and* space complexity which is polylogarithmic in the number of nodes in the network. By *robust*, we mean topologies where the failure of a few nodes is insufficient to disconnect the network.

Main Results and Significance. With this paper, we present an efficient self-stabilizing overlay network with desirable practical properties like robustness and low diameter. In particular, we present a self-stabilizing algorithm for the creation of a Chord network which has expected polylogarithmic space and time requirements. Note that this is the first work to present an efficient (in terms of time and space) self-stabilizing overlay network for a *robust* topology. Note that, while our algorithm is deterministic, it depends upon the prior work of Berns [1], which was randomized, and therefore our results are in expectation.

Our second result is the explicit identification of a "design pattern" we call *network scaffolding* for creating self-stabilizing overlay networks. This pattern has been used in several other works, and the success of this approach, both previously as well as in this work for building Chord, leads us to believe it can be used for many other topologies as well. Our work is a first step towards fully defining and analyzing this design pattern. Our goal is that explicit identification of this design pattern can be useful to other researchers and practitioners in the design and implementation of other self-stabilizing overlay networks.

A preliminary version of this work appeared as a brief announcement at SPAA 2021 [2]. In this version, we provide a more detailed discussion of the

Avatar background, an improved analysis showing a better bound on the degree expansion, and also provide an extended discussion about the identified design pattern.

Related Work and Comparisons. The past few decades have seen tremendous growth in both the theory and practice of overlay networks. Some of this work has focused on *unstructured* overlay networks where connections need not satisfy any particular property and there are no constraints on what is considered a "legal" topology, such as Napster and Gnutella [13].

Our work focuses on *structured* overlay networks, where there is exactly one correct topology for any given set of nodes. While constructing and maintaining the correct topology adds additional work for the algorithm designer, common operations such as routing and searching are much more efficient with these structured networks. One example of such an overlay is Chord. Many of these structured networks, however, provided very limited fault tolerance.

To this end, much previous work has focused on improving the fault tolerance of overlay networks. One approach has looked at *self-healing* networks, where the network can maintain certain properties while a limited number of faults occur during a fixed time period. Examples of this work include the Forgiving Tree [8] and DEX [12]. Many of these approaches also use virtual nodes [14] as done in our work. More recently, Gilbert et al. presented DConstructor [6] which is able to build a correct topology from any initial topology and maintain this in the face of some joins and leaves. Götte et al. [7] also presented an algorithm for transforming a constant-degree network into a tree in $\mathcal{O}(\log n)$ rounds. The key difference between these works and ours, however, is we use a stronger fault model, requiring our algorithm to build the correct configuration regardless of the initial topology *or* the initial state of the nodes. This is a paradigm called self-stabilization, which we discuss next.

Self-stabilizing overlay networks are those that guarantee a legal configuration will be automatically restored by program actions after *any* transient fault so long as the network is not disconnected. This is often modeled as the ability for the network to form a correct topology when starting from an arbitrary weakly-connected state. One of the first such examples of a self-stabilizing overlay network was for the simple structured Linear topology [11] where nodes were arranged in a "sorted list". Since then, there have been several other self-stabilizing structured overlay networks created, including Skip+ [9] and Re-Chord [10]. Unlike the simple Linear topology, Skip+ and Re-Chord maintain several desirable properties for client applications, including low node degree and low diameter. Unfortunately, their worst-case time (in the case of Re-Chord) or space (in the case of Skip+) complexity is linear in the number of nodes.

To date, we are aware of only one self-stabilizing overlay network that is efficient in terms of both time and space. Berns presented the Avatar network framework as a mechanism for ensuring a faulty configuration is detectable for a wide variety of networks, and also gave a self-stabilizing algorithm for the construction of a binary search tree [1]. Our current work builds upon this AVATAR network.

A goal of our current work is to identify a general "design pattern" which can be used for building self-stabilizing overlay networks. There has been little work done on identifying these general patterns for overlay network construction. One exception to this is the *Transitive Closure Framework* [4] (TCF), which provides a way to build any locally-checkable topology by detecting a fault, forming a clique, and then deleting those edges which are not required in the correct configuration. While TCF can create any locally-checkable topology quickly, it requires node degrees to grow to $\mathcal{O}(n)$ during convergence and is therefore not practical for large networks.

2 Preliminaries

2.1 Model of Computation

We model our distributed system as an undirected graph $G = (V, E)$, with processes being the n nodes of V and the communication links being the edges E. Each node u has a unique identifier $u.id \in \mathbb{N}$, which is stored as immutable data in u. Where clear from the context, we will use u to represent the identifier of u.

Each node $u \in V$ has a *local state* consisting of a set of variables and their values, along with its immutable identifier $u.id$. A node may execute *actions* from its *program* to modify the values of the variables in its local state. All nodes execute the same program. Besides modifying its local state, a node can also communicate with its neighbors. We use the *synchronous message passing* model of computation with bounded communication channels, where computation proceeds in synchronous rounds. During each round, a node may receive messages sent to it in the previous round from any of its neighbors, execute program actions to update its state, and send messages to any node in its neighborhood $N(u) = \{v \in V : (u, v) \in E\}$. We assume the communication channels are reliable with bounded delay so that a message is received in some round i if and only if it was sent in round $i - 1$.

In the overlay network model, nodes communicate over logical links that are part of a node's state, meaning a node may execute actions to create or delete edges in G. In particular, in any round a node may delete any edge incident upon it, as well as create any edge to a node v which has been "introduced" to it from some neighbor w, such that (u, w) and (w, v) are both in E. Said in another way, in a particular round a node may connect its neighbors to one another by direct logical links.

The goal for our computation is for nodes to execute actions to update their state (including modifying the topology by adding and deleting edges) until a legal configuration is reached. A *legal configuration* can be represented as a predicate over the state of the nodes in the system, and as links are part of a node's state for overlay networks, a legal configuration is defined at least in part by the network topology. The *self-stabilizing overlay network problem* is to design an algorithm $\mathcal{A}$ such that when executing $\mathcal{A}$ on a connected network with nodes in an arbitrary state, eventually a legal configuration is reached. This means that a self-stabilizing overlay network will always automatically restore a legal configuration after any transient fault so long as the network remains connected.

2.2 Performance Metrics

We analyze the performance of our self-stabilizing overlay network algorithms in terms of both time and space. For time, we are interested in how quickly the network will be able to recover from a transient fault. Specifically, we measure the maximum number of (synchronous) rounds that may be required in the worst case to take any set of n nodes from an arbitrary connected configuration to a legal configuration. This is called the *convergence time*.

The space complexity measure of interest for us is related to the maximum number of neighbors a node might have during convergence that were not present in the initial configuration and are not required in the final configuration. Said in another way, we are interested in the number of "extra" neighbors a node may acquire due to the algorithm during convergence. More specifically, we use the *degree expansion* metric [1], which is the ratio of the maximum node degree of any node during convergence over the maximum node degree from the initial or final configuration.

Finally we note that, as with many distributed algorithms, we consider an efficient algorithm one which keeps these measures polylogarithmic in the number of nodes in the network.

3 Avatar

Our algorithm for creating the Chord network builds upon the work on the Avatar overlay network framework by Berns [1]. We present a brief summary and discussion below to provide the necessary background to understand our new contributions.

3.1 The Avatar Overlay Network

The Avatar overlay network framework can be used to define a variation of any particular network topology. The general idea behind the Avatar framework is to create a dilation-1 embedding of a particular N node *guest network* (with node identifiers in the range of $[0, N)$) onto the n node *host network* (with $n \leq N$). More specifically, in Avatar each node $u \in V$ from the host network (except the two nodes with the smallest and largest identifiers) simulates or "hosts" all nodes from the guest network with identifiers in u's *responsible range*, defined as the range of $[u.id, v.id)$, where $v.id$ is the smallest identifier greater than $u.id$ taken from all nodes in V, which we call the *successor* of u. The node u_0 with the smallest identifier has a responsible range of $[0, v.id)$ (where again v is the successor of u_0), while the node with the largest identifier u_{n-1} has a responsible range of $[u_{n-1}.id, N)$. To ensure a dilation-1 embedding, for every edge (a, b) in the guest network there exists an edge between the host nodes of a and b in the host network, or the host node for a and b is the same – that is, either both $a.id$ and $b.id$ are in the responsible range of the same host node, or there exists the edge $(host(a), host(b))$ in the host network such that $a.id$ is in the responsible range of host node $host(a)$ and $b.id$ is in the responsible range of host node $host(b)$. The definition of some N node guest network GUEST(N) along with the constraints on the corresponding edges in the host network define the legal AVATAR(GUEST(N)) network.

The use of a guest and host network provides two advantages. First, the requirement that there is exactly one correct configuration for any given N (meaning the guest network uses nodes $[0, N)$), along with the fact that the successor relationship used in the host network to define the responsible ranges can easily be determined from a node's local state, ensures that any topology is *locally checkable.* Second, we can design our algorithms (for both stabilization and end-user applications) to execute on the guest network, which has a single predictable configuration for a given N, regardless of the node set V. This simplifies both the design and analysis of our algorithms.

Note that AVATAR does require all nodes to know N, the upper bound on the number of nodes in the network, and that our analysis of convergence time and degree expansion is in terms of this N. Given that all of our algorithms have polylogarithmic time and space requirements, in practice one could easily select an N which was large enough to accommodate any possible node additions while still having a time and space complexity less than many existing algorithms which have complexity at least linear in the actual number of nodes n. Said in another way, even when N is much larger than n, our efficient algorithms may still require fewer resources if $\log N \ll n$. If we consider IPv6, for instance, $\log N$ would be only 128.

3.2 AVATAR(CBT)

Beyond defining the AVATAR framework, Berns also defined the CBT guest network and a self-stabilizing algorithm for building the AVATAR(CBT) network in a polylogarithmic number of rounds with a polylogarithmic degree expansion (both in expectation). The CBT topology is simply a complete binary search tree of the specified N nodes.

Stabilization. Our work building the CHORD network depends upon the existence of a "scaffold" CBT network. To be self-stabilizing, we need a way to build this CBT in a self-stabilizing manner. This is exactly what is provided by the earlier AVATAR work of Berns. The full description of the self-stabilizing algorithm for AVATAR(CBT) can be found in the original work [1]. We present a short informal summary of the algorithm's operation here to assist in understanding and verifying the correctness of our approach.

The general idea for the Avatar(Cbt) algorithm can be described using three components:

1. *Clustering*: The first step in the algorithm is for nodes to form clusters. These clusters begin as a single host node hosting a full N virtual node Cbt network. In the initial configuration, nodes may not be a part of a cluster, but since Avatar is locally checkable, all faulty configurations of Avatar(Cbt) contain at least one node which detects the faulty configuration and will begin forming the single-node clusters. This fault detection and cluster creation will propagate through the network until eventually all nodes are members of N virtual node Cbt clusters.
2. *Matching*: The second step of the algorithm is to match together clusters so that they may merge together. To do this, the root node of the binary tree repeatedly polls the nodes of its cluster, asking them to either find neighboring clusters that are looking for merge partners (called the leader role), or to look at neighboring clusters that can assign them a merge partner (called the follower role). The role of leader or follower is randomly selected. Leader clusters will match together all of their followers for merging, creating a matching between clusters that may not be direct neighbors. This ability to create edges to match non-neighboring clusters allows more matches to occur, and thus more merges, and thus a faster convergence time.
3. *Merging*: The algorithm then deals with the merging of matched clusters. To prevent degrees from growing too large, a cluster is only allowed to merge with at most one other cluster at a time. Once two clusters have matched from the previous step, the roots of the clusters connect as "partners" and update their successor pointers based upon the identifier of the host of the root of the other cluster. One node will have its responsible range become smaller, and this node will send all guest nodes that were in its old responsible range to its partner in the other cluster. The children of the root nodes are connected, and then they repeat the process of updating successor pointers and passing along guest nodes outside their new responsible range. Eventually this process reaches the leaves, at which point all nodes in both clusters have updated their responsible ranges and now form a new legal Cbt cluster.

This process of matching and merging continues until eventually only a single cluster is left, which is the correct Avatar(Cbt) network. We restate the following theorem from the original work and offer a brief sketch of the proof's intuition.

Theorem 1. *The self-stabilizing algorithm for* Avatar(Cbt) *by Berns [1] has a convergence time of* $\mathcal{O}(\log^2 N)$ *rounds in expectation, and a degree expansion of* $\mathcal{O}(\log^2 N)$ *in expectation.*

Intuition: A cluster has a constant probability of being matched and merged with another cluster in $\mathcal{O}(\log N)$ rounds, meaning the number of clusters is reduced by a constant fraction every $\mathcal{O}(\log N)$ rounds in expectation. This matching and merging only needs to happen $\mathcal{O}(\log N)$ times until we have a single cluster, giving us a time complexity of $\mathcal{O}(\log^2 N)$ rounds in expectation. The degree of a node can grow during a merge or during the matching process. However, during a merge a node's degree will grow to at most $\mathcal{O}(\log^2 N)$, and the node's degree will increase by only a constant amount during each match and there are only $\mathcal{O}(\log N)$ such matches in expectation, meaning the degree expansion of the algorithm is also $\mathcal{O}(\log^2 N)$.

Communication. The original work on AVATAR(CBT) also defined a communication mechanism to execute on the guest CBT network. We will also use this mechanism in our algorithm to ensure edges are added systematically and thus limiting unnecessary degree growth. In particular, we will use a variant of a *propagation of information with feedback* (PIF) algorithm [5] which will execute on the (guest) nodes of CBT. While the original work was snap-stabilizing, this would not be a requirement in our work. We are instead simply interested in an organized way to communicate information in waves in a tree.

In PIF, communication happens in waves that are initiated by the root of the binary tree. The root executes a *propagate* action which sends information down the tree level by level until it reaches the leaves, at which point the leaves begin a *feedback* action, performing some operation and then signalling to their parent that the message has been received by all descendants in the tree. Once the root receives the feedback wave, it knows the message was successfully received and acted upon by all nodes in the tree, and the root may continue with further PIF waves if necessary.

We will use this communication mechanism to add edges to our network to build CHORD. As the PIF process itself is previously defined, we only need to provide the actions each node will perform for each part, as well as any data that is sent. In particular, we will say that a tree T executes a $PIF(X)$ wave, meaning the root of tree T will signal to its children that a propagation wave has begun with the $PIF(X)$ message. Furthermore, we will specify the *propagate* action of a, which is what each non-root node a should do when it receives the propagation message $PIF(X)$. We also specify the *feedback* action of a, which is the actions each node a should take when it receives acknowledgements from its children that the most-recent propagation wave has completed and the corresponding feedback wave is underway.

4 AVATAR(CHORD)

In this section we discuss how we can use the existing AVATAR(CBT) self-stabilizing overlay network as the starting point for the efficient creation of a variant of the CHORD overlay network.

4.1 Overview of Our Approach

Arguably one of the major barriers to the practical implementation for self-stabilizing overlay networks is the complexity that must be managed when designing and analyzing these networks, particularly when we desire efficient self-stabilization. For instance, TCF [4] is simple and works with any locally-checkable topology, but it requires $\Theta(n)$ space. One could imagine a simple "design pattern" which simply suggests that in every round, a node computes their ideal neighborhood given the information available to them from their state and the state of their neighbors, and then add and delete edges to form this ideal neighborhood. Unfortunately, analyzing this algorithm in terms of *both* correctness and efficiency is quite difficult as one must consider the implications of a variety of actions on a variety of possible initial configurations.

One approach to managing complexity is to start by building smaller or simpler structures, and then using these to continue towards the final goal. Consider, for instance, the construction of a large building. One common approach is to erect a simple scaffold and use this scaffold to build the more complex permanent structure. As another example, consider the prior work on using *convergence stairs* for analyzing general self-stabilizing algorithms. In this technique, one first must show the system converges to some weaker predicate A_0 from an arbitrary initial configuration, then show it converges to A_1 provided it is in A_0, then show it converges to A_2 provided it is in A_1, and so on until you have reached the correct configuration. These patterns of design and analysis are similar in that they take a complex set of required actions, decompose them into smaller distinct steps, and then rely on prior solutions to the smaller steps to move to the next ones.

In the remainder of this section, we discuss our approach for efficiently creating a self-stabilizing version of the CHORD network based upon this idea of scaffolding. In particular, we shall define the CHORD topology, and then discuss how we can use AVATAR(CBT) as a starting point for constructing CHORD. We then show how nodes can determine in a short amount of time whether they should be building the "scaffold" (CBT) or the target topology (CHORD).

4.2 Chord(N)

Our target network aims to resolve the lack of robustness of the CBT scaffold network. In particular, our target network is an N-node CHORD network defined as follows:

Definition 1. *For any $N \in \mathbb{N}$, let* CHORD(N) *be a graph with nodes $[N]$ and edge set defined as follows. For every node i, $0 \leq i < N$, add to the edge set (i, j), where $j = (i+2^k) \mod N$, $0 \leq k < \log N - 1$. When $j = (i+2^k) \mod N$, we say that j is the k-th finger of i.*

It is worth again noting that our use of the AVATAR framework results in a locally-checkable version of the CHORD network. CHORD as defined on an arbitrary set of nodes is actually *not* locally checkable, particularly because of the "ring" edges (in a legal configuration, exactly one node should have two immediate neighbors with smaller identifiers, but which node this should be cannot be determined if the node set is arbitrary). Unlike prior approaches, then, our stabilizing CHORD network is *silent*, meaning no messages or "probes" need to be continuously exchanged between nodes in a legal configuration.

Our goal, then, is to use the N-node topology of CBT to add edges to the guest nodes (and to the corresponding host network as required to maintain a dilation-1 embedding) until we have formed the correct N-node CHORD network.

4.3 Building CHORD from CBT

Figure 1 elaborates on the algorithm which uses our guest CBT network as a scaffold for creating the guest CHORD network. The algorithm uses the fact that CHORD edges can be created inductively. That is, assuming all fingers from 0 to k are present, the $k+1$ finger can be created in a single round. Specifically, if node b is the $(i-1)$ finger of c_0, and c_1 is the $(i-1)$ finger of b, the ith finger of c_0 is c_1. The algorithm begins by correctly building finger 0, then recursively adds the first finger, then the second, and so on. This adding of edges is done in a metered fashion, however, to prevent unnecessary degree growth from faulty initial configurations.

Once the scaffold network has been built, we can begin the process of constructing our final target topology. We design our algorithm to execute on the N guest nodes of CBT, with the goal being to add edges to the nodes of CBT until they have formed the N guest node CHORD network. For now, we shall assume that the network is in the legal CBT configuration. We will relax this assumption and consider an arbitrary initial configuration shortly.

The algorithm begins with the root of CBT initiating a *PIF* wave which connects each guest node with its 0th finger. Notice that, with the exception of one node, the edges in the host network realizing every guest node's 0th finger are already present. For any guest node $b \neq N-1$, the 0th finger of b is either (i) a guest node with the same host as b, or (ii) a guest node which is hosted by the successor of $host_b$. Edges to guest nodes 0 and $N-1$ are forwarded up the tree during the feedback wave, allowing the root of the tree to connect them at the completion of the wave, thus forming the base ring and completing every guest node's 0th finger. The root then executes $\log N - 1$ additional *PIF* waves, with wave k correctly adding the kth finger for all guest nodes. After $\mathcal{O}(\log^2 N)$ rounds, we have built the correct AVATAR(CHORD) network.

4.4 Phase Selection

The final piece for our self-stabilizing CHORD network is to create a mechanism by which nodes can know which algorithm they should be executing: either executing the steps required to build the AVATAR(CBT) network, or the steps required to build the CHORD target network from an existing CBT network). We assume each host node u maintains a phase variable $phase_u$ whose value is from the set $\{CBT, CHORD, DONE\}$. When $phase_u = CBT$, a node is executing the algorithm for the AVATAR(CBT) network. If $phase_u = CHORD$, then the PIF waves in Algorithm 1 are executed. If $phase_u = DONE$, then a

```
    // Execute when phase_u = CHORD; If phase_u = CBT, then execute
    // the original AVATAR(CBT) algorithm [1].
    // As part of each round, nodes exchange their local state, including
    // LastWave, and check for faulty configurations as described in Section 4.4.
1.  Tree T executes a PIF(MakeFinger(0)) wave:
2.     Propagate Action for a: LastWave_a = 0
3.     Feedback Action for a:
          // Let b be the 0th finger of a.
4.        if LastWave_a = LastWave_b = 0 then
5.           Create the edge (a, b)
6.           Forward an edge to node 0
                or N − 1 (if present) to parent
7.        else phase_u = CBT (where u is host_a) fi
8.  for k = 1, 2, ..., log N − 1 do
9.     Tree T executes a PIF(MakeFinger(k)) wave:
10.       Propagate Action for a: LastWave_a = k
11.       Feedback Action for a:
             // Let b_0, b_1 be the k − 1 fingers of a.
12.          if LastWave_a = LastWave_b0 = LastWave_b1 = k then
13.             Create edge (b_0, b_1), the kth finger of b_0.
14.          else phase_u = CBT (where u is host_a) fi
15. od
```

Fig. 1. Algorithm 1: PIF for Chord Target from Cbt Scaffold

node will take no actions provided its local neighborhood is consistent with a legal Avatar(Chord) network.

Determining which algorithm to execute requires a node be able to determine if the configuration they are in now is either completely correct or consistent with one reached by building Chord from Cbt. We define a subset of states under which Algorithm 1 will converge, and then define a predicate which nodes can use to determine if the network is in one of these states.

Definition 2. *A graph G with node set V is in a* scaffolded Chord configuration *if G is reachable by executing the PIF waves defined by Algorithm 1 on a correct* Avatar(Cbt) *network.*

Thanks to the predictability of the Cbt scaffold network, nodes can determine if their state is consistent with that of a scaffolded Chord configuration. Informally, each guest node can determine this by simply checking to see if its neighborhood is a superset of Cbt but a subset of Chord, with the first k fingers from Chord present, for some $k \in [0, \log N)$. We define the predicate a node can use for this operation below.

Definition 3. *Let $scaffolded_b$ be a predicate defined over the local state of a guest node b, as well as the state of nodes $b' \in N(b)$. The value of $scaffolded_b$ is the conjunction of the following conditions.*

1. *Node b has all neighbors from* CBT*, each with the proper host and tree identifier (a value set as part of a legal* CBT *scaffold network).*
2. *Node b has last executed the kth feedback wave of a* *PIF(MakeFinger(k), ⊥) wave, for some* $0 \leq k < \log N$
3. *All neighbors of b have either all k fingers present, or* $k+1$ *fingers (if a child has just processed a feedback wave), or* $k-1$ *(if parent has not yet processed the current feedback wave), where k is the last feedback wave b has executed*
4. *Node b's parent has last executed the kth feedback wave, and has the first k* CHORD *fingers, or* $k-1$ *fingers if b has just completed the feedback transition and b's parent has not.*

In every round of computation, all nodes are checking their local state and the state of their neighbors to determine if a faulty configuration is found. This check for faults, along with the $scaffolded_b$ predicate, is used to set the $phase_u$ variable as follows. If a fault is detected and $scaffolded_b = false$, then $u = host_b$ sets $phase_u = CBT$. Furthermore, if any neighbor v has a different value for $phase_v$, then $phase_u = CBT$. Notice that once the correct configuration is built, nodes can execute a final *PIF* wave to set $phase_u = DONE$. If any node detects *any* fault during this process, it simply sets $phase_u = CBT$. Since AVATAR(CHORD) is locally checkable, at least one node will not set $phase_u = DONE$ during the final *PIF* wave, and the AVATAR(CBT) algorithm will begin.

5 Analysis

We sketch the proofs for our main results below. Full proofs of convergence and degree expansion can be found in the full version of this paper [3].

Theorem 2. *Algorithm 1, when combined with the self-stabilizing algorithm for* AVATAR(CBT) *from Berns [1], is a self-stabilizing algorithm for the network* AVATAR(CHORD) *with convergence time* $\mathcal{O}(\log^2 N)$ *in expectation.*

Proof Sketch. To prove the convergence time of our algorithm, we first show that if the configuration is not a scaffolded CHORD configuration, within $\mathcal{O}(\log N)$ rounds, all nodes are executing the algorithm to build the scaffold CBT network. We then show that nodes will have built the correct CBT network within an additional $\mathcal{O}(\log^2 N)$ rounds in expectation, at which point all nodes begin building the target CHORD network. We will then show that this process succeeds in $\mathcal{O}(\log^2 N)$ rounds. Putting these together, we get an overall convergence time of $\mathcal{O}(\log^2 N)$ in expectation.

Theorem 3. *Algorithm 1, when combined with the self-stabilizing algorithm for* AVATAR(CBT) *from Berns [1], is a self-stabilizing algorithm for the network* AVATAR(CHORD) *with degree expansion of* $\mathcal{O}(\log^2 N)$ *in expectation.*

Proof Sketch. By design, any edge that is added to the network when building CHORD from CBT is an edge that will remain in the final correct configuration and therefore does not affect the degree expansion. Furthermore, we know from the original AVATAR paper that the expected degree expansion is $\mathcal{O}(\log^2 N)$ when all nodes are executing the CBT algorithm.

The only new piece we need to consider, then, is to analyze the actions nodes might take when they incorrectly believe, based on their local state, that they are building the CHORD network from the CBT scaffold (a "false CHORD" phase), which we show can only happen for $\mathcal{O}(\log N)$ rounds. Since adding CHORD edges is coordinated with a PIF wave, each guest node b can only increase its degree by one during this time. At most, then, a node may increase its degree by a factor of 2 during this time, leading to the initial degree growth of 2 during the "false CHORD" phase.

6 Generalizing Our Approach

Above we have provided an algorithm for using one self-stabilizing overlay network to create another self-stabilizing overlay network. While we are not the first to use this general idea in the construction of overlay networks, we are the first to explicitly define and discuss this approach, which we call *network scaffolding.* To use the network scaffolding approach, one must define several components. In particular, we must define:

- The *scaffold network*, an intermediate topology which we can construct from any initial configuration.
- The *target network*, the network topology that we wish to build for use with our final application.
- A self-stabilizing algorithm for constructing the scaffold network.
- An algorithm for building the target network when starting from the correct scaffold network.
- A local predicate allowing nodes to determine whether they should be building the scaffold network or the target network.

Our self-stabilizing algorithm from above used AVATAR(CBT) as the scaffold network to build a AVATAR(CHORD) target network. To do was relatively straightforward: we defined a way to build CHORD from CBT, and then proved nodes would quickly determine which network they were building.

This network scaffolding approach has been used in some form by other previous work, and we hope it will be extended in future work as well. Our approach heavily depends upon the scaffold network selected. The CBT network has many desirable properties for a scaffold network when compared to other examples in prior work. These properties include:

Efficient Self-stabilization: If the scaffold itself is inefficient to build, we cannot expect the target topology to be built efficiently. TCF [4] can be thought of as an inefficient scaffold network that requires $\mathcal{O}(n)$ space. AVATAR(CBT) is a logical choice as, prior to this work, it is the only self-stabilizing overlay network we are aware of with efficient stabilization in terms of both time and space.

Low Node Degree: Unlike a real scaffold, we maintain the scaffold edges after the target network is built. Therefore, the scaffold network must have low degree if we wish our final configuration to be so. Again, the suitability of AVATAR(CBT) is apparent, as it requires only a few edges per virtual node (and a logarithmic number of edges per real node).

Low Diameter: Low diameter allows (relatively) fast communication for adding the target network's edges one at a time. A previous work, RE-CHORD [10], used a "scaffold" of the LINEAR network, whose $\mathcal{O}(n)$ diameter contributed to the $\mathcal{O}(n \log n)$ convergence time of their algorithm.

Predictable Routing: The predictable routing, particularly for communication, allows us to add edges in a metered and checkable fashion. This predictability helps with both design and analysis. It would be interesting to see if a semi-structured overlay network could be used as a scaffold, as semi-structured overlays may be easier to build. To date, little work has been done on self-stabilizing semi-structured overlay networks, but there are several examples of efficient creation of semi-structured networks in non-self-stabilizing settings [7] which may be interesting starting points for future work.

Local Checkability: To be able to determine which phase of the algorithm should be executed quickly, without "wasting" time and resources adding edges from a faulty configuration, the scaffold should ideally be locally checkable. Some previous overlay networks have used a "probing" approach where messages were circulated continuously to try and detect faulty configurations. The risk of this approach in network scaffolding is that nodes may spend too long adding edges from an incorrect scaffold, or take too long to detect a faulty configuration.

7 Concluding Thoughts

In this paper, we have presented the first time- and space-efficient algorithm for building a CHORD network using a technique we call *network scaffolding*. We discussed considerations for expanding this technique, in particular pointing out considerations and implications for various properties of the scaffold network.

An obvious extension to our work would be to consider building other target topologies using AVATAR(CBT) as a scaffold network. For instance, networks with good load balancing properties or with high resilience to churn could be converted into self-stabilizing variants using AVATAR to define the network topology and the CBT scaffold to build this correct topology. It would also be interesting to investigate the correctness and complexity of this approach when using a more realistic asynchronous communication model.

References

1. Berns, A.: Avatar: a time- and space-efficient self-stabilizing overlay network. In: Pelc, A., Schwarzmann, A.A. (eds.) SSS 2015. LNCS, vol. 9212, pp. 233–247. Springer, Cham (2015). https://doi.org/10.1007/978-3-319-21741-3_16

2. Berns, A.: Network scaffolding for efficient stabilization of the chord overlay network. In: Proceedings of the 33rd ACM Symposium on Parallelism in Algorithms and Architectures, SPAA 2021, pp. 417–419. Association for Computing Machinery, New York (2021). https://doi.org/10.1145/3409964.3461827
3. Berns, A.: Network scaffolding for efficient stabilization of the chord overlay network (2021). https://arxiv.org/abs/2109.14126
4. Berns, A., Ghosh, S., Pemmaraju, S.V.: Building self-stabilizing overlay networks with the transitive closure framework. In: Défago, X., Petit, F., Villain, V. (eds.) SSS 2011. LNCS, vol. 6976, pp. 62–76. Springer, Heidelberg (2011). https://doi.org/10.1007/978-3-642-24550-3_7
5. Delaët, S., Devismes, S., Nesterenko, M., Tixeuil, S.: Snap-stabilization in message-passing systems. J. Parallel Distrib. Comput. **70**(12), 1220–1230 (2010)
6. Gilbert, S., Pandurangan, G., Robinson, P., Trehan, A.: Dconstructor: Efficient and robust network construction with polylogarithmic overhead. In: Proceedings of the 39th Symposium on Principles of Distributed Computing, PODC 2020, pp. 438–447. Association for Computing Machinery, New York (2020). https://doi.org/10.1145/3382734.3405716
7. Götte, T., Hinnenthal, K., Scheideler, C., Werthmann, J.: Time-optimal construction of overlay networks. In: Proceedings of the 2021 ACM Symposium on Principles of Distributed Computing, PODC 2021, pp. 457–468. Association for Computing Machinery, New York (2021). https://doi.org/10.1145/3465084.3467932
8. Hayes, T., Rustagi, N., Saia, J., Trehan, A.: The forgiving tree: a self-healing distributed data structure. In: Proceedings of the Twenty-Seventh ACM Symposium on Principles of Distributed Computing, PODC 2008, pp. 203–212. Association for Computing Machinery, New York (2008). https://doi.org/10.1145/1400751.1400779
9. Jacob, R., Richa, A., Scheideler, C., Schmid, S., Täubig, H.: A distributed polylogarithmic time algorithm for self-stabilizing skip graphs. In: PODC '09: Proceedings of the 28th ACM symposium on Principles of distributed computing, pp. 131–140. ACM, New York (2009). http://doi.acm.org/10.1145/1582716.1582741
10. Kniesburges, S., Koutsopoulos, A., Scheideler, C.: Re-chord: a self-stabilizing chord overlay network. In: Proceedings of the 23rd ACM symposium on Parallelism in algorithms and architectures, SPAA 2011, pp. 235–244. ACM, New York (2011). https://doi.org/10.1145/1989493.1989527. http://doi.acm.org/10.1145/1989493.1989527
11. Onus, M., Richa, A.W., Scheideler, C.: Linearization: locally self-stabilizing sorting in graphs. In: ALENEX. SIAM (2007)
12. Pandurangan, G., Robinson, P., Trehan, A.: Dex: self-healing expanders. Distrib. Comput. **29**(3), 163–185 (2016). https://doi.org/10.1007/s00446-015-0258-3
13. Saroiu, S., Gummadi, K.P., Gribble, S.D.: Measuring and analyzing the characteristics of Napster and Gnutella hosts. Multimedia Syst. **9**(2), 170–184 (2003). https://doi.org/10.1007/s00530-003-0088-1
14. Trehan, A.: Self-healing using virtual structures. CoRR abs/1202.2466 (2012). http://arxiv.org/abs/1202.2466

The Agreement Power of Disagreement

Quentin Bramas[1(✉)], Anissa Lamani[1], and Sébastien Tixeuil[2]

[1] ICUBE, Strasbourg University, CNRS, Strasbourg, France
bramas@unistra.fr
[2] Sorbonne University, CNRS, LIP6, Paris, France

Abstract. We consider the rendezvous problem of two autonomous robots with very weak capacities. This problem is notoriously impossible to solve in the semi-synchronous execution model when robots are deterministic, oblivious, and their ego-centered coordinate system is fully symmetric.

We show that if the robots disagree on the unit distance of their coordinate system, it becomes possible to solve rendezvous and agree on a final common location, without additional assumptions.

1 Introduction

We consider swarms of mobile robots that must coordinate to solve a given task. More precisely, we consider robots modeled by dimensionless points that evolve in a Euclidean bidimensional space according to the Look-Compute-Move (LCM) model introduced by Suzuki and Yamashita [11]. In the LCM model, robots repeatedly execute cycles of Look-Compute-Move phases. In the Look phase, the robot obtains an ego-centered view of the position of the other robots (in its own coordinate system). In the Compute phase, the robot decides where it should move next (still in its own coordinate system). Finally, in the Move phase, the robot simply moves toward its destination.

The vast majority of the research effort in the LCM model [5] focuses on understanding the exact hypotheses that make a task solvable. In most cases, those hypotheses are tightly coupled with the amount of synchronization between robots. Three main synchronization models have been considered: the fully synchronous (FSYNC) model mandates *all* robots to execute their LCM cycles simultaneously, the semi-synchronous (SSYNC) model allows that only a non-empty subset of robots executes its LCM cycle simultaneously, while the asynchronous (ASYNC) model makes no hypothesis about the relative speed of each robot or each phase.

A benchmarking problem in this context is that of *rendezvous*, where two robots have to meet in finite time at the exact same location, not known beforehand. Despite its apparent simplicity, this problem triggered interest from the research community as in FSYNC, it is solvable [11], while in SSYNC, it is unsolvable [3,11] deterministically, without additional assumptions. One of the

This work was partially funded by the ANR project SAPPORO, ref. 2019-CE25-0005-1.

C. Johnen et al. (Eds.): SSS 2021, LNCS 13046, pp. 273–288, 2021.
https://doi.org/10.1007/978-3-030-91081-5_18

key reasons for impossibility is that the two robots may have initial symmetric views (and hence make symmetric moves when operated synchronously), but when only one robot is scheduled for execution (as is possible in SSYNC), only half of the symmetric algorithm is performed at any time, preventing the robots from actually meeting (they only converge toward one another).

Related Works. To circumvent the aforementioned impossibility result, several options for breaking the initial symmetry were considered.

One line of work considers adding extra capacities to the robots. The seminal paper by Suzuki and Yamashita [11] provides a probabilistic solution to the problem (and each robot makes an constant expected number of coin tosses). The rest of the literature focused on deterministic solutions. Another series of papers considers robot that are endowed with some variant of persistent memory. In more details, it was proposed to endow each robot with a *light* [4], that is, a robot is capable of emitting one color among a fixed number of available colors, visible to all other robots. This additional capacity allows to solve rendezvous in the most general ASYNC model, provided that lights of robots are capable to emit at least *four* colors [4]. In the SSYNC model, Viglietta [12] proved that being able to emit two colors is sufficient to solve rendezvous. In the same paper [12], Viglietta proves [12] that three colors are sufficient in ASYNC. Both solutions in ASYNC [4,12] and SSYNC [12] output a correct behavior independently of the initial value of the lights' colors. Then, Okumura *et al.* [9] presented a rendezvous algorithm with two colors in ASYNC assuming *rigid* moves (that is, the move of every robot is never stopped by the scheduler before completion), or assuming non-rigid moves but robots being aware of δ (the minimum distance before which the scheduler cannot interrupt their move). Also, the solution of Okumura *et al.* [9] requires lights to have a specific color in the initial configuration. Finally, Heriban *et al.* [7] prove that two colors are necessary and sufficient in ASYNC without extra assumptions.

Another line of work, most related to the current paper, considers restricting the amount of symmetry that can occur in (an persist from) the initial configuration. One set of papers relates to the directions and orientations of the coordinate systems that are given to each robot by the adversary, to prevent a symmetric situation to occur. This is abstracted by the notion of *compass*, that supposedly points to the "North" of the local coordinate system of each robot. When compasses are perfect (i.e., the two robots have the same "North"), SSYNC rendezvous can be achieved [6]. Also, even if compasses are only eventually perfect (there is a time t, unknown to the robots, after which compasses are perfect), SSYNC rendezvous is also feasible [10]. Finally, a complete SSYNC characterization of rendezvous solvability with respect to compasses is due to Izumi et al. [8]: *(i)* if compasses are fixed (they do not change throughout the execution), rendezvous is solvable if and only if the two compasses angle difference ϕ is smaller than $\frac{\pi}{2}$, and *(ii)* if compasses are dynamic (their direction may vary between two LCM cycles), rendezvous is solvable if and only if ϕ is smaller than $\frac{\pi}{4}$. The case of ASYNC is only partially solved: while the results for static compasses extend to ASYNC, rendezvous with dynamic compasses is feasible if $\phi < \frac{\pi}{6}$ and impossible if $\phi \geq \frac{\pi}{4}$, but the interval $\left[\frac{\pi}{6}, \frac{\pi}{4}\right)$ remains unknown.

Finally, Bramas et al. [1] showed recently that, when robots agree only on the North direction but not on the East (*i.e.*, they might not have the same chirality), then the rendezvous is solvable, which might seems surprising because robots can start with symmetric views (and keep their views symmetric until the rendezvous is achieved).

Our Contribution. We observe that the coordinate system that is given to each robot also includes a *unit distance* (that may be different for each robot as it is given by an adversary). Yet, to our knowledge, this unit distance was never considered as a tool to break initial symmetry, but rather as an assumption (all unit distances are equal) when designing impossibility results.

In this paper, we investigate the possibility to include the two robots' unit distance in the analysis of rendezvous solvability in SSYNC. In more details, we consider two robots that are arbitrarily disoriented (so, the angle difference ϕ may be equal to $\frac{\pi}{2}$), but have a different unit distance. Then, ρ is the ratio between the largest and the smallest unit distance of the robots. In this setting, we show that for any two real numbers ρ_{min} and ρ_{max}, known to the robots, such that $1 < \rho_{min} < \rho_{max}$, if $\rho \in [\rho_{min}, \rho_{max}]$, then rendezvous is solvable in SSYNC, without any additional assumption (robots are deterministic, oblivious, and their compasses are arbitrary). The extreme case $\rho = 1$ (both unit distances are equal) is known to render the problem unsolvable.

The rest of the paper is organized as follows. Section 2 describes the execution model. To warm up, Sect. 3 considers the simple case when $\rho = 2$. Then, the general solution is described in Sect. 4. Concluding remarks are provided in Sect. 5.

2 Preliminaries

We consider two robots, evolving in a Euclidean two dimensional space. Robots are modeled as points and are assumed to be uniform (they execute the same algorithm), and oblivious (they cannot remember past actions).

Let Z be a global coordinate system. A *configuration* at time t, denoted C_t, is a set $\{r_1, r_2\}$ containing the positions of both robots in Z at time t. Notice that r_i, $i = 1, 2$, denotes at the same time a robot and its position in $\mathbb{R}^2$ in the coordinate system Z. Robots do not know Z, instead, each robot r_i has its own coordinate system Z_{r_i} centered at the current position of r_i. We assume *disoriented* robots (they do not agree on any axis) that have different unit distance. Let ρ be the ratio between the largest and the smallest unit distance of the robots *i.e.*, $unit_2 = \rho \cdot unit_1$, with $unit_2 > unit_1$. For a robot r, d_r denotes the distance between the two robots in its own coordinate system. Thus, if r and r' are the two robots, we have $d_r = \rho d_{r'}$ or $d_r = \frac{d_{r'}}{\rho}$. For simplicity, in the remaining of the paper, r_1 denotes the robot with the largest unit distance and r_2 the other robot *i.e.*, $d_1 < d_2$ (where we abusively write d_i instead of d_{r_i}, for $i = 1, 2$). Of course, a robot is *not* aware of it being the robot with the largest unit distance.

Robots operate in cycles that comprise three phases: Look, Compute and Move. More precisely, at each time instant, an activated robot first takes a snapshot to see the position of the other robot in its ego-centered coordinate system. Based on this snapshot, the robot either computes a destination or decides to remain idle. Finally, the robot moves towards the computed destination (if any) following a straight path. We assume *non-rigid* movements *i.e.* a robot can be stopped anywhere along the path to its destination after traveling at least a fixed positive distance δ. The value of δ is common to the two robots but it is unknown. Its value can be arbitrarily small but it is fixed and never changes.

In configuration C, the *local view* of a robot r_i, denoted $\mathcal{V}_{r_i}$ is the output of the look phase. More precisely, when a robot r_i takes a snapshot, it observes the position of the other robot in its own coordinate system Z_{r_i} (translated by $-r_i$ so that r_i is always at the center). An algorithm A is a function mapping local views to destinations. When r_i is activated at time t, algorithm A outputs r_i's destination p in its local coordinate system Z_{r_i} based on $\mathcal{V}_{r_i}$.

We consider the SSYNC model where at each time instant, a non-empty subset of robots is activated by an external entity called scheduler. The activated robots execute their Look-Compute-Move cycle synchronously. We assume that the scheduler is fair *i.e.* each robot is activated infinitely often. An execution $\mathcal{E} = (C_0, C_1, \dots)$ of an algorithm A is a sequence of configurations, where C_0 is an initial configuration, and every configuration C_{t+1} is obtained from C_t by applying A to the robots scheduled for execution by the scheduler.

3 An Algorithm When $\rho = 2$

As an introduction, we show a simple algorithm solving the problem when $\rho = 2$. In this case, the *level* $l_i \in \mathbb{Z}$ of a robot r_i is the unique integer such that $d_i \in [2^{-l_i}, 2^{-l_i+1})$. By construction, we know that $l_2 = l_1 - 1$, because $d_2 = 2d_1$. Then, Algorithm 1 solves the rendezvous with $\rho = 2$. Indeed, by construction, only one robot remains idle and one robot moves to the other.

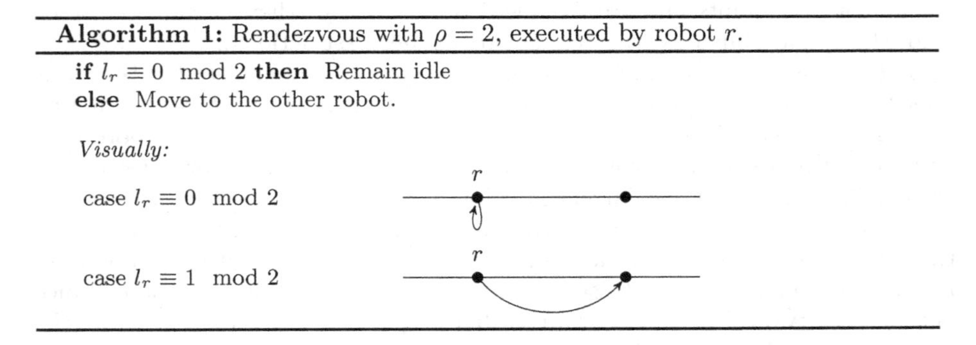

Algorithm 1: Rendezvous with $\rho = 2$, executed by robot r.

if $l_r \equiv 0 \mod 2$ **then** Remain idle
else Move to the other robot.

Visually:

case $l_r \equiv 0 \mod 2$

case $l_r \equiv 1 \mod 2$

Theorem 1. *Algorithm 1 solves the rendezvous problem in SSYNC when $\rho = 2$.*

Proof. As $\rho = 2$, at each time instant, $|l_1 - l_2| = 1$ where for any $i \in \{1, 2\}$, $l_i \in \mathbb{Z}$ is the level of robot r_i. That is, by Algorithm 1, at each time instant, a single robot is allowed to move. Its destination is the other robot's position. Let d_t be the distance between the two robots at time t. Two cases are possible:

1. $d_t \leq \delta$. As the scheduler is assumed to be fair, the robot allowed to move is eventually activated. When it moves, as the other robot remains idle by Algorithm 1, the rendezvous is achieved.
2. $d_t > \delta$. First observe that the distance between the two robots never increases. Indeed, at each time instant, a robot either moves towards the other robot along the straight line connecting them or remains idle. As the scheduler is assumed to be fair, the robot allowed to move is eventually activated. When it moves, the distance between the two robots decreases by at least δ. That is, at each time a robot moves, the distance between the two robots decreases by at least δ. Hence, we can deduce that there exists a time $t' > t$ such that $d_{t'} \leq \delta$ and we retrieve Case 1.

From Cases 1 and 2 we can deduce that the rendezvous is eventually achieved. Hence, the theorem holds. □

4 An Algorithm When $\rho \in [\rho_{\min}, \rho_{\max}]$

In this section, we assume that the robots know an upper and a lower bound on the value ρ *i.e.*, $\rho \in [\rho_{\min}, \rho_{\max}]$. In this case, the intervals defining the level of a robot are more complex.

We define two infinite sequences of intervals as follows:

$$\forall i \in \mathbb{Z} \qquad S_i = [\rho_{\min}^{-i}\rho_{\max}^{-i}, \rho_{\min}^{-(i-1)}\rho_{\max}^{-i}) \tag{1}$$

$$M_i = [\rho_{\min}^{-i}\rho_{\max}^{-(i+1)}, \rho_{\min}^{-i}\rho_{\max}^{-i}) \tag{2}$$

The sets S_i and M_i are called *levels*. We consider that the levels are ordered by the inverse of their length *i.e.*, for all $i \in \mathbb{Z}$, we say level M_i, resp. S_i, is greater than level $M_{i'}$, resp. $S_{i'}$, when $i > i'$. Moreover, level M_i is greater than level S_i.

First, notice that

$$\bigcup_{i \in \mathbb{Z}} S_i \cup M_i = \mathbb{R}_+^*$$

and the intervals are pairwise disjoints, so the sequences form a partition of $\mathbb{R}_+^*$. Figure 1 illustrates this partition. We can see that when the distance d_r seen by a robot r decreases, its level increases. For simplicity, we say a robot r is in a set X, or *has level* X, if its distance d_r is in X.

We now prove a lemma that states that both robots cannot have a level of type S, and the levels of the robots are not too far away.

Lemma 1. *For every $i \in \mathbb{Z}$, if a robot r has level S_i, then the other robot r' has level M_{i-1} or M_i.*

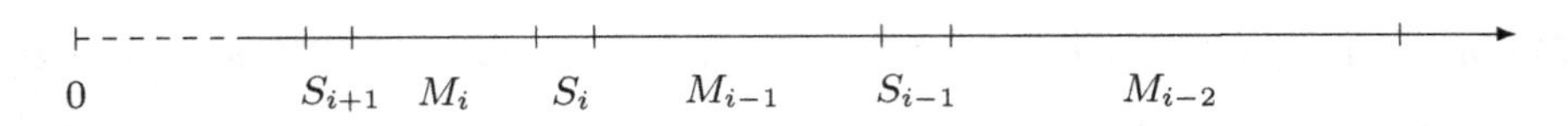

Fig. 1. Partition of the line $\mathbb{R}_+^*$ into levels

Proof. We have $d_r \in [\rho_{\min}^{-i}\rho_{\max}^{-i}, \rho_{\min}^{-(i-1)}\rho_{\max}^{-i})$. If $d_r < d_{r'}$, then

$$d_{r'} = \rho d_r \in \left[\rho_{\min} \times \rho_{\min}^{-i}\rho_{\max}^{-i}, \rho_{\max} \times \rho_{\min}^{-(i-1)}\rho_{\max}^{-i}\right) = \mathcal{M}_{i-1}$$

If $d_r > d_{r'}$, then

$$d_{r'} = \frac{d_r}{\rho} \in \left[\frac{1}{\rho_{\max}} \times \rho_{\min}^{-i}\rho_{\max}^{-i}, \frac{1}{\rho_{\min}} \times \rho_{\min}^{-(i-1)}\rho_{\max}^{-i}\right) = \mathcal{M}_i$$

□

For simplicity, let

$$\mathcal{M}_0 = \bigcup_{i\equiv 0 \mod 2} M_i$$
$$\mathcal{M}_1 = \bigcup_{i\equiv 1 \mod 2} M_i$$
$$\mathcal{S}_0 = \bigcup_{i\equiv 0 \mod 2} S_i$$
$$\mathcal{S}_1 = \bigcup_{i\equiv 1 \mod 2} S_i$$

Also, consider the indexes of those sets modulo 2 *eg.*, $\mathcal{M}_{-1} = \mathcal{M}_3 = \mathcal{M}_1$.

Let $\mathfrak{s}(d)$ be the smallest value defined as follows:

$$d\frac{1-\rho_{\min}^{-1}}{2\rho_{\min}^2\rho_{\max}^2} \leq \mathfrak{s}(d) \leq d\frac{1-\rho_{\min}^{-1}}{2} \text{ such that } \mathfrak{s}(d) \in \mathcal{S}_0 \tag{3}$$

Lemma 2. *$\mathfrak{s}(d)$ is well defined*

Proof. We have to prove that, for any $d > 0$, we have

$$\left[d\frac{1-\rho_{\min}^{-1}}{2\rho_{\min}^2\rho_{\max}^2}, d\frac{1-\rho_{\min}^{-1}}{2}\right] \cap \mathcal{S}_0 \neq \emptyset$$

Assume, for the sake of contradiction, that the intersection is empty for a given $d > 0$. Let a be the smallest number in $\mathcal{S}_0$ such that

$$d\frac{1-\rho_{\min}^{-1}}{2} < a \tag{4}$$

a is well defined because each interval S_i is closed to the left. Then $a \in S_i$ for some $i \in \mathbb{Z}$, $i \equiv 0 \mod 2$, and it is clear that $a = \rho_{\min}^{-i}\rho_{\max}^{-i}$ (*i.e.*, a is the lower bound of the interval S_i). Since by assumption

$$S_{i+2} \cap \left[d\frac{1-\rho_{\min}^{-1}}{2\rho_{\min}^2\rho_{\max}^2}, d\frac{1-\rho_{\min}^{-1}}{2}\right) = \emptyset$$

and by the minimality of a, we have

$$\begin{aligned} & \rho_{\min}^{-(i+2)}\rho_{\max}^{-(i+2)} \leq d\frac{1-\rho_{\min}^{-1}}{2\rho_{\min}^2\rho_{\max}^2} \\ \Rightarrow \quad & a = \rho_{\min}^{-i}\rho_{\max}^{-i} \leq d\frac{1-\rho_{\min}^{-1}}{2} \end{aligned}$$

The last inequality contradicts Eq. (4). □

We say a robot r executes $\texttt{Move}(\mathfrak{s})$ if it moves a distance $\mathfrak{s}(d_r)$ towards the other robot. The first inequality ensures that if both robots execute $\texttt{Move}(\mathfrak{s})$, then one of them eventually reaches the next level. The second part of the definition ensures that if one robot executes $\texttt{Move}(\mathfrak{s})$ and the other executes $\texttt{Move}(Other)$, the one that executes $\texttt{Move}(\mathfrak{s})$ is now in $\mathcal{S}_0$.

Algorithm 2: The movement of a robot r depends on the distance between the two robots d (seen by robot r), and on where the robot r sees itself in the line (on the right or on the left)

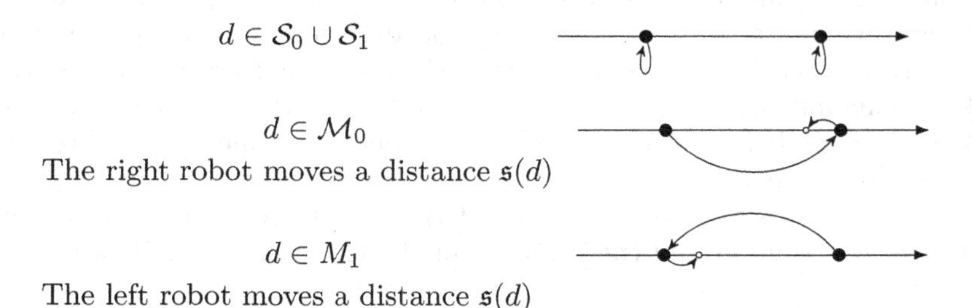
$d \in \mathcal{S}_0 \cup \mathcal{S}_1$
$d \in \mathcal{M}_0$
The right robot moves a distance $\mathfrak{s}(d)$
$d \in M_1$
The left robot moves a distance $\mathfrak{s}(d)$

Our algorithm is defined in Fig. 2. First we notice that, by Lemma 1, robots cannot both stay stationary. Indeed, if r_1 is in S_i, then $d_2 \in M_{i-1}$ and if r_2 is in S_i, then $r_1 \in M_i$ (recall r_1 and r_2 denotes the two robots such that $d_1 < d_2$).

Let $\texttt{Conf}(X, Y)$ be the set of configurations where $r_1 \in X$ and $r_2 \in Y$. Recall that r_1 and r_2 are such that the distance d_1 is smaller than d_2, so r_1 has a greater level than (or the same as) r_2.

We directly have the following lemma to reduce the number of cases we handle in the sequel.

Lemma 3. *For every* $i, j \in \{0, 1\}$,

$$\textit{Conf}(\mathcal{S}_i, \mathcal{M}_i) = \textit{Conf}(\mathcal{M}_{i+1}, \mathcal{S}_i) = \textit{Conf}(\mathcal{S}_i, \mathcal{S}_j) = \emptyset$$

Proof. Since r_1 has a greater level compared to r_2 so, for every $i \in \mathbb{Z}$, if $r_1 \in S_i$, then $r_2 \in M_{i-1}$, by Lemma 1, hence $\texttt{Conf}(\mathcal{S}_i, \mathcal{M}_i) = \emptyset$. Similarly, if $r_2 \in S_i$, then $r_1 \in M_i$ (because M_i is the level right above S_i). Finally, we saw that, by construction, both robots cannot be in an $\mathcal{S}$ level. □

This means $\texttt{Conf}(\mathcal{M}_0, \mathcal{S}_0)$, $\texttt{Conf}(\mathcal{M}_0, \mathcal{M}_0)$, $\texttt{Conf}(\mathcal{S}_1, \mathcal{M}_0)$, $\texttt{Conf}(\mathcal{M}_1, \mathcal{M}_0)$, $\texttt{Conf}(\mathcal{M}_1, \mathcal{S}_1)$, and $\texttt{Conf}(\mathcal{M}_1, \mathcal{M}_1)$ are the only non-empty set of configurations.

From Flexible to Rigid Movements. The following lemma shows that, the robots eventually are, and remain, at distance at most δ from each other. Using this lemma, we can now assume in the remaining of the paper that movements are rigid.

Lemma 4. *If the distance d between the two robots is greater than δ (in the global coordinate system), then, after two rounds, the distance between robots decreases by at least a constant C (that depends only on δ, $\rho_{\min}$ and $\rho_{\max}$).*

Proof. First, it is clear that robots cannot increase the distance between them.

If one or two robots execute $\texttt{Move}(\mathfrak{s})$, then the distance between the robots decreases by at least

$$\min\left(\delta, d\frac{1-\rho_{\min}^{-1}}{2\rho_{\min}^2\rho_{\max}^2}\right) \geq C \qquad \text{with } C = \delta\frac{1-\rho_{\min}^{-1}}{2\rho_{\min}^2\rho_{\max}^2}$$

If one robot remains idle and the other robot execute $\texttt{Move}(\mathfrak{s})$, then the distance decreases by at least δ.

The last remaining case is when both robots are in $\mathcal{M}$ and execute $\texttt{Move}(\mathit{Other})$ at time t. It is possible that the distance does not decrease at all (if both robots reach their destination) or the distance decreases by an arbitrarily small amount. In the next round, at time $t+1$, either (a) the robots are in the same level as before, (b) the level of only one robot increases, or (c) the level of both robots increases.

In case (a), the positions of the robots at time $t+1$ are exchanged, so they now both execute $\texttt{Move}(\mathfrak{s})$ and the property of the lemma is obtained after one more round.

In case (b), one robot reaches $\mathcal{S}$ while the other remains in $\mathcal{M}$, at time $t+1$, so the property is obtained after one more round as well (since one robot remains idle).

In case (c), since both robots cannot reach $\mathcal{S}$ (by Lemma 1) so one robot must increases from M_i to $M_{i'}$ with $i < i'$. we observe that, to do so, the distance must decrease by at least $\frac{d}{\rho_{\min}} \geq \frac{\delta}{\rho_{\min}}$, which is greater than C defined above, and the Lemma is proved. □

When a Single Robot is Activated. First, we compute the configurations that are eventually reached when only one robot is activated.

Lemma 5. *If a single robot is activated and executes $\texttt{Move}(\mathit{Other})$, then the robots gather in one round.*

Lemma 6. *From* $\mathit{Conf}(\mathcal{M}_i, \mathcal{M}_i)$*, if a single robot is activated and executes* $\mathit{Move}(\mathfrak{s})$*, then eventually we reach either a configuration in* $\mathit{Conf}(\mathcal{S}_{i+1}, \mathcal{M}_i)$.

Proof. Consider $C \in \mathtt{Conf}(M_i, M_i)$, with $i \in \mathbb{Z}$, and d_1 and d_2 the distances seen by r_1 and r_2 respectively in C. We have $d_1 \geq \rho_{\min}^{-i} \rho_{\max}^{-(i+1)}$. After a single robot, say r_1, executes $\mathtt{Move}(\mathfrak{s})$, let d'_1 and d'_2 the distances seen by r_1 and r_2 respectively. So:

$$d'_1 = d_1 - \mathfrak{s}(d_1) \geq d_1 - d_1 \frac{1 - \rho_{\min}^{-1}}{2} \geq d_1 \rho_{\min}^{-1} \geq \rho_{\min}^{-(i+1)} \rho_{\max}^{-(i+1)}$$

Where the first inequality comes from the definition of $\mathfrak{s}(d)$, in Eq. 3. Hence, $d'_1 \notin M_{i+1}$. Similarly, $d'_2 \notin M_{i+1}$. The same is true if only r_2 executes $\mathtt{Move}(\mathfrak{s})$.

This implies that, if a single robot comes closer by executing $\mathtt{Move}(\mathfrak{s})$, the level of the robots can increase by at most one, so we reach a configuration in $\mathtt{Conf}(M_i \cup S_{i+1}, \mathcal{M}_i \cup S_{i+1})$.

Then, observe that both robots cannot increase simultaneously their level because $\mathtt{Conf}(S_{i+1}, S_i) = \emptyset$, by Lemma 3. Also, observe that r_2 cannot increase its level alone because $\mathtt{Conf}(M_{i+1}, S_{i+1}) = \emptyset$. Hence, eventually r_1 enters level $\mathcal{S}_{i+1}$ and we reach configuration $\mathtt{Conf}(\mathcal{S}_{i+1}, \mathcal{M}_i)$. □

Lemma 7. *From* $\mathit{Conf}(\mathcal{M}_i, \mathcal{M}_j)$*, with* $i \neq j$*, if a single robot is activated and executes* $\mathit{Move}(\mathfrak{s})$*, then eventually we reach either a configuration in* $\mathit{Conf}(\mathcal{M}_i, \mathcal{S}_i)$.

Proof. We know that $C \in \mathtt{Conf}(M_i, M_{i-1})$, for some $i \in \mathbb{Z}$ (because the level of r_2 is smaller than the one of r_1). Similarly to the previous lemma, a single robot increases its level and it cannot be r_1 because $\mathtt{Conf}(S_{i+1}, M_{i-1}) = \emptyset$. Hence, eventually r_2 enters level S_i and we reach configuration $\mathtt{Conf}(M_i, S_i) \subset \mathtt{Conf}(\mathcal{M}_i, \mathcal{S}_i)$. □

When Both Robots are Activated. The nine following Lemmas consider all the possible cases, when both robots are activated, depending on the level of each robot. Lemma 8–12 consider the cases where both robots are in $\mathcal{M}_*$, depending on which move the robots are executing (both $\mathtt{Move}(\mathfrak{s})$ – Lemma 8–9 –, both $\mathtt{Move}(\mathit{Other})$ – Lemma 10 – or only one $\mathtt{Move}(s)$ – Lemma 11–12), Lemma 13–15 consider the case where one robot is in $\mathcal{S}_*$ (depending on whether the moving robot executes $\mathtt{Move}(\mathit{Other})$ – Lemma 13 – or $\mathtt{Move}(\mathfrak{s})$ – Lemma 14–15) and Lemma 3 proves that the remaining cases cannot occur.

Lemma 8. $\forall i \in \{0, 1\}$*, if* $C \in \mathit{Conf}(\mathcal{M}_i, \mathcal{M}_i)$ *and both robots execute* $\mathit{Move}(\mathfrak{s})$*, then eventually we reach a configuration in* $\mathit{Conf}(\mathcal{S}_{i+1}, \mathcal{M}_i)$*. The same is true if a single robot is activated.*

Proof. Consider $C \in \mathtt{Conf}(M_i, M_i)$, with $i \in \mathbb{Z}$, and d_1 and d_2 the distances seen by r_1 and r_2 respectively in C. We have $d_1 \geq \rho_{\min}^{-i} \rho_{\max}^{-(i+1)}$. After both robots execute $\mathtt{Move}(\mathfrak{s})$, let d'_1 and d'_2 the distances seen by r_1 and r_2 respectively. So:

$$d'_1 = d_1 - \mathfrak{s}(d_1) - \mathfrak{s}(d_2) \geq d_1 - 2d_1 \frac{1 - \rho_{\min}^{-1}}{2} = d_1 \rho_{\min}^{-1} \geq \rho_{\min}^{-(i+1)} \rho_{\max}^{-(i+1)}$$

Where the first inequality comes from the definition of $\mathfrak{s}(d)$, in Eq. 3, and from the fact that $d_1 < d_2$ (the same inequality is true if a single robot is activated). Hence, $d'_1 \notin M_{i+1}$. Similarly, $d'_2 \notin M_{i+1}$. But since robots come closer, robots cannot remain in M_i infinitely and eventually one robot reaches S_{i+1} (both cannot reach S_{i+1} simultaneously because $\texttt{Conf}(S_{i+1}, S_{i+1}) = \emptyset$). Since r_1 have a level greater than r_2, eventually we must reach $\texttt{Conf}(S_{i+1}, M_i) \subset \texttt{Conf}(\mathcal{S}_{i+1}, \mathcal{M}_i)$. □

Lemma 9. *$\forall i \in \{0,1\}$, if $C \in$ Conf$(\mathcal{M}_{i+1}, \mathcal{M}_i)$, and both robots execute Move$(\mathfrak{s})$, then eventually we reach a configuration in Conf$(\mathcal{M}_{i+1}, \mathcal{S}_{i+1})$. The same is true if a single robot is activated.*

Proof. Consider $C \in \texttt{Conf}(M_{i+1}, M_i)$, for some $i \in \mathbb{Z}$. Using the same proof as in the previous Lemma, we obtain that $d'_1 \notin M_{i+2}$ and $d'_2 \notin M_{i+1}$. But since robots come closer, robots cannot remain at the same level infinitely and eventually reaches a level S. Since r_1 have a level greater than r_2, and r_1 cannot reach S_{i+2} while r_2 is still in M_i (Lemma 1), eventually we must reach $\texttt{Conf}(M_{i+1}, S_{i+1}) \subset \texttt{Conf}(\mathcal{M}_{i+1}, \mathcal{S}_{i+1})$. □

Lemma 10. *$\forall i,j \in \{0,1\}$, if $C \in$ Conf$(\mathcal{M}_i, \mathcal{M}_j)$, and both robots execute Move$(Other)$, then after one round, the configuration is still in Conf$(\mathcal{M}_i, \mathcal{M}_j)$ and the robots have reversed their position, so they execute Move$(\mathfrak{s})$ in the next round.*

Proof. If both robots execute $\texttt{Move}(Other)$, then both robots exchange their position and the distance between them remain the same so that the lemma follows. □

Lemma 11. *$\forall i,j \in \{0,1\}$, if $C \in$ Conf$(\mathcal{M}_i, \mathcal{M}_j)$ and only r_1 executes Move$(\mathfrak{s})$ (r_2 execute Move$(Other)$), then we reach a configuration in Conf$(\mathcal{S}_0, \mathcal{M}_1)$ (and the robots have reversed their position).*

Proof. If r_2 execute $\texttt{Move}(Other)$ and r_1 executes $\texttt{Move}(\mathfrak{s})$, then, by definition of $\mathfrak{s}(d_1)$, robot $r_1 \in \mathcal{S}_0$. Then, using Lemma 3, we know that $r_2 \in M_1$. □

Lemma 12. *$\forall i,j \in \{0,1\}$, if $C \in$ Conf$(\mathcal{M}_i, \mathcal{M}_j)$ and only r_2 executes Move$(\mathfrak{s})$ (r_1 execute Move$(Other)$), then we reach a configuration in Conf$(\mathcal{M}_0, \mathcal{S}_0)$ (and the robots have reversed their position).*

Proof. If r_1 execute $\texttt{Move}(Other)$ and r_2 executes $\texttt{Move}(\mathfrak{s})$, then, by definition of $\mathfrak{s}(d_2)$, robot $r_2 \in \mathcal{S}_0$. Then, using Lemma 3, we know that $r_1 \in M_0$. □

Lemma 13. *$\forall i \in \{0,1\}$, if $C \in$ Conf$(\mathcal{S}_i, \mathcal{M}_{i-1}) \cup$ Conf$(\mathcal{M}_i, \mathcal{S}_i)$ and the moving robot executes Move$(Other)$, then the robots gather in one round.*

Proof. Clearly if one robot remains idle while the other executes $\texttt{Move}(Other)$, then the robots gather. □

Lemma 14. *$\forall i \in \{0,1\}$, if $C \in$ **Conf**$(\mathcal{S}_i, \mathcal{M}_{i-1})$ and r_2 executes **Move**$(\mathfrak{s})$, then eventually we reach either a configuration in **Conf**$(\mathcal{M}_i, \mathcal{M}_{i-1})$ or a configuration in **Conf**$(\mathcal{M}_i, \mathcal{S}_i)$.*

Proof. If r_1 remains idle and r_2 comes closer by executing Move$(\mathfrak{s})$, then one or both robots eventually increase their levels. By definition of $\mathfrak{s}(d_2)$ (using the same proof as in Lemma 8), r_2 cannot go from level $\mathcal{M}_{i-1}$ to level $\mathcal{M}_i$ directly. Hence, either r_2 enters level $\mathcal{S}_i$ (in this case, r_1 simultaneously enters $\mathcal{M}_i$), and we reach configuration Conf$(\mathcal{M}_i, \mathcal{S}_i)$, or only r_1 enters $\mathcal{M}_i$ and we reach configuration Conf$(\mathcal{M}_i, \mathcal{M}_{i-1})$. $\square$

Lemma 15. *$\forall i \in \{0,1\}$, if $C \in$ **Conf**$(\mathcal{M}_i, \mathcal{S}_i)$ and r_1 executes **Move**$(\mathfrak{s})$, then eventually we reach either a configuration in **Conf**$(\mathcal{M}_i, \mathcal{M}_i)$ or a configuration in **Conf**$(\mathcal{S}_{i+1}, \mathcal{M}_i)$.*

Proof. If r_2 remains idle and r_1 comes closer by executing Move$(\mathfrak{s})$, then one or both robots eventually increase their levels. By definition of $\mathfrak{s}(d_1)$ (using the same proof as in Lemma 8), r_1 cannot go from level $\mathcal{M}_i$ to level $\mathcal{M}_{i+1}$ directly. Hence, either r_1 enters level $\mathcal{S}_{i+1}$ (in this case, r_2 simultaneously enters $\mathcal{M}_i$), and we reach configuration Conf$(\mathcal{S}_{i+1}, \mathcal{M}_i)$, or only r_2 enters $\mathcal{M}_i$ and we reach configuration Conf$(\mathcal{M}_i, \mathcal{M}_i)$. $\square$

Main Proof of Correctness. We can characterize a configuration by only looking at where is located r_1 (right or left), whether r_2 has the same orientation has r_1, and the level of each robot (in $\mathcal{S}_0$, $\mathcal{S}_1$, $\mathcal{M}_0$ or $\mathcal{M}_1$). For simplicity, we use the notation (X ● ● Y) to denote the configuration where the left robot is in level $X \in \{\mathcal{S}_0, \mathcal{S}_1, \mathcal{M}_0, \mathcal{M}_1\}$ and the right robot has level Y. We add a line over the level of r_1 and a line under the level of r_2. Finally, if both robots have different orientations, we add a minus in front of r_2's level. To help the reader, we also draw the destination of each robot in the configuration.

For instance a configuration $C \in$ ($\underline{\mathcal{M}_1}$ ●→ ● $\overline{\mathcal{S}_0}$) is a configuration where r_1 is located on the right, r_2 on the left (they have the same orientation of the line), r_1 is in $\mathcal{S}_0$ and r_2 in $\mathcal{M}_1$. Recall that, since r_1 has a level greater than r_2, if r_1 is in S_i for some $i \in \mathbb{Z}$ ($i \equiv 0 \mod 2$), then r_2 must be in M_{i-1}. Since $C \in$ Conf$(\mathcal{S}_0, \mathcal{M}_1)$ and r_2 executes Move$(\mathfrak{s})$, by Lemma 14, we eventually reach either a configuration in ($\underline{\mathcal{M}_1}$ ●→ ←● $\overline{\mathcal{M}_0}$) or in ($\underline{\mathcal{S}_0}$ ● ←● $\overline{\mathcal{M}_0}$) (the same is true if a single robot is activated, since r_1 remains idle).

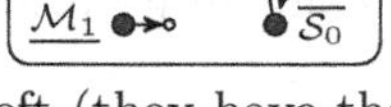

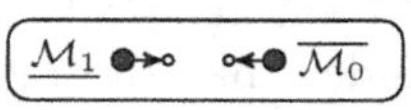

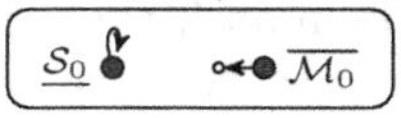

As a second example, $C \in$ ($\overline{\mathcal{M}_0}$ ● ←● $-\underline{\mathcal{M}_1}$) is a configuration where r_1 is on the left and r_2 sees itself on the left as well. Again, if r_1 is in M_i for some $i \in \mathbb{Z}$

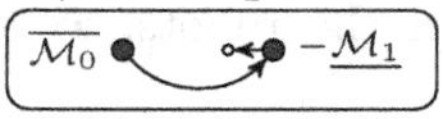

($i \equiv 0 \mod 2$), then r_2 is in M_{i-1}. If both robots are activated, by Lemma 12 the next configuration is in $-\underline{\mathcal{S}_0}$ $\overline{\mathcal{M}_0}$ $-\underline{\mathcal{S}_0}$ $\overline{\mathcal{M}_0}$. If a single robot is activated, either the robots gather in one round (if r_1 is activated) or we can reach configuration $\overline{\mathcal{M}_0}$ $-\underline{\mathcal{S}_0}$ (if only r_2 is activated).

The information of where r_1 is located (indicated by an over-line) does not impact the movement of the robots, however, it limits the possibilities for the reached configuration. For instance if both robots are in $\mathcal{M}_i$ and make a `Move`($\mathfrak{s}$) then we are sure that r_1 is the first to reach $\mathcal{S}_{i+1}$.

Using these notations, we can construct the a graph depicting the transitions between the different sets of configurations. Each arc is proved by one of the previous Lemmas, whose number is indicated on the arc. It is easy to see that if robots agree, resp. do not agree, on the orientation of the line, then the same is true in any reached set of configurations. This implies that we can split the graph in two, one that consider only sets of configuration where the robots agree on the orientation of the line, Fig. 2, and one when they do not, Fig. 3. The dashed arcs correspond to the transitions that can occur when a single robot is activated. Of course, when a single robot is dictated to move (for instance in $\underline{\mathcal{M}_1}$ $\overline{\mathcal{S}_0}$) then activating only this robot results in the same configuration. But with a fair scheduler, the other robot is eventually activated. Also, in this case, activating only the moving robot is similar to activating both robot, so when this happen we only draw the plain arc. Finally, for clarity, we do do represent the dashed arcs corresponding to the case where a single robot is activated and executes `Move`(Other), as robots gather in one round in this case.

Given the previous lemmas that proves the possible transitions between set of configuration, the graphs have been generated by an algorithm (available online [2]). It is easy to check that both graphs are in fact Directed Acyclic Graphs (DAG) with a single sink, the gathered configuration. This means that regardless of the starting configuration, we eventually reach the gathered configuration.

Theorem 2. *When* $\rho \in [\rho_{\min}, \rho_{\max}]$, *Algorithm 2 solves the rendezvous in SSYNC.*

Proof. Using Lemma 4, we know that eventually robots are and remains at distance at most δ so that we can consider that the movements are rigid. From there, we showed in Figs. 2 and Fig. 3 that regardless of the configuration and regardless of the orientation of the robots, we eventually reach the gathered configuration. □

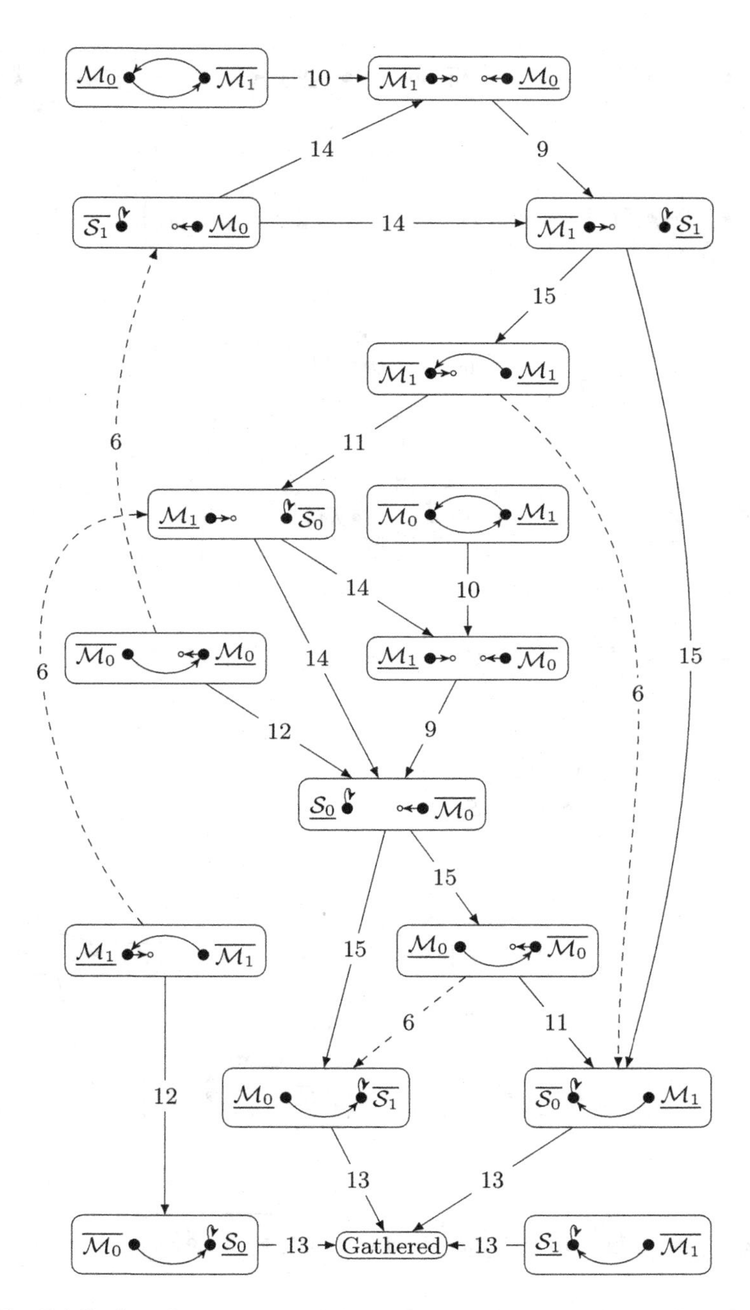

Fig. 2. The DAG of configurations and the transitions between them, when the robots have the same orientation. The number on the edges are the numbers of the Lemmas proving the transition

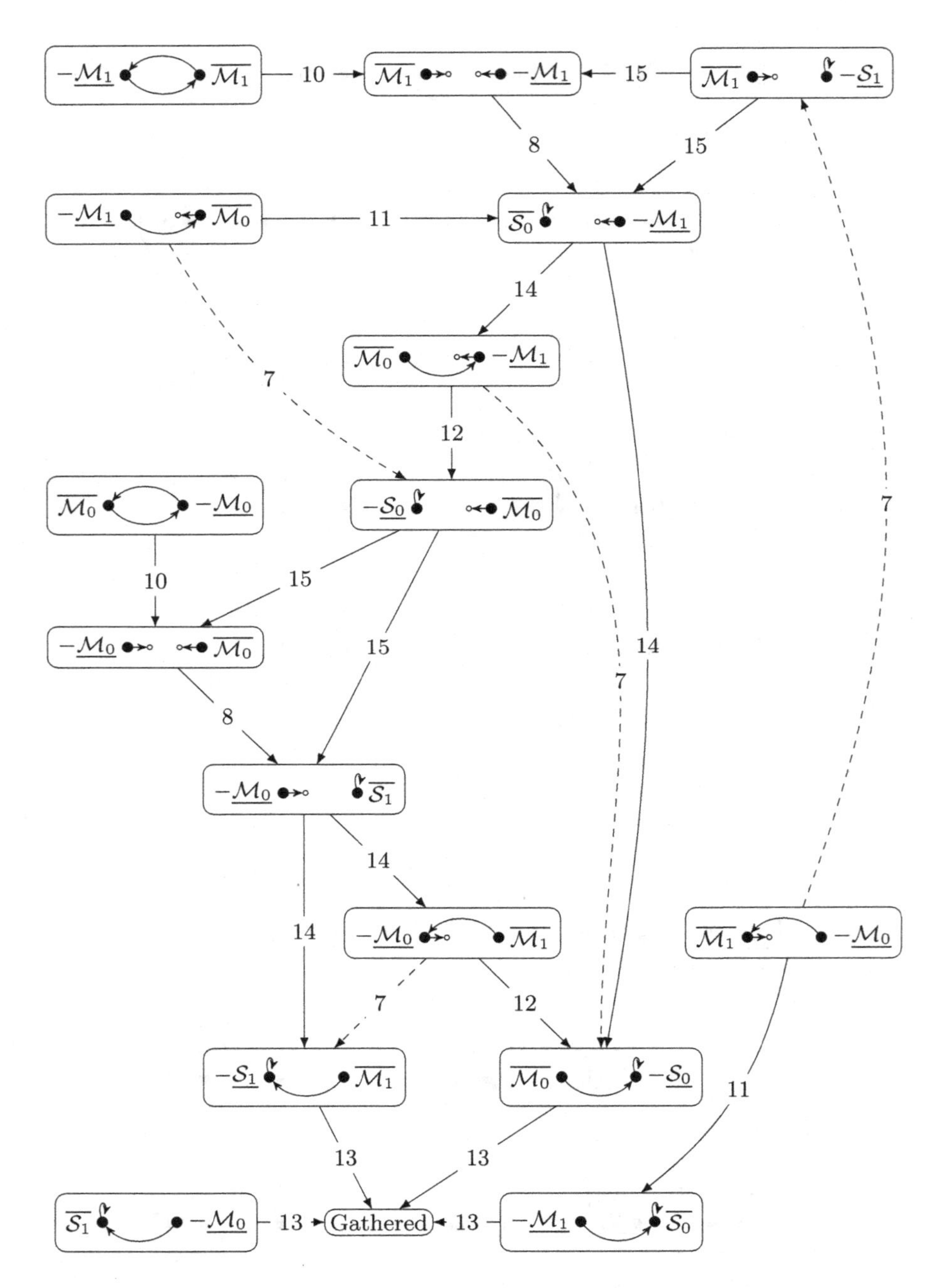

Fig. 3. The DAG of configurations and the transitions between them, when the robots have the opposite orientations. The number on the edges are the numbers of the Lemmas proving the transition.

5 Concluding Remarks

We introduced the possibility to use different unit distances to break symmetries for networks consisting on deterministic oblivious robots that operate in the Look-Compute-Move model. As a case study, we considered the rendezvous problem, that is notoriously impossible to solve in the semi-synchronous execution model, when robots share the same notion of unit distance. By contrast, we proved that when robots have different unit distances (and are unaware of the actual ratio between the unit distances), rendezvous becomes possible in the same model.

A natural open question is to consider the completely asynchronous model (ASYNC). Is it possible to solve rendezvous in ASYNC without any additional capability (no access to a randomness source, nor persistent memory) or constraints (compasses may be fully symmetric) other that the difference in unit distance? Observe that even if $\rho = 2$, the problem seems difficult, as our semi-synchronous algorithm for this special case does not solve rendezvous in ASYNC (one can construct an infinite execution where robots observe one another alternatively as they are moving, and thus never actually reach the other robot destination).

References

1. Bramas, Q., Lamani, A., Tixeuil, S.: Stand up indulgent rendezvous. In: Devismes, S., Mittal, N. (eds.) SSS 2020. LNCS, vol. 12514, pp. 45–59. Springer, Cham (2020). https://doi.org/10.1007/978-3-030-64348-5_4
2. Bramas, Q., Lamani, A., Tixeuil, S.: The agreement power of disagreement: graph generation, September 2021. https://doi.org/10.5281/zenodo.5541136
3. Courtieu, P., Rieg, L., Tixeuil, S., Urbain, X.: Impossibility of gathering, a certification. Inf. Process. Lett. **115**(3), 447–452 (2015)
4. Das, S., Flocchini, P., Prencipe, G., Santoro, N., Yamashita, M.: Autonomous mobile robots with lights. Theor. Comput. Sci. **609**, 171–184 (2016)
5. Flocchini, P., Prencipe, G., Santoro, N. (eds.): Distributed Computing by Mobile Entities-Current Research in Moving and Computing. Lecture Notes in Computer Science, vol. 11340. Springer, Heidelberg (2019). https://doi.org/10.1007/978-3-030-11072-7
6. Flocchini, P., Prencipe, G., Santoro, N., Widmayer, P.: Gathering of asynchronous robots with limited visibility. Theor. Comput. Sci. **337**(1–3), 147–168 (2005)
7. Heriban, A., Défago, X., Tixeuil, S.: Optimally gathering two robots. In: Proceedings of 19th International Conference on Distributed Computing and Networking, ICDCN, pp. 3:1–3:10, January 2018
8. Izumi, T., et al.: The gathering problem for two oblivious robots with unreliable compasses. SIAM J. Comput. **41**(1), 26–46 (2012)
9. Okumura, T., Wada, K., Katayama, Y.: Brief announcement: optimal asynchronous rendezvous for mobile robots with lights. In: Spirakis, P., Tsigas, P. (eds.) SSS 2017. LNCS, vol. 10616, pp. 484–488. Springer, Cham (2017). https://doi.org/10.1007/978-3-319-69084-1_36

10. Souissi, S., Défago, X., Yamashita, M.: Using eventually consistent compasses to gather memory-less mobile robots with limited visibility. ACM Trans. Auton. Adapt. Syst. **4**(1), 9:1–9:27 (2009)
11. Suzuki, I., Yamashita, M.: Distributed anonymous mobile robots: formation of geometric patterns. SIAM J. Comput. **28**(4), 1347–1363 (1999)
12. Viglietta, G.: Rendezvous of two robots with visible bits. In: Flocchini, P., Gao, J., Kranakis, E., Meyer auf der Heide, F. (eds.) ALGOSENSORS 2013. LNCS, vol. 8243, pp. 291–306. Springer, Heidelberg (2014). https://doi.org/10.1007/978-3-642-45346-5_21

The Max-Line-Formation Problem

And New Insights for Gathering and Chain-Formation

Jannik Castenow(✉), Thorsten Götte, Till Knollmann, and Friedhelm Meyer auf der Heide

Heinz Nixdorf Institute and Computer Science Department, Paderborn University, Paderborn, Germany
{jannik.castenow,thorsten.goette,till.knollmann,fmadh}@upb.de

Abstract. We consider n robots with *limited visibility*: each robot can observe other robots only up to a constant distance denoted as the *viewing range*. The robots operate in discrete rounds that are either fully synchronous ($\mathcal{F}$SYNC) or semi-synchronized ($\mathcal{S}$SYNC). Most previously studied formation problems in this setting seek to bring the robots closer together (e.g., GATHERING or CHAIN-FORMATION). In this work, we introduce the MAX-LINE-FORMATION problem, which has a contrary goal: to arrange the robots on a straight line of maximal length.

First, we prove that the problem is impossible to solve by robots with a constant sized *circular* viewing range. The impossibility holds under comparably strong assumptions: robots that agree on *both* axes of their local coordinate systems in $\mathcal{F}$SYNC. On the positive side, we show that the problem is solvable by robots with a constant *square* viewing range, i.e., the robots can observe other robots that lie within a constant-sized square centered at their position. In this case, the robots need to agree on only *one* axis of their local coordinate systems. We derive two algorithms: the first algorithm considers oblivious robots ($\mathcal{OBLOT}$) and converges to the optimal configuration in time $\mathcal{O}(n^2 \cdot \log(n/\varepsilon))$ under the $\mathcal{S}$SYNC scheduler (ε is a convergence parameter). The other algorithm makes use of locally visible lights ($\mathcal{LUMI}$). It is designed for the $\mathcal{F}$SYNC scheduler and can solve the problem exactly in optimal time $\Theta(n)$.

Afterward, we show that both the algorithmic and the analysis techniques can also be applied to the GATHERING and CHAIN-FORMATION problem: we introduce an algorithm with a reduced viewing range for GATHERING and give new runtime bounds for CHAIN-FORMATION.

Keywords: Mobile robots · Runtime · Chain-formation · Gathering · Max-line-formation · Max-chain-formation

1 Introduction

Robot formation tasks aim to arrange n mobile robots in a specific formation. The robots are modeled as points in the Euclidean plane, and usually, the robot

This work was partially supported by the German Research Foundation (DFG) under the project number 453112019; ME 872/14-1. A full version can be found online [4].

C. Johnen et al. (Eds.): SSS 2021, LNCS 13046, pp. 289–304, 2021.
https://doi.org/10.1007/978-3-030-91081-5_19

capabilities are very restricted. Robots are assumed to be externally *identical* (all robots have the same appearance), *anonymous* (no identifiers), *autonomous* (no central control) and *homogeneous* (all robots execute the same algorithm). Furthermore, the robots operate in discrete rounds denoted as `LCM` cycles. Each `LCM` cycle consists of three operations: `Look`, `Compute` and `Move`. During the `Look` operation, each robot takes a snapshot of its surroundings. Afterward, the robot computes a target point during `Compute` and finally moves there in the `Move` operation. With the additional assumptions that robots are *silent* (no communication) and *oblivious* (no memory of previous `LCM` cycles), this is known as the $\mathcal{OBLOT}$ model. The $\mathcal{LUMI}$ model, on the contrary, does not demand the robots to be silent and oblivious. Instead, each robot is equipped with a light that nearby robots (as well as the robot itself) can perceive. The light can have different colors, and thus, the robots obtain a constant-sized memory and can communicate state information to their neighbors. In addition to these core features, both models have a variety of freedom in some other assumptions; for instance, the `LCM` cycles might be fully synchronous ($\mathcal{F}$SYNC), semi-synchronous ($\mathcal{S}$SYNC) or completely asynchronous ($\mathcal{A}$SYNC). All schedulers are assumed to be fair such that each robot can execute its `LCM` cycle infinitely often. Time is measured in *epochs*, i.e., the smallest number of rounds such that each robot has executed its `LCM` cycle at least once. In $\mathcal{F}$SYNC, an epoch is equal to one round. See [15] for an overview about the mentioned models.

Our focus lies on robots with limited visibility, i.e., each robot cannot perceive the entire swarm but only nearby robots. The terms *connectivity range* and *viewing range* are distinguished (see e.g., [6,20]). Robots are connected to all robots up to a distance equal to their connectivity range and can see all robots within their viewing range (the viewing range is at least as large as the connectivity range). Initial configurations are connected w.r.t. the connectivity range and algorithms typically maintain this connectivity. The larger viewing range enhances the local information of the robots. Additionally, viewing and connectivity range can be *circular* or *square*. More precisely, a circular connectivity range of c means that a robot is connected to all robots in the distance at most c. In contrast, the square connectivity range of sc connects a robot r to all other robots located within an axis aligned $2sc \times 2sc$-sized square centered at r. Similarly, circular and square viewing ranges are defined. In many applications, the connectivity and the viewing range are identical. The literature especially focusing on the runtime of formation algorithms often benefits from a viewing range that is larger than the connectivity range, see e.g., [1,3,5,20].

Typical well-studied benchmark problems for robots with limited visibility are the GATHERING and the CHAIN-FORMATION problem. GATHERING demands the robots to gather at a single, not predefined, position. CHAIN-FORMATION considers a chain of robots between two stationary outer robots: each inner robot has two identifiable neighbors (the neighborhoods are predefined and fixed). Robots can observe only the positions of their predefined neighbors and nothing more. The goal is to arrange the robots on the line segment connecting the outer robots. Both GATHERING and CHAIN-FORMATION can be characterized as *contracting*: the robots move closer together. Much less is known about formation tasks for robots with limited visibility that aim to achieve a contrary goal: to *expand* the robots' positions. One example is the UNIFORM-CIRCLE-FORMATION problem in which n robots are to move such that their positions form a regular polygon [12,18]. Another, very recent example and the main inspiration for this work is the MAX-CHAIN-FORMATION problem [6]. The MAX-CHAIN-FORMATION problem is a variant of the CHAIN-FORMATION problem. The difference is that MAX-CHAIN-FORMATION gives the outer robots the ability to move. The new goal is to transform the chain of robots with connectivity and viewing range c into a straight line of length $(n-1) \cdot c$.

In this work, we introduce the MAX-LINE-FORMATION problem. The goal is similar to the MAX-CHAIN-FORMATION problem: to move the robots with connectivity range c such that their positions form a straight line of length $(n-1) \cdot c$. The difference is that MAX-LINE-FORMATION does not consider predefined chain neighborhoods. Instead, robots can observe the positions of all robots within their viewing range and do not have any fixed neighbors. We investigate under which robot capabilities the problem is solvable, derive algorithms, and analyze their runtime.

Related Work We focus on robots that operate in the LCM model and results about GATHERING, CHAIN-FORMATION and MAX-CHAIN-FORMATION with a particular focus on research that considers a runtime analysis of the proposed algorithms. For a very recent and comprehensive overview of different robot formation algorithms, we refer the reader to [15]. Oblivious and disoriented robots ($\mathcal{OBLOT}$), can solve GATHERING in $\mathcal{O}(n^2)$ rounds ($\mathcal{F}$SYNC) with the GTC algorithm. GTC moves robots in each round towards the center of the smallest enclosing circle of their neighborhood [2,10]. GTC achieves the currently best-known runtime for disoriented and oblivious robots in the Euclidean plane. Faster algorithms for disoriented robots could so far only be designed under the $\mathcal{LUMI}$ model. There are two algorithms for robots located on a two-dimensional grid [1,8]. Another algorithm for robots in the Euclidean plane that are connected in a closed chain topology [5] exists. When assuming the $\mathcal{OBLOT}$ model and one axis agreement, an asymptotically optimal algorithm with runtime $\mathcal{O}(\Delta)$ has been introduced in [20]. The algorithm assumes a square connectivity range of 1 and a circular viewing range of $\sqrt{10}$.

CHAIN-FORMATION has been initially introduced in [13]. The authors introduce the GTM algorithm that moves each robot to the midpoint between its neighbors. For the $\mathcal{F}$SYNC scheduler, a runtime of $\mathcal{O}(n^2 \cdot \log(n/\varepsilon))$ rounds has been proven. Later on, an almost matching lower bound (for the algorithm) of $\Omega(n^2 \cdot \log(1/\varepsilon))$ rounds has been derived [16]. Algorithms with stronger assumptions, e.g., the $\mathcal{LUMI}$ model, are able to achieve better runtimes [14,17].

Very recently, the MAX-CHAIN-FORMATION problem has been introduced [6]. Started in one-dimensional configurations, the MAX-GTM algorithm has a runtime of $\mathcal{O}(n^2 \cdot \log(n/\varepsilon))$ and $\Omega(n^2 \cdot \log(1/\varepsilon))$ rounds under the $\mathcal{F}$SYNC scheduler. However, a specific class of input configurations does not converge to the optimal configuration. For two-dimensional configurations, only a convergence result is known. Additionally, for Gathering, Chain-Formation and Max-Chain-Formation, it is known that the problems can be solved optimally in a continuous time model [6,9].

Our Contribution. We introduce the Max-Line-Formation problem. The goal is to arrange n robots with connectivity range c on a straight line of length $(n-1) \cdot c$. We start with an impossibility result and prove that there are initial configurations for which the problem cannot be solved deterministically by robots with constant sized *circular* viewing and connectivity ranges. In addition, also no algorithm that converges to the optimal solution can exist for these configurations. The impossibility result even holds under strong assumptions: fully synchronized robots ($\mathcal{F}$sync) that agree on *both* axes of their local coordinate systems. On the positive side, we show that the problem becomes solvable for robots with identical *square* connectivity and viewing ranges. While *square* connectivity and viewing ranges already have been proven to be useful to derive an efficient Gathering algorithm [20], the Max-Line-Formation is the first known problem that can be solved under *square* viewing ranges but not under *circular* viewing ranges. Our algorithms require the robots to agree on only *one* axis of their local coordinate systems. We introduce two algorithms: The first algorithm considers the $\mathcal{OBLOT}$ model and converges to the optimal configuration in $\mathcal{O}(n^2 \cdot \log(n/\varepsilon))$ epochs under the $\mathcal{S}$sync scheduler. The analysis idea is based on the *sample variance* of time *inhomogeneous* Markov chains (a concept similar to the *mixing time* of the time homogeneous case) inspired by [19]. Afterward, we show that enhancing the robots with the $\mathcal{LUMI}$ model allows us to derive an improved algorithm, i.e., the algorithm solves the problem exactly while simultaneously improving the runtime. The algorithm considers the $\mathcal{F}$sync scheduler and solves the problem in $\Theta(n)$ epochs. The runtime is asymptotically optimal. Moreover, the algorithm can be implemented with 9 colors. Note that, with some additional synchronization, a combination of the two algorithms can solve the problem exactly with the help of lights in $\mathcal{O}(n^2)$ epochs under the $\mathcal{S}$sync scheduler. For more details see [4].

Our results compare to the Max-GTM algorithm for Max-Chain-Formation (which has the same goal but considers predefined and fixed chain neighborhoods) problem as follows: the runtime of our $\mathcal{OBLOT}$ algorithm holds under the $\mathcal{S}$sync scheduler. For Max-GTM, only runtimes in $\mathcal{F}$sync are known [6]. Additionally, our results about Max-Line-Formation hold for *every* input configuration in which robots have distinct initial positions. For Max-GTM, only a convergence result for a large class of configurations is known. Additionally, certain classes of configurations do not converge to the optimal configuration [6].

Moreover, we identify an interesting relation to Gathering and Chain-Formation. We first show that we can apply the main algorithmic idea of the $\Theta(n)$ algorithm to the Gathering problem. More precisely, we derive an algorithm for the $\mathcal{OBLOT}$ model that solves Gathering of n robots that agree on one axis of their local coordinate systems in $\Theta(\Delta)$ epochs under the $\mathcal{F}$sync scheduler, where Δ denotes the maximum distance of two robots in the initial configuration.[1] The algorithm uses a square viewing and connectivity range of 1. Up to now, the best-known algorithm achieving the same runtime uses a square connectivity range of 1 and a circular viewing range of $\sqrt{10}$ [20]. Thus, our algorithm closes the gap between viewing and connectivity range. Furthermore, we show how the analysis technique of the first algorithm (based on time inhomogeneous Markov chains) can also be applied for the CHAIN-FORMATION problem. In this context, *disoriented* robots (no agreement on the local coordinate systems) that are connected in a chain topology are assumed as well as a circular connectivity range and viewing range of 1. We prove that the GTM algorithm [7,14], in which each robot moves to the midpoint between its two direct neighbors in every round, converges to the optimal configuration in $\mathcal{O}(n^2 \cdot \log(n/\varepsilon))$ epochs assuming the $\mathcal{S}$SYNC scheduler. For one-dimensional configurations (all robots are initially collinear) this is a significant improvement over the so far best known runtime bound of $\mathcal{O}(n^5 \cdot \log(n/\varepsilon))$ epochs for this algorithm [7]. For two-dimensional configurations, our result is the first runtime bound for the CHAIN-FORMATION problem derived for the $\mathcal{S}$SYNC scheduler.

2 Model and Notation

Time Model: Robots operate in discrete `LCM` (`Look`, `Compute`, `Move`) cycles, denoted as *rounds*. Each robot takes a snapshot of its neighborhood during `Look`, computes a target point in `Compute`, and moves to this point in `Move`. We assume a *rigid* movement, robots always reach their target points during `Move`. The timing of the executions of the `LCM` cycles is either fully synchronous ($\mathcal{F}$SYNC) or semi-synchronous ($\mathcal{S}$SYNC), i.e., the cycles are synchronous, but only a subset of all robots participates. The executions are always fair: All robots execute their cycles infinitely often. Time is measured in epochs, i.e., the smallest number of rounds until each robot processes one complete `LCM` cycle. The execution starts in round t_0, and we denote the first round of the k-th epoch by t_{e_k}. Thus, $t_{e_1} = t_0$.

Robot Model: We consider n robots $r_1, \ldots, r_n$ positioned in $\mathbb{R}^2$. Initially, the robots are located at pairwise distinct locations. We assume a square connectivity and viewing range of 1, i.e., two robots r_i and r_j are neighbors if r_j is located inside of the 2×2-sized square centered at r_i and vice versa. Note that 1 is only chosen for simplicity; it can be replaced by any constant c. The neighborhood of a robot r_i (the set of all visible robots) in round t is denoted by $N_i(t)$. The square connectivity graph in which two robots share an edge if they are neighbors is initially connected. Robots are assumed to be *transparent* and thus do not block the views between other robots. Moreover, the robots agree on one axis of their local coordinate systems. W.l.o.g., we assume that the robots agree on the x-axis. Thus, the robots have a common understanding of left and right, while up and down can be inverted. However, the robots agree on unit distance and can

[1] $\Omega(\Delta)$ is a trivial lower bound since at least one of the robots forming the diameter Δ must cover a distance of at least $\frac{\Delta}{2}$ to obtain GATHERING. Since the robots have limited visibility, this requires $\Omega(\Delta)$ rounds.

measure distances precisely. When considering the $\mathcal{OBLOT}$ model, the robots are also silent and oblivious.

For one algorithm, we consider the $\mathcal{LUMI}$ model. Each robot is equipped with a constant number of lights $\ell_1, \ldots, \ell_k$ with color sets $C_1, \ldots, C_k$ and at every point in time, each light can have a single color out of its color set.[2] Robots can perceive their light and the lights of their neighbors during `Look` and can manipulate their light during `Compute`. Hence, if a robot r_i decides to change its light color in round t, its neighbors can see this earliest in round $t+1$.

Notation: The position of a robot r_j in round t is denoted by $p_j(t) = (x_j(t), y_j(t))$ in a global coordinate system and by $p_j^i(t) = (x_j^i(t), y_j^i(t))$ in the local coordinate system of r_i. Each robot lies in the center of its local coordinate system and thus $p_i^i(t) = (0,0)$. For a robot r_i, $r_\ell^i(t)$ denotes the leftmost robot of its neighborhood in round t. The position of $r_\ell^i(t)$ in the local coordinate system of r_i in round t is denoted by $p_\ell^i(t) = (x_\ell^i(t), y_\ell^i(t))$. In case there are multiple such robots, $r_\ell^i(t)$ represents an arbitrary robot of all leftmost robots. Similarly, $r_r^i(t)$ and $p_r^i(t)$ are defined for the rightmost neighbor. Additionally, define $r_+^i(t)$ and $p_+^i(t)$ to be the *closest* neighbor above of r_i and its position. Analogously, $r_-^i(t)$ and $p_-^i(t)$ is defined as the *closest* neighbor below and its position. In case no such robot exists, $r_+^i(t) = r_i$ and $r_-^i(t) = r_i$. For a vector v, we denote by $\widehat{v}$ the normalized vector $\frac{1}{\|v\|}v$.

Problem Statement: Max-Line-Formation demands to move n robots with connectivity range c such that their positions form a straight line of length $(n-1) \cdot c$. We say that an $(1-\varepsilon)$-approximation of the optimal configuration is reached if the positions form a straight line of length at least $(1-\varepsilon) \cdot (n-1) \cdot c$. In every round, the connectivity graph has to remain connected.

3 Impossibility Result and Intuition About Square Ranges

This section proves that Max-Line-Formation is unsolvable with constant-sized circular viewing and connectivity ranges. Afterward, we give an intuition on how square ranges circumvent the impossibility.

3.1 Impossibility with Circular Ranges

Theorem 1. *In the $\mathcal{OBLOT}$ model, for every constant* circular *connectivity and viewing range, there exists an initial configuration with robots located at distinct positions such that* Max-Line-Formation *is unsolvable. Furthermore, no convergence algorithm can exist for these configurations. This holds for robots that agree on both axes and the $\mathcal{F}$sync scheduler.*

[2] In the classical $\mathcal{LUMI}$ model [11] each robot is equipped with a single light and color set. Our assumption of multiple lights and color sets can be transferred to the classical setting by choosing a single light with a color set of size at most $2^{\sum_{i=1}^{k} |C_i|}$.

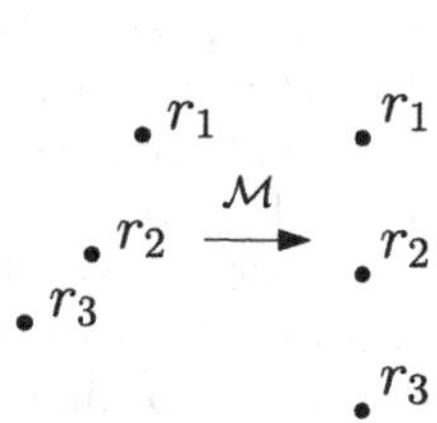

Fig. 1. The config. C_1 transformed by $\mathcal{M}$.

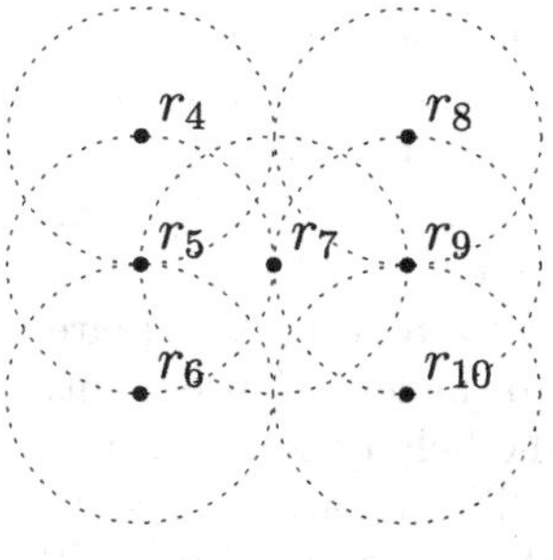

Fig. 2. The configuration C_2.

Proof. Initially, we assume an identical viewing and connectivity range of c. The arguments for viewing ranges that are larger than the connectivity range are analogous and can be found in [4]. We prove the claim by contradiction. We assume that there is an algorithm $\mathcal{M}$ that can solve the MAX-LINE-FORMATION problem. Next, we derive a combination of 2 initial configurations C_1 and C_2 and prove that if $\mathcal{M}$ is able to solve the problem starting in C_1, it cannot solve it starting in C_2. The configuration C_1 consists of three robots r_1, r_2 and r_3 at arbitrary (connected) positions. Since $\mathcal{M}$ can solve the problem, there is a time step t_f such that the MAX-LINE-FORMATION problem is solved. W.l.o.g. we assume that r_1 and r_3 are located at the end of the line and $p_1(t_f), p_2(t_f)$ and $p_3(t_f)$ form a line parallel to the y-axis (otherwise we could rename the robots and rotate the following configuration C_2 accordingly). More precisely, $p_1(t_f) - p_2(t_f) = p_2(t_f) - p_3(t_f) = (0, c)$. See Fig. 1 for a depiction of the effects of $\mathcal{M}$ started in C_1.

The configuration C_2 consists of 7 robots, $r_4, \ldots, r_{10}$ located at the following positions in a global coordinate system (not known to the robots): $p_4(t) = (-c, c), p_5(t) = (-c, 0), p_6(t) = (-c, -c), p_7(t) = (0, 0), p_8(t) = (c, c), p_9(t) = (c, 0)$, and $p_{10}(t) = (c, -c)$. See Fig. 2 for a visualization of the configuration. In C_2, r_4 can only see r_5 and is located in distance c of r_5. Moreover, it holds $p_4(t) - p_5(t) = p_1(t_f) - p_2(t_f)$. Thus, $\mathcal{M}$ is not allowed to move r_4 since $\mathcal{M}$ cannot distinguish r_1 in configuration C_1 after time t_f and r_4 in configuration C_2. By similar arguments, $\mathcal{M}$ is also not allowed to move r_6, r_8 and r_{10}. Hence, the only remaining robots that could be moved by $\mathcal{M}$ are r_5, r_7 and r_9. However, also these robots are not allowed to move. Consider the robot r_5 which is located in maximum distance to r_4, r_6 and r_7. No matter where r_5 moves, it loses the connectivity to either r_4 or r_6 as these robots remain at their position. The same arguments hold for r_7 and r_9. It follows that $\mathcal{M}$ cannot solve the problem started in C_2, which contradicts the assumption. □

3.2 Intuition About Square Ranges

Next, we argue why the proof of Theorem 1 does not hold when considering square viewing and connectivity ranges. Assume that the algorithm $\mathcal{M}$ transforms the configuration C_1 into a line that is parallel to the y-axis. Then, also the configuration C_2 is aligned with the y-axis. Still, the robots r_4, r_6, r_8 and r_{10} are not allowed to move. The robots r_5 and r_9, however, gain the possibility to move horizontally. More precisely, r_5 is allowed to move to the right without losing the connectivity to r_4 and r_6 since the complete line segment connecting r_5 and r_7 is contained in the square viewing range of both r_4 and r_6. Similarly, r_9 can move to the left. Consequently, an algorithm solving the MAX-LINE-FORMATION with the help of square ranges should arrange the robots on a line parallel to the y-axis. The square ranges are only beneficial in case the local coordinate systems have the same orientation. In case the robots are disoriented, the impossibility result of Sect. 3.1 also holds with square ranges.

4 $\mathcal{OBLOT}$ Algorithm

Based on the results of Sect. 3, MAX-LINE-FORMATION is unsolvable with *circular* viewing and connectivity ranges. In this section, we show that equipping the robots with *square* connectivity and viewing ranges allows us to design an algorithm that converges to the optimal solution. More precisely, we give an algorithm that converges to the optimal configuration assuming the $\mathcal{OBLOT}$ model and a square viewing and connectivity range of 1.

4.1 Intuition

The algorithm works in two phases. In the first phase, the positions of all robots are arranged on a straight line parallel to the y-axis. Afterward, the line is stretched in the second phase. Since the robots are oblivious and have limited visibility, robots cannot distinguish the phases and act upon their local view. Nevertheless, we will show that there is a round t' such that all robots have joined the second phase and will remain there for the rest of the execution.

Phase 1: A robot r_i whose neighborhood has not yet formed a line parallel to the y-axis moves only if its position is rightmost in its neighborhood. Then, r_i moves horizontally to the x-coordinate of its leftmost neighbor. If another robot already occupies this position, r_i executes a slight vertical movement into the positive (from its local view) y-direction to avoid a collision. More precisely, if the robot is located topmost in its neighborhood, it moves a constant distance upwards. If the robot is not topmost, it determines the value y^i_{min}, the y-coordinate of its closest neighbor to the top. Afterwards, it moves $\frac{1}{3}y^i_{min}$ upwards. The factor of $\frac{1}{3}$ is essential since the robot with y-coordinate y^i_{min} might do the same movement while having an inverted understanding of up and down. Hence, a collision of the two robots is avoided.

Phase 2: In the second phase, all robots are located on the same line parallel to the y-axis, which can be seen as a particular case of the MAX-CHAIN-FORMATION problem. Thus, the robots execute the MAX-GTM algorithm designed for MAX-CHAIN-FORMATION [6]: each inner robot (robots that have neighbors in each direction) move to the midpoint between their closest northern and their closest southern neighbor. Outer robots (at the end of the line) have to stretch the line and move as far as possible away from their closest neighbor without losing connectivity. Concretely, outer robots move as follows. Let r_1 be an outer robot and r_2 its closest neighbor and $v(t) = p_1(t) - p_2(t)$. Then, r_1 imagines a virtual robot r_v at the position $p_v(t) = p_1(t) + \widehat{v}(t)$ and moves to $\frac{1}{2}p_v(t) + \frac{1}{2}p_2(t)$.

4.2 Algorithm

We define the following set of possibly colliding robots. For a robot r_i, define $C_i(t) = \{r_j \in N_i(t) | x_j^i(t) = 0 \text{ or } x_j^i(t) = x_\ell^i(t)\}$. Now, $r_{min}^i \in C_i(t)$ is the robot with minimal $y_{min}^i(t)$ among all robots with $y_{min}^i(t) > 0$. Thus, r_{min}^i represents the robot lying above of r_i (from r_i's view) that has the smallest y-coordinate among all robots in $C_i(t)$. If no such robot exists, define $y_{min}^i = \frac{1}{10}$. Algorithm 1 describes the movement of a robot r_i.

Algorithm 1. $\mathcal{OBLOT}$ MAX-LINE-FORMATION

1: **if** $x_r^i(t) = 0$ and $x_\ell^i(t) < 0$ **then** ▷ Check if r_i is rightmost but not leftmost
2: **if** no robot is located on $(x_\ell^i(t), 0)$ **then**
3: $p_i(t+1) \leftarrow (x_\ell^i(t), 0)$ ▷ r_i can move safely to the left
4: **else**
5: $p_i(t+1) \leftarrow (x_\ell^i(t), \frac{1}{3} \cdot y_{min}^i)$ ▷ r_i avoids a collision with a vertical movement
6: **else**
7: **if** $x_r^i(t) = 0$ and $x_\ell^i(t) = 0$ **then** ▷ Check if neighbors are collinear
8: **if** $y_+^i(t) = 0$ and $y_-^i(t) < 0$ **then** ▷ Check if r_i is top most
9: $v_-(t) \leftarrow p_-^i(t) - p_i(t)$; $p_v(t) \leftarrow p_i(t) - \widehat{v}_-(t)$ ▷ Position of virtual robot
10: $p_i(t+1) \leftarrow \frac{1}{2}p_-(t) + \frac{1}{2}p_v(t)$
11: **else if** $y_+^i(t) > 0$ and $y_-^i(t) = 0$ **then** ▷ Check if r_i is bottom most
12: $v_+(t) \leftarrow p_+^i(t) - p_i(t)$; $p_v(t) \leftarrow p_i(t) - \widehat{v}_+(t)$ ▷ Position of virtual robot
13: $p_i(t+1) \leftarrow \frac{1}{2}p_+(t) + \frac{1}{2}p_v(t)$
14: **else**
15: $p_i(t+1) \leftarrow \frac{1}{2}p_-(t) + \frac{1}{2}p_+(t)$
16: r_i moves to $p_i(t+1)$

4.3 Analysis

Next, we introduce the analysis idea to prove the main theorem (Theorem 2) about the $\mathcal{OBLOT}$ algorithm. All proofs can be found in [4].

Theorem 2. *For every $0 < \varepsilon < 1$, after $\mathcal{O}(n^2 \cdot \log(n/\varepsilon))$ epochs, the robots have formed a line of length at least $(1-\varepsilon) \cdot (n-1)$.*

First, we argue that the first phase of the algorithm ends after $\mathcal{O}(n^2)$ rounds.

Lemma 1. *After $\mathcal{O}(n^2)$ epochs, all robots are located on distinct positions on the same vertical line parallel to the y-axis. Moreover, the configuration is connected.*

Now, we can assume that the first phase is completed, and thus all robots are located on the same vertical line. W.l.o.g., we rename the robots such that $y_1(t) \leq y_2(t) \leq ... \leq y_n(t)$. Moreover, define $w_1(t) = 1$ and $w_i(t) = y_i(t) - y_{i-1}(t)$ for $2 \leq i \leq n$. In addition, define $z_i(t) = w_i(t) - w_1(t)$. The algorithm is designed such that $\lim_{t \to \infty} w_i(t) = 1$ for all i. To analyze this behavior, we consider the following function: $\Phi(t) = \sum_{i=2}^{n} z_i(t)^2$. The function $\Phi(t)$ is also known as the *sample variance* [19]. The name comes from a relation to time inhomogeneous Markov chains. Although the algorithm is deterministic, the behavior of the vectors $w_i(t)$ can be interpreted as a time inhomogeneous Markov Chain. The main course of our analysis is based on [19], where the authors analyzed a similar behavior in the context of the distributed averaging consensus problem. In this problem, there are n agents, each having a numerical opinion. Every round, an agent gets to know some other opinions and updates its opinion to the average. Our application has one important difference: the values $w_i(t)$ do not average but converge to the fixed value $w_1(t)$. Hence, many parts of the proof in [19] have to be reworked and adapted to our application. First, we derive a bound on the change of $\Phi(t)$ between two epochs. Define $w_{\pi_1}(t_{e_k}), \ldots, w_{\pi_n}(t_{e_k})$ to be the values $w_i(t_{e_k})$ sorted from largest to smallest with ties broken arbitrarily.

Lemma 2. $\Phi(t_{e_k}) - \Phi(t_{e_{k+1}}) \geq \frac{1}{4} \sum_{i=1}^{n-1} \left(w_{\pi_i}(t_{e_k}) - w_{\pi_{i+1}}(t_{e_k})\right)^2$, *for any epoch* k.

Based on Lemma 2, a lower bound on the relative change is derived.

Lemma 3. *Suppose that* $\Phi(t_{e_k}) > 0$. *Then,* $\frac{\Phi(t_{e_k}) - \Phi(t_{e_{k+1}})}{\Phi(t_{e_k})} \geq \frac{1}{8n^2}$.

A combination of Lemmas 2 and 3 yields the statement of Theorem 2.

5 $\mathcal{LUMI}$ Algorithms

In this section, we derive an algorithm that solves MAX-LINE-FORMATION *exactly* with the help of the $\mathcal{LUMI}$ model under the $\mathcal{F}$SYNC scheduler. The algorithm (Algorithm 2) achieves an optimal runtime of $\Theta(n)$ rounds and works in two phases: In the first phase, all robots are arranged on a straight line parallel to the y-axis, and in the second phase, the line is stretched until it has maximal length. Compared to the $\mathcal{OBLOT}$ algorithm (Sect. 4), the algorithm uses different core ideas in both phases. In the first phase, all robots (instead of only the rightmost ones of their neighborhood) move to the left – this is necessary to achieve a linear speedup of the first phase. The second phase makes use of lights to implement a sequential movement denoted as a *run* inspired by [1,5,8,17]. Due to space constraints, we present a variant of the algorithm in which the robots still move to the left during the second phase. More precisely, after a linear number of rounds, the first phase ends, and the robots form a line parallel to the y-axis that continuously moves a distance of 1 to the left. Simultaneously, the robots stretch the line until it has maximal length. However, the line structure is always maintained such that MAX-LINE-FORMATION is solved finally and remains solved (although the line keeps moving to the left). Moving continuously to the left can be removed from the algorithm with some additional effort; an intuition is given in [4].

Phase 1: *All* robots move as far as possible to the left: each robot r_i moves to the x-coordinate $x_r^i(t) - 1$. Again, collision avoidance has to be ensured. While moving to $x_r^i(t) - 1$, the robot r_i could collide with every robot located on its local x-axis. The robot r_i executes a vertical movement to avoid a collision. Based on the ordering of neighbors on the local x-axis, r_i gets assigned a unique y-coordinate as follows: Define $Y_i(t) = \{r_j \in N_i(t) \mid y_j^i(t) = 0\}$ and let $x_{\pi_1}(t), x_{\pi_2}(t), \ldots, x_{\pi_{|Y_i(t)|}}(t)$ be the x-coordinates of robots in $Y_i(t)$ in increasing order. Additionally, let $k_i(t) \in \{1, \ldots, |Y_i(t)|\}$ denote the position of $x_i(t)$ in the sorted sequence $x_{\pi_1}(t), x_{\pi_2}(t), \ldots, x_{\pi_{|Y_i(t)|}}(t)$. Furthermore, define $y_{min}^i(t)$ to be the minimal $y_j^i(t)$ of all $y_j^i(t) > 0$ of robots $r_j \in N_i(t)$. If no such robot exists, define $y_{min}^i(t) = \frac{1}{10}$ (any constant of size at most 1 works). Then, r_i gets assigned the y-coordinate $\frac{k_i(t)-1}{|Y_i(t)|} \cdot \frac{1}{3} y_{min}^i(t)$. The factor $\frac{k_i(t)-1}{|Y_i(t)|}$ is unique for every robot on the local x-axis and the factor of $\frac{1}{3}$ is needed to prevent a collision with other robots that execute the same collision avoidance.

Phase 2: For the second phase, lights are used. Assume w.l.o.g. that the robots are ordered along the y-axis, i.e., $y_1(t) \geq \cdots \geq y_n(t)$. The core idea is a sequential movement started at r_1 and r_n implemented with lights. Such a movement is called a *run* [1,5,8,17]. Assume that a run starts in round t. Then, only r_1 and r_n move. In round $t+1$, only r_2 and r_{n-1} move and so on. A new run is started every three rounds. Runs are realized with lights as follows. The first required light ℓ_c with color set $C_c = \{0, 1, 2\}$ is used as a round counter. Every round, all robots increment their light ℓ_c. Whenever $\ell_c = 2$ holds, both r_1 and r_n activate a light ℓ_{mov} with $C_{mov} = \{0, 1\}$ (the light is either active or inactive). Thus, in the next round, it holds $\ell_c = 0$ and both r_1 and r_n detect an active light ℓ_{mov}. Both r_1 and r_n now execute a movement (see below). Additionally, they deactivate the light ℓ_{mov} and activate a light ℓ_{prev} with color set $C_{prev} = \{0, 1\}$ to remember the movement. Simultaneously, the robots r_2 and r_{n-1} observe a neighbor on the y-axis with active light ℓ_{mov} (r_1 and r_n). Additionally, neither r_2 nor r_{n-1} has activated ℓ_{prev}. Hence, the robots activate ℓ_{mov} to continue the run. In the next round, r_1 and r_n observe a neighbor with active light ℓ_{mov} but do not activate their own light ℓ_{mov} since ℓ_{prev} is active.

Robots that have a run (the light ℓ_{mov} is active) move as follows. In case r_1 has a run and not r_2 ($n > 2$), r_1 moves in distance 1 vertically away from r_2. More formally, $p_1(t+1) = (x_r^1(t)-1, -\frac{y_2^1(t)}{|y_2^1(t)|})$ (remember that in this variant the robots move also in phase 2 to the left). Similar, r_n moves away from r_{n-1} in distance 1. In case a robot r_i has a run that came from r_{i-1} (r_{i-1} has activated ℓ_{prev} and r_i has activated ℓ_{mov}) and r_{i+1} does not have a run, r_i moves in vertical distance 1 away from $r+1$: $p_i(t+1) = (x_r^i(t) - 1, -\frac{y_{i+1}^i(t)}{|y_{i+1}^i(t)|})$. Lastly, in case two neighboring robots have a run, for instance r_i and r_{i+1} have activated ℓ_{mov} both move only a vertical distance of $\frac{1}{2}$ away from each other: $p_i(t+1) = (x_r^i(t) - 1, -\frac{y_{i+1}^i(t)}{2|y_{i+1}^i(t)|})$. The handling of the lights and the corresponding movement is depicted in Fig. 3.

Algorithm 2. $\mathcal{LUMI}$ Algorithm $\mathcal{F}$SYNC executed from the local view of r_i

1: **if** all neighbors are located on the y-axis **then**
2: **if** $r_i = r_+^i(t)$ or $r_i = r_-^i(t)$ **then**
3: **if** $\ell_{mov} = 1$ **then** ▷ $\ell_{mov} = 1$ implies $\ell_c = 0$
4: $\ell_{mov} \leftarrow 0; \ell_{prev} \leftarrow 1$
5: $r_c \leftarrow$ closest neighbor on y-axis
6: **if** r_c has activated ℓ_{mov} **then** ▷ Special case $n = 2$
7: $p_i(t+1) \leftarrow (x_r^i(t) - 1, -\frac{1}{2 \cdot |y_c(t)|} \cdot y_c(t))$ ▷ Move distance of $\frac{1}{2}$
8: **else**
9: $p_i(t+1) \leftarrow (x_r^i(t) - 1, -\frac{1}{|y_c(t)|} \cdot y_c(t))$ ▷ Move distance of 1
10: **else**
11: **if** $\ell_{prev} = 1$ **then** ▷ Deactivate ℓ_{prev}
12: $\ell_{prev} \leftarrow 0$
13: **else**
14: **if** $\ell_c = 2$ **then** ▷ Start new run
15: $\ell_{mov} \leftarrow 1$
16: $p_i(t+1) \leftarrow (x_r^i(t) - 1, 0))$
17: **else**
18: **if** $\ell_{mov} = 1$ **then**
19: $\ell_{mov} \leftarrow 0$, $\ell_{prev} \leftarrow 1$
20: **if** closest neighbor above and below have set $\ell_{mov} = 0$ **then**
21: $r_c \leftarrow$ closest neighbor with $\ell_{prev} = 0$
22: $p_i(t+1) \leftarrow (x_r^i(t) - 1, -\frac{1}{|y_c(t)|} \cdot y_c(t))$
23: **else**
24: $r_c \leftarrow$ neighbor with $\ell_{mov} = 1$
25: $p_i(t+1) \leftarrow (x_r^i(t) - 1, -\frac{1}{2 \cdot |y_c(t)|} \cdot y_c(t))$
26: **else**
27: **if** $\ell_{prev} = 1$ **then**
28: $\ell_{prev} \leftarrow 0$
29: **else**
30: **if** closest neighbor above or below has set $\ell_{mov} = 1$ **then**
31: $\ell_{mov} \leftarrow 1$
32: $p_i(t+1) \leftarrow (x_r^i(t) - 1, 0))$
33: **else**
34: $\{\ell_{mov}, \ell_{prev}\} \leftarrow 0$ ▷ Deactivate lights if neighborhood is not in phase 2
35: **if** $|Y_i(t)| > 0$ **then**
36: $p_i(t+1) \leftarrow (x_r^i(t) - 1, \frac{k_i(t)-1}{|Y_i(t)|} \cdot \frac{1}{3} y_{min}^i(t))$
37: **else**
38: $p_i(t+1) \leftarrow (x_r^i(t) - 1, 0)$
39: $\ell_c \leftarrow \ell_c + 1$
40: r_i moves to $p_i(t+1)$

Analysis: In the analysis [4], it is proven that after a linear number of rounds, the first phase ends. As a part of the proof, it is proven that no collisions occur, and the connectivity is always maintained. Moreover, it is proven that as soon as phase 2 is reached, the robots remain in phase 2 (following from the algorithm's

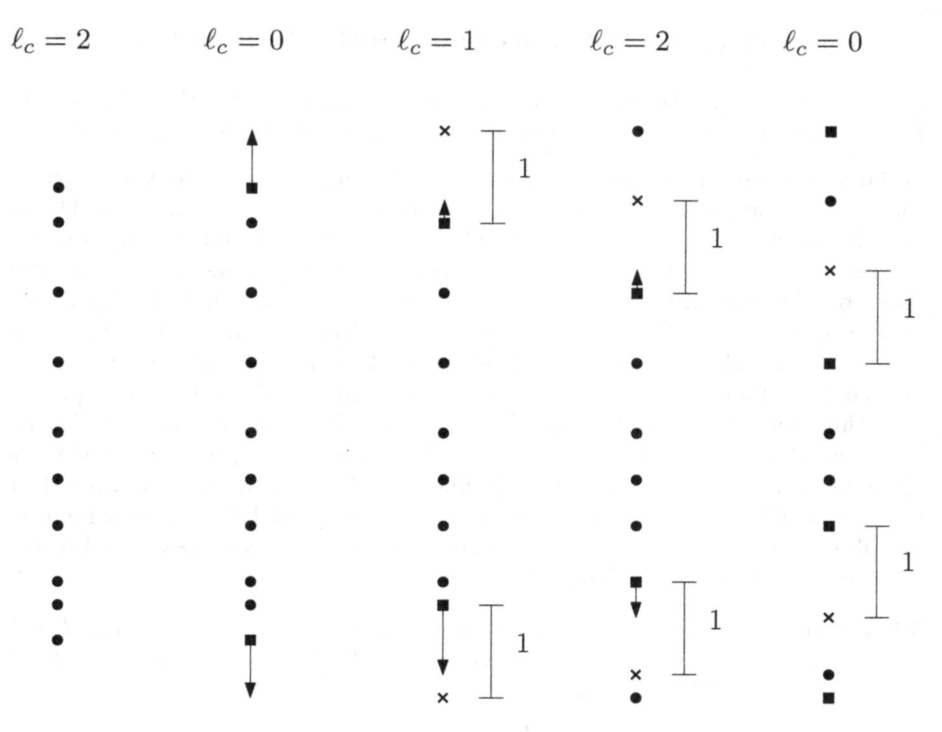

Fig. 3. A square (cross) depicts a robot with active light ℓ_{mov} (ℓ_{prev}). Time proceeds from left to right. In the first line it holds $\ell_c = 2$ for all robots. In this round, the topmost and the bottom-most robot activate ℓ_{mov}. In the next round ($\ell_c = 0$), these two robots move in distance 1 of their neighbor (depicted by an arrow) and additionally deactivate ℓ_{mov} while activating ℓ_{prev}. Afterward, the movement continues.

description). Afterward, the runs of the second phase are analyzed. The first run ensures that after $\mathcal{O}(n)$ rounds, the robots $r_{\lfloor n/2 \rfloor}$ and $r_{\lfloor n/2 \rfloor+1}$ have a vertical distance of 1. The second run ensures the same both for $r_{\lfloor n/2 \rfloor-1}$ and $r_{\lfloor n/2 \rfloor}$ as well as $r_{\lfloor n/2 \rfloor+1}$ and $r_{\lfloor n/2 \rfloor+2}$. Hence, after $\mathcal{O}(n)$ runs, the line reaches maximal length. Since each 3 rounds, a new run is started, the linear runtime follows.

Theorem 3. *After $\mathcal{O}(n)$ epochs, the robots have solved* MAX-LINE-FORMATION.

The algorithm can be implemented in the classical $\mathcal{LUMI}$ model with a single light having 9 colors. Observe that no robot ever activates the lights ℓ_{prev} and ℓ_{mov} at the same time. Thus, for each robot, it always holds: either ℓ_{prev}, ℓ_{mov} or none of both are activated. Additionally, each robot counts rounds with the light ℓ_c requiring 3 colors. Hence, the total number of required colors is 9: 3 colors of ℓ_c, each combined with 3 possible cases for the lights ℓ_{mov} and ℓ_{prev}.

6 Relation to GATHERING and CHAIN-FORMATION

Finally, we apply the main ideas of our algorithms for the MAX-LINE-FORMATION problem in the context of GATHERING and CHAIN-FORMATION.

Gathering: We consider robots in the $\mathcal{OBLOT}$ model that agree on one axis of their local coordinate systems and operate under the $\mathcal{F}$SYNC scheduler. Define Δ to be the maximal distance of two robots in the initial configuration in round t_0. The core idea of the GATHERING algorithm is as follows: to use the first phase of Algorithm 2 to arrange the robots on a vertical line fast. In this phase, every robot moves as far as possible to the left. While in Algorithm 2, collisions have to be avoided, this is not necessary for GATHERING since collisions are desired to gather all robots on a single point. In Sect. 5, it has been proven that this phase requires $\mathcal{O}(n)$ epochs. We show with a slightly more elaborate argument that this phase requires only $\mathcal{O}(\Delta)$ epochs. During the second phase of the algorithm, robots at the end of the line move half the distance towards their farthest neighbor. All other robots move to the midpoint between their farthest neighbor above and below. A pseudocode description, as well as a proof of the following theorem, can be found in [4].

Theorem 4. GATHERING *of* n *robots agreeing on one axis can be solved with a square viewing and connectivity range of* 1 *in* $\mathcal{O}(\Delta)$ *epochs under the* $\mathcal{F}$SYNC *scheduler in the* $\mathcal{OBLOT}$ *model.*

Chain-Formation: Lastly, we study the CHAIN-FORMATION problem that considers *disoriented* robots. The robots are connected in a chain topology: there are $n+2$ robots $r_0, r_1, \ldots, r_{n+1}$; both r_0 and r_{n+1} do not move. Every other robot r_i has exactly two chain neighbors: r_{i-1} and r_{i+1} whose positions it can always observe. The robots have a circular connectivity and viewing range of 1. Define by $w_i(t) = (w_i^x(t), w_i^y(t)) = p_i(t) - p_{i-1}(t)$ the vectors along the chain and $L(t) = \sum_{i=1}^{n+1} \|w_i(t)\|$. Additionally, $D = \|p_0(t) - p_{n+1}(t)\|$. The goal of the CHAIN-FORMATION problem is to move the robots such that $L(t) = D$ and to distribute the robots uniformly along the line segment between r_0 and r_{n+1}. W.l.o.g, assume that r_0 is positioned in the origin of a global coordinate system and r_{n+1} on the positive x-axis in distance D to r_0. Then, in the optimal configuration it holds $w_i(t) = w_\infty = \frac{D}{n+1}$ for all i. We say that an ε-approximation of the optimal configuration is reached in case $\|w_i(t) - w_\infty\| \leq \varepsilon$ for all $1 \leq i \leq n+1$. The GTM algorithm moves each robot in every round to the midpoint between its two direct neighbors [7,14]. The movement is similar to the second phase of the $\mathcal{OBLOT}$ algorithm (Algorithm 1) with the exception that the robots at the end of the line (r_0 and r_{n+1}) do not move instead of stretching the line. Nevertheless, we can apply a very similar analysis idea to the GTM algorithm: We prove convergence independently for $w_i^x(t)$ and $w_i^y(t)$. Since the arguments are identical, we concentrate on $w_i^x(t)$. Define $\overline{x} = \frac{1}{n+1} \cdot \sum_{i=1}^{n+1} w_i^x(t)$. Furthermore, define $z_i(t) = w_i^x(t) - \overline{x}$ and $\Phi_2(t) = \sum_{i=1}^{n+1} z_i(t)^2$. $\Phi_2(t)$ can be analyzed in most parts analogously to $\Phi(t)$ in Sect. 4. See [4] for a proof.

Theorem 5. *For every* $0 < \varepsilon < 1$, GTM *reaches an* ε*-approximation of the optimal configuration in* $\mathcal{O}(n^2 \cdot \log(n/\varepsilon))$ *epochs under the* $\mathcal{S}$SYNC *scheduler.*

References

1. Abshoff, S., Cord-Landwehr, A., Fischer, M., Jung, D., Meyer auf der Heide, F.: Gathering a closed chain of robots on a grid. In: IPDPS, pp. 689–699. IEEE Computer Society (2016)
2. Ando, H., Oasa, Y., Suzuki, I., Yamashita, M.: Distributed memoryless point convergence algorithm for mobile robots with limited visibility. IEEE Trans. Robotics Autom. **15**(5), 818–828 (1999)
3. Castenow, J., Fischer, M., Harbig, J., Jung, D., Meyer auf der Heide, F.: Gathering anonymous, oblivious robots on a grid. Theor. Comput. Sci. **815**, 289–309 (2020)
4. Castenow, J., Götte, T., Jung, D., Knollmann, T., Meyer auf der Heide, F.: The Max-Line-Formation Problem. CoRR abs/2109.11856 (2021). https://arxiv.org/abs/2109.11856
5. Castenow, J., Harbig, J., Jung, D., Knollmann, T., Meyer auf der Heide, F.: Gathering a euclidean closed chain of robots in linear time. In: Gąsieniec, L., Klasing, R., Radzik, T. (eds.) ALGOSENSORS 2021. LNCS, vol. 12961, pp. 29–44. Springer, Cham (2021). https://doi.org/10.1007/978-3-030-89240-1_3
6. Castenow, J., Kling, P., Knollmann, T., Meyer auf der Heide, F.: A discrete and continuous study of the MAX-CHAIN-FORMATION problem. In: Devismes, S., Mittal, N. (eds.) SSS 2020. LNCS, vol. 12514, pp. 65–80. Springer, Cham (2020). https://doi.org/10.1007/978-3-030-64348-5_6
7. Cohen, R., Peleg, D.: Local spreading algorithms for autonomous robot systems. Theor. Comput. Sci. **399**(1–2), 71–82 (2008)
8. Cord-Landwehr, A., Fischer, M., Jung, D., Meyer auf der Heide, F.: Asymptotically optimal gathering on a grid. In: SPAA, pp. 301–312. ACM (2016)
9. Degener, B., Kempkes, B., Kling, P., Meyer auf der Heide, F.: Linear and competitive strategies for continuous robot formation problems. ACM Trans. Parallel Comput. **2**(1), 2:1–2:18 (2015)
10. Degener, B., Kempkes, B., Langner, T., Meyer auf der Heide, F., Pietrzyk, P., Wattenhofer, R.: A tight runtime bound for synchronous gathering of autonomous robots with limited visibility. In: SPAA, pp. 139–148. ACM (2011)
11. Di Luna, G.A., Viglietta, G.: Robots with lights. In: Flocchini, P., Prencipe, G., Santoro, N. (eds.) Distributed Computing by Mobile Entities. LNCS, vol. 11340, pp. 252–277. Springer, Cham (2019). https://doi.org/10.1007/978-3-030-11072-7_11
12. Dutta, A., Gan Chaudhuri, S., Datta, S., Mukhopadhyaya, K.: Circle formation by asynchronous fat robots with limited visibility. In: Ramanujam, R., Ramaswamy, S. (eds.) ICDCIT 2012. LNCS, vol. 7154, pp. 83–93. Springer, Heidelberg (2012). https://doi.org/10.1007/978-3-642-28073-3_8
13. Dynia, M., Kutyłowski, J., Lorek, P., auf der Heide, F.M.: Maintaining communication between an explorer and a base station. In: Pan, Y., Rammig, F.J., Schmeck, H., Solar, M. (eds.) BICC 2006. IIFIP, vol. 216, pp. 137–146. Springer, Boston, MA (2006). https://doi.org/10.1007/978-0-387-34733-2_14
14. Dynia, M., Kutylowski, J., Meyer auf der Heide, F., Schrieb, J.: Local strategies for maintaining a chain of relay stations between an explorer and a base station. In: SPAA, pp. 260–269. ACM (2007)

15. Flocchini, P., Prencipe, G., Santoro, N. (eds.): Distributed Computing by Mobile Entities, Current Research in Moving and Computing. LNCS, vol. 11340. Springer, Cham (2019)
16. Kling, P., Meyer auf der Heide, F.: Convergence of local communication chain strategies via linear transformations. In: SPAA, pp. 159–166. ACM (2011)
17. Kutylowski, J., Meyer auf der Heide, F.: Optimal strategies for maintaining a chain of relays between an explorer and a base camp. Theor. Comput. Sci. **410**(36), 3391–3405 (2009)
18. Mondal, M., Gan Chaudhuri, S.: Uniform circle formation by swarm robots under limited visibility. In: Hung, D.V., D'Souza, M. (eds.) ICDCIT 2020. LNCS, vol. 11969, pp. 420–428. Springer, Cham (2020). https://doi.org/10.1007/978-3-030-36987-3_28
19. Nedic, A., Olshevsky, A., Ozdaglar, A.E., Tsitsiklis, J.N.: On distributed averaging algorithms and quantization effects. IEEE Trans. Autom. Control **54**(11), 2506–2517 (2009)
20. Poudel, P., Sharma, G.: Universally optimal gathering under limited visibility. In: Spirakis, P., Tsigas, P. (eds.) SSS 2017. LNCS, vol. 10616, pp. 323–340. Springer, Cham (2017). https://doi.org/10.1007/978-3-319-69084-1_23

Message Delivery in the Plane by Robots with Different Speeds

Jared Coleman[1(✉)], Evangelos Kranakis[2], Danny Krizanc[3], and Oscar Morales Ponce[4]

[1] Department of Computer Science, University of Southern California, Los Angeles, CA, USA
jaredcol@usc.edu

[2] School of Computer Science, Carleton University, Ottawa, ON, Canada

[3] Department of Mathematics and Computer Science, Wesleyan University, Middletown, CT, USA

[4] Department of Computer Science, California State University, Long Beach, USA

Abstract. We study a fundamental cooperative message-delivery problem on the plane. Assume n robots which can move in any direction, are placed arbitrarily on the plane. Robots each have their own maximum speed and can communicate with each other face-to-face (i.e., when they are at the same location at the same time). There are also two designated points on the plane, S (the *source*) and D (the *destination*). The robots are required to transmit the message from the source to the destination as quickly as possible by face-to-face message passing. We consider both the offline setting where all information (the locations and maximum speeds of the robots) are known in advance and the online setting where each robot knows only its own position and speed along with the positions of S and D.

In the offline case, we discover an important connection between the problem for two-robot systems and the well-known Apollonius circle which we employ to design an optimal algorithm. We also propose a $\sqrt{2}$ approximation algorithm for systems with any number of robots. In the online setting, we provide an algorithm with competitive ratio $\frac{1}{7}\left(5+4\sqrt{2}\right)$ for two-robot systems and show that the same algorithm has a competitive ratio less than 2 for systems with any number of robots. We also show these results are tight for the given algorithm. Finally, we give two lower bounds (employing different arguments) on the competitive ratio of any online algorithm, one of 1.0391 and the other of 1.0405.

Keywords: Delivery · Face-to-Face · Plane · Pony express · Robot · Speed

1 Introduction

We study the problem of delivering a message in minimum time from a source to a destination using autonomous mobile robots with different maximum speeds. We

E. Kranakis—Research supported in part by NSERC Discovery grant.

C. Johnen et al. (Eds.): SSS 2021, LNCS 13046, pp. 305–319, 2021.
https://doi.org/10.1007/978-3-030-91081-5_20

extend the work on this communication problem studied previously on graphs [1, 3,8,9]. In our setting, the robots are initially distributed in arbitrary locations in the plane and the locations of the source and destination are known by all. The robots may move with their own (maximum) speed. Robots cooperate by exchanging information (the message) using face-to-face (F2F) communication. We study message transmission and allow messages to be replicated (as opposed to package delivery). The goal is to give an algorithm which minimizes the time required to deliver the message from the source to the destination through a series of F2F message transfers. In this paper we study how to complete this task efficiently and propose various centralized offline and distributed online algorithms which take into account the knowledge that the robots have about their speeds and initial locations.

1.1 Model, Notation and Terminology

The setup of our pony express problem will be in the Euclidean plane and points will be identified with their cartesian coordinates. We use capital letters to denote points and lower-case letters with subscripts to denote their components (e.g. point $A = (a_1, a_2)$). For any points A, B, C, $|AB|$ denotes the Euclidean distance between A and B, $\angle(ABC)$ denotes the angle formed by A, B, C in this order, and $\triangle(ABC)$ denotes the triangle formed by A, B, C. Finally, $C(A, r)$ denotes a circle centered at A with radius r.

Assume that n robots $r_1, r_2, \ldots, r_n$ are placed at arbitrary positions in the Euclidean plane. The respective speeds of the robots are $v_1, v_2, \ldots, v_n$. The movement of a robot is determined by a well-defined trajectory. A robot trajectory is a continuous function $t \to f(t)$, with $f(t)$ the location of the robot at time t, such that $|f(t) - f(t')| \leq v|t - t'|$, for all t, t', where v is the speed of the robot. A robot can move with its own constant speed and during the traversal of its trajectory it may stop and/or change direction instantaneously and at any time. Robot communication is F2F in that two robots can communicate (instantaneously) with each other only when they are co-located.

Algorithms describe the trajectories robots will follow and we will take into account the time it takes the algorithm to conclude the delivery task from the start, obtaining the message at a given source S, and eventually delivering it to a given destination D. In general, we are interested in offline and online algorithms. In the offline setting, the locations and speeds of all robots are known in advance and are available to a central authority that assigns trajectories to the robots. In the online setting, the robots know only their own initial position and speed, along with the positions of S and D. To measure the performance of our online algorithms, we consider their competitive ratio defined as follows. Let $t^*(I)$ be the optimal delivery time for an instance I of a given problem and $t_A(I)$ be the time needed by some online algorithm A for the same instance. The competitive ratio of A is $\max_I \frac{t_A(I)}{t^*(I)}$. Our goal is to find online algorithms that minimize this competitive ratio.

1.2 Related Work

Communicating mobile robots or agents have been used to address problems such as search, exploration, broadcasting and converge-casting, patrolling, connectivity maintenance, and area coverage (see [11]). For example, [6] addresses the problem of how well a group of collaborating robots with limited communication range is able to monitor a given geographical space. To this end, they study broadcasting and coverage resilience, which refers to the minimum number of robots whose removal may disconnect the network and result in uncovered areas, respectively. Similarly, rendezvous is a relevant communication paradigm and in [13,18] the authors investigate rendezvous in a continuous domain under the presence of spies. A related study on message transmission in a synchronized multi-robot system may be found in [6]. Another application is patrolling whereby mobile robots are required to keep perpetually moving along a specified domain so as to minimize the time a point in the domain is left unvisited by an agent, e.g., see [17] for a related survey.

Data delivery and converge-cast with energy exchange under a centralized scheduler were studied in [16]. A restricted version concerns n mobile agents of limited energy that are placed on a straight line and which need to collectively deliver a single piece of data from a given source point S to a given target point D on the line can be found in [10]. In [12] it is shown that deciding whether the agents can deliver the data is (weakly) NP-complete. Additional research under various conditions and topological assumptions can be found in [4] which studies the game-theoretic task of selecting mobile agents to deliver multiple items on a network and optimizing or approximating the total energy consumption over all selected agents, in [2,5,7] which study data delivery and combine energy and time efficiency, and in [19,20] which are concerned with collaborative exploration in various topologies.

Our problem was previously studied on graphs in [1,3,8,9]. In particular it is shown in [8] that the problem can be solved with k agents on an n-node, m-edge weighted graph in time $O(kn \log n + km)$. We use this algorithm in the development of our approximation algorithm.

Our current work is related to the Pony Express communication problem proposed in [15]. In that paper, the authors provide both optimal offline and online algorithms for the anycast and broadcast problems in the case where the underlying domain was a continuous line segment. To our knowledge, the planar case studied in our paper has not been considered previously.

1.3 Outline and Results of the Paper

In Sect. 2 we propose an optimal offline algorithm for two robots. For ease of exposition, we first consider the case when the slower robot starts at the source and then the general case of arbitrary starting positions. In Sect. 3 we study the offline multirobot case. We propose an algorithm which approximates the optimal delivery time to within a factor of $\sqrt{2}$. Section 4 is dedicated to online algorithms. For two robots we give an algorithm with competitive ratio of $\frac{1}{7}\left(5 + 4\sqrt{2}\right)$ and show that for n robots, this same algorithm has a competitive ratio of at most 2. We also analyze lower bounds for this specific algorithm showing these bounds are tight. In Sect. 5 we prove lower bounds on the competitive ratio of arbitrary online algorithms. We discuss two approaches, one where the position of a robot is unknown and the other where the speed of a robot is unknown. These different approaches provide lower bounds of 1.0391 and 1.0405, respectively. We conclude in Sect. 6 by discussing possibilities for additional research in this area. All missing proofs can be found in the complete version of the paper [14].

2 Optimal Offline Algorithm for Two Robots

In this section, we will consider two robots r_v and r_1 which can move with respective constant speeds v and 1 ($v > 1$) and design optimal offline algorithms with respect to the F2F communication model (observe that by scaling the distances, setting the speed of the slow robot to be 1 yields no loss of generality.) Let L and K be the starting positions of robots r_1 and r_v, respectively. There are three cases to consider:

1. $\frac{|KS|}{v} \leq |LS|$: the fast robot can get to S before the slow robot. In this case, it is clear that in the optimal solution, the fast robot should move to S, acquire the message, and carry it to D.
2. $\frac{|KD|}{v} \geq |LS| + |SD|$: the slow robot can deliver the message to D before the fast robot can even reach D. In this case, the optimal solution is also clear. The slow robot should move to S, acquire the message, and carry it to D.
3. In all other cases, the slow robot can get to S before the fast robot, but the fast robot can get to the destination faster. The optimal solution, then, must involve a handover between the robots at some point M in the plane.

For the first two cases, the optimal solution is trivial. The third case, however, is not as we must find the point at which the robots meet. First, we characterize the optimal meeting point M for Case 3 through a series of lemmas.

Lemma 2.1. *For Case 3, there exists an optimal solution such that if M is the handover, then* $|LS| + |SM| = \frac{|KM|}{v}$.

Intuitively, Lemma 2.1 says that robots must move at their maximum speeds directly towards the location they will acquire the message and then directly toward the location they handover or deliver the message. This restricts the set of feasible meeting points to the set of points in the plane such that both robots, moving in one direction at their maximum speeds, meet at the same time. For the case where the slow robot starts at the source ($L = S$), this is directly related to an ancient theorem by the Greek philosopher Apollonius, which states "the trajectory traced by a point P which moves in such a way that its Euclidean distance from a given point S is a constant multiple of its Euclidean distance from another point K is a circle" [21]. As a consequence, if the robots r_1, r_v start at positions S, K, respectively, then the locus of points at which the two robots may travel directly towards and meet at the same time is the circle of Apollonius (see Fig. 1). This circle, then is the locus of all possible handover points between the two robots. The precise statement in the context of mobile robots is stated in Lemma 2.2.

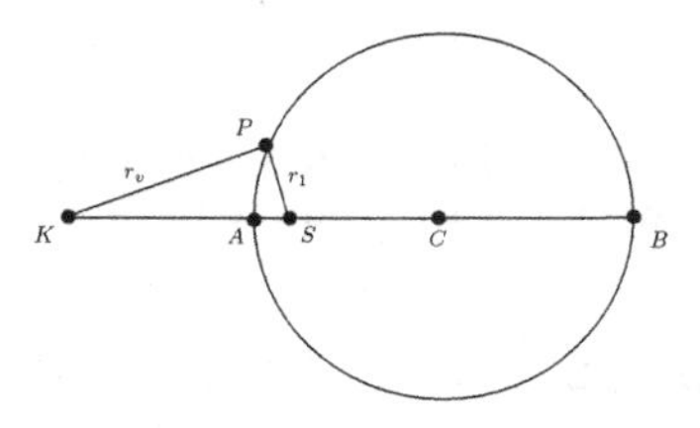

Fig. 1. The Apollonius circle is the locus of points P such that robots r_v and r_1 are equal time away from their starting positions K and S, respectively.

Lemma 2.2. *Two robots r_v and r_1 with speeds v and 1 $(v > 1)$ are initially placed at points K and S, respectively. The locus of points P such that robots r_v and r_1 are equal time away from points K and S, respectively, i.e., $\frac{|PK|}{|PS|} = v$, forms a circle with center C and radius R so that*

$$C = S + \frac{S - K}{v^2 - 1} \text{ and } R = \frac{v|SK|}{v^2 - 1} \tag{1}$$

The following definition of the Apollonius Circle will be used throughout this paper.

Definition 1 (Apollonius Circle). *The circle with center C and radius R given by Eqs. (1) is called the Apollonius circle between robots r_v and r_1 when their respective starting positions are K, S.*

For instances of the problem where $L = S$ and whose optimal solutions do not involve either robot delivering the message by itself, the previous discussion results in the following lemma whose proof follows directly from Lemmas 2.1 and 2.2.

Lemma 2.3. *The optimal meeting point M is the point on the Apollonius circle between robots r_1 and r_v which minimizes the total delivery time $|SM| + \frac{|MD|}{v} = \frac{|KM|+|MD|}{v}$.* □

2.1 Optimal Algorithm When a Robot Starts at the Source

First we give an algorithm in the restricted case where one robot starts at the source where the message is located ($L = S$). Let $S = (0, 0)$ be the source, K be the starting position of the fast robot which we assume to be on the x axis, and $D = (x, y)$ the destination. Without loss of generality, we assume $y \geq 0$ (if $y < 0$, the instance can be reflected about the x axis and solved equivalently, since K

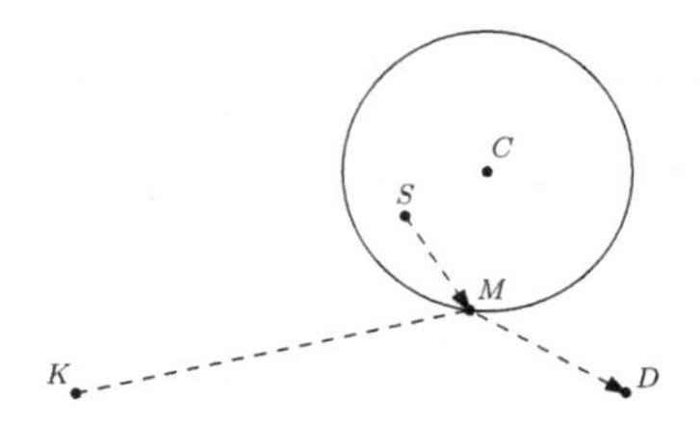

Fig. 2. The two-robot delivery problem with a slow robot at S, a faster robot at K, and an Apollonius circle between the two centered at C.

is on the x axis). By Lemma 2.3, our goal is to find the point M on the robots' Apollonius circle which minimizes the delivery time $|SM| + \frac{|MD|}{v} = \frac{|KM|+|MD|}{v}$ (see Fig. 2).

Consider the following offline algorithm for computing the optimal delivery time.

Algorithm 1. Optimal Two-Robot Algorithm with the Slow Robot Starting at the Source

1: **if** $\frac{|KD|}{v} \geq |SD|$ **then**
2:     **return** $|SD|$
3: $\beta \leftarrow \angle SKD$
4: **if** $\sin(\beta) \leq \frac{1}{v}$ **then**
5:     $\alpha \leftarrow \pi - \beta - \arcsin(v \sin\beta)$
6:     $M \leftarrow \frac{|SK|}{v^2-1}(v\cos\alpha - 1, v\sin\alpha)$
7:     **if** $|KD| < |KM|$ **then**
8:         $M \leftarrow$ point on Apollonius circle such that CM bisects the angle $\angle(DMK)$
9: **else**
10:     $M \leftarrow$ point on Apollonius circle such that CM bisects the angle $\angle(DMK)$
11: **return** $\frac{|KM|+|MD|}{v}$

Theorem 2.1. *Algorithm 1 returns the optimal delivery time for instances with two robots r_1 and r_v with speeds 1 and v, and starting positions S and K, respectively. Algorithm 1 can be implemented using a constant number of operations (including trigonometric functions).*

Proof. First, note that Case 1 (from the three cases at the beginning of the section) is not considered since the slow robot, r_1 is assumed to start at the source. Case 2 is obviously handled by line 1 in the algorithm. Case 3 is divided into two subcases based on whether or not the condition in line 4 is satisfied. First, we consider the case where it is not. Let β be the angle $\angle SKD$. Observe that if KD is tangent to the Apollonius circle (Fig. 3), then $\sin(\beta) = \frac{|SK|v}{v^2-1} \cdot \frac{v^2-1}{|SK|v^2} = \frac{1}{v}$. Clearly for any smaller value for β, KD intersects the Apollonius circle at two points (and for any larger value, KD does not intersect the circle).

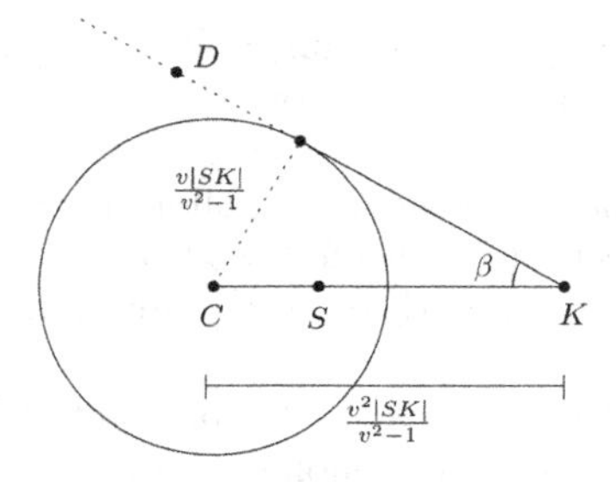

Fig. 3. The maximum β such that KD intersects the Apollonius Circle

Then, let $\alpha = \angle KCM$ and $\gamma = \angle KMC$ (Fig. 4 left). By the law of sines $\frac{(v^2-1)\sin\gamma}{|SK|v^2} = \frac{(v^2-1)\sin\beta}{|SK|v}$ and $\gamma = \arcsin(v\sin\beta)$. Thus $\alpha = \pi - \beta - \arcsin(v\sin\beta)$ and $M = \frac{|SK|}{v^2-1}(v\cos\alpha - 1, v\sin\alpha)$ is just the associated point on the Apollonius circle.

Observe M is the intersection point closest to K. Since the condition in line 4 is satisfied, r_v can move directly toward D and, without veering from a direct path, meet r_1 at M, acquire the message, and continue towards D to deliver the message. We know, since the first case was not satisfied, that r_v can reach the destination before r_1 can, so this is clearly the optimal trajectory. Now, suppose the condition in line 4 is not satisfied. Consider the ellipse with foci K and D whose semi-major axis has length $\frac{1}{2}vt$ for some time $t \geq 0$. Then, by a defining property of an ellipse, the sum of the distances from each foci to a point on the ellipse is equal to a constant value vt. Consequently, a robot starting at K with speed v takes exactly t time to travel to a point on the ellipse and then to D. Observe that if the ellipse and Apollonius circle intersect, then the two robots can meet at one of the intersection points and, by the previous statement, the fast robot can deliver the message in time t. If they intersect at two points,

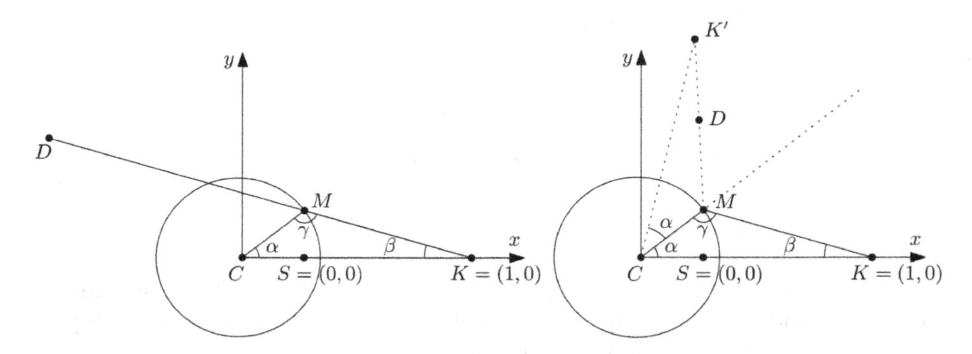

Fig. 4. On the left, D is such that the line-segment KD intersects with the Apollonius Circle. In the general case (right), CM must bisect $\angle(DMK)$, and thus α such that MK' be collinear to MD, where $K' = |CK|(\cos(\alpha), \sin(\alpha))$.

though, then any point on the Apollonius circle between these two intersections would yield a better solution. The solution, then, is to find the minimal t which causes the Apollonius circle and the ellipse to intersect at exactly one point M. Thus CM must be normal to both the Apollonius circle and the ellipse. That CM, therefore, must bisect $\angle(DMK)$ follows from a well-known property of the ellipse, namely that a normal line through a point on an ellipse bisects the angle it forms with the ellipse's foci.

Next, we show the algorithm can be implemented to run using a constant number of operations (including trigonometric functions). The only lines in the algorithm that are not clearly computable with a constant number of operations are lines 8 and 10. To show that M can be computed in constant time, we provide a formulation which can be given as input to Equation Solving tools (e.g., Mathematica) to find a closed-form solution[1]. Let $\alpha = \angle KCM$ and K' be the point given by rotating K 2α around C (into the positive half-plane, Fig. 4 right). Observe that if CM bisects $\angle(DMK)$, then DK' is collinear with MD, or: $\frac{|SK|\cos(2\alpha)-|CS|-x}{|SK|\sin(2\alpha)-y} = \frac{x-\cos\alpha-|CS|}{y-\sin\alpha}$ where $D = (x, y)$. □

2.2 Optimal Algorithm in the General Case

In this subsection we consider the more general case where the slow robot does not start at the source. Let the starting positions of source and destination be $S = (s_1, s_2)$ and $D = (d_1, d_2)$ and let the robots r_v and r_1 start from arbitrary points $K = (k_1, k_2)$ and $L = (l_1, l_2)$ in the plane, respectively. Again, we are interested in finding the point $M = (x_1, x_2)$ for the third case (from the cases at the beginning of the section), since optimal solutions for the first two cases are trivial to find. As depicted in Fig. 5, robot r_v follows a trajectory which first visits a point Q at distance $v|LS|$ from its starting position, then continues along a straight-line trajectory to meet robot r_1 at a suitable point $M = (x_1, x_2)$, and finally delivers the message to the destination D. The main steps of the algorithm are as follows.

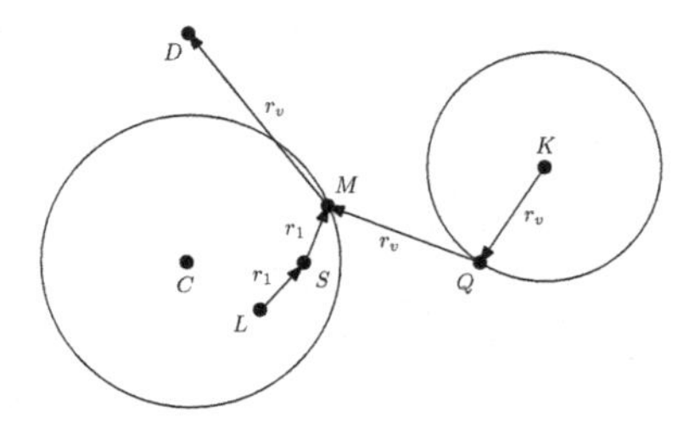

Fig. 5. Trajectories of the robots for message delivery from S to D. Robot r_v starts at the point K and robot r_1 at the point L. Robot r_1 arrives at the source S before r_v does and meets robot r_v at M which then delivers the message to M.

[1] Link to Mathematica solution for Theorem 2.1.

1. If $\frac{|KS|}{v} \leq |LS|$, then r_v reaches S before r_1 and r_v should complete the delivery on its own.
2. Otherwise, if $|LS| + |SD| \leq |KD|$, then r_1 can deliver the message on its own before r_v can even reach the destination.
3. Otherwise r_1 reaches S in time $|LS|$ and, at the same time, robot r_v goes to a specially selected point $Q = (q_1, q_2)$ which lies on the circle centered at K with radius $v|LS|$.
4. Robot r_v meets robot r_1 at a point $M = (x_1, x_2)$ determined by the locus of points which are equal time away from Q and S (by Lemma 2.2, this is the circle with center C and radius R as given in Eq. (1)). Robot r_1 passes message to r_v which delivers it to D.

Observe that by Lemma 2.1, K, Q, and M must be collinear. We can then generalize the result of Sect. 2.1 using the following lemma. Recall that the center of similitude (also known as homothetic center) is a point from which at least two geometrically similar figures can be seen as a dilation or contraction of one another (see [22][Section 1.1.2]).

Lemma 2.4. *Let C be the center of the Apollonius circle between r_1 and r_v when r_1 is at S and r_v is at K. Then, S is the center of similitude of the circles $\mathcal{C}(K, v|LS|)$ and $\mathcal{C}(C, v|LS|/(v^2-1))$. Consider any point Q in the circumference of $\mathcal{C}(K, v|LS|)$. Let β be the angle $\angle(SKQ)$, then $C_\beta = (\frac{v|LS|}{v^2-1}\cos\beta, \frac{v|LS|}{v^2-1}\sin\beta) +$ C is the center of the Apollonius circle of S.*

Consider two points C_β and Q as described in Lemma 2.4, for some β. We can now use Theorem 2.1 to characterize the solution. However, this approach does not lead to a closed-form solution. Instead, in the following lemma, we present another approach using optimization which does.

Lemma 2.5. *Let $a = |LS|$. Then the optimal trajectory is obtained by a point $M = (x_1, x_2)$ which minimizes the objective function*

$$\sqrt{(k_1 - x_1)^2 + (k_2 - x_2)^2} + \sqrt{(x_1 - d_1)^2 + (x_2 - d_2)^2} \tag{2}$$

subject to the condition

$$\left(\frac{(x_1 - k_1)^2 + (x_2 - k_2)^2}{2av^2} - \frac{(x_1 - s_1)^2 + (x_2 - s_2)^2}{2a} - \frac{a}{2}\right)^2 = (x_1 - s_1)^2 + (x_2 - s_2)^2. \tag{3}$$

The resulting optimization problem has two unknowns x_1, x_2 in the objective function (2) and must satisfy the condition of Eq. (3). It can be used to substitute variables and express the final optimization function described in Formula (2) using only a single variable, say x_1, which can then be minimized using standard analytical methods. This is easily seen since Eq. (3) is of degree 4 in the variable x_2 (as well as in the variable x_1, for that matter). Therefore a closed form expression of the variable x_2 in terms of the variable x_1 and the known parameters S, D is easily derived.

There are two symmetries in Eq. (3) which simplify the objective function and make the calculation of the solution easier. They are easily revealed with simple geometric transformations.

For the first symmetry, consider a rotation of the axis and a translation of the entire configuration of points so that K and S lie on the horizontal axis, i.e., $(k_1, k_2) = (0, 0)$ and $(s_1, s_2) = (s_1, 0)$. Then Eq. (3) is transformed to the equation

$$\left(\frac{x_1^2 + x_2^2}{2av^2} - \frac{(x_1 - s_1)^2 + x_2^2}{2a} - \frac{a}{2} \right)^2 = (x_1 - s_1)^2 + x_2^2 \tag{4}$$

The resulting symmetry is along the horizontal x_1-axis in Eq. (4). Namely, if (x_1, x_2) is a solution so is $(x_1, -x_2)$. If we consider Eq. (4) in the unknown x_2 we see that it is of degree 4, but which is also a quadratic in x_2^2. Therefore x_2 can be easily expressed as a function of x_1 using the formula for the roots of the quadratic equation. The second symmetry is obtained in a similar manner. If (x_1, x_2) is a solution so is $(-x_1, x_2)$. One considers a rotation of the axis and a translation of the entire configuration of points so that K and S lie on the vertical axis, i.e., $(k_1, k_2) = (0, 0)$ and $(s_1, s_2) = (0, s_2)$. Details can be completed as above. To sum up we have the following Algorithm 2 which determines the handover point which yields the optimal trajectory.

Algorithm 2. Optimal Two-Robot Algorithm

1: **if** $\frac{|KS|}{v} \leq |LS|$ **then**
2:     **return** $\frac{|KS|+|KD|}{v}$
3: **else if** $\frac{|KD|}{v} \geq |SD|$ **then**
4:     **return** $|SD|$
5: **else**
6:     $M^* \leftarrow M$ which minimizes Formula (2)
7:     **return** $\frac{|KM^*|+|M^*D|}{v}$

Theorem 2.2. *Algorithm 2 returns the optimal delivery time for two robots r_1 and r_v with speeds 1 and v, respectively, and can be implemented using a constant number of operations (including trigonometric functions).*

3 Offline $\sqrt{2}$ Approximation for Multiple Robots

In principle, the equations derived in the previous section can be generalized to solve the problem optimally for n robots. Unfortunately, we are not able to solve the resulting set of equations. We do not speculate on the complexity of the general problem here. Instead, in this section we provide a $\sqrt{2}$-approximation algorithm, The robots know the location of the source S and destination D but

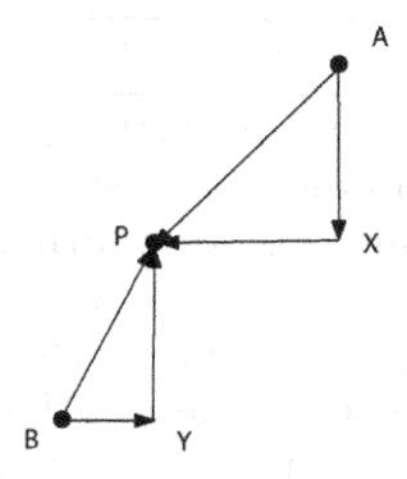

Fig. 6. Replacing Euclidean movements with rectilinear movements.

also all robots know the initial locations and speeds of all other robots. The basic idea of our proof is contained in the following observation depicted in Fig. 6. Suppose that during the execution of an optimal "Euclidean" algorithm (i.e., optimal in the sense of the Euclidean distance) two robots placed at A and B, follow the straight-line trajectories $A \to P$ and $B \to P$, respectively, and meet at the point P.

Now we replace the Euclidean trajectories $A \to P$ and $B \to P$ with the rectilinear trajectories $A \to X \to P$ and $B \to Y \to P$, respectively. Elementary geometry implies that

$$|AX| + |XP| \le \sqrt{2}|AP| \text{ and } |BY| + |YP| \le \sqrt{2}|BP|. \tag{5}$$

This observation leads to the following lemma.

Lemma 3.1. *Consider the pony express problem for n robots, a source S and a destination D in the plane. Then $Opt_{Rect} \le \sqrt{2} \cdot Opt_{Eucl}$, where Opt_{Rect}, Opt_{Eucl} are the delivery costs of the optimal trajectories of the pony express problem for delivering from a source to a destination measured in the rectilinear and Euclidean metrics, respectively.*

Consider n robots in the plane with starting positions $p_1, \ldots, p_n$. Without loss of generality assume the slowest robot has speed 1. Further, let the source of a message be located at a point S and the destination at a point D and assume, without loss of generality, that the line segment SD is horizontal. Enclose the points S, D and $p_1, \ldots, p_n$ in a $\Delta \times \Delta$ square with sides parallel to the $x-, y-$axis, where Δ is a positive real proportional to the diameter of the set $\{S, D\} \cup \{p_1, \ldots, p_n\}$. For $\epsilon > 0$ arbitrarily small, partition the $\Delta \times \Delta$ square with parallel vertical and horizontal lines with consecutive distances $\epsilon > 0$, respectively, so as to form a $\frac{\Delta}{\epsilon} \times \frac{\Delta}{\epsilon}$ grid. Without loss of generality we may assume that S and D are vertices in this grid graph (This is easy to accomplish by choosing ϵ to be an integral fraction of the distance $|SD|$ between S and D.) Clearly, this forms a grid graph with $\left(\frac{\Delta}{\epsilon}\right)^2$ vertices so that S, D are also vertices and $\left(\frac{\Delta}{\epsilon}\right)^2$ edges. Now consider the following algorithm.

Algorithm 3. Grid Algorithm (S source, D destination, $\epsilon > 0$)

1: In phase 1, each robot moves from its starting position p_i to one vertex p'_i of the $\epsilon \times \epsilon$ square in which it is contained; all the robots synchronize so that they can start the next phase at the same time by waiting for time at most ϵ;
2: In phase 2, run the optimal algorithm in [8] on the $\frac{\Delta}{\epsilon} \times \frac{\Delta}{\epsilon}$ grid to provide trajectories for the n robots with starting positions $p'_1, \ldots, p'_n$ in optimal time in order to deliver the message from the course S to the destination D; when a robot meets another robot for a message handover the robot that arrives first waits for the arrival of the second robot;

Theorem 3.1. *For any $\epsilon' > 0$ arbitrarily small, there exists an algorithm that finds trajectories for n robots to deliver the message from the source S to the destination D in time $O\left(n^3 \left(\frac{\Delta}{\epsilon'}\right)^2 \log\left(n\frac{\Delta}{\epsilon'}\right)\right)$ whose delivery time is at most $\sqrt{2}$ multiplied by the delivery time of the optimal Euclidean algorithm plus the additional additive overhead ϵ', where Δ is the diameter of the point set.*

4 Online Upper Bounds

In this section we discuss online algorithms. In Subsect. 4.1 we give an online algorithm with competitive ratio $\frac{1}{7}(5+4\sqrt{2})$ for two robots with knowledge only of the source S and destination D. We show this bound is tight for the given algorithm. In Subsect. 4.2 we show that the same algorithm when generalized to n robots has competitive ratio at most 2. Further, we show that for any $n > 2$, the competitive ratio of our algorithm is at least $2 - \frac{2}{2^n-1}$.

4.1 Two Robot Algorithm with Competitive Ratio $\frac{1}{7}(5 + 4\sqrt{2})$

Consider the following Algorithm 4 for multiple robots.

Algorithm 4. Online Algorithm (S source, D destination)

1: Move toward S
2: Acquire the message at S
3: Move toward and deliver the message to D

Observe that in this algorithm the robots act independently. In particular no attempt is made to co-ordinate the action of the robots and if two robots meet they ignore each other. This is not required in order to achieve the upper bounds below. For our lower bounds on this algorithm, we assume that the robots do not interact even if it may improve the time to complete the task.

Theorem 4.1. *For the case of two robots, Algorithm 4 delivers the message from the source S to the destination D in at most $\frac{1}{7}(5+4\sqrt{2})$ times the optimal offline time.*

Example 1. Now we give a tight lower bound on the competitive ratio of Algorithm 4 for two robots. Consider the following example input. One robot is placed at the source S which is the point $(0,0)$ and has speed $\frac{1}{1+\sqrt{2}}$. The destination D is placed at the point $(1,0)$. The second robot has speed 1 and is placed at the point $(\sqrt{2},0)$. The robots are initially placed at distance $\sqrt{2}$. In the optimal algorithm the robots meet in time $\frac{\sqrt{2}}{1+\frac{1}{1+\sqrt{2}}} = 1$ at the point $x = \frac{1}{1+\sqrt{2}}$. The faster robot picks up the message at x and delivers it to D in additional time $1 - x = 1 - \frac{1}{1+\sqrt{2}}$. Therefore the delivery time of the optimal algorithm is equal to $1 + 1 - x = 2 - \frac{1}{1+\sqrt{2}}$. It follows that the competitive ratio satisfies $c_2 \geq \frac{1+\sqrt{2}}{2-\frac{1}{1+\sqrt{2}}} = \frac{1}{7}(5+4\sqrt{2}) \approx 1.522407\ldots$.

Remark 1. Note that in Example 1 if we parametrize the speed of the slow robot to $\frac{1}{1+y}$, and place the fast robot at position $y > 1$ then similar calculations show that $c_2 \geq \frac{y^2+3y+2}{y^2+y+2}$. Further, it is easy to see that the lower bound $\frac{y^2+3y+2}{y^2+y+2}$ is maximized for $y = \sqrt{2}$.

4.2 Multi Robot Algorithm with Competitive Ratio ≤ 2

Theorem 4.2. *Algorithm 4 has competitive ratio at most* 2 *for any* $n > 2$ *robots.*

Theorem 4.3. *Given* $n > 2$ *robots, there is a robot deployment such that Algorithm 4 has competitive ratio at least* $2 - \frac{2}{2^n-1}$.

Remark 2. Observe that for any $\epsilon > 0$ by taking $n > \log(1 + 2/\epsilon)$ we have the competitive ratio of Algorithm 4 is at least $2 - \epsilon$.

5 Online Lower Bounds for Two Robots

In this section we prove two lower bounds on the competitive ratio for arbitrary online algorithms. Our lower bounds require only two mobile robots. In the first lower bound (Theorem 5.1) we assume that the speed of one of the robots is unknown and in the second (Theorem 5.2) we assume that the starting position of one of the robots is unknown. We provide both bounds (even though the second is slightly better) as the arguments are somewhat different and it seems plausible that an improved lower bound may come from combining the two approaches.

Theorem 5.1. *The lower bound for the competitive ratio when the fast robot does not know whether the speed of the slow robot is one or zero is at least* 1.0391.

Theorem 5.2. *The lower bound for the competitive ratio when the slow robot does not know the position of the fast robot is at least* 1.04059.

6 Conclusion

In this paper we studied the pony express communication problem for delivering a message from a source to a destination in the plane. We gave an optimal offline algorithm for the case of two robots and a $\sqrt{2}$ approximation for n robots. We studied a particular simple online algorithm and provided tight bounds for its performance in both the two robot and n robot case. Finally, we gave two distinct arguments for lower bounds on the competitive ratio of any online algorithm.

Our investigations leave a number of open problems. Of special interest is the complexity of the offline problem for the case of n robots. While it seems unlikely, we can not even be sure the question of deciding if an instance can be solved in a given time bound is decidable as the equations we derive involve trigonometric functions. In the online setting, there remains a gap between the upper bounds and lower bounds in both the case of just two robots and the case of n robots. Based upon this preliminary investigation, it would appear the exact bound will vary with the number of robots considered.

Additional questions arise when one attempts to solve well-known communication tasks such as broadcast and converge-cast. The delivery task considered in this paper made use of our ability to freely replicate the message which is to be delivered from the source to the destination. However, the problem is also interesting if the message to be delivered can be replicated only a given fixed number of times or even if it cannot be replicated at all (e.g., it is a physical object) in which case the problem resembles a transportation problem. Another interesting question would be to study the pony express communication problem in a setting where the robots may be subject to faults.

References

1. Anaya, J., Chalopin, J., Czyzowicz, J., Labourel, A., Pelc, A., Vaxès, Y.: Convergecast and broadcast by power-aware mobile agents. Algorithmica **74**(1), 117–155 (2016)
2. Bärtschi, A., et al.: Energy-efficient delivery by heterogeneous mobile agents. arXiv preprint arXiv:1610.02361 (2016)
3. Bärtschi, A., Graf, D., Mihalák, M.: Collective fast delivery by energy-efficient agents. In: Potapov, I., Spirakis, P.G., Worrell, J. (eds.) 43rd International Symposium on Mathematical Foundations of Computer Science, MFCS 2018, Liverpool, UK, 27–31 August 2018. LIPIcs, vol. 117, pp. 56:1–56:16. Schloss Dagstuhl - Leibniz-Zentrum für Informatik (2018)
4. Bärtschi, A., Graf, D., Penna, P.: Truthful mechanisms for delivery with mobile agents. arXiv preprint arXiv:1702.07665 (2017)
5. Bärtschi, A., Tschager, T.: Energy-efficient fast delivery by mobile agents. In: Klasing, R., Zeitoun, M. (eds.) FCT 2017. LNCS, vol. 10472, pp. 82–95. Springer, Heidelberg (2017). https://doi.org/10.1007/978-3-662-55751-8_8
6. Bereg, S., Brunner, A., Caraballo, L.-E., Díaz-Báñez, J.-M., Lopez, M.A.: On the robustness of a synchronized multi-robot system. J. Comb. Optim. **39**(4), 988–1016 (2020). https://doi.org/10.1007/s10878-020-00533-z

7. Bilò, D., Gualà, L., Leucci, S., Proietti, G., Rossi, M.: New approximation algorithms for the heterogeneous weighted delivery problem. In: Jurdziński, T., Schmid, S. (eds.) SIROCCO 2021. LNCS, vol. 12810, pp. 167–184. Springer, Cham (2021). https://doi.org/10.1007/978-3-030-79527-6_10
8. Carvalho, I.A., Erlebach, T., Papadopoulos, K.: An efficient algorithm for the fast delivery problem. In: Gąsieniec, L.A., Jansson, J., Levcopoulos, C. (eds.) FCT 2019. LNCS, vol. 11651, pp. 171–184. Springer, Cham (2019). https://doi.org/10.1007/978-3-030-25027-0_12
9. Carvalho, I.A., Erlebach, T., Papadopoulos, K.: On the fast delivery problem with one or two packages. J. Comput. Syst. Sci. **115**, 246–263 (2021)
10. Chalopin, J., Das, S., Mihal'ák, M., Penna, P., Widmayer, P.: Data delivery by energy-constrained mobile agents. In: Flocchini, P., Gao, J., Kranakis, E., Meyer auf der Heide, F. (eds.) ALGOSENSORS 2013. LNCS, vol. 8243, pp. 111–122. Springer, Heidelberg (2014). https://doi.org/10.1007/978-3-642-45346-5_9
11. Chalopin, J., Godard, E., Métivier, Y., Ossamy, R.: Mobile agent algorithms versus message passing algorithms. In: Shvartsman, M.M.A.A. (ed.) OPODIS 2006. LNCS, vol. 4305, pp. 187–201. Springer, Heidelberg (2006). https://doi.org/10.1007/11945529_14
12. Chalopin, J., Jacob, R., Mihalák, M., Widmayer, P.: Data delivery by energy-constrained mobile agents on a line. In: Esparza, J., Fraigniaud, P., Husfeldt, T., Koutsoupias, E. (eds.) ICALP 2014. LNCS, vol. 8573, pp. 423–434. Springer, Heidelberg (2014). https://doi.org/10.1007/978-3-662-43951-7_36
13. Chuangpishit, H., Czyzowicz, J., Killick, R., Kranakis, E., Krizanc, D., Morales-Ponce, O.: Optimal rendezvous on a line by location-aware robots in the presence of spies. Discret. Math. Algorithms Appl. (2021, to appear)
14. Coleman, J., Kranakis, E., Krizanc, D., Morales-Ponce, O.: Message delivery in the plane by robots with different speeds. arXiv preprint arXiv:2109.12185 (2021)
15. Coleman, J., Kranakis, E., Krizanc, D., Morales-Ponce, O.: The pony express communication problem. In: Proceedings of IWOCA21; also Extended Version as arXiv preprint arXiv:2105.03545 (2021)
16. Czyzowicz, J., Diks, K., Moussi, J., Rytter, W.: Communication problems for mobile agents exchanging energy. In: Suomela, J. (ed.) SIROCCO 2016. LNCS, vol. 9988, pp. 275–288. Springer, Cham (2016). https://doi.org/10.1007/978-3-319-48314-6_18
17. Czyzowicz, J., Georgiou, K., Kranakis, E.: Patrolling. In: Flocchini, P., Prencipe, G., Santoro, N. (eds.) Distributed Computing by Mobile Entities. LNCS, vol. 11340, pp. 371–400. Springer, Cham (2019). https://doi.org/10.1007/978-3-030-11072-7_15
18. Czyzowicz, J., Killick, R., Kranakis, E., Krizanc, D., Morales-Ponce, O.: Gathering in the plane of location-aware robots in the presence of spies. Theor. Comput. Sci. **836**, 94–109 (2020)
19. Das, S., Dereniowski, D., Karousatou, C.: Collaborative exploration by energy-constrained mobile robots. In: Scheideler, C. (ed.) SIROCCO 2014. LNCS, vol. 9439, pp. 357–369. Springer, Cham (2015). https://doi.org/10.1007/978-3-319-25258-2_25
20. Das, S., Dereniowski, D., Karousatou, C.: Collaborative exploration of trees by energy-constrained mobile robots. Theory Comput. Syst. **62**(5), 1223–1240 (2018)
21. Ogilvy, C.S.: Excursions in Geometry. Dover Publications (1990)
22. Yiu, P.: Introduction to the Geometry of the Triangle. Version 4.0510, Florida Atlantic University Lecture Notes (2004)

Exploring a Dynamic Ring Without Landmark

Archak Das[1], Kaustav Bose[2](✉), and Buddhadeb Sau[1]

[1] Department of Mathematics, Jadavpur University, Kolkata, India
{archakdas.math.rs,buddhadeb.sau}@jadavpuruniversity.in
[2] Advanced Computing and Microelectronics Unit, Indian Statistical Institute, Kolkata, India

Abstract. Consider a group of autonomous mobile computational entities, called agents, arbitrarily placed at some nodes of a dynamic but always connected ring. The agents neither have any knowledge about the size of the ring nor have a common notion of orientation. We consider the EXPLORATION problem where the agents have to collaboratively explore the graph and terminate, with the requirement that each node has to be visited by at least one agent. It has been shown by Di Luna et al. [Distrib. Comput. 2020] that the problem is solvable by two anonymous agents if there is a single observably different node in the ring called landmark node. The problem is unsolvable by any number of anonymous agents in absence of a landmark node. We consider the problem with non-anonymous agents (agents with distinct identifiers) in a ring with no landmark node. The assumption of agents with distinct identifiers is strictly weaker than having a landmark node as the problem is unsolvable by two agents with distinct identifiers in absence of a landmark node. This setting has been recently studied by Mandal et al. [ALGOSENSORS 2020]. There it is shown that the problem is solvable in this setting by three agents assuming that they have edge crossing detection capability. Edge crossing detection capability is a strong assumption which enables two agents moving in opposite directions through an edge in the same round to detect each other and also exchange information. In this paper we give an algorithm that solves the problem with three agents without the edge crossing detection capability.

Keywords: Multi-agent systems · Mobile agent · Dynamic network · Exploration · Meeting

1 Introduction

Consider a team of autonomous computational entities, usually called agents or robots, located at the nodes of a graph. The agents are able to move from a node to any neighboring node. The EXPLORATION problem asks for a distributed algorithm that allows the agents to explore the graph, with the requirement that each node has to be visited by at least one agent. Being one of the fundamental

C. Johnen et al. (Eds.): SSS 2021, LNCS 13046, pp. 320–334, 2021.
https://doi.org/10.1007/978-3-030-91081-5_21

problems in the field of autonomous multi-agent systems, the problem has been extensively studied in the literature. However, the majority of existing literature studies the problem for *static* graphs, i.e., the topology of the graph does not change over time. Recently within the distributed computing community, there has been a surge of interest in highly dynamic graphs: the topology of the graph changes continuously and unpredictably. In highly dynamic graphs, the topological changes are not seen as occasional anomalies (e.g., link failures, congestion, etc.) but rather integral part of the nature of the system [20,27]. We refer the readers to [4] for a compendium of different models of dynamic networks considered in the literature. If time is discrete, i.e., changes occur in rounds, then the evolution of a dynamic graph can be seen as a sequence of static graphs. A popular assumption in this context is *always connected* (Class 9 of [4]), i.e., the graph is connected in each round.

In the dynamic setting, the EXPLORATION problem was first studied in [21]. In particular, the authors studied the EXPLORATION problem in a dynamic but always connected ring by a set of autonomous agents. They showed that EXPLORATION is solvable by two anonymous agents (agents do not have unique identifiers) under fully synchronous setting (i.e., all agents are active in each round) if there is a single observably different node in the ring called *landmark* node. They also proved that in absence of a landmark node, two agents cannot solve EXPLORATION even if the agents are non-anonymous and they have chirality, i.e., they agree on clockwise and counterclockwise orientation of the ring. The impossibility result holds even if we relax the problem to EXPLORATION with partial termination. As opposed to the standard explicit termination setting where all agents are required to terminate, in the partial termination setting at least one agent is required to detect exploration and terminate. If the agents are anonymous, then EXPLORATION with partial termination with chirality remains unsolvable in absence of a landmark node even with arbitrary number of agents. Then in [23], the authors considered the EXPLORATION problem (without chirality and requiring explicit termination) with no landmark node. Since the problem cannot be solved even with arbitrary number of anonymous agents, they considered non-anonymous agents, in particular, agents with unique identifiers. Since the problem is unsolvable by two non-anonymous agents, they considered the question that whether the problem can be solved by three non-anonymous agents. They showed that the answer is yes if the agents are endowed with edge crossing detection capability. Edge crossing detection capability is a strong assumption which enables two agents moving in opposite directions through an edge in the same round to detect each other and also exchange information. In collaborative tasks like exploration, the agents are often required to meet at a node and exchange information. However, the edge crossing detection capability allows two agents to exchange information even without meeting at a node. The assumption is particularly helpful when the agents do not have chirality where it is more difficult to ensure meeting. Even if we do not allow exchange of information, simple detection of the swap can be useful in deducing important information about the progress of an algorithm. In [23], it was also shown that the assumption of edge crossing detection can be removed with the help of randomness. In particular, without assuming edge crossing detection capability, they gave a randomized algorithm that solves EXPLORATION with explicit termination with probability at least $1 - \frac{1}{n}$ where n is the size of the ring. Therefore this leaves the open question that whether the problem can be solved by a deterministic algorithm by three non-anonymous agents without edge crossing detection capability. In this paper, we answer this question affirmatively. In particular we give a deterministic algorithm that solves EXPLORATION in absence of chirality by three non-anonymous agents without edge-crossing detection capability. As basic ingredients of our algorithm, we also solve two problems called MEETING and CONTIGUOUS AGREEMENT. These problems may be of independent interest and useful for solving other problems in similar settings.

1.1 Related Work

The problem of EXPLORATION by mobile agents in static anonymous graph has been studied extensively in the literature [2,6,8,9,11,13,26]. Prior to [21], there have been a few works on EXPLORATION of dynamic graphs, but under assumptions such as complete a priori knowledge of location and timing of topological changes (i.e., *offline* setting) [10,16,18,24] or periodic edges (edges appear periodically) [12,17] or δ-recurrent edges (each edge appears at least once every δ rounds) [18] etc. In the *online* or *live* setting where the location and timing of the changes are unknown, distributed EXPLORATION of graphs without any assumption other than being always connected was first considered in [21]. In particular, they considered the problem on an always connected dynamic ring. They proved that without any knowledge of the size of the ring and without landmark node, EXPLORATION with partial termination is impossible by two agents even if the agents are non-anonymous and have chirality. They also proved that if the agents are anonymous, have no knowledge of size, and there is no landmark node then EXPLORATION with partial termination is impossible by any number of agents even in the presence of chirality. On the positive side the authors showed that under fully synchronous setting, if an upper bound N on the size of the ring is known to two anonymous agents, then EXPLORATION with explicit termination is possible within $3N - 6$ rounds. They then showed that for two anonymous agents, if chirality and a landmark node is present, then exploration with explicit termination is possible within $O(n)$ round, and in the absence of chirality with all other conditions remaining the same, EXPLORATION with explicit termination is possible within $O(n \log n)$ rounds, where n is the size of the ring. They have also proved a number of results in the semi-synchronous setting (i.e., not all agents may be active in each round) under different assumptions. Then in [23], the authors considered agents with unique identifiers and edge crossing detection capability in a ring without any landmark node. They showed that EXPLORATION with explicit termination is impossible in the absence of landmark node and the knowledge of n by two agents with access to randomness, even in the presence of chirality, unique identifiers and edge-crossing detection capability. In the absence of randomness even EXPLO-

RATION with partial termination is impossible in the same setting. With three agents under fully synchronous setting, the authors showed that EXPLORATION with explicit termination is possible by three non-anonymous agents with edge-crossing detection capability in absence of any landmark node. Removing the assumption of edge-crossing detection and replacing it with access to randomness, the authors gave a randomized algorithm for EXPLORATION with explicit termination with success probability at least $1 - \frac{1}{n}$. EXPLORATION of an always connected dynamic torus was considered in [15]. In [14] the problem of PERPETUAL EXPLORATION (i.e., every node is to be visited infinitely often) was studied in temporally connected (i.e., may not be always connected but connected over time) graphs. Other problems studied in dynamic graphs include GATHERING [3,22,25], DISPERSION [1,19], PATROLLING [7] etc.

1.2 Our Contributions

We consider a dynamic but always connected ring of size n. A team of three agents are operating in the ring under a fully synchronous scheduler. Each agent has a k bit unique identifier where k is a fixed constant. The agents do not have any knowledge of n and they do not have chirality. Furthermore, they do not have edge crossing detection capability. In this setting, we give a deterministic algorithm for EXPLORATION with explicit termination. Recall that in [21], the problem was solved by two agents in a ring with a landmark node. The overall idea behind our algorithm is to reduce the problem to a setting similar to [21] by artificially creating a landmark node. For this, we need to solve the problem MEETING where any two agents in the team are required to meet each other at a node. When two agents meet, one of them (say, the one with smaller ID) will settle at that node and act as landmark. Although the situation now becomes similar to [21], they are not exactly the same. This is because, unlike in [21], the landmark agent needs to detect the completion of exploration and terminate. Therefore, several modifications are required in order to employ the strategy from [21] in this setting. However, the main difficult part of the algorithm is to ensure that two agents meet at a node, i.e., to solve the MEETING problem. Solving MEETING in the current setting is a challenging problem and may be of independent interest. It can be useful for solving other problems as well, especially where the agents may need to communicate or exchange information. Recall that in our setting, the only way two agents can communicate is by meeting each other at a node. We first give an algorithm that solves MEETING in the presence of chirality. Then we define and solve a new problem called CONTIGUOUS AGREEMENT which requires the agents to agree on some common direction for some number of consecutive rounds. Then we solve MEETING in the absence of chirality by using these two algorithms as subroutine. In particular, the main idea is to simulate the MEETING algorithm with chirality in the period when the agents agree on a common direction. Our CONTIGUOUS AGREEMENT protocol can be useful for solving other problems as it can be used as a tool to transform certain algorithms that functions in presence of chirality to algorithms that work without chirality. After meeting, one of the agents will become landmark as planned. From there we solve EXPLORATION with termination using a strategy which is partly similar to [21]. Overall, the round complexity of our algorithm is $\Theta(n)$, where n is the size of the ring. This is asymptotically optimal as there are n nodes to be explored and in each round, three agents can visit at most three nodes. Furthermore, our algorithm solves the problem with optimum number of agents as in view of the impossibility results of [21], the problem cannot be solved with two agents in this setting. A comparison of the results obtained in this paper with previous works is given in Table 1.

Table 1. Comparison of our results with previous works. All the above algorithms works without chirality and takes $O(n)$ rounds where n is the size of the ring

Paper	Number of agents	Agents	Landmark node	Edge cross. detection	Algorithm
[21]	2	Anonymous	Yes	No	Deterministic
[23]	3	Have unique identifiers	No	Yes	Deterministic
[23]	3	Have unique identifiers	No	No	Randomized
This paper	3	Have unique identifiers	No	No	Deterministic

1.3 Outline of the Paper

In Sect. 2, we describe the model and terminology used in the paper. In Sect. 3, we give an algorithm for EXPLORATION in the simpler setting where the agents have chirality. In Sect. 4, we use the techniques used in Sect. 3 to give an algorithm for EXPLORATION in the absence of chirality.

2 Model and Terminology

We consider a dynamic ring of size n. All nodes of the ring are identical. Each node is connected to its two neighbors via distinctly labeled ports. The labeling of the ports may not be globally consistent and thus might not provide an orientation. We consider a discrete temporal model i.e., time progresses in rounds. In each round at most one edge of the ring may be missing. Thus the ring is connected in each round. Such a network is known in the literature as a 1-interval connected ring.

We consider a team of three agents operating in the ring. The agents do not have any knowledge of the size of the ring. Each agent is provided with memory and computational capabilities. An agent can move from one node to a neighbouring node if the edge between them is not missing. Two agents moving in opposite direction on the same edge are not able to detect each other. An agent can only detect an active agent co-located at the same node i.e., if an agent terminates it becomes undetectable by any other agent. Two agents can communicate with each other only when they are present at the same node. Each agent has a unique identifier which is a bit string of constant length $k > 1$, i.e., the length k of the identifier is the same for each agent. For an agent r, its unique identifier will be denoted by $r.ID$. Also $val(r.ID)$ will denote the numerical value of $r.ID$. For example $val(00110) = 6$, $val(10011) = 19$, etc. Hence for any agent r, $val(r.ID) < 2^k$.

Each agent has a consistent private orientation of the ring, i.e., a consistent notion of left or right. If the left and right of all three agents are the same then we say that the agents have chirality. By clockwise and counterclockwise we shall refer to the orientations of the ring in the usual sense. These terms will be used only for the purpose of description and the agents are unaware of any such global orientation if they do not have chirality. For two agents r_1 and r_2 on the ring, $d^{\circlearrowright}(r_1, r_2)$ and $d^{\circlearrowleft}(r_1, r_2)$ denotes respectively the clockwise and counterclockwise distance from r_1 to r_2.

We consider a fully synchronous system, i.e., all three agents are active in each round. In each round, the agents perform the following sequence of operations:

Look: If other agents are present at the node, then the agent exchanges messages with them.

Compute: Based on its local observation, memory and received messages, the agent performs some local computations and determines whether to move or not, and if yes, then in which direction.

Move: If the agent has decided to move in the COMPUTE phase, then the agent attempts to move in the corresponding direction. It will be able to move only if the corresponding edge is not missing. An agent can detect if it has failed to move.

During the execution of algorithm, two agents can meet each other in two possible ways: (1) two agents r_1 and r_2 moving in opposite direction come to the same node, or, (2) an agent r_1 comes to a node where there is a stationary agent r_2. In the second case we say that r_1 *catches* r_2. If two agents r_1, r_2 are moving in opposite direction on the same edge in the same round, then we say that r_1 and r_2 *swaps over an edge.*

3 Exploration by Agents with Chirality

In this section, we shall assume that the agents have chirality. Since the agents have agreement in direction we shall use the terms clockwise and counterclockwise instead of right and left respectively. In Sect. 3.1 we present an algorithm for MEETING where at least two agents are required to meet at a node. Then in Sect. 3.2 we shall use this algorithm as a subroutine to solve EXPLORATION.

3.1 Meeting by Agents with Chirality

We have three agents placed arbitrarily at distinct nodes of the ring. Our objective is that at least two of the agents should meet. The algorithm works in several phases. The lengths of the phases are 2^{j+k}, $j = 0, 1, 2, \ldots$. In phase j, an agent

r tries to move clockwise for the first $val(r.ID)2^j$ rounds, and then remains stationary for $(2^k - val(r.ID))2^j$ rounds.

Notice that in each phase, the agent with the smallest ID value stops trying to move first. The main idea behind the algorithm is that if the remaining agents keep trying to move for long enough then a meeting should take place. This is stated in Lemma 1. This lemma will be used several times in the proofs throughout the paper. The intuition behind the lemma is the following. Once the first agent stops moving, the remaining two agents are trying to move towards it. If the closer one is not blocked by edge removals in too many rounds then it will be able to catch the static agent within a certain time. Otherwise if the agent is blocked for too long, it will get caught by the third agent. This is because in the rounds where the agent is blocked, the third agent is able to make progress as at most one edge may disappear in each round. A formal proof of the lemma is given in the full version [5]. Using this lemma, we can show that a meeting is guaranteed to take place at or before the pth phase where $p = \lceil \log 2n \rceil$. So the algorithm solves the problem within $2^k \sum_{i=0}^{p} 2^i < 2^{k+\lceil \log n \rceil + 2}$ rounds. This is formally stated in Theorem 1. The proof of the theorem is given in [5].

Lemma 1. *Let r_1, r_2 and r_3 be three agents in the ring such that at round t, $0 \leq d^{\circlearrowright}(r_1, r_3) < d^{\circlearrowright}(r_1, r_2)$. If r_1 remains static and both r_2 and r_3 try to move clockwise for the next $2n$ rounds, then within these $2n$ rounds either r_2 meets r_1 or r_3 meets r_2.*

Theorem 1. *The above algorithm solves* Meeting *for three agents with chirality. The algorithm ensures that the meeting takes place within $2^{k+\lceil \log n \rceil + 2}$ rounds.*

3.2 Exploration with Termination by Agents with Chirality

We consider three agents in the ring having chirality. For simplicity assume that the agents are initially placed at distinct nodes of the ring. We shall later remove this assumption. Our plan is to first bring two of the agents at the same node using the Meeting algorithm described in Sect. 3.1. Then one of them will settle at that node and play the role of landmark node. Then the situation reduces to a setting similar to [21]. However we cannot use the same algorithm from [21] in our case. This is because unlike in [21] we have to ensure that the agent acting as landmark also terminates. However our algorithm uses some ideas from [21]. We shall now give a brief description algorithm. The detailed pseudocode description of the algorithm is given in [5].

Initially all the agents start with their *state* variable set to `search`. Until an agent meets another agent, it executes the Meeting algorithm described in Sect. 3.1. Now according to Theorem 1, two agents are guaranteed to meet within $2^{k+\lceil \log 2n \rceil + 2}$ rounds from the start of the algorithm. On meeting the agents compare their IDs and the one with smaller ID changes its *state* to `settled` and stops moving. The other agent changes its *winner* variable to *True* and henceforth abandons its phase-wise movement and attempts to move clockwise in each round. Let us now describe the case when an agent with *state* `search` meets the `settled` agent. If an agent with $winner = False$ encounters the `settled` agent it also abandons its phase-wise movement and henceforth tries to move in the clockwise direction in every round. If an agent with $winner = True$ meets the `settled` agent r, then it indicates that it is meeting the `settled` agent for the second time and hence all nodes of the ring have been explored. The agent can also calculate the size of the ring as it is equal to the number of successful moves between the two meetings. The agent assigns this value to the variable $RSize$ and also informs the `settled` agent about it. Then the agent will continue to move in the clockwise direction for $2n$ more rounds. Both these agents will terminate after the completion of these $2n$ rounds. Now consider the case when an agent with $state =$ `search` and $winner = True$ meets an agent with $state =$ `search` and $winner = False$. If the agent with $winner = True$ already knows n, i.e., it has visited the `settled` agent twice, then both of them terminates immediately. If the agent with $winner = True$ does not already know n, then it changes its *state* to `forward` and continues to move in the clockwise direction every round. On the other hand, the agent with $winner = False$ changes its *state* to `bounce` and starts moving in the counterclockwise direction. This phenomenon is called the formation of `settled-forward-bounce` triplet. In this case, both the agents initiate a variable $TTime$ to keep track of the number of rounds elapsed after triplet formation.

After the triplet is created, the agent with *state* `forward` will continue to move in clockwise direction. The agent with *state* `bounce` will move counterclockwise and then on fulfillment of certain conditions, it may change its *state* to `return` and start moving clockwise. Then it may again change its *state* to `bounce` and start moving counterclockwise. The period between any two such *state* changes will be called a *run*. While moving in the clockwise direction with *state* `forward`, the agent keeps count of the number of successful steps with state `forward` in the variable $FSteps$. The variable $BSteps$ (resp. $RSteps$) is used to keep count of the number of successful steps with state `bounce` (resp. `return`) in the current run. Also while moving in the counterclockwise direction with *state* `bounce`, the variable $BBlocked$ counts the number of unsuccessful attempts to move in that run. An agent r with *state* `bounce` will change its *state* to `return` if one of the following takes place: 1) $r.BBlocked$ exceeds $r.BSteps$ or 2) the agent r encounters the `settled` agent twice in the same run. An agent r with *state* `return` will change its *state* to `bounce` if r meets with the agent with *state* `forward` and $r.RSteps > 2r.BSteps$, where $BSteps$ was counted in the last run with *state* `bounce`. Here the main idea is that the agents will try to gauge the size of the ring. An agent may be able to find the size n exactly or calculate an upper bound of n. An agent can exactly find n only if it visits the static `settled` agent twice in the same direction. In this case it will also inform the `settled` agent about n. Clearly when this happens the ring has been explored completely. However the two agents cannot terminate immediately because the third agent is not aware of this. So the agents will remain active till $TTime = 16n$, i.e.,

$16n$ rounds from the time when the triplet was created. It should be noted here that the `settled` agent initially did not know the time when the triplet was created. It came to know about this from the $TTime$ value of some agent that it met and initiated its own $TTime$ counter accordingly. Now it can be shown that within these $16n$ rounds the third agent will meet one of the two agents that already know n. These two agents will terminate immediately upon meeting. Now consider the case where an agent is able to find an upper bound of n. This happens when one of the following three takes place: 1) the `forward` agent meets the agent with *state* `bounce`, 2) the `forward` agent catches the agent with *state* `return`, 3) the agent with *state* `return` catches the `forward` agent with $RSteps \leq 2BSteps$. It can be shown in each of the cases, these two agents will be able to correctly calculate an upper bound $SBound$ of n. Furthermore these cases imply that the ring has been already explored completely. However the two agents cannot terminate immediately because the `settled` agent is not aware of this. Therefore in order to acknowledge the `settled` agent, these two agents will start moving in opposite directions for $SBound$ more rounds and then terminate. Clearly one of them will be able to meet the `settled` agent.

It can be shown that this algorithm solves EXPLORATION with explicit termination in $2^{k+\lceil \log n \rceil+3} + 23n = O(n)$ rounds. This is formally stated in Theorem 2. The proof of this quite involved and is deferred to [5] due to space constraints.

Theorem 2. EXPLORATION *with explicit termination is solvable by three agents with chirality in* $2^{k+\lceil \log n \rceil+3} + 23n = O(n)$ *rounds.*

4 Exploration by Agents Without Chirality

4.1 Contiguous Agreement

In this section we define a new problem called CONTIGUOUS AGREEMENT. Three agents with unique identifiers are placed at three different nodes in the ring. In each round, each agent chooses a direction according to a deterministic algorithm based on its ID and current round. The requirement of the problem is that the agents have to choose the same direction for some N consecutive rounds where N is a constant unknown to the agents.

Before presenting the algorithm, we describe the construction of modified identifiers which will be used in the algorithm. Recall that $r.ID$ is a binary string of length k. We now describe the construction of the corresponding modified identifier $r.MID$ which is a binary string of length $\frac{k(k-1)}{2} + k + 1$. We shall first concatenate a string of length $\frac{k(k-1)}{2}$ at end of $r.ID$. Let us write $\frac{k(k-1)}{2} = l$. To define the string, we shall identify each position of the string as, instead of an integer from $[l] = \{1, \ldots, l\}$, a 2-tuple from the set $S = \{(u, v) \in [k] \times [k] \mid u < v\}$. In order to formally describe this, let us define a bijection $\phi : S \to [l]$ in the following way. Notice that $|S| = l$. Arrange the elements of S in lexicographic order. For any $(u, v) \in S$, we define $\phi((u, v))$ to be the position of (u, v) in this arrangement. For example, if $k = 4$, then the elements of S, arranged in lexicographic order, are (1,2), (1,3), (1,4), (2,3), (2,4), (3,4). Therefore, we have $\phi((1,2)) = 1$, $\phi((1,3)) = 2$, $\phi((1,4)) = 3$, $\phi((2,3)) = 4$, $\phi((2,4)) = 5$ and $\phi((3,4)) = 6$. Now we define the string of length l that will be concatenated with $r.ID$. The ith bit of the string is the $\mathbb{Z}_2$ sum of the uth and vth bit of $r.ID$ where $(u,v) = \phi^{-1}(i)$. After the concatenation, we get a string of length $k + l = k + \frac{k(k-1)}{2}$. Finally we append 0 at the beginning of this string to obtain $r.MID$ of length $\frac{k(k-1)}{2} + k + 1$ (c.f. Fig. 1).

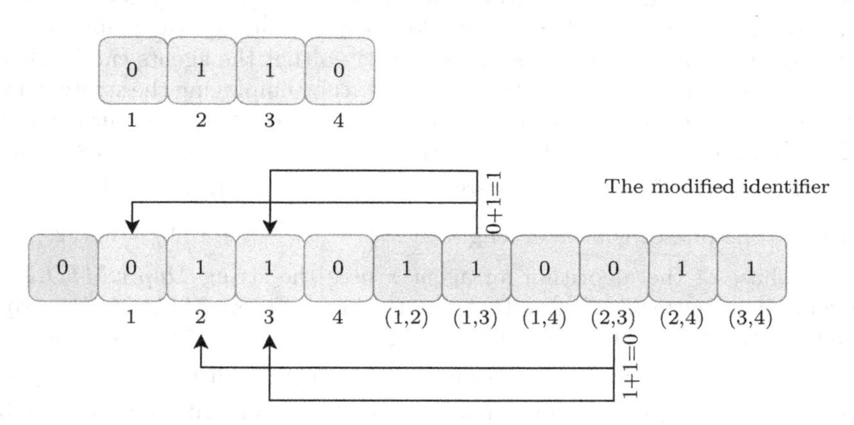

Fig. 1. The construction of the modified identifier.

We now present the algorithm that solves CONTIGUOUS AGREEMENT. The algorithm works in phases with the length of the phases being $2^j\left(\frac{k(k-1)}{2} + k + 1\right)$, $j = 0, 1, 2, \ldots$. For a string S and a positive integer t, let $Dup(S,t)$ denote the string obtained by repeating each bit of string S t times. For example, $Dup(101,3) = 111000111$. For the jth phase, the agent r computes $Dup(r.MID, 2^j)$. Notice that the length of the jth phase is equal to the length of $Dup(r.MID, 2^j)$. In the ith round of the jth phase, r moves left if the ith bit of $Dup(r.MID, 2^j)$ is 0 and otherwise moves right.

The idea behind the algorithm is the following. Let us first look at the 0th phase. If the local orientations of all three agents are the same (say left = counterclockwise for each agent) then all three agents will choose the same direction in the first round. This is because MIDs of all agents start with 0 and in the first round all agents choose left which is the counterclockwise direction. Now if the local orientations of all three agents are not the same still two of the agents will be in agreement. There are three possible cases based on which two agents have the same local orientation. Notice if there is an index where the bit of the MIDs of those two agents are equal, but is different from that of third agent, then all agents will choose the same direction in the round corresponding to that index.

It can be shown that such indices exist for each of the three possible cases. This implies that there is a round in the 0th phase in which all three agents choose the same direction. Hence for $j = \lceil \log N \rceil$, the agents will choose the same direction for N consecutive rounds in the jth phase. The formal proof of correctness of the algorithm is given in [5].

Theorem 3. *The algorithm described above solves* CONTIGUOUS AGREEMENT.

4.2 Meeting by Agents Without Chirality

In Sect. 3.1, we describe an algorithm that solves MEETING by agents with chirality. In the current setting where agents do not have chirality, the main idea is to use the strategy of CONTIGUOUS AGREEMENT so that the agents can implicitly agree on a common direction and solve MEETING by employing the strategy from Sect. 3.1. Similar to the algorithm for CONTIGUOUS AGREEMENT, our algorithm for MEETING also works in phases. In the algorithm for CONTIGUOUS AGREEMENT the length of the phases were $2^j\left(\frac{k(k-1)}{2}+k+1\right)$, $j = 0, 1, 2, \ldots$. For MEETING, the phases will be of length $2^{j+k}\left(\frac{k(k-1)}{2}+k+1\right)$, $j = 0, 1, 2, \ldots$. In the jth phase of the algorithm an agent r uses the string $Dup(r.MID, 2^{j+k})$ to decide its movement. Notice that the length of $Dup(r.MID, 2^{j+k})$ is equal to the length of the jth phase. The string $Dup(r.MID, 2^{j+k})$ is a concatenation of $\left(\frac{k(k-1)}{2}+k+1\right)$ blocks of length 2^{j+k} where each block consists of all 0's or all 1's. Our plan is to simulate the MEETING algorithm from Sect. 3.1. So, in the 2^{j+k} rounds corresponding to each block, the agent r will (try to) move in the first $val(r.ID)2^j$ rounds and will be stationary for the remaining $\left(2^k - val(r.ID)\right)2^j$ rounds. If the block consists of 0's, then the movement will be towards left and otherwise towards right. It can be shown that the algorithm solves the problem within $k^2 2^{k+\lceil \log n \rceil+3}$ rounds as stated in Theorem 4. Formal proof of this is given in [5].

Theorem 4. *The above algorithm solves* MEETING *for three agents without chirality. The algorithm ensures that the meeting takes place within* $k^2 2^{k+\lceil \log n \rceil+3}$ *rounds.*

4.3 Exploration with Termination by Agents Without Chirality

For simplicity assume that the agents are placed arbitrarily at distinct nodes of the ring. We shall remove this assumption at the end of this section. Initially the *state* variable of all three agents are set to `search`. We shall adopt a strategy similar to the one used in Sect. 3. In fact, we only need to make some modifications in the algorithms to be executed by the agents with *state* `search` and `settled`.

The agents will execute the MEETING algorithm described in Sect. 4.2 until another agent is encountered. The agents will keep count of the number of rounds since the beginning in the variable $STime$. Now by Theorem 4, two of the agents are guaranteed to meet within $k^2 2^{k+\lceil \log n \rceil+3}$ rounds. Upon meeting the agents will agree on a common direction, say the right direction of the agent with larger ID. Without loss of generality assume that the agreed direction is the clockwise direction. The agent with smaller ID, say r_1, will become the `settled` agent and the one with larger ID, say r_2, will continue moving in the clockwise direction. The agent r_1 will save the port number leading to the agreed direction, i.e., clockwise. Let r_3 denote the third agent which is still executing the MEETING algorithm. It can be shown that a second meeting is guaranteed to take place on or before $\lceil \log 2n \rceil + 1$th phase, i.e., within $k^2 2^{k+\lceil \log n \rceil+4}$ rounds from the start of the algorithm. However unlike in Sect. 3 where the agents had chirality, here the second meeting may also take place between r_1 and r_2. This is because r_3 is moving in clockwise direction in some rounds and counterclockwise in other rounds. Hence there is a possibility that r_2 and r_3 may swap over an edge and r_1 meets r_2 first. However even then it is not difficult to see that r_3 is guaranteed to meet r_1 or r_2 on or before $(\lceil \log 2n \rceil + 1)$th phase i.e., within $k^2 2^{k+\lceil \log n \rceil+4}$ rounds from the start of the algorithm. To see this, observe that r_2 and r_3 will try to move in the same direction for some $4n$ consecutive rounds in the $(\lceil \log 2n \rceil + 1) = \lceil \log 4n \rceil$th phase. Within the first $2n$ rounds a meeting should take place by Lemma 1. If this meeting involves r_3 then we are done. Otherwise r_1 and r_2 meet each other and then by again applying by Lemma 1, r_3 will meet one of them within the following $2n$ rounds.

Now consider the following cases depending on which of the two robots meet on the second meeting.

1. Suppose that the second meeting takes place between r_1 and r_2. In this case the ring has been explored and r_2 finds out n and informs r_1 about it. Then r_2 will continue moving in the clockwise direction. Both agents will terminate after the round when $STime = k^2 2^{k+\lceil \log n \rceil+4}$. Recall that r_3 is guaranteed to meet one of them in the meantime and will terminate immediately.
2. Suppose that the second meeting takes place between r_2 and r_3. Then r_2 informs r_3 about the agreed direction. Hence the case reduces to the setting of Sect. 3. Therefore r_2 and r_3 will change their *state* to `forward` and `bounce` respectively and execute the algorithms as before.
3. Now suppose that the second meeting takes place between r_1 and r_3. In this case r_3 will come to know about the agreed direction and again the case reduces to the setting of Sect. 3. So r_3 will continue to move in the agreed direction i.e., clockwise.

It follows from the above discussions and the results in Sect. 3 that the agents will terminate after exploring the ring within $k^2 2^{k+\lceil \log n \rceil+4} + 23n = O(n)$ rounds.

Theorem 5. EXPLORATION *with explicit termination is solvable by three agents without chirality in* $k^2 2^{k+\lceil \log n \rceil+4} + 23n = O(n)$ *rounds.*

Recall that we assumed that the agents are placed initially at distinct nodes of the ring. We now show that this assumption is not necessary if the initial configuration has two agents r_1, r_2 at the same node and the third agent r_3 at a different node. Then the case reduces to the situation when the first meeting takes place. Then r_1 and r_2 will change their *state* to `settled` and `forward` while r_3 will execute the MEETING algorithm with *state* `search`. The algorithm will progress as before and achieve exploration with termination.

If all three agents are in the same node in the initial configuration then the three agents will compare their identifiers and will change their *state* to `settled`, `forward` and `bounce` accordingly. Again, the algorithm will progress as before and achieve exploration with termination.

5 Concluding Remarks

We showed that EXPLORATION (with explicit termination) in a dynamic always connected ring without any landmark node is solvable by three non-anonymous agents without chirality. This is optimal in terms of the number of agents used as the problem is known to be unsolvable by two agents. Our algorithm takes $\Theta(n)$ rounds to solve the problem where n is the size of the ring. However, the dependency on k, the bit length of the identifiers is exponential. An interesting question is whether the problem can be solved in $O(poly(k)n)$ rounds. A challenging problem that remains open is EXPLORATION in a dynamic network of arbitrary underlying topology. Except for some bounds on the number of agents obtained in the recent work [14], almost nothing is known.

Acknowledgement. The first author is supported by UGC, Govt. of India. We would like to thank the anonymous reviewers for their valuable comments which helped us to improve the quality and presentation of the paper.

References

1. Agarwalla, A., Augustine, J., Moses, W.K., Jr., Madhav, S.K., Sridhar, A.K.: Deterministic dispersion of mobile robots in dynamic rings. In: Proceedings of the 19th International Conference on Distributed Computing and Networking, ICDCN 2018, Varanasi, India, 4–7 January 2018, pp. 19:1–19:4. ACM (2018). https://doi.org/10.1145/3154273.3154294
2. Albers, S., Henzinger, M.R.: Exploring unknown environments. SIAM J. Comput. **29**(4), 1164–1188 (2000). https://doi.org/10.1137/S009753979732428X
3. Bournat, M., Dubois, S., Petit, F.: Gracefully degrading gathering in dynamic rings. In: Izumi, T., Kuznetsov, P. (eds.) SSS 2018. LNCS, vol. 11201, pp. 349–364. Springer, Cham (2018). https://doi.org/10.1007/978-3-030-03232-6_23
4. Casteigts, A., Flocchini, P., Quattrociocchi, W., Santoro, N.: Time-varying graphs and dynamic networks. Int. J. Parallel Emergent Distrib. Syst. **27**(5), 387–408 (2012). https://doi.org/10.1080/17445760.2012.668546
5. Das, A., Bose, K., Sau, B.: Exploring a dynamic ring without landmark. CoRR abs/2107.02769 (2021). https://arxiv.org/abs/2107.02769
6. Das, S., Flocchini, P., Kutten, S., Nayak, A., Santoro, N.: Map construction of unknown graphs by multiple agents. Theor. Comput. Sci. **385**(1–3), 34–48 (2007). https://doi.org/10.1016/j.tcs.2007.05.011

7. Das, S., Di Luna, G.A., Gasieniec, L.A.: Patrolling on dynamic ring networks. In: Catania, B., Královič, R., Nawrocki, J., Pighizzini, G. (eds.) SOFSEM 2019. LNCS, vol. 11376, pp. 150–163. Springer, Cham (2019). https://doi.org/10.1007/978-3-030-10801-4_13
8. Deng, X., Papadimitriou, C.H.: Exploring an unknown graph. J. Graph Theory **32**(3), 265–297 (1999)
9. Dieudonné, Y., Pelc, A.: Deterministic network exploration by anonymous silent agents with local traffic reports. ACM Trans. Algorithms **11**(2), 10:1–10:29 (2014). https://doi.org/10.1145/2594581
10. Erlebach, T., Hoffmann, M., Kammer, F.: On temporal graph exploration. In: Halldórsson, M.M., Iwama, K., Kobayashi, N., Speckmann, B. (eds.) ICALP 2015. LNCS, vol. 9134, pp. 444–455. Springer, Heidelberg (2015). https://doi.org/10.1007/978-3-662-47672-7_36
11. Flocchini, P., Kellett, M., Mason, P.C., Santoro, N.: Map construction and exploration by mobile agents scattered in a dangerous network. In: 23rd IEEE International Symposium on Parallel and Distributed Processing, IPDPS 2009, Rome, Italy, 23–29 May 2009, pp. 1–10. IEEE (2009). https://doi.org/10.1109/IPDPS.2009.5161080
12. Flocchini, P., Mans, B., Santoro, N.: On the exploration of time-varying networks. Theor. Comput. Sci. **469**, 53–68 (2013). https://doi.org/10.1016/j.tcs.2012.10.029
13. Fraigniaud, P., Ilcinkas, D., Peer, G., Pelc, A., Peleg, D.: Graph exploration by a finite automaton. Theor. Comput. Sci. **345**(2–3), 331–344 (2005). https://doi.org/10.1016/j.tcs.2005.07.014
14. Gotoh, T., Flocchini, P., Masuzawa, T., Santoro, N.: Exploration of dynamic networks: tight bounds on the number of agents. J. Comput. Syst. Sci. **122**, 1–18 (2021). https://doi.org/10.1016/j.jcss.2021.04.003
15. Gotoh, T., Sudo, Y., Ooshita, F., Kakugawa, H., Masuzawa, T.: Group exploration of dynamic tori. In: 38th IEEE International Conference on Distributed Computing Systems, ICDCS 2018, Vienna, Austria, 2–6 July 2018, pp. 775–785. IEEE Computer Society (2018). https://doi.org/10.1109/ICDCS.2018.00080
16. Ilcinkas, D., Klasing, R., Wade, A.M.: Exploration of constantly connected dynamic graphs based on cactuses. In: Halldórsson, M.M. (ed.) SIROCCO 2014. LNCS, vol. 8576, pp. 250–262. Springer, Cham (2014). https://doi.org/10.1007/978-3-319-09620-9_20
17. Ilcinkas, D., Wade, A.M.: On the power of waiting when exploring public transportation systems. In: Fernàndez Anta, A., Lipari, G., Roy, M. (eds.) OPODIS 2011. LNCS, vol. 7109, pp. 451–464. Springer, Heidelberg (2011). https://doi.org/10.1007/978-3-642-25873-2_31
18. Ilcinkas, D., Wade, A.M.: Exploration of the T-interval-connected dynamic graphs: the case of the ring. In: Moscibroda, T., Rescigno, A.A. (eds.) SIROCCO 2013. LNCS, vol. 8179, pp. 13–23. Springer, Cham (2013). https://doi.org/10.1007/978-3-319-03578-9_2
19. Kshemkalyani, A.D., Molla, A.R., Sharma, G.: Efficient dispersion of mobile robots on dynamic graphs. In: 40th IEEE International Conference on Distributed Computing Systems, ICDCS 2020, Singapore, 29 November–1 December 2020, pp. 732–742. IEEE (2020). https://doi.org/10.1109/ICDCS47774.2020.00100
20. Kuhn, F., Oshman, R.: Dynamic networks: models and algorithms. SIGACT News **42**(1), 82–96 (2011). https://doi.org/10.1145/1959045.1959064
21. Di Luna, G., Dobrev, S., Flocchini, P., Santoro, N.: Distributed exploration of dynamic rings. Distrib. Comput. **33**(1), 41–67 (2018). https://doi.org/10.1007/s00446-018-0339-1

22. Luna, G.A.D., Flocchini, P., Pagli, L., Prencipe, G., Santoro, N., Viglietta, G.: Gathering in dynamic rings. Theor. Comput. Sci. **811**, 79–98 (2020). https://doi.org/10.1016/j.tcs.2018.10.018
23. Mandal, S., Molla, A.R., Moses, W.K.: Live exploration with mobile robots in a dynamic ring, revisited. In: Pinotti, C.M., Navarra, A., Bagchi, A. (eds.) ALGOSENSORS 2020. LNCS, vol. 12503, pp. 92–107. Springer, Cham (2020). https://doi.org/10.1007/978-3-030-62401-9_7
24. Michail, O., Spirakis, P.G.: Traveling salesman problems in temporal graphs. Theor. Comput. Sci. **634**, 1–23 (2016). https://doi.org/10.1016/j.tcs.2016.04.006
25. Ooshita, F., Datta, A.K.: Brief announcement: feasibility of weak gathering in connected-over-time dynamic rings. In: Izumi, T., Kuznetsov, P. (eds.) SSS 2018. LNCS, vol. 11201, pp. 393–397. Springer, Cham (2018). https://doi.org/10.1007/978-3-030-03232-6_27
26. Panaite, P., Pelc, A.: Exploring unknown undirected graphs. J. Algorithms **33**(2), 281–295 (1999). https://doi.org/10.1006/jagm.1999.1043
27. Santoro, N.: Time to change: on distributed computing in dynamic networks (Keynote). In: 19th International Conference on Principles of Distributed Systems, OPODIS 2015, Rennes, France, 14–17 December 2015. LIPIcs, vol. 46, pp. 3:1–3:14. Schloss Dagstuhl - Leibniz-Zentrum für Informatik (2015). https://doi.org/10.4230/LIPIcs.OPODIS.2015.3

Loosely-Stabilizing Maximal Independent Set Algorithms with Unreliable Communications

Rongcheng Dong[1(✉)], Yuichi Sudo[2], Taisuke Izumi[1], and Toshimitsu Masuzawa[1]

[1] Osaka University, Osaka, Japan
{r-dong,t-izumi,masuzawa}@ist.osaka-u.ac.jp
[2] Hosei University, Tokyo, Japan
sudo@hosei.ac.jp

Abstract. Self-stabilization is a promising paradigm to design highly adaptive distributed systems. However, it cannot be realized when the communication between processes is unreliable with some constant probability. To circumvent such impossibility, this paper adopts the concept of *loose-stabilization* for the first time, which is a practical alternative of self-stabilization, and proposes three systematic approaches to realize loose-stabilization in the atomic-state model with *probabilistically erroneous communications*, namely, the *redundant-state* approach, the *step-up* approach, and the *repetition* approach. Further, we apply these approaches to design three corresponding loosely-stabilizing algorithms for the maximal independent set problem.

Keywords: Loose-stabilization · Maximal independent set · Probabilistically erroneous communications

1 Introduction

Self-stabilization [4] is a promising paradigm to design distributed systems that can autonomously adapt to dynamics caused by transient faults and topology changes of networks. A self-stabilizing system is characterized by two properties called *convergence* and *closure*. The convergence allows the system to eventually reach legitimate configurations (i.e., configurations satisfying the problem specification) regardless of the initial configuration, and the closure makes the system stay in legitimate configurations *forever*.

An inherent limitation of conventional self-stabilizing algorithms is that it requires the system to be fault-free during its convergence. Thus, the design of self-stabilizing algorithms under the threat of perpetual faults is often recognized as a challenging problem. The adversarial message corruption is one of

This work was partially supported by JSPS KAKENHI Grant Numbers 19H04085, 19K11824, 21H05854, and 20H04140.

C. Johnen et al. (Eds.): SSS 2021, LNCS 13046, pp. 335–349, 2021.
https://doi.org/10.1007/978-3-030-91081-5_22

the popular and strong models of perpetual faults, where at each time step the adversary chooses a set of links (whose size is typically constrained) and modifies the messages transferred through the chosen edges maliciously. Even if the number of corrupted edges is bounded, the adversarial message corruption model can preclude self-stabilizing solutions for most of the non-trivial problems. It can be easily proved by the standard partition-based argument: consider the maximal independent set (MIS) problem in a network consisting of two processes (A and B) and one link between the processes. Briefly speaking, a set S of processes in a graph (or a network) is an independent set if any two processes in S are not adjacent to each other, and if S is not a proper subset of any other independent set, we say S is a maximal independent set. In a legitimate configuration where A is independent (i.e., a member of an MIS S) and B is dominated (i.e., a non-member of S), when A sends a message to inform B that A is an independent process, such message may be corrupted in the link by the adversary and B cannot get the correct information, hence B decides to change its state to become an independent process, which leads to an illegitimate configuration.

The above observation yields the interest in exploring a reasonably relaxed model of message corruption. Probabilistic error models, where message corruption is modeled as a stochastic event that the adversary cannot control, are widely accepted as a reasonable assumption. It is not only standard in information theory but also popular in distributed computing. Self-stabilizing solutions are, however, still ruled out even in most of the probabilistic error-models because it still admits the partition-based argument. More precisely, it allows an execution starting from any legitimate configuration to some non-legitimate one with a non-zero probability.

To circumvent the impossibility of self-stabilization in probabilistic error models, this paper focus on *loose-stabilization* [13], which is a relaxed variant of conventional self-stabilization. While keeping the same convergence property with self-stabilization, loose-stabilization relaxes the closure property: the system is allowed to deviate from legitimate configurations after being legitimate for a long time. Loose-stabilization is practically equivalent to self-stabilization if the duration when the system stays in legitimate configurations (called *holding time*) is much longer (e.g., exponentially longer) than the time required to reach a legitimate configuration (called *convergence time*). Therefore, even if some unexpected error occurs and causes the system to become illegitimate, in a relatively short time the system can converge to legitimate configurations again and keeps being legitimate for a sufficiently long time with high probability.

1.1 Related Work

The maximal independent set problem has been well studied by many researchers for decades because it is a very fundamental problem and has many applications in networks. Luby [11] proposed a classic randomized distributed MIS algorithm with round complexity $O(\log n)$ based on Monte Carlo algorithms, where n is the number of processes. It is first improved by Barenboim et al. [3], who proposed an MIS algorithm running in $O(\log^2 \Delta + 2^{O(\sqrt{\log\log n})})$ rounds where Δ is the maximum degree, while the message size is up to $poly(\Delta \log n)$ bits. Rozhon et al. [12] improved the round complexity to $O(\log \Delta) + poly(\log \log n)$, which is the fastest randomized MIS algorithm currently. In their paper they also proposed a deterministic MIS algorithm that runs in $poly(\log n)$ rounds while the message size is only $O(\log n)$.

Hedetniemi et al. [8] proposed a simple self-stabilizing MIS algorithm with constant round complexity while assuming a centralized scheduler. Arapoglu et al. [2] proposed a self-stabilizing MIS algorithm under the fully distributed scheduler running in $max\{3n - 6, 2n - 1\}$ moves (the total number of process actions). Turau [15] considered randomization and proposed a randomized self-stabilizing MIS algorithm in the synchronous model that converges in $O(\log n)$ rounds.

Self-stabilization over unreliable communications has been studied for decades. Afek et al. [1] proposed a self-stabilizing alternating bit protocol while considering message loss. Dolev et al. [5] proposed a self-stabilizing data-link protocol that emulates a reliable FIFO communication channel over unreliable non-FIFO channels, where messages could be lost, duplicated, created or reordered.

The notion of loose-stabilization was first proposed by Sudo et al. [13] to circumvent the impossibility that the closure of self-stabilizing leader election in the probabilistic population protocol (PPP) model is impossible to satisfy, and it is mainly used on leader election problem in the PPP model such as [10,13,14]. Feldmann et al. [6] adopted the idea of loose-stabilization and applied it in the message passing model to design a randomized congestion control algorithm. Other variants of self-stabilization are also studied by researchers, such as probabilistic stabilization [9] and weak stabilization [7].

1.2 Our Contributions

In this paper, self-stabilization tolerant to erroneous communication is considered for the first time, and the concept of loose-stabilization is applied to circumvent the impossibility of closure property in networks with probabilistically erroneous communication. Specifically, we present three systematic approaches for designing loose-stabilizing algorithms in probabilistic error models: the *redundant-state*, the *step-up*, and the *repetition* approaches. All the approaches assume the atomic-state model (where each process can atomically change its state depending on its state and the neighbors' current states) with the stochastic scheduler and follow a certain kind of error-correction/detection mechanism. Two different models of message corruption are considered: *uniformly distributed error* and *arbitrarily distributed error* models. Both models assume that each message is corrupted with a constant probability ρ $(0 < \rho < 1)$. *Uniformly distributed error* model assumes that the values of the corrupted message are uniformly distributed on the message's domain, while in the *arbitrarily distributed error* model there are no such constraints.

We apply these approaches to design three corresponding loosely-stabilizing algorithms for the *maximal independent set* (MIS) problem, and the performance of the algorithms are summarized in Table 1, where n denotes the number of the processes, d denotes a sufficiently large constant, and Δ denotes the maximum degree among all processes. In the following, we briefly describe these approaches.

Redundant-State Approach. The *redundant-state* approach is applicable to anonymous networks of bounded degree (i.e., $\Delta = O(1)$) and the uniformly distributed error model with error probability $\rho = 1 - \epsilon$ for an arbitrarily small constant ϵ. The key idea of this approach is to enlarge the domain of the message by introducing a large amount of redundancy so that even if some message is corrupted, it becomes a meaningless message (i.e., the bit sequence is invalid in the correct behavior of the algorithm) with high probability. In other words, the erroneous value in such a message can be detected and does not cause any incorrect effect to the receiver with high probability. This mechanism allows processes to greatly confine the influences from erroneous communications.

Step-Up Approach. To circumvent the constraint of bounded degree in the *redundant-state* approach, the *step-up* approach makes better use of the redundant values of the messages. In the previous approach, we only use the redundant values to reduce the effect of corrupted messages, and we do not care about the exact values in the corrupted messages, while in this approach processes confine the effect of corrupted messages by utilizing the redundant values as the buffer, so that even if a process takes a bad action due to some corrupted message, with high probability the result of such action can be corrected before the neighbors of the process are influenced.

Repetition Approach. The *repetition* approach assumes the arbitrarily distributed error model with error probability $\rho = 1/2 - \epsilon$ for an arbitrarily small constant ϵ. Different from the previous two approaches, the *repetition* approach does not use the redundant values. The key idea of this approach is the error-correction by a very simple repetition code with majority decoding. The assumption of $\rho < 1/2 - \epsilon$ admits an error-correction with high probability for a sufficiently large number of repetitions. It should be emphasized that this approach is not so trivial: to utilize block codes (including most of the linear codes with strong error-correction features), some synchronization mechanism is necessary, but our system assumption does not equip with full (round-based) synchrony. It is not clear how one can develop such a mechanism in the loosely-stabilizing manner under the probabilistic scheduler with unreliable communication. Fortunately, the repetition code works without any synchronization mechanism, except for the period around the time when the encoded value changes. Thus it keeps a long holding time after the states of all processes stabilize. In the convergence period, however, the change of processes' states would cause the failure of error correction with an unexpectedly high probability. To prevent such failure, we install an additional mechanism of priority-based monotone fixing of processes' states. Compared with the error-free case, it yields a longer convergence time, but it is still polynomial of n.

Table 1. Performance of three loosely-stabilizing MIS algorithms

	Redundant-State	Step-Up	Repetition
Process ID	Unavailable	Unavailable	Available
Maximum Degree	Constant	Arbitrary	Arbitrary
Error Distribution	Uniform	Uniform	Arbitrary
$maxECT$	$O(\log n)$	$O(n \log^3 n)$	$O(n^2 \log n)$
$minEHT$	$\Omega(n^d)$	$\Omega(n^d)$	$e^{\Omega(n)}$
Space Complexity	$O(\log n)$	$O(\log n)$	$O(n\Delta \log n)$

2 Preliminaries

A distributed system comprises a set of autonomous processes and the communication links that connect the processes. We abstract such system as an undirected graph $G = (V, E)$: the vertex set $V = \{p_0, p_1, ..., p_{n-1}\}$ represents the set of n processes where $n \geq 1$, and the edge set $E \subseteq V \times V$ represents the set of communication links. If $(p_i, p_j) \in E$, we say the process p_i and the process p_j are neighbors and can communicate with each other. The set of neighbors of process p_i is denoted by $Neigh(p_i) = \{p_j \mid (p_i, p_j) \in E\}$, and p_i can distinguish each of its neighbors by some local labeling mechanism. Denote the degree of process p_i by $\Delta_i = |Neigh(p_i)|$ and the maximum degree among all processes by $\Delta = \max\{\Delta_i \mid i \in \{0, 1, \ldots, n-1\}\}$. In this paper, we have no assumptions on the topology of G. In other words, G has an arbitrary topology.

The computational model used in this paper is the atomic-state model: each process p_i has some local variables, and the values of variables of p_i define the state of p_i. A configuration of G is a function that specifies the states of all processes. Each process can read its state and all of its neighbors' states but can update only its state. Since we consider the unreliable communication links in this paper, we assume that each time a process p_i tries to read the value of one variable from process $p_j \in Neigh(p_i)$, with a constant probability ρ $(0 < \rho < 1)$, process p_i obtains an incorrect result.

With respect to the unreliable communication, we introduce two kinds of error model: the *uniformly distributed error* model and the *arbitrarily distributed error* model. In the *uniformly distributed error* model, when the process p_i tries to read the state of its neighbor $p_j \in Neigh(p_i)$ and the erroneous communication occurs, p_i gets the value s as p_j's current state with probability $1/(|S|-1)$ for each $s \in S \setminus \{s_j\}$, where S is the set of all process-states and s_j is the true state of p_j. In this model, we assume $0 < \rho < 1 - \epsilon$ for some positive constant ϵ. In the *arbitrarily distributed error* model, when the process p_i tries to read the state of its neighbor $p_j \in Neigh(p_i)$ and the erroneous communication occurs, then the resultant state p_i reads from p_j is an arbitrary value (i.e., chosen from the adversary). In this model, we assume $0 < \rho < 1/2 - \epsilon$.

The activation of processes is determined by a scheduler. We assume a uniformly distributed scheduler U in this paper: at each *time step* (*step* for short), each process is independently activated by U, with a constant probability ϕ, to make a *move*. If p_i is activated at step t, it reads the states of all its neighbors and its own, then does some local computations, and updates its state if necessary. Given an initial configuration C_0 and the scheduler U, an execution $E_{Alg}(C_0, U)$ of an algorithm Alg is an infinite sequence of configurations $C_0, C_1, ...$ where C_{i+1} is obtained by taking a step from C_i for all $i \geq 0$.

2.1 Loosely-Stabilizing MIS

We first define the specification of the MIS problem and then define the loosely-stabilizing MIS algorithm.

A set of processes $S \subseteq V$ of a graph $G = (V, E)$ is called an independent set if $\forall p, q \in S : (p, q) \notin E$. Further, if S is not a proper subset of any other independent set, we call S the maximal independent set (MIS). We assume that every algorithm specifies its *output function* that maps each process-state to an output value. Given an *output function* ξ mapping each process-state to 0 or 1, the specification of the MIS problem, denoted by SP_{MIS}, is defined as the following predicate on configurations:

Definition 1. *For any configuration C, define $f(C)$ as the set of processes that is in state s such that $\xi(s) = 1$. Then, we define SP_{MIS} as the following: $SP_{MIS}(C) = true$ holds if and only if $f(C)$ is a maximal independent set in G.*

Denote by $\mathcal{C}$ the set of all possible configurations of G. For any $\mathcal{S} \subseteq \mathcal{C}$ and $C \in \mathcal{C}$, define $ECT_{Alg}(C, \mathcal{S})$ as the expected time until an execution of an algorithm Alg reaches a configuration in $\mathcal{S}$ starting from $C \in \mathcal{C}$. Define $EHT_{Alg}(C, \xi)$ as the expected time until an execution of an algorithm Alg that starts from a configuration C changes the output of some process for the first time, i.e., the expected time until the execution reaches a configuration C' such that $\xi(C(v)) \neq \xi(C'(v))$ for some agent v.

Definition 2. *An algorithm Alg is an (α, β)-loosely-stabilizing MIS algorithm if there exists a set $\mathcal{S}$ of configurations satisfying:*

- $\forall C \in \mathcal{S} : SP_{MIS}(C) = true$
- $\max_{C \in \mathcal{C}} ECT_{Alg}(C, \mathcal{S}) \leq \alpha$
- $\min_{C \in \mathcal{S}} EHT_{Alg}(C, \xi) \geq \beta$

We call α and β the maximum expected convergence time and the minimum expected holding time of Alg respectively, and $\mathcal{S}$ the set of safe configurations.

3 Redundant-State Approach

3.1 Description of Algorithm $\mathcal{RS}$

The algorithm $\mathcal{RS}$ is based on the redundant-state approach, thus it assumes a constant maximum degree (i.e. $\Delta = \Theta(1)$) and the uniformly distributed error model with error probability $\rho = 1 - \epsilon$ for an arbitrarily small constant ϵ. Processes are anonymous, that is, identifiers are not available to processes. Each process p_i has one variable $v_i \in \{0, 1, ..., c\}$ where the parameter $c = \Theta(n^{d+1})$ and d is a sufficiently large constant. A process sets value c (resp. 0) when it is a member (resp. non-member) of the MIS the algorithm constructs, and values from 1 to $c-1$ are treated as the redundant values. The output function $\xi_{\mathcal{RS}}$ of $\mathcal{RS}$ is defined as follows: for each process $p_i \in V$, if $v_i = c$ then $\xi_{\mathcal{RS}}(v_i) = 1$, otherwise $\xi_{\mathcal{RS}}(v_i) = 0$.

The algorithm $\mathcal{RS}$ (given in Algorithm 1) works as follows. Each time a process p_i is activated, it reads the states of its neighbors. Next, if v_i is maximal around its neighbors and v_i is not c, it is set to c; if there exists a neighbor of p_i such that v_i is smaller than that of the neighbor, then v_i is set to 0; if v_i is c and there exists a neighbor of p_i such that the state of the neighbor is also c, then v_i is set to 0.

Algorithm 1: $\mathcal{RS}$

Variables in p_i:
$v_i \in \{0, 1, ..., c\}$ where $c = \Theta(n^{d+1})$ and d is a sufficiently large constant

```
1 if ∀p_j ∈ Neigh(p_i) : v_j ≤ v_i ∧ v_i ≠ c then
2 |   v_i ← c
3 if ∃p_j ∈ Neigh(p_i) : v_j > v_i then
4 |   v_i ← 0
5 if ∃p_j ∈ Neigh(p_i) : v_j = c ∧ v_i = c then
6 |   v_i ← 0
```

3.2 Analysis of $\mathcal{RS}$

At first, we divide processes into four types according to the states of their own and their neighbors'.

Definition 3. *(1) independent process: a process p_i is an independent process if $v_i = c$ and $\forall p_j \in Neigh(p_i) : v_j < c$.*
(2) dominated process: a process p_i is a dominated process if $v_i = 0$ and $\exists p_j \in Neigh(p_i) : p_j$ is an independent process.
(3) pseudo-dominated process: a process p_i is a pseudo-dominated process if $v_i = 0$, $\exists p_j \in Neigh(p_i) : v_j = c$, but p_i is not dominated (i.e., there is no independent neighbor).
(4) illegal process: a process p_i is illegal if p_i is not independent, dominated, or pseudo-dominated. That is, (i) $0 < v_i < c$, (ii) $v_i = 0 \wedge \forall p_j \in Neigh(p_i) : v_j < c$, or (iii) $v_i = c \wedge \exists p_j \in Neigh(p_i) : v_j = c$.

Based on Definition 3, we define the safe configurations of $\mathcal{RS}$:

Definition 4. *The set of safe configurations $\mathcal{S}_1$ of $\mathcal{RS}$ is the set of configurations where only independent and dominated processes exist.*

Maximum Expected Convergence Time. In general, the algorithm $\mathcal{RS}$ can be considered to have two phases. In the first phase, we eliminate all the values of processes' states from 1 to $c-1$ by setting them to 0 or c. See Lines 1 to 4 in the pseudo-code. Since once a process makes a move, it can only change its value to 0 or c, and by the union bound every process makes at least one move in $O(\log n)$ steps with high probability, we can conclude that the first phase ends in $O(\log n)$ steps with high probability. Moreover, at the end of this phase, all illegal processes are either in the second or third category of the illegal process.

The second phase is where the redundant states start to work. Due to the existence of the redundant states, the algorithm $\mathcal{RS}$ keeps the independent (or dominated respectively) processes staying independent (or dominated respectively) with high probability. Along with the assumption of constant degree, illegal (or pseudo-dominated, respectively) processes become independent (or dominated, respectively) with at least a constant probability. If the degree is unbounded, the probability that an illegal process p_i in the second or third category becomes independent is exponentially small in the worst case when the values of all p_i's neighbors are not 0, which yields an exponentially long convergence time. Therefore, the constraint of bounded degree is needed in this algorithm. By also taking $O(\log n)$ steps there are no illegal or pseudo-dominated processes in the configuration with high probability. Combining with the first phase we can get the maximum expected convergence time of $\mathcal{RS}$.

Theorem 1. $\max_{C \in \mathcal{C}_1} ECT_{\mathcal{RS}}(C, \mathcal{S}_1) = O(\log n)$ *in terms of steps, where* $\mathcal{C}_1$ *is the set of all possible configurations of* $\mathcal{RS}$.

Minimum Expected Holding Time. Next we analyze the minimum expected holding time of $\mathcal{RS}$. Again, thanks to the redundant states, the probability that at least one process becomes illegal in any step from a safe configuration is at most $O(n/c)$, hence the minimum expected holding time is $\Omega(c/n)$ steps, which is $\Omega(n^d)$ steps by the choice of the parameter c.

Theorem 2. $\min_{C \in \mathcal{S}_1} EHT_{\mathcal{RS}}(C, \xi_{\mathcal{RS}}) = \Omega(n^d)$ *in terms of steps.*

Space Complexity. The space complexity of $\mathcal{RS}$ is straightforward: each process p_i has one variable $v_i \in \{0, 1, ..., c\}$, hence the space complexity is $O(\log c)$, which is $O(\log n)$ bits by the definition of the parameter c.

4 Step-Up Approach

4.1 Description of Algorithm $\mathcal{SU}$

The algorithm $\mathcal{SU}$ is based on the step-up approach that has same setting as the redundant-state approach, and circumvents the constraint of bounded degree, thus we consider the graphs that $\mathcal{SU}$ works on have no restriction on the degree of processes. The output function $\xi_{\mathcal{SU}}$ of $\mathcal{SU}$ is defined as follows: for each process $p_i \in V$, if $v_i = c$ then $\xi_{\mathcal{SU}}(v_i) = 1$, otherwise $\xi_{\mathcal{SU}}(v_i) = 0$.

The algorithm $\mathcal{SU}$ (given in Algorithm 2) works as follows. Each time a process p_i is activated, it reads the states of its neighbors. Next, if p_i has at least one neighbor whose state is c, then v_i is set to 0; in the other case v_i takes a step up: it is increased by $c/\log^2 c$ unless v_i exceeds the domain, and if that happens v_i is set to c. Notice that in $\mathcal{SU}$ we do not treat the states from 1 to $c-1$ as the contaminated values, but as the buffer that the processes can utilize to confine the effect of erroneous communications in the following way: even if a process misrecognizes that no neighbor of it is a member of the MIS due to some erroneous communication, the process gradually approaches to the state of an MIS member instead of instantly becoming an MIS member. This enables the process to correct its state with high probability before becoming an MIS member.

Algorithm 2: $\mathcal{SU}$

Variables in p_i:
$v_i \in \{0, 1, ..., c\}$ where $c = \Theta(n^{d+1})$ and d is a sufficiently large constant

```
7  if ∃p_j ∈ Neigh(p_i) : v_j = c then
8  └ v_i ← 0
9  if ∀p_j ∈ Neigh(p_i) : v_j ≠ c then
10 └ v_i ← min{c, v_i + c/log² c}
```

4.2 Analysis of $\mathcal{SU}$

Processes in $\mathcal{SU}$ are divided into three types according to the states of their own and their neighbors'.

Definition 5. *(1) independent process: a process p_i is an independent process if $v_i = c$ and $\forall p_j \in Neigh(p_i) : v_j \neq c$. Further, process p_i is a safely independent process if $v_i = c$ and $\forall p_j \in Neigh(p_i) : |v_j - c| = \omega(c/\log c)$.*
(2) dominated process: a process p_i is a dominated process if $v_i \neq c$ and $\exists p_j \in Neigh(p_i) : v_j = c$.
(3) illegal process: a process is an illegal process if it does not belong to any of the above two types of processes.

Based on Definition 5, we define a potential set $Pot(C)$ in each configuration C, and a set of safe configurations $\mathcal{S}_2$ of $\mathcal{SU}$ as follows.

Definition 6. *The potential set $Pot(C)$ concerning the configuration C is the set of safely independent processes and all their neighbors in the configuration C.*

Definition 7. *The set of safe configurations $\mathcal{S}_2$ of $\mathcal{SU}$ is the set of configurations C where $Pot(C) = V$.*

Minimum Expected Holding Time. In a safe configuration, there only exist safely independent processes and dominated processes whose values are low enough by Definition 5, thus these are the processes that we need to concern about now. Similar to the algorithm $\mathcal{RS}$, each safely independent process in $\mathcal{SU}$ does not change its values with probability $1 - O(\Delta/c)$ due to the existence of the redundant states. On the other hand, although a dominated process may incorrectly think that the values of all the neighbors are not c and then increases its value with a constant probability, its value is low enough so that $O(\log c)$ such moves are needed to make it illegal, as well as its safely independent neighbors. Such gap (the difference between the values of a safely independent process and a dominated process) works as a buffer that makes dominated processes illegal only with an extremely small probability of $\Theta(1)^{\Omega(\log c)} = e^{-\Omega(\log c)}$. Therefore, the minimum expected holding time of $\mathcal{SU}$ is $\min\{\Omega(c/\Delta), e^{\Omega(\log c)}\} = \Omega(c/\Delta)$ steps, which is $\Omega(n^{d+1}/\Delta) = \Omega(n^d)$ steps by the choices of the parameter c and d.

Theorem 3. $\min_{C \in \mathcal{S}_2} EHT_{\mathcal{SU}}(C, \xi_{\mathcal{SU}}) = \Omega(n^d)$ *in terms of steps.*

Maximum Expected Convergence Time. Next, we analyze the expected convergence time of $\mathcal{SU}$. To do so, we define a set of processes that have *high* values as $High(C)$ in each configuration C. Processes in this set can be seen as good candidates to become safely independent processes in subsequent configurations.

Definition 8. *Define $High(C) \subseteq V - Pot(C)$ to be the set of processes p_i which have values v_i in configuration C such that $|v_i - c| = O(c/\log c)$.*

Algorithm $\mathcal{SU}$ consists of multiple iterations. In each iteration, we can guarantee that $High(C)$ is not empty for long. In other words, there are always some good candidates to become safely independent processes during the iteration with high probability. Consider the worst case where all the values of processes are 0 at the beginning of the iteration, thus all processes now are illegal processes. Each time a process p_i does not obtain value c from any of its neighbors, it can increase its value by $c/\log^2 c$. Even if the degree of processes is unbounded, such case happens with at least a constant probability. By the Chernoff bound and the property of binomial distribution, we can prove that $p_i \in High(C)$ in $O(\log^2 c)$ steps with high probability.

Now we consider the processes in $High(C)$. Processes in $High(C)$ are merely influenced by those that are not in $High(C)$ due to the existence of the buffer between them, hence we focus on the processes that are neighboring with those that are also in $High(C)$. Although such processes may set their values to 0 with constant probability, it can be proved that there exists at least one process in $High(C)$ that reaches value c and keeps this value in $O(\log c)$ steps with at least a constant probability. Therefore, in $O(\log n \cdot \log^2 c)$ steps from the beginning of the iteration, the number of safely independent processes is incremented by at least one with high probability, which means that after $O(n)$ iterations a safe configuration S is reached, i.e., $Pot(S) = V$, with high probability.

Compared with the previous algorithm $\mathcal{RS}$, illegal processes in $\mathcal{SU}$ need to take a non-negligible number of "steps" to become independent. Consequently, the convergence time is longer than that of $\mathcal{RS}$. However, by using this mechanism, we can remove the constraint of bounded degree in $\mathcal{RS}$. By the definition of the parameter c, we have the following theorem.

Theorem 4. $\max_{C \in \mathcal{C}_2} ECT_{\mathcal{SU}}(C, \mathcal{S}_2) = O(n \log^3 n)$ *in terms of steps, where* $\mathcal{C}_2$ *is the set of all possible configurations of* $\mathcal{SU}$.

Space Complexity. Processes in $\mathcal{SU}$ have the same variables as those in $\mathcal{RS}$, thus the space complexity of $\mathcal{SU}$ is also $O(\log n)$ bits.

5 Repetition Approach

5.1 Description of Algorithm $\mathcal{RE}$

The algorithm $\mathcal{RE}$ is based on the repetition approach, thus it assumes the arbitrarily distributed error model with error probability $\rho = 1/2 - \epsilon$ for an arbitrarily small constant ϵ. A unique identifier ID_i for each process p_i is needed in this algorithm to break symmetry. Each process p_i has a binary variable $v_i \in \{0, 1\}$. The output function $\xi_{\mathcal{RE}}$ of $\mathcal{RE}$ is defined as follows: for each process $p_i \in V$, if $v_i = 1$ then $\xi_{\mathcal{SU}}(v_i) = 1$, otherwise $\xi_{\mathcal{SU}}(v_i) = 0$.

The algorithm $\mathcal{RE}$ (given in Algorithm 3) works as follows. For each process p_i, if the counter $counter_i$ that represents the number of readings p_i makes so far is less than $2k+1$ where the parameter $k = \Theta(n)$, it obtains ID_j and v_j from each of its neighbor p_j, stores the ID_j in a dictionary $id\_cand_i(j)$, increments another counter $counter\_one_i(j)$ if $v_j = 1$, which represents the number of value 1 p_i obtains from p_j, and increments $counter_i$. If $counter_i = 2k + 1$, process p_i updates v_i in the following way: at first, for each $p_j \in Neigh(p_i)$, process p_i regards the key with the largest value in the dictionary $id\_cand_i(j)$ as true ID_j, and stores each ID_j in a list id_table_i; besides, process p_i regards $v_j = 1$ if $counter\_one_i(j) > k$. Next, if $\exists p_j \in Neigh(p_i)$ such that $ID_j > ID_i$ and $v_j = 1$, process p_i sets v_i to 0; in other cases p_i sets v_i to 1. At last, process p_i resets all the counters and dictionaries id_cand_i.

5.2 Analysis of $\mathcal{RE}$

Since each process p_i regards the majority value among the results of $2k + 1$ reads from each neighbor p_j, we expect that such majority is the true value of p_j with high probability. However, it does not hold if p_j changes its state during the communication period. Therefore, we try to *fix* the state of a process gradually by introducing the notion of a *maximal process*.

Algorithm 3: $\mathcal{RE}$

```
Constant in p_i:
  ID_i
Variables in p_i:
  v_i ∈ {0,1}
  counter_i ∈ {1,2,...,2k+1}
  counter_one_i(j) ∈ {1,2,...,2k+1}
  id_cand_i(j): a dictionary where keys are ID_js read from p_j ∈ Neigh(p_i)
  id_table_i: a list indexed on the neighbors of p_i
Notation:
  counter_i counts the number of times that p_i communicates with its neighbors.
  counter_one_i(j) counts the number of times that p_i gets v_j = 1 from p_j.
  id_cand_i(j)[ID_j] represents the number of times that p_i gets ID_j from p_j.
  id_table_i[j] represents the ID of p_j that p_i regards.
11 if counter_i < 2k + 1 then
12     counter_i ← counter_i + 1
13     for each p_j ∈ Neigh(p_i) do
14         if ID_j ∈ id_cand_i(j) then
15             id_cand_i(j)[ID_j] ← id_cand_i(j)[ID_j] + 1
16         else
17             create key ID_j in id_cand_i(j)
18             id_cand_i(j)[ID_j] ← 1
19         if v_j = 1 then
20             counter_one_i(j) ← counter_one_i(j) + 1
21 if counter_i = 2k + 1 then
22     for each p_j ∈ Neigh(p_i) do
23         id_table_i[j] ← arg max_key id_cand_i(j)[key]
24     if ∃p_j ∈ Neigh(p_i) : id_table_i[j] > ID_i ∧ counter_one_i(j) > k then
25         v_i ← 0
26     else
27         v_i ← 1
28     counter_i ← 0
29     for each p_j ∈ Neigh(p_i) do
30         counter_one_i(j) ← 0
31         id_cand_i(j) ← empty
```

Definition 9. *Process p_i is a maximal process in G if $\forall p_j \in Neigh(p_i) : ID_j < ID_i$ holds in G.*

Based on Definition 9, we inductively define some related sets of processes.

Definition 10. *Define $M(1)$ to be the set of maximal processes in G, and $N(1)$ to be the set of processes in G that are neighboring to any process in $M(1)$. For* *any integer* $x \geq 2$, *define* $M(x)$ *to be the set of maximal processes in the induced sub-graph* $G_x = G\left[V - \bigcup_{y=1}^{x-1}(M(y) \cup N(y))\right]$, *and* $N(x)$ *to be the set of processes in* G_x *that are neighboring to any process in* $M(x)$.

To define the safe configurations of $\mathcal{RE}$, we first define the set of independent and dominated processes and then introduce the notion of *legitimate initial state* for each process.

Definition 11. *Process* p_i *is in the legitimate initial state if* $counter_i = 0$ *and* $\forall p_j \in Neigh(p_i) : counter\_one_i(j) = 0$ *and* $id\_cand_i(j) = \emptyset$.

Definition 12. *(1) independent process: a process* p_i *is an independent process if* $v_i = 1$ *and* $\forall p_j \in Neigh(p_i) : v_j = 0$. *(2) dominated process: a process* p_i *is a dominated process if* $v_i = 0$ *and* $\exists p_j \in Neigh(p_i) : v_j = 1$.

Based on the above two definitions, we define the safe configurations of $\mathcal{RE}$.

Definition 13. *The set of safe configurations* $\mathcal{S}_3$ *of* $\mathcal{RE}$ *is the set of configurations where* $\forall p \in \bigcup_{y=1}^{n} M(y) : p$ *is an independent process,* $\forall q \in \bigcup_{y=1}^{n} N(y) : q$ *is a dominated process, and each process has reached its legitimate initial state at least once.*

The reason why we require that each process has reached its legitimate initial state at least once in safe configurations is that, if we do not require this, in some initial configurations maybe there only exist independent and dominated processes, but the states of the processes are arbitrary and in the worst case some of them may become illegitimate in one step with a non-negligible probability.

Maximum Expected Convergence Time. Obviously, for any integer $x \geq 2$, if $V - \bigcup_{y=1}^{x-1}(M(y) \cup N(y))$ is not empty, then $M(x) \cup N(x) \neq \emptyset$. Therefore, we have $\bigcup_{y=1}^{n}(M(y) \cup N(y)) = V$. As a consequence, all processes are gradually fixed into the corresponding sets based on the IDs. At the same time, the state of each process is also gradually fixed with high probability by Algorithm $\mathcal{RE}$, starting from the processes in $M(1)$ and expanding into $N(1), M(2), N(2), \cdots$. In this manner, a *fixed* process p_i can change its state only if the majority result of p_i's reading from one of its *fixed* neighbors, say p_j, becomes a wrong value, and such event happens with a negligible probability.

Theorem 5. $\max_{C \in \mathcal{C}_3} ECT_{\mathcal{RE}}(C, \mathcal{S}_3) = O(n^2 \log n)$ *in terms of steps, where* $\mathcal{C}_3$ *is the set of all possible configurations of* $\mathcal{RE}$.

Minimum Expected Holding Time. From a safe configuration, the worst case that makes the specification SP_{MIS} violated is that a process p_i obtains an incorrect value of a variable from $p_j \in Neigh(p_i)$, and the probability that such case happens is $e^{-O(k)}$. Together with the choice of the parameter k, the maximum expected holding time is $\Theta(k)/e^{-O(k)} = e^{\Omega(n)}$

Theorem 6. $\min_{C \in \mathcal{S}_3} EHT_{\mathcal{RS}}(C, \xi_{\mathcal{RS}}) = e^{\Omega(n)}$ *in terms of steps.*

Space Complexity. For each process p_i, the most space consuming variable is the dictionary id_cand_i. For each $p_j \in Neigh(p_i)$, there is a corresponding $id\_cand_i(j)$; each $id\_cand_i(j)$ stores at most n keys, and the value for each key is at most $2k + 1$. Therefore, id_cand_i requires $O(n\Delta \log k)$ bits, and the space complexity is also $O(n\Delta \log k)$ bits, which is $O(n\Delta \log n)$ bits by the definition of the parameter k.

6 Summary

We proposed three systematic approaches to deal with erroneous communications in distributed systems and applied these approaches to design three loosely-stabilizing MIS algorithms.

Loose-stabilization is not widely used for now. To our knowledge, it is only used in the leader election problem in the PPP model and the congestion control problem in the message passing model. Although many problems like leader election have self-stabilizing solutions, if we alter the ideal setting and introduce some probabilistic fault such as unreliable communications, self-stabilizing solutions may become impossible to achieve. At this moment, we can try to apply loose-stabilization and get a solution that may not be perfect but is totally enough for practical use.

References

1. Afek, Y., Brown, G.M.: Self-stabilization over unreliable communication media. Distrib. Comput. **7**(1), 27–34 (1993)
2. Arapoglu, O., Akram, V.K., Dağdeviren, O.: An energy-efficient, self-stabilizing and distributed algorithm for maximal independent set construction in wireless sensor networks. Comput. Standards Interfaces **62**, 32–42 (2019)
3. Barenboim, L., Elkin, M., Pettie, S., Schneider, J.: The locality of distributed symmetry breaking. J. ACM (JACM) **63**(3), 1–45 (2016)
4. Dijkstra, E.W.: Self-stabilizing systems in spite of distributed control. Commun. ACM **17**(11), 643–644 (1974)
5. Dolev, S., Dubois, S., Potop-Butucaru, M., Tixeuil, S.: Stabilizing data-link over non-fifo channels with optimal fault-resilience. Inf. Process. Lett. **111**(18), 912–920 (2011)
6. Feldmann, M., Götte, T., Scheideler, C.: A loosely self-stabilizing protocol for randomized congestion control with logarithmic memory. In: Ghaffari, M., Nesterenko, M., Tixeuil, S., Tucci, S., Yamauchi, Y. (eds.) SSS 2019. LNCS, vol. 11914, pp. 149–164. Springer, Cham (2019). https://doi.org/10.1007/978-3-030-34992-9_13

7. Gouda, M.G.: The theory of weak stabilization. In: Datta, A.K., Herman, T. (eds.) WSS 2001. LNCS, vol. 2194, pp. 114–123. Springer, Heidelberg (2001). https://doi.org/10.1007/3-540-45438-1_8
8. Hedetniemi, S.M., Hedetniemi, S.T., Jacobs, D.P., Srimani, P.K.: Self-stabilizing algorithms for minimal dominating sets and maximal independent sets. Comput. Math. Appl. **46**(5–6), 805–811 (2003)
9. Israeli, A., Jalfon, M.: Token management schemes and random walks yield self-stabilizing mutual exclusion. In: Proceedings of the Ninth Annual ACM Symposium on Principles of Distributed Computing, pp. 119–131 (1990)
10. Izumi, T.: On space and time complexity of loosely-stabilizing leader election. In: Scheideler, C. (ed.) SIROCCO 2014. LNCS, vol. 9439, pp. 299–312. Springer, Cham (2015). https://doi.org/10.1007/978-3-319-25258-2_21
11. Luby, M.: A simple parallel algorithm for the maximal independent set problem. SIAM J. Comput. **15**(4), 1036–1053 (1986)
12. Rozhoň, V., Ghaffari, M.: Polylogarithmic-time deterministic network decomposition and distributed derandomization. In: Proceedings of the 52nd Annual ACM SIGACT Symposium on Theory of Computing, pp. 350–363 (2020)
13. Sudo, Y., Nakamura, J., Yamauchi, Y., Ooshita, F., Kakugawa, H., Masuzawa, T.: Loosely-stabilizing leader election in a population protocol model. Theor. Comput. Sci. **444**, 100–112 (2012)
14. Sudo, Y., Ooshita, F., Kakugawa, H., Masuzawa, T.: Loosely stabilizing leader election on arbitrary graphs in population protocols without identifiers or random numbers. IEICE Trans. Inf. Syst. **103**(3), 489–499 (2020)
15. Turau, V.: Making randomized algorithms self-stabilizing. In: Censor-Hillel, K., Flammini, M. (eds.) SIROCCO 2019. LNCS, vol. 11639, pp. 309–324. Springer, Cham (2019). https://doi.org/10.1007/978-3-030-24922-9_21

On Regenerating Codes and Proactive Secret Sharing: Relationships and Implications

Karim Eldefrawy[1], Nicholas Genise[1](✉), Rutuja Kshirsagar[2], and Moti Yung[3,4]

[1] SRI International, Menlo Park, USA
{karim.eldefrawy,nicholas.genise}@sri.com
[2] Virginia Tech, Blacksburg, USA
rutujak@vt.edu
[3] Google, Menlo Park, USA
motiyung@google.com
[4] Columbia University, New York, USA

Abstract. We look at two basic coding theoretic and cryptographic mechanisms developed separately and investigate relationships between them and their implications. The first mechanism is Proactive Secret Sharing (PSS), which allows randomization and repair of shares using information from other shares. PSS enables constructing secure multi-party computation protocols that can withstand mobile dynamic attacks. This self-recovery and the redundancy of uncorrupted shares allows a system to overcome recurring faults throughout its lifetime, eventually finishing the computation (or continuing forever to maintain stored data). The second mechanism is Regenerating Codes (RC) which were extensively studied and adopted in distributed storage systems. RC are error correcting (or erasure handling) codes capable of recovering a block of a distributedly held codeword from other servers' blocks. This self-healing nature enables more robustness of a code distributed over different machines. Given that the two mechanisms have a built-in self-healing (leading to stabilizing) and that both can be based on Reed Solomon Codes, it is natural to formally investigate deeper relationships between them. We prove that a PSS scheme can be converted into an RC scheme, and that under some conditions RC can be utilized to instantiate a PSS scheme. This allows us, in turn, to leverage recent results enabling more efficient polynomial interpolation (due to Guruswami and Wooters) to improve the efficiency of a PSS scheme. We also show that if parameters are not carefully calibrated, such interpolation techniques (allowing partial word leakage) may be used to attack a PSS scheme over time. Secondly, the above relationships give rise to extended (de)coding notions. Our first example is mapping the generalized capabilities of adversaries (called generalized adversary structures) from the PSS realm into the RC one. Based on this we define a new variant of RC we call Generalized-decoding Regenerating Code (GRC) where not all network servers have a uniform sub-codeword (motivated by non-uniform probability of attacking different servers case). We finally highlight several

C. Johnen et al. (Eds.): SSS 2021, LNCS 13046, pp. 350–364, 2021.
https://doi.org/10.1007/978-3-030-91081-5_23

interesting research directions due to our results, e.g., designing new improved GRC, and more adaptive RC re-coding techniques.

1 Introduction

Many times in the past, it was found that different areas and concepts in computing are related in some fashion. Uncovering such relations, in turn, led to information flowing and ideas transferring across areas. In this work we deal with two areas that are fundamental to major aspects of safe (and secured) self-healing distributed computing. The first area, *Coding Theory*, is one of the oldest and most fundamental branches of modern computing and communication, dealing with keeping integrity of data in various situations (originally communication channels, then storage media, and finally distributed fault tolerant storage). In the context of modern distributed computing (clusters, storage systems, cloud computing, etc.), a fundamental mechanism to perform fault-tolerant distributed storage is to encode the data to be stored using regenerating codes (RC) [1] (our first area's concentration). RC are a class of error correcting codes (ECC) invented in 2010, which, besides the traditional task of ensuring reliability of data recovery, also provide for efficient repair (regeneration) of failed nodes (holding blocks of the codeword) in a distributed storage system, where recovery is from information stored at other nodes, assuring better maintenance capability of the distributed codeword. We note that most RC work is in the erasure codes model, where servers can fail-stop, but some are in the error correcting mode where servers' memory can be maliciously modified.

The second area is *Secret Sharing Schemes*: given that data in distributed cryptographic systems may be sensitive (e.g., a redundant storage of cryptographic keys), these systems may be the target of byzantine/malicious corruption (not only random or fail-stop faults) which attempt to violate confidentiality of the stored data and its availability. To that end, (threshold) secret sharing (SS) was originally proposed as a fundamental cryptographic defense technique in the late 70s. It distributes a secret with redundancy into shares, where, in order to recover the secret, one needs a threshold of shares, while less than the threshold of shares reveals no information about the secret; the mechanism was employed heavily in security protocols. The security of the standard, and first, SS technique (known as Shamir's SS [2]) is based on polynomial interpolation, and is closely related to Reed-Solomon codes [3]. This relation was the first hint regarding certain connections between coding theoretic methods and sharing secrets procedures, yet this, early uncovered relationship did not continue to attract much attention since then.

In 1991, motivated by malicious adversaries being mobile as in the spread of malware, SS was extended to what became known as *Proactive Secret Sharing (PSS)* [4]. PSS (our second area's concentration), in fact, protects against a mobile adversary that can change the subset of corrupted nodes overtime and thus may eventually compromise all involved nodes over a long period of time (while the standard SS notion only assumes that a subset of the nodes can be corrupted, even over a long period of time). PSS, in fact, adds to SS the ability to t-wise randomize the information held by the sharing servers, and allows for the recovery of the current state held by a server in case a share has been previously destroyed by the adversary which moved away. Since then, PSS was extended [5,6] to more generic settings, beyond threshold adversaries, i.e., what is often called general adversary structures. PSS has also been employed in developing secure multi-party computations against mobile adversaries [6], and for building threshold cryptosystems, say, for supporting distributed certification authorities, and in recent years in various Blockchain based protocols, e.g., [7].

In this paper, renewing connections between coding theory and distributed secret sharing methods, we explore fundamental connections between RC and PSS. It is a natural question to ask, due to the fact that both notions involve reconstruction of information (shares or codewords) held by servers using information (shares or codewords) held at other servers, and that some schemes in both areas are, in fact, related to Reed Solomon (or other algebraic) codes. We also suggest (and demonstrate) utilization of the observed connections to imply new useful paradigms and extensions in one area (RC), building on related paradigms existing in the second area (PSS).

Our Contributions: This paper is the first systematic study of the close relation between *proactive* secret sharing (PSS) and regenerating codes (RC), and makes the following concrete contributions:

1. We show how security of common PSS schemes (treating/restricting leakage as/to full shares only leakage) fails to hold in simple generalized leakage models – models allowing leaks of smaller pieces of shares. This is accomplished by developing a new generalized model (Sect. 4) to reason about PSS and analyze its relation to RC. This new model takes into account partial leakage of information about shares of non-compromised nodes as opposed to complete information leaked (all or nothing) from compromised nodes.
2. We demonstrate a (conditional[1]) equivalence between PSS and RC. We provide two main theorems (Theorems 2 and 3 in Sect. 5) as a simple starting point demonstrating the a conditional equivalence between PSS and RC. This allows for a flow of ideas and constructions between the two areas.
3. As a first demonstration of a flow of constructions and ideas from RC to PSS, we show that due to our result proving the equivalence between the two notions, recent techniques for efficient polynomial interpolation due to Guruswami and Wootters [8] may improve efficiency of several bottleneck subprotocols in PSS. We also show that such efficient interpolation may cause a threat if parameters are not carefully calibrated, i,e., such interpolation technique may be used to attack PSS over a long period of time, (Corollary 1 in Sect. 5); further studying such attacks may be of independent interest.

[1] Our condition is that an RC code is MDS. This is to simplify this first treatment of the topic, we note that there is more work required to understand what conditions on the RC side imply certain types of security on the PSS side.

4. Conversely (i.e., considering ideas flowing in the other direction), we map (in Sect. 6) adversarial capabilities – called general adversary structures – from the PSS realm into the RC realm by defining a new notion of RC, called Generalized-decoding Regenerating Code (GRC). In a GRC, the decoding structure is captured by a collection of specific subsets of nodes that may decode, as opposed to the usual case where any set larger than a given threshold can decode. We also show how to construct a GRC scheme based on PSS and Theorem 2. We emphasize a general decoding structure is needed in many applications since network nodes often differ greatly in reliability, trust, and connections.

Paper Outline: Section 2 provides the necessary background and notation, whereas Sect. 3 overviews related work in PSS and RC; Sect. 4 describes a new generalized model to reason about PSS and which allows us to analyze its relation to RC; Sect. 5, in turn, contains the main technical results mapping RC to PSS and PSS to RC, while Sect. 6 discusses how relationships between the PSS and RC notions can further give rise to new extended notions in the RC realm. Our first example of such implications is mapping the generalizations of the capabilities of adversaries (called general adversary structures) from the PSS realm into the RC one, and define a new notion of RC which we call Generalized-decoding Regenerating Code (GRC).

2 Background and Notation

In this section we define the underlying mathematical and protocol notions employed in this work. Due to space constraints, we have further background material in the Appendix.

Finite Fields. We denote finite fields as F, L always such that F is a finite extension of L ($L \leq F$). Let $GF(p)$ be the underlying prime field and let $t = [F : L]$ be $F's$ degree over L. The ring of polynomials with coefficients in F is denoted as $F[x]$ and the subset of polynomials with degree at most $k-1$ is denoted as $F[x]^{\leq k-1}$. The latter is a k-dimensional vector space over F. Whenever we need the fields' cardinalities, we say $F = GF(q^t)$ and $L = GF(q)$ (where $q = p^d$ is some positive integer d).

Definition 1. *Let $F = GF(q^t)$ be a field extension of $L = GF(q)$ with degree t. Then, the* field trace *is defined as*

$$tr_{F/L}(\alpha) = \alpha + \alpha^q + \alpha^{q^2} + \cdots + \alpha^{q^{t-1}}.$$

2.1 Reed-Solomon Codes

Definition 2. *An (n, k)* Reed-Solomon code *with distinct evaluation points $A = \{\alpha_1, \ldots, \alpha_n\} \subset F$ is the subspace of F^n defined as*

$$RS(A, k) := \{(f(\alpha_1), \ldots, f(\alpha_n)) | f \in F[x]^{\leq k-1}\}.$$

We call n the *block-length* and k the *dimension* of the code. Reed-Solomon codes are *Maximally Separable Distance* (MDS) codes since they achieve the Singleton bound. That is, their minimum distance is $n-k+1$ and any collection of k code symbols, $f(\alpha_i)$, can be used to efficiently recover the original message, f. One can also efficiently decode Reed Solomon codes in the presence of $k < n/3$ errors using the Berlekamp-Welch algorithm [9].

2.2 Regenerating Codes

An (n, k, d, α, β) regenerating code [1] distributes a file, represented as a polynomial f in $F[x]^{\leq k-1}$, by encoding it and sending elements of the encoding to n nodes where each node stores α bits of data. A failed node can recover (repair) its share by accessing size β data from d surviving nodes, and we denote the *repair bandwidth* as $\gamma = d\beta$. Any k nodes are able to reconstruct the original file f when using their collective stored data. For example, an (n, k) Reed-Solomon code gives $\alpha = \log_2 |F|$ at each node and using the trivial reconstruction to repair a node yields $d = k$, $\beta = \log_2 |F|$, and $\gamma = k \log_2 |F|$. We give the formal definition in Definition 3 for completeness.

Definition 3. *Let $\mathcal{P}_1, \ldots, \mathcal{P}_n$ be labeled as nodes and $\mathcal{S}$ be labeled as a share-generator. An* (n, k, d, α, β) regenerating code *is a tuple of three protocols (Encode, Repair, Decode) defined as follows:*

- *Encode: This protocol has $\mathcal{S}$ take a file/string, represented by $f \in F[x]^{\leq k-1}$, as input and distributes a codeword $\mathbf{c} = (c_1, \ldots, c_n) \in F^n$ to the storage nodes according to their index (c_i to $\mathcal{P}_i$).*
- *Repair: Here, a failed node $\mathcal{P}_j$ contacts d other nodes, each of which sends it β bits of data. Then, the node computes a function on their sent data $(\delta_1, \ldots, \delta_d)$ to generate its new local storage.*
- *Decode: This protocol accesses any k nodes to reconstruct the original file/string.*

It is clear from the construction that each failed node can be reconstructed by accessing full data from at least k other nodes. However, this does not provide the optimum bandwidth. Regenerating codes facilitate the failed nodes to access fewer bits (β) from more than k surviving nodes for reconstruction. The following theorem summarizes the result for Reed-Solomon codes over general field extensions.

Theorem 1 (Lemma 4 in [10], Implicit in [8]). *Let F be a t-degree extension of a finite field L, let f be a polynomial of degree at most $k-1$ over F, and let $f(\alpha_1), \ldots, f(\alpha_n)$ be evaluations of f on n distinct points $\alpha_1, \ldots, \alpha_n$. Let α_0 be an element in F and let $g_1(x), \ldots, g_t(x)$ be t distinct polynomials over F of degree at most $n-k$ such that $\{g_i(\alpha_0)\}_{i\in[t]}$ is a basis for F over L. Then, it suffices to know the set of values $\bigcup_{i\in[n]}\{tr_{F/L}(g_j(\alpha_i)f(\alpha_i))\}_{j\in[t]}$ in order to recover $f(\alpha_0)$.*

2.3 Secret Sharing Schemes

Here we assume that all secret sharing schemes operate over finite fields. In general, we use the term *secret sharing (SS) scheme* to denote the following:

Definition 4. *Let F be a finite field. A (k, n, F)* information-theoretically secure secret threshold sharing scheme *over F is a pair of protocols used between servers labeled as the, unique and fixed, sharing node $\mathcal{S}$ and the set of storage nodes $\mathcal{A} := \{\mathcal{P}_1, \dots, \mathcal{P}_n\}$:*

Share(s_0): on input $s_0 \in F$, $\mathcal{S}$ randomly generates n shares $x_1, \dots, x_n \in F$ and returns x_i to server $\mathcal{P}_i$.
Reconstruct($\mathbf{s}$): Any $k+1$ nodes combine their shares, represented as a vector $\mathbf{x} \in F^{k+1}$, to reconstruct the secret $s' \in F$.

Let $H(\cdot)$ be the classical Shannon entropy function. For information-theoretic security, we assume s_0 is a non-trivial random variable over F. Then, the scheme's correctness and security is defined as follows:

Security: if $\mathbf{x} \in F^k$ is any (k)-sized subset of shares $x_{j_1}, \dots, x_{j_k}$, then $H(s_0|\mathbf{x}) = H(s_0) > 0$.
Correctness: if $\mathbf{x} \in F^{k+1}$ is any $(k+1)$-sized subset of shares $x_{j_1}, \dots, x_{j_{k+1}}$, then $H(s_0|\mathbf{x}) = 0$.

In the following definition, we break up the timeline into distinct phases once the shares are distributed. Each phase is represented by a positive integer σ.

Definition 5. *A* proactive secret sharing scheme *(PSS) is a secret sharing scheme as in definition 4 with the following additional algorithms:*

Refresh: All storage nodes $\mathcal{P}_1, \dots, \mathcal{P}_n$ use their respective shares from phase t to generate new random shares (for the same secret), $x_1^{(t+1)}, \dots, x_n^{(t+1)}$. Then, it distributes $x_i^{(t+1)}$ to $\mathcal{P}_i$.
Recover: A corrupted node, $\mathcal{P}_r$, contacts d uncorrupted nodes which combine their shares to compute (potentially with new randomness)

Next, we define Shamir's secret sharing scheme and its accompanying proactive protocols [2, 11][2]. We assume the finite field is at least the size of the number of storage nodes plus one, $|F| \geq n+1$. We denote the set of nodes needing to recover their share as $\mathcal{B}$, with $|\mathcal{B}| \leq k$, and the non-corrupted nodes as $\mathcal{D} := \mathcal{A} \backslash \mathcal{B}$. The set of evaluation points $A = \{\alpha_1, \dots, \alpha_n\} \subset F$ is fixed beforehand and known to all nodes.

Definition 6. *The* proactive (n, k) Shamir secret sharing scheme *over a finite field F with evaluation points $A = \{\alpha_1, \cdots, \alpha_n\} \subseteq F$ is defined as follows:*

Share(s_0): on input $s_0 \in F$, $\mathcal{S}$ randomly generates a degree k polynomial f over F conditioned on $f(0) = s_0$. Then, $\mathcal{S}$ sends $x_i^{(0)} := f(\alpha_i) \in F$ to $\mathcal{P}_i$.

[2] There are more efficient PSS schemes, [12] for example, but we describe the scheme in [11] for its clear relation to Reed-Solomon codes.

Reconstruct($\mathbf{x}$): *Any $k+1$ nodes interpolate their shares, represented as a vector $\mathbf{x} \in F^{k+1}$, to reconstruct the secret $s' \in F$.*
Refresh: Each storage node $\mathcal{P}_i$ generates a random polynomial δ_i of degree k conditioned on $\delta_i(0) = 0$ and sends $\delta_i(\alpha_j)$ to $\mathcal{P}_j$ for all $j \neq i$. Then, each storage node $\mathcal{P}_i$ updates their share as $x_i^{t+1} \leftarrow x_i^t + \sum_j \delta_j(\alpha_i)$ (and erases all intermediate values used to compute x_i^{t+1}).
Recover: For each corrupted node $\mathcal{P}_r \in \mathcal{B}$, each $\mathcal{P}_i \in \mathcal{D}$ does the following. Generate a uniformly random polynomial of degree k, ξ_i, such that $\xi_i(\alpha_r) = 0$. Then, send $\xi_i(\alpha_j)$ to $\mathcal{P}_j$ for all $\mathcal{P}_j \in \mathcal{D}$. Each $\mathcal{P}_j \in \mathcal{D}$ updates their share as $x_j^t \leftarrow x_j^t + \sum_{i,\mathcal{P}_i \in \mathcal{D}} \xi_i(\alpha_j)$. Finally, each $\mathcal{P}_i \in \mathcal{D}$ sends its updated share x_i^t to $\mathcal{P}_r$ and $\mathcal{P}_r$ interpolates them to get its original share, x_r^t.

2.4 Leakage Model

Leakage is fundamental to modeling secret sharing schemes. Here we define leakage functions using the terminology of [13]. We restrict the output of the leakage function to field elements, either in some large field F or some subfield $L \leq F$.

Definition 7. *Let $L \leq F$ such that $t = [F : L]$, and fix a secret sharing scheme over F, denoted by $\mathsf{Share} : F \to F^n$. We denote $\mathsf{Leak}^L = (\mathsf{Leak}_1^L, \ldots, \mathsf{Leak}_n^L)$ for a* length l L-leakage function *with $l < t$, where $\mathsf{Leak}_i : F \to L^l$ is a leakage function, possibly randomized, defined for each node. When $(x_1, \ldots, x_n) \leftarrow \mathsf{Share}(s_0)$, we denote the collection of leakage outputs as $(b_1, \ldots, b_n) \leftarrow \mathsf{Leak}^L(x_1, \ldots, x_n)$ where $b_i = \mathsf{Leak}^L(x_i)$. In addition, for any $S \subseteq [n]$ we define $\mathsf{Reveal}_S : F^n \to F^n$ as revealing an entire share at node i if $i \in S$. That is, $\mathsf{Reveal}_S(i) = \mathbb{1}_{i \in S} \cdot \mathsf{Share}_i$ where $\mathbb{1}_{i \in S}$ is the indicator function for S.*

For epoch i, we denote $\mathcal{F}^{(i)} \subset [n]$ as the set of nodes which leak full shares and $\mathcal{L}^{(i)}$ as the set of nodes which leak partial shares (elements in L^l).

3 Related Work

Now that we have introduced the basic notions and mechanisms we need and employ, let us review further related earlier work.

Proactive Secret Sharing. The notion of proactive security was first proposed by Ostrovsky and Yung [4], and subsequently utilized to protect cryptographic keys by secret sharing them and computing RSA signatures in a distributed manner [11]. Specifically, Proactive Secret Sharing (PSS) aims to protect against a mobile adversary that can change the subset of corrupted parties over time and thus may eventually compromise all involved parties over a long period of time; the model assumes that such a mobile adversary is limited to simultaneously corrupting no more than t parties during the same period though. PSS initially only considered static groups and for settings with honest majorities, Dynamic Proactive Secret Sharing (DPSS) schemes are both proactively secure and allow the set of parties to dynamically change over time. The dynamic group problem has been addressed [14–20], but mostly for the honest majority and non-proactive settings, and only in [21] in the proactive setting.

In the dishonest majority setting most of the PSS literature [22,23] assumes a static group of parties, i.e., unchanged during the secret lifetime. PSS protocols for *dynamic* groups with dishonest majorities were only recently constructed [6,24]. As for any secret sharing against dishonest majorities, security is only computational. In addition to efficiently handling dynamic groups, recent work [24] introduces a notion of batched PSS that retain fairness against mixed (passive and active) adversaries and reduces the communication complexity of DPSS from $O(n^4)$ to $O(n^2)$ when batching is used and $O(n^3)$ in the single secret setting.

Regenerating Codes. A large file can divided into pieces, each of which is stored at a different nodes using distributed storage. Server corruption can lead to loss of information. Error-correction coding[3] techniques allow the recovery of information stored in a corrupted node using the information stored in other servers. Regenerating codes were first introduced by Dimakis et al. in [1] to improve the repair bandwidth for distributed storage systems. The aim is to recreate the information stored in a corrupted node without recreating the entire encoded information. Given a file of size $\mathcal{M}$, it can be divided into k pieces of size $\mathcal{M}/k$ which are stored in n nodes using an (n, k) MDS code. Each node stores α symbols. Information stored in a corrupted node can be recovered by accessing β sub-symbols from d surviving nodes. This can be done in the following three ways.

- *Exact repair:* the encoded block is regenerated exactly as before.
- *Functional repair:* the corrupted nodes are regenerated such that the new system represents an MDS code of length n.
- *Exact repair of systematic parts:* this is a hybrid between the above two repair schemes. The code contains exactly one replica of the information. Systematic parts of the code are regenerated using the exact repair scheme and the non-systematic part is regenerated using the functional repair scheme.

The trade-off between storage efficiency and repair bandwidth is a point of interest. Two special cases of regenerating codes are given by the optimal cases: minimum-storage regenerating (MSR) codes and minimum-bandwith regenerating (MBR) codes.

Other Works. Though there are other works combining regenerating codes and secret sharing, we emphasize that this paper is the first to do so in the *proactive* secret sharing setting. First, Huang and Bruck [10] apply the GW paradigm to threshold secret sharing schemes and prove optimality. They stay in the simplest

[3] Regenerating codes are studied from the point of view of erasure recovery in coding theory literature. However, here we refer to it as a subset of error-correction because we also care about corruption of nodes along with node failures.

security model, ignoring leakage, and are not concerned with proactive schemes. A previous work by Huang et al. [25] studies the communication complexity of threshold secret schemes and presents a scheme with optimal decoding bandwidth based on Reed-Solomon codes. However, [25] does not concern repairing in secret sharing schemes, leakage, randomizing shares (proactive schemes), or generalized decoding structures.

4 Leakage and Reconstruction: Old Models, New Lens

This section is to show how considering a simple leakage model affects the security of proactive secret sharing (PSS) schemes across epochs. We show the connection between repairing Reed-Solomon codes (Subsection 2.1) and PSS schemes using the leakage model described in (Subsection 2.4). In other words, we show how the algorithms comprised in [8] can be used to attack an incorrectly-implemented PSS scheme.

Model. Let $(\mathsf{Share}, \mathsf{Reconstruct}, \mathsf{Refresh}, \mathsf{Recover})$ be a Shamir-based (n, k, F) PSS scheme (Definition 6) with evaluation points $A' \subset F$. Our model is simple. During each epoch of a PSS scheme an adversary $\mathcal{A}$ receives either nothing, a collection of $l < t$ small field elements ($\alpha \in L^l$), or a full field element from each node. Let $l^{(i)}$ be the number of subfield elements leaked in epoch i, and let $f^{(i)}$ be the number of full field elements leaked during epoch i. The key notion throughout this section is that the linear transformations used in [8]'s reconstruction algorithms are *independent* of the polynomial f representing the distributed data (potentially secret). In the notation of [8, Algorithm 1], the polynomials $\mu_{\zeta,\alpha}(x)$ only depend on A[4]. Lastly, we extend the evaluation points to include 0 in order to bridge Definitions 6 and a linear exact repair scheme [8, Algorithm 1], $A := A' \cup \{0\}$.

4.1 Static Leakage

Here we look at the case to where the leakage function Leak^L is static between epochs.

Proposition 1. *Let b be the repair bandwidth of the Reed-Solomon linear exact repair scheme also being used as a Shamir-based PSS scheme. There is an efficient adversary which receives $l^{(i)}$ leaked subfield elements and $f^{(i)}$ leaked full field elements which needs $b-(tf^{(i)}+l^{(i)})$ subfield elements during a single epoch to reconstruct the secret $s_0 \in F$ for a static leakage function between epochs.*

The above proposition shows how an adversary can be under the security threshold for whole shares, $f^{(i)} < k$, during each epoch but can still reconstruct the secret!

[4] For example, Corollary 9 in [8] constructs these polynomials as $\mu_{\zeta,\alpha}(\alpha^*) = p(\alpha) \cdot \frac{\prod_{\beta\in A\setminus\{\alpha^*\}}(\beta-\alpha^*)}{\prod_{\beta\in A\setminus\{\alpha\}}(\beta-\alpha)}$ where p is a polynomial dependent only on the evaluation points A.

4.2 Dynamic Leakage

Here we consider a leakage function that changes between epochs. This setting represents the case where storage nodes fail to completely erase some data used in computing the Refresh protocol.

Proposition 2. *Let b be the repair bandwidth of the Reed-Solomon linear exact repair scheme also being used as a Shamir-based PSS scheme. Then, there is an efficient adversary which needs b leaked subfield elements in order to reconstruct the secret across epochs assuming the leaked nodes store a non-zero value.*

Note that this is a strong leakage model. However, updating a leakage function via multiplying and inverting finite field elements at each node is a plausible scenario which should be known to those utilizing and implementing PSS schemes.

5 On the Equivalence of Regenerating Codes and Proactive Secret Sharing

In this section we prove the equivalence between Regenerating Codes (RC) and threshold Proactive Secret Sharing (PSS) under certain conditions. Theorem 2 treats the PSS to RC direction, while Theorem 3 treats the reverse one, restricted to linear, MDS codes. These properties on the RC are required for threshold security and to construct a simple Refresh protocol.

Theorem 2. *For each (t, n) proactive secret sharing scheme represented as a tuple of algorithms (Share, Reconstruct, Recover, Refresh) which stores α bits at each node and contacts d nodes in the $\mathsf{Recover}_{PSS}$ protocol (Definition 5), each sending β bits of data to the failed node, there is an erasure $(n' = n, k' = t+1, d' = d, \alpha' = \alpha, \beta' = \beta)$ regenerating code represented as a tuple of algorithms (Encode, Decode, Repair), as in Definition 3.*

Remark 1. We note that the rate given implicitly in Theorem 2 is only $1/n$. However, there are clear ways to achieve a better rate. The first is when the PSS scheme is linear. Here, the encoding procedure usually involves a linear code of dimension $t+1$ and t of the input symbols are uniformly random elements in F. This randomness is only for security so we can replace these t symbols with data elements. The second is the case of batching [21], employing a basic technique from [26], where the PSS scheme takes a secret as an element in F^l. The proof of Theorem 2 for the batching case is the same except $\mathbf{s}, \mathbf{s}' \in F^l$.

The other direction, from RC to PSS only makes sense if we are able to show threshold security. This was proven for non-proactive secret sharing schemes with repair in [10, Theorems 1 and 2]. Here we extend this result to the proactive setting.

Theorem 3. *For every $(n+1, k+1, d, \alpha, \beta)$ MDS linear regenerating code over F, there is a (k, n) proactive secret sharing scheme whose $\mathsf{Recover}_{PSS}$ protocol contacts d nodes, each sending β bits to the damaged node.*

Theorem 3 and the result of Guruswami and Wootters [8], Theorem 1 in the special case where the number of parties equals the degree of the extension, together imply the existence of an alternative $\mathsf{Recover}_{PSS}$ protocol that only receives symbols in a subfield for the Shamir-based PSS scheme. This may be advantageous in settings with restricted communication during the recovery phase. This is summarized in the following Corollary.

Corollary 1. *The Shamir-based PSS scheme, Definition 6 [2,11], over a finite field F with subfield $L \leq F$ with degree $t = [F : L]$ has an alternative $\mathsf{Recover}_{PSS}$ protocol which contacts the remaining $n-1$ nodes and receives t symbols in L from each node in order to recover the lost share.*

We emphasize that Corollary 1's efficient recover protocol is information-theoretically secure in the setting illustrated by Theorem 1, without leakage. This is because Theorem 1's repair algorithm only sends a user t subfield elements and its secret is exactly comprised of t subfield elements.

6 From General Adversary Structures to General Decoding Structures

In this section we demonstrate how, due to the relationships between the notions, an interesting useful paradigm in one area (PSS) can induce an interesting new notion in the other area (RC). Secret sharing (SS) schemes, and proactive SS (PSS) in particular, can be extended [6,21,23,24] to accommodate for different adversaries beyond a threshold one (i.e., more general secure subsets of shares), and to deal with dynamic change of the servers holding the shares (dynamic groups). Such extensions make sense in the case of a network of nodes/processors, holding shares jointly. Since RC is a coding theoretic technique for a network of nodes/processors to hold sub-codewords, the extended notions make sense in storage-oriented RC codes. Here we demonstrate such translation of a relevant extension.

The most generic adversarial capabilities are captured (since PODC'97 [5]) using the *general adversary structure (GAS)* in the secure computation literature. The GAS notion is a more general (and practically motivated via different availability of different types of servers in a network) and more flexible one when modeling adversaries, compared to only the threshold limitation on corruptions. GAS applies to various scenarios, for example when only special combinations of nodes is required to reconstruct a secret, when some nodes are authorized by some authority and others by another authority and combination of authorities is required, etc.

Let $2^{\mathcal{P}}$ denote the set of all the subsets of nodes ($\mathcal{P}$) involved in a secret sharing scheme. A subset of $2^{\mathcal{P}}$ is *qualified* if nodes in the subset can reconstruct the secret, while a subset of $2^{\mathcal{P}}$ that nodes in the set obtain no information about the secret is called *ignorant*. Every subset of $\mathcal{P}$ is either qualified or ignorant[5]. The secrecy condition is stronger: even if any ignorant set of nodes hold any kind of partial information about the shared value, they must not obtain any additional information about the shared value.

The *access structure* Γ is the set of all qualified subsets of $\mathcal{P}$ and the *secrecy structure* Σ is the set of all ignorant subsets of $\mathcal{P}$. Naturally, Γ includes all supersets of each element in it (so often called *monotone access structure*), while Σ includes all subsets of each element in it. We call such minimum or maximum sets as *basis structure*, and denote it with $\tilde{\cdot}$. i.e., the basis access structure $\tilde{\Gamma}$ is the set of all minimal subsets in Γ, and the basis secrecy structure $\tilde{\Sigma}$ is the set of all maximal subsets in Σ.

The *adversary structure* $\Delta \subseteq \Sigma$ is a set of subsets of nodes that can be potentially corrupted. The adversary can choose a set in Δ and corrupt all the nodes in the set. Note that the adversary structure in t-threshold SS is the set of all subsets of $\mathcal{P}$ of at most t nodes and GAS extends this to non-threshold models. A GAS includes all of these structures, (Γ, Σ, Δ). We define a family of security properties on the sets Δ by a covering condition: $\Delta \in Q^k(\mathcal{P}, \Delta)$ if for all distinct $A_1, \ldots, A_k \in \Delta$, $A_1 \cup \cdots \cup A_k \neq \mathcal{P}$ [5,28]. That is, no k different adversary patterns can cover the set of nodes in the protocol. Further, we are focused on the $Q^2(\mathcal{P}, \Delta)$ setting: $\forall\, A, B \in \Delta, A \cup B \neq \mathcal{P}$.

One can extend the definition of secret sharing for the threshold structure (see Definition 4) to the GAS as follows:

Definition 8 (Secret Sharing for GAS). *Let F be a finite set. An (n, F)* information-theoretically secure secret threshold sharing scheme *over F for a GAS (Γ, Σ, Δ) is a pair of protocols used between servers labeled as the, unique and fixed, sharing node $\mathcal{S}$ and the set of storage nodes $\mathcal{A} := \{\mathcal{P}_1, \ldots, \mathcal{P}_n\}$:*

> Share(s_0): *on input $s_0 \in F$, $\mathcal{S}$ randomly generates n shares $x_1, \ldots, x_n \in F$ and returns x_i to server $\mathcal{P}_i$.*
> Reconstruct$(\mathbf{s})$: *Any set of nodes in Γ can combine their shares, represented as a vector of elements $\mathbf{x}$, to reconstruct the secret $s' \in F$.*

For information-theoretic security, we assume s_0 is a non-trivial random variable over F. Then, the scheme's correctness and security is defined as follows:

> *Security: if $\mathbf{x}$ is any subset of shares corresponding to a set in Σ, then $H(s_0|\mathbf{x}) = H(s_0) > 0$.*
> *Correctness: if $\mathbf{z}$ is any subset of shares with indices corresponding to a set of nodes in Γ then $H(s_0|\mathbf{z}) = 0$.*

When extending the definition to the proactive setting, we break up the timeline into distinct phases once the secret shares are distributed. Each phase is represented by a positive integer σ (analogous to Definition 5).

[5] That is, we only consider *perfect* secret sharing schemes [27, Definition 11.59].

Definition 9 (Proactive Secret Sharing for GAS). *A* proactive secret-sharing scheme *for a GAS* (Γ, Σ, Δ) *is a secret-sharing scheme as in definition 8 with the following additional algorithms:*

Refresh: All storage nodes $\mathcal{P}_1, \ldots, \mathcal{P}_n$ *use their respective shares from phase t to generate new random shares (for the same secret),* $x_1^{(t+1)}, \ldots, x_n^{(t+1)}$. *Then, it distributes* $x_i^{(t+1)}$ *to* $\mathcal{P}_i$.
Recover: A corrupted node, $\mathcal{P}_r$, *contacts d uncorrupted nodes which combine their shares to compute (potentially with new randomness) a new share for the corrupted node.* $\mathcal{P}_r$ *receives their new (recovered) share.*

Lemma 1. *There exists a proactive secret sharing (PSS) scheme for general adversary structures (GAS) as defined in Definition 9 with* $Q^2(\mathcal{P}, \Delta)$ *adversaries.*

Proof. This follows a constructive proof, from a PSS for a GAS given in [6].

6.1 Generalized Decoding in Regenerating Codes

Let $\mathcal{P}$ be a set of all servers involved in distributed storage and $2^{\mathcal{P}}$ all its subsets. A subset of $2^{\mathcal{P}}$ is available if servers in the subset can be accessed to reconstruct the original data, while a subset of $2^{\mathcal{P}}$ that cannot be used for data reconstruction is called unavailable. Every subset of $\mathcal{P}$ is either available or unavailable. Therefore, let Γ_{RC} be the sets of servers which can decode when available. We call Γ_{RC} the *decoding structure.* Any decoding structure is monotone, i.e., it is closed under taking supersets.

The analogy between Γ on the secret sharing side and Γ_{RC} on the regenerating code side is clear: these are the sets which can decode the secret. So, there must remain one which is not tampered with, or, equivalently, an erasure pattern, υ, can come from $\mathcal{P}$ as long as there exists $B \in \Gamma$ such that $\upsilon \cap B = \varnothing$. On the other hand, the coding analogy of the sets of tolerable active adversaries, Δ, is more subtle. It is, however, the sets of error patterns which the code can withstand and still decode. In addition, we define the error property $Q^k_{RC}(\mathcal{P}, \Delta)$ analogously in the coding setting: For all distinct $A_1, \ldots, A_k \in \Delta_{RC}$, $A_1 \cup \cdots \cup A_k \neq \mathcal{P}$.

Definition 10 (Generalized-decoding Regenerating Code (GRC)). *Let* $\mathcal{P}_1, \ldots, \mathcal{P}_n$ *be labeled as servers and* $\mathcal{S}$ *be labeled as a share-generator. Let F be a finite set. An* (n, k, d, α, β) *general error and erasure regenerating code for a decoding structure for* $(\Gamma_{RC}, \Delta_{RC})$, $\varnothing \neq \Gamma_{RC}, \Delta_{RC} \subset 2^{\mathcal{P}}$ *and* $\Gamma_{RC} \cap \Delta_{RC} = \varnothing$, *is a tuple of three protocols (Encode, Repair, Decode) which are defined as follows:*

Encode: This protocol has $\mathcal{S}$ *take a file, represented by* $f \in F[x]^{\leq k-1}$, *as input and distributes a codeword* $\mathbf{c} = (c_1, \ldots, c_n) \in F^n$.
Repair: Here, a failed node $\mathcal{P}_j$ *contacts d other uncorrupted, each of which sends it* β *bits of data. Then, the node computes a function on their sent data* $(\delta_1, \ldots, \delta_d)$ *to generate its new local storage.*

Decode: This protocol accesses any set of servers in Γ_{RC} to reconstruct the original file.

Moreover, we require the following correctness constraints:

Errors: for all error patterns in Δ_{RC}, Decode correctly decodes to the original message.
Erasures: for all erasure patterns υ such that there exists a $B \in \Gamma_{RC}$ such that $\upsilon \cap B = \varnothing$, Decode correctly decodes to the original message.

Now we show the existence of any GRC with Q^2_{RC} error patterns via the mapping applied in Theorem 2's proof. We emphasize that the following theorem is an existence result and not optimized for parameters. We leave optimization open for future works.

Theorem 4. *For every $(\Gamma_{RC}, \Delta_{RC})$ with property $Q^2_{RC}(\mathcal{P}, \Delta_{RC})$, there exists a linear GRC as defined in Definition 10 over a finite field F. Moreover, all erasure patterns, υ, such that there is a $B \in \Gamma_{RC}$ such that $\upsilon \cap B = \varnothing$ can be correctly decoded.*

References

1. Dimakis, A.G., Godfrey, B., Wu, Y., Wainwright, M.J., Ramchandran, K.: Network coding for distributed storage systems. IEEE Trans. Inf. Theory **5**, 4539–4551 (2010)
2. Shamir, A.: How to share a secret. Commun. ACM **22**, 612–613 (1979)
3. McEliece, R.J., Sarwate, D.V.: On sharing secrets and Reed-Solomon codes. Commun. ACM **24**, 583–584 (1981)
4. Ostrovsky, R., Yung, M.: How to withstand mobile virus attacks (extended abstract). In: PODC (1991)
5. Hirt, M., Maurer, U.M.: Complete characterization of adversaries tolerable in secure multi-party computation (extended abstract). In: PODC. IEEE (1997)
6. Eldefrawy, K., Hwang, S., Ostrovsky, R., Yung, M.: Communication-efficient (proactive) secure computation for dynamic general adversary structures and dynamic groups. In: Galdi, C., Kolesnikov, V. (eds.) SCN 2020. LNCS, vol. 12238, pp. 108–129. Springer, Cham (2020). https://doi.org/10.1007/978-3-030-57990-6_6
7. Maram, S.K.D., et al.: CHURP: dynamic-committee proactive secret sharing. In: CCS 2019 (2019)
8. Guruswami, V., Wootters, M.: Repairing Reed-Solomon codes. In: STOC (2016)
9. Berlekamp, E.R.: Bounded distance+1 soft-decision Reed-Solomon decoding. IEEE Trans. Inf. Theory **42**, 704–720 (1996)
10. Huang, W., Bruck, J.: Secret sharing with optimal decoding and repair bandwidth. In: ISIT. IEEE (2017)
11. Herzberg, A., Jarecki, S., Krawczyk, H., Yung, M.: Proactive secret sharing or: how to cope with perpetual leakage. In: Coppersmith, D. (ed.) CRYPTO 1995. LNCS, vol. 963, pp. 339–352. Springer, Heidelberg (1995). https://doi.org/10.1007/3-540-44750-4_27
12. Baron, J., Eldefrawy, K., Lampkins, J., Ostrovsky, R.: How to withstand mobile virus attacks, revisited. In: PODC. ACM (2014)

13. Nielsen, J.B., Simkin, M.: Lower bounds for leakage-resilient secret sharing. In: Canteaut, A., Ishai, Y. (eds.) EUROCRYPT 2020. LNCS, vol. 12105, pp. 556–577. Springer, Cham (2020). https://doi.org/10.1007/978-3-030-45721-1_20
14. Damgård, I., Nielsen, J.B.: Scalable and unconditionally secure multiparty computation. In: Menezes, A. (ed.) CRYPTO 2007. LNCS, vol. 4622, pp. 572–590. Springer, Heidelberg (2007). https://doi.org/10.1007/978-3-540-74143-5_32
15. Damgård, I., Ishai, Y., Krøigaard, M., Nielsen, J.B., Smith, A.: Scalable multiparty computation with nearly optimal work and resilience. In: Wagner, D. (ed.) CRYPTO 2008. LNCS, vol. 5157, pp. 241–261. Springer, Heidelberg (2008). https://doi.org/10.1007/978-3-540-85174-5_14
16. Damgård, I., Ishai, Y., Krøigaard, M.: Perfectly secure multiparty computation and the computational overhead of cryptography. In: Gilbert, H. (ed.) EUROCRYPT 2010. LNCS, vol. 6110, pp. 445–465. Springer, Heidelberg (2010). https://doi.org/10.1007/978-3-642-13190-5_23
17. Desmedt, Y., Jajodia, S.: Redistributing secret shares to new access structures and its applications. Technical Report ISSE TR-97-01, George Mason University, July 1997
18. Schultz, D.: Mobile proactive secret sharing. Ph.D. thesis, Massachusetts Institute of Technology (2007)
19. Wong, T.M., Wang, C., Wing, J.M.: Verifiable secret redistribution for archive system. In: IEEE Security in Storage Workshop (2002)
20. Zhou, L., Schneider, F.B., van Renesse, R.: APSS: proactive secret sharing in asynchronous systems. ACM Trans. Inf. Syst. Secur. **8**, 259–286 (2005)
21. Baron, J., Defrawy, K.E., Lampkins, J., Ostrovsky, R.: Communication-optimal proactive secret sharing for dynamic groups. In: Malkin, T., Kolesnikov, V., Lewko, A.B., Polychronakis, M. (eds.) ACNS 2015. LNCS, vol. 9092, pp. 23–41. Springer, Cham (2015). https://doi.org/10.1007/978-3-319-28166-7_2
22. Dolev, S., ElDefrawy, K., Lampkins, J., Ostrovsky, R., Yung, M.: Proactive secret sharing with a dishonest majority. In: Zikas, V., De Prisco, R. (eds.) SCN 2016. LNCS, vol. 9841, pp. 529–548. Springer, Cham (2016). https://doi.org/10.1007/978-3-319-44618-9_28
23. Eldefrawy, K., Ostrovsky, R., Park, S., Yung, M.: Proactive secure multiparty computation with a dishonest majority. In: Catalano, D., De Prisco, R. (eds.) SCN 2018. LNCS, vol. 11035, pp. 200–215. Springer, Cham (2018). https://doi.org/10.1007/978-3-319-98113-0_11
24. Eldefrawy, K., Lepoint, T., Leroux, A.: Communication-efficient proactive secret sharing for dynamic groups with dishonest majorities. In: Conti, M., Zhou, J., Casalicchio, E., Spognardi, A. (eds.) ACNS 2020. LNCS, vol. 12146, pp. 3–23. Springer, Cham (2020). https://doi.org/10.1007/978-3-030-57808-4_1
25. Huang, W., Langberg, M., Kliewer, J., Bruck, J.: Communication efficient secret sharing. IEEE Trans. Inf. Theory **62**, 7195–7206 (2016)
26. Franklin, M.K., Yung, M.: Communication complexity of secure computation (extended abstract). In: STOC (1992)
27. Cramer, R., Damgård, I., Nielsen, J.B.: Secure Multiparty Computation and Secret Sharing. Cambridge University Press, Cambridge (2015)
28. Hirt, M., Maurer, U.M.: Player simulation and general adversary structures in perfect multiparty computation. J. Cryptol. **13**, 31–60 (2000)

Extending Lattice Linearity for Self-stabilizing Algorithms

Arya Tanmay Gupta(✉) and Sandeep S. Kulkarni

Computer Science and Engineering, Michigan State University, East Lansing, USA
{atgupta,sandeep}@msu.edu

Abstract. In this article, we focus on extending the notion of lattice linearity to self-stabilizing programs. Lattice linearity allows a node to execute its actions with old information about the state of other nodes and still preserve correctness. It increases the concurrency of the program execution by eliminating the need for synchronization among its nodes.

The extension –denoted as eventually lattice linear algorithms– is performed with an example of the service-demand based minimal dominating set (SDDS) problem, which is a generalization of the dominating set problem; it converges in $2n$ moves. Subsequently, we also show that the same approach could be used in various other problems including minimal vertex cover, maximal independent set and graph coloring.

Keywords: Eventually lattice linear algorithms · Self-stabilization · Dominating set · Vertex cover · Graph coloring

1 Introduction

In a distributed program, a node cooperates with other nodes to solve the problem at hand such as leader election, mutual exclusion, tree construction, dominating set, independent set, etc. There are several models for such distributed programs. These can be broadly classified as message passing programs or shared-memory programs. In message passing programs, nodes do not share memory. Rather, they communicate with each other via messages. On the other hand, the shared-memory model allows a node to read the memory of other nodes to solve the given problem.

Implementation of such shared memory programs introduces several challenges to allow a node to read the state of its neighbors in a consistent fashion. One solution in this context is that the nodes only execute in a coordinated manner where when a node is activated by the scheduler, it reads the variables of other nodes and updates its own state. Furthermore, the scheduler needs to ensure that any *conflicting* nodes are not activated at the same time. This approach, however, is expensive and requires synchronization among nodes.

To alleviate the issue of consistency while reading remote variables, Garg (2020) [4] introduced lattice linear predicate detection in combinatorial optimization problems. In [4], it is shown that when an algorithm exploits lattice

C. Johnen et al. (Eds.): SSS 2021, LNCS 13046, pp. 365–379, 2021.
https://doi.org/10.1007/978-3-030-91081-5_24

linearity of the underlying problem, it preserves correctness even if nodes execute with old information. However, this work assumes that the algorithm begins in a specific initial state and, hence, is not applicable for self-stabilizing algorithms since a self-stabilizing algorithm guarantees that starting from an arbitrary state, the algorithm reaches a legitimate state (*invariant*) and remains there forever. With this intuition, in this work, we extend the results in [4] to self-stabilizing algorithms.

We proceed as follows. We begin with the problem of service-demand based minimal dominating set (SDDS) which is a generalization of the dominating set problem. We devise a self-stabilizing algorithm for SDDS. We scrutinize this algorithm and disassemble it into two parts, one of which satisfies the lattice linearity property of [4] if it begins in a *feasible* state. Furthermore, we show that the second part of the algorithm ensures that the algorithm reaches a *feasible* state. We show that the resulting algorithm is self-stabilizing, and the algorithm has *limited-interference* property (to be discussed in Sect. 5.5) due to which it is tolerant to the nodes reading old values of other nodes.

We also demonstrate that this approach is generic. It applies to various other problems including vertex cover, independent set and graph coloring.

1.1 Contributions of the Article

- We present a self-stabilizing algorithm for the minimal SDDS problem. The algorithm can be modified to solve other generalizations of dominating set present in the literature. The algorithm converges in $2n$ moves, which is an improvement over the other algorithms in the literature.
- We extend the notion of *lattice linear predicate detection* from [4] to introduce the class of *lattice linear self-stabilizing algorithms* and *eventually lattice linear self-stabilizing algorithms*. Such algorithms allow the program to converge even when the nodes read old values. This is unlike the algorithms presented in [4] where it is required that (1) the problems have only one optimal state, and (2) the program needs to start in specific initial states.
- Our solution to SDDS can be extended to other problems including minimal vertex cover, maximal independent set and graph coloring problems. The resulting algorithms are eventually lattice linear and can be modified to lattice linear self-stabilizing algorithms.

1.2 Organization of the Article

In Sect. 2, we discuss the related work in the literature. In Sect. 3, we discuss some notations and definitions that we use in the article. In Sect. 4, we describe the algorithm for the service-demand based dominating set problem. In Sect. 5, we analyze the characteristics of the algorithm and show that it is eventually lattice linear. In Sect. 6, we use the structure of eventually lattice linear self-stabilizing algorithms to develop algorithms for vertex cover, independent set and graph coloring problems. Finally, we conclude the article in Sect. 7.

2 Literature Study and Discussion on Our Contribution

Self-stabilizing algorithms for the minimal dominating set problem have been proposed in several works in the literature, for example, in [2,6,9,16,19]. The best convergence time among these works is $4n$ moves.

Other variations of the dominating set problem are also studied. Fink and Jacobson (1985) [3] proposed the minimal k-dominating set problem; here, the task is to compute a minimal set of nodes $\mathcal{D}$ such that for each node $v \in V(G)$, $v \in \mathcal{D}$ or there are at least k neighbors of v in $\mathcal{D}$. When $k = 1$, the definition of $\mathcal{D}$ here is same as that in the general dominating set problem. Kamei and Kakugawa (2003) [10] proposed self-stabilizing algorithm for tree networks under central and distributed schedulers for the minimal k-dominating set; the converge time is n^2 moves. Kamei and Kakugawa (2005) [11] have proposed a self-stabilizing distributed algorithm which converges in $2n + 3$ rounds; their algorithm runs on synchronous daemon.

A generalization of the dominating set problem is described in Kobayashi et al. (2017) [13]. This article assumes the input to include wish sets (of nodes) for every node. For each node i, either i should be in the dominating set $\mathcal{D}$ or at least one of its wish set must be a subset of $\mathcal{D}$. In this case, the input size may be exponential. The nodes require to read the latest values of other nodes.

Self-stabilizing algorithms for the vertex cover problem has been studied in Kiniwa (2005) [12], Astrand and Suomela (2010) [15], and Turau (2010) [17]. A survey of self-stabilization algorithms on independence, domination and coloring problems can be found in Guellati and Kheddouci (2010) [7].

Garg (2020) [4] studied the exploitation of lattice linear predicates in several problems to develop parallel processing algorithms. Lattice linearity ensures convergence of the system to an optimal solution while the nodes perform executions parallely and are allowed to do so without coordination, and are allowed to perform executions based on the old values of other nodes. Garg (2021) [5] introduces lattice linearity to dynamic programming problems such as longest increasing subsequences and knapsack problem. In this approach, the lattice arises from the computation. We are going to pursue a similar goal for self-stabilizing problems.

Our SDDS algorithm uses local checking to determine if it is in an inconsistent state and local correction to restore it. Thus, it differs from [18] where global correction in the form of reset is used. Local detection and correction is also proposed in [1,14]. The key difference with our work is that we are focusing on scenarios where local detection can be performed without requiring coordination with other nodes.

3 Preliminaries

3.1 Modeling Algorithms

Throughout the article, we denote G to be an arbitrary graph on which we apply our algorithms. $V(G)$ is the vertex-set and $E(G)$ is the edge-set of G. In G, for any node i, Adj_i is the nodes connected to i in G, and $N_i = Adj_i \cup \{i\}$. $deg(i)$ denotes the degree of node i.

Each node i is associated with a set of variables. The algorithm is written in terms of rules, where each *rule* for process i is of the form $guard \longrightarrow action$ where *guard* is a proposition over variables of some nodes which may include the variables of i itself along with the variables of other nodes. If any of the guards hold true for some node, we say that the node is *enabled.* As the algorithm proceeds, we define a *move* with reference to a node to be an action in which it changes its state. A *round* with reference to a scheduler is a minimum time-frame where each node is given a chance to evaluate its guards and take action (if some guard evaluates to true) at least once.

An algorithm is *silent* if no node is enabled when G reaches an optimal state (we describe the respective optimal states as we discuss the problems in this article).

Scheduler/Daemon. A *central scheduler/daemon* is a scheduler which chooses only one node to evaluate its guards in a time-step and execute the corresponding action. A *distributed scheduler/daemon* chooses an arbitrary subset of nodes of $V(G)$ in a time-step to evaluate their guards and execute the corresponding actions respectively. A *synchronous scheduler/daemon* chooses all the nodes in $V(G)$ in each time-step to evaluate their guards and execute the corresponding actions respectively together.

Read/Write Model. In the read/write model, we partition the variables of a node as *public* variables that can be read by others, and *private* variables that are only local to that node. In this model, the rules of the node are allowed to be either:

(a) *read rules*, where any node i is allowed to read the public variables of one or more or all of its neighbors and copies them into a private variable of i, or
(b) *write rules*, where i reads only its own variables to update its public variables.

3.2 Lattice Linear Predicates

Lattice Linearity [4] of a problem is a phenomenon by which all the state vectors of a global state of a system G form a distributive lattice. The predicate which defines an optimal state of the problem (under which such a lattice forms) is called a *lattice linear predicate.* In such a lattice, if the state of a system G is false according to the predicate, then at least one node $i \in V(G)$ can be identified such that it is *forbidden*, that is, in order for G to reach an optimal state, i must change its state. Since this article studies self-stabilization problems, we define the predicate to be an optimal state with respect to the respective problems.

3.3 The Communication Model

The nodes of a graph communicate via shared memory. In each action, a node reads the values of its distance-k neighbors (where value of k depends upon the specific algorithm) and updates its own state. We make no assumptions about atomicity with respect to reading the variables. In other words, while one node is in the middle of updating its state, its neighbors may be updating their owns state as well. In turn, this means that when node i changes its state (based on state of node j) it is possible that the state of node j has changed. In other words, i is taking an action based on an old value of node j. Therefore, our algorithms will run equally well in a message passing model (with distance-k flooding), without the requirement of synchronization or locks.

4 Service-Demand Based Dominating Set

In this section, we introduce a generalization of the dominating set problem, the service-demand based dominating set problem and describe an algorithm to solve it.

Definition 1. ***Service-demand based dominating set problem (SDDS).*** *In the minimal service-demand based dominating set problem, the input is a graph G and a set of services S_i and a set of demands D_i for each node i in G; the task is to compute a minimal set $\mathcal{D}$ such that for each node i,*

1. *either $i \in \mathcal{D}$, or*
2. *for each demand d in D_i, there exists at least one node j in Adj_i such that $d \in S_j$ and $j \in \mathcal{D}$.*

In the following subsection, we present a self-stabilizing algorithm for the minimal SDDS problem. Each node i is associated with variable $st.i$ with domain $\{IN, OUT\}$. $st.i$ defines the state of i. We define $\mathcal{D}$ to be the set $\{i \in V(G) : st.i = IN\}$.

4.1 Algorithm for SDDS Problem

The list of constants stored in each node is described in the following table. For a node i, D_i is the set of demands of i, S_i is the set of services that i can provide to its neighbors. D_i and S_i are provided as part of the input.

Constant	What it stands for
D_i	the set of demands of node i
S_i	the set of services provided by node i

The list of macros stored in each node is described in the following table. Recall that $\mathcal{D}$ is the set of nodes which currently have the state as IN. SATISFIED(i) is true if $i \in \mathcal{D}$ or each demand d in D_i is being served by some node j in Adj_i. If REMOVABLE(i) is true, then $\mathcal{D} \setminus \{i\}$ is also a dominating set given that $\mathcal{D}$ is a dominating set. DOMINATORS-OF(i) is the set of nodes that are (possibly) dominating node i: if some node j is in DOMINATORS-OF(i), then there is at least one demand $d \in D_i$ such that $d \in S_i$. We also defined FORBIDDEN(i) to capture the notion of *forbidden* in [4] (discussed in Sect. 3).

Macro	What it stands for
$\mathcal{D}$	$\{i \in V(G) : st.i = IN\}$
SATISFIED(i)	$st.i = IN \vee (\forall d \in D_i, \exists j \in Adj_i : d \in S_j \wedge st.j = IN)$
UNSATISFIED-DS(i)	$\neg$SATISFIED(i)
REMOVABLE-DS(i)	$(\forall d \in D_i : (\exists j \in Adj_i : d \in S_j \wedge st.j = IN)) \wedge$ $(\forall j \in Adj_i, \forall\, d \in D_j : d \in S_i \implies$ $(\exists k \in Adj_j, k \neq i : (d \in S_k \wedge st.k = IN)))$
DOMINATORS-OF(i)	$\{j \in Adj_i, st.j = IN : \exists d \in D_i : d \in S_j\} \cup \{i\}$ if $st.i = IN$ $\{j \in Adj_i, st.j = IN : \exists d \in D_i : d \in S_j\}$ otherwise
FORBIDDEN-DS(i)	$st.i = IN \wedge$ REMOVABLE-DS(i)$\wedge$ $(\forall j \in Adj_i, \forall\, d \in D_j : d \in S_i \implies$ $((\forall k \in$ DOMINATORS-OF(j), $k \neq i : (d \in S_k \wedge st.k = IN)) \implies$ $(id.k < id.i \vee \neg$REMOVABLE-DS(k))))

The general idea our algorithm is as follows.

1. We allow a node to enter the dominating set unconditionally if it is unsatisfied, i.e., SATISFIED(i) is false. This ensures that G enters a feasible state (where $\mathcal{D}$ is a dominating set) as quickly as possible.
2. While entering the dominating set is not coordinated with others, leaving the dominating set is coordinated with neighboring nodes. Node i can leave the dominating set only if it is removable. But before it does that, it needs to coordinate with others so that too many nodes do not leave, creating a race condition. Specifically, if i serves for a demand d in D_j where $j \in Adj_i$ and the same demand is also served by another node k ($k \in Adj_j$) then i leaves only if (1) $id.k < id.i$ or (2) k is not removable. This ensures that if some demand d of D_j is satisfied by both i and k both of them cannot leave the dominating set simultaneously. This ensures that j will remain dominated.

Thus, the rules for Algorithm 1 are as follows:

Algorithm 1 *Rules for node i.*

$$\text{FORBIDDEN-DS}(i) \longrightarrow st.i = OUT.$$
$$\text{UNSATISFIED-DS}(i) \longrightarrow st.i = IN.$$

We decompose Algorithm 1 into two parts: (1) Algorithm 1.1, that only consists of first guard and action of Algorithm 1 and (2) Algorithm 1.2, that only consists of the second guard and action of Algorithm 1. We use this decomposition in some of the following parts of this article section to relate the algorithm to eventual lattice linearity.

5 Lattice Linear Characteristics of Algorithm 1

In this section, we analyze the characteristics of Algorithm 1 to demonstrate that it is eventually lattice linear. We proceed as follows. In Sect. 5.1, we state the propositions which define the feasible and optimal states of the SDDS problem, along with some other definitions. In Sect. 5.2, we show that G reaches a state where it manifests a (possibly non-minimal) dominating set. In Sect. 5.3, we show that after when G reaches a feasible state, Algorithm 1 behaves like a lattice linear algorithm. In Sect. 5.4, we show that when $\mathcal{D}$ is a minimal dominating set, no nodes are enabled. In Sect. 5.5, we argue that because there is a bound on interference between Algorithm 1.1 and 1.2 even when the nodes read old values, Algorithm 1 is an Eventually Lattice Linear Self-Stabilizing (ELLSS) algorithm. In Sect. 5.6, we study the time and space complexity attributes of Algorithm 1.

5.1 Propositions Stipulated by the SDDS Problem

Notice that the SDDS problem stipulates that the nodes whose state is IN must collectively form a dominating set. Formally, we represent this proposition as $\mathcal{P}'_d$ which is defined as follows.

$$\mathcal{P}'_d(\mathcal{D}) \equiv \forall i \in V(G) : (i \in \mathcal{D} \vee (\forall d \in D_i, \exists j \in Adj_i : (d \in S_j \wedge j \in \mathcal{D}))).$$

The SDDS problem stipulates an additional condition that $\mathcal{D}$ should be a minimal dominating set. We formally describe this proposition $\mathcal{P}_d$ as follows.

$$\mathcal{P}_d(\mathcal{D}) \equiv \mathcal{P}'_d(\mathcal{D}) \wedge (\forall i \in \mathcal{D}, \neg\mathcal{P}'_d(\mathcal{D} \setminus \{i\})).$$

If $\mathcal{P}'_d(\mathcal{D})$ is true, then G is in a feasible state. And, if $\mathcal{P}_d(\mathcal{D})$ is true, then G is in an optimal state.

Based on the above definitions, we define two scores with respect to the global state, $RANK$ and $BADNESS$. $RANK$ determines the number of nodes needed to be added to $\mathcal{D}$ to change $\mathcal{D}$ to a dominating set. $BADNESS$ determines the number of nodes that are needed to be removed from $\mathcal{D}$ to make it a minimal dominating set, given that $\mathcal{D}$ is a (possibly non-minimal) dominating set. Formally, we define $RANK$ and $BADNESS$ as follows.

Definition 2. $RANK(\mathcal{D}) \equiv \min\{|\mathcal{D}'| - |\mathcal{D}| : \mathcal{P}'_d(\mathcal{D}') \wedge \mathcal{D} \subseteq \mathcal{D}'\}$.

Definition 3. $BADNESS(\mathcal{D}) \equiv \max\{|\mathcal{D}| - |\mathcal{D}'| : \mathcal{P}'_d(\mathcal{D}') \wedge \mathcal{D}' \subseteq \mathcal{D}\}$.

5.2 Guarantee to Reach a Feasible State by Algorithm 1.2

In this subsection, we show that if the nodes execute Algorithm 1.2 only, then G is guaranteed to reach a *feasible* state where $\mathcal{D}$ is a (possibly non-minimal) dominating set.

Lemma 1. *Let $t.\mathcal{D}$ be the value of $\mathcal{D}$ at the beginning of round t. If $t.\mathcal{D}$ is not a dominating set then $(t+1).\mathcal{D}$ is a dominating set.*

Proof. Let i be a node such that $i \in t.\mathcal{D}$ and $i \notin (t+1).\mathcal{D}$, i.e., i leaves the dominating set in round t. This means that i will remain satisfied and each node in Adj_i is satisfied, even when i is removed. This implies that i will not reduce the feasibility of $t.D$; it will not increase the value of $RANK$.

Now let ℓ be a node such that $\ell \notin t.\mathcal{D}$ which is not satisfied when it evaluates its guards in round t. This implies that $\exists\ d \in D_\ell$ such that d is not present in S_j for any $j \in Adj_\ell$. According to the algorithm, the guard of the second action is true for ℓ. This implies that $st.\ell$ will be set to IN.

It can also be possible for the node ℓ that it is satisfied when it evaluates its guards in round t. This may happen if some other nodes around ℓ already decided to move to $\mathcal{D}$, and as a result ℓ is now satisfied. Hence $\ell \notin (t+1).\mathcal{D}$ and we have that ℓ is dominated at round $t+1$.

Therefore, we have that $(t+1).\mathcal{D}$ is a dominating set, which may or may not be minimal. □

By Lemma 1, we have that if the G is in a state where $RANK > 0$ then by the next round, $RANK$ will be 0.

5.3 Lattice Linearity of Algorithm 1.1

In the following lemma, we show that Algorithm 1.1 is lattice linear.

Lemma 2. *If $t.\mathcal{D}$ is a non-minimal dominating set then according to Algorithm 1 (more specifically, Algorithm 1.1), there exists at least one node such that G cannot reach a minimal dominating set until that node is removed from the dominating set.*

Proof. Since $\mathcal{D}$ is a dominating set, we have that the second guard is not true for any node in G.

Since $\mathcal{D}$ is not minimal, there exists at least one node that must be removed in order to make $\mathcal{D}$ minimal. Let S' be the set of nodes which are removable. Let M be some node in S'. If M is not serving any node, then Forbidden(M) is trivially true. Otherwise there exists at least one node j which is served by M, that is, $\exists d \in D_j : d \in S_M$. We study two cases which are as follows: (1) for some node j served by M, there does not exist a node $b \in S'$ which serves j, and (2) for any node $b \in S'$ such that M and b serve some common node j, $id.b < id.M$.

In the first case, M cannot be removed because Removable(M) is false and, hence, M cannot be in S', thereby leading to a contradiction. In the second case, Forbidden(M) is true and Forbidden(b) is false since $id.b < id.M$. Thus, node b cannot leave the dominating set in that b cannot leave until M leaves. In both the cases, we have that j stays dominated.

Since ID of every node is distinct, we have that there exists at least one node M for which Forbidden(M) is true. For example, Forbidden(M) is true for the node with the highest ID in S'; G cannot reach a minimal dominating set until M is removed from the dominating set. □

From Lemma 2, it follows that Algorithm 1.1 satisfies the condition of lattice linearity defined in Sect. 3. It follows that if we start from a state where $\mathcal{D}$ is a (possibly non-minimal) dominating set and execute Algorithm 1.1 then it will reach a state where $\mathcal{D}$ is a minimal dominating set even if nodes are executing with old information about others. Next, we have the following result which follows from Lemma 2.

Lemma 3. *Let $t.\mathcal{D}$ be the value of $\mathcal{D}$ at the beginning of round t. If $t.\mathcal{D}$ is a non-minimal dominating set then $|(t+1).\mathcal{D}| \leq |t.\mathcal{D}| - 1$, and $(t+1).\mathcal{D}$ is a dominating set.*

Proof. From Lemma 2, at least one node M (including the maximum ID node in S' from the proof of Lemma 2) would be removed in round t. Furthermore, since $\mathcal{D}$ is a dominating set, UNSATISFIED(i) is false at every node i. Thus, no node is added to $\mathcal{D}$ in round t. Thus, the $|(t+1).\mathcal{D}| \leq |t.\mathcal{D}| - 1$.

For any node M that is removable, FORBIDDEN(i) is true only if any node j which is (possibly) served by M has other neighbors (of a lower ID) which serve the demands which M is serving to it. This guarantees that j stays dominated and hence $(t+1).\mathcal{D}$ is a dominating set. □

5.4 Termination of Algorithm 1

The following lemma studies the action of Algorithm 1 when $\mathcal{D}$ is a minimal dominating set.

Lemma 4. *Let $t.\mathcal{D}$ be the value of $\mathcal{D}$ at the beginning of round t. If $\mathcal{D}$ is a minimal dominating set, then $(t+1).\mathcal{D} = t.\mathcal{D}$.*

Proof. Since $\mathcal{D}$ is a dominating set, SATISFIED(i) is true for every node in $V(G)$. Hence, the second action is disabled for every node in $V(G)$. Since $\mathcal{D}$ is minimal, REMOVABLE(i) is false for every node in $\mathcal{D}$. Hence, the first action is disabled at every node i in $\mathcal{D}$. Thus, $\mathcal{D}$ remains unchanged. □

5.5 Eventual Lattice Linearity of Algorithm 1

Lemma 2 showed that Algorithm 1.1 is lattice linear. In this subsection, we make additional observations about Algorithm 1 to generalize the notion of lattice linearity to eventually lattice linear algorithms. We have the following observations.

1. From Lemma 1, starting from any state, Algorithm 1.2 will reach a feasible state even if a node reads old information about the neighbors. This is due to the fact that Algorithm 1.2 only adds nodes to $\mathcal{D}$.
2. From Lemma 2, if we start G in a feasible state where no node has incorrect information about the neighbors in the initial state then Algorithm 1.1 reaches a minimal dominating set. Note that this claim remains valid even if the nodes execute actions of Algorithm 1.1 with old information about the neighbors as long as the initial information they use is correct.
3. Now, we observe that Algorithm 1.1 and Algorithm 1.2 have very limited interference with each other, and so an arbitrary graph G will reach an optimal state even if nodes are using old information. In this case, observe that any node i can execute the action of the first guard incorrectly at most once. After going OUT incorrectly, when it reads the correct information about other nodes, then it will execute the guard of the second action and change its state to IN, after which, it can go out only if it evaluates that FORBIDDEN(i) is true.

From the above observations, if we allow the nodes to read old values, then the nodes can violate the feasibility of G finitely many times and so G will eventually reach a feasible state and stay there forever. We introduce the term Eventually Lattice Linear Self-stabilizing algorithms (ELLSS). Before defining ELLSS algorithms, we define the class of Lattice Linear Algorithms (LL) as follows.

Definition 4. ***Lattice Linear Algorithms.*** *LL algorithms are the algorithms under which a system G is forced to traverse a lattice of states and proceed to reach an optimal solution.*

Note that LL is a generalization of the notion of lattice linearity introduced in [4]. In [4], the existence of the lattice arises from the problem at hand. In LL, it arises by constraints imposed by the algorithm.

The class of ELLSS algorithms can be defined as follows.

Definition 5. ***Eventually Lattice Linear Self-Stabilizing Algorithms.*** *An algorithm is ELLSS if its rules can be split into F_1 and F_2 and there exists a predicate $\mathcal{R}$ such that*

(a) Any computation of $F_1[]F_2$, that is, the union of the actions in F_1 and F_2, eventually reaches a state where $\mathcal{R}$ is stable in $F_1[]F_2$, i.e., $\mathcal{R}$ is true and remains true subsequently (even if the nodes read old values of other nodes),
(b) F_2 is an LL algorithm, given that it starts in a state in $\mathcal{R}$,
(c) Actions in F_1 are disabled once the program reaches $\mathcal{R}$.

In Algorithm 1, F_1 corresponds to Algorithm 1.2 and F_2 corresponds to Algorithm 1.1. And, the above discussion shows that this algorithm satisfies the properties of Definition 5 and $\mathcal{R}$ corresponds to the predicate that $\mathcal{D}$ is a dominating set, i.e., $\mathcal{R} \equiv \mathcal{P}'_d(\mathcal{D})$.

Finally, we note that in Algorithm 1, we chose to be aggressive for a node to enter the dominating set but cautious to leave the dominating set. A similar ELLSS SDDS algorithm is feasible where a node is cautious to enter the dominating set but aggressive to leave it.

5.6 Analysis of Algorithm 1: Time and Space Complexity

We make the following observations about the time and space complexities of Algorithm 1. For reasons of space, the proofs are provided in [8].

1. Algorithm 1 converges in $2n$ moves.
2. $\mathcal{D}$ will be feasible within 1 round. After entering a feasible state, $\mathcal{D}$ will be optimal within n moves.
3. Algorithm 1 is self-stabilizing and silent.
4. At any time-step, we have that a node will take $O((\Delta)^4 \times (max_d)^2)$ time, where (1) Δ is the maximum degree of any node in $V(G)$, and (2) max_d is the total number of distinct demands made by all the nodes in $V(G)$.
5. The space required to store the services and demands in each node is $O(max_d)$, where max_d is the total number of distinct demands made by all the nodes in $V(G)$.

6 Other Examples

The sequence of states of G under Algorithm 1 is essentially divided into two phases: (1) the system entering a feasible state (reduction of $RANK$ to zero), and then (2) the system entering an optimal state (reduction of $BADNESS$ to zero). Algorithm 1 first takes the system to a feasible state where $RANK = 0$ and then it takes the system to an optimal state where $RANK = 0 \wedge BADNESS = 0$.

This notion was used to define the concept of ELLSS algorithms. The notion of ELLSS algorithms can be extended to numerous other problems where the optimal global state can be defined in terms of a minimal (or maximal) set $\mathcal{S}$ of nodes. This includes the vertex cover problem, independent set problem and their variants. Once the propositions regarding $\mathcal{S}$ is defined where its structure depends on some relation of nodes with their neighbours, the designed algorithm can decide which node to put IN the set and which nodes to take OUT.

In this section, we describe algorithms for vertex cover and independent set, along with graph coloring, which follow from the structure that we have laid for the SDDS problem. Thus, the algorithms we describe for these problems are also ELLSS algorithms. The proofs of correctness follow from the proofs of correctness described above for the SDDS problem.

6.1 Vertex Cover

In the *vertex cover* (VC) problem, the input is an arbitrary graph G, and the task is to compute a minimal set $\mathcal{V}$ such that for any edge $\{i, j\} \in E(G)$, $(i \in \mathcal{V}) \vee (j \in \mathcal{V})$. If a node i is in $\mathcal{V}$, then $st.i = IN$, otherwise $st.i = OUT$. To develop an algorithm for VC, we utilize the macros in the following table.

REMOVABLE-VC(i)	$(\forall j \in Adj_i, st.j = IN)$.
UNSATISFIED-VC(i)	$(st.i = OUT) \wedge (\exists j \in Adj_i : st.j = OUT)$.
FORBIDDEN-VC(i)	$(st.i = IN) \wedge (\text{REMOVABLE-VC}(i)) \wedge$ $(\forall j \in Adj_i : (id.j < id.i) \vee \neg\text{REMOVABLE-VC}(j))$.

The proposition $\mathcal{P}'_v$ defining a feasible state and the proposition $\mathcal{P}_v$ defining the optimal state can be defined as follows.

$$\mathcal{P}'_v(\mathcal{V}) \equiv \forall i \in V(G) : ((i \in \mathcal{V}) \vee (\forall j \in Adj_i, j \in \mathcal{V})).$$
$$\mathcal{P}_v(\mathcal{V}) \equiv \mathcal{P}'_v(\mathcal{V}) \wedge (\forall i \in \mathcal{V}, \neg\mathcal{P}'_v(\mathcal{V} - \{i\})).$$

Based on the definitions above, the algorithm for VC is described as follows.

Algorithm 2 *Rules for node i.*

FORBIDDEN-VC(i) $\longrightarrow st.i = OUT$.
UNSATISFIED-VC(i) $\longrightarrow st.i = IN$.

Once again, this is an ELLSS algorithm in that it satisfies the conditions in Definition 5, where F_1 corresponds to the second action of Algorithm 2, F_2 corresponds to its first action, and $\mathcal{R} \equiv \mathcal{P}'_v(\mathcal{V})$. Thus, starting from any arbitrary state, the algorithm eventually reaches a state where $\mathcal{V}$ is a minimal vertex cover.

Note that in Algorithm 2, the definition of REMOVABLE relies only on the information about distance-1 neighbors. Hence, the evaluation of guards take $O(\Delta^3)$ time. In contrast, (the standard) dominating set problem requires information of distance-2 neighbors to evaluate REMOVABLE. Hence, the evaluation of guards in that would take $O(\Delta^4)$ time.

6.2 Independent Set

In VC and SDDS problems, we tried to reach a minimal set. Here on the other hand, we have to obtain a maximal set. In the *independent set* (IS) problem, the input is an arbitrary graph G, and the task is to compute a maximal set $\mathcal{I}$ such that for any two nodes $i \in \mathcal{I}$ and $j \in \mathcal{I}$, if $i \neq j$, then $\{i, j\} \neq E(G)$.

The proposition $\mathcal{P}'_i$ defining a feasible state and the proposition $\mathcal{P}_i$ defining the optimal state can be defined as follows.

$$\mathcal{P}'_i(\mathcal{I}) \equiv \forall i \in V(G) : ((i \notin \mathcal{I}) \vee (\forall j \in Adj_i : j \notin \mathcal{I})).$$
$$\mathcal{P}_i(\mathcal{I}) \equiv \mathcal{P}'_i(\mathcal{I}) \wedge (\forall i \in V(G) \setminus \mathcal{I}, \neg\mathcal{P}'_i(\mathcal{I} \cup \{i\})).$$

If a node i is in $\mathcal{I}$, then $st.i = IN$, otherwise $st.i = OUT$. To develop the algorithm for independent set, we define the macros in the following table.

ADDABLE(i)	$(\forall j \in Adj_i, st.j = OUT)$.
UNSATISFIED-IS(i)	$(st.i = IN) \wedge (\exists j \in Adj_i : st.j = IN)$.
FORBIDDEN-IS(i)	$st.i = OUT \wedge$ ADDABLE$(i) \wedge$ $(\forall j \in Adj_i : ((id.j < id.i) \vee (\neg$ADDABLE$(j)))$.

Based on the definitions above, the algorithm for IS is described as follows.

Algorithm 3 *Rules for node i.*

FORBIDDEN-IS(i) $\longrightarrow$	$st.i = IN$.
UNSATISFIED-IS(i) $\longrightarrow$	$st.i = OUT$.

This algorithm is an ELLSS algorithm as well: as per Definition 5, F_1 corresponds to the second action of Algorithm 2, F_2 corresponds to its first action, and $\mathcal{R} \equiv \mathcal{P}'_i(\mathcal{I})$. Thus, starting from any arbitrary state, the algorithm eventually reaches a state where $\mathcal{I}$ is a maximal independent set.

In Algorithm 3, the definition of ADDABLE relies only on the information about distance-1 neighbors. Hence, the evaluation of guards take $O(\Delta^3)$ time.

6.3 Coloring

In this section, we extend ELLSS algorithms to graph coloring. In the *graph coloring* (GC) problem, the input is a graph G and the task is to assign colors to all the nodes of G such that no two adjacent nodes have the same color.

Unlike vertex cover, dominating set or independent set, coloring does not have a binary domain. Instead, we correspond the equivalence of changing the state to IN to the case where a node increases its color. And, the equivalence of changing the state to OUT corresponds to the case where a node decreases its color. With this intuition, we define the macros as shown in the following table.

$\text{CONFLICTED}(i)$	$(\exists j \in Adj_i : (color.j = color.i))$
$\text{SUBTRACTABLE}(i)$	$\exists c \in [1 : color.i - 1] : ((\forall j \in Adj_i : color.j \neq c))$
$\text{UNSATISFIED-GC}(i)$	$\text{CONFLICTED}(i)$
$\text{FORBIDDEN-GC}(i)$	$\neg\text{CONFLICTED}(i) \wedge \text{SUBTRACTABLE}(i) \wedge$ $(\forall j \in Adj_i : (id.j < id.i \vee \neg\text{SUBTRACTABLE}(j)))$

The proposition $\mathcal{P}'_c$ defining a feasible state and the proposition $\mathcal{P}_c$ defining an optimal state is defined below. $\mathcal{P}_c$ is true when all the nodes have lowest available color, that is, for any node i and for all colors c in $[1 : deg(i) + 1]$, either c should be greater than $color.i$ or c should be equal to the color of one of the neighbors j of i.

$$\begin{aligned}
&\mathcal{P}'_c \equiv \forall i \in V(G), \forall j \in Adj_i : color.i \neq color.j.\\
&\mathcal{P}_c \equiv \mathcal{P}'_c \wedge (\forall i \in V(G) : (\forall c \in [1 : color.i - 1] :\\
&(c < color.i \implies (\exists j \in Adj_i : color.j = c)))).
\end{aligned}$$

Unlike SDDS, VC and IS, in graph coloring (GC), each node is associated with a variable *color* that can take several possible values (the domain can be as large as the set of natural numbers). As mentioned above, the action of increasing the color is done whenever a conflict is detected. However, decreasing the color is achieved only with coordination with others. Thus, the actions of the algorithm are shown in Algorithm 4.

Algorithm 4 *Rules for node i.*

$\text{FORBIDDEN-GC}(i) \longrightarrow color.i = \min_c\{c \in [1 : color.i - 1] : (\forall j \in Adj_i : color.j \neq c)\}.$
$\text{UNSATISFIED-GC}(i) \longrightarrow color.i = color.i + id.i.$

Algorithm 4 is an ELLSS algorithm: according to Definition 5, F_1 corresponds to the second action of Algorithm 2, F_2 corresponds to its first action, and $\mathcal{R} \equiv \mathcal{P}'_c$. Thus, starting from any arbitrary state, the algorithm eventually reaches a state where no two adjacent nodes have the same color and no node can reduce its color.

7 Conclusion

We extended lattice linear algorithms from [4] to the context of self-stabilizing algorithms. The approach in [4] relies on the assumption that the algorithm starts in specific initial states and, hence, it is not directly applicable in self-stabilizing algorithms. A key benefit of lattice linear algorithms is that correctness is preserved even if nodes are reading old information about other nodes. Hence, they allow a higher level of concurrency.

We began with the service-demand based dominating set (SDDS) problem and designed a self-stabilizing algorithm for the same. Subsequently, we observed that it consists of two parts: One part is a lattice linear algorithm that constructs a minimal dominating set if it starts in some valid initial states, say $\mathcal{R}$. The second part makes sure that it gets the program in a state where $\mathcal{R}$ becomes true and stays true forever. We also showed that these parts can only have bounded interference thereby guaranteeing that the overall program is self-stabilizing even if the nodes read old values of other nodes.

We introduced the notion of eventually lattice-linear self-stabilization to capture such algorithms. We demonstrated that it is possible to develop eventually lattice linear self-stabilizing (ELLSS) algorithms for vertex cover, independent set and graph coloring.

We note that Algorithms 1–4 could also be designed to be lattice linear self-stabilizing algorithms (LLSS) if we change the second action of these algorithms to account for the neighbors in the same fashion as done for the second action.

The Algorithms 1–4 converge under central, distributed, or synchronous daemon. Due to the property of ELLSS, its straightforward implementation in read/write model is also self-stabilizing. Intuitively, in a straightforward translation, each remote variable is replaced by a local copy of that variable and this copy is updated asynchronously. Normally, such straightforward translation into read/write atomicity does not preserve self-stabilization. However, the ELLSS property of the self-stabilization ensures correctness of the straightforward translation.

As future work, an interesting direction can be to study which class of problems can the paradigm of LL and ELLSS algorithms be extended to. Also, it is interesting to study if we can implement approximation algorithms under these paradigms.

References

1. Arora, A., Gouda, M.: Closure and convergence: a foundation of fault-tolerant computing. IEEE Trans. Software Eng. **19**(11), 1015–1027 (1993)
2. Chiu, W.Y., Chen, C., Tsai, S.-Y.: A 4n-move self-stabilizing algorithm for the minimal dominating set problem using an unfair distributed daemon. Inf. Process. Lett. **114**(10), 515–518 (2014)
3. Fink, J.F., Jacobson, M.S.: N-Domination in Graphs, pp. 283–300. Wiley, Hoboken (1985)

4. Garg, V.K.: Predicate detection to solve combinatorial optimization problems. In: Proceedings of the 32nd ACM Symposium on Parallelism in Algorithms and Architectures. SPAA 2020, pp. 235–245. Association for Computing Machinery, New York (2020)
5. Garg, V.K.: A lattice linear predicate parallel algorithm for the dynamic programming problems. CoRR abs/2103.06264 (2021)
6. Goddard, W., Hedetniemi, S.T., Jacobs, D.P., Srimani, P.K., Xu, Z.: Self-stabilizing graph protocols. Parallel Process. Lett. **18**(01), 189–199 (2008)
7. Guellati, N., Kheddouci, H.: A survey on self-stabilizing algorithms for independence, domination, coloring, and matching in graphs. J. Parallel Distrib. Comput. **70**(4), 406–415 (2010)
8. Gupta, A.T., Kulkarni, S.S.: Extending lattice linearity for self-stabilizing algorithms. CoRR abs/2109.13216 (2021)
9. Hedetniemi, S., Hedetniemi, S., Jacobs, D., Srimani, P.: Self-stabilizing algorithms for minimal dominating sets and maximal independent sets. Comput. Math. Appl. **46**(5), 805–811 (2003)
10. Kamei, S., Kakugawa, H.: A self-stabilizing algorithm for the distributed minimal k-redundant dominating set problem in tree networks. In: Proceedings of the Fourth International Conference on Parallel and Distributed Computing, Applications and Technologies, pp. 720–724 (2003)
11. Kamei, S., Kakugawa, H.: A self-stabilizing approximation algorithm for the distributed minimum k-domination. IEICE Trans. Fundam. Electron. Commun. Comput. Sci. **E88-A**(5), 1109–1116 (2005)
12. Kiniwa, J.: Approximation of self-stabilizing vertex cover less than 2. In: Tixeuil, S., Herman, T. (eds.) SSS 2005. LNCS, vol. 3764, pp. 171–182. Springer, Heidelberg (2005). https://doi.org/10.1007/11577327_12
13. Kobayashi, H., Kakugawa, H., Masuzawa, T.: Brief announcement: a self-stabilizing algorithm for the minimal generalized dominating set problem. In: Spirakis, P., Tsigas, P. (eds.) SSS 2017. LNCS, vol. 10616, pp. 378–383. Springer, Cham (2017). https://doi.org/10.1007/978-3-319-69084-1_27
14. Leal, W., Arora, A.: Scalable self-stabilization via composition. In: 2004 Proceedings of the 24th International Conference on Distributed Computing Systems, pp. 12–21 (2004)
15. Åstrand, M., Suomela, J.: Fast distributed approximation algorithms for vertex cover and set cover in anonymous networks. In: Proceedings of the Twenty-Second Annual ACM Symposium on Parallelism in Algorithms and Architectures. SPAA 2010, pp. 294–302. Association for Computing Machinery, New York (2010)
16. Turau, V.: Linear self-stabilizing algorithms for the independent and dominating set problems using an unfair distributed scheduler. Inf. Process. Lett. **103**(3), 88–93 (2007)
17. Turau, V.: Self-stabilizing vertex cover in anonymous networks with optimal approximation ratio. Parallel Process. Lett. **20**(02), 173–186 (2010)
18. Varghese, G.: Self-stabilization by local checking and correction. Ph.D thesis, Massachusetts Institute of Technology, October 1992
19. Xu, Z., Hedetniemi, S.T., Goddard, W., Srimani, P.K.: A synchronous self-stabilizing minimal domination protocol in an arbitrary network graph. In: Das, S.R., Das, S.K. (eds.) IWDC 2003. LNCS, vol. 2918, pp. 26–32. Springer, Heidelberg (2003). https://doi.org/10.1007/978-3-540-24604-6_3

Information Exchange in the Russian Cards Problem

Zoe Leyva-Acosta[1], Eduardo Pascual-Aseff[1], and Sergio Rajsbaum[2](✉)

[1] Instituto de Investigaciones en Matemáticas Aplicadas y en Sistemas, Universidad Nacional Autónoma de México (UNAM), 04510 Mexico City, Mexico
[2] Instituto de Matemáticas, Universidad Nacional Autónoma de México (UNAM), 04510 Mexico City, Mexico
rajsbaum@im.unam.mx

Abstract. Alice and Bob wish to privately exchange information by public announcements overheard by Cath. To do so by a deterministic protocol, their inputs must be correlated. Dependent inputs are represented using a deck of cards. There is a publicly known *signature* $(\mathbf{a}, \mathbf{b}, \mathbf{c})$, meaning that A gets $\mathbf{a}$ cards, B gets $\mathbf{b}$ cards, and C gets $\mathbf{c}$ cards, out of the deck of n cards, $n = \mathbf{a} + \mathbf{b} + \mathbf{c} + \mathbf{r}$. We use a perspective inspired by distributed computing that considers colorings of a generalization of Johnson graphs, together with techniques based on Singer difference sets and shifting, to study the classic *Russian cards* problem $\mathbf{a} = \mathbf{b} = 3$, $\mathbf{c} = 1$, and $\mathbf{r} = 0$. We consider also a novel variant where they wish to learn *something* about each other's hands. We focus on the number of bits that Alice and Bob need to exchange to solve either the classic or the minimally informative version of the problem.

1 Introduction

Peter Winkler [24] motivated a long research line e.g. [12,15,18], by "the surprising discovery that information can be passed both covertly and legally between bridge partners". It inspired Fischer and Wright e.g. [12] to consider *card games*, where A, B, C draw cards from a deck D of n cards, as specified by a *signature* $(\mathbf{a}, \mathbf{b}, \mathbf{c})$, with $n = \mathbf{a}+\mathbf{b}+\mathbf{c}$, meaning that A gets $\mathbf{a}$ cards, B gets $\mathbf{b}$ cards, and C gets $\mathbf{c}$ cards; they thought of the cards as representing correlated random initial local variables for the players, that have a simple structure. Their protocols are information-theoretic secure, but they are not concerned with keeping the cards of A and B secret from C.

Another research line started with an in-depth, combinatorial and epistemic logic study of van Ditmarsch [8] of the *Russian cards* problem, presented at the Moscow Mathematics Olympiad in 2000, where the cards of A and B should be kept secret from C. Here A, B and C draw $(3, 3, 1)$ cards, respectively, from a deck of 7 cards. First A makes an announcement, and then B makes an announcement. The goal is for A and B to learn each other cards, while ensuring that C,

Supported by the UNAM-PAPIIT project IN106520.

C. Johnen et al. (Eds.): SSS 2021, LNCS 13046, pp. 380–394, 2021.
https://doi.org/10.1007/978-3-030-91081-5_25

who listens to both announcements, cannot deduce a single card of A (nor of B). The problem has received a fair amount of attention since then, including its *generalized* form $(\mathbf{a}, \mathbf{b}, \mathbf{c})$, and other variants, including multiround, multiplayer, and different security requirements, using techniques from modular arithmetic, combinatorial designs, and epistemic logic e.g. [1,2,4–6,9,10,16,22,23].

There is yet another long research line on communication complexity based on correlated inputs in information theory e.g. [19], but the focus is not in correlated inputs from card games nor preserving the privacy of the inputs; our work takes inspiration from this line, as well as from fault-tolerant distributed computing e.g. [7].

Motivation and the New Approach. In a recent work Rajsbaum [20] considered the generalized Russian cards problem $(\mathbf{a}, \mathbf{b}, \mathbf{c})$, for a deck D, with the possibility that nobody gets $\mathbf{r}$ cards, $n = \mathbf{a} + \mathbf{b} + \mathbf{c} + \mathbf{r}$, and studied the case of *one-way* information transfer, where A sends a public message using a deterministic protocol P_A. Given a hand a, she sends a public message $P_A(a)$ to B, from which B can deduce a using his own hand b, secretly against C. It showed that the possibility of $\mathbf{r} > 0$ introduces interesting new aspects to the problem. In fact, even in the one-way information transfer case, a full characterization of the signatures for which the problem is solvable is still open.

The set of possible hands of A is $\mathscr{P}_{\mathbf{a}}(D)$, all subsets of size $\mathbf{a}$ of D. The vertices of the *d-distance Johnson graph* $J^d(n, \mathbf{a})$ are $\mathscr{P}_{\mathbf{a}}(D)$. Vertices a, a' of $J^d(n, \mathbf{a})$ are adjacent whenever $\mathbf{a} - d \leq |a \cap a'|$. In particular, we have a Johnson graph when $d = 1$. Let $\mathcal{M}_A$ be the set of messages that A may send. The protocol P_A can be viewed as a coloring of the vertices of $J^d(n, \mathbf{a})$, $d = \mathbf{c} + \mathbf{r}$, a function

$$P_A : \mathscr{P}_{\mathbf{a}}(D) \to \mathcal{M}_A.$$

It was noticed in [20] that B can deduce the hand of a if and only if P_A is a proper coloring of $J^d(n, \mathbf{a})$, $d = \mathbf{c} + \mathbf{r}$. Thus, the chromatic number of $J^d(n, \mathbf{a})$ determines the number of messages needed for P_A to be informative. There are many interesting open questions concerning the chromatic number even when $d = 1$, i.e., Johnson graphs [13].

Contributions. We extend the one-way approach of [20] to *two-step protocols*, (P_A, P_B). Using her protocol P_A, A makes a public announcement, $M = P_A(a)$, based on her hand, a. Using his own hand, b, B responds using his protocol P_B with a public announcement $P_B(b, M)$. In the language of e.g. [4,6,9], the protocol is *informative*, if A and B learn each other hands, and a protocol is *safe* if C does not learn any of the cards of A and B. We consider additionally *minimally informative* protocols, where A and B should learn *something* about each other hands.

We start by modeling formally the setting using ideas from distributed computing [14], dealing with the additional difficulties of allowing $\mathbf{r} > 0$, and also, those of the new notion of minimally informative protocol. We hope this general formulation is interesting in itself, although for the following results, we use it only in special cases.

We introduce several combinatorial techniques, and apply them to the classic Russian cards problem, where $\mathbf{a}, \mathbf{b} = 3$. We are interested in communication complexity, namely, if $\mathcal{M}_A, \mathcal{M}_B$ are the sets of messages that A and B can send, we study how small can $\mathcal{M}_A, \mathcal{M}_B$ be, to implement either an informative or a minimally informative safe protocol.

Remarkably, A and B can implement safe, minimal information transmission, by exchanging just one bit. We show that when $\mathbf{c} + \mathbf{r} = 1$, minimal information transmission is equivalent to learning one of each other cards. The protocol is designed using Singer difference sets [21].

Regarding informative protocols, when $\mathbf{a}, \mathbf{b} = 3$, since there is no informative and safe two-step protocol if $\mathbf{c} + \mathbf{r} \geq 2$, we consider the case of $\mathbf{c} + \mathbf{r} = 1$. We show that there is a safe informative solution using 6 messages for the *weak Russian cards* problem, when $\mathbf{c} = 0$ and $\mathbf{r} = 1$. In the Russian cards case, $\mathbf{c} = 1, \mathbf{r} = 0$, there is a well-known informative and safe solution with 7 messages[1], known since [17]. But there are solutions with only 6 messages [22], and this is optimal, since the chromatic number of $J(7, 3)$ is known to be 6. However, as opposed to the 7 message solution, these solutions are not *uniform;* color classes are of different sizes. We prove that there is no uniform safe and informative solution with 6 messages. This result provides evidence that usual uniform solutions based on modular arithmetic or combinatorial design techniques would not work with 6 messages.

Organization. In Sect. 2 we formalize the problem for general signatures, then rephrase it in terms of Johnson graphs, and discuss some consequences. In Sect. 3 we present the results about the minimal information variant of the problem when $\mathbf{a} = \mathbf{b} = 3$, and in Sect. 4 we present the results about informative six-message solutions. The conclusions are in Sect. 5. For lack of space some proofs have been omitted from this extended abstract.

2 Secure Information Exchange

The problem is defined here, adapting the distributed computing formalization of [14] to the case of an eavesdropper.

2.1 Correlated Inputs

Let $D = \{0, \ldots, n-1\}$, $n > 1$, be the *deck* of n distinct cards. An element in the deck is a *card.* A subset x of cards is a *hand,* $x \in \mathscr{P}(D)$. We may say for short that x, $|x| = m$, is an m-set or m-hand, namely, if $x \in \mathscr{P}_m(D)$, the subsets of D of size m. A *deal* $= (a, b, c)$ consists of three disjoint hands, meaning that cards in a are dealt to A, cards in b to B, and cards in c to C. We call $\gamma = (\mathbf{a}, \mathbf{b}, \mathbf{c})$ the *signature* of the *deal* $= (a, b, c)$ if $|a| = \mathbf{a}$, $|b| = \mathbf{b}$ and $|c| = \mathbf{c}$, following the

[1] First A sends the sum of her cards modulo 7. Using this information, B can deduce the hand of A, and thus responds by announcing the card of C.

notation by Fischer and Wright [12]. We assume that A, B and C are aware of the deck and the signature. It has been usually assumed that $n = \mathbf{a}+\mathbf{b}+\mathbf{c}$, but following [20] we consider the case where nobody gets $\mathbf{r}$ cards, $n = \mathbf{a}+\mathbf{b}+\mathbf{c}+\mathbf{r}$. While A and B get at least one card, $\mathbf{a}, \mathbf{b} \geq 1$, C may get none $\mathbf{c} \geq 0$. However, we assume $\mathbf{c}+\mathbf{r} \geq 1$. Otherwise, if $\mathbf{c} = \mathbf{r} = 0$, without any communication A and B know each other hands.

All possible deals are represented by sets of the form $I = \{(A, a), (B, b), (C, c)\}$, which define an input configuration. The sets of all such deals I, together with all their subsets, form the *input complex* $\mathcal{I}$. The vertices are of the form (Y, y), $Y \in \{A, B, C\}$, and y a hand. Such a vertex is called a Y-vertex. We say that the hand y is the *input* of X. Notice that the A-vertices of $\mathcal{I}$ are in a one-to-one correspondence with $\mathscr{P}_{\mathbf{a}}(D)$, the B-vertices with $\mathscr{P}_{\mathbf{b}}(D)$, and the C-vertices with $\mathscr{P}_{\mathbf{c}}(D)$. Indeed, when $\mathbf{c} = 0$, there is a single vertex for C in $\mathcal{I}$.

2.2 Informative, Minimally Informative and Safe Protocols

In a *one-step protocol,* A announces a message M, defined by a deterministic function $P_A(a) = M$, for each input vertex $(A, a) \in \mathcal{I}$, where M belongs to $\mathcal{M}_A$, the domain of possible messages that A may send.

In a *two-step protocol,* after listening to A's announcement, B responds with his own announcement, M', using his protocol, also a deterministic function, $P_B(b, M) = M'$, $M' \in \mathcal{M}_B$. We say that P_A is the *protocol of* A, respectively. P_B is the *protocol of* B, and (P_A, P_B) is simply a *protocol.* The third player C listens to the conversation, without sending any messages[2].

For a facet $I = \{(A, a), (B, b), (C, c)\} \in \mathcal{I}$, and a protocol (P_A, P_B), we denote by $\alpha(I)$ the *execution* starting with I, where A announces $P_A(a)$ and B responds with $P_B(b, P_A(a))$, and C observes both announcements. For a one-step protocol P_A, an execution is defined only by A announcing $P_A(a)$. For an execution α, denote by $input_X(\alpha)$ the hand of player $X \in \{A, B, C\}$.

For an input simplex $I' \in \mathcal{I}$, $|I'| < 3$, defining inputs for either one or two players, $\alpha(I') = \{\alpha(I) : I' \subset I \in \mathcal{I}, |I| = 3\}$. For example, if $I' = \{(C, c)\}$ then $\alpha(I')$ denotes all executions where A and B start with disjoint hands of size $\mathbf{a}, \mathbf{b}$, respectively, that are disjoint from c.

The *view* of a player in an execution consists of its input and the sequence of messages announced in the execution. Two executions α, α' are *indistinguishable* to $X \subseteq \{A, B, C\}$, if the players in X have the same view in both, denoted $\alpha \stackrel{X}{\sim} \alpha'$ (e.g. [3]).

Definition 1 (Informative and minimally informative). *Let $X, X' \in \{A, B\}$, be two different agents, and (X', x') by any vertex in $\mathcal{I}$.*

- *A protocol P_X is* informative *for X' if any two executions $\alpha_1, \alpha_2 \in \alpha(X', x')$, such that $input_X(\alpha_1) \neq input_X(\alpha_2)$, are distinguishable to X'.*

[2] Notice that when C listens to a message M, say from A, she knows that the hand of A could be any a such that $P_A(a) = M$. Thus, M can be seen as an encoding of all such hands a, as in some previous papers e.g. [4].

- *A protocol* P_X *is* minimally informative *for* X' *if there exist two executions* $\alpha_1, \alpha_2 \in \alpha(X', x')$, *such that* $input_X(\alpha_1) \neq input_X(\alpha_2)$, *which are distinguishable to* X'.
- *A protocol* (P_A, P_B) *is* (minimally) informative *if it is* (minimally) informative *for both* A *and* B.

This definition is based on the graph which is the subcomplex of $\mathcal{I}$ induced by the A-vertices and the B-vertices. The C-vertices appear now; following previous papers, we require that C cannot tell who holds any individual card.

Definition 2 (Safe protocol). *A protocol* P_A *(or a two-step protocol* (P_A, P_B)*) is* safe *if for any* C*-vertex* (C, c) *of* $\mathcal{I}$*, any protocol execution* α *in* $\alpha(C, c)$*, and any pair of cards* x, y *held by* A *and* B*, respectively, in* α*, there are two other executions of the protocol* $\alpha_1, \alpha_2 \in \alpha(C, c)$*, with* $y \in input_A(\alpha_1)$ *and* $x \in input_B(\alpha_2)$*, such that* $\alpha \overset{C}{\sim} \alpha_1$ *and* $\alpha \overset{C}{\sim} \alpha_2$.

Notice that this is a generalization to $\mathbf{r} \geq 0$ of the definition used in previous papers for the case $\mathbf{r} = 0$. When $\mathbf{r} > 0$, C may learn cards that neither A nor B hold, but for any other pair of cards (x, y), she does not learn which party holds which card. That is, C might be able to learn a set of cards of size $\mathbf{c} + \mathbf{r}$ (including her own).

2.3 Protocols and Johnson Graphs

Recall the *d-distance Johnson graph* $J^d(n, \mathbf{a})$, consisting of vertices $a, a' \in \mathscr{P}_{\mathbf{a}}(D)$, adjacent whenever $\mathbf{a} - d \leq |a \cap a'|$. In particular, we have a Johnson graph when $d = 1$.

The vertices of A that B considers possible with input b, are the A-neighbors of (B, b) in $\mathcal{I}$. They are denoted $K_p(\bar{b})$, where $\bar{b} = D - b$. They induce a clique in $J^d(n, \mathbf{a})$, $d = \mathbf{c} + \mathbf{r}$ (overloading notation the clique itself is also sometimes denoted by $K_p(\bar{b})$). The vertices in $K_p(\bar{b})$ are all $a \subseteq \bar{b}$ with $|a| = \mathbf{a}$. Thus, when B has input b, B considers possible that A has any input a, $a \in K_p(\bar{b})$. Notice that $K_p(\bar{b})$ is of size $p = \binom{n-\mathbf{b}}{\mathbf{a}}$, consisting of all $a \in \mathscr{P}_{\mathbf{a}}(D)$, such that $a \subset \bar{b}$, and is in fact a maximal clique[3].

We have, for one-step protocols, the following reformulation of Definitions 1 and 2.

Theorem 1. *Let* $P_A : \mathscr{P}_{\mathbf{a}}(D) \to \mathcal{M}_A$ *be a protocol.*

- P_A *is informative for* B *if and only if* P_A *is a proper vertex coloring of* $J^{\mathbf{c}+\mathbf{r}}(n, \mathbf{a})$
- *When* $\mathbf{c} + \mathbf{r} \geq 1$, P_A *is minimally informative for* B *if and only if for each* $b \in \mathscr{P}_{\mathbf{b}}(D)$ *there is some edge* $\{a, a'\}$ *in the clique* $K_p(\bar{b})$ *of* $J^{\mathbf{c}+\mathbf{r}}(n, \mathbf{a})$ *such that* $P_A(a) \neq P_A(a')$

[3] The set of maximal cliques in $J(n, m)$ have been well-studied, they are of size $n-m+1$ and $m + 1$ e.g. [13].

– *Consider any* $c \in \mathscr{P}_{\mathbf{c}}(D)$*, and any* $y \in \bar{c}$*.* P_A *is safe if and only if for each* $M \in P_A(K_p(\bar{c}))$*, there exist* $a, a' \in K_p(\bar{c})$ *with* $P_A(a) = P_A(a') = M$ *such that* $y \in a \triangle a'$*.*

One of the interesting implications is the following observation [20]. It is a generalization of the well-known duality of Johnson graphs: the graphs $J(n, m)$ and $J(n, n - m)$ are isomorphic. In the following, the protocol $\bar{P}_A : \mathscr{P}_{n-\mathbf{a}}(D) \to \mathcal{M}_A$ is defined by $\bar{P}_A(a) = P_A(\bar{a})$.

Lemma 1 (duality). *Assume* $\mathbf{c} + \mathbf{r} = 1$*, so* $n = \mathbf{a} + \mathbf{b} + 1$*. A protocol* P_A *is informative and safe for* $(\mathbf{a}, \mathbf{b}, \mathbf{c})$ *if and only if the protocol* $\bar{P}_A$ *is informative and safe for* $(\mathbf{b} + 1, \mathbf{a} - 1, \mathbf{c})$*.*

3 Minimal Information Exchange

We consider here minimal informative solutions for the classic signature $(3, 3, 1)$, $\mathbf{r} = 0$, $n = 7$. We present first a two-message minimally informative protocol P_A, in Sect. 3.1. We then show in Sect. 3.2 that for general signatures with $\mathbf{c} + \mathbf{r} = 1$, any two-message minimally informative protocol is also safe. This implies that the previous protocol P_A for $(3, 3, 1)$ is also safe. Furthermore, in Sect. 3.3 we show how this protocol can be used in a two-step minimally informative and safe solution.

A natural two-message protocol P_A sends the sum of the cards modulo 2. However, it is not minimally informative for $(3, 3, 1)$, as was proved in [20].

3.1 Two-Message Minimally Informative Protocol from Singer Sets

The construction we present here is based on Singer difference sets [21] and is inspired by the *good announcement* construction proposed in [2, Theorem 3]. A set S of size $m + 1$ is a *Singer difference set* if the differences $s_i - s_j$ modulo $m(m + 1) + 1$, with $i \neq j$, $s_i, s_j \in S$, are all the different integers from 1 to $m(m + 1)$. In the following, the notation $x + S$ for a set S stands for the set $\{x + s \mod v \mid s \in S\}$.

Lemma 2. *Let* S *be a Singer difference set of size* $m+1$ *and* $v = m(m+1)+1$*, then for any two distinct elements* $l_1, l_2 \in \{x + S \mid x \in \mathbb{Z}_v\}$*, it holds that* $|l_1 \cap l_2| = 1$*.*

For a prime power m there is a Singer difference set of size $m + 1$, with all elements between 0 and $m(m+1)$ [21]. Thus, we know there is a Singer difference set S of size 3, such that $S \subseteq \mathbb{Z}_7$ which is what we need for the following protocol construction.

Let S be a Singer difference set of size 3 and S' a 3-*set* such that $S' \subseteq D - S$. Let L and L' be defined as follows:

$$L = \{x + S \mid x \in \mathbb{Z}_7\}, \quad L' = \{x + S' \mid x \in \mathbb{Z}_7\} \tag{1}$$

Then, the protocol $\chi_S : \mathscr{P}_3(D) \to \mathbb{Z}_2$ is defined by,

$$\chi_S(0)^{-1} = L \cup L', \quad \chi_S(1)^{-1} = \mathscr{P}_3(D) - \chi_S(0)^{-1}.$$

Lemma 3. *The cliques $K_p(\bar{a})$ of $J(7,3)$, with $a \in L$, partition $\mathscr{P}_3(D) - L$.*

The main result of this section is the following:

Theorem 2. *Let S be a Singer difference set of size 3 and S' a 3-set such that $S' \subseteq D - S$. The protocol $P_A = \chi_S$ is minimally informative for B, for $(3,3,1)$.*

Proof. By Lemma 3, for any $b \in \mathscr{P}_3(D)$ we have two cases, namely $b \in L$ or $b \in K_p(\bar{a})$ for some $a \in L$.

Suppose $b \in L$, then there is $x \in \mathbb{Z}_7$, such that $b = x + S$. Therefore $x + S' \in K_p(\bar{b})$, otherwise if $x + S$ and $x + S'$ were to have common elements, it would mean that S and S' are not disjoint. Thus, as $|L| = |L'|$, for any $b \in L$ there is exactly one element $a \in L'$ such that $a \in K_p(\bar{b})$, given that all the cliques $K_p(\bar{b})$, $b \in L$ are disjoints by Lemma 3. Finally, let $a \in K_p(\bar{b})$ be $x + S'$ and a' be any element in $K_p(\bar{b}) - a$, then $\chi_S(a) = 0$ and $\chi_S(a') = 1$.

Now suppose $b \in K_p(\bar{a})$ for some $a \in L$. Then $a \in K_p(\bar{b})$ and $\chi_S(a) = 0$. Let a' be any element in $K_p(\bar{b}) - \{a\}$, then $dist(a,a') = 1$, i.e. $|a \cap a'| = 2$ and therefore $a' \notin L$, otherwise it would contradict Lemma 2. Now let a_1, a_2, a_3 be the three elements in $K_p(\bar{b}) - \{a\}$, and assume for contradiction that $\chi_S(a_1) = \chi_S(a_2) = \chi_S(a_3) = 0$, i.e. $a_1, a_2, a_3 \in L'$. Let $\bar{b} = \{x,y,z,k\}$, then w.l.o.g. $a_1 = \{x,y,z\}$, $a_2 = \{x,y,k\}$ and $a_3 = \{x,k,z\}$. Thus, there are only three different ways in which the elements of these sets could be obtained according to (1), from $S' = \{s'_1, s'_2, s'_3\}$ and different $i,j,l \in \mathbb{Z}_7$:

	s'_1	s'_2	s'_3
$+i$	x	y	z
$+j$	y	x	k
$+l$	-	-	-

	s'_1	s'_2	s'_3
$+i$	x	y	z
$+j$	k	x	y
$+l$	z	k	x

	s'_1	s'_2	s'_3
$+i$	x	y	z
$+j$	y	k	x
$+l$	z	x	k

Thus, for any of the previous scenarios to hold we would need resp. that

$$x - y \equiv y - x, \quad x - z \equiv z - x, \quad k - x \equiv x - k \mod 7.$$

Given that 7 is prime, any of the previous statements is clearly impossible. Then, we have arrived to a contradiction, and the theorem follows.

Regarding the final argument from the previous proof, the reader can verify that we can always arrive to a contradiction of the form $x - y \equiv y - x \mod 7$, for different $x, y \in \mathbb{Z}_7$ or the more trivial $x + i \equiv x + j \mod 7$, for different $x, i, j \in \mathbb{Z}_7$ (which we discarded from the beginning in the proof).

3.2 Safety for Two-Message Minimally Informative Protocols

The two-message protocol P_A of the previous section is safe, in light of the following theorem.

Theorem 3. *Let* $\mathbf{c}+\mathbf{r}=1$, $\mathbf{b}\geq 2$. *If a two-message protocol* P_A *for* $(\mathbf{a},\mathbf{b},\mathbf{c})$ *is minimally informative then it is safe.*

Proof. Let $\chi : \mathscr{P}_{\mathbf{a}}(D) \rightarrow \mathbb{Z}_2$ be a minimally informative coloring for $J(n,\mathbf{a})$. Assume for contradiction that χ is not safe. That is, according to the safety characterization Theorem 1 there $\exists c \in D, \exists M \in \mathbb{Z}_2, \exists x \in \bar{c}$, such that for any $\mathbf{a}$-sets $a, a' \subseteq \bar{c}$, it holds $\neg(\chi(a)=\chi(a')=M) \vee x \notin a \triangle a'$. Thus, for such c, M and x we have $\chi(a)=\chi(a')=M \Rightarrow x \in a \cap a' \vee x \notin a \cup a'$; so, if we consider any $a, a' \subseteq \bar{c}$ such that $a, a' \in \chi^{-1}(M)$, if $x \in a$ then $x \in a'$ or else if $x \notin a$ then $x \notin a'$. This means, for this c, M and x one of the following should hold:

1. for any $a \subseteq \bar{c}$, if $\chi(a)=M$ then $x \notin a$
2. for any $a \subseteq \bar{c}$, if $\chi(a)=M$ then $x \in a$

Let $\{M'\}=\mathbb{Z}_2-\{M\}$, then the previous is equivalent to:

1. for any $a \subseteq \bar{c}$, if $x \in a$, then $\chi(a)=M'$
2. for any $a \subseteq \bar{c}$, if $x \notin a$, then $\chi(a)=M'$

Suppose case 1 holds. Let $a' \subseteq \bar{c}$ be an $\mathbf{a}$-set such that $x \notin a'$, then $\chi(a') \neq M'$, otherwise if $b = D - a' - \{x\}$ (so that $\bar{b} = a' \cup \{x\}$), for any $a_1 \in K_p(\bar{b})$, we have $\chi(a_1)=M'$ (given that $x \in a_1$ or $a_1 = a'$), and then χ is not minimally informative. Thus, let b be a $\mathbf{b}$-set such that $c, x \in b$, then for all $\mathbf{a}$-set $a' \in K_p(\bar{b})$, we have $\chi(a') \neq M'$. Thus, all $\mathbf{a}$-set $a' \in K_p(\bar{b})$ are equally colored by χ (given that there are only two colors in $\mathbb{Z}_2$), a contradiction to χ being a minimally informative coloring for $J(n,\mathbf{a})$, so χ is safe.

Suppose case 2 holds. Let b be a $\mathbf{b}$-set such that $c, x \in b$, then for all $\mathbf{a}$-set $a' \in K_p(\bar{b})$, we have $\chi(a')=M'$. Thus all $\mathbf{a}$-set $a' \in K_p(\bar{b})$ are equally colored by χ, thus we arrived to the same contradiction as before.

Notice that $\mathbf{c}+\mathbf{r}=1$ is necessary for the arguments in the proof. Thus, for example, the result does not hold for the case $(3,3,2)$.

3.3 Two-Step Minimally Informative Solution for (3, 3, 1)

A two-step solution, for signature $(3,3,1)$ is obtained when both A and B send just one bit, using the same protocol, based on the construction from Sect. 3.1, denoted (χ_S, χ_S). We remark that the message sent by B does not depend on the one by A, so they could both send their announcements concurrently.

Theorem 4. *Let* S *be a Singer difference set of size* 3 *and* S' *a* 3*-set such that* $S' \subseteq D-S$. *The two-step protocol* (χ_S, χ_S) *is a minimally informative and safe for* $(3,3,1)$, *with* $|\mathcal{M}_A|=|\mathcal{M}_B|=2$.

Proof. By Theorem 2, it is straightforward that (χ_S, χ_S) is minimally informative for $(3, 3, 1)$. Regarding the safety property, according to Definition 2, we must consider any card $c \in D^4$ that C might hold, any execution $\alpha \in \alpha(C, c)$, and any pair of cards x, y held by A and B, respectively, in α. Let M and M' be the messages announced by A and B in α, respectively, i.e., $M = \chi_S(input_A(\alpha))$ and $M' = \chi_S(input_B(\alpha))$. We must show that there are two other executions of the protocol $\alpha_1, \alpha_2 \in \alpha(C, c)$, with $y \in input_A(\alpha_1)$ and $x \in input_B(\alpha_2)$, such that $\alpha \,{}^C\!\!\sim \alpha_1$ and $\alpha \,{}^C\!\!\sim \alpha_2$.

First, assume that $M = M'$. Now, consider the execution α' for the deal $(input_B(\alpha), input_A(\alpha), c)$. Since both A and B use the same protocol, $\alpha' \,{}^C\!\!\sim \alpha$ and $y \in input_A(\alpha')$ and $x \in input_B(\alpha')$. Thus, we are done.

Assume $(M, M') = (0, 1)$. Notice that, there are exactly three elements in $L' = \{x + S' \mid x \in \mathbb{Z}_7\}$ that contain card c. Let us say these elements are $t + S', u + S'$ and $v + S'$. Then, the elements $t + S, u + S$ and $v + S$ from L are in $K_p(\bar{c}) \cap \chi_S^{-1}(0)$. Let $t + S$, $u + S$ and $v + S$ be a_1, a_2 and a_3 respectively. Consider the 3-sets $b_1 = D - a_1 - c$, $b_2 = D - a_2 - c$ and $b_3 = D - a_3 - c$, so that $b_1 \in K_p(\bar{a_1})$, $b_2 \in K_p(\bar{a_2})$ and $b_3 \in K_p(\bar{a_3})$. Let α_1, α_2 and α_3 be the executions for the deals (a_1, b_1, c), (a_2, b_2, c) and (a_3, b_3, c), respectively. Notice that, by Lemma 2, $b_1, b_2, b_3 \notin L$, since $a_1 \cap b_1 = \emptyset$, $a_2 \cap b_2 = \emptyset$ and $a_3 \cap b_3 = \emptyset$. Assume for contradiction that there are at least two distinct elements in $K_p(\bar{a_1}) \cap L'$. Then, one of these elements must be $t + S'$, and other must be $k + S'$ for some $k \in \mathbb{Z}_7$, with $k \neq t$. This means, that $K_p(\overline{t + S})$ and $K_p(\overline{k + S})$ are not disjoint, a contradiction with Lemma 3. Then, there is exactly one element in $K_p(\bar{a_1}) \cap L'$, which is $t + S'$. But, since $c \in t + S'$, it follows that $b_1 \notin K_p(\bar{a_1}) \cap L'$, i.e., $b_1 \notin L'$. Thus, $b_1 \notin \chi_S^{-1}(0)$, which means $\chi_S(b_1) = 1$. Similarly, we can prove that $b_2 \notin L'$ and $b_3 \notin L'$ then, $b_2 \notin \chi_S^{-1}(0)$ and $b_3 \notin \chi_S^{-1}(0)$, therefore $\chi_S(b_2) = 1$ and $\chi_S(b_3) = 1$. It follows that $\alpha \,{}^C\!\!\sim \alpha_1$, $\alpha \,{}^C\!\!\sim \alpha_2$ and $\alpha \,{}^C\!\!\sim \alpha_3$. By Lemma 2, the intersection of any pair of elements from L is exactly one. Then, since $c \notin a_1 \cup a_2 \cup a_3$, from the inclusion-exclusion principle, it follows that $|a_1 \cup a_2 \cup a_3| = 6$ and $|a_1 \cap a_2 \cap a_3| = 0$. Then, there must be two elements in $\{a_1, a_2, a_3\}$ such that one contains the card y, and other that do not contains the card x. W.l.o.g, let's say these elements are a_1 and a_2. Thus, we have that $y \in input_A(\alpha_1)$ and $x \in input_B(\alpha_2)$.

Assume $(M, M') = (1, 0)$. Given that $\mathbf{a} = \mathbf{b}$, A and B use the same protocol, this follows from the previous case, i.e., they are symmetric.

4 Russian Cards Problems

We study here informative two-step protocols for the case $\mathbf{a} = \mathbf{b} = 3$. We consider the case of $\mathbf{c} + \mathbf{r} = 1$, since there are no informative and safe solutions when $\mathbf{c} + \mathbf{r} \geq 2$. In the classic Russian cards problem $\mathbf{c} = 1, \mathbf{r} = 0$. In the *weak Russian cards problem* $\mathbf{c} = 0, \mathbf{r} = 1$, namely, C gets no cards at all.

[4] We also denote the singleton set with card c as c, as it is always clear from the context which case it is.

First, by [20, Theorem 5], we have the following impossibility, justifying why we consider only the case of $n = 7$, for both the classic signature $(3, 3, 1)$ and for the weak variant $(3, 3, 0)$.

Theorem 5. *There is no informative and safe two-step protocol for either of the two following cases: signature* $(3, 3, 1)$ *and* $\mathbf{r} \geq 1$ *and signature* $(3, 3, 0)$ *and* $\mathbf{r} \geq 2$.

The classic Russian cards problem with signature $(3, 3, 1)$ and $n = 7$ has been thoroughly studied, an exhaustive analysis can be found in [8]. It is well-known that there is a simple, modular arithmetic solution sending seven messages, $|\mathcal{M}_A| = |\mathcal{M}_B| = 7$.

4.1 Solutions with Six Messages

Considering Theorem 1, recall that the chromatic number of Johnson graphs has been well studied due to its importance in combinatorics [13] and coding theory [11]. In general, determining the chromatic number of a Johnson graph is an open problem [13, Chapter 16]. It is known that $\chi(J(7, 3)) = 6$, and hence there is an informative protocol with 6 messages, and no less[5]. We present an explicit solution below, which is informative, but not safe for the weak version, and then a solution that is informative and safe for the weak version, but not for the classic version. At the end we prove there is no uniform solution with 6 messages for the classic version.

While $J(7, 3)$ is the same for the Russian cards problem and for its weak version, for the protocol to be safe one needs to consider the possible inputs of C. In the Russian cards problem, the C-vertices are $\mathscr{P}_{\mathbf{c}}(D)$, $\mathbf{c} = 1$, while in the weak version, there is a single C-vertex, $(C, \emptyset)$. It is not hard to see that in both cases there is a two-step protocol with $|\mathcal{M}_A| = 6$ (which is informative and safe), see the technical report of [20]. While this is the smallest number of messages for A, it is possible for B to send only 4 different messages, if safety is not required.

Theorem 6. *There is an informative (non-safe) two-step protocol for the Russian cards problem and also for the weak version, that is optimal for both A and B, with* $|\mathcal{M}_A| = 6$ *and* $|\mathcal{M}_B| = 4$.

4.2 Impossibility of Uniform Solutions

Here we discuss a new technique to study the structure of six message protocols P_A to the Russian cards problem. In a six message solution, every color class must be of size at least 5, or at most 7. Given that there are 35 possible hands of A, there are only three possibilities for the classes' size distribution: the *uniform* solutions $5, 6, 6, 6, 6, 6$ and $5, 5, 6, 6, 6, 7$, and the *non uniform* one $5, 5, 5, 6, 7, 7$. This is optimal for a six message *uniform* solution, namely, two color classes

[5] The same lower bound is [22, Theorem 4], proved by reduction to a combinatorial design theorem.

must be of size 7. Thus, there is no solution with six messages with classes of sizes 5, 6, 6, 6, 6, 6 nor 5, 5, 6, 6, 6, 7. This is stated in the following theorem.

Theorem 7. *There is no uniform protocol P_A for the Russian cards problem with six messages.*

Proof. Assume for contradiction that there is such a protocol P_A, which partitions all the 35 possible hands of A into 6 color classes. One class must have 5 hands, by a counting argument, not all can have at least 6, and it is not hard to check that a color class cannot have only 4 hands. Also, a color class cannot have more than 7 hands (as observed in [8]). Thus, the most uniform solution induces a partition of sizes 5, 6, 6, 6, 6, 6. And the less-uniform solutions are either of sizes 5, 5, 6, 6, 6, 7, or 5, 5, 5, 6, 7, 7.

A partition with 5 hands must have a single card, say 0, that appears in 3 hands. All other cards appear twice. There are 15 hands containing 0. Consider all remaining 12 hands containing 0 in the other color classes, say 2 through 6.

In the remaining 5 classes there must be 3 with two hands containing 0, and 2 classes with three hands containing 0. Recall that each card must appear at least twice in a color class, [20]. Also, no color class can have 4 hands containing 0, because then two hands would have an intersection of 2 cards (and share an edge of $J(7, 3)$, violating the properness of the coloring).

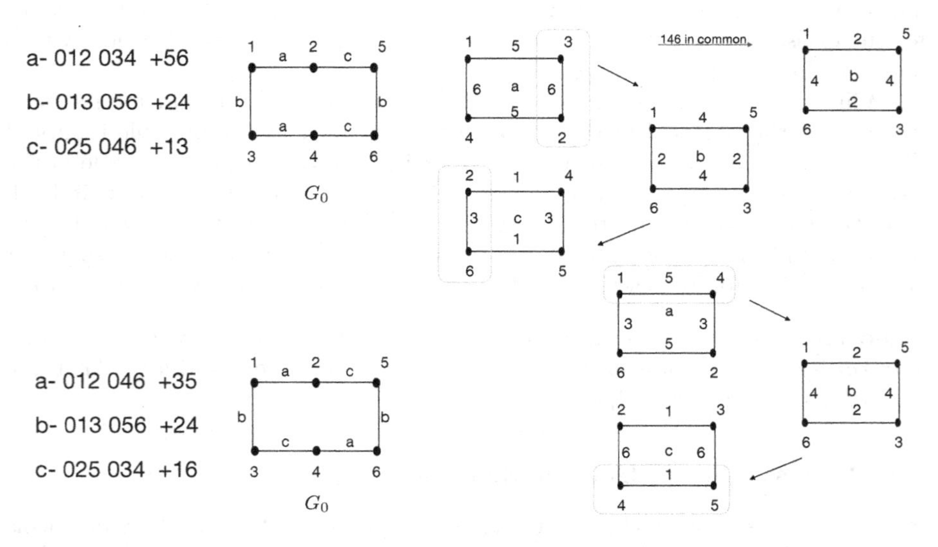

Fig. 1. First configuration on top 12, 34; 13, 56; 25, 46. Second configuration on bottom 12, 46; 13, 56; 25, 34. On the right part of the trees of possible ways of completing them.

Consider three color classes of size 6, denoted a, b, c, each one has exactly two hands containing 0. The case where one of these classes is of size 7, and hence it has three hands containing 0, is similar; it will be discussed at the end.

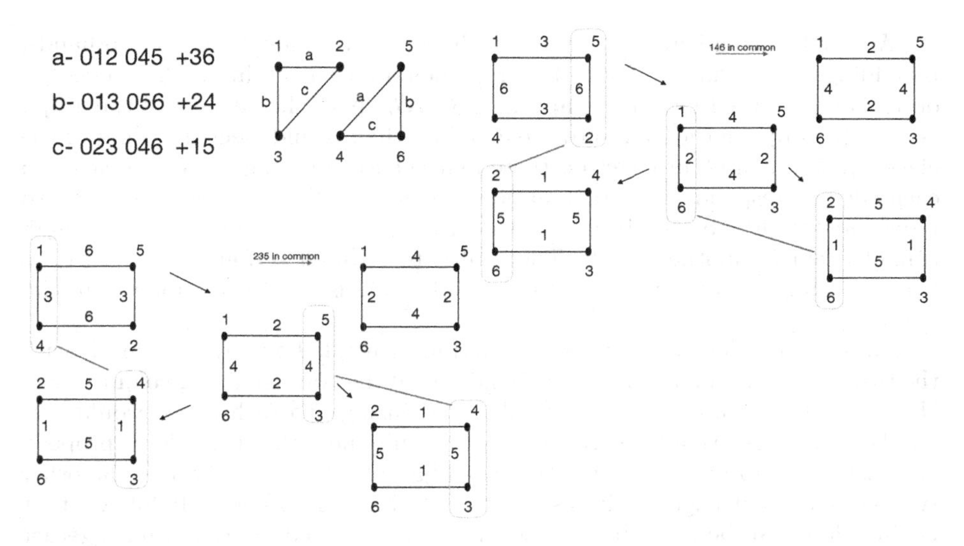

Fig. 2. The a, b, c classes define two triangles. On the right part are the trees of possible ways of completing first a, then b and then c, to have each 6 hands. Each hand is represented by an edge.

The 3 color classes a, b, c with two hands containing 0 define a graph G_0 on the vertices $D \backslash 0 = \{1, 2, 3, 4, 5, 6\}$, each vertex representing a card. An edge of this graph is colored with an element from $\{a, b, c\}$, meaning that if an edge x, y is colored i, then the hand $0xy$ is in class $i \in \{a, b, c\}$.

Since two hands in a class cannot have an intersection of more than one card, it follows that the edges of the same color are independent in G_0.

Now, assume for contradiction that a vertex, say 1, has degree 3. The three edges $\{1, v_1\}, \{1, v_2\}, \{1, v_3\}$ are colored with different elements from $\{a, b, c\}$. As we shall see, this implies that 1 appears in three hands of each class, a, b, c. Therefore, it appears in two hands, of each of the remaining classes, d, e, f. We can thus consider the graph G_1 on the vertices $D \backslash 1$, with edges colored with elements from $\{d, e, f\}$, meaning that if an edge x, y is colored i, then the hand $1xy$ is in class i. The vertex 0 of G_1 must then have degree 3, because as we shall see, this is needed for 0 to appear three times in each class d, e, f. But this implies that 0 is incident to one of v_1, v_2, v_3, say v_i, since the graph has only 6 vertices. Namely, $\{0, v_i\}$ is and edge of G_1, and $\{1, v_i\}$ is and edge of G_0, so the hand $01v_i$ appears twice, in a class of $\{a, b, c\}$ and a class of $\{d, e, f\}$, a contradiction to the assumption that a vertex has degree three in G_0.

Thus, the edges of G_0 either they form a cycle or two triangles. There are two types of cyclic configurations for the three classes a, b, c with two hands containing 0: either for each $i \in \{a, b, c\}$, the edges colored i are opposite in the cycle or not. For instance, 12, 34; 13, 56; 25, 46 (all plus 0) or else 12, 46; 13, 56; 25, 34 (all plus 0). See Fig. 1 for these two cyclic configurations, and Fig. 2 for the triangles case. These figures illustrate the case where 0 appears in exactly two hands, and the color classes are of size 6.

We need to complete each set of two hands to form a color class of 6 hands, by adding 4 more hands. These 4 more hands do not contain 0. The process to do it, is represented by three graphs, G_a, G_b, G_c. Now the vertices of the graph G_i, $i \in \{a, b, c\}$ are the four cards spanned by the two independent edges of the class G_i. There are four edges on these four vertices forming a cycle in each G_i; each edge corresponds to a combination that *does not* appear in one of the two independent edges of G_i (because two cards that already appeared in a hand, cannot occur in another hand). The goal is to color these four edges, with the two remaining colors (0 is no longer available, because it already appears in two hands).

Notice that a loop on a vertex x could in principle be used, coloring it with the two remaining colors, giving the hand xyz, if the two remaining colors are yz. However, at most one such loop can be used (using two such loops, would give hands with intersection yz, with violets the requirement that the color is proper). And using a loop prevents using the two adjacent edges, leaving only the other two, non-adjacent edges to be used, i.e., coloring only 3 edges. It follows that no such loop can be used, because we need to color 4 edges, to obtain together with the 2 hands containing 0, the total number of hands which is 6 in the color class.

Consider all 4 combinations of taking one card from each pair (of 2 values different from 0). Then add each of the two remaining cards to complementary pairs, as illustrated in the figures. For example, in Fig. 1, for the pairs (a) 12,34 one most add values 56. And there are only two options of getting independent edges. Add 5 to 13 and to 24; add 6 to 14 and to 23, as in the figure. Or else add 6 to 13 and to 24; add 5 to 14 and to 23.

Once 5 is added to 13 and to 24, and 6 to 14 and to 23, the next move is determined, to complete class (b). In the figure a blue arrow shows that 146 would be in common to the next class, if we added 2 to 15 and 36; and 4 to 16 and 35. Thus, the only option is the complementary choice. But then, either way, it is not possible to add 1 and 3 to class (c). In the figure one choice is shown, where 236 is repeated in classes (a) and (c). The reader can verify that in either of the two types of configurations, this process cannot be completed. The figure of the full tree for the first configuration and the proof for the class of size 7 are in the technical report of [20].

5 Conclusions

We have presented an indistinguishability-based formalization of the problem of A and B communicating to each other information about their hands, by public announcements, without C learning for any of their cards who holds it. We described two variants: one where A and B learn each other hands, and one where they learn *something* about each other hands. For concreteness, the formalization was done for two-step protocols, where A makes the first announcement, and then B makes the second announcement, but can be used also for multi-round protocols. We then viewed the formalization as vertex colorings of distance d

Johnson graphs. This allows to prove general properties about when there exists a solution, in terms of the number of cards n and the signature $(\mathbf{a}, \mathbf{b}, \mathbf{c})$, and exposing a relation with coding theory.

We studied in detail the case where $\mathbf{a} = \mathbf{b} = 3$, introducing new ideas both for the informative and for the minimally informative case, that we hope can be useful for more general signatures. But already in this specific case there are interesting combinatorial problems. We presented the two following main technical results. First, the surprising result that exploits the correlated inputs of A and B to inform one of their cards to each other privately, using a single bit announcement (Theorem 4). Second, there are no uniform six-message solutions to the Russian cards problem, and hence no corresponding colorings of $J(7, 3)$ (Theorem 7).

Many interesting questions remain open. While for $\mathbf{c} = 1$, $\mathbf{r} = 0$, there is a characterization of the signatures $(\mathbf{a}, \mathbf{b}, \mathbf{c})$ that allow for a two-step informative and safe solution [20], namely, for graphs $J(n, \mathbf{a})$, the general case $J^{\mathbf{c}+\mathbf{r}}(\mathbf{a}, \mathbf{b})$ remains open. Even more cases remain open, for the minimally informative problem. We followed the definition of security of most previous papers e.g. [1,2,4–6,8,9], that requires privacy for individual cards only, but it would be interesting to investigate stronger requirements, in particular where C should learn nothing about the hands of A and B, such as in [16,22], which become more subtle in the case of $\mathbf{r} > 0$, and have not been studied at all for the minimally informative problem. It would also be interesting to study the case of four players, with signature $(\mathbf{a}, \mathbf{b}, \mathbf{c}, \mathbf{r})$, $n = \mathbf{a} + \mathbf{b} + \mathbf{c} + \mathbf{r}$, which seems similar to the case of three players we considered here where no one gets $\mathbf{r}$ cards, and relate it to previous multi-players work e.g. [10].

Acknowledgement. We thank Hans van Ditmarsch for his comments.

References

1. Albert, M., Cordón-Franco, A., van Ditmarsch, H., Fernández-Duque, D., Joosten, J.J., Soler-Toscano, F.: Secure communication of local states in interpreted systems. In: Abraham, A., Corchado, J.M., González, S.R., De Paz Santana, J.F. (eds.) International Symposium on Distributed Computing and Artificial Intelligence. Advances in Intelligent and Soft Computing, vol. 91, pp. 117–124. Springer, Heidelberg (2011). https://doi.org/10.1007/978-3-642-19934-9_15
2. Albert, M.H., Aldred, R.E.L., Atkinson, M.D., van Ditmarsch, H., Handley, C.C.: Safe communication for card players by combinatorial designs for two-step protocols. Australas. J. Comb. **33**, 33–46 (2005)
3. Attiya, H., Rajsbaum, S.: Indistinguishability. Commun. ACM **63**(5), 90–99 (2020)
4. Cordón-Franco, A., van Ditmarsch, H., Fernández-Duque, D., Joosten, J.J., Soler-Toscano, F.: A secure additive protocol for card players. Australas. J. Comb. **54**, 163–176 (2012)
5. Cordón-Franco, A., Ditmarsch, H., Fernández-Duque, D., Soler-Toscano, F.: A geometric protocol for cryptography with cards. Des. Codes Cryptogr. **74**(1), 113–125 (2015)

6. Cordón-Franco, A., Van Ditmarsch, H., Fernández-Duque, D., Soler-Toscano, F.: A colouring protocol for the generalized Russian cards problem. Theor. Comput. Sci. **495**, 81–95 (2013). https://doi.org/10.1016/j.tcs.2013.05.010
7. Delporte-Gallet, C., Fauconnier, H., Rajsbaum, S.: Communication complexity of wait-free computability in dynamic networks. In: Richa, A.W., Scheideler, C. (eds.) SIROCCO 2020. LNCS, vol. 12156, pp. 291–309. Springer, Cham (2020). https://doi.org/10.1007/978-3-030-54921-3_17
8. van Ditmarsch, H.: The Russian cards problem. Stud. Log. **75**, 31–62 (2003)
9. van Ditmarsch, H., Soler-Toscano, F.: Three steps. In: Proceedings of CLIMA XII. Lecture Notes in Computer Science, vol. 6814, pp. 41–57. Springer, New York (2011)
10. Duan, Z., Yang, C.: Unconditional secure communication: a Russian cards protocol. J. Comb. Optim. **19**(4), 501–530 (2010)
11. Etzion, T., Bitan, S.: On the chromatic number, colorings, and codes of the Johnson graph. Discret. Appl. Math. **70**(2), 163–175 (1996)
12. Fischer, M.J., Wright, R.N.: An efficient protocol for unconditionally secure secret key exchange. In: Proceedings of 4th Symposium Discrete Algorithms (SODA). p. 475–483. SIAM, USA (1993)
13. Godsil, C., Meagher, K.: Erdős-Ko-Rado Theorems: Algebraic Approaches. Cambridge Studies in Advanced Mathematics, Cambridge University Press, Cambridge (2015)
14. Herlihy, M., Kozlov, D., Rajsbaum, S.: Distributed Computing Through Combinatorial Topology. Elsevier-Morgan Kaufmann (2013)
15. Koizumi, K., Mizuki, T., Nishizeki, T.: A revised transformation protocol for unconditionally secure secret key exchange. Theory Comput. Syst. **42**(2), 187–221 (2008)
16. Landerreche, E., Fernández-Duque, D.: A case study in almost-perfect security for unconditionally secure communication. Des. Codes Cryptogr. **83**(1), 145–168 (2017)
17. Makarychev, Y.S., Makarychev, K.: The importance of being formal. Math. Intelli. **23**(1) (2001)
18. Mizuki, T., Shizuya, H., Nishizeki, T.: A complete characterization of a family of key exchange protocols. Int. J. Inf. Secur. **1**(2), 131–142 (2002)
19. Orlitsky, A., Viswanathan, K.: One-way communication and error-correcting codes. IEEE Trans. Inf. Theory **49**(7), 1781–1788 (2003)
20. Rajsbaum, Sergio: A distributed computing perspective of unconditionally secure information transmission in Russian cards problems. In: Jurdziński, Tomasz, Schmid, Stefan (eds.) SIROCCO 2021. LNCS, vol. 12810, pp. 277–295. Springer, Cham (2021). https://doi.org/10.1007/978-3-030-79527-6_16. Full preliminary version in arXiv:2009.13644, September 2020
21. Singer, J.: A theorem in finite projective geometry and some applications to number theory. Trans. Am. Math. Soc. **43**(3), 377–385 (1938)
22. Swanson, C.M., Stinson, D.R.: Combinatorial solutions providing improved security for the generalized Russian cards problem. Des. Codes Cryptogr. **72**(2), 345–367 (2014)
23. Swanson, C.M., Stinson, D.R.: Additional constructions to solve the generalized Russian cards problem using combinatorial designs. Electron. J. Combin. **21**(3) (2014)
24. Winkler, P.: The advent of cryptology in the game of bridge. Cryptologia **7**(4), 327–332 (1983)

Compact Distributed Interactive Proofs for the Recognition of Cographs and Distance-Hereditary Graphs

Pedro Montealegre[1(✉)], Diego Ramírez-Romero[2,3], and Ivan Rapaport[3]

[1] Facultad de Ingeniería y Ciencias, Universidad Adolfo Ibáñez, Santiago, Chile
p.montealegre@uai.cl

[2] Departamento de Ingeniería Matemática, Universidad de Chile, Santiago, Chile
dramirez@dim.uchile.cl

[3] DIM-CMM (UMI 2807 CNRS), Universidad de Chile, Santiago, Chile
rapaport@dim.uchile.cl

Abstract. We present compact distributed interactive proofs for the recognition of two important graph classes, well-studied in the context of centralized algorithms, namely *complement reducible graphs* and *distance-hereditary graphs*. Complement reducible graphs (also called *cographs*) are defined as the graphs not containing a four-node path P_4 as an induced subgraph. Distance-hereditary graphs are a super-class of cographs, defined as the graphs where the distance (shortest paths) between any pair of vertices is the same on every induced connected subgraph.

First, we show that there exists a distributed interactive proof for the recognition of cographs with two rounds of interaction. More precisely, we give a dAM protocol with a proof size of $\mathcal{O}(\log n)$ bits that recognizes cographs with high probability. Moreover, our protocol can be adapted to verify any Turing-decidable predicate restricted to cographs in dAM with certificates of size $\mathcal{O}(\log n)$.

Second, we give a three-round, dMAM interactive protocol for the recognition of distance-hereditary graphs, still with a proof size of $\mathcal{O}(\log n)$ bits.

Finally, we show that any one-round (denoted dM) or two-round, dMA protocol for the recognition of cographs or distance-hereditary graphs requires certificates of size $\Omega(\log n)$ bits. Moreover, we show that any constant-round dAM protocol using shared randomness requires certificates of size $\Omega(\log \log n)$.

Keywords: Distributed interactive proofs · Distributed recognition of graph classes · Cographs · Distance hereditary graph

Partially supported by CONICYT via PIA/ Apoyo a Centros Científicos y Tecnológicos de Excelencia AFB 170001 (P.M. and I.R.), FONDECYT 1170021 (D.R. and I.R.) and FONDECYT 11190482 (P.M.) and PAI + Convocatoria Nacional Subvención a la Incorporación en la Academia Año 2017 + PAI77170068 (P.M.).

C. Johnen et al. (Eds.): SSS 2021, LNCS 13046, pp. 395–409, 2021.
https://doi.org/10.1007/978-3-030-91081-5_26

1 Introduction

The study of graph classes provides important insights to address basic graph problems such as coloring, maximum independent set, dominating set, etc. Indeed, as such problems are hard in general, restricting the input to a particular graph-class is a natural approach in order to exploit structural properties for designing efficient algorithms.

A well-known example is the class of perfect graphs [20], i.e., the class of graphs satisfying that the chromatic number equals the size of the largest clique of every induced subgraph. Many NP-complete problems on general graphs, such as coloring, maximum clique, maximum independent set, etc., can be solved in polynomial-time when the input is known to be a perfect graph [22].

The design of efficient algorithms for particular graph-classes has also interest in the context of distributed algorithms. Besides the classes of sparse and bounded degree graphs, there are many examples of efficient distributed algorithms specially designed to run on planar graphs [19], interval graphs [23], chordal graphs [32] and others. It is therefore very important to efficiently *check the membership* of a graph to a given class. Through this checking procedure we make sure that the execution is performed in the right type of input, in order to avoid erroneous computations or even the lack of termination.

Distributed Interactive Proofs. Distributed decision refers to the task in which the nodes of a connected graph G have to collectively decide (whether G satisfies) some graph property [38]. For performing any such task, the nodes exchange messages through the edges of G. The input of distributed decision problems may also include labels given to the nodes and/or to the edges of G. For instance, the nodes could decide whether G is properly colored, or decide whether the graph belongs to a given graph-class.

Acceptance and rejection are defined as follows. If G satisfies the property, then all nodes must accept; otherwise, at least one node must reject. This type of algorithm could be used in distributed fault-tolerant computing, where the nodes, with some regularity, must check whether the current network configuration is in a legal state for some Boolean predicate [33]. Then, if the configuration becomes illegal at some point, the rejecting node(s) raise the alarm or launch a recovery procedure.

Deciding whether a given coloring is proper can be done locally, by exchanging messages between neighbors. These types of properties are called *locally decidable*. Nevertheless, some other properties, such as deciding whether G is a simple path, are not. As a remedy, the notion of *proof-labeling scheme* (PLS) was introduced [33]. Similar variants were also introduced: non-deterministic local decisions [16], locally checkable proofs [21], and others.

Roughly speaking, in all these models, a powerful prover gives to every node v a certificate $c(v)$. This provides G with a global distributed proof. Then, every node v performs a local verification using its local information together with $c(v)$. PLSs can be seen as a distributed counterpart to the class NP, where, thanks to nondeterminism, the power of distributed algorithms increases.

Just as it happened in the centralized framework a natural step forward is to consider a model where the nodes are allowed to have *more than one interaction round* with the prover. Interestingly, there is no gain when interactions are all deterministic. When there is no randomness, the prover, from the very beginning, has all the information required to simulate the interaction with the nodes. Then, in just one round, the prover could simply send to each node the transcript of the whole communication, and the nodes simply verify that the transcript is indeed consistent. A completely different situation occurs when the nodes have access to some kind of randomness [2,18]. In that case, the exact interaction with the nodes is unknown to the prover until the nodes communicate the realization of their random variables. Adding a randomized phase to the non-deterministic phase gives more power to the model [2,18].

The notion of *distributed interactive protocols* was introduced by Kol, Oshman, and Saxena in [31] and further studied in [10,17,36,37]. In such protocols, a centralized, untrustable prover with unlimited computation power, named Merlin, exchanges messages with a randomized distributed algorithm, named Arthur. Specifically, Arthur and Merlin perform a sequence of exchanges during which every node queries Merlin by sending a random bit-string, and Merlin replies to each node by sending a bit-string called proof. Neither the random strings nor the proofs need to be the same for each node. After a certain number of rounds, every node exchanges information with its neighbors in the network, and decides (i.e., it outputs accept or reject). For instance, a dMAM protocol involves three interactions: Merlin provides a certificate to Arthur, then Arthur queries Merlin by sending a random string. Finally, Merlin replies to the query by sending another certificate. Recall that this series of interactions is followed by a phase of distributed verification performed between every node and its neighbors.

When the number of interactions is k we refer to $\mathsf{dAM}[k]$ protocols (if the last player is Merlin) and $\mathsf{dMA}[k]$ protocols (otherwise). For instance, $\mathsf{dAM}[2] = \mathsf{dAM}$, $\mathsf{dMA}[3] = \mathsf{dAMA}$, etc. Also, the scenario of distributed verification, where there is no randomness and only Merlin interacts, corresponds dM. In other words, dM is the PLS model.

In distributed interactive proofs, Merlin tries to convince the nodes that G satisfies some property in a small number of rounds and through short messages. We say that an algorithm uses $\mathcal{O}(f(n))$ bits if the messages exchanged between the nodes (in the verification round) and also the messages exchanged between the nodes and the prover are upper bounded by $\mathcal{O}(f(n))$. We include this *bandwidth bound* in the notation, which becomes $\mathsf{dMA}[k, f(n)]$ and $\mathsf{dAM}[k, f(n)]$ for the corresponding protocols.

It is known that all Turing-decidable predicates on graphs admit a PLS with certificates of size $\mathcal{O}(n^2)$ bits [33]. Interestingly, some distributed problems are hard, even when a powerful prover provides the nodes with certificates. It is the case of SYMMETRY, the language of graphs having a non-trivial automorphism (i.e., a non-trivial one-to-one mapping from the set of nodes to itself preserving edges). Any PLS recognizing SYMMETRY requires certificates of size $\Omega(n^2)$ [21]. However, many problems requiring $\Omega(n^2)$-bit certificates in any PLS, such as SYMMETRY, admit distributed interactive protocols with small certificates, and very few interactions. In fact, SYMMETRY is in both $\mathsf{dMAM}[\log n]$ and $\mathsf{dAM}[n \log n]$ [31].

Local Certification of Graph Classes. Regarding local certification of graph classes, there exist PLSs (with logarithmic-sized certificates) for the recognition of many graph classes such as acyclic graphs [33], planar graphs [15], graphs with bounded genus [14], etc. More recently, Busquet et al. [3] tackle the problem of locally certifying graphs classes defined by a finite set of minors.

Recently, Naor, Parter and Yogev defined in [37] a *compiler* which (1) turns any problem solved in NP in time $\tau(n)$ into a dMAM protocol using private randomness and bandwidth $\tau(n) \log n/n$ and; (2) turns any problem which can be solved in NC into a dAM protocol with private randomness, poly $\log n$ rounds of interaction and bandwidth poly $\log n$. This result has implications in the recognition of graph-classes. For example, it implies that any class of sparse graphs that can be recognized in linear time, can also be recognized by a dMAM protocol with logarithmic-sized certificates. This raises automatically the question of whether one can design, for the recognition of a given graph class, a distributed interactive proof based on fewer interactions than the interactions given by directly applying the compiler (while keeping the certificates as small as possible).

A graph-class is hereditary if the class is closed under vertex and edge deletion. Examples of hereditary graph classes include planar graphs, forests, bipartite graphs, perfect graphs, etc. Interestingly, all graph properties that are known to require large certificates (e.g. small diameter [6], non-3- colorability [21], having a non-trivial automorphism [21]), are non-hereditary.

Therefore, natural question is whether all hereditary graph-classes admit a distributed interactive proof with a constant number of interactions, and logarithmic-sized certificates. In this work we address the problem of the distributed recognition of two hereditary graph classes (which are in fact perfect graphs), namely *complement reducible graphs* and *distance-hereditary graphs.*

Cographs and Distance-Hereditary Graphs. The class of complement reducible graphs, or simply *cographs*, has several equivalent definitions, as it has been re-discovered in many different contexts [8,28,39,40]. A graph is a cograph if it does not contain a four-node path P_4 as an induced subgraph. Equivalently, a graph is a cograph if it can be generated recursively from a single vertex by complementation and disjoint-union. A graph is a *distance-hereditary graph* if the distance between any two vertices is the same on every connected induced subgraph [25]. An equivalent definition is that every path between two vertices is a shortest path. It is known that every cograph is a distance-hereditary graph.

Many NP-complete problems are solvable in polynomial-time, or even linear time, when restricted to cographs and distance-hereditary graphs. For instance, maximum clique, maximum independent set, coloring (as distance-hereditary graphs are perfect [25]), hamiltonicity [27], Steiner tree and connected domination [13], computing the tree-width and minimum fill-in [5], among others. By a result of Courcelle, Makowsky and Rotics [9], every decision problem expressible in a type of monadic second order logic can be solved in linear time on distance-hereditary graphs. Observe that all these results also apply to cographs. Other problems, like graph isomorphism, can be solved in linear time on cographs [8].

In the centralized setting, both cographs and distance-hereditary graphs can be recognized in linear time [12].

Respect to algorithms in the distributed setting, both recognition problems have also been addressed in the One-Round Broadcast Congested Clique Model (1BCC), also known as the Distributed Sketching Model [1]. In this model, the nodes of a graph send a single message to a *referee*, which initially has no information about the graph and, only using the received messages, has to decide a predicate of the input graph. 4 In [29] and [35], randomized protocols recognizing both classes of graphs in the 1BCC model are given. Interestingly, these protocols not only recognize the classes but *reconstruct them*, meaning that the referee learns all the edges of the input graph.

In this work, we focus on the recognition of cographs and distance-hereditary graphs in the model of distributed interactive proofs. We show that both classes can be recognized with compact certificates and constant (two or three) rounds of interaction.

Our Results. We show that the recognition of cographs is in $\mathsf{dAM}[\log n]$. Our result consists of adapting an algorithm given in [29,35], originally designed for the 1BCC model. In this regard, we exploit the natural high connectivity of this class, combined with the use of non-determinism in order to route all messages in the network to a leader node, which is delegated to act as a *referee*. In fact, our protocol allows this leader to learn all the edges of the input graph. We use this fact to show that any Turing-decidable predicate restricted to cographs is decidable in $\mathsf{dAM}[\log n]$.

Interestingly, our results imply that any one-round deterministic protocol in the 1BCC model recognizing cographs, would immediately imply a dM (i.e. a PLS) protocol for the recognition of cographs. Unfortunately, up to our knowledge, it is not known whether recognizing cographs can be done through a deterministic 1BCC protocol.

Then, we adapt the protocol for the recognition of cographs and we combine it with a set of tools related to the structure of distance-hereditary graphs in order to show that the recognition of this class is in $\mathsf{dMAM}[\log n]$. In this case, we are not able to simulate the 1BCC protocol by gathering all the information in a single node representing the referee. Instead, we find a way to verify each step of the computation of the referee in a distributed manner, by choosing nodes that can receive (with the help of the prover) all necessary messages for performing the task.

We remark that our protocols beat the performance of the compiler of Naor, Parter and Yogev. In fact, both graph classes can have $\Theta(n^2)$ edges and, therefore, the use of the compiler shows that the recognition of these classes is in $\mathsf{dMAM}[n \log n]$ and in $\mathsf{dAM}[\operatorname{poly}\log n, \operatorname{poly}\log n]$ (note that cographs and distance-hereditary graphs can be recognized in NC [11] and in linear time in the centralized setting [12], see the Related Work section).

Finally, we give some lower-bounds. More precisely, we show that any dM or dMA protocol for the recognition of cographs or distance-hereditary graphs, requires messages of size at least $\Omega(\log n)$. Our results are obtained extending a lower-bound technique described in [21], for the detection of a single leader in the context of locally checkable proofs. We note that our protocols use shared randomness. In that sense we prove that, any dAM protocol using shared randomness for the previous problems, requires messages of size at least $\Omega(\log \log n)$.

Related Work. The recognition of cographs and distance-hereditary graphs has been studied thoroughly in the parallel setting, where both problems have been shown to be in NC [11,24,30]. The currently best algorithms for the recognition of both classes run in time $\mathcal{O}(\log^2 n)$ and using a linear number of processors in a CREW-PRAM [11]. There also exist fast-parallel algorithms for NP-hard problems restricted to cographs and distance-hereditary graphs [26,34].

Unfortunately, there are no much research regarding distributed algorithms specially designed for cographs and distance-hereditary. Nevertheless, the structural properties of distance-hereditary graphs have been used in the design of compact routing tables for interconnection networks [7].

Structure of the Article. Section 2 is the preliminary section, where we give some graph-theoretic background, including the formal definitions of cographs and distance-hereditary graphs. We also give the precise definition of distributed interactive proofs. In Sect. 3 we give the results regarding cographs, and in Sect. 4 we give the results regarding distance-hereditary graphs. Finally, in Sect. 5, we provide some lower-bounds. Due the lack of space, the results are only outlined in their corresponding sections, while the full proofs are detailed in the appendix

2 Preliminaries

Background on Cographs and Distance-Hereditary Graphs. All the graphs in this paper are simple and undirected. Let $G = (V, E)$ be a graph. For a set $U \subseteq V$, we define $G[U]$, the *induced subgraph* of $G = (V, E)$ according to U, as the graph $H = (U, E(U))$, where $E(U) = E \cap \binom{U}{2}$. We denote $H \subseteq G$ when H is s an induced subgraph of G. If, instead, we have a graph with vertex set U such that its edges are only contained in $E(U)$, we simply call it a *subgraph* of G. A *spanning subgraph* of G is a subgraph H with $V(H) = V(G)$. Given two nodes u, v of a connected graph H, the distance between them, denoted by $d_H(u, v)$, is defined as the length of the shortest path between u and v in H. A P_4 is an induced path of length four.

Given two graphs $G_1 = (V_1, E_1)$ and $G_2 = (V_2, E_2)$, we define the *union* between both graphs, denoted by $G_1 \cup G_2$ as the graph $\tilde{G} = (\hat{V}, \tilde{E})$, with $\tilde{V} = V_1 \cup V_2$ and $\tilde{E} = E_1 \cup E_2$. Given two graphs $G_1 = (V_1, E_1)$ and $G_2 = (V_2, E_2)$, we define the *join* between both graphs, denoted by $G_1 * G_2$ as the graph $\hat{G} = (\hat{V}, \hat{E})$, with $\hat{V} = V_1 \cup V_2$ and $\hat{E} = E_1 \cup E_2 \cup \{v_1 v_2 \text{ such that } v_1 \in V_1, v_2 \in V_2\}$.

The set of neighbors of a node u is denoted $N(u)$, and the closed neighborhood $N[u]$ is the set $N(u) \cup \{u\}$. A node v is said to be a *pending node* if it has a unique neighbor in the graph. A pair of nodes $u, v \in V$ are said to be *twins* if their neighborhoods are equal. That is, $N(u) = N(v)$ or $N[u] = N[v]$. In the case that u and v are adjacent ($N[u] = N[v]$) we refer to them as *true twins* and, otherwise, we refer to them as *false twins*.

As we mentioned in the introduction, a cograph is a graph that does not contain a P_4 as an induced subgraph (i.e. it is P_4-*free*). Another equivalent definition states that cographs are the graphs which can be obtained recursively following three rules: (1) A single vertex is a cograph, (2) the disjoint union between two cographs is a cograph and (3) the join of two cographs is a cograph. An advantage of cographs is that they admit other characterizations that may be useful for local verification. First, we define a *twin ordering* as an ordering $(v_i)_{i=1}^n$ of the nodes of V such that, for each $j \geq 2$, v_j has a twin in $G[\{v_1, \ldots v_j\}]$ (Fig. 1).

Proposition 1 ([29]). *Given a graph G the following are equivalent:*

1. *G is a cograph.*
2. *Each non trivial induced subgraph of G has a pair of* twins.
3. *G is* P_4-free.
4. *G admits a* twin ordering.

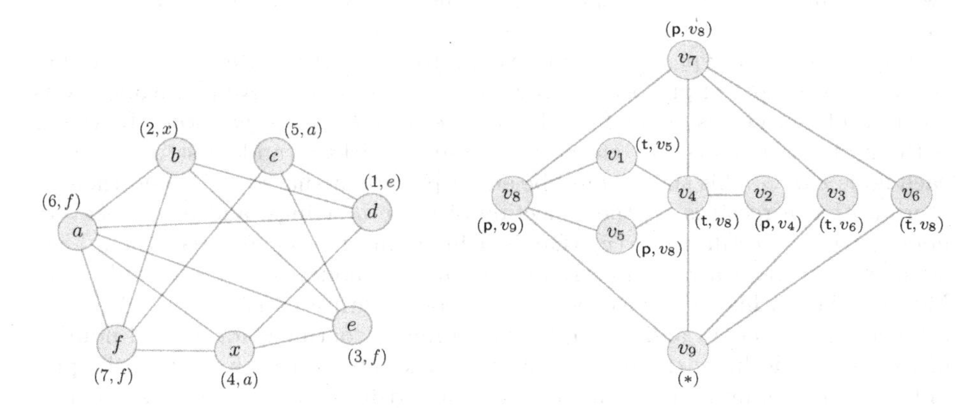

Fig. 1. Left: A cograph with labels according to a twin ordering. The first entry represents the step at which they are removed, while the second entry indicates the node's twin at such step. Right: A distance-hereditary graph with labels according to its ordering. The first entry indicates whether it is removed as a *true twin* (t̄), as a *false twin* (t) or as a pending node (p).

A graph G is said to be distance-hereditary if for any induced subgraph $H \subseteq G$ and any pair $u, v \in H$ satisfy that $d_H(u, v) = d_G(u, v)$. That is, any induced path between a pair of nodes is a shortest path. A relevant characterization for this class is the following.

Proposition 2 ([4]). *An n-node graph G is distance hereditary iff there exists an ordering $(v_i)_{i=1}^n$ such that, for any $i \in [n]$, either there exists $j < i$ such that v_i and v_j are* twins *in $G_i = G[\{v_1, ..v_i]\}$ or v_i is a pending node at G_i.*

Model Definitions. Let G be a simple connected n-node graph, let $I : V(G) \to \{0,1\}^*$ be an input function assigning labels to the nodes of G, where the size of all inputs is polynomially bounded on n. Let $\mathsf{id} : V(G) \to \{1, \ldots, \text{poly}(n)\}$ be a one-to-one function assigning identifiers to the nodes. A *distributed language* $\mathcal{L}$ is a (Turing-decidable) collection of triples (G, id, I), called *network configurations.*

A distributed interactive protocol consists of a constant series of interactions between a *prover* called Merlin, and a *verifier* called Arthur. The prover Merlin is centralized, has unlimited computing power and knows the complete configuration (G, id, I). However, he cannot be trusted. On the other hand, the verifier Arthur is distributed, represented by the nodes in G, and has limited knowledge. In fact, at each node v, Arthur is initially aware only of his identity $\mathsf{id}(v)$, and his label $I(v)$. He does not know the exact value of n, but he knows that there exists a constant c such that $\mathsf{id}(v) \leq n^c$. Therefore, for instance, if one node v wants to communicate $\mathsf{id}(v)$ to its neighbors, then the message is of size $\mathcal{O}(\log n)$.

Given any network configuration (G, id, I), the nodes of G must collectively decide whether (G, id, I) belongs to some distributed language $\mathcal{L}$. If this is indeed the case, then all nodes must accept; otherwise, at least one node must reject (with certain probabilities, depending on the precise specifications we are considering).

There are two types of interactive protocols: Arthur-Merlin and Merlin-Arthur. Both types of protocols have two phases: an interactive phase and a verification phase. Let us define first *Arthur-Merlin interactive protocols*. If Arthur is the party that starts the interactive phase, he picks a random string $r_1(v)$ at each node v of G (this string could be either private or shared) and send them to Merlin. Merlin receives r_1, the collection of these n strings, and provides every node v with a certificate $c_1(v)$ that is a function of v, r_1 and (G, id, I). Then again Arthur picks a random string $r_2(v)$ at each node v of G and sends r_2 to Merlin, who, in his turn, provides every node v with a certificate $c_2(v)$ that is a function of v, r_1, r_2 and (G, id, I). This process continues for a fixed number of rounds. If Merlin is the party that starts the interactive phase, then he provides at the beginning every node v with a certificate $c_0(v)$ that is a function of v and (G, id, I), and the interactive process continues as explained before. In Arthur-Merlin protocols, the process ends with Merlin. More precisely, in the last, k-th round, Merlin provides every node v with a certificate $c_{\lceil k/2 \rceil}(v)$. Then, the verification phase begins. This phase is a one-round deterministic algorithm executed at each node. More precisely, every node v broadcasts a message M_v to its neighbors. This message may depend on $\mathsf{id}(v)$, $I(v)$, all random strings generated by Arthur at v, and all certificates received by v from Merlin. Finally, based on all the knowledge accumulated by v (i.e., its identity, its input label, the generated random strings, the certificates received from Merlin, and all the messages received from its neighbors), the protocol either accepts or rejects at node

v. Note that Merlin knows the messages each node broadcasts to its neighbors because there is no randomness in this last verification round.

A *Merlin-Arthur interactive protocols* of k interactions is an Arthur-Merlin protocol with $k-1$ interactions, but where the verification round is randomized. More precisely, Arthur is in charge of the k-th interaction, which includes the verification algorithm. The protocol ends when Arthur picks a random string $r(v)$ at every node v and uses it to perform a (randomized) verification algorithm. In other words, each node v randomly chooses a message M_v from a distribution specified by the protocol, and broadcast M_v to its neighbors. Finally, as explained before, the protocol either accepts or rejects at node v. Note that, in this case, Merlin does not know the messages each node broadcasts to its neighbors (because they are randomly generated). If $k = 1$, a distributed Merlin-Arthur protocol is a (1-round) randomized decision algorithm; if $k = 2$, it can be viewed as the non-deterministic version of randomized decision, etc.

Definition 1. *Let $\mathcal{V}$ be a verifier and $\mathcal{M}$ a prover of a distributed interactive proof protocol for languages over graphs of n nodes. If $(\mathcal{V}, \mathcal{M})$ corresponds to an Arthur-Merlin (resp. Merlin Arthur) k-round, $\mathcal{O}(f(n))$ bandwidth protocol, we write $(\mathcal{V}, \mathcal{M}) \in \mathsf{dAM}_{\mathsf{prot}}[k, f(n)]$ (resp. $(\mathcal{V}, \mathcal{M}) \in \mathsf{dMA}_{\mathsf{prot}}[k, f(n)]$).*

Definition 2. *Let $\varepsilon \leq 1/3$. The class $\mathsf{dAM}_\varepsilon[k, f(n)]$ (resp. $\mathsf{dMA}_\varepsilon[k, f(n)]$) is the class of languages $\mathcal{L}$ over graphs of n nodes for which there exists a verifier $\mathcal{V}$ such that, for every configuration (G, id, I) of size n, the two following conditions are satisfied.*

Completeness. *If $(G, \mathsf{id}, I) \in \mathcal{L}$ then, there exists a prover $\mathcal{M}$ such that $(\mathcal{V}, \mathcal{M}) \in \mathsf{dAM}_{\mathsf{prot}}[k, f(n)]$ (resp. $(\mathcal{V}, \mathcal{M}) \in \mathsf{dMA}_{\mathsf{prot}}[k, f(n)]$) and*

$$\mathbf{Pr}\Big[\mathcal{V} \text{ accepts } (G, \mathsf{id}, I) \text{ in every node given } \mathcal{M}\Big] \geq 1 - \varepsilon.$$

Soundness. *If $(G, \mathsf{id}, I) \notin \mathcal{L}$ then, for every prover $\mathcal{M}$ such that $(\mathcal{V}, \mathcal{M}) \in \mathsf{dAM}_{\mathsf{prot}}[k, f(n)]$ (resp. $(\mathcal{V}, \mathcal{M}) \in \mathsf{dMA}_{\mathsf{prot}}[k, f(n)]$),*

$$\mathbf{Pr}\Big[\mathcal{V} \text{ rejects } (G, \mathsf{id}, I) \text{ in at least one nodes given } \mathcal{M}\Big] \geq 1 - \varepsilon.$$

We also denote $\mathsf{dAM}[k, f(n)] = \mathsf{dAM}_{1/3}[k, f(n)]$ and $\mathsf{dMA} = \mathsf{dMA}_{1/3}[k, f(n)]$, and omit the subindex ε when its value is obvious from the context.

In this paper, we are interested mainly in two languages, that we call COGRAPH and DIST-HEREDITARY which are the languages of graphs that are cographs and distance-hereditary graphs, respectively. Formally,

- COGRAPH $= \{\langle G, \mathsf{id}\rangle$ s.t. G is a cograph$\}$.
- DIST-HEREDITARY $= \{\langle G, \mathsf{id}\rangle$ s.t. G is distance-hereditary$\}$.

Also, for a distributed language $\mathcal{L}$, the *restriction of $\mathcal{L}$ to cographs*, denoted $\mathcal{L}_{\text{COGRAPH}}$ is the subset of network configurations $(G, \mathsf{id}, I) \in \mathcal{L}$ such that G is a cograph.

3 Cographs

We first show a way to distribute the proofs received by the network in such a way that we can centralize the verification process.

Lemma 1. *Given a cograph G, it is possible to construct a spanning tree T of* depth *two, such that each node at depth one, has at most one child.*

By the previous lemma, we know that for any two round protocol $\mathcal{P}$ over a cograph G with cost $\Omega(\log n)$ bits, we may assume, without loss of generality, that there is a root ρ with access to all coins and messages received by the whole network. We simply construct the spanning tree given by Lemma 1, by choosing the root ρ as follows. First, a bipartite graph can be easily verified with two colors, and ρ can be chosen to be the node in G_2 with the smallest identifier. Then, it suffices to assign to each node u of depth one in the spanning tree, both its proof and the proof received by its child w, along with the random coin it drew. Then, the nodes can locally verify the consistency of this message and the root will have received all the messages in the network.

Lemma 2. *Given any* dM *(resp.* dAM*) protocol with bandwidth L that runs over a cograph, we can construct a* dM *(resp.* dAM*) protocol with bandwidth cost $L + \mathcal{O}(\log n)$ and where there exists a node ρ which has access to all messages (resp. all messages and coins) in the network.*

An advantage of this procedure is that we may simulate any protocol in the (non-deterministic) One-Round Broadcast Congested Clique model (by using the root ρ as referee) by either using one round of interaction (if the simulated protocol is deterministic) or two rounds (when the simulated protocol is randomized). From here it follows that we can use the protocol by [29] to recognize cographs, therefore constructing a protocol for cograph detection in two rounds of interaction and $\mathcal{O}(\log n)$ bits. That is, COGRAPH $\in$ dAM$[\log n]$. For the sake of completeness, we now describe the protocol of [29].

Definition 3. *Given a cograph $G = (V, E)$, we can define its* canonical order *as follows. We start by choosing the smallest pair of* twins *(those with the smallest identifiers in lexicographic order) which we know to exist by Proposition 1. From there we choose and remove the smallest node from this pairing. Then, we repeat this process by finding another pair and removing one of its members until we end up with a single node.*

Let p be a prime and $\phi = (\phi_w)_{w \in V}$ be a family of linearly independent polynomials in $\mathbb{Z}_p[x]$. Given $w \in V$ we define, $q_w = \sum_{w' \in N(w)} \phi_{w'}$ and $\bar{q}_w = q_w + \phi_w$. We also define the *derived polynomials* of ϕ as the collection

$$\alpha_{u,v} = \phi_u - \phi_v \qquad \beta_{u,v} = q_u - q_v, \qquad \gamma_{u,v} = \bar{q}_u - \bar{q}_v, \quad u.v \in V$$

Now, given a pair of *twins* u and v, we assign to $G - v$ the polynomials $\{\phi'_w\}_{w \in V-v}$ defined as

$$\phi'_w = \begin{cases} \phi_w & \text{if } w \neq u \\ \phi_u + \phi_v & \text{if } w = u \end{cases}$$

With this construction, from $\phi_u(x) = x^{\mathsf{id}(u)}$ it is possible to construct a sequence of polynomials ϕ_u^i for $i \in [n]$ according to the *canonical order* $\{v_i\}_{i=1}^n$ and u in the graph $G - \{v_j\}_{j=i+1}^n$. We call these functions the *basic* polynomials of G. And so the *canonical family* of polynomials of G is defined as the union between its basic and derived polynomials. It follows that this family of functions has at most $3n^3$ elements.

Definition 4. *Let G be a cograph. We say that a vector $m = ((a_w, b_w))_{w\in V} \in (\mathbb{Z}_p)^{2n}$ is* valid *for G in $t \in \mathbb{Z}_p$ if there exists a family of linearly independent polynomials $(\phi_w)_{w\in V}$ in $\mathbb{Z}_p[X]$ such that $a_w = \phi_w(t)$ and $b_w = q_w(t)$ for each $w \in V$.*

Lemma 3. *Let $m = ((a_w, b_w))_{w\in V} \in (\mathbb{Z}_p)^{2n}$ be a valid vector for G in t. Consider u, v to be a pair of twins in G such that $a_u \neq a_v$. Then, the vector $m' = ((a'_w, b'_w))_{w\in V-v} \in (\mathbb{Z}_p)^{2n-2}$ is valid for $G - v$ in t, where its coordinates are given by*

$$(a'_w,\, b'_w) = \begin{cases} (a_w,\, b_w) & \text{if } w \in V - \{u, v\} \\ (a_u + a_v,\, b_u - a_v\delta_{uv}) & \text{if } w = u \end{cases}$$

with δ_{uv} equals one if and only if $a_u + b_u = a_v + b_v$

With this lemma now we can proceed to describe the protocol.

Theorem 1. *There is a distributed interactive proof with two rounds for the recognition of cographs, i.e., COGRAPH $\in$ $\mathsf{dAM}[\log n]$. Moreover, the protocol uses shared randomness and gives the correct answer with high probability.*

Proof. Let $G = (V, E)$ be an n-node graph. Without loss of generality we may assume the graph has identifiers in $[n]$ as, following Lemma 1, it is possible to implement a PERMUTATION protocol in a single round: Merlin sends to each node v an identifier $\bar{\mathsf{id}} : V \to [n]$ and the root, by receiving all proofs, can see that they all received distinct identifiers which are consistent with their original ones.

Let p be a prime such that $3n^{c+4} \leq p \leq 6n^{c+4}$. The protocol is the following: All nodes collectively generate a seed $t \in \mathbb{F}_p$ uniformly at random. Then Merlin sends to each node w a message m_w such that $m = (m_w)_{w\in V}$ is a valid vector for G at t. Each node then computes such message by defining $\phi_w(x) = x^{\bar{\mathsf{id}}(w)}$.

After the nodes exchange messages, following Lemma 1 we obtain that the root ρ owns a vector $m \in \mathbb{F}_p^{2n}$. From here, the root repeats the following procedure at most $n-1$ times trying to construct a canonical ordering $\{v_i\}_{i=1}^n$ for G. At step i, it starts at graph G^i and a vector $m_i \in \mathbb{F}_p^{2(n-i+1)}$ (where $G^1 = G$ and $m^1 = m$) and looks for a pair of nodes u, v in G^i such that $a_u^i \neq a_v^i$ and either $b_u^i = b_v^i$ or $a_u^i + b_u^i = a_v^i + b_b^i$. Then it chooses, among all pairs it has found, the first in lexicographic order. If no such pair exists, then he rejects. On the contrary, he defines $G^{i+1} = G^i - v$, and setting $v_{n-i+1} = v$ (without loss of generality we assume that $\mathsf{id}(v) < \mathsf{id}(u)$). Then the root computes m^{i+1} from the previous vector m^i following Lemma 3. If the root reaches step $n-1$ then it accepts.

Completeness and **Soundness.** It follows then that as the messages depend on the original identifiers and the root ρ has access to all messages, then both acceptance errors depend solely in the 1BCC construction. Now, by Lemma 3 it follows that the only point at which the protocol might fail is if the chosen t turns out to be a root for any of the polynomials in the canonical family from Definition 3. As there are at most $3n^3$ such polynomials, each of degree at most n , we have that the acceptance error is at most $3n^4/3n^{c+4} = 1/n^c$ and the theorem follows. □

As we mentioned in the Introduction, the result obtained in [29] is much stronger than just recognizing cographs. In fact, the referee not only can recognize a cograph but actually can *reconstruct* it. In other words, when the input graph is a cograph, after the communication round, the referee learns all the edges. In our context, this implies that the root ρ not only recognizes cographs, but also can recognize any distributed language restricted to them.

Theorem 2. *For every distributed language $\mathcal{L}$, there is a distributed interactive proof with two rounds for its restriction to cographs, i.e. $\mathcal{L}_{\text{COGRAPH}} \in \mathsf{dAM}[\log n]$. Moreover, the protocol uses shared randomness and gives the correct answer with high probability.*

Proof. It is sufficient to notice that the tree-root ρ in the construction from Lemma 1 has access to all proofs in the network. In particular, the id's and positions for each node in the twin-ordering π. As such, ρ has knowledge of the entire topology of the network and its inputs (provided that these are of size $\mathcal{O}(\log n)$) and can compute any property related to them, with the acceptance error matching that of the verification procedure in Theorem 1. As for the rest of the nodes, they simply accept and delegate this decision to the root. □

4 Distance-Hereditary Graphs

Following the protocol described for cographs, it is possible to derive an interactive protocol for distance-hereditary graphs, which admit a similar construction. Indeed, as described before, any distance-hereditary graph can be constructed by sequentially adding twins or pending nodes. Notice that for the protocol in Theorem 1, the verification process is done by the root as it prunes the graph in $n-1$ steps. This leads to an order by which the nodes were selected, and we call it *canonical ordering.* While we cannot delegate the verification routine to a single node (as distance-hereditary graphs can have arbitrarily large diameter), we can distribute the verification process by letting different nodes check different steps of the computation. As the rule described in Lemma 3 for pruning the graph involves only the pair of twins at each step, we only need to find nodes that, for a fixed node v, can receive all the proofs sent by v, its twins and its pending nodes.

In order to prune the graph in this new setting, we need a rule for pruning pending nodes from a graph and updating the vectors of each node accordingly. Here, we use the definition of a *valid vector*as described in Sect. 3.

Lemma 4. *Let $m = ((a_w, b_w))_{w \in V} \in (\mathbb{Z}_p)^{2n}$ be a valid vector for G at some point t. If $u \in G$ has v as a pending node adjacent to it, then, the vector $m' = ((a'_w, b'_w))_{w \in V-v} \in (\mathbb{Z}_p)^{2n-2}$ is valid for $G - v$ in t, where the coordinates of m' are given by*

$$(a'_w, b'_w) = \begin{cases} (a_w, b_w) & \text{if } w \in V - \{u, v\} \\ (a_w, b_w - a_v) & \text{if } w = u \end{cases}$$

In order to distribute the verification procedure, for any fixed v we wish to set a node to compute the correctness of the vectors of all nodes assigned as twins of v. Indeed, for a fixed ordering π for pruning the graph and a node v with $\pi_v < n$, consider the *predecessor* of v, denoted by $\mathsf{ant}(v)$, to be v's neighbor whose value for $\pi(\cdot)$ is immediately after that of v among its neighbors. As all previous nodes in the order which are twins of v have the same neighborhood, it follows that all these nodes must be adjacent to $\mathsf{ant}(v)$. In case that no such a node exists, by assuming that G is connected, it follows that the last node according to π which is assigned as a twin of v must be a *true* twin and, therefore, be adjacent to him. And the same reasoning holds.

Thus, the main strategy of our protocol is that, given an initial vector (a_v, b_v) for a node v in the graph, each node $\mathsf{ant}(v)$ has the task of updating this vector until it obtains the vector v that the referee should have at the time the node is pruned from the graph, which we denote by (a_v^π, b_v^π). Then, each node u which is a twin of v provides its vector (a_u^π, b_u^π) (which is proved to be correct by some other node) and so the predecessor of v compares and updates v's vector according to the rules from Lemmas 3 and 4.

Theorem 3. *There is a distributed interactive proof with three rounds of interaction for the recognition of distance-hereditary graphs, i.e.,* DIST-HEREDITARY $\in$ $\mathsf{dMAM}[\log n]$. *Moreover, the protocol uses shared randomness and gives the correct answer with high probability.*

5 Lower Bounds

In this section, we provide lower-bounds on the certificate size of distributed interactive proofs for COGRAPH or DIST-HEREDITARY. Due the lack of space, the proof of the result on this section are detailed in the appendix. The following result is based on a construction by [21].

Theorem 4. *If* COGRAPH *or* DIST-HEREDITARY *belongs to the class* $\mathsf{dM}[f(n)]$, *then* $f(n) = \Omega(\log n)$. *Moreover, for any fixed k, if* COGRAPH *or* DIST-HEREDITARY *belongs to* $\mathsf{dAM}^{\mathsf{pub}}[k, g(n)]$, *then* $g(n) = \Omega(\log \log n)$.

Following an approach introduced by Fraigniaud et al. [17], we obtain that the graph constructions used in the proof of Theorem 4 can be adapted in order to obtain lower-bounds for the models dMA.

Corollary 1. *If any of the problems* COGRAPH *or* DIST-HEREDITARY *belongs to* $\mathsf{dMA}_{1/7}[f(n)]$, *then* $f(n) = \Omega(\log n)$.

References

1. Assadi, S., Kol, G., Oshman, R.: Lower bounds for distributed sketching of maximal matchings and maximal independent sets. In: Emek, Y., Cachin, C. (eds.) PODC '20: ACM Symposium on Principles of Distributed Computing, Virtual Event, Italy, 3–7 August 2020, pp. 79–88. ACM (2020)
2. Baruch, M., Fraigniaud, P., Patt-Shamir, B.: Randomized proof-labeling schemes. In: Symposium on Principles of Distributed Computing, pp. 315–324 (2015)
3. Bousquet, N., Feuilloley, L., Pierron, T.: Local certification of graph decompositions and applications to minor-free classes. arXiv preprint arXiv:2108.00059 (2021)
4. Brandstadt, A., Spinrad, J.P., et al.: Graph Classes: A Survey, vol. 3. SIAM (1999)
5. Broersma, H., Dahlhaus, E., Kloks, T.: A linear time algorithm for minimum fill-in and treewidth for distance hereditary graphs. Discret. Appl. Math. **99**(1–3), 367–400 (2000)
6. Censor-Hillel, K., Paz, A., Perry, M.: Approximate proof-labeling schemes. Theoret. Comput. Sci. **811**, 112–124 (2020)
7. Cicerone, S., Di Stefano, G., Flammini, M.: Compact-port routing models and applications to distance-hereditary graphs. J. Parallel Distrib. Comput. **61**(10), 1472–1488 (2001)
8. Corneil, D., Lerchs, H., Burlingham, L.: Complement reducible graphs. Discret. Appl. Math. **3**(3), 163–174 (1981)
9. Courcelle, B., Makowsky, J.A., Rotics, U.: Linear time solvable optimization problems on graphs of bounded clique-width. Theory Comput. Syst. **33**(2), 125–150 (2000). https://doi.org/10.1007/s002249910009
10. Crescenzi, P., Fraigniaud, P., Paz, A.: Trade-offs in distributed interactive proofs. In: 33rd International Symposium on Distributed Computing (DISC 2019). Schloss Dagstuhl-Leibniz-Zentrum fuer Informatik (2019)
11. Dahlhaus, E.: Efficient parallel recognition algorithms of cographs and distance hereditary graphs. Discret. Appl. Math. **57**(1), 29–44 (1995)
12. Damiand, G., Habib, M., Paul, C.: A simple paradigm for graph recognition: application to cographs and distance hereditary graphs. Theoret. Comput. Sci. **263**(1–2), 99–111 (2001)
13. D'Atri, A., Moscarini, M.: Distance-hereditary graphs, Steiner trees, and connected domination. SIAM J. Comput. **17**(3), 521–538 (1988)
14. Feuilloley, L., Fraigniaud, P., Montealegre, P., Rapaport, I., Rémila, É., Todinca, I.: Local certification of graphs with bounded genus. arXiv preprint arXiv:2007.08084 (2020)
15. Feuilloley, L., Fraigniaud, P., Montealegre, P., Rapaport, I., Rémila, É., Todinca, I.: Compact distributed certification of planar graphs. Algorithmica **83**, 1–30 (2021). https://doi.org/10.1007/s00453-021-00823-w
16. Fraigniaud, P., Korman, A., Peleg, D.: Towards a complexity theory for local distributed computing. J. ACM (JACM) **60**(5), 1–26 (2013)
17. Fraigniaud, P., Montealegre, P., Oshman, R., Rapaport, I., Todinca, I.: On distributed Merlin-Arthur decision protocols. In: Censor-Hillel, K., Flammini, M. (eds.) SIROCCO 2019. LNCS, vol. 11639, pp. 230–245. Springer, Cham (2019). https://doi.org/10.1007/978-3-030-24922-9_16
18. Fraigniaud, P., Patt-Shamir, B., Perry, M.: Randomized proof-labeling schemes. Distrib. Comput. **32**(3), 217–234 (2019). https://doi.org/10.1007/s00446-018-0340-8
19. Ghaffari, M., Haeupler, B.: Distributed algorithms for planar networks I: planar embedding. In: Proceedings of the 2016 ACM Symposium on Principles of Distributed Computing, pp. 29–38 (2016)

20. Golumbic, M.C.: Algorithmic Graph Theory and Perfect Graphs (Annals of Discrete Mathematics, vol 57). North-Holland Publishing Co., NLD (2004)
21. Göös, M., Suomela, J.: Locally checkable proofs in distributed computing. Theory Comput. **12**(1), 1–33 (2016)
22. Grötschel, M., Lovász, L., Schrijver, A.: Geometric Algorithms and Combinatorial Optimization. Springer, Heidelberg (1993). https://doi.org/10.1007/978-3-642-78240-4
23. Halldórsson, M.M., Konrad, C.: Distributed algorithms for coloring interval graphs. In: Kuhn, F. (ed.) DISC 2014. LNCS, vol. 8784, pp. 454–468. Springer, Heidelberg (2014). https://doi.org/10.1007/978-3-662-45174-8_31
24. He, X.: Parallel algorithm for cograph recognition with applications. J. Algorithms **15**(2), 284–313 (1993)
25. Howorka, E.: A characterization of distance-hereditary graphs. Q. J. Math. **28**(4), 417–420 (1977)
26. Hsieh, S.Y., Ho, C.W., Hsu, T.S., Ko, M.T., Chen, G.H.: Efficient parallel algorithms on distance hereditary graphs. Parallel Process. Lett. **09**(01), 43–52 (1999)
27. Hung, R.W., Chang, M.S.: Linear-time algorithms for the Hamiltonian problems on distance-hereditary graphs. Theoret. Comput. Sci. **341**(1–3), 411–440 (2005)
28. Jung, H.: On a class of posets and the corresponding comparability graphs. J. Comb. Theory Ser. B **24**(2), 125–133 (1978)
29. Kari, J., Matamala, M., Rapaport, I., Salo, V.: Solving the INDUCED SUBGRAPH problem in the randomized multiparty simultaneous messages model. In: Scheideler, C. (ed.) SIROCCO 2014. LNCS, vol. 9439, pp. 370–384. Springer, Cham (2015). https://doi.org/10.1007/978-3-319-25258-2_26
30. Kirkpatrick, D.G., Przytycka, T.: Parallel recognition of complement reducible graphs and cotree construction. Discret. Appl. Math. **29**(1), 79–96 (1990)
31. Kol, G., Oshman, R., Saxena, R.R.: Interactive distributed proofs. In: ACM Symposium on Principles of Distributed Computing, pp. 255–264. ACM (2018)
32. Konrad, C., Zamaraev, V.: Brief announcement: distributed minimum vertex coloring and maximum independent set in chordal graphs. In: Proceedings of the 2018 ACM Symposium on Principles of Distributed Computing, pp. 159–161 (2018)
33. Korman, A., Kutten, S., Peleg, D.: Proof labeling schemes. Distrib. Comput. **22**(4), 215–233 (2010). https://doi.org/10.1007/s00446-010-0095-3
34. Lin, R., Olariu, S.: Fast parallel algorithms for cographs. In: Nori, K.V., Veni Madhavan, C.E. (eds.) FSTTCS 1990. LNCS, vol. 472, pp. 176–189. Springer, Heidelberg (1990). https://doi.org/10.1007/3-540-53487-3_43
35. Montealegre, P., Perez-Salazar, S., Rapaport, I., Todinca, I.: Graph reconstruction in the congested clique. J. Comput. Syst. Sci. **113**, 1–17 (2020)
36. Montealegre, P., Ramírez-Romero, D., Rapaport, I.: Shared vs private randomness in distributed interactive proofs. LIPIcs, vol. 181, pp. 51:1–51:13. Schloss Dagstuhl - Leibniz-Zentrum für Informatik (2020)
37. Naor, M., Parter, M., Yogev, E.: The power of distributed verifiers in interactive proofs. In: ACM-SIAM Symposium on Discrete Algorithms, pp. 1096–115. SIAM (2020)
38. Naor, M., Stockmeyer, L.: What can be computed locally? SIAM J. Comput. **24**(6), 1259–1277 (1995)
39. Seinsche, D.: On a property of the class of n-colorable graphs. J. Comb. Theory Ser. B **16**(2), 191–193 (1974)
40. Sumner, D.P.: Dacey graphs. J. Aust. Math. Soc. **18**(4), 492–502 (1974)

Asynchronous Gathering Algorithms for Autonomous Mobile Robots with Lights

Rikuo Nakai[1(✉)], Yuichi Sudo[2], and Koichi Wada[3]

[1] Graduate School of Science and Engineering, Hosei University, Koganei City, Japan
rikuo.nakai.7r@stu.hosei.ac.jp

[2] Faculty of Computer and Information Sciences, Hosei University, Koganei City, Japan

[3] Faculty of Science and Engineering, Hosei University, Koganei City, Japan

Abstract. We consider a *Gathering* problem for n autonomous mobile robots with persistent memory called *light* in an asynchronous scheduler (ASYNC). Gathering is well known to be impossible when robots have no lights in basic standard models if the system is semi-synchronous (SSYNC) or even centralized (only one robot is active at each time). It is known that robots can solve Gathering with 10 colors of lights in ASYNC. This result is obtained by combining the following results. (1) The simulation of SSYNC robots with k colors by ASYNC robots with $5k$ colors [7], and (2) Gathering is solved by SSYNC robots with 2 colors [28].

In this paper, we improve the result by reducing the number of colors and show that Gathering can be solved by ASYNC robots with 3 colors of lights. We also show that we can construct a simulation algorithm of any *unfair* SSYNC algorithm using k colors by ASYNC robots with $3k$ colors, where unfairness does not guarantee that every robot is activated infinitely often. Combining this simulation and the Gathering algorithm by SSYNC robots with 2 colors [28], we obtain a Gathering algorithm by ASYNC robots with 6 colors. Our main result can be obtained by reducing the number of colors from 6 to 3.

1 Introduction

1.1 Background and Motivation

The computational issues of autonomous mobile entities have been the object of much research in distributed computing. In this paper, we focus on mobile objects operating on a two-dimensional Euclidean space but there are several research on three-dimensional spaces and graphs [16]. Each robot operate in *Look-Compute-Move* (*LCM*) cycles. In the *Look* phase, an entity, viewed as a

This research was supported in part by JSPS KAKENHI No. 20H04140, 20KK0232, 20K11685, 21K11748, and by Japan Science and Technology Agency (JST) SICORP Grant#JPMJSC1806.

C. Johnen et al. (Eds.): SSS 2021, LNCS 13046, pp. 410–424, 2021.
https://doi.org/10.1007/978-3-030-91081-5_27

point and usually called *robot*, obtains a snapshot of the space; in the *Compute* phase it executes its algorithm (the same for all robots) using the snapshot as input; it then moves towards the computed destination in the *Move* phase. Repeating these cycles, the robots are able to collectively perform some tasks and solve some problems. The research interest has been on determining the impact that *internal* capabilities (e.g., memory, communication) and *external* conditions (e.g. synchrony, activation scheduler) have on the solvability of a problem.

We also explore such weakest capabilities to solve the task. The problem considered in this paper is *Gathering*, which is one of the most fundamental tasks of autonomous mobile robots. Gathering is the process of n mobile robots, initially located on arbitrary positions, meeting within finite time at a location, not known a priori. When there are only two robots, this task is called *Rendezvous*. Since Gathering and Rendezvous are simple but essential problems, they have been intensively studied, and a number of possibility and/or impossibility results have been shown under the different assumptions [1,2,4,6,9,10,14,19–24,26]. The solvability of Gathering and Rendezvous depends on the activation schedule and the synchronization level. Usually three basic types of schedulers are identified, the fully synchronous (FSYNC), the semi-synchronous (SSYNC) and the asynchronous (ASYNC)[1] Gathering and Rendezvous are trivially solvable in FSYNC and the basic model. However, these problems cannot be solved in SSYNC without any additional assumptions [15], and the same is true in ASYNC.

In [7], persistent memory called *light* has been introduced to reveal relationship between ASYNC and SSYNC and they show asynchronous robots with lights equipped with a constant number of colors, are strictly more powerful than semi-synchronous robots without lights: for any algorithm $\mathcal{A}$ designed for semi-synchronous robots (without colors), they give a simulation algorithm by which asynchronous robots with 5 colors simulate an execution of $\mathcal{A}$. Rendezvous can be solved by robots with lights without any other additional assumptions [7,17,29]. Gathering is also solvable by robots with lights and it can be solved by robots with 2 colors of lights in SSYNC [28]. The power of lights to solve other problems are discussed in [11–13].

1.2 Our Contribution

In this paper, we study Gathering algorithms by robots with lights in the most realistic schedulers, ASYNC and some of the weakest conditions in term of computational power. As for Gathering algorithms in ASYNC, the following results are known; Cielieback et al. [6] solves the distinct Gathering for more than two robots with weak multiplicity detection, where the distinct gathering means all robots are initially placed in different positions, and weak multiplicity detection helps a robot to identify multiple occurrences of robots at a single point. Bhagat et al. [3] solves a gathering problem for five or more robots under the additional constraint to minimize the maximum distance traversed by any robots, with

[1] In addition to these basic models, the new model semi-asynchronous (SAsync) [5] is recently proposed to reveal the gap between SSYNC and ASYNC.

weak multiplicity detection or 4 colors of lights. Both algorithms work without any extra assumptions like agreements of coordinate systems, unit distance and chirality and rigidity of movement, but they solve some constrained gathering problem. Gathering can be solved by ASYNC robots with 10 colors of lights by using the following two results;

(a) Any algorithm in SSYNC using k colors of light can be simulated by ASYNC robots with $5k$ colors of lights [7], and
(b) A Gathering algorithm is constructed by SSYNC robots with 2 colors of lights [28].

Since the algorithm shown in [28] needs the chirality assumption but no any other extra assumptions, the obtained algorithm works only with chirality.[2]

This paper improves the result just stated above by reducing the number of colors and shows that Gathering can be solved by ASYNC robots with 3 colors of lights in no any other extra assumptions except chirality as follows;

1. We construct a simulation algorithm of any *unfair* SSYNC algorithm using k colors by ASYNC robots with $3k$ colors of lights, where unfair SSYNC is that the adversary makes enabled robots (changing its color or moving a different location) active in SSYNC. We have reduced the number of colors used in the simulation to $3k$ from $5k$, although the simulated algorithms are limited to ones working in unfair SSYNC. Since many robots algorithms seem to work in unfair SSYNC if it works in (fair) SSYNC, this simulation is interesting in itself and can be used to reduce the number of colors used in algorithms working in ASYNC.
2. We show that the Gathering algorithm with 2 colors of light shown in [28] can still work in unfair SSYNC. Hence we obtain that Gathering can be solved by ASYNC robots with 6 colors of lights in no any other extra assumptions except chirality. The Gathering algorithm of [28] is divided into two sub-algorithms. The first one makes a configuration that all robots are located on one straight line from any initial configuration and the second one is a Gathering algorithm from any initial configuration such that all robots are located on the straight line. The first one needs no lights but assumption of chirality and the second one need no extra assumptions and uses 2 colors. We show that the both algorithms can work in unfair SSYNC by defining a potential function for each algorithm and showing that each function becomes monotonically decreasing for any behaviour of each algorithm.
3. We improve the number of colors used in the algorithm into 3. Since the second algorithm uses 2 colors in SSYNC, the resultant algorithm have 6 colors. Thus in order to reduce the number of colors, we directly construct a 3-color Gathering algorithm from any configuration such that all robots are located on the straight line in ASYNC. Combining with the simulation of the first algorithm, we have obtained an ASYNC Gathering algorithm with 3 colors of light.

---

[2] Note that Rendezvous and Gathering cannot be solved by SSYNC and ASYNC robots without light even if the chirality is assumed [15].

2 Model and Preliminaries

2.1 The Basics

The systems considered in this paper consist of a team $R = \{r_0, \cdots, r_{n-1}\}$ of computational entities moving and operating in the Euclidean plane $\mathbb{R}^2$. Viewed as points, and called *robots*, the entities can move freely and continuously in the plane. Each robot has its own local coordinate system and it always perceives itself at its origin; there might not be consistency between these coordinate systems. A robot is equipped with sensorial devices that allows it to observe the positions of the other robots in its local coordinate system.

The robots are *identical*: they are indistinguishable by their appearance and they execute the same protocol. The robots are *autonomous*, without a central control.

At any point in time, a robot is either *active* or *inactive*. Upon becoming active, a robot r_i executes a *Look-Compute-Move* (*LCM*) cycle performing the following three operations:

1. *Look*: The robot activates its sensors to obtain a snapshot of the positions occupied by robots with respect to its own coordinate system[3]. The snapshot of r_i is denoted as $\mathcal{SS}_i$.
2. *Compute*: The robot executes its algorithm using the snapshot as input. The result of the computation is a destination point.
3. *Move*: The robot moves in a straight line toward the computed destination but the robot may be stopped by an adversary before reaching the computed destination. In this case, the movement is called *non-rigid*. Otherwise, it is called *rigid*. When stopped before reaching its destination in the non-rigid movement, a robot moves at least a minimum distance $\delta > 0$. If the distance to the destination is at most δ, the robot can reach it. We assume non-rigid movement throughout the paper. If the destination is the current location, the robot stays still.

When inactive, a robot is idle. All robots are initially idle. The amount of time to complete a cycle is assumed to be finite, and the *Look* operation is assumed to be instantaneous.

There might not be consistency between the local coordinate systems and their unit of distance. The absence of any a-priori assumption on consistency of the local coordinate systems is called *disorientation*.

The robots are said to have *chirality* if they share the same circular orientation of the plane (i.e., they agree on "clockwise" direction). If there is chirality, then there exists a unique circular ordering of locations occupied robots [27]. Thus, for each edge of the convex hull obtained by locations of n robots ($n \geq 3$), all robots can agree with the right vertex of the edge.

---

[3] This is called the *full visibility* (or unlimited visibility) setting; restricted forms of visibility have also been considered for these systems.

2.2 The Models

Different models, based on the same basic premises defined above, have been considered in the literature and we will use the following two models.

In the most common model, $\mathcal{OBLOT}$, the robots are *silent*: they have no explicit means of communication; furthermore they are *oblivious*: at the start of a cycle, a robot has no memory of observations and computations performed in previous cycles.

In the other common model, $\mathcal{LUMI}$, each robot r_i is equipped with a persistent visible state variable ℓ_i, called *light*, whose values are taken from a finite set C of states called *colors* (including the color that represents the initial state when the light is off). The colors of the lights can be set in each cycle by r_i at the end of its *Compute* operation. A light is *persistent* from one computational cycle to the next: the color is not automatically reset at the end of a cycle; the robot is otherwise oblivious, forgetting all other information from previous cycles. In $\mathcal{LUMI}$, the *Look* operation produces a colored snapshot; i.e., it returns the set of pairs $(position, color)$ of the other robots[4]. Note that if $|C| = 1$, then the light is not used; thus, this case corresponds to the $\mathcal{OBLOT}$ model.

We denote by $\ell_i(t)$ the color of light r_i has at time t and $p_i(t) \in \mathbb{R}^2$ the position occupied by robot r_i at time t represented in some global coordinate system. A *configuration* $\mathcal{C}(t)$ at time t is a multi-set of n pairs $(\ell_i(t), p_i(t))$, each defining the color of light and the position of robot r_i at time t. When no confusion arises, $\mathcal{C}(t)$ is simply denoted by $\mathcal{C}$.

If a configuration is that robots are located on a line segment connecting p and q (denoted as pq), this configuration is denoted by a regular-expression-like sequence of colors robots have from the endpoint p to the other endpoint q. Formally we define *color-configurations* for a configuration of line segment pq as follows; Let $\mathcal{C}(t)$ be the line-segment configuration at time t. *Color-configurations* for $\mathcal{C}(t)$ are defined as (0)–(3) as follows;

(0) Factor f is defined as either α, $(\alpha|\beta)$ or $(\alpha|\beta|\gamma)$, where α, β and γ are colors and $(\alpha|\beta)$ and $(\alpha|\beta|\gamma))$ denote α or β and α, β or γ, respectively. Color(s) which robots at a point have are denoted as f. Let f, g and h be factors.
(1) fg denotes a configuration that all robots at p have colors f, all robots at q have colors g, and there are no robots inside the segment.
(2) fg^+h denotes a configuration that all robots at p have colors f, all robots at q have colors h, and there exists at least one point inside the segment that all robots located there have colors g.
(3) fg_mh, if all robots at p have colors f, all robots at q have colors h, and all robots at the mid-point of the segment have colors g and there are no robots except on the three locations.

If a color-configuration is one of (1)–(3), it is denoted as fg^*h. Let $dis(\mathcal{C}(t))$ denote the length of the segment in the configuration $\mathcal{C}(t)$. The color-configuration for $\mathcal{C}(t)$ is denoted as $cc(\mathcal{C}(t))$ and the number of points which have

[4] If (strong) multiplicity detection is assumed, the snapshot is a multi-set.

color α in $\mathcal{C}(t)$ is denoted as $\#_\alpha(\mathcal{C}(t))$. We also use this notation for a snapshot of robot.

In Sect. 4, color-configurations defined here are used, and in addition, one abuse of notation is used as follows. Letting f, g, and h be factors, f^+gh^* denotes that all robots at p have colors f, all robots at q have colors g or h, and all robots inside the segment have colors f, g, and h.

2.3 The Schedulers

With respect to the activation schedule of the robots, and the duration of their LCM cycles, the fundamental distinction is between the *asynchronous* and *synchronous* settings.

In the *synchronous* setting (SSYNC), also called semi-synchronous, time is divided into discrete intervals, called *rounds*; in each round some robots are activated simultaneously, and perform their LCM cycle in perfect synchronization.

A popular synchronous setting which plays an important role is the *fully-synchronous* setting (FSYNC), where every robot is activated in every round; that is, the activation scheduler has no adversarial power.

In the *asynchronous* setting (ASYNC), there is no common notion of time, each robot is activated independently of the others, the duration of each phase is finite but unpredictable and might be different in different cycles. In this paper, we are concerned with ASYNC and we assume the following; In a *Look* operation, a snapshot of the environment is taken at some time t_L and we say that the *Look operation is performed at time* t_L. Each *Compute* operation of r_i is assumed to be done at time t_C and the color of its light $\ell_i(t)$ and its pending destination des_i are both set to the computed values for any time greater than t_C[5]. When the movement in a *Move* operation begins at time t_B and ends at t_E, we say that it is performed during interval $[t_B, t_E]$, and the beginning (resp. ending) of the movement is denoted by $Move_{BEGIN}$ (resp. $Move_{END}$) occurring at time t_B (resp. t_E). In the following, *Compute*, $Move_{BEGIN}$ and $Move_{END}$ are abbreviated as *Comp*, M_B and M_E, respectively. When a cycle has no actual movement (i.e., robots only change color and their destinations are the current positions), we can equivalently assume that the *Move* operation in this cycle is omitted, since we can consider the *Move* operation to be performed just before the next *Look* operation.

Without loss of generality, we assume the set of time instants at which the robots start executions of *Look*, *Comp*, M_B and M_E to be $\mathbb{N}$. We also assume the followings for each operation.

1. *Comp* operation is performed instantaneously at integer time t_C and if some robot performs a *Look* operation at time t_C, then it observes the former color and if it does at time $t_C + 1$, then it observes the newly computed color.

[5] Note that if some robot performs a *Look* operation at time t_C, thenit observes the former color and if it does at time $t_C + \epsilon(\forall\epsilon > 0)$, thenit observes the newly computed color.

2. When the movement in a *Move* operation begins at t_B and ends at $t_E \geq t_B + 1$ and if a robot performs a *Look* operation at time t_B then it observes the location before moving and it does at time $t(t_B + 1 \leq t \leq t_E)$, then it observes any location on the half-open line segment between one before moving (inclusive) and the destination (exclusive) satisfying the following condition, letting p_t be the location of the moving robot at time t, for times t and t' such that $t_B+1 \leq t < t' \leq t_E$, it holds that $dis(p_{t_B}, p_t) < dis(p_{t_B}, p_{t'})$, where $dis(p, q)$ denotes the distance between p and q. The selected location is assumed to be determined by adversary. Also if it does at time $t_E + 1$, it observes the destination.

In SSYNC and ASYNC settings, the selection of which robots are activated is made by an adversarial scheduler, whose only limit is that every robot must be activated infinitely often (i.e., it is a *fair* scheduler). We also consider an *unfair* scheduler. When a robot becomes active and performs the LCM cycle, the robot is *enabled* if it changes its color and/or the computed destination is different from the current position at each activation. The unfair scheduler does not guarantee that every robot is activated infinitely often. It is only guaranteed that if there is one or more enabled-robots at a time t, at least one enabled-robot will be activated or become non-enabled at some time $t' > t$. Note that in a computation under this scheduler, an enabled-robot may not be activated until it becomes the only enabled robot.

3 Simulating Algorithms in Unfair SSYNC by ASYNC $\mathcal{LUMI}$ Robots

In this section, we show that any $\mathcal{OBLOT}$ algorithm working in unfair SSYNC can be simulated by $\mathcal{LUMI}$ robots with 3 colors in ASYNC. In what follows, all proofs are omitted due to lack of space and are shown in the full version [25].

3.1 Simulation in ASYNC for Algorithms in Unfair SSYNC

We show an algorithm in ASYNC that simulates algorithms in unfair SSYNC. The algorithm is shown in Algorithm 1. The algorithm in square brackets indicates that it is given as input of a robot and simulated. Let A_{unfair} be a simulated algorithm in unfair SSYNC. When a robot r is enabled at a configuration $\mathcal{C}(t)$ in algorithm A, we say that *r is A-enabled at $\mathcal{C}(t)$*. Our simulating algorithm uses light with 3 colors, S(tay), M(ove), and E(nd). We use the notation $\forall col$ for a color col denoting a set of configuration such that all robots have color col. We also use the notation $\forall col_1, col_2$ for colors col_1 and col_2 denoting a set of configurations such that each robot has color col_1 or col_2 and there exists at least one robot with color col_1 and there exists at least one robot with color col_2. Initial configuration is in $\forall S$. This algorithm repeats a *color-cycle*, that is, the transition of $\forall S \rightarrow \forall M \rightarrow \forall E$. When the configuration is in $\forall S$, since A_{unfair}-enabled-robots exist, some A_{unfair}-enabled-robots that become active among

Algorithm 1. SIM-for-Unfair(r_i)[A_{unfair}]

Assumptions: non-rigid, $\mathcal{LUMI}$, ℓ_i has 3 colors(S, M, and E), initially $\ell_i = S$;
Input: A_{unfair} : algorithm working in unfair SSYNC, snapshot $\mathcal{SS}_i$ of r_i;

```
1:  case cc(SS_i) of
2:  ∈ ∀S:
3:      if r_i is A_unfair-enabled then
4:          r_i executes A_unfair
5:          des_i ← the computed destination of A_unfair
6:          l_i ← M
7:  ∈ ∀S, M:
8:      l_i ← M
9:  ∈ ∀M or ∀M, E:
10:     l_i ← E
11: ∈ ∀E or ∀S, E:
12:     l_i ← S
13: endcase
```

those execute A_{unfair}, change their colors to M, and move to the computed destination. While they move after changing their colors to M, other robots change their colors to M until the configuration becomes one in $\forall M$. Note that when the configuration is in $\forall M$, some robots may be still moving. After the robots reach a configuration in $\forall M$, the robots change their colors to E until the configuration becomes one in $\forall E$. In the same way, the configuration changes from a configuration in $\forall E$ to a configuration in $\forall S$. This cycle is repeated until the robots reach a configuration where no robot is A_{unfair}-enabled and all robots are colored S. Each time a configuration in $\forall S$ where one or more robots are A_{unfair}-enabled is reached, at least one of them becomes active and performs A_{unfair} observing the same configuration. In consequence, this algorithm can simulate algorithms in unfair SSYNC.

Lemma 1. *Let the configuration be in $\forall S$ at time t_S and let R_e be a set of A_{unfair}-enabled-robots at t_S. After t_S, the followings hold for SIM-for-Unfair(r_i) if $R_e \neq \emptyset$.*

(1) There is a time $t_M > t_S$ at which the configuration is in $\forall M$.
(2) There are a time $t_f (t_S < t_f < t_M)$ and a non-empty subset R'_e of R_e such that all robots in R'_e perform A_{unfair} observing the same configuration and all robots in $R_e - R'_e$ do nothing between t_S and t_f.

Lemma 2. *If the configuration is in $\forall M$ at time t_M, there is a time t_E at which it is in $\forall E$.*

Note that since robots with M stay when changing their colors from M to E, the configuration at t_E is unchanged until some robots are activated after t_E.

Lemma 3. *If the configuration is in $\forall E$ at time t_E, there is a time t_S at which it is in to $\forall S$.*

Using Lemmas 1, 2 and 3, we can verify that algorithm SIM-for-Unfair(r_i) simulates $\mathcal{OBLOT}$-algorithm A_{unfair} in unfair SSYNC correctly in ASYNC with 3 colors of $\mathcal{LUMI}$-light, and if A_{unfair} uses k colors of $\mathcal{LUMI}$-light, SIM-for-Unfair(r_i) uses $3k$ colors. Then the following theorem is obtained.

Theorem 1. *SIM-for-Unfair in ASYNC with $\mathcal{LUMI}$ of $3k$ colors simulates algorithms in unfair SSYNC with $\mathcal{LUMI}$ of k colors.*

3.2 Gathering Algorithm with Simulation

In order to show that an algorithm A works in unfair SSYNC, we use a concept of potential function for A, which represents how close current configuration is to the final configuration. *A potential function* f_A for algorithm A is a function from time t ($\in \mathbb{N}$) to feature value obtained from configuration $\mathcal{C}(t)$ for the algorithm A, which is taken from a total ordered set. If the potential function f_A for algorithm A working in SSYNC is monotonically decreasing, that is, $f_A(t) > f_A(t+1)$ for any t, we can show that the algorithm A can work in unfair SSYNC.

We show that the Gathering algorithm with two colors of light shown in SSYNC in [28] can still work in unfair SSYNC by constructing potential functions.

The Gathering algorithm [28] is divided into two sub-algorithms. The first one (called ElectOneLDS) obtains a configuration that all robots are located on one straight line segment (called onLDS) from any initial configuration, and the second one (called $\mathcal{LUMI}$-Gather) is a Gathering algorithm from any initial configuration of onLDS.

We obtain an algorithm with SIM-for-Unfair, ElectOneLDS, and $\mathcal{LUMI}$-Gather by replacing the line 4 in Algorithm 1 with the line

if not onLDS **then** ElectOneLDS(r_i) **else** $\mathcal{LUMI}$-Gather(r_i).

This algorithm simulates ElectOneLDS(r_i) until onLDS is attained and once onLDS is obtained it simulates $\mathcal{LUMI}$-Gather and Gathering is completed. If ElectOneLDS and $\mathcal{LUMI}$-Gather can work in unfair SSYNC, we can show that the synthesized algorithm solves Gathering in ASYNC. We can show that potential functions are defined for ElectOneLDS and $\mathcal{LUMI}$-Gather and these functions are monotonically decreasing. However, due to lack of space, the details are stated in the full version [25].

We obtain the following theorem by Theorem 1 indicating that an algorithm with k colors in unfair SSYNC can be simulated by ASYNC robots with $3k$ color, and the result that ElectOneLDS and $\mathcal{LUMI}$-Gather work in unfair SSYNC.

Theorem 2. *Gathering can be solved in ASYNC by $\mathcal{LUMI}$ robots having 6 colors under non-rigid movement and agreement of chirality.*

4 Gathering Algorithm in ASYNC with 3 Colors

In this section, we give a Gathering algorithm called 3-color-Gather-in-ASYNC working in ASYNC with 3 colors. The pseudocode is shown in Algorithm 2. This algorithm consists of two algorithms, where one is to make onLDS and uses the simulation of ElectOneLDS (SIM-for-Unfair[ElectOneLDS]), and the other is a Gathering algorithm from onLDS and does not use the simulation and is newly developed (called $\mathcal{LUMI}$-Gather-in-ASYNC). As we will show in Corollary 1, in SIM-for-Unfair[ElectOneLDS], once a configuration becomes onLDS, it remains onLDS forever. Therefore, the algorithm works in ASYNC with 3 colors and therefore 3-color-Gather-in-ASYNC attains Gathering in ASYNC with 3 colors.

Algorithm 2. 3-color-Gather-in-ASYNC(r_i)

Assumptions: non-rigid, ASYNC,
Subroutine: SIM-for-Unfair(r_i), ElectOneLDS(r_i), $\mathcal{LUMI}$-Gather-in-ASYNC(r_i);
1: **if not** onLDS **then** SIM-for-Unfair(r_i)[ElectOneLDS]
2: **else** $\mathcal{LUMI}$-Gather-in-ASYNC(r_i)

4.1 Configurations Becoming OnLDS

In 3-color-Gather-in-ASYNC, it is switched to $\mathcal{LUMI}$-Gather-in-ASYNC from the simulation when the configuration becomes onLDS. We consider configurations which become onLDS when ElectOneLDS is simulated by SIM-for-Unfair.

Since we are concerned with ASYNC, $\mathcal{C}(t)$ contains moving robots[6] and/or robots having performed *Look* but not performing *Compute*. The former robots are said to be in *pending move* at $\mathcal{C}(t)$ and the latter robots are said to be in *pending color* at $\mathcal{C}(t)$ [8]. Then the following notations are introduced in color-configurations. In factor f of a color-configuration for $\mathcal{C}(t)$, if some robots have the possibility to be in pending move or pending color at a position represented by f, the factor is denoted by $f[pm]$ and $f[pc \rightarrow \alpha]$, respectively. If there is possibility of robots being in pending move and in pending color, the factor is denoted by $f[pm, pc]$, where $pc \rightarrow \alpha$ shows that the color is changed to α when performing *Compute*. When robots in pending move with color α move to the destination d in the factor $\alpha[pm]$, we say that $\alpha[pm]$ *has destination* d.

Lemma 4. *If the configuration becomes onLDS at t when SIM-for-Unfair simulates ElectOneLDS. It holds that*

(1) $cc(\mathcal{C}(t)) = SS^*S$,
(2) $cc(\mathcal{C}(t)) = (S|S[pc \rightarrow M]|M|M[pm])(S|S[pc \rightarrow M]|M|M[pm])^*(S|S[pc \rightarrow M]|M|M[pm])$ *with at least one* M,
(3) $cc(\mathcal{C}(t)) = (M|M[pm, pc \rightarrow E]|E)(M|M[pm, pc \rightarrow E]|E)^*(M|M[pm, pc \rightarrow E]|E)$ *with at least one* M,
(4) $cc(\mathcal{C}(t)) = (S|M)$ *with at least one* M, *or*
(5) $cc(\mathcal{C}(t)) = (M|E)$ *with at least one* M.

[6] Robots having performed *Compute* and not finishing *Move* yet.

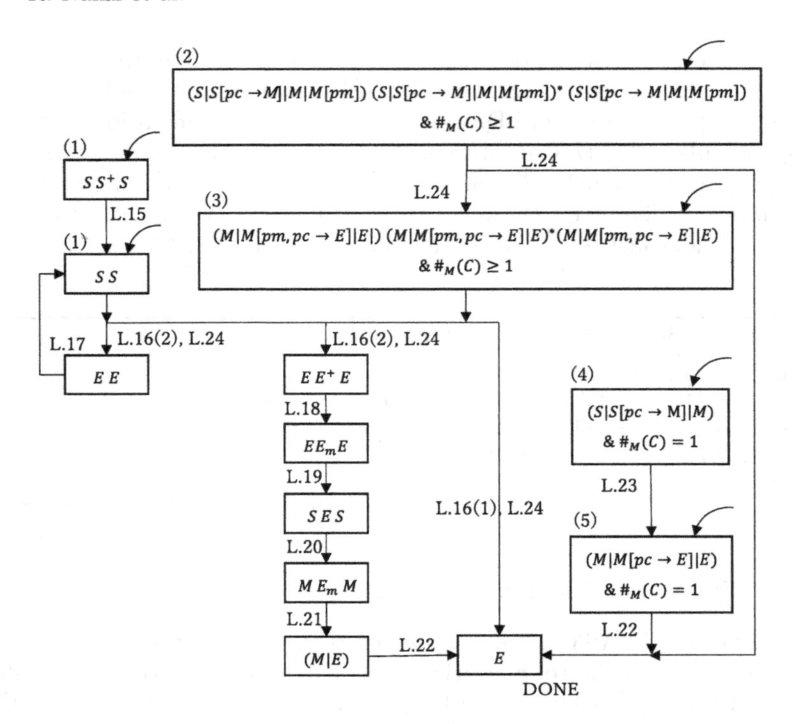

Fig. 1. Transition graph for $\mathcal{LUMI}$-Gather-in-ASYNC from $\mathcal{C}$ in Lemma 4.

In (2) and (3), all $M[pm]$ has a destination of a point on the straight line through onLDS in $\mathcal{C}(t)$.

Corollary 1. *In SIM-for-Unfair[ElectOneLDS], if the configuration becomes onLDS from non-onLDS at time t, destination of any moving robot at t is a point on the straight line through onLDS in $\mathcal{C}(t)$.*

We will show that $\mathcal{LUMI}$-Gather-in-ASYNC can work from the configurations shown in Lemma 4.

4.2 Correctness of $\mathcal{LUMI}$-Gather-in-ASYNC

$\mathcal{LUMI}$-Gather-in-ASYNC (Algorithm 3) is an extension of Algorithm $\mathcal{LUMI}$-Gather [28] so that it can work in ASYNC, and uses color-cycles similar to that of Algorithm 1. This algorithm uses 3 colors S, M, and E and its color-cycle repeats $\forall S(SS) \rightarrow \forall M \rightarrow \forall E(EE) \rightarrow \forall S(SS)$. Notations in parentheses indicate that the configuration is limited to two points.

In the algorithm, robots gather at the midpoint of some onLDS, or *Gathering point*, where configuration $\mathcal{C}$ has a Gathering point p_G if and only if $\mathcal{C}$ is in $\forall M, E$, $\#_E(\mathcal{C}) = 1$ and p_G has E. Thus, the aim of this algorithm to create Gathering point during color-cycles.

Algorithm 3. $\mathcal{LUMI}$-Gather-in-ASYNC(r_i)

Assumptions: non-rigid, $\mathcal{LUMI}$, 3 colors(S, M and E).
Input: configuration onLDS and configuration satisfying Lemma 4.

1: (Let p_n be the nearest endpoint to p_i, and let p_f be the furthest endpoint to p_i)
2: **case** $cc(\mathcal{SS}_i)$ **of**
3: $\in \forall S$:
4: **if** $cc(\mathcal{SS}_i) = SS$ **then**// $\rightarrow SMS$, SM^+S, MM^*M, SMM, or SM^*M
5: $l_i \leftarrow M$
6: $des_i \leftarrow (p_n + p_f)/2$
7: **else** //$cc(\mathcal{SS}_i) = SS^+S$
8: **if** $p_i \neq p_n$ **then** $des_i \leftarrow p_n$ //$\rightarrow SS$
9: $\in \forall S, M$:
10: **if** $cc(\mathcal{SS}_i) = M^+(S|M)M^*$, $M^*(S|M)M^+$, or $(S|M)$ **then**
11: **if** $l_i = S$ **then** $l_i \leftarrow E$ //$M^+(S|M)M^* \rightarrow M^+EM^*$ or $(S|M) \rightarrow (M|E)$
12: **else if** $cc(\mathcal{SS}_i) = (S|M)M^*(S|M)$ **and** $\#_S(\mathcal{SS}_i) = 2$ **and** $l_i = S$ **then**
13: $l_i \leftarrow M$ //$\rightarrow SM^*M$ or MM^*M
14: $des_i \leftarrow (p_n + p_f)/2$
15: **else if** $\#_S(\mathcal{SS}_i) \geq 2$ **and** $l_i = S$ **then** $l_i \leftarrow M$ //The number of S decreases.
16: $\in \forall S, E$:
17: **if** $cc(\mathcal{SS}_i) = (S|E)(S|E)$ **and** $l_i = E$ **then** $l_i \leftarrow S$ //$\rightarrow SS$
18: **else** //$(S|E)E(S|E)$
19: **if** $cc(\mathcal{SS}_i) = (S|E)E(S|E)$ **and** $\#_E(\mathcal{SS}_i) > 1$ **and** $p_i = p_n$ **and** $l_i = E$ **then**
20: $l_i \leftarrow S$ //$\rightarrow SES$
21: **else if** $cc(\mathcal{SS}_i) = SES$ **and** $l_i = S$ **then** //$\rightarrow (S|M)E(S|M)$
22: $l_i \leftarrow M$
23: $\in \forall M$:
24: $l_i \leftarrow E$ //$MM^*M \rightarrow (M|E)(M|E)^*(M|E)$
25: $\in \forall M, E$:
26: **if** $cc(\mathcal{SS}_i) = M^+(E|M)M^*$ or $M^*(E|M)M^+$ **then**
27: (Let p_E be a point with E)
28: **if** $p_i \neq p_E$ **then** $des_i \leftarrow p_E$ //Possibly the color-configuration becomes $(M|E)$.
29: **else**// $\#_E(\mathcal{SS}_i) \geq 2$ or $cc(\mathcal{SS}_i) = (M|E)$
30: **if** $l_i = M$ **then** $l_i \leftarrow E$ //$(M|E)(M|E)^*(M|E) \rightarrow EE^*E$ or $(M|E) \rightarrow E$
31: $\in \forall E$:
32: **if** $cc(\mathcal{SS}_i) = E$ **then** do nothing //Gather
33: **else if** $cc(\mathcal{SS}_i) = EE$ **then** $l_i \leftarrow S$ //$\rightarrow (S|E)(S|E)$
34: **else if** $cc(\mathcal{SS}_i) = EEE$ **then**
35: **if** $p_i = p_n$ **then**
36: $l_i \leftarrow S$ //$\rightarrow (S|E)E(S|E)$
37: **else** //$cc(\mathcal{SS}_i) = EE^+E$
38: **if** $p_i \neq p_n$ **then** $des_i \leftarrow (p_n + p_f)/2$ //$EE^+E \rightarrow EEE$
39: $\in \forall S, M, E$:
40: **if** $cc(\mathcal{SS}_i) = M^+(S|M|E)M^*$, $M^*(S|M|E)M^+$ or $(S|M|E)$ **and** $l_i = S$ **then**
41: $l_i = E$ //$M^+(S|M|E)M^* \rightarrow M^+(E|M)M^*$ or $(S|M|E) \rightarrow (M|E)$
42: **else if** $cc(\mathcal{SS}_i) = (S|M)E(S|M)$ **and** $p_i = p_n$ **and** $l_i = S$ **then**
43: $l_i = M$ //$\rightarrow MEM$
44: **endcase**

In $\forall S(SS) \rightarrow \forall M$, robots on the two points with S change their colors to M and move to the midpoint. Note that robots at the endpoints move to the midpoint only if the endpoints have S and the color of the robots is S. Gathering point is created during transitions in color-cycles for the following cases;

(1) During $\forall S(SS) \rightarrow \forall M$, robots with S look configuration $\mathcal{C}$ such that $\#_S(\mathcal{C}) = 1$.
(2) configuration $\mathcal{C}$ during $\forall M \rightarrow \forall E$.
(3) After configuration becomes $\forall E$ configuration $\mathcal{C}$ such that $\#_E(\mathcal{C}) \geq 3$.

We show how to create a Gathering point from each of (1), (2), and (3).

For (1), a robot r_i with S changes its color to E if r_i looks the configuration with $\#_S(\mathcal{C}) = 1$. From the configuration it will make a configuration such that $\#_E(\mathcal{C}) = 1$ and $\forall M, E$, and then a Gathering point is created.

For (2), Gathering point is lost if there are more than one point with E. On the other hands, Gathering point is confirmed if there is only one point with E. Let t_M be a time at which the configuration becomes $\forall M$, let $t_E(> t_M)$ be the first time at which some robot changes its color to E, and let p_E be the location having robots with E. If there are not activated robots at points except p_E between $t_M + 1$ and t_E, robots activated after t_E observe a configuration with $\#_E(\mathcal{C}) = 1$ and $\forall M, E$. Therefore, there are no robots that change their colors to E, and the Gathering point p_E is confirmed.

In (3), robots with E do not move, and the both endpoints are fixed and robots at points except the endpoints move to the midpoint. Thus the configuration becomes EEE. The transition of the configuration becomes $EEE \rightarrow SES \rightarrow MEM$, where the last configuration satisfies $\#_E(\mathcal{C}) = 1$ and in $\forall M, E$. This case also determines a Gathering point.

If the configuration becomes (1), (2) or (3), Gathering point is made. If the configuration does not become (1), (2) and (3), it becomes EE, will change SS and again begins the next color-cycle $\forall S(SS) \rightarrow \forall M \rightarrow \forall E(EE) \rightarrow \forall S(SS)$. If color-cycles are repeated, the distance of the endpoints is reduced by at least 2δ in one cycle. Therefore the distance will become less than 2δ when the color configuration becomes $\forall S(SS)$. Then, if the configuration is in $\forall M$, robots with M reach the midpoint, and Gathering is achieved.

Transitions between color configurations in Algorithm 3 are shown in Fig. 1. In this figure, boxes with numbers and $\curvearrowleft$ are starting configurations and the number corresponds to that in Lemma 4. Arrow labelled with $L.n(i)$ means it is proved in Lemma $n(i)$ which will be shown in Appendix. "DONE" means Gathering is attained.

Since we can show that Algorithm 3 can work from the configurations in Lemma 4, we have the following theorem and obtain our main result.

Theorem 3. *$\mathcal{LUMI}$-Gather-in-ASYNC solves Gathering from onLDS for $\mathcal{LUMI}$ robots having 3 colors, under non-rigid movement.*

Theorem 4. *Gathering can be solved in ASYNC by $\mathcal{LUMI}$ robots having 3 colors under non-rigid movement and agreement of chirality.*

5 Concluding Remarks

We have shown a Gathering algorithm in non-rigid and ASYNC with $\mathcal{LUMI}$ of three colors. In order to obtain the algorithm, we have shown a simulating algorithm of any algorithm in unfair SSYNC by $\mathcal{LUMI}$ of three colors in ASYNC.

We have reduced the number of colors used in the simulation to three from five, although the simulated algorithms are ones in unfair SSYNC.

The method by combining the simulation of SSYNC robots by ASYNC ones and algorithms working in SSYNC not only reduces the number of colors used in the resultant algorithm but also simplifies the proof of correctness of it. As is known from an example of ElectOneLDS, it seems to be very complicated to extend ElectOneLDS such that it can work in ASYNC and prove the correctness. However, about correctness of the synthesized algorithm, it is enough to prove the correctness of the simulation working in ASYNC because the correctness of ElectOneLDS working in SSYNC has been obtained.

One of the interesting open questions is the number of colors to solve Gathering in ASYNC, although two colors are enough to solve Rendezvous in ASYNC [18], we conjecture that three colors are necessary to solve Gathering in ASYNC.

References

1. Agmon, N., Peleg, D.: Fault-tolerant gathering algorithms for autonomous mobile robots. SIAM J. Comput. **36**(1), 56–82 (2006)
2. Ando, H., Osawa, Y., Suzuki, I., Yamashita, M.: A distributed memoryless point convergence algorithm for mobile robots with limited visibility. IEEE Trans. Robot. Autom. **15**(5), 818–828 (1999)
3. Bhagat, S., Mukhopadhyaya, K.: Optimum gathering of asynchronous robots. In: Gaur, D., Narayanaswamy, N.S. (eds.) CALDAM 2017. LNCS, vol. 10156, pp. 37–49. Springer, Cham (2017). https://doi.org/10.1007/978-3-319-53007-9_4
4. Bouzid, Z., Das, S., Tixeuil, S.: Gathering of mobile robots tolerating multiple crash faults. In: The 33rd International Conference on Distributed Computing Systems, pp. 334–346 (2013)
5. Cicerone, S., Stefano, G.D., Navarra, A.: "semi-asynchronous": a new scheduler in distributed computing. IEEE Access **9**, 41540–41557 (2021)
6. Cieliebak, M., Flocchini, P., Prencipe, G., Santoro, N.: Distributed computing by mobile robots: gathering. SIAM J. Comput. **41**(4), 829–879 (2012)
7. Das, S., Flocchini, P., Prencipe, G., Santoro, N., Yamashita, M.: Autonomous mobile robots with lights. Theor. Comput. Sci. **609**, 171–184 (2016)
8. Défago, X., Hériban, A., Tixeuil, S., Wada, K.: Using model checking to formally verify rendezvous algorithms for robots with lights in euclidean space. In: 2020 International Symposium on Reliable Distributed Systems (SRDS), pp. 113–122 (2020)
9. Défago, X., Potop-Butucaru, M.G., Clément, J., Messika, S., Parvédy, P.R.: Fault and byzantine tolerant self-stabilizing mobile robots gathering - feasibility study. CoRR abs/1602.05546, arXiv (2016)
10. Degener, B., Kempkes, B., Langner, T., auf der Heide, F.M., Pietrzyk, P., Wattenhofer, R.: A tight run-time bound for synchronous gathering of autonomous robots with limited visibility. In: 23rd ACM SPAA, pp. 139–148 (2011)
11. D'Emidio, M., Frigioni, D., Navarro, A.: Synchronous robots vs asynchronous lights-enhanced robots on graphs. Electr. Notes Theor. Comput. Sci. **322**, 169–180 (2016)
12. D'Emidio, M., Stefano, G.D., Frigioni, D., Navarra, A.: Characterizing the computational power of mobile robots on graphs and implications for the euclidean plane. Inf. Comput. **263**, 57–74 (2018)

13. Di Luna, G., Flocchini, P., Chaudhuri, S., Poloni, F., Santoro, N., Viglietta, G.: Mutual visibility by luminous robots without collisions. Inf. Comput. **254**(3), 392–418 (2017)
14. Dieudonné, Y., Petit, F.: Self-stabilizing gathering with strong multiplicity detection. Theor. Comput. Sci. **428**(13), 47–57 (2012)
15. Flocchini, P., Prencipe, G., Santoro, N.: Distributed Computing by Oblivious Mobile Robots. Morgan & Claypool (2012)
16. Flocchini, P., Prencipe, G., Santoro, N. (eds.): Distributed Computing by Mobile Entities, Current Research in Moving and Computing. Lecture Notes in Computer Science, vol. 11340. Springer, Heidelberg (2019). https://doi.org/10.1007/978-3-030-11072-7
17. Flocchini, P., Santoro, N., Viglietta, G., Yamashita, M.: Rendezvous with constant memory. Theor. Comput. Sci. **621**, 57–72 (2016)
18. Hériban, A., Défago, X., Tixeuil, S.: Optimally gathering two robots. In: Proceedings of 19th International Conference on Distributed Computing and Networking (ICDCN), pp. 1–10 (2018)
19. Izumi, T., Katayama, Y., Inuzuka, N., Wada, K.: Gathering autonomous mobile robots with dynamic compasses: an optimal result. In: Pelc, A. (ed.) DISC 2007. LNCS, vol. 4731, pp. 298–312. Springer, Heidelberg (2007). https://doi.org/10.1007/978-3-540-75142-7_24
20. Izumi, T., et al.: The gathering problem for two oblivious robots with unreliable compasses. SIAM J. Comput. **41**(1), 26–46 (2012)
21. Kamei, S., Lamani, A., Ooshita, F., Tixeuil, S.: Asynchronous mobile robot gathering from symmetric configurations without global multiplicity detection. In: Kosowski, A., Yamashita, M. (eds.) SIROCCO 2011. LNCS, vol. 6796, pp. 150–161. Springer, Heidelberg (2011). https://doi.org/10.1007/978-3-642-22212-2_14
22. Klasing, R., Markou, E., Pelc, A.: Gathering asynchronous oblivious mobile robots in a ring. Theoret. Comput. Sci. **390**(1), 27–39 (2008)
23. Lin, J., Morse, A., Anderson, B.: The multi-agent rendezvous problem. Parts 1 and 2. SIAM J. Comput. **46**(6), 2096–2147 (2007)
24. Prencipe, G.: Impossibility of gathering by a set of autonomous mobile robots. Theor. Comput. Sci. **384**(2–3), 222–231 (2007)
25. Rikuo, R., Sudo, Y., Wada, K.: Asynchronous gathering algorithms for autonomous mobile robots with lights. arXiv.org cs (ArXiv:2109.12289) (2021)
26. Souissi, S., Défago, X., Yamashita, M.: Using eventually consistent compasses to gather memory-less mobile robots with limited visibility. ACM Trans. Auton. Adapt. Syst. **4**(1), 1–27 (2009)
27. Suzuki, I., Yamashita, M.: Distributed anonymous mobile robots: formation of geometric patterns. SIAM J. Comput. **28**, 1347–1363 (1999)
28. Terai, S., Wada, K., Katayama, Y.: Gathering problems for autonomous mobile robots with lights. arXiv.org cs (ArXiv:1811.12068) (2018)
29. Viglietta, G.: Rendezvous of two robots with visible bits. In: Flocchini, P., Gao, J., Kranakis, E., Meyer auf der Heide, F. (eds.) ALGOSENSORS 2013. LNCS, vol. 8243, pp. 291–306. Springer, Heidelberg (2014). https://doi.org/10.1007/978-3-642-45346-5_21

Synchronization Modulo k in Dynamic Networks

Louis Penet de Monterno[1(✉)], Bernadette Charron-Bost[2], and Stephan Merz[3]

[1] École polytechnique, IP Paris, 91128 Palaiseau, France
penetdemonterno@lix.polytechnique.fr
[2] DI ENS, CNRS, École Normale Supérieure, 45 rue d'Ulm, 75005 Paris, France
[3] Université de Lorraine, CNRS, Inria, LORIA, 54000 Nancy, France

Abstract. We define the mod k-synchronization problem as a weakening of the Firing Squad problem, where all nodes fire not at the same round, but at rounds that are all equal modulo k. We propose an algorithm that achieves mod k-synchronization in any dynamic network where there exist – possibly several – fixed spanning stars within each period of Δ consecutive rounds. In other words, we require that there always exists a temporal path of length at most Δ between some fixed node γ and every other node. As opposed to the perfect synchronization achieved in the Firing Squad problem, mod k-synchronization thus does not require any strong connectivity property in the network. In our algorithm, all the nodes "know" Δ, but they ignore what nodes are the centers of the spanning stars. We also prove that if the bound Δ for guaranteeing fixed spanning stars exists but is unknown to the agents, then mod k-synchronization is impossible.

All nodes in our algorithm fire in less that $6kn + 4k$ rounds after all nodes become active, but unfortunately uses unbounded counters. We then propose a refinement of this algorithm so that it becomes finite state while maintaining the same time complexity. The correctness of our first algorithm has been formally established in the proof assistant Isabelle.

1 Introduction

Distributed algorithms are often designed in a synchronous computing model, in which computation is divided into *communication-closed rounds*: any message sent at some round can be received only at that round. In this model, it is usually assumed that each run of an algorithm is started by all nodes simultaneously, i.e., at the same round, or even at round one. For instance, most synchronous consensus algorithms (e.g., [8,13,14]), as well as many distributed algorithms for dynamic networks (e.g., [10,11]) require synchronous starts.

This assumption makes the sequential composition of two distributed algorithms $A; B$ – in which each node starts executing B when it has completed the execution of A – quite problematic. Indeed, nodes start the algorithm B asynchronously when the algorithm A terminates asynchronously, and the properties of B are no more guaranteed in this context of asynchronous starts.

C. Johnen et al. (Eds.): SSS 2021, LNCS 13046, pp. 425–439, 2021.
https://doi.org/10.1007/978-3-030-91081-5_28

This leads to the problem of simulating synchronous starts, classically referred to as the *firing squad problem*: Each node is initially *passive* and then becomes *active* at an unpredictable round. The goal is to guarantee that the nodes, when all active, eventually synchronize by *firing* – i.e., entering a designated state for the first time – at the same round.

Unfortunately, the impossibility result in [4] demonstrates that the firing squad problem is not solvable without a strong connectivity property of the network, namely, there exists some positive integer Δ such that the communication graph within every period of Δ consecutive rounds is strongly connected and the bound on the delay Δ is "known"[1]. In many situations, this connectivity property is not guaranteed: as an example, in the dynamic graphs corresponding to the Heard-Of models for benign failures, a node that suffers permanent and complete send omissions is constantly a sink in the communication graph.

However, looking more closely at many distributed algorithms designed in the round-based model, we see that these algorithms actually do not require perfectly synchronous starts, and still work under the weaker condition that all the nodes start executing the algorithms in rounds with numbers that are equal modulo k, for some positive integer k. The corresponding synchronization problem, that we call mod k-*synchronization*, is formally specified as follows:

Termination. If all nodes are eventually active, then every node eventually fires.

mod k-simultaneity. If two nodes fire at round t and t', then $t' \equiv t \mod k$.

Indeed, let A be an algorithm organized into regular *phases* consisting of a fixed number k of consecutive rounds: the sending and transition functions of every node at round t are entirely determined by the value of t modulo k. Moreover, assume that A has been proved correct (with respect to some given specification) when all nodes start A synchronously (at round one), but with any dynamic graph in a family $\mathcal{G}$ that is stable under the addition of arbitrary finite prefixes. For instance, the *ThreePhaseCommit* algorithm for non-blocking atomic commitment [1], as well as the consensus algorithms in [9] or the *LastVoting* algorithm [6] – corresponding to the consensus core of *Paxos* – fulfill all the above requirements for phases of length $k = 3$ and $k = 4$, respectively, and the family $\mathcal{G}$ of dynamic graphs in which there exists an infinite number of "good" communication patterns (e.g., a sequence of $2k$ consecutive communication graphs in which a majority of nodes is heard by all nodes in each graph). The use of a mod k-synchronization algorithm prior to the algorithm A yields a new algorithm that executes exactly like A does, after a finite preliminary period during which every node becomes active and fires. The above property on the set of dynamic graphs $\mathcal{G}$ then guarantees this variant of A to be correct with asynchronous starts and dynamic graphs in $\mathcal{G}$.

Another typical example for which perfect synchronization can be weakened to synchronization modulo k is the development of the basic *rotating coordinator* strategy in the context of asynchronous starts. Roughly speaking, this

[1] in a sense that will be detailed in Sect. 3.5.

strategy consists in the following: if nodes have unique identifiers in $\{1, \ldots, n\}$, the coordinator at round t is the node whose identifier is t modulo n. For that, each node u maintains a local counter c_u whose current value is the number of rounds in which it has been active. At each round, the coordinator of u is the node with the identifier that is equal to the current value of c_u modulo n. Since there may be only one coordinator per round, such a selection rule requires synchronous starts. Clearly, with the use of a mod n-synchronization algorithm in a preliminary phase and a counter for each node that now counts the number of rounds elapsed since the node fired, the above scheme implements the rotating coordinator strategy from the first round where all nodes have fired.

A natural question is then whether synchronization modulo k may be achieved without strong connectivity. In this paper, we address this issue and show that this problem is solvable under the sole assumption of a fixed center γ with a fixed and "known" bound on the delay Δ, that is, every node receives a message from γ (possibly indirectly) in every period of Δ consecutive rounds. In fact, we exhibit an algorithm, denoted by $SynchMod_k$, that achieves synchronization modulo k in any dynamic graph with a fixed center and a delay at most equal to k. The case where $\Delta > k$ will be covered separately, in Sect. 3.5. Interestingly, our algorithm requires no node identifiers. In particular, nodes are not assumed to "know" what node is the center of the graph. Provided that the communication graph is centered with delay at most Δ, no other assumption is made on the dynamic graph.

The correctness proof of our algorithm relies on a series of preliminary lemmas that consider all the possible cases for the respective values of the variables in the algorithm. In order to increase our confidence in the correctness and remove any doubts on such combinatorial proofs, we have developed a formal proof of the correctness of our algorithm in the interactive theorem prover Isabelle/HOL [12].

2 Preliminaries

2.1 The Computational Model

We consider a networked system with a *fixed* set V of n nodes. We assume a round-based computational model in the spirit of the Heard-Of model [6], in which point-to-point communications are organized into *synchronized rounds*: each node sends messages to all nodes and receives messages sent by *some* of the nodes. Rounds are communication closed in the sense that no node receives messages in round t that are sent in a round different from t. The collection of communications (which nodes receive messages from which nodes) at each round t is modelled by a directed graph (digraph, for short) with a set of nodes equal to V. The digraph at round t is denoted by $\mathbb{G}(t) = (V, E_t)$, and is called the *communication graph at round t*. The set of u's incoming neighbors in the digraph $\mathbb{G}(t)$ is denoted by $In_u(t)$.

We assume a self-loop at each node in all these digraphs since every node can communicate with itself instantaneously. The sequence of such digraphs $\mathbb{G} = (\mathbb{G}(t))_{t \geq 1}$ is called a *dynamic graph* [3].

In round t ($t = 1, 2, \ldots$), each node u successively (a) broadcasts messages determined by its state at the beginning of round t, (b) receives *some* of the messages sent to it, and finally (c) performs an internal transition to a successor state. A *local algorithm* for a node is given by a *sending function* that determines the messages to be sent in step (a) and a *transition function* for state updates in step (c). An *algorithm* for the set of nodes V is a collection of local algorithms, one per node.

We also introduce the notion of *start schedules*, represented as collections $\mathbb{S} = (s_u)_{u \in V}$, where each s_u is a positive integer or is equal to ∞.

The execution of an algorithm A with the dynamic graph $\mathbb{G}$ and the start schedule $\mathbb{S}$ then proceeds as follows: Each node u is initially *passive*. If $s_u = \infty$, then the node u remains passive forever. Otherwise, s_u is a positive integer, and u becomes *active* at the beginning of round s_u, setting up its local variables. In round t ($t = 1, 2 \ldots$), a passive node sends only heartbeats, corresponding to *null* messages, and cannot change its state. An active node applies its sending function in A to its current state to generate the messages to be sent, then it receives the messages sent by its incoming neighbors in the directed graph $\mathbb{G}(t)$, and finally applies its transition function $\mathcal{T}_u$ in A to its current state and the list of messages it has just received (including the null messages from passive nodes), to compute its next state. Since each local algorithm is deterministic, an execution of the algorithm A is entirely determined by the initial state of the network, the dynamic graph $\mathbb{G}$, and the start schedule $\mathbb{S}$.

The states "passive" and "active" do not refer to any physical notion, and are relative to the algorithm under consideration: as an example, if two algorithms A and B are sequentially executed according to the order "A followed by B", then at some round, a node may be active w.r.t. A while it is passive w.r.t. B. In such a situation, the node is integrally part of the system and can send messages, but these messages are empty with respect to the semantics of the algorithm B.

2.2 Network Model and Start Model

Let us first recall the notion of *product* of two digraphs $G_1 = (V, E_1)$ and $G_2 = (V, G_2)$, denoted by $G_1 \circ G_2$ and defined as follows [5]: $G_1 \circ G_2$ has V as its set of nodes, and (u, v) is an edge if there exists $w \in V$ such that $(u, w) \in G_1$ and $(w, v) \in G_2$. For any dynamic graph $\mathbb{G}$ and any integer $t' > t \geq 1$, we let $\mathbb{G}(t : t') = \mathbb{G}(t) \circ \mathbb{G}(t+1) \circ \cdots \circ \mathbb{G}(t')$. By extension, we let $\mathbb{G}(t : t) = \mathbb{G}(t)$.

The set of incoming neighbors of u in $\mathbb{G}(t : t')$ is noted as $In_u(t : t')$. The set $In_u(t : t)$ is simply noted $In_u(t)$.

Each edge (u, v) in the digraph $\mathbb{G}(t : t')$ corresponds to a $u \triangleright v$ *path in the interval* $[t, t']$, i.e., a finite sequence of nodes $u = w_{t-1}, w_t, \ldots, w_{t'} = v$ such that each pair (w_i, w_{i+1}) is an edge of $\mathbb{G}(t + i)$. This path is said to be *active* if each node $w_{t-1}, w_t, \ldots, w_{t'}$ is active in rounds $t - 1, t, \ldots t'$, respectively.

A *network model* is any non-empty set of dynamic graphs. We will focus on those network models $\mathcal{G}^*_\Delta$ of dynamic graphs $\mathbb{G}$ where each digraph $\mathbb{G}(t : t+\Delta-1)$ contains a *fixed* star graph, namely,

$$\exists \gamma \in V, \forall t \in \mathbb{N}, \forall u \in V, \gamma \in In_u(t : t + \Delta - 1).$$

The dynamic graph $\mathbb{G}$ is said to be *centered at* the node γ with delay Δ, and γ is called a Δ-*center of* $\mathbb{G}$.

The network model $\mathcal{G}^*_\Delta$ contains some dynamic graphs which are partitionned during less than Δ consecutive rounds. If the network model containing dynamic graphs which are rooted in each round is denoted by $\mathcal{G}^{rooted}$, we can easily check that, because of self-loops, if a node γ is a root of each digraph $\mathbb{G}(t)$, then the dynamic graph $\mathbb{G}$ is centered at γ with delay $|V| - 1$. Then $\mathcal{G}^{rooted} \subseteq \mathcal{G}^*_{|V|-1}$. Similarly, if the network model containing dynamic graphs which are strongly connected in each round is denoted by $\mathcal{G}^{strong}$, we get $\mathcal{G}^{strong} \subseteq \mathcal{G}^{rooted} \subseteq \mathcal{G}^*_{|V|-1}$.

We also define a *start model* as a non-empty set of start schedules. A start schedule $\mathbb{S} = (s_u)_{u \in V}$ is *complete* if every s_u is finite, i.e., no node is passive forever. Synchronous starts correspond to complete start schedules where all s_u are finite and equal. The point of this paper is to simulate mod k*-synchronous starts* defined by $s_u \equiv s_v \bmod k$ for every pair of nodes u and v, with any complete start schedule.

The algorithm we introduce in the next section requires the existence of a Δ-center. By comparison, the firing squad problem is solvable with some strong connectivity hypothesis. In other words, every node must be a Δ-center.

3 The Algorithm

In this section, we present simultaneously the pseudo-code of our algorithm, and its formal definition in the Isabelle framework. The correctness of the algorithm has been formally verified in Isabelle.[2] The proof that we present in this article closely follows our formal proof.

3.1 Pseudo-code and Formal Definition

The state of each node is represented by five variables whose initial value is given below.

record *locState* =
  x :: nat
  synch :: bool
  ready :: bool
  force :: nat — *force* $\in \{0, 1, 2\}$
  level :: nat — *level* $\in \{0, 1, 2\}$
definition *initState* **where**
  initState ≡ ⦇ *x = 0, synch = False, ready = False, force = 0, level = 0* ⦈

We define a datatype for messages sent between two nodes u and v: messages either carry a value of some type *'msg*, or are equal to *Null* if u is passive, or to *Void* if u is not an incoming neighbor of v.

datatype *'msg message = Content 'msg | Null | Void*

[2] The complete Isabelle development is available at https://github.com/louisdm31/asynchronous_starts_HO_model/tree/master/proof/sync-mod.

Algorithm 1: The $SynchMod_k$ algorithm

1 **Initialization:**
2 $c_u \in \mathbb{N}$, initially 0
3 $synch_u \leftarrow false$
4 $ready_u \leftarrow false$
5 $force_u \in \{0, 1, 2\}$, initially 0
6 $level_u \in \{0, 1, 2\}$, initially 0
7 **At each round:**
8 send $\langle c_u, synch_u, force_u, ready_u \rangle$ to all
9 receive incoming messages: let In^a be the set of nodes from which a non-null message is received.
10 **if** *all received messages are non-null* **then**
11 $synch_u \leftarrow \bigwedge_{v \in In^a} synch_v \wedge c_v \equiv c_u \mod k$
12 **end**
13 **else**
14 $synch_u \leftarrow false$
15 **end**
16 $ready_u \leftarrow \bigwedge_{v \in In^a} ready_v$
17 $force_u \leftarrow \max_{v \in In^a} force_v$
18 $c_u \leftarrow 1 + \min_{\substack{v \in In^a \\ force_v = force_u}} c_v$
19 **if** $c_u \equiv 0 \mod k$ **then**
20 **if** $level_u = 0 \wedge synch_u$ **then**
21 $level_u \leftarrow 1$
22 **if** $force_u < 2$ **then**
23 $force_u \leftarrow 1$
24 $c_u \leftarrow 0$
25 **end**
26 **end**
27 **else if** $level_u = 1 \wedge ready_u \wedge synch_u$ **then**
28 $level_u \leftarrow 2$ /* the node u fires */
29 $force_u \leftarrow 2$
30 $c_u \leftarrow 0$
31 **end**
32 $synch_u \leftarrow true$
33 $ready_u \leftarrow level_u > 0$
34 **end**

3.2 Informal Description of the Algorithm

We fix some $k > 2$. In this algorithm, the nodes hold a *level* variable. When they become active, they move from passive state to level 0. They later move to level 1, then to level 2. Each time a node moves from some level to the next, this constitutes a *level-up event*. From now on, the level reached during this level-up event will be called the *strength* of this event. Reaching level 2 means firing. The conditional statements at lines 20 and 27 of Algorithm 1 are executed when the node reaches level 1 and 2 respectively. The intuition of the algorithm can be summarized by two simple ideas.

Firstly, *each node keeps track of the most recent strongest level-up event.* Only the strongest level-up events are considered: if some node "knows" about a level-up event from level 1 to level 2, it will not record any level-up event from level 0 to level 1, nor any level-up event from passive state to level 0. Among the strongest level-up events, the nodes keep track of the age of the most recent one. For that purpose, they hold two variables c_u and $force_u$. At any round, node u knows that c_u rounds ago, some node reached a level equal to $force_u$ from the previous level (as proved in Lemma 6), and the node does not know any node which reached a level equal to $force_u$ (or higher) in any more recent round (as proved in Lemma 7). With lines 17 and 18, they update their c_u and $force_u$ variables using those of their incoming neighbors. The presence of self-loops implies that, in these lines, the minima and maxima are well-defined.

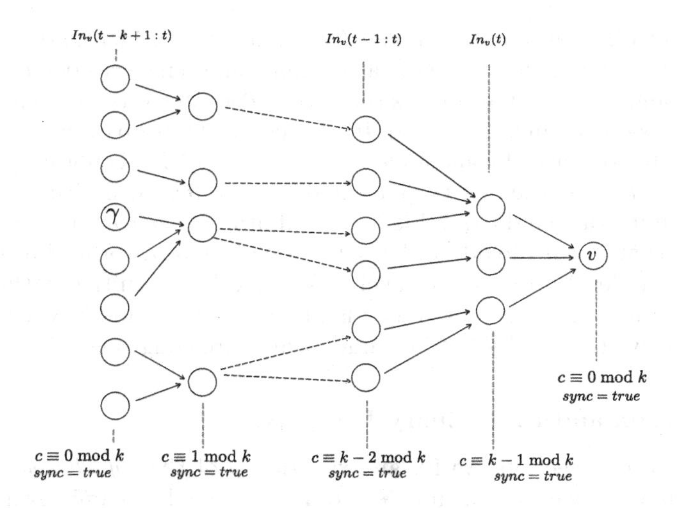

Fig. 1. Evolution of the incoming neighbors of u between round $t-k$ and t: case where every c_u is congruent to 0 in round $t-k$

Secondly, *a node may level up in round t only if its counter c_u is congruent to 0 and the counter of γ was also congruent to 0 k rounds ago.* Since the nodes do not "know" a fixed Δ-center, they conservatively level up only if all of their incoming neighbors $v \in In_u(t-k+1:t)$ were congruent to 0 k rounds ago. The assumption $\Delta \leq k$ guarantees that γ is one of these incoming neighbors. For that purpose, they use a Boolean variable *synch*. When the counter of some node v becomes congruent to 0 in some round $t-k$, it sets its $synch_v$ variable to *true* in line 32. During the next $k-1$ rounds, it will check whether the counters of its incoming neighbors are all congruent to its own counter (line 11). In case they are not, the node will set its $synch_u$ variable to *false*. This *false* value will

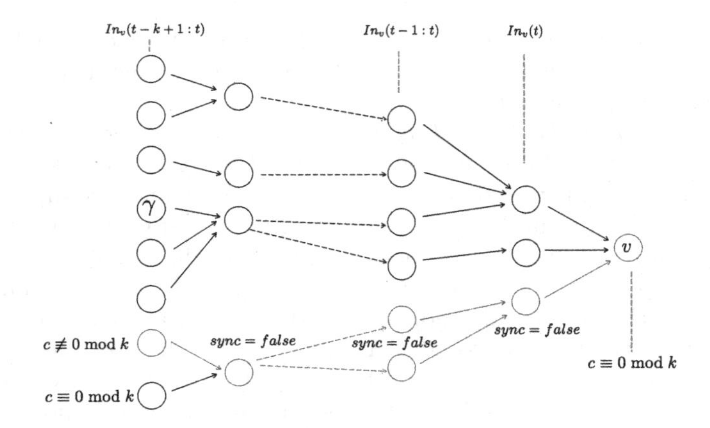

Fig. 2. Evolution of the incoming neighbors of u between round $t-k$ and t: case where some c_u are not congruent to 0 in round $t-k$

disseminate to its outgoing neighbors (also line 11). If, in round t, its $synch_u$ variable is still true, node u knows that no non-congruence was detected between round $t-k$ and round t. This means that every Δ-center was congruent with 0 in round $t-k$ (as proved in Lemma 3.c). In that case, a level-up event will take place (see Fig. 1). In contrast, if some node $v \in In_u(t-k+1:t)$ is not congruent with 0 in round $t-k$, then the line 11 guarantees that $synch_u$ will ultimately be *false* at the beginning of round t (see Fig. 2). In addition to *synch*, the *ready* variable makes sure that γ was already in level 1 k rounds ago (as proved in Lemma 4). Otherwise, the level-up event to level 2 is forbidden. Intuitively, the round t_γ in which γ reaches level 1 is used as a landmark for the mod k-synchronization: Lemma 9 shows that nodes fire in rounds which are congruent to t_γ modulo k.

3.3 Notation and Preliminary Lemmas

In the rest of this section, we fix an execution ρ of the $SynchMod_k$ algorithm for a complete activation schedule $\mathbb{S}$ and a Δ-centered dynamic graph $\mathbb{G} \in \mathcal{G}^*_\Delta$ with $\Delta \leq k$. Let $s^{\max} = \max_{u \in V} s(u)$ (note that $s^{\max} < \infty$) and let γ denote any Δ-center of $\mathbb{G}$.

If the node u is active in round t, the value of any u's variable x_u just before u executes line 19 at round t and at the very end of round t are denoted by $x_u^{pre}(t)$ and $x_u(t)$ respectively. By extension, $x_u(t)$ refers to the initial state if $t = s_u - 1$. In our formal proof, these values are encapsulated in a *rho* variable: for any round t, for any node u, *rho t u* returns either *Passive* or *Active s*, where *s :: locState* contains $c_u(t), synch_u(t) \ldots$. We now prove that this execution satisfies both properties of the mod k-synchronization problem.

definition *liveness* **where** — *termination*
  liveness rho $\equiv \forall$ *u.* $\exists$ *t s. rho t u = Active s* $\wedge$ *level s = 2*
definition *safety* **where** — mod *k-simultaneity*
  safety rho $\equiv \exists$ *c.* $\forall$ *u t s ss.*

rho t u = Active s $\longrightarrow$ *level s* $< 2 \longrightarrow$
rho (Suc t) u = Active ss $\longrightarrow$ *level ss* $= 2 \longrightarrow$ *t mod k = c*

We proved these propositions under the following assumptions:

assumes $\forall$ *u t. path In gamma u t k* — *gamma is a k-center*
and $\forall$ *u t.* $u \in$ *In t u* — *the graph contains self-loops*
and *HORun (HOMachine k) rho In* — *rho is an execution*
and $\forall$ *p.* $\exists$ *t. rho t p* $\neq$ *Asleep* — *the schedule is complete*
and $k > 2$

The *HORun* term above is defined in [7] and characterizes executions of an algorithm. Since this definition was first written for synchronous starts, we adapted it to describe asynchronous starts.

We denote $In_u^a(t)$ the subset of nodes in $In_u(t)$ which are active in round $t-1$ in this execution. Some simple claims follow immediately from the definition of the transition function, regardless of the connectivity properties of $\mathbb{G}$. We consider some node $u \in V$ and some round t in which u is active (i.e., $t \geq s_u$).

Lemma 1.

(a) $level_u(t+1) \in \{level_u(t), level_u(t)+1\}$
(b) If $c_u(t) \neq 0$*, then* $force_u(t) = force_u^{pre}(t)$ *and* $c_u(t) = c_u^{pre}(t)$.
(c) $c_u(t) \equiv c_u^{pre}(t) \mod k$.
(d) If $synch_u^{pre}(t) = true$ *holds, then each node* $v \in In_u(t)$ *is active at round* $t-1$ *with:* $c_v^{pre}(t-1)+1 \equiv c_u^{pre}(t) \mod k$.
(e) If $c_u^{pre}(t) \not\equiv 1 \mod k$ *and* $synch_u^{pre}(t)$ *holds, then each node* $v \in In_u(t)$ *is active in round* $t-1$ *with* $synch_v^{pre}(t-1)$.
(f) If $c_u^{pre}(t) \not\equiv 1 \mod k$ *and* $synch_u^{pre}(t) = ready_u^{pre}(t) = true$*, then for every node* $v \in In_u^a(t)$*, it holds that* $ready_v^{pre}(t-1) = true$.
(g) For every $v \in In_u^a(t)$*, we have:*
$$force_v^{pre}(t-1) \leq force_v(t-1) \leq force_u^{pre}(t) \leq force_u(t).$$
(h) $\forall v \in In_u^a(t)$, $force_v^{pre}(t-1) = force_u^{pre}(t) \Rightarrow$
$$c_u^{pre}(t) \leq 1 + c_v(t-1) \leq 1 + c_v^{pre}(t-1).$$
(i) $level_u(t) \leq force_u(t)$.

Lemma 2. *No node can perform a level-up event action in round* $k-1$ *or earlier.*

We now show a few properties on the incoming neighbors of nodes that reach level 1 or 2. This situation is illustrated in Fig. 1.

Lemma 3. *Let* i *be an integer,* $0 \leq i < k$*, and let* u *and* v *be two nodes such that* $u \in In_v(t-k+i+1:t)$*. If* v *is active in round* t*, if* $c_v^{pre}(t) \equiv 0 \mod k$ *and* $synch_v^{pre}(t) = true$ *hold, then*

(a) $t \geq k$.
(b) u *is active in round* $t-k+i$.
(c) $c_u^{pre}(t-k+i) \equiv i \mod k$.
(d) If $ready_v^{pre}(t)$ *is true and* $i > 0$*, then* $ready_u^{pre}(t-k+i)$ *is true as well.*

Lemma 4. *If some node* u *reaches level 2 in round* t_u*, then* γ *is already in level 1 in round* t_u.

Lemma 5. *If γ reaches level 1 in round t_γ, no node can reach level 1 or 2 in any of the rounds $t_\gamma + 1, \ldots, t_\gamma + k - 1$.*

Lemma 6. *Let u be some node, and t be some round in which u is active. There exists some node w which reached a level equal to $force_u^{pre}(t)$ in round $t - c_u^{pre}(t)$. Moreover, an active $w \rhd u$ path exists in the interval $[t - c_u^{pre}(t) + 1, t]$.*

We consider the set $Z = \{(f, t), \exists u \in V, level_u(t) = f \wedge level_u(t-1) \neq f\}$. This set is the finite set of level-up events. Using Lemma 6, any node u satisfies $z_u(t) = (force_u^{pre}(t), t - c_u^{pre}(t)) \in Z$ in every round $t \geq s_u$ in which u is active. We order Z lexicographically. The following two lemmas prove that $z_u(t)$ is the most recent strongest level-up event "known" by u in round t.

Lemma 7. *For every node u and v, if u leveled up in round t, then for every $i > 0$ such that there exists an active $u \rhd v$ path in the interval $[t+1, t+i]$,*

$$\begin{aligned} & level_u(t) \leq force_v^{pre}(t+i) \\ \wedge\ & level_u(t) = force_v^{pre}(t+i) \Rightarrow c_v^{pre}(t+i) \leq i. \end{aligned}$$

Lemma 8. *If there exists an active $u \rhd v$ path between two nodes u and v in the interval $[t+1, t']$, then $z_u(t) \leq z_v(t')$.*

Lemma 9. *If γ reached level 1 in some round t_γ, whereas some u reaches level 1 or 2 in some round $t_u \geq t_\gamma$, then $t_u \equiv t_\gamma \mod k$.*

Proof. By contradiction, we consider the earliest node u which levels up in some round $t_u \geq t_\gamma$ with $t_u \not\equiv t_\gamma \mod k$. By Lemma 2, $t_\gamma \geq k$. The Lemma 6 implies the existence of a node v which reached a level equal to $force_u^{pre}(t_u)$ in some round $t_v = t_u - c_u^{pre}(t_u)$.

In the case $force_u^{pre}(t_u) = 2$, from Lemma 4, we obtain $t_v \geq t_\gamma$.

In the case $force_u^{pre}(t_u) = 1$, Lemma 5 tells us that $t_u - t_\gamma \geq k$. Using self-loops and $\mathbb{G} \in \mathcal{G}_\Delta^*$ respectively, there exists a $\gamma \rhd \gamma$ path in the interval $[t_\gamma + 1, t_u - k]$ and a $\gamma \rhd u$ path in the interval $[t_u - k + 1, t_u]$. By concatenation, we obtain a $\gamma \rhd u$ path in the interval $[t_\gamma + 1, t_u]$ Using Lemma 3.b, this path is active. From Lemma 7, $c_u^{pre}(t_u) \leq t_u - t_\gamma$. We also get $t_v \geq t_\gamma$.

The case $force_u^{pre}(t_u) = 0$ is impossible: we have $force_\gamma(t) \geq level_\gamma(t) \geq 1$ by Lemma 1.i. Using Lemma 1.g, we get $1 \leq force_\gamma(t) \leq force_{w_{t+1}}^{pre}(t+1) \leq \cdots \leq force_u^{pre}(t_u)$, where $w_t, w_{t+1}, \ldots, w_{t_u}$ is the $\gamma \rhd u$ path constructed above.

In both possible cases, we have $t_v \geq t_\gamma$. By line 19, we have $c_u^{pre}(t) \equiv 0 \mod k$. Recalling $t_v = t_u - c_u^{pre}(t_u)$, we obtain $t_v \equiv t_u \not\equiv t_\gamma \mod k$. This contradicts the fact that u was the earliest such node. □

We say that the system is *monovalent* in round t if every node u is active and the values in the family $(c_u^{pre}(t))_{u \in V}$ are mutually congruent modulo k. Moreover, we denote $\bar{c}^{pre}(t)$ some integer which is congruent to every value $(c_u^{pre}(t))_{u \in V}$.

Lemma 10. *If the system is monovalent in round t, it is monovalent in any round $t + i$. Moreover, $\bar{c}^{pre}(t+i) \equiv \bar{c}^{pre}(t) + i \mod k$.*

Lemma 11. *If, in some round t, the system is monovalent, then every node u is in level 1 in round $t + 2k$ and in level 2 in round $t + 3k$.*

3.4 Correctness Proof

Lemma 12. *Under the assumption of a Δ-centered dynamic graph with $\Delta \leq k$, any execution of the SynchMod$_k$ algorithm satisfies the* mod k*-simultaneity property.*

Proof. We fix some node u, and we assume that u reaches level 2 in round t_u. From Lemma 4, we obtain $t_u \geq t_\gamma$, where t_γ is the round in which γ reaches level 1. By Lemma 9, $t_u \equiv t_\gamma \mod k$. That proves the mod k-simultaneity property. □

Lemma 13. *Under the assumptions of a complete activation schedule and of a Δ-centered dynamic graph with $\Delta \leq k$, any execution of the SynchMod$_k$ algorithm terminates.*

Proof. For every node u, the sequence $(z_u(t))_{t \geq s_u}$ belongs to the finite set Z. Moreover, by Lemma 8, this sequence is non-decreasing. Then it eventually stabilizes to some value z_u^{max}. Let z^{min} be $min\{z_u^{max}, u \in V\}$. We consider the round t^0 in which every node is active, and every sequence $(z_u(t))_{t \geq s_u}$ has stabilized to z_u^{max}. We consider the subset $V_{min} = \{u \in V, z_u^{max} = z^{min}\}$. We claim that $\forall t > t^0, \forall u \in V_{min}, In_u(t) \subseteq V_{min}$:

By contradiction, if in some round $t > t^0$, some $w \notin V_{min}$ belongs to $In_u(t)$, we would obtain $z_u^{max} = z^{min} < z_w(t-1) \leq z_u(t)$, using $u \in V_{min}$, $w \notin V_{min}$ and Lemma 8.

We apply Lemma 11 to the subsystem consisting of V_{min}. Since for all $t > t^0$ and $u \in V_{min}$, $In_u(t) \subseteq V_{min}$, this subsystem behaves like an independent system. Then, in round $t^0 + 3k$, every node in V_{min} is in level 2. By Lemma 1.i, every node $u \in V_{min}$ satisfies $force_u^{pre}(t^0 + 3k) = 2$. By definition of V_{min}, every node $u \in V$ has $force_u^{pre}(t^0 + 3k) = 2$. Now, we prove that in round $t^0 + 3k$, the entire system is monovalent:

Let us consider two nodes u_1 and u_2. By Lemma 6, we obtain two nodes w_1 and w_2 which reached level 2 in round $t^0 + 3k - c_{u_1}^{pre}(t^0 + 3k)$ and $t^0 + 3k - c_{u_2}^{pre}(t^0 + 3k)$ respectively. By Lemma 12, we obtain $c_{u_1}^{pre}(t^0 + 3k) \equiv c_{u_2}^{pre}(t^0 + 3k) \mod k$. That proves monovalence.

The termination property now follows from Lemma 11. □

The previous two lemmas yield the following correctness theorem:

Theorem 1. *Under the assumption of a Δ-centered dynamic graph with $\Delta \leq k$, and a complete activation schedule, the SynchMod$_k$ algorithm solves the* mod k*-synchronization problem for any integer k greater than 2.*

3.5 Solvability Results

We show that the mod k-synchronization problem is always solvable, regardless of the value of k, if the bound Δ on the delay is known: for each possible Δ, we can exhibit an algorithm which solves mod k-synchronization in any Δ-centeed dynamic graph.

Corollary 1. *For any positive integer k, the* mod k*-synchronization problem is solvable in each network model $\mathcal{G}^*_\Delta$ in any complete activation schedule.*

Proof. Depending on the relative values of k and Δ, we consider the following cases:

1. $k = 1$. The problem is trivially solvable in any network model, in particular $\mathcal{G}^*_\Delta$.
2. $\Delta \leq k$ and $k > 2$. By Theorem 1, the *SynchMod*$_k$ algorithm solves the mod k-synchronization problem in $\mathcal{G}^*_\Delta$ if $k > 2$.
3. $\Delta \leq k = 2$. Theorem 1 shows that the *SynchMod*$_4$ algorithm achieves mod 4-synchronization in $\mathcal{G}^*_2$, and hence achieves mod 2-synchronization in $\mathcal{G}^*_2$.
4. $\Delta > k$. We have $\Delta \leq \lceil \frac{\Delta}{k} \rceil \cdot k$. By Theorem 1, the mod $\lceil \frac{\Delta}{k} \rceil \cdot k$-synchronization problem is solvable in $\mathcal{G}^*_\Delta$ using *SynchMod*$_{\lceil \frac{\Delta}{k} \rceil \cdot k}$. The mod k-synchronization problem is also solvable in $\mathcal{G}^*_\Delta$, *a fortiori*. □

In contrast, we show that the mod k-synchronization problem is not solvable if the delay Δ is unknown to the nodes.

Theorem 2. *If $k > 1$, then the* mod k*-synchronization problem is not solvable in the network model $\bigcup_{i \in \mathbb{N}} \mathcal{G}^*_i$.*

Proof. By contradiction, assume that an algorithm A solves the problem in the above-mentioned network model. We consider any system and we fix two nodes u and v in this system. We denote I the digraph only containing self-loops. We denote C_u and C_v the digraphs only containing self-loops and a star centered in u and v respectively. We construct four executions of A:

1. Every node starts in round 1. The dynamic graph is equal to C_u at each round. This dynamic graph belongs to $\mathcal{G}^*_1$. Using the termination of A, u fires in some round f_u.
2. Every node starts in round 1. The dynamic graph is equal to C_v at each round. This dynamic graph belongs to $\mathcal{G}^*_1$. Using the termination of A, v fires in some round f_v.
3. Every node starts in round 1. During the first $f_u + f_v$ rounds, the communication graph is equal to I. In every subsequent round, the communication graph is equal to C_u. This dynamic graph belongs to $\mathcal{G}^*_{1+f_u+f_v}$.
4. the node u starts in round 1, whereas every other node starts in round 2. During the first $f_u + f_v$ rounds, the communication graph is equal to I. In every subsequent round, the communication graph is equal to C_u. This dynamic graph belongs to $\mathcal{G}^*_{1+f_u+f_v}$.

From the point of view of u, the third execution is indistinguishable from the first execution. Then u fires in round f_u in the third execution. From the point of view of v, the third execution is indistinguishable from the second execution during the first f_v rounds. Then v fires in round f_v in the third execution. Using the mod k-simultaneity of A in the third execution, we obtain:

$$f_u \equiv f_v \bmod k.$$

Similarly, u fires in round f_u and v fires in round $1+f_v$ in the forth execution. Using the mod k-simultaneity of A in the forth execution, we obtain:

$$f_u \equiv f_v + 1 \bmod k.$$

We obtain a contradiction if $k > 1$. □

4 Complexity Analysis

4.1 Time Complexity Analysis

Theorem 3. *There are at most* $6kn + 4k$ *rounds between the activation of all nodes and the firing of all nodes.*

Proof. We now bound the number of rounds between the activation of all nodes (noted s^{max}) and the firing of all nodes. Let t_γ be the round in which γ reaches level 1. First, we try to bound $t_\gamma - s^{max}$. We consider the non-decreasing series $(z_\gamma(t))_{t \geq s_\gamma}$. By Lemma 4, no node can reach level 2 before round t_γ. Then, for any $t \in \{s_\gamma, \ldots, t_\gamma\}$, we have $z_\gamma(t) \in Z^- = \{(f,t) \in Z, f < 2\}$. This set $Z^- \subseteq Z$ is the set of level-up events of strength 0 or 1. Since nodes can reach level 0 and 1 only once, the cardinality of Z^- is bounded by $2n$, where n is the total number of nodes. We can show that γ is in level 1 in round t if $(z_\gamma(t))_{t\in\mathbb{N}}$ remains stable between rounds $t - 3k$ and t. Then the worst case scenario happens if $z_\gamma(s_\gamma)$ starts with the lowest value of Z^-, and every $3k$ rounds, $z_\gamma(t)$ moves to the closest greater element of Z^-. Then $t_\gamma - s^{max}$ is bounded by $2n \times 3k = 6kn$.

Second, if γ is in level 1 in round t_γ, then every node u satisfies $z_u(t_\gamma + k) \geq z_\gamma(t_\gamma)$. By Lemma 9, the system is monovalent in round $t_\gamma + k$. By Lemma 11, every node is in level 2 in round $t_\gamma + 4k$. We finally obtain that there is at most $6kn + 4k$ rounds between the activation of all nodes and the firing of all nodes. □

4.2 Reducing Memory Usage

For all nodes u, for all rounds t, we have $(force_u^{pre}(t), t - c_u^{pre}(t)) \in Z$ by Lemma 6. Since Z is finite, $c_u^{pre}(t)$ tends to infinity as t tends to infinity. We present below a idea (inspired by [2]) which can alleviate this issue: in each execution of Algorithm 1, total memory usage increases forever, whereas in each execution of Algorithm 2, total memory usage grows during some arbitrarly-long initial period, and then drops and remains bounded forever. The idea is as follows:

As soon as $force_u(t) = 2$, the node u "knows" that some node v fired in round $t - c_u(t)$ (see Lemma 6). Then u may fire in any round $t' \equiv t - c_u(t) \bmod k$. At this point, the transition function can thus be simplified as in Algorithm 2. This simplified version uses a constant amount of memory.

Algorithm 2: The $OptSynchMod_k$ algorithm

```
1  Initialization:
2    initialize with SynchMod_k's initial state
3  At each round:
4    if force_u = 2 then
5      send ⟨c_u, true, 2, true⟩ to all
6      c_u ← 1 + c_u mod k
7      if level_u < 2 ∧ c_u = 0 then
8        level_u ← 2
9      end
10   end
11   else
12     apply SynchMod_k's transition function
13   end
```

Theorem 4. *Under the assumption of a Δ-centered dynamic graph with $\Delta \leq k$ and a complete activation schedule the Algorithm 2 solves the* $\bmod k$*-synchronization problem. Moreover, in each execution of Algorithm 2, the memory usage of each node is finite.*

5 Conclusion and Future Work

In this paper, we presented the $\bmod k$-synchronization problem, and we introduced an algorithm solving this problem. We provided an optimized version of this algorithm to tackle large memory usage. We also provided an upper-bound on the number of rounds between the start of all nodes and the firing of all nodes. This bound is linear in both k and n, which is not bad. However, this bound is deteriorated by a few nasty worst-case scenarios. We believe that some additional assumptions could provide a much tighter bound, which would not depend on n. That would be especially useful in very large systems. This consitutes a possible topic for a future work.

References

1. Bernstein, P.A., Hadzilacos, V., Goodman, N.: Concurrency Control and Recovery in Database Systems. Addison-Wesley, Boston (1987)
2. Boldi, P., Vigna, S.: Universal dynamic synchronous self-stabilization. Distrib. Comput. **15**(3), 137–153 (2002)
3. Casteigts, A., Flocchini, P., Quattrociocchi, W., Santoro, N.: Time-varying graphs and dynamic networks. In: Frey, H., Li, X., Ruehrup, S. (eds.) ADHOC-NOW 2011. LNCS, vol. 6811, pp. 346–359. Springer, Heidelberg (2011). https://doi.org/10.1007/978-3-642-22450-8_27
4. Charron-Bost, B., Moran, S.: The firing squad problem revisited. In: 35th Symposium on Theoretical Aspects of Computer Science (STACS 2018). Leibniz International Proceedings in Informatics (LIPIcs), vol. 96, pp. 20:1–20:14 (2018)

5. Charron-Bost, B., Moran, S.: Minmax algorithms for stabilizing consensus. CoRR abs/1906.09073 (2019). http://arxiv.org/abs/1906.09073
6. Charron-Bost, B., Schiper, A.: The heard-of model: computing in distributed systems with benign faults. Distrib. Comput. **22**(1), 49–71 (2009)
7. Debrat, H., Merz, S.: Verifying fault-tolerant distributed algorithms in the heard-of model. Archive of Formal Proofs (2012). http://isa-afp.org/entries/Heard_Of.html. Formal proof development
8. Dolev, D., Strong, H.R.: Authenticated algorithms for Byzantine agreement. SIAM J. Comput. **12**(4), 656–666 (1983)
9. Dwork, C., Lynch, N.A., Stockmeyer, L.: Consensus in the presence of partial synchrony. J. ACM **35**(2), 288–323 (1988)
10. Kuhn, F., Lynch, N., Oshman, R.: Distributed computation in dynamic networks. In: Proceedings of 42nd ACM Symposium on Theory of Computing (STOC 2010), pp. 513–522. ACM, New York (2010). https://doi.org/10.1145/1806689.1806760
11. Kuhn, F., Moses, Y., Oshman, R.: Coordinated consensus in dynamic networks. In: Proceedings of 30th ACM Symposium on Principles of Distributed Computing (PODC). ACM (2011)
12. Nipkow, T., Paulson, L., Wenzel, M.: Isabelle/HOL. A Proof Assistant for Higher-Order Logic. Lecture Notes in Computer Science, vol. 2283. Springer, Heidelberg (2002). https://doi.org/10.1007/3-540-45949-9
13. Pease, M., Shostak, R., Lamport, L.: Reaching agreement in the presence of faults. J. ACM **27**(2), 228–234 (1980)
14. Srikanth, T.K., Toueg, S.: Simulating authenticated broadcasts to derive simple fault-tolerant algorithms. Distrib. Comput. **2**(2), 80–94 (1987)

Partial Gathering of Mobile Agents in Dynamic Rings

Masahiro Shibata[1(✉)], Yuichi Sudo[2], Junya Nakamura[3], and Yonghwan Kim[4]

[1] Kyushu Institute of Technology, Kitakyushu, Fukuoka, Japan
shibata@cse.kyutech.ac.jp

[2] Hosei University, Chiyoda, Tokyo, Japan
sudo@hosei.ac.jp

[3] Toyohashi University of Technology, Toyohashi, Aichi, Japan
junya@imc.tut.ac.jp

[4] Nagoya Institute of Technology, Nagoya, Aichi, Japan
kim@nitech.ac.jp

Abstract. In this paper, we consider the partial gathering problem of mobile agents in synchronous dynamic bidirectional rings. The partial gathering problem is a generalization of the (well-investigated) total gathering problem, which requires that all k agents distributed in the network terminate at a non-predetermined single node. The partial gathering problem requires, for a given positive integer $g\,(< k)$, that agents terminate in a configuration such that either at least g agents or no agent exists at each node. The requirement for the partial gathering problem is strictly weaker than that for the total gathering problem, and thus it is interesting to clarify the difference in the move complexity between them. So far, partial gathering has been considered in static graphs. In this paper, we consider this problem in 1-interval connected rings, that is, one of the links in the ring may be missing at each time step. In such networks, we aim to clarify the solvability of the partial gathering problem and the move complexity, focusing on the relationship between values of k and g. First, we consider the case of $3g \le k \le 8g - 2$. In this case, we show that our algorithm can solve the problem with the total number of $O(kn)$ moves, where n is the number of nodes. Since $k = O(g)$ holds when $3g \le k \le 8g - 2$, the move complexity $O(kn)$ in this case can be represented also as $O(gn)$. Next, we consider the case of $k \ge 8g - 3$. In this case, we show that our algorithm can also solve the problem and its move complexity is $O(gn)$. These results mean that, when $k \ge 3g$, the partial gathering problem can be solved also in dynamic rings. In addition, agents require a total number of $\Omega(gn)$ (resp., $\Omega(kn)$) moves to solve the partial (resp., total) gathering problem. Thus, the both proposed algorithms can solve the partial gathering problem with the asymptotically optimal total number of $O(gn)$ moves, which is strictly smaller than that for the total gathering problem.

Keywords: Mobile agent · Partial gathering problem · Dynamic ring

C. Johnen et al. (Eds.): SSS 2021, LNCS 13046, pp. 440–455, 2021.
https://doi.org/10.1007/978-3-030-91081-5_29

1 Introduction

1.1 Background and Related Work

A *distributed system* comprises a set of computing entities (*nodes*) connected by communication links. As a promising design paradigm of distributed systems, (mobile) agents have attracted much attention [6]. The agents can traverse the system, carrying information collected at visited nodes, and execute an action at each node using the information to achieve a task. In other words, agents can encapsulate the process code and data, which simplifies design of distributed systems [10].

The *total gathering problem* (or the rendezvous problem) is a fundamental problem for agents' coordination. When a set of k agents are arbitrarily placed at nodes, this problem requires that all the k agents terminate at a non-predetermined single node. By meeting at a single node, all agents can share information or synchronize their behaviors. The total gathering problem has been considered in various kinds of networks such as rings [4,8,9], trees [1], tori [7], and arbitrary networks [3].

Recently, a variant of the total gathering problem, called the *g-partial gathering problem* [13], has been considered. This problem does not require all agents to meet at a single node, but allows agents to meet at several nodes separately. Concretely, for a given positive integer g $(< k)$, this problem requires that agents terminate in a configuration such that either at least g agents or no agent exists at each node. From a practical point of view, the g-partial gathering problem is still useful especially in large-scale networks. That is, when g-partial gathering is achieved, agents are partitioned into groups each of which has at least g agents, each agent can share information and tasks with agents in the same group, and each group can partition the network and then patrol its area that it should monitor efficiently. The g-partial gathering problem is interesting also from a theoretical point of view. Clearly, if $k < 2g$ holds, the g-partial gathering problem is equivalent to the total gathering problem. On the other hand, if $k \geq 2g$ holds, the requirement for the g-partial gathering problem is strictly weaker than that for the total gathering problem. Thus, there exists possibility that the g-partial gathering problem can be solved with strictly smaller total number of moves (i.e., lower costs) compared to the total gathering problem.

As related work, in case of $k \geq 2g$, Shibata et al. considered the g-partial gathering problem in rings [13,14,18], trees [16], and arbitrary networks [15]. In [13,14], they considered it in unidirectional ring networks with whiteboards (or memory spaces that agents can read and write) at nodes. They mainly showed that, if agents have distinct IDs and the algorithm is deterministic, or if agents do not have distinct IDs and the algorithm is randomized, agents can achieve g-partial gathering with the total number of $O(gn)$ moves (in expectation), where n is the number of nodes. Notice that in the above results agents do not have any global knowledge such as n or k. In [18], they considered g-partial gathering for another mobile entity called *mobile robots* that have no memory but can observe all nodes and robots in the network. In case of using mobile robots, they also showed that g-partial gathering can be achieved with the total number

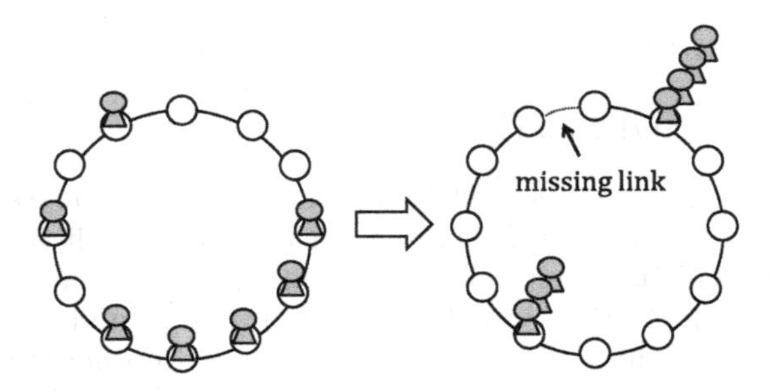

Fig. 1. An example of the g-partial gathering problem in a dynamic ring ($g = 3$).

of $O(gn)$ moves. In addition, the g-partial (resp., the total) gathering problem in ring networks requires a total number of $\Omega(gn)$ (resp., $\Omega(kn)$) moves. Thus, the above results are asymptotically optimal in terms of the total number of moves, and the total number $O(gn)$ of moves is strictly smaller than that for the total gathering problem when $g = o(k)$. In tree and arbitrary networks, they also proposed algorithms to solve the g-partial gathering problem with strictly smaller total number of moves compared to the total gathering problem for some settings, but we omit the details in this paper.

Although all the above work on the total gathering problem and the g-partial gathering problem are considered in *static graphs* where a network topology does not change during an execution, recently many problems involving agents have been studied in *dynamic graphs*, where a topology changes during an execution. For example, the total gathering problem [12], the exploration problem [5,11], the compact configuration problem [2] and the uniform deployment problem [17] are considered in dynamic graphs. However, to the best of our knowledge, there is no work for g-partial gathering in dynamic graphs, and hence in this paper we consider it in dynamic rings as a first step.

1.2 Our Contribution

In this paper, we consider the g-partial gathering problem of mobile agents in synchronous dynamic bidirectional rings with whiteboards at nodes. In this paper, we consider *1-interval connected rings* [2,11,12,17], that is, one of the links may be missing at each time step. An example is given in Fig. 1. In such networks, we aim to clarify the solvability of the g-partial gathering problem and the move complexity, focusing on the relationship between values of k and g.

In this paper, we assume that agents have distinct IDs, chirality, and knowledge of n and k. In Table 1, we compare our contributions with the result for agents with distinct IDs in static rings. We also analyze the time complexity for solving the problem. First, we consider the case of $3g \leq k \leq 8g - 2$. In this case, we show that our algorithm can solve the problem with $O(n)$ time and the total number of $O(kn)$ moves. Next, we consider the case of $k \geq 8g - 3$. In this case, we show that our algorithm can also solve the problem and the time complexity and the move complexity are $O(n)$ and $O(gn)$, respectively. These results mean

Table 1. Results of g-partial gathering for agents with distinct IDs in ring networks (n: #nodes, k: #agents).

	Result in [13]	Results of this paper	
		Result 1 (Sect. 3)	Result 2 (Sect. 4)
Static/Dynamic ring	Static	Dynamic	Dynamic
Knowledge of n and k	No	Available	Available
Relation between k and g	$k \geq 2g$	$3g \leq k \leq 8g-2$	$k \geq 8g-3$
Time complexity	$\Theta(n)$	$\Theta(n)$	$\Theta(n)$
Total number of agent moves	$\Theta(gn)$	$O(kn)(=O(gn))$	$\Theta(gn)$

that, although it is open that whether or not the g-partial gathering problem can be solved in dynamic rings when $2g \leq k < 3g$, it can be solved when $k \geq 3g$ and the time complexity $O(n)$ of our algorithms is asymptotically optimal. In addition, since $k = O(g)$ holds when $3g \leq k \leq 8g-2$ holds like the first case, the both proposed algorithms can achieve g-partial gathering also with the asymptotically the total number of $O(gn)$ moves, which is strictly smaller than the move complexity for the total gathering problem. Furthermore, it is worthwhile to mention that, while total gathering (i.e., all agents gather at a single node) cannot be solved in dynamic rings and it needs to relax the requirement so that agents stay at either of two nodes connected by a link [12], g-partial gathering can be achieved without relaxing the requirement.

Due to the page limitation, we omit several pseudocodes and proofs of theorems and lemmas.

2 Preliminaries

2.1 System Model

We basically follow the model defined in [12]. A *dynamic bidirectional ring* R is defined as 2-tuple $R = (V, E)$, where $V = \{v_0, v_1, \ldots, v_{n-1}\}$ is a set of n nodes and $E = \{e_0, e_1, \ldots, e_{n-1}\}$ (e_i=$\{v_i$, $v_{(i+1) \bmod n}\}$) is a set of links. For simplicity, we denote $v_{(i+j) \bmod n}$ (resp., $e_{(i+j) \bmod n}$) by v_{i+j} (resp., $e_{(i+j)}$) for any integers i and j. We define the direction from v_i to v_{i+1} (resp., v_i to v_{i-1}) as the *forward* or *clockwise* (resp., *backward* or *counterclockwise*) direction. In addition, one of links in the ring may be missing at each time step, and which link is missing is controlled by an *adversarial scheduler*. Such a dynamic ring is known as a *1-interval connected ring*. The *distance* from node v_i to v_j is defined to be $(j-i) \bmod n$. Note that this definition of the distance is correct when any of the links from v_i to v_j is not missing. Moreover, we assume that nodes are anonymous, i.e., they do not have IDs. Every node $v_i \in V$ has a whiteboard that agents at node v_i can read from and write on.

Let $A = \{a_0, a_1, \ldots, a_{k-1}\}$ be a set of $k\,(\leq n)$ agents. Agents can move through directed links, that is, they can move from v_i to v_{i+1} (i.e., move forward) or from v_i to v_{i-1} (i.e., move backward) for any i. Agents have distinct IDs and knowledge of n and k[1]. Agents have *chirality*, that is, they agree on the orientation of clockwise and counterclockwise direction in the ring. In addition, agents cannot detect whether other agents exist at the current node or not. An agent a_i is defined as a deterministic finite automaton (S, W, δ, $s_{initial}$, s_{final}, $w_{initial}$, $w'_{initial}$). The first element S is the set of all states of an agent, including two special states, initial state $s_{initial}$ and final state s_{final}. The second element W is the set of all states (contents) of a whiteboard, including two special initial states $w_{initial}$ and $w'_{initial}$. We explain $w_{initial}$ and $w'_{initial}$ in the next paragraph. The third element $\delta : S \times W \mapsto S \times W \times M$ is the state transition function that decides, from the current states of a_i and the current node's whiteboard, the next states of a_i and the whiteboard, and whether a_i moves to its neighboring node or not. The last element $M = \{-1, 0, 1\}$ in δ represents whether a_i makes a movement or not. The value 1 (resp., -1) means moving forward (resp., backward) and 0 means staying at the current node. We assume that $\delta(s_{final}, w_j) = (s_{final}, w_j, 0)$ holds for any state $w_j \in W$, which means that a_i never changes its state, updates the contents of a whiteboard, or leaves the current node once it reaches state s_{final}. We say that an agent *terminates* when its state changes to s_{final}. Notice that $S, \delta, s_{initial}$, and s_{final} can be dependent on the agent's ID.

In an agent system, (global) *configuration* c is defined as a product of the states of all agents, the states (whiteboards' contents) of all nodes, and the locations (i.e., the current nodes) of all agents. We define C as a set of all configurations. In an initial configuration $c_0 \in C$, we assume that agents are deployed arbitrarily at mutually distinct nodes, (or no two agents start at the same node), and the state of each whiteboard is $w_{initial}$ or $w'_{initial}$ depending on the existence of an agent. That is, when an agent exists at node v in the initial configuration, the initial state of v's whiteboard is $w_{initial}$. Otherwise, the state is $w'_{initial}$.

During an execution of the algorithm, we assume that agents move instantaneously, that is, they always exist at nodes (do not exist on links). Each agent executes the following four operations in an *atomic action*: 1) reads the contents of its current node's whiteboard, 2) executes local computation (or changes its state), 3) updates the contents of the current node's whiteboard, and 4) moves to its neighboring node or stays at the current node. If several agents exist at the same node, they take atomic actions interleavingly in an arbitrary order. In addition, when an agent tries to move to its neighboring node (e.g., from node v_j to v_{j+1}) but the corresponding link (e.g., link e_j) is missing, we say that the agent is *blocked*, and it still exists at v_j at the beginning of the next atomic action.

In this paper, we consider a *synchronous execution*, that is, in each time step called *round*, all agents perform atomic actions. Then, an *execution* starting from c_0 is defined as $E = c_0, c_1, \ldots$ where each c_i $(i \geq 1)$ is the configuration reached

[1] The knowledge of k is used for agents to decide which proposed algorithm they apply by comparing it with the value of g.

from c_{i-1} by atomic actions of all agents. An execution is infinite, or ends in a *final configuration* where the state of every agent is s_{final}.

2.2 The Partial Gathering Problem

The requirement for the g-partial gathering problem is that, for a given integer g, agents terminate in a configuration such that either at least g agents or no agent exists at each node. Formally, we define the problem as follows.

Definition 1. *An algorithm solves the g-partial gathering problem in dynamic rings when the following conditions hold:*

- *Execution E is finite (i.e., all agents terminate in state s_{final}).*
- *In the final configuration, at least g agents exist at any node where an agent exists.*

In this paper, we evaluate the proposed algorithms by the time complexity (the number of rounds for agents to solve the problem) and the total number of agents moves. In [13], the lower bound on the total number of agent moves for static rings is shown to be $\Omega(gn)$. This theorem clearly holds also in dynamic rings.

Theorem 1. *A lower bound on the total number of agent moves required to solve the g-partial gathering problem in dynamic rings is $\Omega(gn)$ if $g \geq 2$.*

On the time complexity, the following theorem holds. Intuitively, this is because there exist an initial configuration and link-missings such that the distance between some agent a_i and its nearest agent is $\Omega(n)$, which requires $\Omega(n)$ rounds for a_i to meet with other agents.

Theorem 2. *A lower bound on the time complexity required to solve the g-partial gathering problem in dynamic rings is $\Omega(n)$.*

3 The Case of $3g \leq k \leq 8g - 2$

In this section, when $3g \leq k \leq 8g - 2$, we propose a naive algorithm to solve the g-partial gathering problem in dynamic rings with $O(n)$ rounds and the total number of $O(kn)$ moves. Since $k = O(g)$ holds in this case, this algorithm is asymptotically in terms of both the time and move complexities, similar to the second algorithm explained in Sect. 4. In this algorithm, all agents try to travel once around the ring to get IDs of all agents, and then determine a single common node where all agents should gather. However, it is possible that some agent cannot travel once around the ring and get IDs of all agents due to missing links. Agents treat this by additional behaviors explained by the following subsections. The algorithm comprises two phases: the selection phase and the gathering phase. In the selection phase, agents move in the ring and determine the *gathering node* where they should gather. In the gathering phase, agents try to stay at the gathering node.

3.1 Selection Phase

The aim of this phase is that each agent achieves either of the following two goals: (i) It travels once around the ring and gets IDs of all agents, or (ii) it detects that all agents stay at the same node. To this end, we use an idea similar to [12] which considers total gathering in dynamic rings. First, each agent a_i writes its ID on the current whiteboard, and then tries to move forward for $3n$ rounds. During the movement, a_i memorizes values of observed IDs to array $a_i.ids[]$. After the $3n$ rounds, the number $a_i.nVisited$ of nodes that a_i has visited is (a) at least n or (b) less than n due to missing links. In case (a), a_i must have completed traveling once around the ring. Thus, a_i can get IDs of all k agents (goal (i) is achieved). Then, a_i (and the other agents) select the gathering node v_{gather} as the node where the minimum ID min is written.

In case (b) (i.e., a_i has visited less than n nodes during the $3n$ rounds), we show in Lemma 1 that all k agents stay at the same node (goal (ii) is achieved). This situation means that agents already achieve g-partial (or total) gathering, and they terminate the algorithm execution.

Concerning the selection phase, we have the following lemma.

Lemma 1. *After finishing the selection phase, each agent achieves either of the following two goals: (i) It travels once around the ring and gets IDs of all agents, or (ii) it detects that all agents stay at the same node.*

3.2 Gathering Phase

In this phase, agents aim to achieve g-partial (or total) gathering by trying to visit the gathering node v_{gather}. Concretely, for $3n$ rounds from the beginning of this phase, each agent a_i tries to move forward until it reaches v_{gather}. If agents are blocked few times, all agents can reach v_{gather} and they achieve g-partial (or total) gathering. However, it is possible that some agent cannot reach v_{gather} due to link-missings. To treat this, we introduce a technique called *splitting*. Intuitively, in this technique, when at least $2g$ agents exist at some node, from there an agent group with at least g agents tries to move forward and another agent group with at least g agents tries to move backward. In addition, when an agent group with at least g agents visits a node where less than g agents exist, the less than g agents join the agent group and try to move to the same direction as that of the group. By this behavior, it does not happen that all agents are blocked, and agents can eventually terminate in a configuration such that at least g agents exist at each node where an agent exists.

Concretely, after the $3n$ round from when agents tried to move forward to reach v_{gather}, by the similar discussion of Lemma 1, all agents that do not reach v_{gather} stay at the same node. Let v' be the node. Then, there are at most two nodes v_{gather} and v' where agents exist after the movement. If between g and $2g-1$ agents exist at v_{gather} or v' (or both), the agents staying there terminate the algorithm execution. On the other hand, if less than g agents exist at v_{gather} (resp., v'), at least $2g$ agents exist at v' (resp., v_{gather}) since we consider the

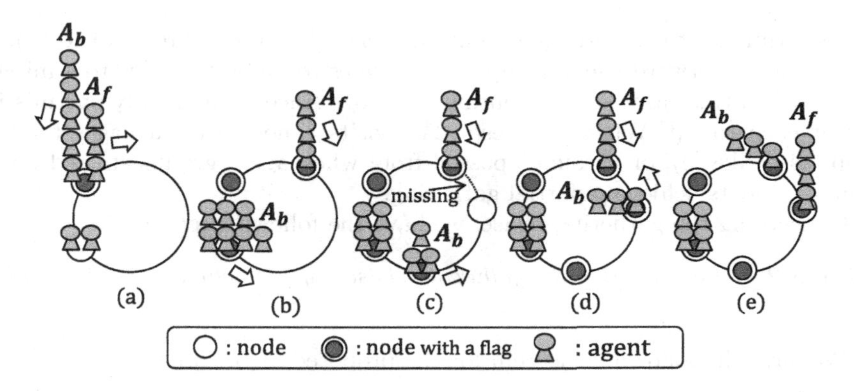

Fig. 2. An execution example of the gathering phase ($g = 3$).

case of $k \geq 3g$. We call the node with at least $2g$ agents v_{more}. Notice that it is possible that at least $2g$ agents exist at both v_{gather} and v'. Let $k'(\geq 2g)$ be the number of agents staying at v_{more}. Then, each agent a_i at v_{more} calculates how small its ID is among the k' agents. We denote the ordinal number by $a_i.rank$. Then, if $1 \leq a_i.rank \leq g$ holds, it belongs to the *forward agent group* A_f and tries to move forward. Else if $(k' < 3g) \vee (g+1 \leq a_i.rank \leq 2g)$ holds, it belongs to the *backward agent group* A_b and tries to move backward. If a_i does not satisfy any of the above conditions, it terminates the algorithm execution because there still exist at least g agents even after A_f and A_b leave v_{more}.

While A_f and A_b move in the ring, if A_f (resp., A_b) visits a new node v_j, it sets a flag $v_j.fMarked$ (resp., $v_j.bMarked$) representing that v_j is visited by A_f (resp., A_b). These flags are used for an agent group A to check whether or not the current node is visited by another agent group and A can stop moving in the ring. In addition, if A_f (resp., A_b) visits a node with less than g agents, the less than g agents join A_f (resp., A_b) and try to move forward (resp., backward). However, it is possible that the number *num* of agents in the updated group is more than $2g$. In this case, using their IDs, only g agents continue to try moving and the remaining $num - g$ agents terminate the algorithm execution at the current node. By this behavior, each link is passed by at most $2g$ agents and the total number of moves for agent groups can be reduced to $O(gn)$ (this technique is used in Sect. 4). Moreover, since A_f or A_b can visit a next node at each round even when some link is missing, A_f (resp., A_b) repeats such a behavior for n rounds or until it visits some node v_j with $v_j.bMarked = true$ (resp., $v_j.fMarked = true$), which implies that all the remaining nodes that A_f (resp., A_b) should visit are already visited by another agent group A_b (resp., A_f).

An example is given in Fig. 2 (we omit nodes unrelated to the example). From (a) to (b), a backward group A_b visits a node with two ($< g$) agents, and the two agents join A_b. Then, since the number of agents in the updated A_b is $7, (> 2g)$, only three agents continue to try moving and the remaining four agents terminate the algorithm execution there ((b) to (c)). From (c) to (d), we assume that a forward agent group A_f continues to be blocked due to a missing link. Even in this case, A_b can continue to move since there is only one missing link at each round. When A_f (resp., A_b) visits a node with a flag set by A_b (resp., A_f) like (e), or n rounds passed from when agent groups started trying to move, agents achieve g-partial gathering.

Concerning the gathering phase, we have the following lemma.

Lemma 2. *After finishing the gathering phase, agents achieve g-partial gathering.*

We have the following theorem for the proposed algorithm.

Theorem 3. *When $3g \leq k \leq 8g - 2$ holds, the proposed algorithm solves the g-partial gathering problem in dynamic rings with $O(n)$ rounds and the total number of $O(kn)$ moves.*

4 The Case of $k \geq 8g - 3$

In this section, when $k \geq 8g - 3$, we propose an algorithm to solve the problem with $O(n)$ rounds and the total number of $O(gn)$ (i.e., optimal) moves. Since the move complexity is not $O(kn)$ but $O(gn)$, it is not possible that all agents try to travel once around the ring as in Sect. 3. Hence, in this section agents aim to reduce the total number of moves using distinct IDs and the fact of $k \geq 8g - 3$. The algorithm comprises three phases: the semi-selection phase, the semi-gathering phase, and the achievement phase. In the semi-selection phase, agents select a set of *gathering-candidate nodes* each of where at least $2g$ agents may gather. In the semi-gathering phase, agents try to stay at a gathering-candidate node. As a result, at least $2g$ agents gather at some node (the node may not be a gathering-candidate node due to link-missings). In the achievement phase, agents achieve g-partial gathering by the same method as that for the gathering phase in Sect. 3.2.

4.1 Semi-selection Phase

The aim of this part is to select a set of gathering-candidate nodes each of where at least $2g$ agents may gather. A possible approach is that each agent a_i moves forward and backward for getting IDs of its 1-st, 2-nd, $\ldots, (2g-1)$-st forward agents and IDs of its 1-st, 2-nd, $\ldots, (2g-1)$-st backward agents, and then returns to its initial node. Here, the i-th ($i \neq 0$) forward (resp., backward) agent a' of agent a represents the agent such that $i-1$ agents exist between a and a' in a's forward (resp., backward) direction in the initial configuration. Thereafter, a_i compares its ID and the obtained $4g-2$ IDs. If its ID is the minimum, a_i selects its initial node as a gathering-candidate node v_{candi}. Then, the $2g-1$ agents existing in a_i's backward direction try to move forward to stay at v_{candi} and eventually $2g$ agents may gather at v_{gather}. However, since we consider 1-interval connected rings, there are two problems: (1) it is possible that no gathering-candidate node is selected since some agent may not be able to collect $4g-2$ IDs due to link-missings, and (2) even if a gathering-candidate node v_{candi} is selected, it is possible that some agent cannot reach v_{candi} due to link-missings and only less than $2g$ agents gather at each node.

To treat these problems, each agent a_i in this phase keeps trying to move forward, tries to observe more than $4g-2$ IDs, and considers some observed ID as its own ID when it observed the necessary number of IDs. Concretely, for $3n$ rounds, each agent a_i tries to move forward until it observes $10g-4$ IDs or at least $2g$ agents exist at the current node. Thereafter, a_i determines its behavior depending on whether it observed at least $8g-3$ IDs or not. If a_i did not observe at least $8g-3$ IDs, we show in Lemma 3 that at least $2g$ agents exist at some node v_j and then a flag $v_j.candi$ is set to true to represent that v_j is a gathering-candidate node (problem (1) is solved). Intuitively, this is because a_i does not observe at least $(10g-4)-(8g-4)=2g$ IDs and this means that at least $2g-1$ agents existing in a_i's backward direction also do not observe the necessary number of IDs and they eventually stay at a_i's node.

On the other hand, if a_i observed at least $8g-3$ IDs, it uses the first $8g-3$ IDs for comparison and considers the $(4g-1)$-st ID as its own ID. Then, this situation is similar to one that a_i compares its ID with $4g-2$ forward IDs and $4g-2$ backward IDs. Hence, if the $(4g-1)$-st ID is the minimum among the $8g-3$ IDs, a_i sets $v_j.candi = true$ at the current node v_j. Then, since $k \geq 8g-3$ holds, all the $8g-3$ IDs are distinct and thus $4g-2$ agents existing in a_i's backward direction can recognize a_i's staying node as the nearest gathering-candidate node v_{candi} in the forward direction when they observed at least $8g-3$ IDs. Thus, the $4g-1$ agents in total (a_i and the $4g-2$ agents) try to move forward and stay at v_{candi} (the detail is explained in the next subsection). Then, when some link continues to be missing, the $4g-1$ agents are partitioned into two groups and at least one group has $2g$ agents (problem (2) is solved).

The pseudocode of the semi-selection phase is described in Algorithm 1. Global variables used in the algorithm is summarized in Table 2 (several variables are used in other sections). Concerning the semi-selection phase, we have the following lemma.

Lemma 3. *After finishing the semi-selection phase, there exists at least one node v_j with $v_j.candi = true$.*

Proof. Let a_{min} be the agent with minimum ID among all agents and a_i be the $(4g-2)$-nd backward agent of a_{min}. We consider the cases that the value of $a_i.nIDs$ after executing Algorithm 1 is (a) less than $8g-3$ and (b) at least $8g-3$ in this order. First, (a) if $a_i.nIDs < 8g-3$ holds, let $g' = (10g-4) - a_i.nIDs$ be the number of IDs that a_i could not observe and $a_{i-1}, a_{i-2}, \dots, a_{i-(g'-1)}$ be the 1-st, 2-nd, ..., $(g'-1)$-st backward agents of a_i. Then, since $(g'-1)+a_i.nIDs = ((10g-4)-a_i.nIDs)-1+a_i.nIDs = 10g-5 < 10g-4$, agent $a_{i-(g'-1)}$ does not observe the required number $10g-4$ of IDs. Thus, $a_{i-1}, a_{i-2}, \dots, a_{i-(g'-1)}$ also observed less than $10g-4$ IDs and they stay at the same node (a_i's node) by the similar discussion of Lemma 1. Since $g'-1 \geq (10g-4)-(8g-4)-1 = 2g-1$

Table 2. Global variables used in the proposed algorithm.

Variables for agent a_i			
Type	Name	Meaning	Initial value
int	$a_i.rounds$	number of rounds from some round	1
int	$a_i.nIDs$	number of different IDs that a_i has observed from some round	0
int	$a_i.nVisited$	number of nodes that a_i has ever visited	0
int	$a_i.rank$	ordinal number of how its ID is small among IDs of agents at the same node	0
array	$a_i.ids[]$	sequence of IDs that a_i has observed	$\perp$
Variables for node v_j			
Type	Name	Meaning	Initial value
int	$v_j.id$	ID stored by v_j	$\perp$
int	$v_j.nAgents$	number of agents staying at v_j	0
boolean	$v_j.fMarked$	whether v_j is visited by a forward group or not	false
boolean	$v_j.bMarked$	whether v_j is visited by a backward group or not	false
boolean	$v_j.candi$	whether v_j is a gathering-candidate node or not	false

Algorithm 1. The behavior of agent a_i in the semi-selection phase (v_j is the current node of a_i.)

Main Routine of Agent a_i

1: $v_j.id := a_i.id$, $a_i.ids[a_i.nIDs] := v_j.id$
2: $a_i.nIDs := a_i.nIDs + 1$, $v_j.nAgents := v_j.nAgents + 1$,
3: **while** $a_i.rounds < 3n$ **do**
4: **if** $(a_i.nIDs < 10g - 4) \wedge (v_j.nAgents < 2g)$ **then**
5: $v_j.nAgents := v_j.nAgents - 1$
6: Try to move from the current node v_j to the forward node v_{j+1}
7: **if** (a_i reached v_{j+1} (that becomes new v_j)) $\wedge$ $(v_j.id \neq \perp)$ **then**
8: $a_i.ids[a_i.nIDs] := v_j.id$, $a_i.nIDs := a_i.nIDs + 1$
9: **end if**
10: $v_j.nAgents := v_j.nAgents + 1$, $a_i.rounds := a_i.rounds + 1$
11: **end if**
12: **end while**
13: **if** $(v_j.nAgents \geq 2g) \vee ((a_i.nIDs \geq 8g-3) \wedge (\forall h \in [0, 8g-2] \setminus \{4g-2\}; a_i.ids[4g-2] < a_i.id[h]))$ **then**
14: $v_j.candi := true$
15: Terminate the semi-selection phase and enter the semi-gathering phase
16: **end if**

holds, at least $2g$ agents (including a_i) stay at the same node v_j and thus $v_j.candi$ is set to true. Next, (b) if $a_i.nIDs \geq 8g - 3$ holds, a_i recognizes that a_{min}'s ID is its own ID and the ID is the minimum among the $8g - 3$ IDs. Hence, a_i sets $v_j.candi = true$ at the current node v_j. Therefore, the lemma follows. □

Algorithm 2. The behavior of agent a_i in the semi-gathering phase (v_j is the current node of a_i.)

Main Routine of Agent a_i

1: $a_i.rounds := 1, a_i.nIDs := 1$
2: **while** $(a_i.rounds < 3n) \wedge (a_i.nIDs \neq 4g - 1)$ **do**
3:   **if** $v_j.candi = false$ **then**
4:     $v_j.nAgents := v_j.nAgents - 1$
5:     Try to move from the current node v_j to the forward node v_{j+1}
6:     **if** (a_i reached v_{j+1} (that becomes new v_j)) $\wedge$ ($v_j.id \neq \perp$) **then** $a_i.nIDs := a_i.nIDs + 1$
7:     $v_j.nAgents := v_j.nAgents + 1$
8:     **if** $v_j.nAgents \geq 2g$ **then** $v_j.candi := true$
9:   **end if**
10:   $a_i.rounds = a_i.rounds + 1$
11: **end while**
12: Terminate the semi-gathering phase and enter the achievement phase

4.2 Semi-gathering Phase

In this phase, agents aim to make a configuration such that at least $2g$ agents exist at some node. By Lemma 3, there exists at least one gathering-candidate node v_j with $v_j.candi = true$ at the end of the semi-selection phase. In the following, we call such a candidate node v_{candi}. Then, if less than $2g$ agents exist at v_{candi}, $4g - 2$ agents in total that already stay at v_{candi} and exist in v_{candi}'s backward direction try to stay at v_{candi}. Concretely, in this phase, for $3n$ rounds each agent tries to move forward until it stays v_{candi} or at least $2g$ agents exist at the current node. Then, due to link-missings, it is possible that only less than $2g$ agents gather at v_{candi} after the movement. In this case, we can show by the similar discussion of Lemma 1 that all the agents that do not reach v_{candi} among the $4g - 2$ agents stay at the same node. Then, the $4g - 1$ agents (the $4g - 2$ agents and the agent originally staying at v_{candi}) are partitioned into two groups and at least one group has at least $2g$ agents in any partition. Thus, agents can make a configuration such that at least $2g$ agents exist at some node.

The pseudocode of the semi-gathering phase is described in Algorithm 2. Note that, during the movement, when agents are blocked few times and they do not stay at a node with at least $2g$ agents, agents may require the total number of more than $O(gn)$ moves. To avoid this, each agent stop moving when it observed $4g - 1$ IDs even if it does not stay at a node with at least $2g$ agents (line 2).

Concerning the semi-gathering phase, we have the following lemma.

Lemma 4. *After finishing the semi-gathering phase, there exists at one node v_j with $v_j.nAgents \geq 2g$.*

Proof. We consider a configuration such that there exists no node with at least $2g$ agents at the beginning of the semi-gathering phase. By Lemma 3, there exists at least one node v_j with $v_j.candi = true$, and $4g - 2$ agents in total that already stay at v_j and exist in v_j's backward direction try to stay at v_j by Algorithms 1 and 2. Then, by the similar discussion of the proof of Lemma 1, after executing Algorithm 2 for $3n$ rounds, all agents among the $4g-2$ agents that do not reach v_j stay at the same node. Thus, the $4g-1$ agents (the $4g-2$ agents and the agent originally staying at v_j) are partitioned into two groups and at least one group has at least $2g$ agents in any partition. Therefore, the lemma follows. □

4.3 Achievement Phase

In this phase, agents aim to achieve g-partial gathering. By Lemma 4, there exists at least one node with at least $2g$ agents as in Sect. 3.2. The difference from Sect. 3.2 is that there may exist more than two nodes with agents and there may exist several nodes each of which has at least $2g$ agents. Also from this situation, agents can achieve g-partial gathering using the same method as that in Sect. 3.2, that is, (1) agents staying at a node with at least $2g$ agents are partitioned into a forward group and a backward group and they try to move forward and backward respectively, and (2) when a forward group (resp., a backward group) visits a node with less than $2g$ agents, the less than $2g$ agents join the forward group (resp., a backward group).

An example is given in Fig. 3. In Fig. 3 (a), there exist two nodes v_p and v_q each of which has $6\,(=2g)$ agents. Hence, a forward group A_{f_p} and a backward group A_{b_p} (resp., A_{f_q} and A_{b_q}) start moving from node v_p (resp., from node v_q). From (a) to (b), A_{b_p} reaches node v_ℓ with less than $2g$ agents, and the less than $2g$ agents join A_{b_p} and try to move backward. From (b) to (e), we assume that A_{f_q} continues to be blocked by a missing link. From (b) to (c), A_{f_p} and A_{b_q} crossed and they recognize the fact by the existence of flags, and they terminate the algorithm execution. From (c) to (d), A_{b_p} reaches node v_m with less than $2g$ agents, and the less than $2g$ agents join A_{b_p} and try to move backward. Then, since the number *num* of agents in the updated A_{b_p} is $7\,(>2g)$, by using their IDs, only g agents continue to try moving backward and the remaining $num-g$ agents terminate the algorithm execution there ((d) to (e)). By this behavior, during this phase each link is passed by at most $2g$ agents and the achievement phase can be achieved with the total number of $O(gn)$ moves. From (e) to (f), A_{f_q} and A_{b_p} reach some node simultaneously and recognize the fact by flags, and they terminate the algorithm execution and agents achieve g-partial gathering.

Concerning the achievement phase, we have the following lemma.

Lemma 5. *After executing the achievement phase, agents achieve g-partial gathering.*

We have the following theorem for the proposed algorithm.

Theorem 4. *When $k \geq 8g-3$ holds, the proposed algorithm solves the g-partial gathering problem in dynamic rings with $O(n)$ rounds and the total number of $O(gn)$ moves.*

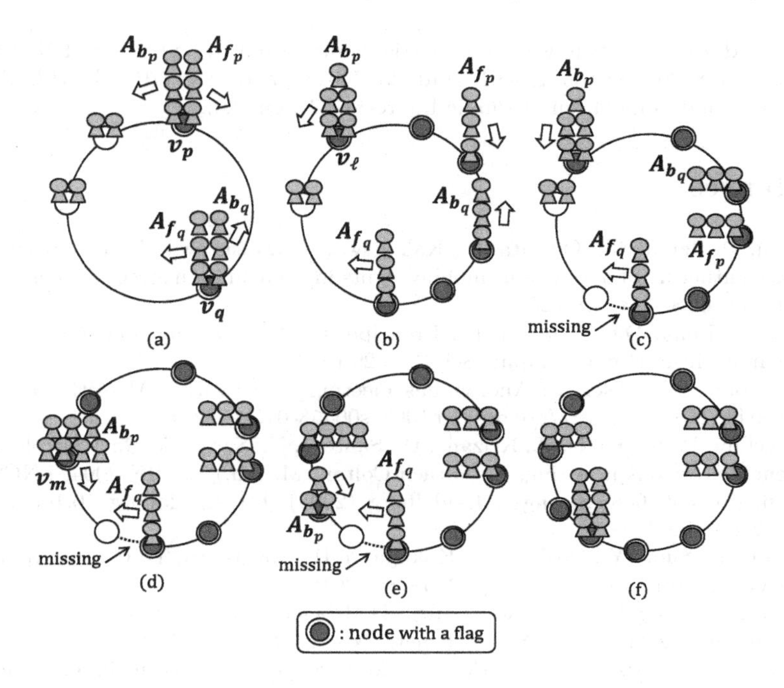

Fig. 3. An execution example of the achievement phase ($g = 3$).

5 Conclusion

In this paper, we considered the g-partial gathering problem in bidirectional dynamic rings and considered the solvability of the problem and the move complexity, focusing on the relationship between values of k and g. First, when $3g \leq k \leq 8g - 2$, we showed that the proposed algorithm can solve the problem with $O(n)$ rounds and the total number of $O(kn)$ moves. Next, when $k \geq 8g - 3$, we showed that the proposed algorithm can solve the problem with $O(n)$ rounds and the total number of $O(gn)$ moves. These results show that, when $k \geq 3g$, the g-partial gathering problem can be solved also in dynamic rings. In addition, since $k = O(g)$ holds when $3g \leq k \leq 8g - 2$ holds like the first case, the both proposed algorithms can achieve g-partial gathering with the asymptotically optimal total number of agent moves.

Future works are as follows. First, we consider the solvability in case of $2g \leq k < 3g$. Second, when $3g \leq k < 8g - 3$, we consider whether agents can achieve g-partial gathering with the total number of moves smaller than $O(kn)$ or not. Finally, we will consider agents with weaker capabilities, e.g., agents without distinct IDs, without chirality, or agents that behave semi-synchronously or asynchronously. In any of the above cases, we conjecture that agents cannot solve the problem or require more total number of moves than the proposed algorithms.

Acknowledgement. This work was partially supported by JSPS KAKENHI Grant Number 18K18029, 18K18031, 20H04140, 20KK0232, and 21K17706; the Hibi Science Foundation; and Foundation of Public Interest of Tatematsu.

References

1. Baba, D., Izumi, T., Ooshita, F., Kakugawa, H., Masuzawa, T.: Linear time and space gathering of anonymous mobile agents in asynchronous trees. Theoret. Comput. Sci. **478**, 118–126 (2013)
2. Das, S., Luna, D.G., Mazzei, D., Prencipe, G.: Compacting oblivious agents on dynamic rings. PeerJ Comput. Sci. **7**, 1–29 (2021)
3. Dieudonné, Y., Pelc, A.: Anonymous meeting in networks. Algorithmica **74**(2), 908–946 (2016). https://doi.org/10.1007/s00453-015-9982-0
4. Flocchini, P., Kranakis, E., Krizanc, D., Santoro, N., Sawchuk, C.: Multiple mobile agent rendezvous in a ring. In: Farach-Colton, M. (ed.) LATIN 2004. LNCS, vol. 2976, pp. 599–608. Springer, Heidelberg (2004). https://doi.org/10.1007/978-3-540-24698-5_62
5. Gotoh, T., Sudo, Y., Ooshita, F., Kakugawa, H., Masuzawa, T.: Group exploration of dynamic tori. In: ICDCS, pp. 775–785 (2018)
6. Gray, R.S., Kotz, D., Cybenko, G., Rus, D.: D'agents: applications and performance of a mobile-agent system. Softw. Pract. Exper. **32**(6), 543–573 (2002)
7. Kranakis, E., Krizanc, D., Markou, E.: Mobile agent rendezvous in a synchronous torus. In: Correa, J.R., Hevia, A., Kiwi, M. (eds.) LATIN 2006. LNCS, vol. 3887, pp. 653–664. Springer, Heidelberg (2006). https://doi.org/10.1007/11682462_60
8. Kranakis, E., Krozanc, D., Markou, E.: The mobile agent rendezvous problem in the ring. Syn. Lect. Distrib. Comput. Theory **1**, 1–122 (2010)
9. Kranakis, E., Santoro, N., Sawchuk, C., Krizanc, D.: Mobile agent rendezvous in a ring. In: ICDCS, pp. 592–599 (2003)
10. Lange, D., Oshima, M.: Seven good reasons for mobile agents. CACM **42**(3), 88–89 (1999)
11. Di Luna, G., Dobrev, S., Flocchini, P., Santoro, N.: Distributed exploration of dynamic rings. Distrib. Comput. **33**(1), 41–67 (2018). https://doi.org/10.1007/s00446-018-0339-1
12. Luna, D.G., Flocchini, P., Pagli, L., Prencipe, G., Santoro, N., Viglietta, G.: Gathering in dynamic rings. Theoret. Comput. Sci. **811**, 79–98 (2018)
13. Shibata, M., Kawai, S., Ooshita, F., Kakugawa, H., Masuzawa, T.: Partial gathering of mobile agents in asynchronous unidirectional rings. Theoret. Comput. Sci. **617**, 1–11 (2016)
14. Shibata, M., Kawata, N., Sudo, Y., Ooshita, F., Kakugawa, H., Masuzawa, T.: Move-optimal partial gathering of mobile agents without identifiers or global knowledge in asynchronous unidirectional rings. Theoret. Comput. Sci. **822**, 92–109 (2020)
15. Shibata, M., Nakamura, D., Ooshita, F., Kakugawa, H., Masuzawa, T.: Partial gathering of mobile agents in arbitrary networks. IEICE Trans. Inf. Syst. **102**(3), 444–453 (2019)
16. Shibata, M., Ooshita, F., Kakugawa, H., Masuzawa, T.: Move-optimal partial gathering of mobile agents in asynchronous trees. Theoret. Comput. Sci. **705**, 9–30 (2018)

17. Shibata, M., Sudo, Y., Nakamura, J., Kim, Y.: Uniform deployment of mobile agents in dynamic rings. In: Devismes, S., Mittal, N. (eds.) SSS 2020. LNCS, vol. 12514, pp. 248–263. Springer, Cham (2020). https://doi.org/10.1007/978-3-030-64348-5_20
18. Shibata, M., Tixeuil, S.: Partial gathering of mobile robots from multiplicity-allowed configurations in rings. In: Devismes, S., Mittal, N. (eds.) SSS 2020. LNCS, vol. 12514, pp. 264–279. Springer, Cham (2020). https://doi.org/10.1007/978-3-030-64348-5_21

Optimal Protocols for 2-Party Contention Resolution

Dingyu Wang(✉)

University of Michigan, Ann Arbor MI, USA
wangdy@umich.edu

Abstract. *Contention Resolution* is a fundamental symmetry-breaking problem in which n devices must acquire temporary and exclusive access to some *shared resource*, without the assistance of a mediating authority. For example, the n devices may be sensors that each need to transmit a single packet of data over a broadcast channel. In each time step, devices can (probabilistically) choose to acquire the resource or remain idle; if exactly one device attempts to acquire it, it succeeds, and if two or more devices make an attempt, none succeeds. The complexity of the problem depends heavily on what types of *collision detection* are available. In this paper we consider *acknowledgement-based protocols*, in which devices only learn whether their own attempt succeeded or failed; they receive no other feedback from the environment whatsoever, i.e., whether other devices attempted to acquire the resource, succeeded, or failed.

Nearly all work on the Contention Resolution problem evaluated the performance of algorithms *asymptotically*, as $n \to \infty$. In this work we focus on the simplest case of $n = 2$ devices, but look for *precisely* optimal algorithms. We design provably optimal algorithms under three natural cost metrics: minimizing the expected average of the waiting times (AVG), the expected waiting time until the first device acquires the resource (MIN), and the expected time until the last device acquires the resource (MAX). We first prove that the optimal algorithms for $n = 2$ are *periodic* in a certain sense, and therefore have finite descriptions, then we design optimal algorithms under all three objectives.

AVG. The optimal contention resolution algorithm under the AVG objective has expected cost $\sqrt{3/2} + 3/2 \approx 2.72474$.

MIN. The optimal contention resolution algorithm under the MIN objective has expected cost 2. (This result can be proved in an ad hoc fashion, and may be considered folklore.)

MAX. The optimal contention resolution algorithm under the MAX objective has expected cost $1/\gamma \approx 3.33641$, where $\gamma \approx 0.299723$ is the smallest root of $3x^3 - 12x^2 + 10x - 2$ (We may also express γ in radical form: $\gamma = -\frac{1}{6}\left(1 - i\sqrt{3}\right)\sqrt[3]{13 + i\sqrt{47}} + \frac{4}{3} - \frac{1+i\sqrt{3}}{\sqrt[3]{13+i\sqrt{47}}}$.).

Keywords: Contention resolution · Probabilistic algorithm

C. Johnen et al. (Eds.): SSS 2021, LNCS 13046, pp. 456–468, 2021.
https://doi.org/10.1007/978-3-030-91081-5_30

1 Introduction

The goal of a contention resolution scheme is to allow multiple devices to eventually obtain exclusive access to some shared resource. In this paper we will use often use the terminology of one particular application, namely, wireless devices that wish to broadcast messages on a multiple-access channel. However, contention resolution schemes are used in a variety of areas [13,15,20], not just wireless networking. We consider a model of contention resolution that is distinguished by the following features.

Discrete Time. Time is partitioned into discrete *slots*. It is the goal of every device to obtain exclusive access to the channel for exactly one slot, after which it no longer participates in the protocol. We assume that all n devices begin at the same time, and therefore agree on slot zero. (Other work considers an infinite-time model in which devices are injected adversarially [3,8,10], or according to a Poisson distribution [17,21] with some constant mean.)

Feedback. At the beginning of each time slot each device can choose to either transmit its message or remain idle. If it chooses to idle, it receives no feedback from the environment; if it chooses to transmit, it receives a signal indicating whether the transmission was successful (all other devices remained idle). ("Full sensing" protocols like [3,8,10,17,21], in contrast, depend on receiving ternary feedback at each time slot indicating whether there was no transmission, some successful transmission, or a collision.)

Noiseless operation. The system is errorless; there is no environmental noise.

Anonymity. Devices are indistinguishable and run the same algorithm, but can break symmetry by generating (private) random bits.

There are many ways to measure the time-efficiency of contention resolution protocols. In infinite-time models, we want to avoid deadlock [2,5–8,10], minimize the latency of devices in the system, and generally make productive use of a (large) constant fraction of the slots [5,8,10]. When all n devices begin at the same time [6,7], there are still several natural measures of efficiency. In this paper we consider three: minimizing the time until the *first* successful transmission (MIN), the *last* successful transmission time (MAX, a.k.a. the *makespan*), and the *average* transmission time (AVG).

1.1 Prior Work

Classic infinite-time protocols like ALOHA [1] and binary exponential backoff algorithms [14,15] are simple but suffer from poor worst case performance and eventual deadlock [2,6,7], even under *non*-adversarial injection rates, e.g., Poisson injection rates with arbitrary small means. These are *acknowledgement-based* protocols which do not require constant (ternary) channel feedback. One line of work aimed to achieve deadlock-freeness under Poisson arrivals [9,12,17,21], assuming ternary channel feedback. The maximum channel usage rate is known to be between 0.48776 [17,21] and 0.5874 [16]. A different line of work aimed at achieving deadlock-freeness and constant rate of efficiency under *adversarial* injections and possibly adversarial *jamming*, also assuming ternary feedback. See [3,8,10] for robust protocols that can tolerate a jamming adversary. One problem with both of these lines of work is that all devices must monitor the channel constantly (for the ternary silence/success/collision feedback). Bender et al. [5] considered adversarial injection rates and showed that it is possible to achieve a constant efficiency rate while only monitoring/participating in $O(\log(\log^* n))$ time slots. This was later shown to be optimal [11].

When all n devices start at the same time slot (n unknown), we have a pretty good understanding of the AVG, MIN, and MAX objectives. Here there are still variants of the problem, depending on whether the protocol is full-sensing (requiring ternery feedback) or merely acknowledgement-based. Willard [22] and Nakano and Olariu [18] gave full sensing protocols for the MIN objective when n is unknown that takes time $O(\log\log n + \log f^{-1})$ with probability $1-f$, which is optimal. The *decay* algorithm [4] is an acknowledgement-based protocol for the MIN objective that runs in $O(\log n \log f^{-1})$ time with probability $1-f$, which is also known to be optimal [19]. When n is unknown, binary exponential backoff achieves optimal $O(n)$ time under the AVG objective, but suboptimal $\Theta(n \log n)$ time under the MAX objective [6,7]. The *sawtooth* protocol of Bender et al. [6,7] is optimal $O(n)$ under both AVG and MAX; it is acknowledgement-based.

1.2 New Results

In this paper we consider what seems to be the *simplest non-trivial symmetry breaking problem*, namely, resolving contention among two parties ($n = 2$) via an acknowledgement-based protocol. The *asymptotic* complexity of this problem is not difficult to derive: $O(1)$ time suffices, under any reasonable objective function, and $O(\log f^{-1})$ time suffices with probability $1 - f$. However, our goal is to discover *precisely* optimal algorithms.

We derive the optimal protocols for the AVG, MIN, and MAX objectives, in expectation, which are produced below. The optimal MIN protocol is easy to obtain using *ad hoc* arguments; it has expected cost 2. However, the optimal protocols for AVG and MAX require a more principled, rigorous approach to the problem. We show that the protocol minimizing AVG has expected cost $\sqrt{3/2} + 3/2 \approx 2.72474$, and that the optimal protocol minimizing MAX has expected cost $1/\gamma \approx 3.33641$, where $\gamma \approx 0.299723$ is the unique root of $3x^3 - 12x^2 + 10x - 2$ in the interval $[1/4, 1/3]$.

AVG-Contention Resolution:

Step 1. Transmit with probability $\frac{4-\sqrt{6}}{3} \approx 0.516837$. If successful, halt; if there was a collision, repeat Step 1; otherwise proceed to Step 2.
Step 2. Transmit with probability $\frac{1+\sqrt{6}}{5} \approx 0.689898$. If successful, halt; if there was a collision, go to Step 1; otherwise proceed to Step 3.
Step 3. Transmit with probability 1. If successful, halt; otherwise go to Step 1.

MIN-**Contention Resolution:**

- In each step, transmit with probability 1/2 until successful.

MAX-**Contention Resolution:**

Step 1. Transmit with probability $\alpha \approx 0.528837$, where α is the unique root of $x^3 + 7x^2 - 21x + 9$ in $[0, 1]$. If successful, halt; if there was a collision, repeat Step 1; otherwise proceed to Step 2.
Step 2. Transmit with probability $\beta \approx 0.785997$, where β is the unique root of $4x^3 - 8x^2 + 3$ in $[0, 1]$. If successful, halt; if there was a collision, go to Step 1; otherwise proceed to Step 3.
Step 3. Transmit with probability 1. If successful, halt; otherwise go to Step 1.

One may naturally ask: what is the point of understanding Contention Resolution problems with $n = O(1)$ devices? The most straightforward answer is that in some applications, contention resolution instances between $n = O(1)$ devices are commonplace.[1] However, even if one is only interested in the asymptotic case of $n \to \infty$ devices, understanding how to resolve $n = O(1)$ optimally is essential. For example, the protocols of [9,12,17,21] work by repeatedly isolating subsets of the n' active devices, where n' is Poisson distributed with mean around 1.1, then resolving conflicts within this set (if $n' > 1$) using a near-optimal procedure. The channel usage rate of these protocols (≈ 0.48776) depends critically on the efficiency of Contention Resolution among n' devices, where $\mathbb{E}[n'] = O(1)$. Moreover, *improving* these algorithms will likely require a much better understanding of $O(1)$-size contention resolution.

Organization. In Sect. 2 we give a formal definition of the model and state Theorem 1 on the *existence* of an optimal protocol for any reasonable objective function. In Sect. 3 we prove another structural result on optimal protocols for $n = 2$ devices under the AVG, MIN, and MAX objectives (Theorem 2), and use it to characterize what the optimal protocols for AVG (Theorem 3), MIN (Theorem 4), and MAX (Theorem 5) should look like. Corollary 1 derives that AVG-**Contention Resolution** is the optimal protocol under the AVG objective, and Corollary 2 does the same for MAX-**Contention Resolution** under MAX. The proofs of Theorems 1 and 2 and Corollaries 1 and 2 appear in the Appendix in the full version of this paper[2].

2 Problem Formulation

After each time step the channel issues responses to the devices from the set $\mathcal{R} = \{0, 1, 2_+\}$. If the device idles, it always receives 0. If it attempts to transmit,

[1] For a humorous example, consider the Canadian Standoff problem https://www.cartoonstock.com/cartoonview.asp?catref=CC137954.

[2] The full version is available at http://www.ancientwang.com/document/Optimal_Protocols_for_2_Party_Contention_Resolution%20(1).pdf.

it receives 1 if successful and 2_+ if unsuccessful. A *history* is word over $\mathcal{R}^*$. We use exponents for repetition and $*$ as short for $\mathcal{R}^*$; e.g., the history $0^3 2_+^2$ is short for 0002_+2_+ and $*1*$ is the set of all histories containing a 1. The notation $a \in w$ means that symbol a has at least one occurrence in word w.

Devices choose their action (transmit or idle) at time step $t \in \mathbb{N}$ and receive feedback at time $t + 0.5$. A *policy* is a function f for deciding the probability of transmitting. Define $\mathcal{F} = \{f : \mathcal{R}^* \to [0,1] \mid \forall w \in \mathcal{R}^*, 1 \in w \implies f(w) = 0\}$ to be the set of all proper policies, i.e., once a device is successful ($1 \in w$), it must halt ($f(w) = 0$).[3] Every particular policy $f \in \mathcal{F}$ induces a distribution on decisions $\{D_{k,t}\}_{k\in[n],t\in\mathbb{N}}$ and responses $\{R_{k,t}\}_{k\in[n],t\in\mathbb{N}}$, where $D_{k,t} = 1$ iff the kth device transmits at time t and $R_{k,t} \in \mathcal{R}$ is the response received by the kth device at time $t + 0.5$. In particular,

$$\mathbb{P}(D_{k,t} = 1 \mid R_{k,0}R_{k,1}\cdots R_{k,t-1} = h) = f(h), \tag{1}$$

$$R_{k,t}(w) = \begin{cases} 0, & D_{k,t}(w) = 0 \\ 1, & (D_{k,t}(w) = 1) \wedge (\forall j \neq k, D_{j,t}(w) = 0) \\ 2^+, & (D_{k,t}(w) = 1) \wedge (\exists j \neq k, D_{j,t}(w) = 1) \end{cases} \tag{2}$$

Define X_i to be the random variable of the number of time slots until device i succeeds. Note that since we number the slots starting from zero,

$$X_i = 1 + \min\{t \geq 0 \mid R_{i,t} = 1\}.$$

Note that $\{X_i\}_{i\in[n]}$ are identically distributed but not independent. For example, minimizing the average of $\{X_i\}_{i\in[n]}$ is equivalent to minimizing X_1 since:

$$\mathbb{E}\frac{\sum_{i=1}^n X_i}{n} = \frac{\sum_{i=1}^n \mathbb{E}X_i}{n} = \frac{n\mathbb{E}X_1}{n} = \mathbb{E}X_1.$$

2.1 Performance Metrics and Existence Issues

For our proofs it is helpful to assume the existence of an *optimal protocol* but it is not immediate that there *exists* such an optimal protocol. (Perhaps there is just an infinite succession of protocols, each better than the next.) In the full version, we prove that optimal protocols exist for all "reasonable" objectives. A *cost function* $T : \mathbb{Z}_+^n \to \mathbb{R}_+$ is one that maps the vector of device latencies to a single (positive) cost. The objective is to minimize $\mathbb{E}T(X_1, \ldots, X_n)$.

Definition 1 (Informal). *A function* $T : \mathbb{Z}_+^n \to \mathbb{R}^+$ *is* reasonable *if for any* $s > 0$ *there exists some* $N > 0$ *such that* $T(x_1, \ldots, x_n) < s$ *can be known if each of* $x_1, x_2, \ldots, x_n$ *is either known or known to be greater than* N.

For example, $T_1(x_1, \ldots, x_n) = \frac{\sum_{k=1}^n x_k}{n}$ (AVG), $T_2(x_1, \ldots, x_n) = \min(x_1, \ldots, x_n)$ (MIN), and $T_3(x_1, \ldots, x_n) = \max(x_1, \ldots, x_n)$ (MAX) are all reasonable, as are all ℓ_p norms, etc.

[3] A policy may have no finite representation, and therefore may not be an *algorithm* in the usual sense.

Theorem 1. *Given the number of users n and a reasonable objective function T, there exists an optimal policy $f^* \in \mathcal{F}$ that minimizes $\mathbb{E}T(X_1, X_2, \ldots, X_n)$.*

3 Contention Resolution Between Two Parties

In this section we restrict our attention to the case $n = 2$. One key observation that makes the $n = 2$ case special is that whenever one device receives 2_+ (collision) feedback, it knows that its history and the other device's history are identical. For many reasonable objective functions the best response to a collision is to restart the protocol. This is proved formally in Theorem 2 for a class of objective functions that includes AVG, MIN, and MAX. See the full version for proof.

Theorem 2. *Let $n = 2$, T be a reasonable objective function, and f be an optimal policy for T. Another policy f^* is defined as follows.*

$$\begin{aligned} f^*(0^k) &= f(0^k), \quad \forall k \in \mathbb{N} \\ f^*(*2_+0^k) &= f(0^k), \quad \forall k \in \mathbb{N} \\ f^*(*1*) &= 0 \end{aligned}$$

If $T(x + c, y + c) = T(x, y) + c$ for any c (scalar additivity), then f^* is also an optimal policy for T.

Theorem 2 tells us that for the objectives that are scalar additive (including AVG, MIN, and MAX), we can restrict our attention to policies $f \in \mathcal{F}$ defined by a vector of probabilities $(p_i)_{i\geq 0}$, such that $f(w0^k) = p_k$, where w is empty or ends with 2_+, i.e., the transmission probability cannot depend on anything that happened *before* the last collision.

3.1 AVG: Minimizing the Average Transmission Time

Let $(p_k)_{k\geq 0}$ be the probability sequence corresponding to an optimal policy f for AVG. We first express our objective $\mathbb{E}X_1$ in terms of the sequence (p_k). Then, using the optimality of f, we deduce that $(p_k)_{k\geq 0}$ must take on the special form described in Theorem 3. This Theorem does not completely specify what the optimal protocol looks like. Further calculations (Corollary 1) show that choosing $N = 2$ is the best choice, and that AVG-**Contention Resolution** (see Sect. 1) is an optimal protocol.

Theorem 3. *There exists an integer $N > 0$ and $a_0, a_1, a_2 \in \mathbb{R}$ where $a_0 - a_1 + a_2 = 1$ and $a_0 + a_1N + a_2N^2 = 0$ such that the following probability sequence*

$$p_k = 1 - \frac{a_0 + a_1k + a_2k^2}{a_0 + a_1(k-1) + a_2(k-1)^2}, \quad 0 \leq k \leq N,$$

induces an optimal policy that minimizes $\mathbb{E}X_1$.

Remark 1. Note that defining $p_0, \dots, p_N$ is sufficient, since $p_N = 1$ induces a certain collision if there are still 2 devices in the system, which causes the algorithm to reset. In the next time slot both devices would transmit with probability p_0.

Proof. Assume we are using an optimal policy f^* induced by a probability sequence $(p_i)_{i=1}^{\infty}$. Define $S_1, S_2 \geq 0$ to be the random variables indicating the *index* of the first slot in which devices 1 and 2 first transmit. Observe that S_1 and S_2 are i.i.d. random variables, where $\mathbb{P}(S_1 = k) = \mathbb{P}(S_2 = k) = p_k \prod_{i=0}^{k-1}(1-p_i)$.[4] We have

$$\mathbb{E}X_1 = \sum_{k=0}^{\infty} [\mathbb{P}(S_1 = k)(k + 1 + \mathbb{P}(S_2 = k) \cdot \mathbb{E}X_1)]$$

$$\iff \mathbb{E}X_1 = \sum_{k=0}^{\infty} \left[p_k \left(\prod_{i=0}^{k-1} (1 - p_i) \right) \cdot \left(k + 1 + p_k \left(\prod_{i=0}^{k-1} (1 - p_i) \right) \cdot \mathbb{E}X_1 \right) \right]. \tag{3}$$

Define $m_k = \prod_{i=0}^{k}(1 - p_i)$ to be the probability that a device idles in time steps 0 through k, where $m_{-1} = 1$. Note that $p_k m_{k-1} = m_{k-1} - m_k$ is true for all $k \geq 0$. We can rewrite Eq. (3) as:

$$\mathbb{E}X_1 = \sum_{k=0}^{\infty} [(m_{k-1} - m_k)(k + 1 + (m_{k-1} - m_k) \cdot \mathbb{E}X_1)]$$

$$\iff \mathbb{E}X_1 = \mathbb{E}X_1 \cdot \sum_{k=0}^{\infty} (m_{k-1} - m_k)^2 + \sum_{k=0}^{\infty} (m_{k-1} - m_k)(k + 1)$$

$$\iff \mathbb{E}X_1 = \frac{\sum_{k=0}^{\infty} m_{k-1}}{1 - \sum_{k=0}^{\infty}(m_{k-1} - m_k)^2}. \tag{4}$$

By definition, $(m_k)_{k=-1}^{\infty}$ is a non-increasing sequence with $m_{-1} = 1$ and $m_k \geq 0$. There is no optimal policy with $m_{k-1} = m_k \neq 0$ (meaning $p_k = 0$), since otherwise we can delete m_k from the sequence, leaving the denominator unchanged but reducing the numerator. This implies $(m_k)_{k=-1}^{\infty}$ is either an infinite, positive, strictly decreasing sequence or a finite, positive, strictly-decreasing sequence followed by a tail of zeros. Pick any index $k_0 \geq 0$ such that $m_{k_0} > 0$. We know $m_{k_0-1} > m_{k_0} > m_{k_0+1}$. By the optimality of f^*, m_{k_0} must, holding all other parameters fixed, be the optimal choice for this parameter in its neighborhood. In other words, we have $\frac{\partial \mathbb{E}X_1}{\partial m_{k_0}} = 0$, which implies

$$\frac{1 - \sum_{k=0}^{\infty}(m_{k-1} - m_k)^2 + \sum_{k=0}^{\infty} m_{k-1}(-2(m_{k_0-1} - m_{k_0}) + 2(m_{k_0} - m_{k_0+1}))}{(1 - \sum_{k=0}^{\infty}(m_{k-1} - m_k)^2)^2} = 0$$

Therefore we have for any $k_0 \geq 0$ such that $m_{k_0} > 0$,

$$2m_{k_0} - m_{k_0+1} - m_{k_0-1} = C$$

$$\iff m_{k_0} - m_{k_0+1} = m_{k_0-1} - m_{k_0} + C \tag{5}$$

[4] We use the convention that $\prod_{i=0}^{-1} a_k = 1$, where $(a_k)_{k=0}^{\infty}$ is any sequence.

where $C = \frac{\sum_{k=0}^{\infty}(m_{k-1}-m_k)^2-1}{2\sum_{k=0}^{\infty} m_{k-1}}$ is a real constant. Note that $C = -\frac{1}{2\mathbb{E}X_1} < 0$. Fix any $k_1 \geq 0$ such that $m_{k_1} > 0$. By summing up Eq. (5) for $k_0 = 0, 1, \ldots, k_1$ and rearranging terms, we have

$$m_{k_1} - m_{k_1+1} = (k_1 + 1)C + m_{-1} - m_0. \tag{6}$$

Fix any $k_2 \geq 0$ such that $m_{k_2} > 0$. By summing up Eq. (6) for $k_1 = 0, 1, \ldots, k_2$, we have

$$m_{k_2+1} = (m_0 - m_{-1})(k_2 + 2) + m_{-1} - \frac{(k_2+1)(k_2+2)}{2}C \tag{7}$$

$$= -\frac{C}{2}k_2^2 + \left(-\frac{3C}{2} + m_0 - m_{-1}\right)k_2 + 2m_0 - m_{-1} - C. \tag{8}$$

Recall that $C < 0$ and $m_k \in [0, 1]$. This rules out the possibility that the sequence $(m_k)_{k=0}^{\infty}$ is an infinite strictly decreasing sequence, since a non-degenerate quadratic function is unbounded as k goes to infinity. As a result, there must be a positive integer $N \geq 1$ for which $m_{N-1} > 0$ and $m_N = 0$. Also note that Eq. (7) is not only true for $k_2 = 0, 1, \ldots N-1$, but also true for $k_2 = -1$ and -2. (This can be checked by directly setting $k_2 = -1$ and -2.) We conclude that it is possible to write (m_k) as

$$m_k = a_0 + a_1 k + a_2 k^2, \quad -1 \leq k \leq N,$$

for some constants a_0, a_1, a_2 satisfying

$$m_{-1} = a_0 - a_1 + a_2 = 1$$
$$m_N = a_0 + a_1 N + a_2 N^2 = 0.$$

Writing $p_k = 1 - \frac{m_k}{m_{k-1}}$ gives the statement of the theorem.

Based on Theorem 3, we can find the optimal probability sequence for each fixed N by choosing the best a_2. It turns out that $N = 2$ is the best choice, though $N = 3$ is only marginally worse. The proof of Corollary 1 is in the full version.

Corollary 1. **AVG-Contention Resolution** *is an optimal protocol for* $n = 2$ *devices under the* AVG *objective. The expected average time is* $\sqrt{3/2} + 3/2 \approx 2.72474$.

3.2 MIN: Minimizing the Earliest Transmission Time

It is straightforward to show $\mathbb{E}\min(X_1, X_2) = 2$ under the optimal policy. Nonetheless, it is useful to have a general closed form expression for $\mathbb{E}\min(X_1, X_2)$ in terms of the (m_k) sequence of an arbitrary (suboptimal) policy, as shown in the proof of Theorem 4. This will come in handy later since $\mathbb{E}\max(X_1, X_2)$ can be expressed as $2\mathbb{E}X_1 - \mathbb{E}\min(X_1, X_2)$.

Theorem 4. *The policy that minimizes* $\mathbb{E}\min(X_1, X_2)$, MIN-**Contention Resolution**, *transmits with constant probability* $1/2$ *until successful. Using the optimal policy,* $\mathbb{E}\min(X_1, X_2) = 2$.

Proof. By Theorem 2 we can consider an optimal policy defined by a sequence of transmission probabilities $(p_k)_{k\geq 0}$. Let $H_{j,k}$ be the transmission/idle history of player $j \in \{1, 2\}$ up to time slot k. Then we have

$$\begin{aligned}
\mathbb{E}\min(X_1, X_2) &= \sum_{k=0}^{\infty} \mathbb{P}(H_{1,k} = H_{2,k} = 0^k 1)(k + 1 + \mathbb{E}\min(X_1, X_2)) \\
&\quad + \sum_{k=0}^{\infty} \mathbb{P}(\{H_{1,k}, H_{2,k}\} = \{0^k 1, 0^k 0\})(k+1) \\
&= \sum_{k=0}^{\infty} \left(\prod_{i=0}^{k-1} (1-p_i)^2 \cdot p_k^2 (k + 1 + \mathbb{E}\min(X_1, X_2)) \right) \\
&\quad + \sum_{k=0}^{\infty} \left(\prod_{i=0}^{k-1} (1-p_i)^2 \cdot 2p_k(1-p_k)(k+1) \right) \\
&= \sum_{k=0}^{\infty} \left(\prod_{i=0}^{k-1} (1-p_i)^2 \cdot ((1-(1-p_k)^2)(k+1) + p_k^2 \cdot \mathbb{E}\min(X_1, X_2)) \right).
\end{aligned}$$

Defining $m_k = \prod_{i=0}^{k} (1-p_i)$ as before, we have

$$= \sum_{k=0}^{\infty} ((m_{k-1}^2 - m_k^2)(k+1) + m_{k-1}^2 p_k^2 \cdot \mathbb{E}\min(X_1, X_2))$$

As $m_{k-1}p_k = m_{k-1} - m_k$, we can write $\mathbb{E}\min(X_1, X_2)$ in closed form as

$$\begin{aligned}
\mathbb{E}\min(X_1, X_2) &= \frac{\sum_{k=0}^{\infty}(m_{k-1}^2 - m_k^2)(k+1)}{1 - \sum_{k=0}^{\infty}(m_{k-1} - m_k)^2} \\
&= \frac{\sum_{k=0}^{\infty} m_{k-1}^2}{1 - \sum_{k=0}^{\infty}(m_{k-1} - m_k)^2} \qquad (9) \\
&= \frac{\sum_{k=0}^{\infty} m_{k-1}^2}{2\sum_{k=0}^{\infty}(m_{k-1} - m_k)m_k} \\
&\geq \frac{\sum_{k=0}^{\infty} m_{k-1}^2}{2\sum_{k=0}^{\infty}(\frac{m_{k-1}}{2})^2} = 2. \qquad (10)
\end{aligned}$$

Thus $\mathbb{E}\min(X_1, X_2)$ attains minimum 2 if and only if for all $k \in \mathbb{N}$, $m_{k-1} - m_k = m_k$, i.e. $m_k = \frac{m_{k-1}}{2}$ and $m_0 = \frac{1}{2}$. Thus $p_k = 1 - \frac{m_k}{m_{k-1}} = \frac{1}{2}$ for all k. This constant probability sequence corresponds to the constant policy with sending probability $\frac{1}{2}$ (i.e., MIN-**Contention Resolution**).

3.3 MAX: Minimizing the Last Transmission Time

Before determining the optimal policy under the MAX objective, it is useful to have a crude estimate for its cost.

Lemma 1. *Let f be the optimal policy for the* MAX *objective and X_1^f, X_2^f be the latencies of the two devices. Then $\mathbb{E}\max(X_1^f, X_2^f) \in [3, 4]$.*

Proof. The optimal policy under the MIN objective, f^*, sends with probability 1/2 until successful. It is easy to see that $\mathbb{E}\max(X_1^{f^*}, X_2^{f^*}) = 4$, so f can do no worse. Under f (or any policy), $\mathbb{E}\max(X_1^f, X_2^f) \geq 1 + \mathbb{E}\min(X_1^f, X_2^f)$. By the optimality of f^* for MIN, $\mathbb{E}\min(X_1^f, X_2^f) \geq \mathbb{E}\min(X_1^{f^*}, X_2^{f^*}) = 2$, so $\mathbb{E}\max(X_1^f, X_2^f) \geq 3$.

Theorem 5. *Let f be the optimal policy for the* MAX *objective and define $1/\gamma = \mathbb{E}\max(X_1^f, X_2^f)$ to be its expected cost. Let x_1, x_2 be the roots of the polynomial*

$$x^2 - (2 - \gamma)x + 1. \tag{11}$$

There exists an integer $N \geq 0$ and reals C_1, C_2 where $C_1 x_1^{-1} + C_2 x_2^{-1} = 0$ and $C_1 x_1^{N+1} + C_2 x_2^{N+1} = -1$, such that the following probability sequence

$$p_k = 1 - \frac{C_1 x_1^k + C_2 x_2^k + 1}{C_1 x_1^{k-1} + C_2 x_2^{k-1} + 1}, \quad 0 \leq k \leq N + 1,$$

induces an optimal policy that minimizes $\mathbb{E}\max(X_1, X_2)$.

Remark 2. Note that $p_{N+1} = 1$, thus it is sufficient to only define $p_0, p_1, \ldots, p_{N+1}$.

Proof. Assume the optimal policy f is characterized by the probability sequence $(p_k)_{k=0}^{\infty}$. Using the derived expressions (Eq. (4) and Eq. (9)) in Theorem 3 and 4, we have

$$\begin{aligned}
\mathbb{E}\max(X_1, X_2) &= 2\mathbb{E}X_1 - \mathbb{E}\min(X_1, X_2) \\
&= 2\frac{\sum_{k=0}^{\infty} m_{k-1}}{1 - \sum_{k=0}^{\infty}(m_{k-1} - m_k)^2} - \frac{\sum_{k=0}^{\infty} m_{k-1}^2}{1 - \sum_{k=0}^{\infty}(m_{k-1} - m_k)^2} \\
&= \frac{2\sum_{k=0}^{\infty} m_{k-1} - \sum_{k=0}^{\infty} m_{k-1}^2}{1 - \sum_{k=0}^{\infty}(m_{k-1} - m_k)^2},
\end{aligned} \tag{12}$$

where $m_k = \prod_{i=0}^{k}(1 - p_i)$ with $m_{-1} = 1$.

The only requirement on the sequence $(m_k)_{k=-1}^{\infty}$ is that it is strictly decreasing with $m_k \in [0, 1]$. First we observe if $m_k = m_{k+1}$, we must have both of them equal to zero. Otherwise, we can remove m_k which will leave the denominator unchanged but reduce the numerator. Therefore, the optimal sequence is either a strictly decreasing sequence or a strictly decreasing sequence followed by a tail of zeros. Fix any $v \geq 0$ for which $m_v > 0$ we have, by the optimality of $(m_k)_{k=-1}^{\infty}$,

$$\frac{\partial \mathbb{E}\max(X_1, X_2)}{\partial m_v} = \frac{(2 - 2m_v)B - 2A(m_{v-1} - m_v - (m_v - m_{v+1}))}{D} = 0, \tag{13}$$

where $B = 1 - \sum_{k=0}^{\infty}(m_{k-1} - m_k)^2$ and $A = 2\sum_{k=0}^{\infty} m_{k-1} - \sum_{k=0}^{\infty} m_{k-1}^2$. Let $\gamma = \frac{B}{A} = \frac{1}{\mathbb{E}\max(X_1,X_2)}$, then we have, from Eq. (13)

$$m_{v+1} = (2-\gamma)m_v - m_{v-1} + \gamma \tag{14}$$

$$\iff (m_{v+1} - 1) = (2-\gamma)(m_v - 1) - (m_{v-1} - 1) \tag{15}$$

Equation (15) defines a linear homogeneous recurrence relation for the sequence $(m_{v+1} - 1)$, whose characteristic roots are $x_1, x_2 = \frac{2-\gamma\pm\sqrt{\gamma^2-4\gamma}}{2}$. One may verify that they satisfy the following identities.

$$x_1 + x_2 = 2 - \gamma \tag{16}$$

$$x_1 x_2 = 1 \tag{17}$$

From Lemma 1 we know $\gamma \in [\frac{1}{4}, \frac{1}{3}]$. Thus we have $\gamma^2 - 4\gamma < 0$ which implies x_1 and x_2 are distinct conjugate numbers and of the same norm $\sqrt{x_1 x_2} = 1$. Then $m_k - 1 = C_1 x_1^k + C_2 x_2^k$ for all k for which at least one of m_{k-1}, m_k or m_{k+1} is greater than zero.

If it were the case that $m_k > 0$ for all k, then by summing (14) up for all $v \in \mathbb{N}$, we have

$$\sum_{k=0}^{\infty} m_{k+1} = (2-\gamma)\sum_{k=0}^{\infty} m_k - \sum_{k=0}^{\infty} m_{k-1} + \gamma \cdot \infty$$

which implies $\sum_{k=0}^{\infty} m_k = \infty$. This is impossible since, by the upper bound of Lemma 1,

$$4 \geq \mathbb{E}\max(X_1, X_2) = \frac{1 + 2\sum_{k=0}^{\infty} m_k - \sum_{k=0}^{\infty} m_k^2}{1 - \sum_{k=0}^{\infty}(m_{k-1} - m_k)^2} \geq \frac{1 + \sum_{k=0}^{\infty} m_k}{1}.$$

Therefore the optimal sequence must be of the form

$$m_k = (C_1 x_1^k + C_2 x_2^k + 1)\mathbb{1}_{k \leq N}$$

for some integer $N \geq 0$, where $C_1 x_1^N + C_2 x_2^N + 1 = 0$ and $C_1 x_1^{-1} + C_2 x_2^{-1} + 1 = 1$. Writing $p_k = 1 - \frac{m_k}{m_{k-1}}$ gives the statement of the theorem.

The proof of Corollary 2 is in the full version.

Corollary 2. MAX-**Contention Resolution** *is an optimal protocol for $n = 2$ devices under the* MAX *objective. The expected maximum latency is $1/\gamma \approx 3.33641$, where γ is the unique root of $3x^3 - 12x^2 + 10x - 2$ in the interval* $[1/4, 1/3]$.

4 Conclusion

In this paper we established the existence of optimal contention resolution policies for any *reasonable* cost metric, and derived the first optimal protocols for resolving conflicts between $n = 2$ parties under the AVG, MIN, and MAX objectives.

Generalizing our results to $n \geq 3$ or to more complicated cost metrics (e.g., the ℓ_2 norm) is a challenging problem. Unlike the $n = 2$ case, it is not clear, for example, whether the optimal protocols for $n = 3$ select their transmission probabilities from a finite set of reals. It is also unclear whether the optimal protocols for $n = 3$ satisfy some analogue of Theorem 2, i.e., that they are "recurrent" in some way.

References

1. Abramson, N.: The ALOHA system: another alternative for computer communications. In: Proceedings of the November 17–19, 1970, Fall Joint Computer Conference, pp. 281–285. ACM (1970)
2. Aldous, D.J.: Ultimate instability of exponential back-off protocol for acknowledgment-based transmission control of random access communication channels. IEEE Trans. Inf. Theory **33**(2), 219–223 (1987). https://doi.org/10.1109/TIT.1987.1057295
3. Awerbuch, B., Richa, A.W., Scheideler, C.: A jamming-resistant MAC protocol for single-hop wireless networks. In: Proceedings of the Twenty-Seventh Annual ACM Symposium on Principles of Distributed Computing (PODC), pp. 45–54 (2008). https://doi.org/10.1145/1400751.1400759
4. Bar-Yehuda, R., Goldreich, O., Itai, A.: On the time-complexity of broadcast in multi-hop radio networks: an exponential gap between determinism and randomization. J. Comput. Syst. Sci. **45**(1), 104–126 (1992)
5. Bender, M., Kopelowitz, T., Pettie, S., Young, M.: Contention resolution with constant throughput and log-logstar channel accesses. SIAM J. Comput. **47**(5), 1735–1754 (2018). https://doi.org/10.1137/17M1158604
6. Bender, M.A., Farach-Colton, M., He, S., Kuszmaul, B.C., Leiserson, C.E.: Adversarial analyses of window backoff strategies. In: Proceedings 18th International Parallel and Distributed Processing Symposium (IPDPS) (2004). https://doi.org/10.1109/IPDPS.2004.1303230, https://doi.org/10.1109/IPDPS.2004.1303230
7. Bender, M.A., Farach-Colton, M., He, S., Kuszmaul, B.C., Leiserson, C.E.: Adversarial contention resolution for simple channels. In: Proceedings of the 17th Annual ACM Symposium on Parallelism in Algorithms and Architectures (SPAA), pp. 325–332 (2005). https://doi.org/10.1145/1073970.1074023
8. Bender, M.A., Fineman, J.T., Gilbert, S., Young, M.: Scaling exponential backoff: constant throughput, polylogarithmic channel-access attempts, and robustness. J. ACM **66**(1), 6:1–6:33 (2019)
9. Capetanakis, J.: Tree algorithms for packet broadcast channels. IEEE Trans. Inf. Theory **25**(5), 505–515 (1979). https://doi.org/10.1109/TIT.1979.1056093
10. Chang, Y., Jin, W., Pettie, S.: Simple contention resolution via multiplicative weight updates. In: Proceedings 2nd Symposium on Simplicity in Algorithms (SOSA), pp. 16:1–16:16 (2019). https://doi.org/10.4230/OASIcs.SOSA.2019.16
11. Chang, Y., Kopelowitz, T., Pettie, S., Wang, R., Zhan, W.: Exponential separations in the energy complexity of leader election. In: Proceedings 49th Annual ACM Symposium on Theory of Computing (STOC), pp. 771–783 (2017). https://doi.org/10.1145/3055399.3055481, http://doi.acm.org/10.1145/3055399.3055481

12. Gallager, R.G.: Conflict resolution in random access broadcast networks. In: Proceedings AFOSR Workshop on Communications Theory Applications, Provincetown, MA, 17–20 September, pp. 74–76 (1978)
13. Jacobson, V.: Congestion avoidance and control. ACM SIGCOMM Comput. Commun. Rev. **18**, 314–329. ACM (1988)
14. Kwak, B.J., Song, N.O., Miller, L.E.: Performance analysis of exponential backoff. IEEE/ACM Trans. Netw. (TON) **13**(2), 343–355 (2005)
15. Metcalfe, R.M., Boggs, D.R.: Ethernet: distributed packet switching for local computer networks. Commun. ACM **19**(7), 395–404 (1976)
16. Mikhailov, V.A., Tsybakov, B.S.: Upper bound for the capacity of a random multiple access system. Problemy Peredachi Informatsii **17**(1), 90–95 (1981)
17. Mosely, J., Humblet, P.A.: A class of efficient contention resolution algorithms for multiple access channels. IEEE Trans. Commun. **33**(2), 145–151 (1985). https://doi.org/10.1109/TCOM.1985.1096261
18. Nakano, K., Olariu, S.: Uniform leader election protocols for radio networks. IEEE Trans. Parallel Distrib. Syst. **13**(5), 516–526 (2002). https://doi.org/10.1109/TPDS.2002.1003864
19. Newport, C.: Radio network lower bounds made easy. In: Proceedings of the 28th International Symposium on Distributed Computing (DISC), pp. 258–272 (2014)
20. Rajwar, R., Goodman, J.R.: Speculative lock elision: enabling highly concurrent multithreaded execution. In: Proceedings of the 34th Annual ACM/IEEE International Symposium on Microarchitecture, pp. 294–305. IEEE Computer Society (2001)
21. Tsybakov, B.S., Mikhailov, V.A.: Slotted multiaccess packet broadcasting feedback channel. Problemy Peredachi Informatsii **14**(4), 32–59 (1978)
22. Willard, D.E.: Log-logarithmic selection resolution protocols in a multiple access channel. SIAM J. Comput. **15**(2), 468–477 (1986). https://doi.org/10.1137/0215032

Computer Aided Formal Design of Swarm Robotics Algorithms

Thibaut Balabonski[1], Pierre Courtieu[2], Robin Pelle[1], Lionel Rieg[3], Sébastien Tixeuil[4(✉)], and Xavier Urbain[5]

[1] Université Paris-Saclay, CNRS, LMF, Gif-sur-Yvette, France
[2] Cédric, Conservatoire des Arts et Métiers, Paris, France
[3] VERIMAG, UMR 5160, Grenoble INP, Univ. Grenoble Alpes, Grenoble, France
[4] Sorbonne University, CNRS, LIP6, Paris, France
Sebastien.Tixeuil@lip6.fr
[5] Université Claude Bernard Lyon 1, LIRIS UMR5205, Lyon, France

Abstract. Previous works on formally studying mobile robotic swarms consider necessary and sufficient system hypotheses enabling to solve theoretical benchmark problems (geometric pattern formation, gathering, scattering, etc.). We argue that formal methods can also help in the early design stage of mobile robotic swarms correct-by-design protocols, even for tasks closer to real-world use cases and not previously studied theoretically. Our position is supported by a concrete case study. Starting from a real-world case scenario, we jointly design the formal problem specification, a family of protocols that are able to solve the problem, and their corresponding proof of correctness, all expressed with the same formal framework. The concrete framework we use for our development is the PACTOLE library based on the COQ proof assistant.

Context. Swarm robotics envisions groups of mobile robots self-organizing and cooperating toward the resolution of common objectives, such as patrolling, exploring and mapping disaster areas, constructing ad hoc mobile communication infrastructures to enable communication with rescue teams, etc. As several of those applications are life-critical, the correctness of the deployed protocols becomes of paramount importance. In turn, correctness reasoning about autonomous moving and computing entities that collaborate to achieve a global objective in a setting where unpredictable hazards may occur is complex and error prone. A first step into more formal reasoning is to use a sound *mathematical model.*

Suzuki & Yamashita [25] introduced such a mathematical model describing the behaviour of robots in this context. The model is targeted at swarms of very weak robots evolving in harsh environments. At its core, the model simply commands individual robots to repetitively *observe* their environment before *computing* a path of actions to pursue and acting on it, usually by *moving* to a

This work was partially supported by Project SAPPORO of the French National Research Agency (ANR) under the reference 2019-CE25-0005-1.

C. Johnen et al. (Eds.): SSS 2021, LNCS 13046, pp. 469–473, 2021.
https://doi.org/10.1007/978-3-030-91081-5_31

specific location. Three different levels of synchronization have been commonly considered. The fully-synchronous (FSYNC) case [25] ensures each phase of each cycle is performed simultaneously by all robots. The semi-synchronous (SSYNC) case [25] allows that only a proper subset of robots execute a cycle at each round. Finally, the asynchronous (ASYNC) case [18] allows each robot to execute its cycle at its own pace. The Look-Compute-Move (LCM) model received a considerable amount of attention from the Distributed Computing community,[1] yielding a large variety of submodels used to assess the solvability of a certain task assuming certain system hypotheses. As such, the Distributed Computing literature about mobile robots so far can be seen as *computability-oriented.*

Alas, the various submodels make it extremely tedious to check whether a particular property of a robot protocol holds in a particular setting. Furthermore, these variants do not behave well regarding proof reusability: checking that a property holding in a given setting also holds in another setting that is not strictly contained in the former often amounts to developing a completely new proof, regardless of the proof arguments similarity. This problem is specially acute because of the great diversity of subtly different models: one may be tempted to simply hand-wave their way around the issue by declaring that the proof in this model is "obviously" also valid in this very close model, even more so as even a careful examination may not always find the most subtle errors. Last but not least, protocols are typically written in an informal high-level language: assessing whether they conform to a particular model setting is particularly cumbersome, and may lead to hard to find mismatches. As a result, sustained research effort was made in the last decade to use *formal methods* in the context of mobile robotic swarms.

Related Work. Formal methods encompass a long-lasting path of research that is meant to overcome errors of human origin. Perhaps the most well known instance in the Distributed Computing community is the *Temporal Logic of Actions* and its companion tools TLA/TLA+[12,22]. Though very expressive, TLA is designed for the shared memory and message passing contexts, thus not well suited to studying mobile robotic swarms. *Model-checking* and its powerful automation proved useful to find bugs in existing literature [7,16,17], and to assess formally published algorithms [7,13,15]. Automatic program synthesis (for the problem of perpetual exclusive exploration in a ring-shaped discrete space) is due to Bonnet *et al.* [9], and can be used to obtain automatically algorithms that are "correct-by-design". The approach was refined by Millet *et al.* [23] for the problem of gathering in a discrete ring network. However, these are limited to instances with few robots. Generalizing them to an arbitrary number of robots with similar models is doubtful as Sangnier *et al.* [24] proved that safety and reachability problems are undecidable in the parameterized case.

The approach on which we focus in this work is *formal proof*, that is proof development mechanically certified by a proof assistant. Mechanical proof assistants are proof management systems where a user can express data, programs,

[1] Yamashita received the "Prize for Innovation in Distributed Computing" for his seminal work on this model.

theorems and proofs. In sharp contrast with automated provers (like model-checkers), they are mostly interactive, and thus require some kind of expertise from their users. Sceptical proof assistants provide an additional guarantee by checking mechanically the soundness of a proof after it has been interactively developed. Formal proofs allow for more generality as this approach is not limited to *particular instances* of algorithms. During the last twenty years, the use of tool-assisted verification has extended to the validation of distributed processes, in contexts such as process algebras [8,19], symmetric interconnection networks [20], message passing settings [21], self-stabilization [1,14], etc. The main approach for mechanized proof dedicated to swarms of mobile entities is the PACTOLE framework (https://pactole.liris.cnrs.fr), based on the COQ proof assistant. Briefly, COQ is a Curry-Howard-based interactive proof assistant that enjoys a trustworthy kernel. Then, a proof development consists in building, interactively and using tactics, a λ-term, the type of which corresponds to the theorem to be proven. Most importantly: *a theorem or a lemma can only be saved/defined in the system if it comes with its type-checked proof.* Designed for mobile entities, and making the most of COQ's assets, PACTOLE allows for working on a given protocol to establish and certify its correctness [5,11], as well as for quantifying over all protocol so as to prove *impossibility* results [2,6,10], with an unspecified number of robots, possibly including a proportion of Byzantine faults, in continuous or discrete spaces. FSYNC/SSYNC and ASYNC modes are all supported, and the framework is expressive enough to state and certify results as theoretical as comparisons between demons or models [3].

Our Contribution. Taking some perspective over aforementioned works mixing formal methods and swarm robotics, one can only notice that the computability-centric approach of the Suzuki and Yamashita model yielded a concentration of efforts towards few benchmark problems that are theoretically interesting (one can get impossibility results or correctness certification) but of little practical relevance, such as perpetual or terminating exploration of a ring-shaped graph, and gathering or concentrating all robots at a particular location.

On the other side, relevant practical problems, such as constructing ad hoc mobile communication infrastructures to enable communication with rescue teams, remain untouched using a formal approach. Yet, their correctness is crucial, and possibly life-critical, so it should be assessed formally and mechanically verified. Overall, for those practical problems, the question is not really to characterize which system hypotheses enable problem solvability, but rather how to design a provably correct solution using hypotheses that correspond to real devices.

This paper is the first step in this direction. In more details, we start from a real-life application scenario to jointly design *(i)* its formal specification, *(ii)* a family of protocols that are able to solve the problem, and *(iii)* their corresponding proof of correctness, all expressed with the same formal framework, PACTOLE. In this process, we illustrate how formal methods and PACTOLE in particular could be used to derive protocols that are correct-by-design before they are deployed to actual devices.

The full version of the paper is available on ArXiv [4]. The developments described in this work, for Coq v8.13, based on the current version of Pactole, are available at https://pactole.liris.cnrs.fr.

References

1. Altisen, K., Corbineau, P., Devismes, S.: A framework for certified self-stabilization. In: Albert, E., Lanese, I. (eds.) FORTE 2016. LNCS, vol. 9688, pp. 36–51. Springer, Cham (2016). https://doi.org/10.1007/978-3-319-39570-8_3
2. Auger, C., Bouzid, Z., Courtieu, P., Tixeuil, S., Urbain, X.: Certified impossibility results for Byzantine-tolerant mobile robots. In: Higashino, T., Katayama, Y., Masuzawa, T., Potop-Butucaru, M., Yamashita, M. (eds.) SSS 2013. LNCS, vol. 8255, pp. 178–190. Springer, Cham (2013). https://doi.org/10.1007/978-3-319-03089-0_13
3. Balabonski, T., Courtieu, P., Pelle, R., Rieg, L., Tixeuil, S., Urbain, X.: Continuous *vs.* discrete asynchronous moves: a certified approach for mobile robots. In: Atig, M.F., Schwarzmann, A.A. (eds.) NETYS 2019. LNCS, vol. 11704, pp. 93–109. Springer, Cham (2019). https://doi.org/10.1007/978-3-030-31277-0_7
4. Balabonski, T., Courtieu, P., Pelle, R., Rieg, L., Tixeuil, S., Urbain, X.: Computer aided formal design of swarm robotics algorithms. CoRR abs/2101.06966 (2021). https://arxiv.org/abs/2101.06966
5. Balabonski, T., Delga, A., Rieg, L., Tixeuil, S., Urbain, X.: Synchronous gathering without multiplicity detection: a certified algorithm. Theory Comput. Syst. **63**(2), 200–218 (2017). https://doi.org/10.1007/s00224-017-9828-z
6. Balabonski, T., Pelle, R., Rieg, L., Tixeuil, S.: A foundational framework for certified impossibility results with mobile robots on graphs. In: Bellavista, P., Garg, V.K. (eds.) Proceedings of the 19th International Conference on Distributed Computing and Networking, ICDCN 2018, Varanasi, India, 4–7 January 2018, pp. 5:1–5:10. ACM (2018). https://doi.org/10.1145/3154273.3154321
7. Bérard, B., Lafourcade, P., Millet, L., Potop-Butucaru, M., Thierry-Mieg, Y., Tixeuil, S.: Formal verification of mobile robot protocols. Distrib. Comput. **29**(6), 459–487 (2016). https://doi.org/10.1007/s00446-016-0271-1
8. Bezem, M., Bol, R., Groote, J.F.: Formalizing process algebraic verifications in the calculus of constructions. Formal Aspects Comput. **9**, 1–48 (1997)
9. Bonnet, F., Défago, X., Petit, F., Potop-Butucaru, M., Tixeuil, S.: Discovering and assessing fine-grained metrics in robot networks protocols. In: 33rd IEEE International Symposium on Reliable Distributed Systems Workshops, SRDS Workshops 2014, Nara, Japan, 6–9 October 2014, pp. 50–59. IEEE (2014). https://doi.org/10.1109/SRDSW.2014.34
10. Courtieu, P., Rieg, L., Tixeuil, S., Urbain, X.: Impossibility of gathering, a certification. Inf. Process. Lett. **115**, 447–452 (2015). https://doi.org/10.1016/j.ipl.2014.11.001
11. Courtieu, P., Rieg, L., Tixeuil, S., Urbain, X.: Certified universal gathering in $\mathbb{R}^2$ for oblivious mobile robots. In: Gavoille, C., Ilcinkas, D. (eds.) DISC 2016. LNCS, vol. 9888, pp. 187–200. Springer, Heidelberg (2016). https://doi.org/10.1007/978-3-662-53426-7_14
12. Cousineau, D., Doligez, D., Lamport, L., Merz, S., Ricketts, D., Vanzetto, H.: TLA$^+$ Proofs. In: Giannakopoulou, D., Méry, D. (eds.) FM 2012. LNCS, vol. 7436, pp. 147–154. Springer, Heidelberg (2012). https://doi.org/10.1007/978-3-642-32759-9_14

13. Défago, X., Heriban, A., Tixeuil, S., Wada, K.: Using model checking to formally verify rendezvous algorithms for robots with lights in Euclidean space. In: International Symposium on Reliable Distributed Systems, SRDS 2020, Shanghai, China, 21–24 September 2020, pp. 113–122. IEEE (2020). https://doi.org/10.1109/SRDS51746.2020.00019
14. Deng, Y., Monin, J.F.: Verifying self-stabilizing population protocols with coq. In: Chin, W.N., Qin, S. (eds.) Third IEEE International Symposium on Theoretical Aspects of Software Engineering (TASE 2009), Tianjin, China, pp. 201–208. IEEE Computer Society, July 2009
15. Devismes, S., Lamani, A., Petit, F., Raymond, P., Tixeuil, S.: Optimal grid exploration by asynchronous oblivious robots. In: Richa, A.W., Scheideler, C. (eds.) SSS 2012. LNCS, vol. 7596, pp. 64–76. Springer, Heidelberg (2012). https://doi.org/10.1007/978-3-642-33536-5_7
16. Doan, H.T.T., Bonnet, F., Ogata, K.: Model checking of a mobile robots perpetual exploration algorithm. In: Liu, S., Duan, Z., Tian, C., Nagoya, F. (eds.) SOFL+MSVL 2016. LNCS, vol. 10189, pp. 201–219. Springer, Cham (2017). https://doi.org/10.1007/978-3-319-57708-1_12
17. Doan, H.T.T., Bonnet, F., Ogata, K.: Model checking of robot gathering. In: Aspnes, J., Bessani, A., Felber, P., Leitão, J. (eds.) 21st International Conference on Principles of Distributed Systems, OPODIS 2017, Lisbon, Portugal, 18–20 December 2017. LIPIcs, vol. 95, pp. 12:1–12:16. Schloss Dagstuhl - Leibniz-Zentrum für Informatik (2017). https://doi.org/10.4230/LIPIcs.OPODIS.2017.12
18. Flocchini, P., Prencipe, G., Santoro, N., Widmayer, P.: Gathering of asynchronous robots with limited visibility. Theor. Comput. Sci. **337**(1–3), 147–168 (2005). https://doi.org/10.1016/j.tcs.2005.01.001
19. Fokkink, W.: Modelling Distributed Systems. EATCS Texts in Theoretical Computer Science, Springer, Heidelberg (2007)
20. Gaspar, N., Henrio, L., Madelaine, E.: Bringing coq into the world of GCM distributed applications, pp. 643–662 (2014)
21. Küfner, P., Nestmann, U., Rickmann, C.: Formal verification of distributed algorithms. In: Baeten, J.C.M., Ball, T., de Boer, F.S. (eds.) TCS 2012. LNCS, vol. 7604, pp. 209–224. Springer, Heidelberg (2012). https://doi.org/10.1007/978-3-642-33475-7_15
22. Lamport, L.: The temporal logic of actions. ACM Trans. Program. Lang. Syst. **16**(3), 872–923 (1994). https://doi.org/10.1145/177492.177726
23. Millet, L., Potop-Butucaru, M., Sznajder, N., Tixeuil, S.: On the synthesis of mobile robots algorithms: the case of ring gathering. In: Felber, P., Garg, V. (eds.) SSS 2014. LNCS, vol. 8756, pp. 237–251. Springer, Cham (2014). https://doi.org/10.1007/978-3-319-11764-5_17
24. Sangnier, A., Sznajder, N., Potop-Butucaru, M., Tixeuil, S.: Parameterized verification of algorithms for oblivious robots on a ring. Formal Methods Syst. Des. (6), 55–89 (2019). https://doi.org/10.1007/s10703-019-00335-y
25. Suzuki, I., Yamashita, M.: Distributed anonymous mobile robots: formation of geometric patterns. SIAM J. Comput. **28**(4), 1347–1363 (1999)

Delta-State JSON CRDT: Putting Collaboration on Solid Ground

Amos Brocco(✉)

University of Applied Sciences and Arts of Southern Switzerland, Lugano, Switzerland
amos.brocco@supsi.ch

Abstract. In this paper we present a framework to support the implementation of offline-first asynchronous collaboration using a variety of data storage and communication backends. In particular, our approach can make use of Solid pods to exchange data between users.

Keywords: Collaborative applications · JSON · CRDT · Solid

1 Introduction

In this paper we present a framework for the development of collaboration features into applications, by combining the advantages of CRDTs [7,10] with the flexibility and safety of decentralized storage. As such, our intent is to exploit Solid pods [1] as a communication channel for exchanging data between users. However, thanks to the simplicity and modularity of our design, storage solutions such as shared folders or cloud file-sharing platforms, as well as physical devices like thumb drives, can also serve the same function as a pod. To ease the integration of our solution, we allow transparent replication of complex JSON documents without explicit editing. It is therefore possible to implement collaborative features into existing programs without altering their data model. The remaining of this paper presents a brief review of the relevant related work in the field of CRDTs, a formal overview of our data structure and some details of its implementation. Finally, an evaluation of the proposed solution and the corresponding results will be discussed.

2 Related Work

CRDTs can be grouped into three different categories, namely operation-based, state-based and delta-state based. Operation-based solutions [3] rely on update operations which are propagated to all replicas, and are best suited for high-frequency updates, such as in the context of real-time collaborative text editors. In the literature it is possible to find examples of operation-based CRDTs which support JSON-like data, for instance Yjs [8] or the *Automerge* [6] library, but require explicit editing of the document in order to keep track of each modification. State-based CRDTs [10] always store the full state of the data, and are

C. Johnen et al. (Eds.): SSS 2021, LNCS 13046, pp. 474–478, 2021.
https://doi.org/10.1007/978-3-030-91081-5_32

therefore better suited for situations where updates are less frequent, communication is unreliable, or operations are not commutative. The main drawback of state-based CRDTs is that the size of the state can become very large [2], consequently delta-state solutions (referred to as δ-CRDTs) have been proposed [2,9]. δ-CRDTs rely on disseminating small updates (changesets) called delta mutations: these updates are idempotent, which means that they can be applied possibly several times to an existing state without compromising its consistency, and can be disseminated over unreliable communication channels. The CRDT discussed in this paper uses delta states and presents a practical architecture for collaborative applications with a modular design supporting different types of communication and storage backends.

3 Overview of the Data Structure

We consider a generic collection C of JSON objects which is generated from an arbitrary JSON document using a reversible data transformation algorithm [5]. In contrast to other solutions, which require explicit editing operations, this algorithm processes an input document (as produced by an application) to automatically extract nested objects and determine their changes. This collection can be replicated on multiple sites and concurrently updated by each participant. Each object o in this collection is identified by a UUID id_o and its value (or content) can be modified independently on each replica. We assume that objects are atomic and immutable. The collection is a grow-only data structure, where deletions are recorded using a *tombstone*. To efficiently compare different versions of an object, the content x is hashed to produce a string digest $H(x)$.

Object Map. On each replica of the data structure, the set O of tuples $\langle H(x), x\rangle$ stores the content of each version of each object. The set O is referred to as the *object map*, and allows for retrieving the value associated with a specific digest.

Revision String. Let x_N represent the content of the N-th version of object x. The *revision string* r_N associated with x_N is defined as N-$H(x_N)\_Tail_N$, where N is a monotonically increasing numerical index (starting at 1), $Tail_N = T(H(r_{N-1}))$, and T is a determistic function (such as the identity function, a hashing function or a simple string transformation). A revision string r_k univocally refers to a specific version x_k in the history of an object, and allows for retrieving the exact content through the embedded digest string $H(x_k)$.

Revision Trees. Modifications made to an object can be recorded as sequences of *revision strings*. The history of modifications made to each object o across all replicas is represented by a *revision tree* rt_o, which can be conveniently stored as a collection of tuples $\langle r_N, r_{N-1}\rangle$ (r_{N-1} being referred to as the *parent* of r_N). The revision tree for a newly created object is $\langle r_1, \emptyset\rangle$. The revision with the highest numerical index is considered the *winning revision* r_W, and determines the contents that shall be returned when querying for the latest version of an object. If multiple revisions share the same index, revision strings are compared in lexicographic sort order.

State Set. Given a replica of a collection of JSON objects C, we define its *state set* (or simply *state*) $X = D \cup O$, where $D = \bigcup_{o \in C} rt_o$, and O is the *object map* as defined above.

3.1 Delta-State Decomposition

According to [2], a δ-CRDT consists of a triple (S, M^δ, Q), where S is a join-semilattice of states, M^δ is a set of delta-mutators, and Q is a set of query functions. The state transition at each replica is given by either joining the current state $X \in S$ with a delta-mutation ($X' = X \sqcup m_\delta(X)$), or by joining the current state with some received delta-group D ($X' = X \sqcup D$). Delta-mutators are defined as functions, corresponding to an update operation, which take a state X in a join-semilattice S as parameter and return a delta-mutation $m_\delta(X) \in S$. Finally, a delta-group is inductively defined as either a delta-mutation or a join of several delta-groups. In the considered scenario, each transition from state X to state X' can be represented by an *update set* $U = X' \setminus X$, with $U \in S$, since S is closed under set difference. Update sets are equivalent to delta-mutations, since $X' = m_\delta(X) = X \sqcup m_\delta(X) = X \sqcup U$, where $X, X' \in S$ and $m_\delta(X) \in S$ is the delta mutation. Delta-mutators m_δ are defined by the relation $m_\delta(X) = X \sqcup U$. Furthermore, update sets also translate into delta-groups D, as the relation $X' = X \sqcup D$ holds when $D = m_\delta(X)$, and by associativity this relation is verified even when considering a join of several delta-groups. Since the join operation is associative, commutative and idempotent, the requirements for convergence (as stated in [2]) are fulfilled.

3.2 Delta-State Serialization and Adapters

The serialization format is derived from the one presented in [4]. Update sets are serialized into two different JSON structures, namely *delta blocks* and *data packs*. The former stores revision tree updates, whereas the latter maintains the actual content of each new object. Both structures are immutable, hence they can be cached locally to reduce network overhead. To cope with the possibility that a *data pack* hasn't yet been delivered to a replica, we redefine *winning revision* as the one with the highest numerical index *whose content is available*. To store and replicate data on different platforms, the low-level task of reading or writing the delta blocks and data packs is fulfilled by means of *adapters*. Adapters can be used to seamlessly support different types of storage, such as main memory, filesystems (where delta blocks and packs are files), databases, decentralized data pods, or cloud sharing platforms (such as Dropbox and Nextcloud).

3.3 Example Architecture of a Collaborative Application

The functionalities of the underlying CRDT are exposed through a high-level API which implements methods to update, read and synchronize replicas. An application can make use of these methods to support asynchronous collaborative editing without reinventing its data model.

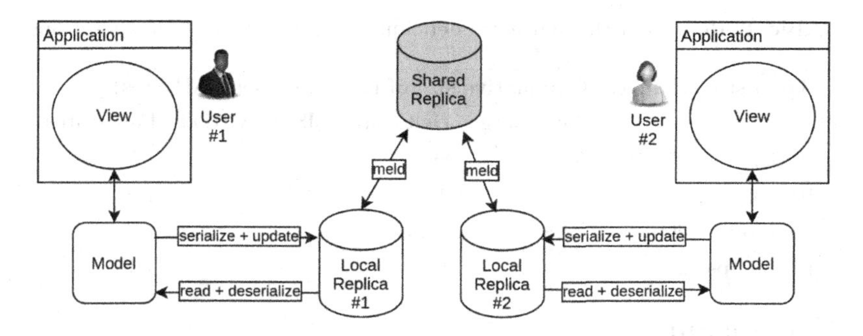

Fig. 1. Example architecture of a collaborative application

Figure 1 shows an example architecture with two local replicas and one shared replica (a simple data store). Each user can work on their data while offline. Replication is achieved by serializing the contents of the data model into a JSON document and subsequently *update* the local replica. By means of the *melding* procedure, changes made to the local replica are propagated to the shared replica through a backend adapter. Afterwards, changes from the shared replica are *melded* into the local one, integrating modifications made by other users. Finally, the local replica can be *read* and deserialized so as to obtain an updated model. This workflow can be used to mimic the co-authoring functionality called *save and refresh* offered by Microsoft Office.

Solid Pod Adapter. The Solid specification [1] provides a standard for building an ecosystem of personal web-accessible data *pods*. Access to a pod is controlled by the owner, and linked data is exploited to promote interoperability between applications and pods. In our framework we store delta blocks and data packs as LDP Resources inside LDP Containers.

4 Evaluation

To evaluate the proposed solution, we consider a synthetic benchmark to determine the space overhead in comparison to Automerge [6]. We employ an editing trace[1] of a large text document with 182 315 single-character insertion operations, and 77 463 single-character deletion operations. To simulate asynchronous collaboration, the editing process is divided into batches of 10, 100, 1 000, and 10 000 single-character operations: when a batch is completed, a *changeset* is generated. As shown in Table 1, as the number of edits in each batch increases, our delta-state CRDT incurs a smaller space overhead compared to Automerge.

[1] https://github.com/automerge/automerge-perf.

Table 1. Results of the synthetic benchmark after 259 778 edit operations

Batch size	Batches	Cumulative size of the changesets (MBytes)		
		Automerge	Delta-State JSON CRDT	Full-states
10 ops	25 978	56.5	86	1 699.8
100 ops	2 598	54	23.7	170
1 000 ops	260	52.8	16.5	17
10 000 ops	26	51.5	15	1.7

5 Conclusion

In this paper, we presented a JSON δ-CRDT solution to support the development of multi-user applications based on asynchronous collaboration. Our design is both simple and modular, and by means of adapters, it allows for seamless interoperation between different storage and communication backends, such as Solid pods. Future work includes performance optimizations, the implementation of additional adapters, and porting the library to other languages and platforms (such as WebAssembly).

Acknowledgments. This work has been financially supported by the Swiss Innovation Agency, Project nr. 42832.1 IP-ICT and by Banana.ch SA.

References

1. Solid technical reports (2021). https://solidproject.org/TR/. Accessed 5 Aug 2021
2. Almeida, P.S., Shoker, A., Baquero, C.: Delta state replicated data types. J. Parallel Distrib. Comput. **111**, 162–173 (2018)
3. Baquero, C., Almeida, P.S., Shoker, A.: Making operation-based CRDTs operation-based. In: Proceedings of the First Workshop on Principles and Practice of Eventual Consistency, PaPEC 2014. Association for Computing Machinery, New York (2014)
4. Brocco, A.: The document chain: a delta-CRDT framework for arbitrary JSON data. In: SEBD: 29th Italian Symposium on Advanced Database Systems (2021)
5. Brocco, A., Ceppi, P., Sinigaglia, L.: libJoTs: JSON that syncs! In: SEBD: 28th Italian Symposium on Advanced Database Systems (2020)
6. Kleppmann, M., Beresford, A.R.: A conflict-free replicated JSON datatype. IEEE Trans. Parallel Distrib. Syst. **28**(10), 2733–2746 (2017)
7. Letia, M., Preguiça, N., Shapiro, M.: Consistency without concurrency control in large, dynamic systems. SIGOPS Oper. Syst. Rev. **44**(2), 29–34 (2010)
8. Nicolaescu, P., Jahns, K., Derntl, M., Klamma, R.: Yjs: a framework for near real-time P2P shared editing on arbitrary data types, June 2015
9. Rinberg, A., Solomon, T., Khazma, G., Lushi, G., Shlomo, R., Ta-Shma, P.: Array CRDTs using delta-mutations. In: 8th Workshop on Principles and Practice of Consistency for Distributed Data, PaPoC 2021. ACM, April 2021
10. Shapiro, M., Preguiça, N., Baquero, C., Zawirski, M.: A comprehensive study of Convergent and Commutative Replicated Data Types. Research Report RR-7506, Inria - Centre Paris-Rocquencourt; INRIA, January 2011

Self-stabilization and Byzantine Tolerance for Maximal Independent Set

Johanne Cohen[1], Laurence Pilard[2], and Jonas Sénizergues[1(✉)]

[1] LISN-CNRS, Université Paris-Saclay, Orsay, France
{Johanne.Cohen,Jonas.Senizergues}@lri.fr
[2] LI-PaRAD, UVSQ, Université Paris-Saclay, Orsay, France
Laurence.Pilard@uvsq.fr

Abstract. We analyze the impact of transient and Byzantine faults on the construction of a maximal independent set in a general network. We adapt the self-stabilizing algorithm presented by Turau [15] for computing such a vertex set. Our algorithm is self-stabilizing, and also works under the more difficult context of arbitrary Byzantine faults.

Byzantine nodes can prevent nodes close to them from taking part in the independent set for an arbitrarily long time.

We give boundaries to their impact using a variation on the notion of containment radius. As far as we know, we present the first algorithm tolerating both transient and Byzantine faults under the fair distributed daemon. We prove that this algorithm converges in $\mathcal{O}(\Delta n)$ rounds with high probability. Additionally, we present a modified version of this algorithm for anonymous systems under the adversarial distributed daemon that converges in $\mathcal{O}(n^2)$ expected number of steps.

Introduction

Maximal independent set has received a lot of attention in different areas. For instance, in wireless networks, the maximum independent sets can be used as a black box to perform communication (to collect or to broadcast information) (see [5,12] for example). An *independent set I* in a graph is a set of vertices such that no two of them form an edge in the graph. It is called *maximal* when it is maximal inclusion-wise (in which case it is also a minimal dominating set). The maximal independent set (MIS) problem has been extensively studied in parallel and distributed settings, following the seminal works of [1,11,13].

Self-stabilizing algorithms for maximal independent set have been designed in vario us models (anonymous network [14,16] or not [6,9]). Up to our knowledge, Shukla *et al.* [14] present the first algorithm designed for finding a MIS in a graph using self-stabilization paradigm for anonymous networks. Some other self-stabilizing works deal with this problem assuming identifiers: with a synchronous daemon [6] or distributed one [9]. These two works require $O(n^2)$ moves to converge. Turau [15] improves these results to $O(n)$ moves under the distributed daemon. Recently, some works improved the results in the synchronous model.

C. Johnen et al. (Eds.): SSS 2021, LNCS 13046, pp. 479–483, 2021.
https://doi.org/10.1007/978-3-030-91081-5_33

For non-anonymous networks, Hedetniemi [8] designed a self-stabilization algorithm for solving the problem related to dominating sets in graphs in particular for a maximal independent set which stabilizes in $O(n)$ synchronous rounds. Moreover, for anonymous networks, Turau [16] designs some Randomized self-stabilizing algorithms for maximal independent set w.h.p. in $O(\log n)$ rounds. See the survey [7] for more details on MIS self-stabilizing algorithms.

We focus on the construction of a MIS handling both transient and Byzantine faults. On one side, transient faults can appear in the whole system, possibly impacting all nodes. However, these faults are not permanent, thus they stop at some point of the execution. Self-stabilization [3] is the classical paradigm to handle transient faults. On the other side, (permanent) Byzantine faults [10] are located on some faulty nodes and so those faults only occur from them.

1 Model

A system consists of a set of processes where two adjacent processes can communicate with each other. The communication relation is represented by a graph $G = (V, E)$ where V is the set of the processes (*nodes*) and E represents the neighborhood relation between them, *i.e.* $uv \in E$ when u and v can communicate. By convention we write $|V| = n$. If u is a node, we note $N(u) = \{v \in V | uv \in E\}$, $deg(u) = |N(u)|$, and $\Delta = \max\{deg(u) | u \in V\}$.

We assume the system to be *anonymous* meaning that a node has no identifier. We use the *state model*, which means that each node has a set of *local variables* which make up the *local state* of the node. A node can read its local variables and all the local variables of its neighbors, but can only rewrite its own local variables. A *configuration* is the value of the local states of all nodes. When u is a node and x a local variable, the *x-value* of u is the value x_u.

An algorithm is a set of *rules*, where each rule is of the form $\langle guard \rangle \rightarrow \langle command \rangle$. The *guard* is a predicate over the variables of a node and its neighbors. The *command* is some actions that may change the values of the node's variables. The *activation* of a rule on a node may only change the value of variables of that specific node, but multiple moves may be performed at the same time, as long as they act on different nodes. A *move* is a couple (u, r) where u is a node, and r a rule.

Then, a *transition* is a triplet $\gamma \xrightarrow{t} \gamma'$ such that γ' is a configuration accessible from γ by performing simultaneously all the moves of t. An *execution* is an alternate sequence of configurations and move sets such that the sequence either is infinite or finishes by a configuration and every factor of length three beginning by a configuration is a transition.

When the execution is finite, the last element of the sequence is the *last configuration* of the execution. An execution is *maximal* if it is infinite, or it is finite and no node is activable in the last configuration (such a configuration is called *stable*). It is called *partial* otherwise.

The *daemon* is the adversary that chooses, from a given configuration, which nodes to activate in the next transition. Two types are used in this paper: the *adversarial distributed daemon* that allows all possible executions and the *fair distributed daemon* that only allows executions where nodes cannot be continuously activable without being eventually activated.

Given a specification and $\mathcal{L}$ the associated set of *legitimate configuration*, *i.e.*, the set of the configurations that verify the specification, a probabilistic algorithm is *self-stabilizing* when these properties are true: (*correctness*) every configuration of an execution starting by a configuration of $\mathcal{L}$ is in $\mathcal{L}$ and (*convergence*) from any configuration, whatever the strategy of the daemon, the resulting execution eventually reaches a configuration in $\mathcal{L}$ with probability 1.

The time complexity of an algorithm that assumes the fair distributed daemon is given as a number of *rounds*. The concept of round was introduced by Dolev *et al.* [4], and reworded by Cournier *et al.* [2] to take into account activable nodes. We quote the following definition from Cournier *et al.* [2]:

Definition 1. *Let $\mathcal{E}$ be an execution. A* round *is a sequence of consecutive transitions in $\mathcal{E}$. The first round begins at the beginning of $\mathcal{E}$; successive rounds begin immediately after the previous round has ended. The current round ends once every node $u \in V$ satisfies at least one of the following two properties: (i) u has been activated in at least one transition during the current round or (ii) u has been non-activable in at least one configuration during the current round.*

Our first algorithm is to be executed in the presence of *Byzantine* nodes: there is a subset $B \subseteq V$ of adversarial nodes. Byzantine nodes are always activable, and when activated, they are free to update or not their local variables. Finally, observe that in the presence of Byzantine nodes all maximal executions are infinite. We denote by $d(u, B)$ the minimal (graph) distance between node u and a Byzantine node, and we define for $i \in \mathbb{N}$: $V_i = \{u \in V | d(u, B) > i\}$. Note that V_0 is exactly the set of non-Byzantine nodes, and that V_{i+1} is exactly the set of nodes of V_i whose neighbors are all in V_i.

2 With Byzantine Nodes Under the Fair Daemon

The algorithm builds a maximal independent set represented by a local variable s. The approach of the state of the art is the following: when two nodes are candidate to be in the independent set, then a local election decides who will remain in the independent set. To perform a local election, the standard technique is to compare the identifiers of nodes. Unfortunately, this mechanism is not robust to the presence of Byzantine nodes.

Keeping with the approach outlined above, when a non-Byzantine node observes that its neighbors are not in (or trying to be in) the independent set, the node decides to join it with a certain probability. The choice of probability should reduce the impact of Byzantine nodes while maintaining the efficiency of the algorithm. $Rand(p)$ with $p \in [0, 1]$ represents the random function that outputs 1 with probability p, and 0 otherwise.

Algorithm 1. *Any node u has two local variables $s_u \in \{\bot, \top\}$ and $x_u \in \mathbb{N}$ and may make a move according to one of the following rules:*
(Refresh) $x_u \neq |N(u)| \rightarrow x_u := |N(u)| \quad (= deg(u))$
(Candidacy?) $(x_u = |N(u)|) \wedge (s_u = \bot) \wedge (\forall v \in N(u), s_v = \bot) \rightarrow$
. $\quad$ *if* $Rand\left(\frac{1}{1+\max(\{x_v | v \in N(u) \cup \{u\}\})}\right) = 1$ *then* $s_u := \top$
(Withdrawal) $(x_u = |N(u)|) \wedge (s_u = \top) \wedge (\exists v \in N(u), s_v = \top) \rightarrow s_u := \bot$

A node joins the MIS with a probability $\frac{1}{1+\max(\{x_v | v \in N(u) \cup \{u\}\})}$. The idea to ask the neighbors about their own number of neighbors (through the use of the x variable) to choose the probability of a candidacy comes from the mathematical property $\forall k \in \mathbb{N}, (1 - \frac{1}{k+1})^k > e^{-1}$, which will allow to have a good lower bound for the probability of the event "some node made a successful candidacy, but none of its neighbors did".

Since Byzantine nodes are not bound to follow the rules, we cannot hope for a correct solution in the entire graph. What we wish to do is to find a solution that works when we are far enough from the Byzantine nodes. One could think about a fixed containment radius around Byzantine nodes, but it is not as simple, and does not work with our approach. We define on any configuration γ the following set of nodes, that represents the already built independent set: $I_\gamma = \{u \in V_1 | (s_u^\gamma = \top) \wedge \forall v \in N(u), s_v^\gamma = \bot\}$ in order to adapt the notion of containment radius.

Theorem 1. *For any $p \in [0, 1[$. From any configuration γ, Algorithm is self-stabilizing for a configuration γ' where $I_{\gamma'}$ is a maximal independent set of $V_2 \cup I_{\gamma'}$, with time complexity $1 + \max\left(-\alpha^2 \ln p, \frac{\sqrt{2}}{\sqrt{2}-1}\frac{n}{\alpha}\right)$ rounds with probability at least $1 - p$, where $\alpha = \frac{1}{(\Delta+1)e}$.*

3 Anonymous System Under the Adversarial Daemon

The previous algorithm works in this case, but as the measure of complexity changes without fairness hypothesis, the above analysis would not give a suitable bound. We keep the idea of having nodes making candidacy, and then withdraw if the situation to be candidate is not right. As we still do not have identifiers, we also need probabilistic tie breaking. But contrary to the Byzantine case, we move the probabilities to the **Withdrawal** rule: a non-candidate node with no candidate neighbor will always become candidate when activated, but a candidate node with a candidate neighbor will only withdraw with probability $\frac{1}{2}$.

Algorithm 2. *Any node u has a single local variable $s_u \in \{\bot, \top\}$ and may make a move according to one of the following rules:*
(Candidacy) $(s_u = \bot) \wedge (\forall v \in N(u), s_v = \bot) \rightarrow s_u := \top$
(Withdrawal?) $(s_u = \top) \wedge (\exists v \in N(u), s_v = \top) \rightarrow$ *if* $Rand(\frac{1}{2}) = 1$ *then* $s_u := \bot$

The idea behind this is that we can give a non-zero lower bound on the probability that a connected component of candidate nodes eventually collapses into at least one definitive member of the independent set. As every transition with **Candidacy** moves makes such sets appear, and **Withdrawal?** moves make those collapse into members of the independent set with non-zero probability, it should converge toward a maximal independent set.

Theorem 2. *From any configuration γ, the expected number of steps to reach a stable configuration is at most $3n^2$.*

References

1. Alon, N., Babai, L., Itai, A.: A fast and simple randomized parallel algorithm for the maximal independent set problem. J. Algorithms **7**(4), 567–583 (1986)
2. Cournier, A., Devismes, S., Villain, V.: Snap-stabilizing PIF and useless computations. In: ICPADS, pp. 39–48 (2006)
3. Dijkstra, E.W.: Self-stabilizing systems in spite of distributed control. Commun. ACM **17**(11), 643–644 (1974)
4. Dolev, S., Israeli, A., Moran, S.: Uniform dynamic self-stabilizing leader election. IEEE Trans. Parallel Distrib. Syst. **8**(4), 424–440 (1997)
5. Gao, X., et al.: A novel approximation for multi-hop connected clustering problem in wireless networks. IEEE/ACM Trans. Netw. **25**(4), 2223–2234 (2017)
6. Goddard, W., Hedetniemi, S.T., Jacobs, D.P., Srimani, P.K.: Self-stabilizing protocols for maximal matching and maximal independent sets for ad hoc networks. In: IPDPS, p. 14-p. IEEE (2003)
7. Guellati, N., Kheddouci, H.: A survey on self-stabilizing algorithms for independence, domination, coloring, and matching in graphs. JPDC **70**, 406–415 (2010)
8. Hedetniemi, S.T.: Self-stabilizing domination algorithms. In: Structures of Dominationin Graphs, pp. 485–520 (2021)
9. Ikeda, M., Kamei, S., Kakugawa, H.: A space-optimal self-stabilizing algorithm for the maximal independent set problem. In: PDCAT, pp. 70–74 (2002)
10. Lamport, L., Shostak, R.E., Pease, M.C.: The byzantine generals problem. ACM Trans. Program. Lang. Syst. **4**(3), 382–401 (1982)
11. Linial, N.: Distributive graph algorithms global solutions from local data. In: 28th Annual Symposium on Foundations of Computer Science, pp. 331–335 (1987)
12. Liu, T., Wang, X., Zheng, L.: A cooperative SWIPT scheme for wirelessly powered sensor networks. IEEE Trans. Commun. **65**(6), 2740–2752 (2017)
13. Luby, M.: A simple parallel algorithm for the maximal independent set problem. SIAM J. Comput. **15**(4), 1036–1053 (1986)
14. Shukla, S.K., Rosenkrantz, D.J., Ravi, S.S., et al.: Observations on self-stabilizing graph algorithms for anonymous networks. In: Proceedings of the Second Workshop on Self-stabilizing Systems, vol. 7, p. 15 (1995)
15. Turau, V.: Linear self-stabilizing algorithms for the independent and dominating set problems using an unfair distributed scheduler. Inf. Process. Lett. **103**(3), 88–93 (2007)
16. Turau, V.: Making randomized algorithms self-stabilizing. In: Censor-Hillel, K., Flammini, M. (eds.) SIROCCO 2019. LNCS, vol. 11639, pp. 309–324. Springer, Cham (2019). https://doi.org/10.1007/978-3-030-24922-9_21

Coordinating Amoebots via Reconfigurable Circuits

Michael Feldmann[1], Andreas Padalkin[1](✉), Christian Scheideler[1], and Shlomi Dolev[2]

[1] Paderborn University, Paderborn, Germany
{michael.feldmann,andreas.padalkin,scheideler}@upb.de
[2] Ben-Gurion University of the Negev, Be'er Sheva, Israel
dolev@cs.bgu.ac.il

Abstract. We consider an extension to the geometric amoebot model that allows amoebots to form so-called *circuits*. Given a connected amoebot structure, a circuit is a subgraph formed by the amoebots that permits the instant transmission of signals. We show that such an extension allows for significantly faster solutions to a variety of problems related to programmable matter. More specifically, we provide algorithms for leader election, consensus, compass alignment, chirality agreement, and shape recognition. Leader election can be solved in $\Theta(\log n)$ rounds, w.h.p., consensus in $O(1)$ rounds, and both, compass alignment and chirality agreement, can be solved in $O(\log n)$ rounds, w.h.p. For shape recognition, the amoebots have to decide whether the amoebot structure forms a particular shape. We show that the amoebots can detect a shape composed of triangles within $O(1)$ rounds. Finally, we show how the amoebots can detect a parallelogram with linear and polynomial side ratio within $\Theta(\log n)$ rounds, w.h.p.

Keywords: Progammable matter · Amoebot model · Reconfigurable circuits

1 Introduction

Programmable matter is a physical substance consisting of tiny, homogeneous robots (also called *particles*) that is able to dynamically change its physical properties like shape or density. Such a substance can be deployed, for example, for minimal invasive surgeries through injection into the human body (detecting cancer cells, repairing bones, closing blood vessels, etc.). Programmable matter has been envisioned for 30 years [7] and is yet to be realized in practice. However, theoretical investigations on various models (such as the self-assembly model [6],

This work has been supported by the DFG Project SCHE 1592/6-1 (PROGMATTER).
A full version of this brief announcement is available online at the following address: https://arxiv.org/abs/2105.05071.

C. Johnen et al. (Eds.): SSS 2021, LNCS 13046, pp. 484–488, 2021.
https://doi.org/10.1007/978-3-030-91081-5_34

the nubot model [8] or the geometric amoebot model [2]) have already started and is still continuing in the distributed computing community.

Shape formation algorithms are of particular interest. Algorithms of polylogarithmic complexity are known for the nubot model [8]. However, these assume particles on the molecular scale since it requires the rotation of entire substructures. Due to the acting forces, this would not be possible on the micro or macro scale. In contrast, many problems for the geometric amoebot model come with a natural lower bound of $\Omega(D)$, where D is the diameter of the structure formed by the amoebots. The main goal of our research is to formulate a model that is able to break this lower bound while still being reasonable on the micro or even macro scale.

Many of the various models for programmable matter take their inspiration from nature. For example, the particles of the nubot model resemble molecules, and the locomotion of the particles of the amoebot model is inspired by amoeba. However, many more fascinating forms of locomotion can be found in nature. Our model is inspired by the muscular system. Muscles are composed of muscle fibers, which can be stimulated to perform coordinated contractions. These contractions (and their counterpart relaxations) allow for fast locomotion. The stimuli are inflicted by the nervous system. The nervous system consists of highly connected nerves. These are able to rapidly transmit primitive signals (the stimuli) over long distances. Our aim is to come up with a model for programmable matter incorporating both concepts: the muscular system and the nervous system.

Instead of proposing an entirely new model, we build our model on top of the geometric amoebot model. This model is tailor-made for our purpose since it already provides contractions (and expansions) on a small scale of single particles. Inspired by the nervous system described above, in this brief announcement, we introduce reconfigurable circuits to the geometric amoebot model. Each particle is allowed to create a constant amount of circuits with a subset of the particle structure. A circuit formed by particles allows for the instantaneous transmission of primitive signals to all of these. Since physical constraints like the maximum force at which particles can contract or expand have to be taken into account, the proper modeling of the muscular system is subject of future work.

2 Reconfigurable Circuit Extension

We build our extension on the canonical amoebot model recently proposed by Daymude et al. [1] and focus on the geometric variant though our circuit extension can also be applied to any other grid graph.

In the geometric amoebot model [1], a set of n amoebots is placed on the infinite regular triangular grid graph $G_{eqt} = (V, E)$ (see Fig. 1a). An amoebot is an anonymous, randomized finite state machine that either occupies one or two adjacent nodes of G_{eqt}, and every node of G_{eqt} is occupied by at most one amoebot. If an amoebot occupies just one node, it is called *contracted* and otherwise *expanded*, and exactly one of its occupied nodes is called its *head*. We assume that initially, the amoebot structure is connected and all amoebots are contracted. Each amoebot has a compass orientation (it defines one of its incident edges as the northern direction) and a chirality (a sense of clockwise or counterclockwise rotation) that it can maintain as it moves, but initially the amoebots might not agree on their compass orientation and chirality. We refer to [1] for more details on the model.

In our reconfigurable circuit extension, each edge between two neighboring amoebots u and v is replaced by k edges called *external links* with endpoints called *pins*, for some constant $k \geq 1$ that is the same for all amoebots. For each of these links, one pin is owned by u while the other pin is owned by v. We assume that the k pins on the side of u resp. v are consecutively numbered from 1 to k, and there are two possible ways by which these pins are matched (i.e., belong to the same link). If u and v have the same chirality, pin i of u is matched with pin $k - i + 1$ of v, and if u and v have different chiralities, pin i of u is matched with pin i of v (see also Fig. 1c).

Each amoebot u can connect its pin set $P(u)$ via an arbitrary set of *internal links* $L(u)$, i.e., its pin set can form an arbitrary undirected graph $H(u) = (P(u), L(u))$. We call $H(u)$ the *pin configuration* of u. Let X be the set of all external links. Moreover, let $P = \bigcup_A P(u)$ be the set of all pins in the system and $L = X \cup (\bigcup_{u \in S} L(u))$ be the set of all links in the system. Then, we call $H = (P, L)$ the *pin configuration* of the system and any connected component C of H a *circuit* (see Fig. 1b). Note that if $\bigcup_{u \in S} L(u)$ is empty, then every circuit of H just connects two neighboring amoebots. However, an external link between the neighboring amoebots u and v can only be maintained as long as both, u and v occupy the incident nodes. If either amoebot leaves the respective node, the external link, its pins, and all internal links attached to those pins are removed from the pin configurations. An amoebot is part of a circuit iff the circuit contains at least one of its pins. A priori, an amoebot u may not know whether two of its pins belong to the same circuit or not since initially it only knows $H(u)$.

Each amoebot can send a primitive signal (a *beep*) via any of its pins p that is received by all amoebots of the circuit belonging to p in the next round. More specifically, an amoebot receives a beep at pin p if at least one amoebot sends a beep on the circuit belonging to p, but the amoebots neither know the origin of the signal nor the number of origins. We have chosen a primitive signal instead of more complex messages to keep our extension as simple as possible. Also, we do not have to worry about interference issues in this case.

We assume the fully synchronous activation model, i.e., the time is divided into synchronous rounds, and every amoebot is active in each round. The time complexity of an algorithm is measured by the number of synchronized rounds required by it. The amoebots operate in synchronized *look-compute-move-send* cycles, where each cycle takes place in one round. In the look phase, the amoebot listens to each of its pins. In the compute phase, it performs any finite number of calculations on its local memory. In the move phase, it may perform a movement. In the send phase, it may reconfigure its pin configuration and send a beep via any of its pins. These beeps are received during the next look phase.

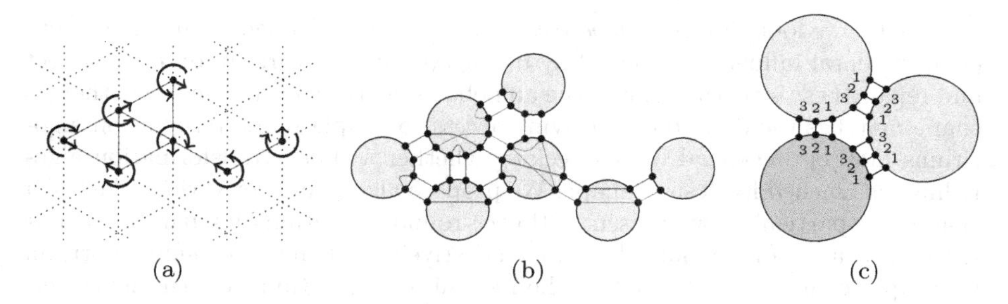

Fig. 1. (a) shows an amoebot structure. (b) shows an amoebot structure with $k = 2$ external links between neighboring amoebots. The amoebots are shown in gray. The pins are the nodes and the internal and external links are the colored edges. Each color indicates another circuit. (c) shows the local labeling of the pins. The yellow amoebot labels its pins counterclockwise, while the gray amoebots label their pins clockwise. The pin labels of the yellow amoebot match the pin labels of the gray amoebots. However, the pin labels between the two gray amoebots do not match. (Color figure online)

3 Problem Statement and Our Contribution

Table 1 summarizes the following results. First, we study the *leader election problem*. Here, the amoebots have to eventually agree on exactly one amoebot, which becomes the leader. Since amoebots only have constant storage capacity and due to the absence of node identifiers, leader election is a non-trivial problem. We propose a protocol which requires $\Theta(\log n)$ rounds, w.h.p.[1] This is a significant improvement on the runtime of previous algorithms that require at least linear complexity. We refer to [3] for an overview of leader election algorithms.

Next, we consider the *consensus problem*. Initially, each amoebot has a constant-sized input value stored in its local memory. The amoebots are tasked to agree on the same value (agreement). However, the agreed on value has to be one of the input values (validity). Also, each amoebot has to eventually decide on the agreed on value (termination). Our algorithm solves the problem after $O(1)$ rounds.

Then, we study the *compass alignment problem*. Initially, the amoebots may not agree on a common orientation. Thus, the goal of the problem is to align the compasses of all amoebots globally. A compass alignment is essential to coordinate synchronized movements. Our algorithm requires $O(\log n)$ rounds, w.h.p. The same approach can also be used to solve the *chirality agreement problem*. Within the standard amoebot model, this problem was solved by Di Luna et al. [5]. Their algorithm requires $O(n)$ rounds. Note that compass alignment and chirality agreement are harder problems than consensus since one is not able to just use a global circuit in order to change the compasses/chiralities of all amoebots at once.

[1] An event holds with high probability (w.h.p.) if it holds with probability at least $1 - 1/n^c$ where the constant c can be made arbitrarily large.

Finally, we look into the *shape recognition problem.* Amoebots are exposed to environmental influences, which may damage their structure. In order to detect and repair these structural flaws, the amoebot structure has to check whether its shape matches the desired one. Having access to simple shape recognition algorithms may be beneficial when checking whether a shape transformation algorithm has reached its desired shape. We propose algorithms for various classes of shapes. In particular, we present $\Theta(\log n)$-round algorithms for parallelograms with linear and polynomial side ratio, respectively, and an $O(1)$-round algorithm for shapes composed of triangles. The recognition algorithms for parallelograms are based on results by Gmyr et al. [4].

Table 1. An overview over our algorithmic results.

Problem	Minimum required pins	Common chirality	Runtime
Leader election	1	No	$\Theta(\log n)$ w.h.p.
Consensus	1	No	$O(1)$
Compass alignment	1	Yes	$O(\log n)$ w.h.p.
Chirality agreement	2	No	$O(\log n)$ w.h.p.
Shape recognition			
Shapes composed of triangles	1	Yes	$O(1)$
Parallelograms	1	No	$O(1)$
Parallelograms with linear side ratio	1	No	$\Theta(\log n)$ w.h.p.
Parallelograms with polynomial side ratio	2	No	$\Theta(\log n)$ w.h.p

References

1. Daymude, J.J., Richa, A.W., Scheideler, C.: The canonical amoebot model: algorithms and concurrency control. In: Gilbert, S. (ed.) 35th International Symposium on Distributed Computing (DISC 2021), LIPIcs, vol. 209, pp. 20:1–20:19. Schloss Dagstuhl – Leibniz-Zentrum für Informatik, Germany (2021). https://doi.org/10.4230/LIPIcs.DISC.2021.20. https://drops.dagstuhl.de/opus/volltexte/2021/14822
2. Derakhshandeh, Z., Dolev, S., Gmyr, R., Richa, A.W., Scheideler, C., Strothmann, T.: Brief announcement: amoebot - a new model for programmable matter. In: SPAA, pp. 220–222. ACM (2014)
3. Dufoulon, F., Kutten, S., Moses Jr., W.K.: Efficient deterministic leader election for programmable matter. In: PODC, pp. 103–113. ACM (2021)
4. Gmyr, R., Hinnenthal, K., Kostitsyna, I., Kuhn, F., Rudolph, D., Scheideler, C.: Shape recognition by a finite automaton robot. In: MFCS. LIPIcs, vol. 117, pp. 52:1–52:15. Schloss Dagstuhl - Leibniz-Zentrum für Informatik (2018)
5. Di Luna, G.A., Flocchini, P., Santoro, N., Viglietta, G., Yamauchi, Y.: Shape formation by programmable particles. Distrib. Comput. **33**(1), 69–101 (2019). https://doi.org/10.1007/s00446-019-00350-6
6. Rothemund, P.W.K., Winfree, E.: The program-size complexity of self-assembled squares (extended abstract). In: STOC, pp. 459–468. ACM (2000)
7. Toffoli, T., Margolus, N.: Programmable matter: concepts and realization. Int. J. High Speed Comput. **5**(2), 155–170 (1993)
8. Woods, D., Chen, H., Goodfriend, S., Dabby, N., Winfree, E., Yin, P.: Active self-assembly of algorithmic shapes and patterns in polylogarithmic time. In: ITCS, pp. 353–354. ACM (2013)

On Optimal Doorway Egress by Autonomous Robots

Rory Hector[1(✉)], Ramachandran Vaidyanathan[1], Gokarna Sharma[2], and Jerry Trahan[1]

[1] Louisiana State University, Baton Rouge, LA 70803, USA
{rhecto1,vaidy,jtrahan}@lsu.edu
[2] Kent State University, Kent, OH 44242, USA
sharma@cs.kent.edu

Abstract. We consider the distributed setting of n autonomous mobile robots operating in Look-Compute-Move (LCM) cycles in the Euclidean plane. In this paper, we introduce and study the problem of exiting n robots positioned initially on one side of a wall to its other side through a door (which we call the *Doorway Egress* problem). This problem is fundamental as it resembles evacuating robots from a working area after they are done with an assigned job. We consider both point (dimensionless) and fat (unit circle) robots. For each of these, we consider three abilities: visibility, lights, and synchronization. We show that without any of these abilities (i.e., obstructed visibility, no lights, and asynchronous setting), both point and fat robots can solve Doorway Egress in $O(n)$ time. We then show that with any one of the three abilities (global visibility or lights or semi-synchronous setting) point robots can solve the problem in optimal $\Theta(1)$ time. Finally, we show that with global visibility, grid-alignment, and any one of the two other abilities (lights or semi-synchronous setting), fat robots can solve the problem in optimal $\Theta\left(\sqrt{n} + \frac{n}{s}\right)$ time, where $2s$ is the width of the door. Our results also point to possible dependencies and trade-offs between these abilities.

Keywords: Doorway egress · Mobile robots · Point and fat robots · Time complexity

1 Introduction

Coordination problems for robot swarms have attracted increasing interest recently. In particular, the literature has focused on the minimal capabilities necessary for a swarm of robots to accomplish a task. While a more powerful swarm (equipped with better sensors, memory, etc.) may function well, distributed problem solving by simple, cheap robots allows for greater scalability. In this paper, we consider only the distributed computing aspect [4].

In the literature, the robots are typically modeled as *autonomous* (no external control), *anonymous* (no unique identifiers), *indistinguishable* (no external identifiers), and *disoriented* (no inherent agreement on local coordinate systems).

C. Johnen et al. (Eds.): SSS 2021, LNCS 13046, pp. 489–494, 2021.
https://doi.org/10.1007/978-3-030-91081-5_35

Each robot operates in *Look-Compute-Move* (LCM) cycles. That is, when a robot is initially activated, it takes a snapshot of its surroundings from its local perspective (*Look*), then it uses the snapshot and the algorithm it is executing to determine an action to take (*Compute*), then it performs the action (*Move*). The robots each execute the same algorithm.

Often, the robots are assumed to be *oblivious* (retaining no memory of prior LCM cycles). Furthermore, the robots are often assumed to be *silent* (no means of direct communication). These assumptions are standard in the *classical* model. A popular variant of the classical model is the *robots with lights* model [3,5], which relaxes the assumptions of obliviousness and silence. In this model, each robot is endowed with an external light that may assume any one color at a time from a constant-sized set of colors. The lights are *persistent* (continuing across LCM cycles until changed by a robot in the "Move" phase of an LCM cycle). These lights can be perceived by others, offering a form of direct communication, and persist, offering a form of memory across LCM cycles. Except for the lights, robots are oblivious as in the classical model.

The literature varies in the notion of visibility and extent of the robots. Visibility typically falls into one of two categories: *global visibility* (the robots are *transparent*, and all robots can determine the positions of all robots at any time) or *obstructed visibility* (the robots are *opaque*, and a robot r_i can only see the position of another robot r_j if there is no robot r_k between them that obstructs r_i's view of r_j and vice-versa). The extent of the robots typically falls into one of two categories: *point* (each robot is dimensionless) or *fat* (each robot is a disc of unit diameter) [1,2]. We assume that fat robots are initially *grid-aligned*. That is, the initial configuration of fat robots places each robot in a cell of an imaginary, infinite 2-dimensional discrete grid. The origin of the grid is at the center of the doorway (defined in the next section) and the grid lines are parallel or perpendicular to the wall (also defined in the next section). Each cell is of unit width (which is the diameter of a robot).

One final, important variation is that of synchronization. In the fully synchronous setting ($\mathcal{FSYNC}$), every robot is activated every LCM cycle. Further, all robots begin and end each phase of each LCM cycle (Look, Compute, and Move) at exactly the same time. Immediately after the conclusion of the "Move" phase of an LCM cycle, the "Look" phase of the next LCM cycle follows. That is, no robots have any period of inactivity. In the semi-synchronous setting ($\mathcal{SSYNC}$), at least one robot is active in every LCM cycle. Over an infinite number of $\mathcal{SSYNC}$ LCM cycles, each robot is activated infinitely many times. While a subset of robots may be inactive, the timing of LCM cycles is as in $\mathcal{FSYNC}$, where each phase of the LCM cycle begins and ends at the same time across all robots. In the asynchronous setting ($\mathcal{ASYNC}$), robots have no common notion of time. There is no consistency between start times or durations of phases of LCM cycles across robots. As in $\mathcal{SSYNC}$, robots may be inactive for periods but each robot is activated infinitely many times over an infinite period. Time is measured in *rounds* in $\mathcal{FSYNC}$ and in *epochs* in $\mathcal{SSYNC}$ and $\mathcal{ASYNC}$, where an epoch is the smallest time interval during which each robot

Table 1. Doorway Egress Algorithms. '–' means the classical model (no lights).

Algorithm	Extent	{Lights, Visibility, Synchronization}	Time
1	Point/Fat	{–, Obstructed, $\mathcal{ASYNC}$}	$O(n)$
2	Point	{Light with 10 Colors, Obstructed, $\mathcal{ASYNC}$}	$\Theta(1)$
3	Point	{–, Global, $\mathcal{ASYNC}$}	$\Theta(1)$
4	Point	{–, Obstructed, $\mathcal{SSYNC}$}	$\Theta(1)$
5	Fat	{–, Global, $\mathcal{SSYNC}$}	$\Theta(\sqrt{n} + \frac{n}{s})$
6	Fat	{Light with 9 Colors, Global, $\mathcal{ASYNC}$}	$\Theta(\sqrt{n} + \frac{n}{s})$

is active at least once. Therefore, in $\mathcal{FSYNC}$, a round is equivalent to an epoch. We will use the term "time" generically to mean rounds for $\mathcal{FSYNC}$ and epochs for $\mathcal{SSYNC}$ and $\mathcal{ASYNC}$.

In this paper, we introduce the problem of *Doorway Egress*. In this problem, the 2-dimensional real plane is bisected by an infinite line or *wall*. Let the robots begin in any arbitrary, but unique, positions on one side of the wall. The goal is to move the robots through a doorway of width $2s$ that is located on the wall, at which time they terminate and disappear. Doorway Egress is fundamental as it simulates the evacuation of robots that have completed their task, or the infiltration of robots into a target space. We investigate this problem across many combinations of the model variants previously discussed. Specifically, we vary the visibility (global or obstructed), lights (robots with lights or oblivious), and synchronization ($\mathcal{SSYNC}$ or $\mathcal{ASYNC}$) of the model. We then provide one linear-time algorithm (for point/fat robots) and five time-optimal algorithms (three for point robots and two for grid-aligned, fat robots) for a variety of these combinations. All six algorithms are collision-free.

Contributions. We establish six results (summarized in Table 1) for Doorway Egress.

Result 1: A $O(n)$ time algorithm for point or fat robots with obstructed visibility in $\mathcal{ASYNC}$ in the classical model (no lights). Therefore, this algorithm runs correctly in $O(n)$ for any of the model variants we discuss in this paper.

Results 2–4: Three $O(1)$ time, time-optimal, algorithms for point robots with each algorithm using exactly one of {lights, global visibility, $\mathcal{SSYNC}$}.

Results 5–6: Two $O(\sqrt{n}+\frac{n}{s})$ time, time-optimal, algorithms for fat robots that are initially grid-aligned, with each algorithm using global visibility and exactly one of {lights, $\mathcal{SSYNC}$}.

2 Doorway Egress

Let W be any line that bisects the real plane $\mathbb{R}^2$. Let $\widehat{W}$ be a finite segment of W that has length $2s$. The real plane can be viewed as one with a "wall," W

with a doorway $\widehat{W}$ of width $2s > 0$. The Doorway Egress Problem models the movement of robots on one side of a wall to the other side through the doorway. More specifically, a set of n robots are placed on one side of the wall. To move through the doorway, each robot first positions itself on the line segment, $\widehat{W}$, representing the doorway, and then it crosses over to the other side immediately or at a subsequent epoch. When robots cross over to the other side of the wall, they "disappear" spontaneously. A solution to Doorway Egress is to cause all robots to disappear across the wall.

We now provide a $O(n)$-time, collision-free algorithm (Algorithm 1) for both point and fat robots with no abilities, i.e. obstructed visibility, $\mathcal{ASYNC}$, and classical model (no lights). This algorithm running in $O(n)$ time moves the innermost robot (closest to c, the center of the doorway) with the smallest angle $\angle(W, c, r_i)$, call it a leader, through the doorway along the line $\overline{cr_i}$.

Theorem 1. *Algorithm 1 solves Doorway Egress in $O(n)$ epochs for both point and fat robots in the classical model in $\mathcal{ASYNC}$ under obstructed visibility.*

3 Doorway Egress by Point Robots

We discuss here Algorithms 2–4. The key challenge for point robots to solve Doorway Egress optimally is how to move $\Omega(n)$ robots to the door during at least one epoch without robots moving to the same location or crossing paths. Note that any $O(1)$-time algorithm is asymptotically optimal for Doorway Egress for point robots since the robots need at least 1 epoch to reach the doorway. Therefore, Algorithms 2–4 we design here are time-optimal.

Algorithm 2 operates in 4 phases. Phase 1 positions special guiding robots called *beacons* beside each line l_i (passing through c and a robot's position) containing multiple robots. In Phase 2, each non-beacon robot moves directly towards its corresponding beacon such that all robots form unique lines from c. In Phase 3, the robots move along their lines to a distance of $\frac{s}{3}$ from c. Then, they lie on the perimeter of one quadrant of a circle. This necessarily positions them in unique heights within the height of the door. In Phase 4, the robots move horizontally from the perimeter of the circle to the door and disappear.

This algorithm uses 10 colors. Algorithms 3 and 4 use the same framework.

Theorem 2. *Algorithm 2 solves Doorway Egress in $O(1)$ epochs for point robots without collisions in the lights model with 10 colors in $\mathcal{ASYNC}$ under obstructed visibility.*

Theorem 3. *Algorithm 3 solves Doorway Egress in $O(1)$ epochs for point robots without collisions in the classical model in $\mathcal{ASYNC}$ under global visibility.*

Theorem 4. *Algorithm 4 solves Doorway Egress in $O(1)$ epochs for point robots without collisions in the classical model in $\mathcal{SSYNC}$ under obstructed visibility.*

4 Doorway Egress by Fat Robots

We discuss here Algorithms 5 and 6. The key challenge for fat robots to solve Doorway Egress optimally is how to move s robots to the door each epoch (after some initial setup epochs) without robots moving to the same location or crossing paths, even partially. Unlike point robots, fat robots are limited in that no more than s can move through simultaneously (on each half-plane). Thus, $\Omega(\frac{n}{s})$ epochs are needed. Further, if the robots are initially placed in a packed $\sqrt{n} \times \sqrt{n}$ square, then $\Omega(\sqrt{n})$ epochs are required for the innermost robots to have room to move, irrespective of the width of the doorway. This explains the $\Omega(\sqrt{n} + \frac{n}{s})$ lower bound. Both Algorithms 5 and 6 match this lower bound.

Algorithms 5 and 6 use the same 3-phase framework. Phase 1 moves all of the robots to unique horizontal lanes. Phase 2 moves all of the robots within their unique horizontal lanes to be in unique vertical lanes as well. In Phase 3, the rightmost s robots move vertically then horizontally through the door before disappearing; this process is repeated until all robots have exited the doorway.

Theorem 5. *Algorithm 5 solves Doorway Egress in $O(\sqrt{n}+\frac{n}{s})$ epochs for grid-aligned, fat robots without collisions in the classical model (no lights) in $\mathcal{SSYNC}$ under global visibility.*

Theorem 6. *Algorithm 6 solves Doorway Egress in $O(\sqrt{n}+\frac{n}{s})$ epochs for grid-aligned, fat robots without collisions in lights model with 9 colors in $\mathcal{ASYNC}$ under global visibility.*

5 Concluding Remarks

In this paper, we have introduced the problem of Doorway Egress and provided five time-optimal algorithms for specific model variants and provided a $O(n)$-time algorithm for any model variant. We conjecture that for either point or fat robots with no abilities, $\Omega(n)$ time is required. It would also be interesting to examine fat robots under obstructed visibility. This model offers unique challenges with respect to visibility (such as complete obstruction of the wall and/or door) and collisions. Finally, it would be of interest to characterize the relationship between capabilities of the robots. Our results seem to point to a relation between abilities and time.

References

1. Agathangelou, C., Georgiou, C., Mavronicolas, M.: A distributed algorithm for gathering many fat mobile robots in the plane. In: PODC, pp. 250–259 (2013)
2. Bose, K., Adhikary, R., Kundu, M.K., Sau, B.: Arbitrary pattern formation by opaque fat robots with lights. In: Changat, M., Das, S. (eds.) CALDAM 2020. LNCS, vol. 12016, pp. 347–359. Springer, Cham (2020). https://doi.org/10.1007/978-3-030-39219-2_28

3. Bose, K., Kundu, M.K., Adhikary, R., Sau, B.: Arbitrary pattern formation by asynchronous opaque robots with lights. In: Theoretical Computer Science, vol. 849, pp. 138–158. Elsevier (2021)
4. Flocchini, P., Prencipe, G., Santoro, N.: Distributed Computing by Mobile Entities. Current Research in Moving and Computing. Springer, Heidelberg (2019). https://doi.org/10.1007/978-3-030-11072-7
5. Sharma, G., Vaidyanathan, R., Trahan, J.L.: Optimal randomized complete visibility on a grid for asynchronous robots with lights. Int. J. Netw. Comput. **11**, 50–77 (2021). IJNC Editorial Committee

Byz-GentleRain: An Efficient Byzantine-Tolerant Causal Consistency Protocol

Kaile Huang[1], Hengfeng Wei[1(✉)], Yu Huang[1], Haixiang Li[2], and Anqun Pan[2]

[1] State Key Laboratory for Novel Software Technology, Nanjing University, Nanjing, China
MG1933024@smail.nju.edu.cn, {hfwei,yuhuang}@nju.edu.cn
[2] Tencent Inc., Shenzhen, China
{blueseali,aaronpan}@tencent.com

Abstract. Causal consistency is a widely used weak consistency model and there are plenty of research prototypes and industrial deployments of causally consistent distributed systems. However, none of them consider Byzantine faults, except Byz-RCM proposed by Tseng et al. Byz-RCM achieves causal consistency in the client-server model with $3f + 1$ servers where up to f servers may suffer Byzantine faults, but assumes that clients are non-Byzantine. In this work, we present Byz-Gentlerain, the first causal consistency protocol which tolerates up to f Byzantine servers among $3f + 1$ servers in each partition and any number of Byzantine clients. Byz-GentleRain is inspired by the stabilization mechanism of GentleRain for causal consistency. To prevent causal violations due to Byzantine faults, Byz-GentleRain relies on PBFT to reach agreement on a sequence of global stable times and updates among servers, and only updates with timestamps less than or equal to such common global stable times are visible to clients. Byz-GentleRain achieves Byz-CC, the causal consistency variant in the presence of Byzantine faults. All reads and updates complete in one round-trip. The preliminary experiments show that Byz-GentleRain is efficient on typical workloads.

Keywords: Causal consistency · Byzantine faults · PBFT · GentleRain

1 Introduction

Causal consistency [1,2,6] is a widely used weak consistency model that allows high availability despite network partitions. It guarantees that *an update does not become visible to clients until all its causality are visible.* There are plenty of research prototypes and industrial deployments of causally consistent distributed

He is also with Software Institute, Nanjing University, China. This work was partially supported by the CCF-Tencent Open Fund (CCF-Tencent RAGR20200124) and the National Natural Science Foundation of China (61772258). A full version of this work is available at https://arxiv.org/abs/2109.14189.

C. Johnen et al. (Eds.): SSS 2021, LNCS 13046, pp. 495–499, 2021.
https://doi.org/10.1007/978-3-030-91081-5_36

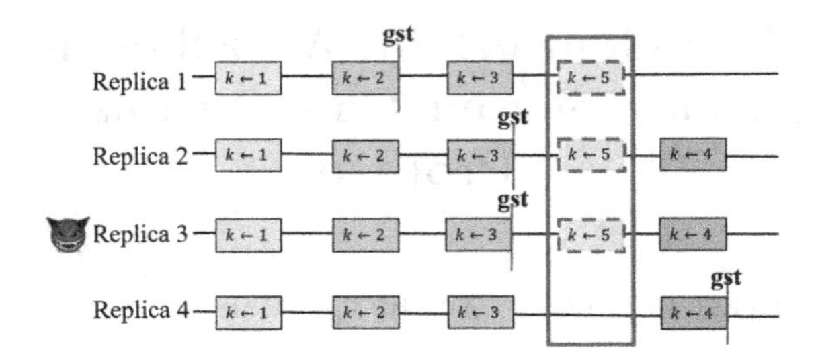

Fig. 1. Why the servers need to synchronize their global stable times.

systems (e.g., COPS [5], GentleRain [4], MongoDB [8], and Byz-RCM [7]). GentleRain in a key-value store uses a stabilization mechanism to make updates visible while respecting causal consistency. It timestamps all updates with the physical clock value of the server where they originate. Each server s periodically computes a *global stable time* gst, which is a lower bound on the physical clocks of all servers. This ensures that no updates with timestamps $\leq \mathsf{gst}$ will be generated. Thus, it is safe to make the updates with timestamps $\leq \mathsf{gst}$ at s visible to clients. A get operation on key k with dependency time dt issued to s will wait until $\mathsf{gst} \geq dt$ and then obtain the latest version of k before gst.

However, none of these causal consistency protocols/systems consider Byzantine faults, except Byz-RCM (Byzantine Resilient Causal Memory) in [7]. Byz-RCM achieves causal consistency in the client-server model with $3f+1$ servers where up to f servers may suffer Byzantine faults, and any number of clients may crash. However, Byz-RCM did not tolerate Byzantine clients, and thus it could rely on clients' requests to identify bogus requests from Byzantine servers [7].

In this work, we present Byz-GentleRain, the first Byzantine-tolerant causal consistency protocol which tolerates up to f Byzantine servers among $3f+1$ servers in each partition *and* any number of Byzantine clients. It uses PBFT [3] to reach agreement among servers on a total order of client requests. The major challenge Byz-GentleRain faces is to ensure that the agreement is consistent with the causal order. To this end, Byz-GentleRain should prevent causality violations caused by Byzantine clients or servers: Byzantine clients may violate the session order by fooling some servers that a request happened before another that was issued earlier. Byzantine servers may forge causal dependencies by attaching arbitrary metadata for causality tracking to the forward messages. To migrate the potential damages of Byzantine servers, we let clients assign totally ordered timestamps to updates in Byz-GentleRain. Utilizing the digital signatures mechanism, Byzantine servers cannot forge causal dependencies.

To preserve causality, Byz-GentleRain uses the stabilization mechanism of GentleRain. As explained above, the timestamps in Byz-GentleRain are generated by clients. However, it is unrealistic to compute a lower bound on physical clock values of an arbitrary number of clients. Therefore, each server s in Byz-GentleRain maintains and periodically computes a global stable time gst which is a lower bound on physical clock values of the clients it is aware of. Simply refusing any updates with timestamps $\leq \mathsf{gst}$ on each server may lead to causality violations. Consider a system of four servers which are replicas all maintaining a single key k, as shown in Fig. 1. Due to asynchrony, these four servers may have different values of gst. Without loss of generality, we assume that $\mathsf{gst}_1 < \mathsf{gst}_2 = \mathsf{gst}_3 < \mathsf{gst}_4$, as indicated by vertical lines. Now suppose that a new update $u : k \leftarrow 5$ with timestamp between gst_3 and gst_4 arrives, and we want to install it on ≥ 3 servers, using quorum mechanism. In this scenario, if each server refuses any updates with timestamps smaller than or equal to its gst, the update u can only be accepted by the first 3 servers, indicated by dashed boxes. Suppose that server 3 is a Byzantine server, which may expose or hide the update u as it will. Consequently, later read operations which read from ≥ 3 servers may or may not see this update u. That is, the Byzantine server 3 may cause causality violations.

To cope with this problem, we synchronize the global stable times of servers. When a server periodically computes its gst, it checks whether no larger global stable time has been or is being synchronized. If so, the server will try to synchronize its gst among all servers, by running PBFT independently in each partition. For each partition, the PBFT leader is also responsible for collecting updates with timestamps $\leq$ gst from $2f+1$ servers, and synchronizing them on all servers. Once successfully synchronized, a global stable time becomes a *common global stable time*, denoted cgst, and in each partition the updates with timestamps $\leq$ cgst on all correct servers are the same. Therefore, each server can safely refuse any updates with timestamps smaller than or equal to its cgst.

Still, the classic PBFT is insufficient, since a Byzantine leader of each partition may propose an arbitrary set of updates. To avoid this, the leader will also include the sets of updates it collects from $2f + 1$ servers in its PROPOSE message. A server will reject the PROPOSE message if it finds the contents of this message have been manipulated by checking hash and signatures.

2 Byzantine Causal Consistency

Byz-GentleRain achieves Byzantine Causal Consistency (Byz-CC) defined as follows. For two events e and f, we say that e *happens before* f, denoted $e \rightsquigarrow f$, if and only if one of the following three rules holds:

- *Session-order.* Events e and f are two operation requests issued by the same *correct* client, and e is issued before f.
- *Read-from relation.* Event e is a PUT request issued by some client and f is a GET request issued by a *correct* client, and f reads the value updated by e. Since a GET of Byzantine clients may return an arbitrary value, we do *not* require read-from relation induced by it.
- *Transitivity.* There is another operation request g such that $e \rightsquigarrow g$ and $g \rightsquigarrow f$.

If $e \rightsquigarrow f$, we also say that f causally depends on e and e is a causal dependency of f. A version vv of a key k causally depends on version vv' of key k', if the update of vv causally depends on that of vv'. A key-value store satisfies Byz-CC if, when a certain version of a key is visible to a client, then so are all of its causal dependencies.

3 The Byz-GentleRain Protocol

We consider a distributed multi-version key-value store. It runs at D data centers, each of which has a full copy of a data. In each data center, the full data is shared in to P partitions. We denote by r_d^p the replica of partition p in data center d, and store_d^p the store at replica r_d^p. Each partition consists of at least $3f+1$ replicas and at most f of them may be Byzantine. Any clients may be Byzantine.

When a client issues an update, it assigns to the update a timestamp which is its current clock value. All updates are totally ordered according to their timestamps, with client identifiers used for tie-breaking. We distinguish between the updates that have been received by a server and those that have been made visible to clients. Byz-GentleRain guarantees that an update can be made visible to clients only if so are all its causal dependencies.

In Byz-GentleRain, both clients and servers maintain a *common global stable time* cgst. We denote the cgst at client c by cgst_c and that at replica r_d^p by cgst_d^p. Replicas get their gst synchronized using PBFT, to obtain a cgst. All the updates with timestamps $\leq \mathsf{cgst}$ issued to each individual partition will be synchronized as well. Byz-GentleRain maintains the following key invariants about cgst:

INV (I): Consider cgst_c at any time σ. All updates issued by correct client c after time σ have a timestamp $> \mathsf{cgst}_c$.
INV (II): Consider cgst_d^p at any time σ. No updates with timestamps $\leq \mathsf{cgst}_d^p$ will be successfully executed at $> f$ correct replicas of partition p after time σ.
INV (III): Consider a $cgst$ value. For any two correct replicas r_d^p and r_i^p (where $i \neq d$) of partition p, if $\mathsf{cgst}_d^p \geq cgst$ and $\mathsf{cgst}_i^p \geq cgst$, then the updates with timestamps $\leq cgst$ in store_d^p and store_i^p are the same.

Each operation returns only when it receives at least $2f+1$ replies from the replicas of the partition it accesses. Byz-GentleRain enforces the following rules for reads and updates:

RULE (I): For a correct replica r_d^p, any updates with timestamps $> \mathsf{cgst}_d^p$ in store_d^p are invisible to any clients (via GET operations).
RULE (II): Any correct replica r_d^p will reject any updates with timestamps $\leq \mathsf{cgst}_d^p$.
RULE (III): For a read operation with timestamp ts issued by client c, any correct replica r_d^p that receives this operation must wait until $\mathsf{cgst}_d^p \geq ts$ before it returns a value to client c.

4 Evaluation

We implement both Byz-GentleRain and Byz-RCM in Java. The key-value stores hold 300 keys in main memory, with each key of size 8 bytes and each value of size 64 bytes. We run all experiments on 4 Aliyun[1] instances running Ubuntu

[1] Alibaba Cloud: https://www.alibabacloud.com/.

16.04. Each instance is configured as a data center, with 1 virtual CPU core, 300 MB memory, and 1G SSD storage. All keys are shared into 3 partitions within each data center, according to their hash values.

We first explore the system throughput and the latency of GET and PUT operations of both Byz-GentleRain and Byz-RCM in failure-free scenarios. First, Byz-GentleRain is quite efficient on typical workloads, especially for read-heavy workloads (2,000 $\sim$ 3,000 operations per second). Second, Byz-RCM performs better than Byz-GentleRain, especially with low GET : PUT ratios. This is because Byz-RCM assumes Byzantine fault-free clients and is signature-free. In contrast, Byz-GentleRain requires clients sign each PUT request. Third, the performance of Byz-GentleRain is closely comparable to that of Byz-RCM, if digital signatures are omitted deliberately from Byz-GentleRain.

We then evaluate the impacts of various Byzantine failures on the system throughput of Byz-GentleRain. Specifically, we consider Byzantine clients that may send GET or PUT requests with incorrect timestamps, and Byzantine replicas that may broadcast different global stable time *cgst* to replicas in different partitions. We find that these Byzantine failures have little impact on throughput. In contrast, frequently sending arbitrary messages in PBFT does hurt throughput. This is probably due to the signatures carried by these messages.

References

1. Ahamad, M., Neiger, G., Burns, J.E., Kohli, P., Hutto, P.W.: Causal memory: definitions, implementation, and programming. Distrib. Comput. **9**(1), 37–49 (1995). https://doi.org/10.1007/BF01784241
2. Burckhardt, S., Gotsman, A., Yang, H., Zawirski, M.: Replicated data types: specification, verification, optimality. In: Proceedings of the 41st ACM Symposium on Principles of Programming Languages, pp. 271–284. POPL 2014 (2014)
3. Castro, M.: Practical Byzantine fault tolerance. Ph.D. thesis, Massachusetts Institute of Technology, Cambridge, MA, USA (2000)
4. Du, J., Iorgulescu, C., Roy, A., Zwaenepoel, W.: Gentlerain: cheap and scalable causal consistency with physical clocks. In: Proceedings of the ACM Symposium on Cloud Computing, pp. 1–13. SoCC 2014 (2014)
5. Lloyd, W., Freedman, M.J., Kaminsky, M., Andersen, D.G.: Don't settle for eventual: scalable causal consistency for wide-area storage with cops. In: Proceedings of the 23rd ACM Symposium on Operating Systems Principles, pp. 401–416. SOSP 2011 (2011). https://doi.org/10.1145/2043556.2043593
6. Perrin, M., Mostefaoui, A., Jard, C.: Causal consistency: beyond memory. In: Proceedings of the 21st ACM Symposium on Principles and Practice of Parallel Programming. PPoPP 2016 (2016). https://doi.org/10.1145/2851141.2851170
7. Tseng, L., Wang, Z., Zhao, Y., Pan, H.: Distributed causal memory in the presence of byzantine servers. In: 18th IEEE International Symposium on Network Computing and Applications, NCA 2019, pp. 1–8 (2019)
8. Tyulenev, M., Schwerin, A., Kamsky, A., Tan, R., Cabral, A., Mulrow, J.: Implementation of cluster-wide logical clock and causal consistency in mongodb. In: Proceedings of the 2019 International Conference on Management of Data, pp. 636–650. SIGMOD 2019 (2019). https://doi.org/10.1145/3299869.3314049

Mitigating Internal, Stealthy DoS Attacks in Microservice Networks

Amr Osman[1(✉)], Jeannine Born[2], and Thorsten Strufe[1]

[1] KIT Karlsruhe, Karlsruhe, Germany
amr.osman@kit.edu
[2] TU Dresden, Dresden, Germany

Abstract. The advent of Microservice (MS) architectures has led to increasingly complex communication patterns between distributed web applications in the cloud. In order to process an incoming request, each MS must invoke multiple remote API calls to the MSes that it is connected to along a service dependency graph. This allows attackers to exploit long-running remote API calls along the performance-critical path to cause application DoS, and potentially amplify subsequent inter-MS communication. This paper focuses on mitigating a class of stealthy, low-volume DDoS attacks that are launched internally from within and exploit this. The attacker uses the MSes under its control to disguise then send and resource-heavy requests to target MSes in a way that is indistinguishable from benign requests. We propose a probabilistic algorithm to proactively identify MSes involved in DDoS, and mitigate the attack in real-time.

1 Introduction

In today's inter-connected microservices, a single incoming request could virtually trigger a chain of hundreds of expensive remote API calls between the involved MSes, forming a complex dependency chain and consuming a lot of CPU and I/O resources. Even a single slow-performing MS may act as a bottleneck to other MSes along the path that is traversed by inter-MS network requests. As a consequence, a high end-to-end delay is experienced.

This opens up new attack vectors to overwhelm MSes with expensive requests in the form of *Stealthy, Internal, Low-volume* DDoS attacks [6] targeted at slowly-performing MSes (SILVDDoS). In such attacks, the adversary disguises resource-consuming requests at low rates using patterns below the detection thresholds of DDoS countermeasures; making them difficult to detect and mitigate as they evade signatures and anomalies observed by traditional Intrusion detection systems [3]. Also as the attacks originate internally and masquerade benign traffic, most countermeasures that rely on perimeter defense [4] and autoscaling [2] are not effective against them.

This work identifies the MSes that participate in SILVDDoS and mitigates the attack on a granular level. Our main contributions are the following: **(1)** A risk metric to evaluate the likelihood of a MS becoming either a target or

C. Johnen et al. (Eds.): SSS 2021, LNCS 13046, pp. 500–504, 2021.
https://doi.org/10.1007/978-3-030-91081-5_37

a source for a SILVDDoS attack. **(2)** A probabilistic algorithm to approximate the identity of the sources SILVDDoS and mitigate it. **(3)** We mount a stealthy, SILVDDoS attack and evaluate the effectiveness of our approach using a popular container-based open-source MS application.

2 Assumptions and Threat Model

We consider a MSes deployment in a container-based cloud environment, e.g. a docker cluster. Each MS is isolated from other MSes via a separate Linux container and communicates with other MSes through a network-exposed API such as REST over HTTPS, or secure RPC. The MSes are exposed to the outside network, i.e. Internet, via an external load-balancing gateway that receives the requests from the end users. MSes may be elastically scaled by internal load balancers. This model is aligned with the vast array of MS deployments and kubernetes clusters today.

Adversary. The main goals of SILVDDoS is to exhaust the CPU and/or I/O resources on target MSes such that the network requests traversing the paths leading to them experience a high end-to-end delay leading to *unavailability*. A side-goal is *financial DoS* as cloud schedulers elastically scale the attacked MSes and cloud customers are charged for the newly provisioned resources, e.g. Yo-yo attack [2]. The adversary remotely controls multiple MSes internally and uses them to initiate its SILVDDoS attack on other target MSes from within. By controlling a MS, the attacker has access to all its resources such as mounted volumes (e.g. Databases), network name-spaces, processes, and user groups. Thus, it also has access to secret keys, and may authenticate itself to the network and other MSes. It may *passively* observe traffic and learn about its neighbours and measure their request-response times, or *actively* replay legitimate user traffic. It aims to remain stealthy by following the same paths and patterns that are followed by benign requests [3,5].

3 Mitigating SILVDDoS

Existing DDoS countermeasures assume that attack traffic is distinguishabe, which is not true in a SILVDDoS [4]. Cluster resource management protects the infrastructure from overutilization but cannot be used in the presence of SILVDDoS where the bottleneck are the MSes themselves [5] [3]. Chaos Engineering [1] is only used for resilience testing and does not assume malicious intent. Critical-Path-Analysis unfortunately requires knowledge about the application, low-level instrumentation and does not assume adversarial presence [7].

Our approach. We propose a risk metric that assesses the susceptibility of MSes and costly APIs of being used in a SILVDDoS attack based on key network performance and graph properties, and then later use this risk metric in an iterative probabilistic estimation algorithm that improves its quality every iteration to identify the attack sources and mitigate the attack. Unlike a greedy Critical-Path-Analysis (CPA), our algorithm also considers cases when the adversary uses MSes that may not lie on the critical path to mount the attack.

we formulate a risk metric that is computed for each API endpoint a for each MS m to determine DDoS sinks as follows:

$$\overset{<}{h}_{m,a} = \frac{w_1(1-T) + w_2(1-R) + w_3E + w_4L + w_5D_{in} + w_6A + w_0}{7} \quad (1)$$

Similarly, we formulate a DDoS source risk metric for each API end point per node as:

$$\overset{>}{h}_{m,a} = \frac{w_1(T) + w_2(R) + w_3(1-E) + w_4(1-L) + w_5D_{out} + w_6A + w_0}{7} \quad (2)$$

where $w_{0..6} \in [0,1] \subset \mathbb{Q}$ are selected weights for each metric, and T, R, E, L, D, A are the arithmetic means in their normalized form in the range $[0,1]$ and are calculated for each API $a \in m$ for each MS m.

These properties used are: *Transfer rate (T)*, *Request rate (R)*, *Error rate (E)*, *Latency (L)*, *Node degree (D)*: The number of possible logical connections each MS has to other MSes in the topology. We distinguish between incoming (D_{in}) and outgoing connections (D_{out}), *Amplification factor (A)*: The ratio between the number of external requests sent to other MSes and a given incoming request. The algorithm to mitigate SILVDDoS on a MS s can be summarized in the following steps:

1. **Pick** $a \in s$ with $j+1^{th}$ highest $\overset{<}{h}_{s,a}$ and $m \in M$ with $i+1^{th}$ highest $\overset{>}{h}_{m,a'}$, where $m \xrightarrow{a} s$
2. **Apply** either rate-limiting or container-restart to m with respect to a and s
3. **Measure** the performance metrics, i.e. health, of s.
4. With probabilies $p1$ and $p2$, **increment** j and i to the next API and microservice respectively.
5. With a probability $p3$, **undo** all rate-limiting and container-restart $\forall_m \forall_a$
6. If the health of s has improved, **go to** step 1. Otherwise, **undo** step 2 and **go to** step 1.
7. After a health threshold for s or num. of iterations is reached, **terminate** and **return** all (m, a) that were used in step 2.

4 Preliminary Evaluation

We deployed an open-source heterogeneous MSes-based application [8] with the topology in Fig. 1. We then used FastNetmon to detect high volume DoS, and Zeek to detect traffic anomalies and low-volume DoS. After that, we initiated a SILVDDoS attack on 'carts' from 'frontend' and 'orders' following the same benign user traffic paths. (See Eq. 1 and 2). *Neither FastNetmon nor zeek with*

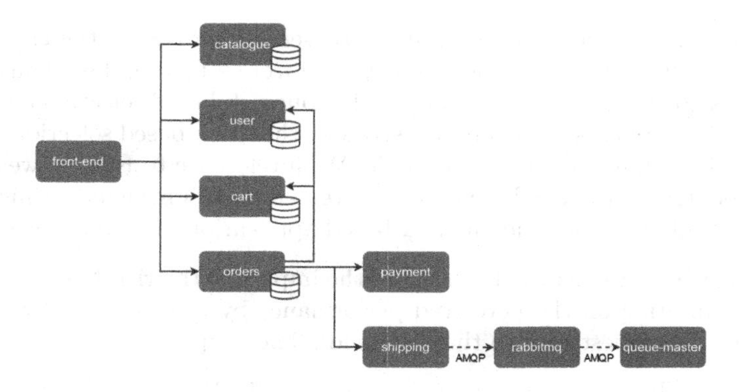

Fig. 1. Sockshop MSes topology

the latest up to date rules were able to detect SILVDDoS and zero alarms were triggered, confirming the stealthiness of our attack.

We then evaluate two main questions: **(1)** How does our risk-based placement of countermeasures compare to critical-path-analysis or a random placement? **(2)** What is the impact of our approach on benign user traffic?

Quality of risk-based selection. We compared the risk-based application of rate-limiting compared to a random placement approach that, based on majority-occurance, rate-limits requests from 'orders' to 'user'. Second, we performed a Critical path analysis with the sink as a root node and applied the rate limiters to the MSes that lie on that critical path. CPA rate-limits requests from 'orders' to 'user' and from 'orders' to 'carts'. The output can be see in Fig. 2a.

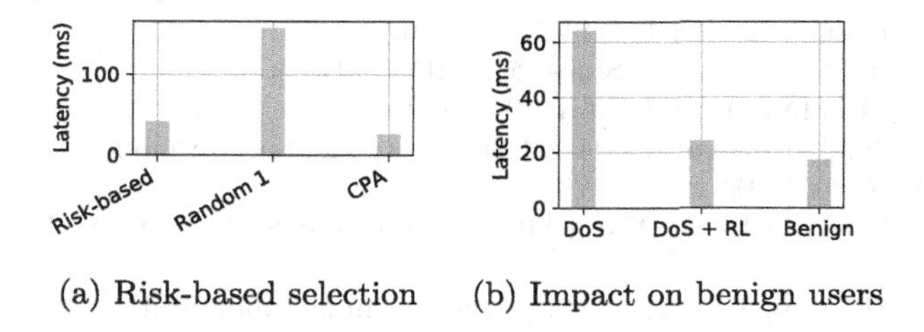

(a) Risk-based selection (b) Impact on benign users

Fig. 2. Latency measurements. Subfig. (a) shows the median response time of 'carts' (under attack). Subfig. (b) shows the median response times experienced by endusers at the frontend. Each experiment was evaluated with a confidence level of 80%

We observe that random 1 does not lead to a performance restoration. The sources of the DoS were not correctly identified and restricted. CPA however, correctly identified and rate-limited only one of the sources, namely, 'orders'. It also rate-limited requests from 'orders' to 'user' which lies on the critical path. That effectively reduced the concurrency level from 2 to 1, and explains why the sink has a slightly lower latency compared to our risk-based selection mechanism. So, while CPA appears to be a better strategy than risk-based selection, it in fact also excessively limits benign user traffic. With reference to Fig. 2b, we observed that the benign user traffic latency was increased to 38.4 ms which is higher than both: the benign setting and the risk-based application of countermeasures.

Impact on benign users. To measure the impact of the risk-based application of rate-limiting on the perceived performance by the end users, we measure the HTTP request rate to the 'frontend'. The output is in Fig. 2b.

During DoS, the user experiences 64.1 ms of HTTP response time instead of 18 ms in the case of benign traffic. The risk-based application of rate-limiting improved the response time and brought the HTTP response time down to 22.7 ms. Hence, the end user temporarily experiences a 26.11 % performance penalty when the risk-based application of rate-limiting is done, instead of a 256.11% performance penalty.

Discussion. Our approach mitigates the attack, but *temporarily* rate-limits some existing benign traffic, until the sources of the attack are identified and replaced with fresh instances. Unlike a greedy CPA, our approach is able to correctly identify attacks that may be off the critical path, and requires neither prior knowledge of the system and application, nor low-level instrumentation. In the future, we would like to optimize the algorithm parameter selection with respect to the properties of multiple MS topologies and compare the effectiveness of different countermeasures other than rate-limiting.

References

1. Blohowiak, A., Basiri, A., Hochstein, L., Rosenthal, C.: A platform for automating chaos experiments. In: IEEE ISSREW (October 2016)
2. Bremler-Barr, A., Brosh, E., Sides, M.: DDoS attack on cloud auto-scaling mechanisms. In: IEEE INFOCOM (May 2017)
3. Ficco, M., Rak, M.: Stealthy denial of service strategy in cloud computing. IEEE TCC **3**(1), 80–94 (2015)
4. Garcia, V.F., et al.: DeMONS: a DDoS mitigation NFV solution. In: IEEE AINA (May 2018)
5. Li, Z., Jin, H., Zou, D., Yuan, B.: Exploring new opportunities to defeat low-rate DDoS attack in container-based cloud environment. IEEE TPDS **31**(3), 695–706 (2020)
6. Payne, B., Behrens, S.: Starting the avalanche: application ddos in microservice architectures (July 2017). https://netflixtechblog.com/starting-the-avalanche-640e69b14a06
7. Qiu, H., et al.: FIRM: An intelligent fine-grained resource management framework for slo-oriented microservices. In: USENIX OSDI 20, pp. 805–825 (November 2020)
8. Weaveworks: Microservice sockshop (June 2021). https://microservices-demo.github.io/

Flat-Combining-Based Persistent Data Structures for Non-volatile Memory

Matan Rusanovsky[1,3(✉)], Hagit Attiya[2], Ohad Ben-Baruch[1], Tom Gerby[1], Danny Hendler[1(✉)], and Pedro Ramalhete[4]

[1] Ben-Gurion University of the Negev, Be'er Sheva, Israel
{matanru,ohadben,tomger}@post.bgu.ac.il, hendlerd@cs.bgu.ac.il
[2] Department of Computer Science, Technion, Haifa, Israel
hagit@cs.technion.ac.il
[3] Israel Atomic Energy Commission, Tel Aviv-Yafo, Israel
[4] Cisco Systems, San Jose, USA

Abstract. In this work, we present the first *persistent* (also called *durable* or *recoverable*) object implementations that employ the *flat-combining* (FC) synchronization paradigm. Specifically, we introduce a detectable FC-based implementation of concurrent LIFO stack, FIFO queue, and double-ended queue. Our empirical evaluation establishes that our novel FC-based implementations require a much smaller number of costly persistence instructions than competing algorithms and are therefore able to significantly outperform them.

1 Introduction

Byte-addressable non-volatile main memory (NVM) combines the performance benefits of conventional (volatile) DRAM-based main memory with the durability of secondary storage. The recent availability of NVM-based systems increased the interest in the development of *persistent* concurrent objects. These are objects that are able to recover from system failures (*crashes*) and ensure consistency by retaining their state in NVM and fixing it, if required, upon recovery. Of particular interest are *detectable* objects [5] that also allow recovery code to infer if a failed operation took effect before the crash and obtain its response.

The correctness condition for persistent objects that we use in this work is *durable linearizability* [8], which, simply stated, requires that linearizability be maintained in spite of crash-failures. Devising durably linearizable recoverable objects in general, and detectable ones in particular, is challenging. Although data stored in main memory will not be lost upon a system crash, with the currently available technology, caches and registers *are* volatile and their content *is* lost if the system fails before they are *persisted* (that is, written to NVM). A system crash may occur in the midst of operations applied to the object and leave it in an inconsistent state that must be fixed upon recovery. Ensuring correctness is further complicated since cache lines are not necessarily evicted in the order in which they were written by the program. Consequently, program

C. Johnen et al. (Eds.): SSS 2021, LNCS 13046, pp. 505–509, 2021.
https://doi.org/10.1007/978-3-030-91081-5_38

stores may be persisted out of order. Persistence order can be guaranteed by explicitly invoking *persistence instructions* such as flushes and fences. However, these instructions are expensive and should be used as sparingly as possible.

Persistent transactional memories (PTMs) (e.g., [1,3,9]) are general-purpose implementations that support persistent memory transactions. Although they make the construction of persistent objects easier, PTMs often incur significant performance overheads. Another shortcoming of PTMs is that none of them provides detectability. Unfortunately, for many key concurrent objects, optimized non-transactional detectable implementations still do not exist.

Flat combining (FC) [6] is a coarse-grained lock-based synchronization technique, in which threads delegate their work to a single *combiner thread*, which combines operations by multiple threads in a manner that exploits the semantics of the implemented concurrent object and then jointly applies them.

This work presents *detectable flat combining (DFC)*, a generic approach for persistent objects that is based on FC, and applies it to derive detectable stacks, queues and double-ended queues. We experimented in our DFC objects with the concept of elimination. In the case of a stack object, pairs of *push* and *pop* are combined by "eliminating" them: each *pop* operation in the pair return the item that is the argument of the *push* operation in the pair. Surplus *push* or *pop* operations are combined by atomically extending or truncating a linked-list stack representation. We employ elimination in our double-ended-queue implementation as well. We compare the performance of our DFC-based algorithms with that of several PTM-based implementations. Our results establish that the new algorithms greatly outperform the competition and the margin increases when a large fraction of operations are eliminated. Our DFC queue, which does not employ elimination, also significantly outperforms PTM-based queues.

System Model. Shared memory holds both non-volatile shared variables (residing in NVM) and volatile shared variables (residing in DRAM). The contents of the cache and processor registers are volatile. Writes to non-volatile variables are persisted to NVM using explicit flush instructions, or when a cache line is evicted. A write-back to persistent storage is triggered by a *persistent write-back (pwb)* instruction. The program order of `pwb` instructions is not necessarily preserved. When ordering is required, a `pfence` instruction orders preceding `pwb` instructions before all subsequent `pwb` instructions. A `psync` instruction waits until all previous `pwb` instructions complete the write back. For each memory location, persistent write-backs preserve program order.

Since in current architectures a `pfence` acts as both `pfence` and `psync`, our pseudocode uses `pfence` to indicate the execution of both. A *system-wide crash-failure* may occur, which resets all volatile variables to their initial values, but preserves the values stored in the NVM. An operation's response is lost if a crash occurs before it was *persisted* to a non-volatile variable. Following a crash, the system resurrects all threads and lets them execute the `Recover` procedure, in order to recover the data-structure by fixing inconsistencies in it, if any. An implementation is *detectable* [2] if `Recover` also finishes p's crashed operation (if there is one) and returns its response.

We only describe our DFC-based stack. The full paper [10] contains additional details for the stack and the detailed algorithms for the other data structures.

2 The DFC Stack

As done in FC-based algorithms, each process *announces* its operation by writing its operation code and arguments to its entry in an *announcement array*. Then, each process attempts to capture a global lock that protects a sequential data structure and become a *combiner*. Processes that fail to capture the lock (*non-combiner* processes) wait for the combiner to apply their operations, whereas the combiner proceeds to traverse the announcement array, collect announced operations, apply them to the data structure, and write operation responses to their corresponding announcement array entries. In our implementation, the stack is represented by a linked list.

Algorithm 1 (left) presents the OP procedure, which implements both PUSH and POP operations. *cEpoch* counts the number of combining phases performed so far (multiplied by 2, for reasons we explain soon). At the beginning of each operation (lines 2–3), a thread t creates a local copy *opEpoch* of the current *cEpoch*. Then, t announces its operation in the next available announcement structure in $tAnn[t]$, which consists of two such announcement structures. A 2-bit variable *valid* indicates which of the announcement structures is the active one and whether it is ready for the combiner to collect. The update of *valid* is done in two stages. This ensures that in case of a crash, the *recovery combiner* will handle the correct announcement structure. Then, a thread attempts to become the combiner in the TAKELOCK procedure (line 13). The combiner returns from TAKELOCK without waiting and proceeds to combine all announced operations by calling the COMBINE procedure in line 17 (the code of this procedure appears in the full paper [10]).

In our implementation, the combiner employs *elimination* [7] to pair concurrent PUSH and POP operations. It applies each operations pair by setting the response of the POP operation to be the input of the PUSH operation without accessing the linked list. This reduces the number of persistency instructions. The combiner modifies the linked list only if the numbers of PUSH and POP operations that it collected differ. COMBINE persists the data in two stages, to ensure that in case of a crash it can find a stable copy of the stack. For this reason, *cEpoch* is incremented twice. A non-combiner t busy-waits for the combiner to increase *cEpoch* by 2 (or more). If this condition is satisfied, it is guaranteed that all combined announcement structures received valid response values. Consequently, if t finds a response value in line 47, it can safely return that value. Otherwise, if *cEpoch* was incremented but there is no response value, then t has arrived late, that is, it has completed announcing its operation only after the combiner checked its announcement structure. In this case, *opEpoch* is incremented by two in line 48 in order to wait for the next combiner to collect the operation. The algorithm guarantees that this scenario can occur only once, since the next combiner will surely collect the thread's announcement.

Algorithm 1. DFC Stack. Op, Recover and auxiliary procedures

```
1:  procedure Op(param)
2:    opEpoch := cEpoch
3:    if opEpoch%2 = 1 then opEpoch + +
4:    nOp := 1 − LSB(tAnn[t].valid)
5:    tAnn[t].ann[nOp].val :=⊥
6:    tAnn[t].ann[nOp].epoch := opEpoch
7:    tAnn[t].ann[nOp].param := param
8:    tAnn[t].ann[nOp].name := Op
9:    pwb(&tAnn[t].ann[nOp]); pfence()
10:   tAnn[t].valid = nOp
11:   pwb(&tAnn[t].valid); pfence()
12:   MSB(tAnn[t].valid) := 1
13:   value := TakeLock(opEpoch)
14:   if value ≠⊥ then
15:     return value
16:   else
17:     Combine()
18:     return tAnn[t].ann[nOp].val
19: procedure TakeLock(opEpoch)
20:   if cLock.Cas (0, 1) = False then
21:     while cEpoch ≤ opEpoch + 1 do
22:       if (cLock = 0 and
              cEpoch ≤ opEpoch + 1) then
23:         return TakeLock (opEpoch)
24:     return TryToReturn(opEpoch)
25:   else return ⊥
26: procedure Recover(param)
27:   if rLock.Cas (0, 1) then
28:     if cEpoch%2 = 1 then
29:       cEpoch + +
30:       pwb(&cEpoch); pfence()
31:     GarbageCollect()
32:     for i = 1 to N do
33:       vOp := tAnn[i].valid
34:       opEpoch := tAnn[i].ann[LSB(vOp)].epoch
35:       if MSB(vOp) = 0 then
36:         MSB(tAnn[i].valid) := 1
37:       if opEpoch = cEpoch then
38:         tAnn[i].ann[LSB(vOp)].val :=⊥
39:     Combine()
40:     rLock = 2
41:   else
42:     while rLock = 1 do spin
43:   return tAnn[t].ann[LSB(tAnn[t].valid)].val
44: procedure TryToReturn(opEpoch)
45:   vOp := LSB(tAnn[t].valid)
46:   val := tAnn[t].ann[vOp].val
47:   if val =⊥ then                         ▷ late arrival
48:     opEpoch := opEpoch + 2
49:     return TakeLock(opEpoch)
50:   else return val
```

The Recovery Procedure. In case of a crash, all threads execute the Recover procedure upon recovery (Algorithm 1 (right)). If required, the recovery function recovers the shared stack by re-executing the last combining phase, thus completing all pending operations and updating their responses. Each thread first attempts to capture the recovery lock *rLock* that protects the critical section of the recovery code. If the thread fails to capture the lock, it simply busy-waits until the lock is freed in line 42. If it succeeds, it becomes the *recovery combiner.* DFC needs to re-collect and re-apply all operations of the last crashed combining phase, even ones for which response values were persisted before the crash. Thus, all announcement structures are traversed again and inconsistencies are fixed, such that in the following combining phase in line 39, all operations from the last crashed combining phase will be collected and applied again. Finally, the recovery combiner releases the lock in line 40 and returns its own response value (as do all other threads) in line 43.

3 Experimental Evaluation

We compared the performance of the DFC stack with stack implementations using Romulus [3], OneFile [9] and PMDK [1] in an experiment in which we executed 1M pairs of push and pop operations, that were distributed equally between the threads. More details regarding these algorithms and the machine we used appear in the full paper [10]. Figure 1 presents (from left to right) the throughput, and the average number of `pwb` and `pfence` instructions. As can be seen, the DFC stack slightly lags behind the Romulus stack for up to 8 threads

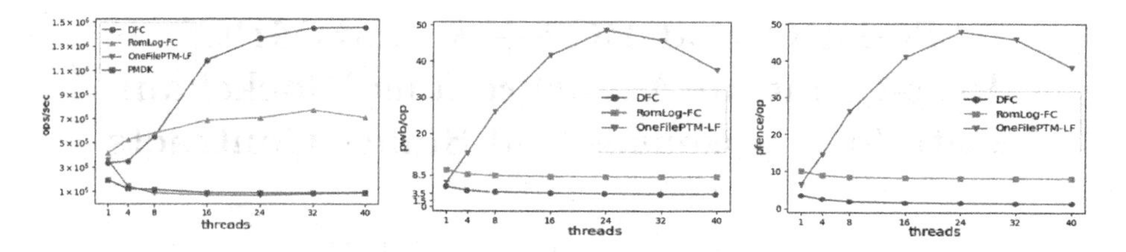

Fig. 1. Throughput, `pwbs`, `pfences` for push-pop benchmark.

but outperforms all algorithms by a wide margin for higher concurrency levels. This is largely due to the fact that its operations perform on average significantly fewer `pwb` and `pfence` instructions.

Very recent work [4] presents detectable implementations of several objects, including a stack. This work employs combining as well, but there are some differences in technical implementation details. We provided our previously-archived code to the authors and the performance evaluation they present shows higher throughput in comparison to our DFC stack for high concurrency levels, on a different machine than ours. As we did not have access to their code, we were unable to fully explore the differences between the two implementations, and could not evaluate the performance of their implementation on our machine.

References

1. Persistent memory development kit (2020). https://pmem.io/pmdk
2. Ben-David, N., Blelloch, G.E., Friedman, M., Wei, Y.: Delay-free concurrency on faulty persistent memory. In: SPAA, pp. 253–264 (2019)
3. Correia, A., Felber, P., Ramalhete, P.: Romulus: efficient algorithms for persistent transactional memory. In: SPAA, pp. 271–282 (2018)
4. Fatourou, P., Kallimanis, N.D., Kosmas, E.: Persistent software combining. CoRR abs/2107.03492 (2021). https://arxiv.org/abs/2107.03492
5. Friedman, M., Herlihy, M., Marathe, V.J., Petrank, E.: A persistent lock-free queue for non-volatile memory. In: SPAA, pp. 28–40 (2018)
6. Hendler, D., Incze, I., Shavit, N., Tzafrir, M.: Flat combining and the synchronization-parallelism tradeoff. In: SPAA, pp. 355–364 (2010)
7. Hendler, D., Shavit, N., Yerushalmi, L.: A scalable lock-free stack algorithm. J. Parallel Distrib. Comput. **70**(1), 1–12 (2010)
8. Izraelevitz, J., Mendes, H., Scott, M.L.: Linearizability of persistent memory objects under a full-system-crash failure model. In: Gavoille, C., Ilcinkas, D. (eds.) DISC 2016. LNCS, vol. 9888, pp. 313–327. Springer, Heidelberg (2016). https://doi.org/10.1007/978-3-662-53426-7_23
9. Ramalhete, P., Correia, A., Felber, P., Cohen, N.: OneFile: a wait-free persistent transactional memory. In: DSN, pp. 151–163. IEEE (2019)
10. Rusanovsky, M., Attiya, H., Ben-Baruch, O., Gerby, T., Hendler, D., Ramalhete, P.: Flat-combining-based persistent data structures for non-volatile memory (2021). https://arxiv.org/abs/2012.12868v3

SodsBC/SodsBC++ & SodsMPC: Post-quantum Asynchronous Blockchain Suite for Consensus and Smart Contracts

Shlomi Dolev[1] and Ziyu Wang[1,2](✉)

[1] Department of Computer Science, Ben-Gurion University of the Negev, Beersheba, Israel
Dolev@cs.bgu.ac.il

[2] School of Cyber Science and Technology, Beihang University, Beijing, China
wangziyu@buaa.edu.cn

Abstract. SodsBC is the first asynchronous permissioned blockchain consensus (asynchronous BFT, aBFT in short) utilizing the concurrent preprocessing model to achieve post-quantum security while keeping high performance simultaneously. SodsBC concurrently preprocesses common random coins (from a global secret sharing pool) for the consensus necessary randomness usage, and also preprocesses symmetric encryption keys for considerable censorship resilience. The finalization of a stage of the global secret sharing pool utilizes the aBFT itself, as a transaction in the new block. SodsBC (and its followed work SodsBC++) is faster than the quantum-sensitive aBFT competitors. SodsMPC is a post-quantum smart contract system, in which all participants execute a contract by secure multi-party computation (MPC) protocols. MPC ensures the contract execution correctness while keeping the data privacy. Moreover, SodsMPC expresses the logic of a contract by a blind polynomial with secret-shared coefficients, and utilizes a finite state machine (FSM) to simplify the blind polynomial for better efficiency. When using MPC to compute this blind polynomial, the contract business logic privacy is obtained. SodsMPC participants also preprocess random permutation matrices to hide the real contract input relation, which protects the contract user anonymous identity.

Keywords: Blockchain · Post-quantum · Asynchrony · Smart Contract

1 Introduction

Shor algorithm can efficiently break the basic mathematical intractabilities like integer factorization and discrete logarithm (Dlog). The recent theoretical or

---

This brief announcement introduces the published work SodsBC [6] (with a full version [8]) and SodsMPC [7].

C. Johnen et al. (Eds.): SSS 2021, LNCS 13046, pp. 510–515, 2021.
https://doi.org/10.1007/978-3-030-91081-5_39

practice developments of blockchain consensus protocols or smart contract systems, heavily depend on the use of quantum-sensitive cryptography tools. A perfect information-theoretical (I.T.) secure algorithm can be *proved* to resist a quantum adversary. Also, some symmetric cryptography tools are *believed* to be post-quantum if the security parameter is long enough.

Honeybadger [12] first makes an aBFT practical by trying to solve a consistent union of all block parts proposed by each participant, which adopts a quantum-sensitive threshold encryption scheme to avoid censorship in an asynchronous common subset (ACS) and instantiates a common random coin source via quantum-sensitive threshold signature. Dumbo [10] breaks the Honeybadger $O(\log n)$ round complexity bottleneck and designs a quantum-sensitive threshold-signature-based external validity predicate when replaying n binary Byzantine agreement (BBA) instances with a multi-value Byzantine agreement (MVBA), which terminates in a constant round. The design of practical post-quantum aBFT based on both BBA-ACS and MVBA-ACS approaches was an open problem which we now close.

Several permissionless blockchain private contract systems [4,5] cope with the contract privacy problem relying on zero-knowledge proof (ZKP). In these systems, users execute the contract (with or without privacy protection) in the off-chain style. Later, the execution result proof is included in the blockchain, which may not reveal the contract executions. However, the deployed ZKP schemes are quantum-sensitive in these systems.

Our Contribution. For consensus, we propose a post-quantum framework for aBFT via concurrent preprocessing, and instantiate this framework to SodsBC. We preprocess *common random coins* and *symmetric encryption keys* utilizes the same aBFT architecture as the aBFT for blocks. We design an aBFT based blockchain consensus, while the consensus itself provides the consensus ability for the aBFT protocol reversely. Common random coins supply the necessary aBFT randomness, while the symmetric encryption keys (and the encryption scheme) provides the considerable aBFT anti-censorship property. The practical performance of the SodsBC implementation demonstrates that it is efficient to deploy concurrent preprocessing to achieve post-quantum security. Moreover, we also propose a scheme to implement a global wait-free bootstrap for SodsBC, utilizing waiting (partial-synchronous) components.

For executing a smart contract, we also arrange permissioned blockchain participants to act as MPC participants. When executing a contract by an MPC protocol, *data privacy* is basically obtained. In SodsMPC, a contract can be expressed by a *blind* polynomial whose coefficients are secretly shared. So that participants execute the contract without knowing the contract *business logic.* SodsMPC specially utilizes a finite state machine (FSM) to encode a contract to state transitions to save the multiplication overhead. When enforcing the "mixing-then-contract" paradigm, the secret-shared inputs of a contract are first mixed and then the mixing results are regarded as contract inputs for *user anonymity.* Our robust online phase has only one matrix-vector multiplication consuming a preprocessed permutation matrix, which keeps the secret share form outputs and achieves a full randomized shuffle.

2 SodsBC: Post-quantum aBFT Consensus

Asynchronous Weak Verifiable Secret Sharing (awVSS). In a weak commitment VSS scheme, participants verify the threshold in a reconstruction stage [1]. In our awVSS scheme, honest participants can detect the malicious behavior and set a shared secret to a default value (e.g., zero) after reconstruction.

A Post-quantum Common Random Coin Design. In a SodsBC online stage, our BBA will consume post-quantum, fresh, and one-time used common random coins, reconstructing from the shared secrets distributed in history preprocessing stages. This online stage is also a preprocessing stage, simultaneously, in which each participant shares secrets for future coins by awVSS. Previous aBFTs [9,10,12] use quantum-sensitive coins.

A SodsBC common random coin encompasses $f+1$ shared random secrets produced by $f+1$ distinct dealers, i.e., $\mathsf{coin} = \mathsf{secret}_1 + \cdots + \mathsf{secret}_{f+1} \mod 2$. Since each coin component is secretly shared, before the first honest participant invocates the coin recovery, the at most f adversaries do not learn any information about the coin value if at least one coin component is well-shared by an honest participant under the $f+1$ threshold (the awVSS secrecy). Every participant will recover the $f+1$ consistent coin components resulting in the consistent coin. Moreover, honest participants consistently set at most f coin components to zero when detecting the malicious behaviors, while one successful reconstruction still keeps a well-defined coin without bias.

Algorithm 1. SodsBC Consensus (for p_i) [6].

1: $\mathsf{AESEnc}(\mathsf{aesKey}_i, \mathcal{B}_{\mathsf{p\_part}_i}) \rightarrow \mathcal{B}_{\mathsf{c\_part}_i}$.
   // Consensus core
2: In RBC_i^*, broadcast $\mathcal{B}_{\mathsf{c\_part}_i}$, share aesKey_i and secrets by piggybacked awVSS messages.
3: Input 1 to BBA_i if RBC_i^* finishes.
4: Input 0 to remained BBAs if $n-f$ BBAs output 1.
   // Reconstruct BBA coins by awVSS.
   // Decryption and output
5: If BBA_j outputs 1, reconstruct aesKey_j and AES decrypts $\mathcal{B}_{\mathsf{c\_part}_j}$. If the decryption fails, or RBC_j is aborted (BBA_j outputs 0), set $\mathcal{B}_{\mathsf{part}_j} = \perp$.
6: Make $\mathcal{B} = \bigcup_{j=1}^{n} \mathcal{B}_{\mathsf{part}_j}$, and assign agreed awVSS batches to n queues.

Different asynchronous participants may have different observations about the secrets shared by participants. We employ SodsBC itself to achieve secret pool consensus. Each SodsBC dealer runs an awVSS batch to sharing some secrets, and n BBAs in SodsBC finalize these n awVSS batches, leading to a global awVSS pool. Then, honest participants can assign $f+1$ secrets (shared from $f+1$ distinct dealers) to one coin, and assign each coin to n BBA queues. From a global view to observe this pool, the assigned object is for each secret. Locally, every honest participant will assign its share from the pool it stores.

A Post-quantum Censorship Resilience Solution. No matter adopting BBA-ACS or MVBA-ACS to implement aBFT, **censorship resilience** cannot be guaranteed since malicious participants can intentionally exclude some particular block parts according to the content of these block parts. We follow the *encryption-consensus-decryption* idea [12] to achieve censorship resilience. Each participant p_i AES encrypts its block part $\mathcal{B}_{\mathsf{p\_part}_i}$, and secretly sharing aesKey_i simultaneously when reliably broadcasting $\mathcal{B}_{\mathsf{c\_part}_i}$. After consensus, participants reconstruct the shared AES keys, then decrypt the agreed encrypted block parts. The previous protocols [9,10,12] use quantum-sensitive threshold encryption schemes [2,13] to further encrypt the AES keys.

The SodsBC Protocol (Algorithm 1)**.** SodsBC participants agree on the termination of n awVSS batches distributed by a specific dealer for future coins ($\mathsf{Pred}_{\mathsf{awVSS\_coin}}$), n awVSSs for AES keys ($\mathsf{Pred}_{\mathsf{awVSS\_aeskey}}$), and n RBC instances for block parts ($\mathsf{Pred}_{\mathsf{RBC}}$). Hence, a SodsBC predicate is $\mathsf{Pred} = \mathsf{Pred}_{\mathsf{RBC}} \wedge \mathsf{Pred}_{\mathsf{awVSS\_coin}} \wedge \mathsf{Pred}_{\mathsf{awVSS\_aeskey}}$, which decides a complex instance having three sub-instances. Since SodsBC keeps the Honeybadger-ACS architecture, Algorithm 1 can satisfy the BFT agreement, total order, and liveness properties.

The SodsBC communication complexity is $O(|\mathcal{B}|n+\lambda n^4 \log n)$, slightly larger than the quantum-sensitive competitors HoneyBadger and BEAT ($O(|\mathcal{B}|n + \lambda n^3 \log n)$)[1]. However, our post-quantum improvements avoid the time expensive cryptographic functions usage, so that the SodsBC prototype can be around 53% faster than Honeybadger when $n = 100$ and $|\mathcal{B}_{\mathsf{part}}| = 20,000$ benchmark (250B) transactions in an AWS LAN network [8].

Extending the Post-quantum Framework to an MVBA-Based aBFT. The bottleneck of a BBA-ACS-based aBFT gradually emerges when the number of participants n is increasing, since the slowest BBA instance of the n BBAs would spend $O(\log n)$ rounds. Dumbo [10] breaks through this bottleneck utilizing an MVBA, which dramatically decreases the BBA usage to a constant time. However, Dumbo heavily relies on quantum-sensitive threshold signatures. SodsBC can be extended to SodsBC++ to make Dumbo post-quantum. Still in a typical $n = 100$ AWS LAN network, SodsBC++ prototype accelerates Dumbo to be faster in 7.5% or so [8].

3 SodsMPC: Post-quantum Smart Contracts

Blind-Coefficient Polynomials and Finite State Machine. A smart contract can be a mutual-execution distributed protocol run by distrusted entities. Roughly speaking, if this contract (a computer program) can be expressed by an arithmetic polynomial, we can use an MPC protocol to compute this contract. If the coefficients of the polynomial are also secret-shared, the contract business logic is also hidden. However, directly computing a contract polynomial may

---

[1] λ is the security parameter for SodsBC symmetric cryptography schemes and Honeybadger/BEAT asymmetric and quantum-sensitive schemes.

not be efficient. For instance, a three-input millionaire problem (find the maximum input index in the field $GF(11)$) can be encoded to a very long polynomial $f(x, y, z)$, which has 909 monomials from 0 to $x^{10}y^{10}z^{10}$ (the coefficients of some terms are zero). Directly solving this polynomial (with all secret-shared coefficients) requires 30 rounds and around 20,000 multiplication gates [3]. Accordingly, our FSM-based comparator with blind transitions spends only 16 rounds and 32 multiplications to achieve the same target.

Besides, we also instantiate an FSM-based adder and a secret sharing base conversion protocol (for integer to binary) in the published version [7]. Using an FSM to design a smart contract also assists a contract programmer to express the correct design logic with fewer bugs [11].

MPC Mixing for the User Anonymity. Our MPC for transaction mixing in the online phase is only one matrix-vector multiplication. The preprocessing work for a permutation matrix lies in the verification and randomness extraction, which is from the permutation matrix definition. Every row or every column of a permutation matrix has and only has one 1-value item, and the remained elements should be 0-value. Moreover, each matrix item should be 1-value or 0-value secret share, so that we extra blindly test whether the shares of each matrix item $[x]$ satisfying $[x^2] - [x] = [0]$ utilizing a preprocessed square tuple. If the reconstruction is 0, then x must be 0 or 1. For extracting a random permutation matrix from $f + 1$ dealers, the direct way is to multiply $f + 1$ valid permutation matrices generated by $f + 1$ distinct dealers, i.e., $\mathbf{M} = \Pi_{i=1}^{i=f+1}\mathbf{M}_i$.

References

1. Backes, M., Kate, A., Patra, A.: Computational verifiable secret sharing revisited. In: Lee, D.H., Wang, X. (eds.) ASIACRYPT 2011. LNCS, vol. 7073, pp. 590–609. Springer, Heidelberg (2011). https://doi.org/10.1007/978-3-642-25385-0_32
2. Baek, J., Zheng, Y.: Simple and efficient threshold cryptosystem from the gap Diffie-Hellman group. In: GLOBECOM 2003, pp. 1491–1495 (2003)
3. Bitan, D., Dolev, S.: Optimal-round preprocessing-MPC via polynomial representation and distributed random matrix (extended abstract). IACR Cryptology ePrint Archive, p. 1024 (2019)
4. Bowe, S., Chiesa, A., Green, M., Miers, I., Mishra, P., Wu, H.: ZEXE: enabling decentralized private computation. In: S&P 2020, pp. 947–964 (2020)
5. Bünz, B., Agrawal, S., Zamani, M., Boneh, D.: Zether: towards privacy in a smart contract world. In: Bonneau, J., Heninger, N. (eds.) FC 2020. LNCS, vol. 12059, pp. 423–443. Springer, Cham (2020). https://doi.org/10.1007/978-3-030-51280-4_23
6. Dolev, S., Wang, Z.: SodsBC: stream of distributed secrets for quantum-safe blockchain. In: IEEE Blockchain 2020, pp. 247–256 (2020)
7. Dolev, S., Wang, Z.: SodsMPC: FSM based anonymous and private quantum-safe smart contracts. In: NCA 2020, pp. 1–10 (2020)
8. Dolev, S., Wang, Z.: SodsBC: a post-quantum by design asynchronous blockchain framework. IACR Cryptology ePrint Archive, p. 205 (2020)
9. Duan, S., Reiter, M.K., Zhang, H.: BEAT: asynchronous BFT made practical. In: CCS 2018, pp. 2028–2041 (2018)

10. Guo, B., Lu, Z., Tang, Q., Xu, J., Zhang, Z.: Dumbo: faster asynchronous BFT protocols. In: CCS 2020, pp. 803–818 (2020) (2020)
11. Mavridou, A., Laszka, A.: Designing secure ethereum smart contracts: a finite state machine based approach. In: Meiklejohn, S., Sako, K. (eds.) FC 2018. LNCS, vol. 10957, pp. 523–540. Springer, Heidelberg (2018). https://doi.org/10.1007/978-3-662-58387-6_28
12. Miller, A., Xia, Y., Croman, K., Shi, E., Song, D.: The honey badger of BFT protocols. In: CCS 2016, pp. 31–42 (2016)
13. Shoup, V., Gennaro, R.: Securing threshold cryptosystems against chosen ciphertext attack. In: Nyberg, K. (ed.) EUROCRYPT 1998. LNCS, vol. 1403, pp. 1–16. Springer, Heidelberg (1998). https://doi.org/10.1007/BFb0054113

Distributed Reconfiguration of Spanning Trees

Yukiko Yamauchi[1(✉)], Naoyuki Kamiyama[1], and Yota Otachi[2]

[1] Kyushu University, 744 Motooka, Nishi-ku, Fukuoka 819-0395, Japan
yamauchi@inf.kyushu-u.ac.jp, kamiyama@imi.kyushu-u.ac.jp
[2] Nagoya University, Furocho, Chikusa-ku, Nagoya 464-8601, Japan
otachi@nagoya-u.jp

Abstract. We introduce a new type of distributed reconfiguration problem, where an initial instance of a combinatorial object is transformed to a goal instance by "local" exchange operations. We present a distributed algorithm that transforms an arbitrary spanning tree to another one through a sequence of spanning trees. We then discuss distributed reconfiguration of hypertrees and maximum bipartite matchings.

Keywords: Distributed reconfiguration problem · Spanning tree · Hypertree · Maximum bipartite matching

1 Introduction

Construction of combinatorial objects such as a spanning tree, maximum matching, maximum independent set, and vertex coloring in spite of local control has been one of the most important problems in distributed computing. As dynamic networks such as P2P networks and mobile ad-hoc networks become widely available, it is expected that a distributed system continuously updates its behavior so that it can adapt to dynamic changes. However, most construction algorithms build an instance from scratch and no instance is available during construction.

Recently, in sequential algorithm theory, *reconfiguration problems* introduced by Ito et al. [4] has attracted much attention. A reconfiguration problem asks when given two feasible solutions F_s and F_t for an instance I of a problem $\mathcal{P}$ and an *adjacency relation* between feasible solutions of I, if there exists a sequence $F_0 = F_s, F_1, F_2, \ldots, F_k = F_t$ of feasible solutions for I such that F_i and F_{i+1} are adjacent for $i = 0, 1, 2, \ldots, k-1$. In other words, the reconfiguration problem asks reachability from a feasible solution to another feasible solution by exchange operations specified by the adjacency relation. It is shown that reconfiguration problems of many NP-complete problems such as independent sets, cliques, and

This work was supported by JSPS KAKENHI Grant Numbers JP20H05793, JP20H05795, JP18H04091, JP18K11168, JP18K11169, and JP21K11752. The authors would like to thank Shuji Kijima for helpful discussions.

C. Johnen et al. (Eds.): SSS 2021, LNCS 13046, pp. 516–520, 2021.
https://doi.org/10.1007/978-3-030-91081-5_40

vertex covers, where an exchange operation is adding or deleting a single vertex, are PSPACE-complete [4]. However, little is known about reconfiguration problems in distributed settings.

Our Contribution. In this paper, we introduce a new type of distributed reconfiguration problem, where adjacency relation conforms to locality in distributed systems. That is, each intermediate object is obtained by an update of local variables at a single process. We first consider distributed reconfiguration of spanning trees. The base exchange property of a graphic matroid guarantees reconfiguration between any pair of spanning trees. However, it does not consider locality. We first present a distributed reconfiguration algorithm for spanning trees. We then discuss distributed reconfiguration of hypertrees and maximum bipartite matchings.

Related Work. The minimum spanning tree reconfiguration problem by edge exchanges was first presented by Ito et al. [4]. They generalized the problem to a reconfiguration of matroid bases with a weight function and showed that the problem is in P. Bousquet et al. considered reconfiguration of spanning trees by edge flips [2]. Mizuta et al. considered reconfiguration of minimum Steiner trees by vertex exchanges on an unweighted graph [7].

Several papers considered distributed reconfiguration problems with a different type of adjacency relations. Bonamy et al. considered distributed reconfiguration of vertex colorings, where adjacent processes (i.e., vertices) cannot change their colors simultaneously [1]. They investigated the number of communication rounds for the processes to agree on a recoloring schedule, the length of the schedule, and the effect of adding extra colors. Censor-Hillel and Rabie investigated distributed reconfiguration of maximal independent sets, where adjacent processes cannot change their membership simultaneously [3]. They relaxed covering requirement during reconfiguration and showed a trade-off between the length of a reconfiguration schedule and the number of communication rounds.

2 Preliminaries

A *distributed system* is represented by an undirected graph $G = (V, E)$, where the set V of vertices is the set of *processes* and the set E of undirected edges is the set of *bidirectional communication links*. The cardinality of V is n. Each process maintains a set of *local variables* and a *state* of a process is an assignment of values to its local variables. Each process $u \in V$ is assigned a unique ID denoted by ID_u. When an edge $\{u, v\} \in E$, we say process u is a neighboring process of v. Let $N_u = \{v \mid \{u, v\} \in E, v \neq u\}$. Each process u can read the local variables at itself and its neighboring processes without any delay, and u can change the values of its own local variables.

An algorithm at process u is a finite set of *guarded actions* in the form of $\langle label \rangle \colon \langle guard \rangle \rightarrow \langle action \rangle$, where $\langle guard \rangle$ is a predicate over the states of processes in $N_u \cup \{u\}$ and $\langle action \rangle$ is a statement that changes u's state. A guarded action is said to be *enabled* if its guard is evaluated to *true* and a process is said to be *enabled* if at least one of its guarded actions is enabled.

A *configuration* is the set of states of all processes. The *central scheduler* selects a single process from enabled processes at each discrete time step, and a selected process non-deterministically chooses one of its enabled guards and executes the corresponding action. We assume that the scheduler is *weakly fair*, that is, any continuously enabled process is selected in finite time.

An *execution* of a distributed algorithm under a given scheduler is a maximal sequence of configurations $\mathcal{E} = \gamma_0, \gamma_1, \ldots$ such that for each $i \geq 0$, γ_{i+1} is obtained from γ_i by executions of local algorithms at a single process. The time complexity of a distributed algorithm is measured by *rounds*.

We assume that a spanning tree in G is represented by a pointer $parent_u \in N_u \cup \{u\}$ at each process $u \in V$. We say $T = \{parent_u \mid u \in V\}$ is a *spanning tree* when $E_T = \{\{u, v\} \mid \{u, v\} \in E, parent_u = v\}$ is a spanning tree. Process r is a *root* if $parent_r = r$. The condition says that (*i*) E_T is a rooted in-tree or (*ii*) E_T consists of two subtrees and the roots of the subtrees point at each other. A *local exchange operation* at process $u \in V$ is an update of $parent_u$.

Definition 1 (Spanning tree reconfiguration problem). *Given two spanning trees T_A and T_B in a distributed system $G = (V, E)$, find a finite sequence of spanning trees $T_A = T_0, T_1, T_2, \ldots, T_k = T_B$ where T_{i+1} is obtained from T_i by a local exchange operation for all $i = 0, 1, 2, \ldots, k-1$.*

3 Proposed Algorithm

In this section, we present a distributed spanning tree reconfiguration algorithm. The proposed algorithm shows that an orientation of a spanning tree helps each process locally decide when and on which pair of incident edges it performs a local exchange operation so that it can avoid cycles and keep connectivity. The proposed algorithm terminates in $O(n)$ rounds and requires $O(\log n)$ bits memory at each process[1].

Let an initial spanning tree T_A and a goal spanning tree T_B be rooted in-trees with a root r_A and r_B, respectively. A local exchange operation at process $u \in V$ is changing its parent to its parent in T_B. The proposed algorithm is based on the following two observations. First, a local exchange operation at a leaf of T_A does not form any cycle. Then, a local exchange operation at process u whose descendants have performed local exchange operations keep connectivity and does not form any cycle. Its new parent does not belong to its current descendants because all the descendants belong to a subtree of T_B rooted at u. The proposed algorithm starts local exchange operations at leaves of T_A and propagates the update to their ancestors.

Secondly, when $r_B \neq r_A$, r_B cannot perform any local exchange operation because it disconnects r_B. Furthermore, this results in a deadlock because the ancestors of r_B cannot perform any local exchange operation. We use "root

[1] An orientation of a tree can be obtained in $O(n)$ rounds by a tree orientation algorithm [5]. When an initial and a goal spanning trees are not oriented, the proposed algorithm together with the orientation algorithm terminates in $O(n)$ rounds.

Algorithm 3.1. Spanning tree reconfiguration algorithm at process $u \in V$

(Local exchange operation at $u \neq r_B$)
S_1: $(update_u = false) \wedge (\forall v \in Child(u) : update_v = true) \wedge \neg NewRoot(u) \wedge \{parent_u = v \wedge parent_v \neq u\} \rightarrow parent_u := new\_parent_u, update_u := true$
(Local exchange operation at r_B)
S_2: $(update_u = false) \wedge (\forall v \in Child(u) : update_u = true) \wedge (NewRoot(u) \wedge Root(u)) \rightarrow update_u := true$
(Root hopping)
S_3: $Root(u) \wedge \neg NewRoot(u) \rightarrow parent_u := NextChild(u, receive_u)$
(Accept root hopping)
S_4: $parent_u = v \wedge v \neq u \wedge parent_v = u \rightarrow parent_u := u, receive_u := v$

hopping," in which a root of the current tree passes the privilege to one of its children so that the root walks along a tree and eventually reaches r_B.

Given an initial spanning tree $T_A = \{parent_u \mid u \in V\}$ and a goal spanning tree $T_B = \{new\_parent_u \mid u \in V\}$, Algorithm 3.1 solves the distributed spanning tree reconfiguration problem. Each process $u \in V$ maintains four local variables, $parent_u$, new_parent_u, $update_u$, and $receive_u$. The pointer $parent_u$ represents u's current parent and the pointer new_parent_u represents u's parent in T_B. The Boolean variable $update_u$ represents whether u has updated its pointer or not, and $receive_u$ is a pointer for root hopping. In an initial configuration, we have $update_u = false$, $receive_u = u$ for all $u \in V$, and $parent_u$ and new_parent_u are set according to T_A and T_B. In a terminal configuration, we have $parent_u = new\_parent_u$ and $update_u = true$ for all $u \in V$.

We use the following predicates and functions at $u \in V$.

- $Root(u)$ returns *true* if $parent_u = u$, otherwise *false*.
- $NewRoot(u)$ returns *true* if $new\_parent_u = u$, otherwise *false*.
- $Child(u)$ returns the set $\{v \mid v \in N_u, parent_v = u\}$ in the current configuration. Any predicate on $Child(u)$ holds if $Child(u)$ is an empty set.

We make the root hopping follow a depth first search (DFS) in T_A. Each process has a list of its parent and children in T_A in some local ordering, whose last element is the parent. Let δ_u be the degree of u in T_A and the *i-th child* of u be the $(i \mod \delta_u)$-th child in u's local list. Function $NextChild(u, v)$ returns the process that (i) is a current child or a parent of u, (ii) has not performed local exchange operation, and (iii) appears after v in the local list at u. If no neighboring process satisfies the above condition or v is not in $N_u \cup \{u\}$, it returns u. We note that the DFS does not need to visit a process that has performed a local exchange operation because r_B is not in the descendants of the process.

Let $\mathcal{E} = \gamma_0, \gamma_1, \ldots$ be an arbitrary execution of Algorithm 3.1 under the central scheduler. In configuration γ_i, let $T_i = \{parent_u \mid u \in V\}$. When process u performs a local exchange operation by S_1, its current descendants are also descendants in T_B. Hence, u does not point at its descendants. No process disconnects itself because r_B waits until it becomes a current root by S_2. Hence, each T_i is a spanning tree in G.

The root walks along a DFS in T_A by S_3 and S_4 and eventually reaches r_B. In the worst case, the root hopping procedure moves the root n times which takes $2n$ rounds. The local exchange operations take place concurrently with the root hopping operations. Hence, $T_0, T_1, \ldots$ is finite and it is a sequence of spanning trees.

4 Discussion and Conclusion

We introduced a new type of distributed reconfiguration problem for spanning trees and presented a distributed algorithm for the problem.

We finally discuss an extension of the proposed algorithm and distributed reconfiguration of related combinatorial objects. We can extend the proposed distributed spanning tree reconfiguration algorithm to hypertrees. A hypertree can be represented by a spanning tree in its graph representation [6]. We can easily show that a sequence of spanning trees that our algorithm outputs corresponds to a sequence of hypertrees.

We can also obtain a distributed reconfiguration algorithm for maximum bipartite matchings. Given two maximum bipartite matchings M_A and M_B, $M_A \cup M_B$ consists of connected components, each of which is either a path or a cycle of even size. We assume that matching edges are represented by pointers of both endpoint processes and a local exchange operation is an update of this pointer at a single process. For a given graph, a *near maximum matching* is a matching whose size is smaller than the maximum matching by one. We can obtain a distributed reconfiguration sequence of near maximum matchings from M_A to M_B by circulating a single token to allow at most one connected component undergo reconfiguration and electing a leader in each connected component to propagate local exchange operations from the leader along the path or cycle.

References

1. Bonamy, M., Ouvrard, P., Rabie, M., Suomela, J., Uitto, J.: Distributed recoloring. In: Proceedings of the DISC 2018, pp. 12:1–12:17 (2018)
2. Bousquet, N., et al.: Reconfiguration of spanning trees with many or few leaves. In: Proceedings of the ESA 2020, pp. 24.1–25.15 (2020)
3. Censor-Hillel, K., Rabie, M.: Distributed reconfiguration of maximal independent sets. In: Proceedings of the ICALP 2019, pp. 135:1–135:14 (2019)
4. Ito, T., et al.: On the complexity of reconfiguration problems. Theor. Comput. Sci. **412**, 1054–1065 (2011)
5. Karaata, M.H., Pemmaraju, S.V., Bruell, S.C., Ghosh, S.: Self-stabilizing algorithms for finding centers and medians of trees. In: Proceedings of the PODC 1994, pp. 374 (1994)
6. Lovász, L.: A generalization of König's theorem. Acta Mathematic Academia Sceientiarum Hungaricae **21**, 443–446 (1970)
7. Mizuta, H., Hatanaka, T., Ito, T., Zhou, X.: Reconfiguration of minimum steiner trees via vertex exchanges. In: Proceedings of the MFCS 2019, pp. 79:1–79:11 (2019)

Correction to: Distributed Computing with the Cloud

Yehuda Afek, Gal Giladi, and Boaz Patt-Shamir

Correction to:
Chapter "Distributed Computing with the Cloud" in: C. Johnen et al. (Eds.): *Stabilization, Safety, and Security of Distributed Systems*, LNCS 13046, https://doi.org/10.1007/978-3-030-91081-5_1

In an older version of this paper, the presentation of Gal Giladi's affiliation was misleading. This has been corrected.

The updated version of this chapter can be found at
https://doi.org/10.1007/978-3-030-91081-5_1

C. Johnen et al. (Eds.): SSS 2021, LNCS 13046, p. C1, 2022.
https://doi.org/10.1007/978-3-030-91081-5_41

Author Index